LEEDS COLLEGE
CLASS NO. 343.
BARCODE T 3

D1485593

LEEDS COLLEGE OF BUILDING
WITHDRAWN FROM STOCK

THE
BUILDING REGULATIONS

EXPLAINED AND ILLUSTRATED

The authors

M. J. Billington, *BSc*, *MRICS*, who has worked on various editions of the book since 1986, is a chartered building surveyor. He was formerly Senior Lecturer in building control and construction at De Montfort University, Leicester, before leaving to join the private sector, where he continued to act as visiting lecturer at a number of universities. Currently he is Managing Director of Construction Auditing Services Ltd, a company that specialises in latent defects insurance technical auditing, and is Managing Director of Certass Ltd, a Building Regulations Competent Person Scheme for replacement windows and doors.

K. T. Bright, *MSc*, *FRICS*, *FBEng*, *MCIOB*, *NRAC (Consultant)*, is Emeritus Professor of Inclusive Environments at the University of Reading and Visiting Professor of Inclusive Environments in the School of Art and Design at the University of Ulster. He is also a Director of KBC Ltd, an independent access consultancy. He has published widely in academic and practice journals and is the author and editor of several highly regarded reference books related to the development of inclusive environments.

J. R. Waters, *BSc*, *MPhil*, *PhD*, *MCIBSE*, *CEng*, is Honorary Research Fellow at Coventry University, where he was formerly Head of Building Services Engineering. He is a consultant on environmental aspects of building design, has written *Energy Conservation in Buildings* and co-authored *Sound Control in Buildings*, both from Blackwell Publishing.

THE BUILDING REGULATIONS

EXPLAINED AND ILLUSTRATED

THIRTEENTH EDITION

M. J. Billington

K. T. Bright

J. R. Waters

Blackwell
Publishing

© 1967, 1968, 1970, 1973, 1978, 1982 Walter S. Whyte & Vincent Powell-Smith
© 1986 Seventh Edition The Estate of Vincent Powell-Smith & Walter S. Whyte
© 1986 New material The Estate of Vincent Powell-Smith & M.J. Billington
© 1990 Eighth Edition The Estate of Vincent Powell-Smith & M.J. Billington
© 1992 Ninth Edition The Estate of Vincent Powell-Smith & M.J. Billington
© 1995 Tenth Edition The Estate of Vincent Powell-Smith & M.J. Billington
© 1999 Eleventh Edition The Estate of Vincent Powell-Smith & M.J. Billington
© 2004 Twelfth Edition M.J. Billington, M.W. Simons, J.R. Waters and Blackwell Publishing Ltd
© 2007 Thirteenth Edition M.J. Billington, J.R. Waters and Blackwell Publishing Ltd

Blackwell Publishing editorial offices:
Blackwell Publishing Ltd, 9600 Garsington Road, Oxford OX4 2DQ, UK
    Tel: +44 (0)1865 776868
Blackwell Publishing Inc., 350 Main Street, Malden, MA 02148-5020, USA
    Tel: +1 781 388 8250
Blackwell Publishing Asia Pty Ltd, 550 Swanston Street, Carlton, Victoria 3053, Australia
    Tel: +61 (0)3 8359 1011

The right of the Author to be identified as the Author of this Work has been asserted in accordance with the Copyright, Designs and Patents Act 1988.

All rights reserved. No part of this publication may be reproduced, stored in a retrieval system, or transmitted, in any form or by any means, electronic, mechanical, photocopying, recording or otherwise, except as permitted by the UK Copyright, Designs and Patents Act 1988, without the prior permission of the publisher.

First edition published by Crosby Lockwood & Son Ltd 1967, Second Edition 1968, Third Edition 1970, Fourth Edition 1973, Fifth Edition published by Granada Publishing Ltd in Crosby Lockwood Staples 1978, Sixth Edition published by Granada Publishing Ltd 1982, Seventh Edition published by Collins Professional Books 1986, Eighth Edition published by BSP Professional Books 1990, Ninth Edition published by Blackwell Scientific Publications 1992, Tenth Edition 1995, Eleventh Edition published by Blackwell Science Ltd 1999, Twelfth Edition published by Blackwell Science, a Blackwell Publishing company 2004, Thirteenth Edition published by Blackwell Publishing Ltd 2007

ISBN: 978-1-4051-5922-7

Library of Congress Cataloging-in-Publication Data
Billington, M. J. (Michael J.)
  The building regulations : explained and illustrated / M.J. Billington, K.T. Bright, J.R. Waters.—13th ed.
      p. cm
  Includes bibliographical references and index.
  ISBN: 978-1-4051-5922-7 (hardback : alk. paper)
  1. Building laws—England. I. Bright, Keith. II. Waters, J. R. III. Title.

KD1140.W48 2007
343.41'07869—dc22
                                                                                          2007002033

A catalogue record for this title is available from the British Library

Set in 10/12.5 pt Times by Sparks, Oxford (www.sparks.co.uk)
Printed and bound in Great Britain by TJ International Ltd, Padstow, Cornwall

The publisher's policy is to use permanent paper from mills that operate a sustainable forestry policy, and which has been manufactured from pulp processed using acid-free and elementary chlorine-free practices. Furthermore, the publisher ensures that the text paper and cover board used have met acceptable environmental accreditation standards.

For further information on Blackwell Publishing, visit our website:
www.blackwellpublishing.com/construction

III Appendix

Preface to Thirteenth Edition

The twelfth edition of this book was published in 2004. The radical changes brought about by amendments to the Building Regulations 2000, which were reflected in the twelfth edition, have been built on: several parts have been substantially revised and control has been extended to cover electrical installations by the introduction of a new Part P covering electrical safety. These regulation changes have also been matched by extension of the Competent Person concept which was necessitated by Part P. Therefore, many parts of the previous edition have been redrafted with the inevitable consequence that the length of this publication has been extended yet again!

In detail the following chapters have been completely rewritten:

Chapter 6: Structural stability – revised to take account of the 2004 edition of Approved Document A structure. This document updated many codes and standards, considerably altered the guidance on disproportionate collapse, provided new extended guidance on the design and construction of domestic garages and removed the timber sizing tables for floors and roofs of traditional house construction. These tables are now to be found in a revised form in a TRADA publication.

Chapter 8: Materials and workmanship, site preparation and resistance to contaminants and moisture – updated to reflect changes brought about by the 2004 edition of Approved Document C and the changes to paragraph C of Schedule 1 to the Building Regulations 2000. In the new edition of AD C much more guidance is given on resistance to contaminants, and interstitial and surface condensation are dealt with for the first time (the previous reference to condensation in roofs which was originally found in Approved Document F has now been relocated in AD C).

Chapter 11: Ventilation – rewritten to take account of changes brought about by the introduction of a new Approved Document F, which came into effect on 6 April 2006 replacing the 1995 edition. The new guidance adopts a mainly performance-based approach due to the necessity for the fabric of buildings to be significantly more airtight than was previously the case. More guidance is given on domestic mechanical and natural ventilation systems and ventilation of basements in dwellings is covered for the first time.

Chapter 16: Conservation of fuel and power – revised to reflect changes brought about by the 2006 edition of Approved Document L, which replaced the 2002 edition. The need to further reduce CO_2 emissions has resulted in extensive amendments to guidance which was in fact only four years old. Part L now has four approved documents providing guidance on:

- new dwellings (AD L1A),
- existing dwellings (AD L1B),
- new buildings other than dwellings (AD L2A), and
- existing buildings other than dwellings (AD L2B).

The most significant changes are the omission of the elemental and target U-value methods for designing new dwellings and their replacement with a single calculation method, and the omission of all the previous methods for designing new buildings other than dwellings and their replacement with a calculated single annual CO_2 emission rate for the completed building (using approved commercial software) which must be compared with a target, set by reference to a notional building.

It is also necessary for all buildings to be tested for air-tightness. The result of these changes is that designers will find it difficult to carry out detailed design without employing the services of specialists to provide the necessary calculations. As with the twelfth edition, we are extremely grateful to Bob Waters for his expertise in rewriting Chapters 11 and 16.

Chapter 17: Access to and use of buildings – rewritten to incorporate changes brought about by the 2004 edition of Approved Document M. The changes to the guidance are mainly concerned with buildings other than dwellings and it is significant that Part M no longer refers to 'disabled people' since the aim of the new Part M and AD M is to foster a more inclusive approach to design to accommodate the needs of all people. A new section has been introduced into AD M to cover audience and spectator facilities, refreshment facilities, sleeping accommodation and switches, outlets and controls. Also included is guidance on educational establishments and clarification of the treatment of purpose-built student accommodation. An interesting development is the introduction, for the first time, of the concept of an Access Statement that can identify the philosophy and approach to inclusive design, and this can be particularly useful when the design approach differs from that contained in the AD M guidance. We are grateful for the services of the well-known independent access consultant and author, Keith Bright, for his help in redrafting this chapter.

A new Chapter 19: Electrical safety has been added to this edition to cover Part P of Schedule 1 to the Building Regulations 2000, which was introduced on 1 January 2005 and was subsequently revised with effect from 6 April 2006. Part P covers design and installation of electrical installations, and its accompanying Approved Document also includes guidance on inspection and testing and on the provision of information. It is clear that most electrical installations will be self-certified by Competent Persons and will not go through the normal local authority or Approved Inspector control routes.

The early chapters, which set out the legal and administrative provisions of the regulations have been revised to take account of further expansion of the Approved Inspector control system and Chapter 5 has been changed in order to provide information on the increasing number of Competent Person Schemes. These are being created to provide building control on work which is either of a specialist nature, such as electrical installations under Part P and the installation of heat-producing appliances under Part J, or would probably be considered too trivial to interest local authorities, such as replacement windows and doors under Part L.

As always, the aim is to provide a convenient and straightforward guide and reference to a complex and constantly evolving subject. It must be stressed that this book is a guide to the Regulations and approved and other documents, and is not a substitute for them. We hope that it may shed light on some of the more obscure and difficult to understand parts of the source documents. It should also be stressed that the guidance in the Approved Documents is not mandatory and differences of opinion can quite legitimately exist between controllers and developers or designers as to whether a particular detail in a building design does actually satisfy the mandatory functional requirements of the Building Regulations.

The intended readers of this book are all those concerned with building work – architects and other designers, building control officers, Approved Inspectors, Competent Persons, building surveyors, clerks of works, services engineers and contractors, etc. – as well as their potential successors, the current generation of students on built environment and architectural courses. This book is designed to be of use to both students and teachers and it is gratifying that successive editions are widely adopted by various academic institutions and professional bodies.

We are grateful to Martin Simons for his original work on Chapter 11: Ventilation in the last edition of the book. As always we are especially grateful to Julia Burden, Publisher at Blackwell Publishing, and her editorial team, for their help, patience, interest and sense of humour during the production of this edition.

The law is stated on the basis of cases reported and other material available to us on 1 October 2006.

M.J. Billington
K.T. Bright
J.R. Waters

STOP PRESS (January 2007)

As this edition went to press, DCLG published a new edition of Approved Document B which came into force on 1 April 2007. The main changes brought about by this new edition are as follows:

General

- The former 2000 edition of Approved Document B has now been split into two volumes. Volume 1 deals with dwelling-houses, whilst Volume 2 deals with all other buildings.
- The use of residential sprinkler systems in accordance with BS 9251:2005 is recognised.
- Suitable certification schemes for passive and active fire safety systems may be accepted by Building Control Bodies as evidence of compliance.
- Reference is made to the code of practice for fire safety in adult placements.
- New guidance is given on the need to ensure that management regimes for premises are realistic.
- The following Alternative Approaches are recommended:
  - HTM 05 'Firecode' for the design of hospitals and similar health care premises.
  - Building Bulletin 100 for the design of schools.

- A new Regulation (16B) has been introduced to ensure that sufficient information is recorded to assist the eventual owner/occupier/employer to meet their statutory duties under the Regulatory Reform (Fire Safety) Order 2005.

Changes to Requirement B1 – Means of warning and escape

- The guidance on smoke alarms has been amended such that alarms should be installed in accordance with BS 5839-6:2004 and a commentary on this standard has now been provided and the guidance for buildings other than dwellings has been updated to take account of the 2002 edition of BS 5839-1.
- All smoke alarms should have a standby power supply and smoke alarms should be provided in circulations spaces where a dwelling-house is extended.
- Additional guidance has been provided in relation to work on existing houses.
- Locks and child resistant safety stays may be provided on escape windows.
- The alternative approach for loft conversions to two storey houses has been removed.
- New guidance has been provided on the provision of galleries and inner-inner rooms.
- An option of providing sprinkler protection instead of alternative escape routes has been included for dwelling-houses with a floor more than 7.5 m above ground level.
- Guidance on the application of B1 to replacement windows has been included.
- Guidance on the use of air circulation systems in houses with protected stairways is given.
- Additional options of providing sprinkler protection and/or a protected stairway instead of alternative escape routes has been included for flats with more than one storey.
- Guidance on the use of air circulation systems in flats with protected entrance halls or stairways is given.
- The provisions for smoke control in the common areas of flats have been changed.
- Guidance on means of escape in buildings with open spatial planning has been included.
- A method has been provided for calculating acceptable final exit widths for merging escape routes at ground floor level.
- Guidance applicable to small premises, previously in BS 5588-11, has been incorporated into the text.
- New guidance on the design of residential care homes has been given, including the use of sprinklers and/or free swing door closing devices.
- Guidance on means of escape for disabled people has been incorporated in the general guidance.
- In tall buildings with phased evacuation, the interaction of firefighters with people attempting to evacuate the building now needs to be considered.

Changes to Requirement B3 – Internal fire spread (structure)

- As an alternative to the 100 mm step between dwelling-houses and integral garages, it is now permissible to have a sloping floor in the garage.
- Window and door frames are only suitable for use as cavity barriers if they are constructed of steel or timber of an appropriate thickness.
- The design of compartment walls should take account of the predicted deflection of a floor, in the event of a fire.
- Sprinkler systems should be provided in blocks of flats exceeding 30 m in height.
- A maximum compartment size now applies to unsprinklered single-storey warehouse buildings.
- Extensive cavities in floor voids should be subdivided with cavity barriers.
- Guidance on the specification and installation of fire dampers has been provided.
- Non-combustible materials should be used in the construction of a car park for it to be regarded as 'open sided' for the purposes of establishing the necessary period of fire resistance. Other car parks should achieve the standard period of fire resistance.

Changes to Requirement B4 – External fire spread

- The guidance on roof coverings incorporates the BS EN 13501-5:2005 system of classification.
- Space separation (including the use of 'Notional Boundaries') should be considered where more than one building is on the same site but operated by different 'organisations'.

Changes to Requirement B5 – Access and facilities for the fire service

- There should be access for a pump appliance to within 45 m of all points within a dwelling-house. For other dwellings (flats etc.), an alternative to this would be the provision of a fire main.
- A building with a compartment of 280 m² or more, constructed more than 100 m from a highway, should be provided with suitable fire hydrants.
- An assembly building with any floor exceeding 900 m² which is over 7.5 m above ground level, should be provided with firefighting shafts.
- In unsprinklered buildings, every part of every storey over 18 m in height should be within 45 m of a fire main outlet.

Appendix B

- In dwelling-houses, fire doors need not be provided with self closing devices (except for doors between a dwelling-house and an integral garage). This also applies to fire doors within flats.

Appendix C

- The floor space factors table used to calculate occupancy factors has been updated and moved to Appendix C.
- The method of measurement for door width has been changed to align with Approved Document M.
- Guidance is given on the measurement of free area for smoke ventilators.

Appendix G

- This new Appendix provides guidance on the new requirement for fire safety information to be recorded and passed on to the 'responsible' person (see also new Regulation 16(B)).

Acknowledgements

All material reproduced from the Approved Documents is Crown copyright and is reproduced with the permission of the Controller of HMSO.

The authors are also grateful to the following for permission to reproduce copyright material:

Extracts from British Standards are reproduced with the permission of BSI under licence number 2006/SK008.

Data from SAP 2005 and BR 443 is published with the permission of the Building Research Establishment.

BRE/CRC and BRESCU publications, plus most others, may be obtained from BRE Bookshop, 151 Rosebery Avenue, London EC1R 4GB.

BSI publications can be obtained from BSI customer services, 389 Chiswick High Road, London W4 4AL (tel: +44(0)20 8996 9001, email: cservices@bsi-global.com).

Legal and Administrative

1 Building control: an overview

1.1 Introduction

The building control system in England and Wales was radically revised in 1985. After a long period of gestation, Building Regulations were laid before Parliament and came into general operation on 11 November 1985. They applied to Inner London from 6 January 1986. Subject to specified exemptions, all building work (as defined in the regulations) in England and Wales is governed by Building Regulations.

The current regulations are the Building Regulations 2000 which came into force on 1 January 2001. The 2000 Regulations have been amended eight times since then, the latest being the Building and Approved Inspectors (Amendment) Regulations 2006, and the provisions of all these amendments are reflected in this book.

A separate system of building control applies in Scotland and in Northern Ireland.

The power to make Building Regulations is vested in the Secretary of State for the Environment by section 1 of the Building Act 1984 which sets out the basic framework. Building Regulations may be made for the following broad purposes:

- Securing the health, safety, welfare and convenience of people in or about buildings and of others who may be affected by buildings or matters connected with buildings.
- Furthering the conservation of fuel and power.
- Preventing waste, undue consumption, misuse or contamination of water.

The 2000 Regulations are very short and contain no technical detail. That is found in a series of Approved Documents and certain other non-statutory guidance, all of which refer to other non-statutory documents such as National Standards or Technical Specifications (e.g. British Standards or Agrément Certificates), with the objective of making the system more flexible and easier to use. The 2000 Regulations implement the final conclusions of a major review of both the technical and procedural requirements.

A significant feature of the system is that there are alternative systems of building control – one by local authorities, and the other a private system of certification which relies on 'approved inspectors' operating under a separate set of regulations called The Building (Approved Inspectors, etc.) Regulations 2000. These set out the detailed procedures for operating the system of private certification and came into effect at the same time as the main regulations. Since April 2002 a further system of approval has been added whereby certain competent persons can self-certify their work as complying with the Building Regulations. This system is fully discussed in Chapter 5.

1.2   The Building Act 1984

The Building Act 1984 received the Royal Assent on 31 October 1984 and the majority of its provisions came into force on 1 December 1984. It consolidated most, but not all, of the primary legislation relating to building which was formerly scattered in numerous other Acts of Parliament.

Part I of the Building Act 1984 is concerned with Building Regulations and related matters, while Part II deals with the system of private certification discussed in Chapter 4. Other provisions about buildings are contained in Part III which, amongst other things, covers drainage, and the local authority's powers in relation to dangerous buildings, defective premises, etc.

The provisions of the 1984 Act are of the greatest importance in practice, and many of them are referred to in this and subsequent chapters.

'Building' is defined in the 1984 Act in very wide terms. A building is 'any permanent or temporary building and, unless the context otherwise requires, it includes any other structure or erection of whatever kind or nature (whether permanent or temporary)'. 'Structure or erection' includes a vehicle, vessel, hovercraft, aircraft or other movable object of any kind in such circumstances as may be prescribed by the Secretary of State. The Secretary of State's opinion is, however, qualified. The circumstances must be those which 'in [his] opinion … justify treating it … as a building'.

The result of this definition is that many things which would not otherwise be thought of as a building may fall under the Act – fences, radio towers, silos, air-supported structures and the like.

In the past, doubt has been cast over the status of structures such as residential park homes and marquees. Provided that a residential park home conforms to the definition given in the Caravan Sites and Control of Development Act 1960 (as augmented by the Caravan Sites Act 1968) it is exempt from the definition of 'building' contained in the regulations, and according to the *Manual to the Building Regulations* a marquee is not regarded as a building.

Happily, as will be seen, there is a more restrictive definition of 'building' for the purposes of the 2000 Regulations, but a comprehensive definition is essential for general purposes, e.g. in connection with the local authority's powers to deal with dangerous structures. Hence the statutory definition is necessarily couched in the widest possible terms. In general usage (and at common law) the word 'building' ordinarily means 'a structure of considerable size intended to be permanent or at least to last for a considerable time' (*Stevens* v. *Gourely* (1859) 7 CBNS 99) and considerable practical difficulties arose as to the scope of earlier Building Regulations which the 1984 definition has removed.

In *Seabrink Residents Association* v. *Robert Walpole Campion and Partners* (1988) (6-CLD-08–13; 6-CLD-08–10; 6-CLD-06–32) for example, the High Court held that walls and bridges on a residential development were not subject to the then Building Regulations 1972 because they were not part of 'a building'. The development was not to be considered as a homogenous whole. The then regulations, said Judge Esyr Lewis QC, were 'concerned with structures which have walls and roofs into which people can go and in which goods can be stored'. Each structure in the development must be looked at separately to see whether the regulations applied. 'Obviously a wall may be part of a building and so, in my view, may be a bridge'.

1.3   The linked powers

Local authorities exercise a number of statutory public health functions in conjunction with the process of building control, although these have been reduced in recent years; for example, controls on construction of drains and sewers. These provisions are commonly called 'the linked powers' because their operation is linked with the local authority's building control functions, both in checking deposited plans or considering a building notice, and under the approved inspector system of control. Many of the former linked powers have been brought under the Building Regulations, but local authorities are responsible for certain functions now found in the 1984 Act. In those cases, the local authority must reject the plans (or building notice) or the approved inspector's initial notice if relevant compliance is not achieved or else must impose suitable safeguards. The relevant provisions are:

(a)   Section 21 – Provision of drainage. Although sub-sections (1) and (2) of this section have been replaced by requirement H1 of Schedule 1 to the Building Regulations 2000 (see Chapter 13, section 13.3), a local authority (or on appeal a magistrates' court) may still require a proposed drain to connect with a sewer where that sewer is within 100 feet of the site of the building. In cases where the sewer is located more than 100 feet from the site of the building, the local authority may still require connection to that sewer if they undertake to bear the additional cost (i.e. for the length of drain in excess of 100 feet) of construction, maintenance and repair. Disputes regarding the cost of the additional work may be referred to the magistrates' court. Additionally, the local authority can insist that the drainage connects to a nearby public sewer. Disputes under section 21 are dealt with by a magistrates' court. A related provision is section 98 of the Water Industry Act 1991 under which owners or occupiers of premises can require the water authority to provide a public sewer for domestic purposes in their area, subject to various conditions which can include in an appropriate case the making of a financial contribution.

(b)   Section 22 – Drainage of building in combination. The powers of a local authority under section 21 are extended by this section so that where two or more buildings are involved, the local authority may require them to be drained in combination (instead of each making a separate connection) into an existing sewer. As for section 21, the drain may be constructed by the owners (or by the local authority on their behalf) and the expenses of construction, maintenance and repair may be proportioned between each owner and the local authority as appropriate. Disputes regarding the cost of the apportionment may be referred to the magistrates' court.

(c)   Section 25 – Provision of water supply. This section requires the local authority to reject plans of a house submitted under the Building Regulations unless they are satisfied with the proposals for providing the occupants with a sufficient supply of wholesome water for domestic purposes, by pipes or otherwise. The water supply can be provided in any of the following ways:

- by connecting the house to a water supply provided by a water undertaker (i.e. a mains supply);

- where it is not reasonable to connect to a mains supply (in remote country districts there may be no mains supply) by taking the water into the house by means of a pipe (e.g. from a well or spring);
- where circumstances exist which make either of the foregoing solutions unreasonable, the supply of water may be located within a reasonable distance of the house.

This last solution is interesting when considered against the requirements of paragraph G2 of Schedule 1 to the Building Regulations. G2 demands that in a dwelling a '*bathroom shall be provided containing either a fixed bath or shower bath, and there shall be a suitable installation for the provision of hot and cold water to the bath or shower bath*'. It is difficult to see how this could be achieved if a water supply is not provided in the dwelling.

The wholesomeness of water is judged by reference to section 67 of the Water Industry Act 1991 (standards of wholesomeness of water) as read with regulations made under that Act (i.e. the Water Supply (Water Quality) Regulations 2001). Disputes are determined by the magistrates' court. A related provision is section 37 of the Water Industry Act 1991, which enables a landowner who proposes to erect buildings to require the water authority to lay necessary mains for the supply of water for domestic purposes to a point which will enable the buildings to be connected to the mains at a reasonable cost, a provision which is of considerable use to developers.

1.4   Building Regulations

The Secretary of State is given power to make comprehensive regulations about the provision of services, fittings and equipment in or in connection with buildings, as well as about the design and construction of buildings. A very comprehensive list of the subject matter of Building Regulations is contained in Schedule 1 of the 1984 Act. The Regulations are supported by Approved Documents, giving 'practical guidance' (see section 2.3).

Building Regulations may include provision as to the deposit of plans of executed, as well as proposed work; for example where work has been done without the deposit of plans or there has been a departure from the approved plans. Broad powers are given to make Building Regulations about the inspection and testing of work, and the taking of samples.

Prescribed classes of buildings, services, etc. may be wholly or partially exempted from regulation requirements. Similarly, the Secretary of State may, by direction, exempt any particular building or buildings at a particular location.

Schedule 1 of the 1984 Act is a flexible provision and covers the application of the regulations to existing buildings. It enables regulations to be made regarding not only alterations and extensions, but also the provision, alteration or extension of services, fittings and equipment in or in connection with existing buildings. It also enables the regulations to be applied on a *material change of use* as defined in the regulations and, very importantly, makes it possible for the regulations to apply where re-construction is taking place, so that the regulations can deal with the whole of the building concerned and not merely with the new work.

The 1984 Act contains enabling powers for the making of regulations on a number of procedural matters.

The regulations made and currently in force are:

- The Building Regulations 2000 (as amended)
- The Building (Approved Inspectors, etc.) Regulations 2000 (as amended)
- The Building (Local Authority Charges) Regulations 1998
- The Building (Inner London) Regulations 1985

Most of these regulations have been amended, in some cases several times, and care should be taken to ensure that the most recent amendments are being used.

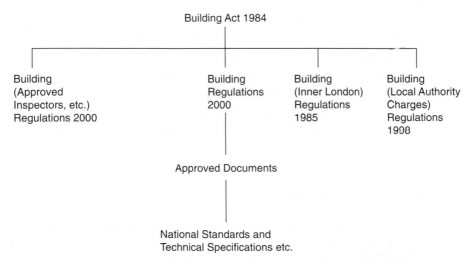

Fig. 1.1   Building control: the legislative scheme.

1.5   Building Regulations – exemptions

1.5.1   Crown immunity

The Building Regulations do not apply to premises which are occupied by the Crown. It is an established rule of statutory interpretation that the Crown is not bound by an Act of Parliament except by express provision or necessary implication. Therefore, an Act of Parliament must specifically state that the Crown is covered by the provisions in order that it be bound by them and, in fact, there is a provision in section 44 of the Building Act 1984 to apply the substantive requirements of the regulations to Crown buildings but this has never been activated.

In practice, it is normal for government department building work to be designed and constructed in accordance with the Building Regulations. In some areas the plans and particulars may even be submitted to the local authority for comment, although it is more usual for these to be scrutinised by specialist companies (replacing the service which was formally given by the Property Services Agency) who will also carry out on-site inspections of the works in progress. Even so, such companies have no legal control over the work and cannot take enforcement action in the event of a breach of the Regulations.

Interestingly, Crown premises are not exempt from control under the Regulatory Reform (Fire Safety) Order 2005, which replaced the former fire certification system under the Fire Precautions Act 1971. However, where independent inspection is needed of such premises, they are inspected *not* by the relevant fire and rescue authority, but by the Crown Premises Inspection Group within the Home Office Fire Service Inspectorate, a bureaucratic anomaly, which has attracted much criticism. Unfortunately, the powers of entry to premises contained in the former 1971 Act (and now contained in the Regulatory Reform (Fire Safety) Order 2005) do not apply to premises occupied by the Crown.

According to the Building Act 1984, a Crown building is defined as '*a building in which there is a Crown interest or a Duchy interest*'. This definition necessitates the following additional definitions:

Crown interest means – '*an interest belonging to Her Majesty in right of the Crown, or belonging to a government department, or held in trust for Her Majesty for the purposes of a government department*'

Duchy interest means – '*an interest belonging to Her Majesty in right of the Duchy of Lancaster, or belonging to the Duchy of Cornwall*'.

Examples of Crown buildings include not only the Royal Palaces, the Houses of Parliament, 10 Downing Street, etc. but also all government offices (such as local Job Centres) across England and Wales.

Over the years a number of bodies have lost Crown immunity. These include:

- Health Service Premises – under the provisions of section 60 of the National Health Service and Community Care Act 1990 health service bodies are no longer regarded as the servant or agent of the Crown in respect of land over which they have powers of disposal or management, or which is otherwise used or occupied by them. Sub-section (7) of section 60 defines *Health Service Bodies* in relation to England and Wales as a Family Health Services Authority, the Dental Practice Board and the Public Health Laboratory Service Board. In practice, this covers regional, district and special health authorities and means that health service buildings are now subject to the full substantive and procedural provisions of building, planning and fire precautions legislation enforceable by local authorities.
- The Metropolitan Police – although no longer regarded as servants or agents of the Crown, the Metropolitan Police Authority has been exempted from having to comply with the procedural requirements of the Building Regulations, using the powers available under section 5 of the Building Act 1984 (exemption of public bodies from the procedural requirements and enforcement of Building Regulations). However, it is still required to comply with the substantive or technical requirements of the Regulations. As an exempt body the Metropolitan Police Authority is also exempt from enforcement procedures by local authorities. Instead, the Metropolitan Police Authority as a '*Public Body*' is bound by the provisions of the Building Act 1984, section 54 (Supervision of their own work by public bodies) and by Part VII (Public Bodies) of the Building (Approved Inspectors, etc.) Regulations 2000 (as amended).

Finally, the reorganisation of the Post Office has meant that, whilst the Royal Mail is still regarded as Crown property, Post Office Counters is not.

1.5.2   Building Act exemptions

Taken together, the Building Act 1984 and the Building Regulations 2000 (as amended), exempt certain uses of buildings and many categories of work from control as is illustrated by the following examples.

- As a result of the repeal of regulation 8 of the Education (Schools and Further and Higher Education) Regulations 1989, maintained schools in England ceased to have exemption from the Building Regulations from 1 April 2000. A similar situation has existed in Wales since 1 January 2002 following the passing of the Education (Schools and Higher and Further Education) (Amendment) (Wales) Regulations 2001. As a result, building works at schools are now treated in the same way as in other user groups and are subject to normal building control procedures. This is a change in the approval process, meaning that building regulation submissions in respect of work to maintained schools now have to be made to the appropriate building control body (local authority or Approved Inspector), but does not affect the standards applicable to schools. The change does not affect in any way the status of the Education (School Premises) Regulations 1999, which continues to apply to all schools. These regulations cover general standards of provision of facilities, such as:
  - in day schools, accommodation for washrooms, medical purposes, staff, cloakrooms, canteens, etc.;
  - in boarding schools, accommodation for sleeping, washing (including bathrooms), living (for study outside school hours and for social purposes), preparing and consuming meals, medical purposes, staff and storage.

  They also cover general constructional requirements such as:
  - structural stability;
  - weather protection;
  - means of escape in case of fire and other health, safety and welfare issues;
  - acoustics, lighting, heating and ventilation;
  - water supplies, and drainage.

  For many years, constructional standards for schools have been set by the Department for Education and Science (DfES) in England and the National Assembly for Wales in Wales, the most recent ones being the 1997 Constructional Standards. Virtually all of the requirements of these standards for school buildings have now been incorporated into the building regulation Approved Documents, which sometimes refer to DfES *Building Bulletins* as alternatives to the normal Approved Document guidance. For example:
  - Ventilation provisions in schools can be made in accordance with the guidance in DfES Building Bulletin 101, *Ventilation of School Buildings*, (see www.teachernet. gov.uk/iaq) and in The Education (School Premises) Regulations 1999. In spaces where noxious fumes may be generated additional provision for ventilation should be made and may require the use of fume cupboards. Fume cupboards in schools should comply with DfES Building Bulletin 88, *Fume cupboards in schools*, 1998.

○ For acoustics, Approved Document E (*Resistance to the passage of sound*) refers the reader to Building Bulletin 93 (*The acoustic design of schools*) for guidance on acoustic conditions and disturbance by noise. (See Chapter 10, section 10.11.)

Care should be taken to use the most recent edition of these alternative sources of guidance since they are regularly updated by DfES.

Further information on alternative sources of guidance can be obtained from the School Premises Team, Department for Education and Skills, Caxton House, Room 762, 6–12 Tothill Street, London SW1H 9NA, tel 020 7273 6023, email: premises.schools@dfes.gsi.gov.uk.

● *Statutory undertakers and other public bodies:* Under section 4 of the Building Act 1984, a building belonging to a statutory undertaker, the United Kingdom Atomic Energy Authority, or the Civil Aviation Authority is exempt from the application of Building Regulations, provided that the building in question is held or used by them for the purposes of their undertaking.

'Statutory undertaker' as defined in section 126 of the Building Act 1984 (as amended) means '*persons authorised by an enactment or statutory order to construct, work or carry on a railway, canal, inland navigation, dock harbour, tramway or other public undertaking but does not include a universal service provider (within the meaning of the Postal Services Act 2000), the Post Office company (within the meaning of Part IV of that Act) or any subsidiary or wholly-owned subsidiary (within the meanings given by section 736 of the Companies Act 1985) of the Post Office company*'.

From this definition it is clear that Post Offices (i.e. the actual high street 'shops' where the public resort for postal services) are no longer regarded as statutory undertakers within the meaning of section 126, consequently they are now required to comply with the Building Regulations, including submissions of work to the appropriate building control body. Interestingly, Royal Mail (which deals with the collection, sorting and delivery of mail) is still regarded as a Crown body and is therefore exempt from compliance with the Building Regulations.

The status of statutory undertakers has been complicated by the fact that a number of former public bodies are now in private hands. This has meant that it has been necessary to pass additional legislation in order to clarify the status of some of these bodies. As a consequence, the following bodies are deemed to be statutory undertakers:

○ public gas suppliers (see sections 67(1), 67(3) and 67(4) of the Gas Act 1986);
○ electricity suppliers (see section 112(4) of the Electricity Act 1989);
○ the National Rivers Authority (see section 190(1) of the Water Act 1989);
○ water and sewerage undertakers (see section 190(1) of the Water Act 1989).

The building of the statutory undertaker must be one which is held or used by them for the purposes of the undertaking; therefore the exemption from the application of Building Regulations granted by virtue of section 4 of the Building Act 1984 is subject to the following exceptions, in respect of which Building Regulations do apply:

(1) a house
(2) a building used as offices or showrooms unless:
    (a) it forms part of a railway station, or
    (b) in the case of the Civil Aviation Authority, it is on an aerodrome owned by the Authority.

A further class which enjoys an exemption under section 4 is 'relevant airport operators' as defined in section 57 of the Airports Act 1986. Section 4 applies in relation to a relevant airport operator as it applies to a statutory undertaker, subject to the following variations:

(1) hotels are not exempt from compliance with the regulations;
(2) offices and showrooms are exempt from compliance with the regulations, if they are on any airport to which Part V of the Airport Act 1986 applies.

It should be noted that local authority buildings are *not* exempt from either the procedural or substantive requirements of the Building Regulations.

1.5.3   Miscellaneous

Under section 16 of the Building Act 1984 there is power to approve the plans of a proposed building by stages. Usually, the initiative will rest with the applicant as to whether to seek approval by stages – subject to the local authority's agreement.

However, local authorities may – of their own initiative – give approval by stages; they might, for example, await further information. In giving stage approval, local authorities will be able to impose a condition that certain work will not start until the relevant information has been produced.

Plans may also be approved subject to agreed modifications, e.g. where there is a minor defect in the plans.

Section 19 of the Building Act 1984 deals with the use of short-lived materials. The provision applies where plans, although conforming to the regulations, include the use of items listed in the regulations for the purpose of section 19. In such circumstances the local authority has discretion:

- to pass the plans;
- to reject the plans; or
- to pass them subject to the imposition of a time limit, whether conditionally or otherwise.

Interestingly, the Building Regulations 2000 (as amended) contain no specific references to any particular materials; however, as will be seen, regulation 7 of the 2000 Regulations requires that building work which must comply with the Schedule 1 requirements must be carried out 'with proper materials which are appropriate for the circumstances in which they are used …', and the supporting approved document deals with the use of short-lived materials.

The local authority may impose a time limit either on the whole of a building or on particular work. Additionally, they may impose conditions as to the use of a building or the particular items concerned. Appeal against the local authority's decision lies to the Secretary of State.

Eventually, section 19 will cease to have effect when section 20, which is wider in scope, is brought into force by the Secretary of State.

Building regulations may impose continuing requirements on the owners and occupiers of buildings, including buildings which were not, at the time of their erection, subject to Building Regulations. These requirements are of two kinds.

Continuing requirements may be imposed *first,* in respect of designated provisions of the regulations to ensure that their purpose is not frustrated, e.g. the keeping clear of fire escapes; and *second,* in respect of services, fittings and equipment, e.g. a requirement for the periodical maintenance and inspection of lifts in flats.

Type relaxations may be granted by the Secretary of State; he may dispense with or relax some regulation requirement generally. A type relaxation can be made subject to conditions or for a limited period only. It should be noted that before granting a type relaxation the Secretary of State must consult such bodies as appear to him to be representative of the interests concerned and he has to publish notice of any relaxations issued.

The Building Act 1984, sections 39 to 43, contains the appeal provisions. The principal appeals to the Secretary of State are:

- appeals against rejection of plans by a local authority; and
- appeals against a local authority's refusal to give a direction dispensing with or relaxing a requirement of the regulations or against a condition attached by them to such a direction.

Interestingly, section 38 of the Building Act 1984 is concerned with civil liability but has yet to be activated. Under this section, breach of duty imposed by the regulations will be actionable at civil law, where damage is caused, except where the regulations otherwise provide. 'Damage' is defined as including the death of, or injury to, any person (including any disease or any impairment of a person's physical or mental condition). The regulations themselves may provide for defences to such a civil action and section 38 will not, when operative, prejudice any right which exists at common law.

1.6   Dangerous structures, etc.

Local authorities have power to deal with a building or structure which is in a dangerous condition or is overloaded. The procedure is for the local authority to apply to the magistrates' court for an order requiring the owner to carry out remedial works or, at his option, to demolish the building or structure and remove the resultant rubbish. The court may restrict the use of the building if the danger arises from overloading. If the owner fails to comply with the order within the time limit specified by the court, the local authority may execute the works themselves and recover the expenses incurred from the owner, who is also liable to a fine (Building Act 1984, section 77).

Under section 78 of the Building Act 1984 the local authority may take immediate action in an emergency so as to remove the danger, e.g. if a wall is in danger of imminent collapse. Where it is practicable to do so, they must give notice of the proposed action to the owner and occupier. The local authority may recover expenses which they have reasonably incurred in taking emergency action, unless the magistrates' court considers that they might reasonably have proceeded under section 77. An owner or occupier who suffers damage as a result of action taken under section 78 may in some circumstances be entitled to recover compensation from the local authority.

Section 79 of the 1984 Act empowers local authorities to deal with ruinous and dilapidated buildings or structures and neglected sites 'in the interests of amenity', which is a term of wider significance than 'health and safety': *Re Ellis and Ruislip* v. *Northwood*

UDC [1920] 1 KB 343. (Section 76 of the Act enables them to deal with defective premises which are 'prejudicial to health or a nuisance'.)

Under section 79, where a building or structure is in such a ruinous or dilapidated condition as to be seriously detrimental to the amenities of the neighbourhood, the local authority may serve notice on the owner requiring him to repair or restore it or, at his option, demolish the building or structure and clear the site.

Demolition is itself subject to control. Section 80 requires a person who intends to demolish the whole or part of a building to notify the local authority, the occupier of any adjacent building and the gas and electricity authorities of his intention to demolish. He must also comply with any requirements which the local authority may impose by notice under section 82.

The demolition notice procedure does not apply to the demolition of:

- an internal part of an occupied building where it is intended that the building should continue to be occupied;
- a building with a cubic content (ascertained by external measurement) of not more than 1750 cubic feet (50 m³) or a greenhouse, conservatory, shed or prefabricated garage which forms part of a larger building;
- an agricultural building unless it is contiguous to a non-agricultural building or falls within the preceding paragraph.

The local authority may by notice require a person undertaking demolition to carry out certain works:

- To shore up any adjacent building.
- To weatherproof any surfaces of an adjacent building exposed by the demolition.
- To repair and make good any damage to any adjacent building caused by the demolition.
- To remove material and rubbish resulting from the demolition and clearance of the site.
- To disconnect and seal and/or remove any sewers or drains in or under the building.
- To make good the ground surface.
- To make arrangements with the gas, electricity and water authorities for the disconnection of supplies.
- To make suitable arrangements with the fire authority (and Health and Safety Executive, if appropriate) with regard to burning of structures or materials on site.
- To take such steps in connection with the demolition as are necessary for the protection of the public and the preservation of public amenity.

1.7   Other legislation

Although the Building Act 1984 attempted to rationalise the main controls over buildings, there are in fact a great many pieces of legislation, in addition to the Building Act and the Building Regulations, which affect the building, its site and environment and the safety of working practices on and within the building. Reference to some of this additional legislation is made throughout this book in subsequent chapters.

2 The Building Regulations and Approved Documents

2.1 Introduction

Although the statutory framework of building control is found in the Building Act 1984, the 2000 Regulations, as amended, contain the detailed rules and procedures. The regulations are comparatively short because the technical requirements have mostly been cast in a functional form.

Each technical requirement is supported by a document approved by the Secretary of State intended to give practical guidance on how to comply with the requirements. The Approved Documents refer to British Standards and other guidance material such as BRE publications and thus give designers and builders a great degree of flexibility.

The 2000 Regulations became effective on 1 January 2000, and have since been amended eight times.

2.2 Division of the Regulations

There are 37 regulations, arranged logically in seven parts. The division is as follows:

PART I: GENERAL

Reg. 1.   Citation and commencement.
Reg. 2.   Interpretation.

PART II: CONTROL OF BUILDING WORK

Reg. 3.   Meaning of building work.
Reg. 4.   Requirements relating to building work.
Reg. 4A.   Requirements relating to thermal elements.
Reg. 4B.   Requirements relating to change of energy status.
Reg. 5.   Meaning of material change of use.
Reg. 6.   Requirements relating to material change of use.
Reg. 7.   Materials and workmanship.
Reg. 8.   Limitation on requirements.
Reg. 9.   Exempt buildings and work.

PART III: EXEMPTION OF PUBLIC BODIES FROM PROCEDURAL REQUIREMENTS

Reg 10.   The Metropolitan Police Authority.

PART IV: RELAXATION OF REQUIREMENTS

Reg. 11.   Power to dispense with or relax requirements.

PART V: NOTICES AND PLANS

Reg. 12.   Giving of a building notice or deposit of plans.
Reg. 13.   Particulars and plans where a building notice is given.
Reg. 14.   Full plans.
Reg. 14A. Consultation with sewerage undertaker.
Reg. 15.   Notice of commencement and completion of certain stages of work.
Reg. 16.   Energy rating.
Reg. 16A. Provisions applicable to self-certification schemes.
Reg. 17.   Completion certificates.

PART VA: (ENERGY PERFORMANCE OF BUILDINGS)

Reg. 17A. Methodology of calculation of the energy performance of buildings.
Reg. 17B. Minimum energy performance requirements for buildings.
Reg. 17C. New buildings.
Reg. 17D. Consequential improvements to energy performance.
Reg. 17E. Interpretation.

PART VI: MISCELLANEOUS

Reg. 18.   Testing of building work.
Reg. 19.   Sampling of material.
Reg. 20.   Supervision of building work otherwise than by local authorities.
Reg. 20A. Sound insulation testing.
Reg. 20B. Pressure testing.
Reg. 20C. Commissioning.
Reg. 20D. CO_2 emission rate calculation.
Reg. 21.   Unauthorised building work.
Reg. 22.   Contravention of certain regulations not to be an offence.
Reg. 23.   Transitional provisions.
Reg. 24.   Revocations.

There are also five schedules:

SCHEDULE 1 – REQUIREMENTS

This contains technical requirements which are almost all expressed in functional terms and grouped in 14 parts set out in tabular form:

PART A: STRUCTURE – Covers loading, ground movement and disproportionate collapse.

PART B: FIRE SAFETY – Covers means of warning and escape, internal and external fire spread, and access and facilities for the fire service.

PART C: SITE PREPARATION AND RESISTANCE TO CONTAMINANTS AND MOISTURE – Covers preparation of site and resistance to contaminants, subsoil drainage, and resistance to weather, interstitial and surface condensation and ground moisture.

PART D: TOXIC SUBSTANCES – Deals with cavity insulation.

PART E: RESISTANCE TO THE PASSAGE OF SOUND – Protection against sound from other parts of a building and adjoining buildings, protection against sound emanating within relevant buildings, reverberation in the common internal parts of relevant buildings and acoustic conditions in schools.

PART F: VENTILATION – Covers means of ventilation in dwellings and buildings other than dwellings.

PART G: HYGIENE – Deals with bathrooms, hot water storage and sanitary conveniences and washing facilities.

PART H: DRAINAGE AND WASTE DISPOSAL – Deals with foul water drainage, wastewater treatment systems and cesspools, rainwater drainage, building over sewers, separate systems of drainage and solid waste storage.

PART J: COMBUSTION APPLIANCES AND FUEL STORAGE SYSTEMS – Covers air supply, discharge of products of combustion, protection of the building, provision of information, protection of liquid fuel storage systems and protection against pollution.

PART K: PROTECTION FROM FALLING, COLLISION AND IMPACT – Covers stairs, ladders and ramps, protection from falling, vehicle barriers and loading bays, protection from collision with open windows, etc. and protection against impact from and trapping by doors.

PART L: CONSERVATION OF FUEL AND POWER – Now divided into four separate documents dealing with conservation of fuel and power in new dwellings, existing dwellings, new buildings other than dwellings and existing buildings other than dwellings.

PART M: ACCESS TO AND USE OF BUILDINGS – Requires that buildings, and the facilities provided within them are reasonably accessible. Specifically mentioned are sanitary conveniences in dwellings and means to ensure that extensions to buildings do not reduce the level of access and use.

PART N: GLAZING – SAFETY IN RELATION TO IMPACT, OPENING AND CLEANING – Deals with reducing the risks associated with glazing in critical locations in buildings and covers the safe operation and cleaning of windows, skylights and ventilators, etc.

PART P: ELECTRICAL SAFETY – Covers design, installation and inspection of electrical installations.

SCHEDULE 2 – EXEMPT BUILDINGS AND WORK

This lists exempt buildings and work in seven classes, and one of its effects is significantly to reduce the extent of control by giving complete exemptions for certain buildings and extensions.

SCHEDULE 2A – SELF-CERTIFICATION SCHEMES AND EXEMPTIONS FROM REQUIREMENT TO GIVE BUILDING NOTICE OR DEPOSIT FULL PLANS

This schedule lists certain types of work (for example, the installation of various kinds of combustion appliances or the installation of replacement windows, doors and rooflights) and gives details of certain classes of people who can carry out the work without giving a building notice or depositing full plans with the local authority. Such individuals will need to be registered under various industry schemes appropriate to the work in order to benefit from the exemption.

SCHEDULE 2B – DESCRIPTIONS OF WORK WHERE NO BUILDING NOTICE OR DEPOSIT OF FULL PLANS REQUIRED

Lists types of work which can be carried out without notifying the local authority. This is mainly concerned with minor electrical work but also includes some work on heating or cooling systems and the replacement of an external door which is not substantially glazed.

SCHEDULE 3 – REVOCATION OF REGULATIONS

This lists the former regulations which are revoked, i.e. the Building Regulations 1991, and parts of other relevant regulations.

2.3   Approved Documents

There are 18 Approved Documents issued by the Department for Communities and Local Government (DCLG) intended to give practical guidance on how the technical requirements of Schedule 1 may be complied with. They are written in straightforward technical terms with accompanying diagrams and the intention is that they will be quickly updated as necessary. Additionally, there are two other Approved Documents covering respectively: *Timber intermediate floors for dwellings* published by the Timber Research and Development Association, and *Basements for dwellings* published by the British Cement Association.

The status and use of Approved Documents is prescribed in sections 6 and 7 of the Building Act 1984. Section 6 provides for documents giving 'practical guidance with respect to the requirements of any provision of Building Regulations' to be approved by the Secretary of State or some body designated by him. The documents so far issued have been approved by the Secretary of State, although they refer to other non-statutory material.

The legal effect of 'Approved Documents' is specified in section 7. Their use is not mandatory, and failure to comply with their recommendations does not involve any civil or criminal liability, but they can be relied upon by either party in any proceedings about an alleged contravention of the requirements of the regulations. If the designer or contractor proves that he has complied with the requirements of an Approved Document, in any proceedings which are brought against him he can rely upon this 'as tending to negative liability'. Conversely, failure to comply with an Approved Document may be relied on by the local authority 'as tending to establish liability'. In other words, the onus will be upon the designer or contractor to establish that he has met the functional requirements in some other way.

The position is illustrated by *Richards* v. *Kerrier District Council* (1987) CILL 345, 4-CLD-04–26 where it was held that if the local authority proved that the works did not comply with the Approved Document, it was then for the appellant to show compliance with the regulations. If the designer fails to follow an Approved Document, it is for him to prove (if prosecuted) that he used an equally effective method or practice.

All the Approved Documents are in a common format, and their provisions are considered in subsequent chapters. They may be summarised as follows:

A: STRUCTURE – This supports Schedule 1, A1, A2 and A3. Section 1 gives details of codes and standards that can be used for all building types and emphasises certain basic principles which must be taken into account if other approaches are adopted. Section 2, which deals with houses and other small buildings, contains guidance on the sizing of timber members, wall thicknesses, masonry chimneys and concrete foundations. Sections

3 and 4 cover wall claddings and roof coverings respectively and section 5 deals with disproportionate collapse and is relevant to all types of building.

B: FIRE SAFETY – This supports Schedule 1, B1, B2, B3, B4 and B5 and is probably the most complex part of the Regulations. B1 deals with means of warning and escape (fire alarm systems and the design of buildings to permit rapid evacuation in the event of fire). B2 and B3 cover the ability of the building to resist fire spread over the surfaces of internal walls and ceilings, the ability of a building to stand up to the effects of a fire so that it will not collapse before people have had a chance to escape and the way a building can be designed so that fire is prevented from spreading through its internal structure (in floor, ceiling and wall voids and past party walls, etc.). B4 deals with the prevention of external fire spread across an open space where it might affect a neighbouring building and B5 covers ways of making buildings accessible for fire fighters when they need to save lives.

C: SITE PREPARATION AND RESISTANCE TO CONTAMINANTS AND MOISTURE – Read in conjunction with Schedule 1, Part C, it deals with the necessary basic requirements. Section 1 covers clearance or treatment of unsuitable materials. Section 2 deals with contaminants, including the erection of buildings on sites affected by radon gas or the landfill gases, methane and carbon dioxide. Additionally, it covers any substances in the ground which might cause a danger to health, and its provisions effectively replace those of the repealed section 29 of the Building Act 1984. Section 3 deals with sub-soil drainage and sections 4 to 6 describe the measures necessary in order to prevent the passage of moisture to the inside of the building including resistance to damage from the effects of interstitial and surface condensation.

D: TOXIC SUBSTANCES – This supports Schedule 1, Part D and is very short. It gives advice on guarding against fumes from urea formaldehyde foam.

E: RESISTANCE TO THE PASSAGE OF SOUND – This supports Schedule 1, Part E and deals with the ability of a building to prevent the passage of unwanted sound from internal sources (sound penetration through external walls is covered by planning legislation, not Building Regulations). The details apply to dwellings and to rooms in other buildings which are used for residential purposes (like hotel bedrooms and similar rooms in hostels and residential homes for the elderly). This means, for example, that it is now necessary to insulate walls between hotel bedrooms and also to apply lining materials to wall surfaces of common access stairs and corridors in such buildings. E4 gives guidance on how to improve acoustic conditions in schools.

F: VENTILATION – Supporting Part F of Schedule 1, this covers means of ventilation and applies to all building types. Some guidance is also given on work to existing buildings caused by the addition of extensions.

G: HYGIENE – Supporting Part G of Schedule 1, it includes the requirements of certain repealed sections (sections 26 to 28) of the Building Act 1984 dealing with water closets and bathrooms as well as covering unvented hot-water systems.

H: DRAINAGE AND WASTE DISPOSAL – This supports Part H of Schedule 1 and covers above and below ground drainage, wastewater treatment systems and cesspools. Certain new sections deal with building over sewers and separate systems of drainage, thereby replacing the repealed section 18 and some sub-sections of section 21 of the Building Act 1984; and solid-waste storage.

J: COMBUSTION APPLIANCES AND FUEL STORAGE SYSTEMS – Supporting Part J of Schedule 1, this deals with gas appliances up to 60 kW and solid and oil fuel appliances up to 45 kW, as well as protection of liquid fuel storage systems and protection against pollution caused by heating oil leakage.

K: PROTECTION FROM FALLING, COLLISION AND IMPACT – This supports Part K of Schedule 1 and covers the design and construction of stairs, ramps and guarding. It has been extended in the 1998 edition to cover vehicle loading bays, protection from collision with open windows, skylights and ventilators, and protection against impact from and trapping by doors.

L: CONSERVATION OF FUEL AND POWER – Supporting Part L, Approved Document L is now divided into four separate documents dealing with conservation of fuel and power in new dwellings, existing dwellings, new buildings other than dwellings and existing buildings other than dwellings. This is mainly concerned with making sure that buildings are reasonably efficient in their use of energy and that carbon dioxide emissions are kept to a minimum.

M: ACCESS TO AND USE OF BUILDINGS – Supporting Part M of Schedule 1, this gives practical guidance on means of access, use of buildings, sanitary conveniences, passenger lifts and common stairs and accessible switches and socket outlets.

N: GLAZING – SAFETY IN RELATION TO IMPACT, OPENING AND CLEANING – Supporting Part N, this covers safe operation and access for cleaning windows, etc., in addition to measures designed to reduce the risks of accidents caused by contact with glazing.

P: ELECTRICAL SAFETY – Supporting Part P, this Approved Document gives guidance on design, installation, inspection and testing of electrical installations.

There is a further Approved Document – MATERIALS AND WORKMANSHIP to support regulation 7 – and it is phrased in very general terms.

Relaxations of the mandatory requirements may be given only by local authorities in appropriate cases, with the possibility of an appeal against refusal to the Secretary of State. An approved inspector cannot grant a relaxation.

2.4   Definitions in the Regulations

Regulation 2 provides a number of general definitions, but not all of them are equally important or helpful. In this section full definitions are given for purposes of ease of reference, although the various special definitions will be referred to again in later chapters. The definitions are:

THE ACT – This means the Building Act 1984.

AMENDMENT NOTICE – This is a notice given by an approved inspector under section 51A of the Building Act 1984 where the scope of the work has changed to such an extent that the original Initial Notice no longer truly reflects the work actually being carried out.

BUILDING – The regulations apply only to buildings as defined. There is a narrow definition of 'building' for the purposes of the regulations:

- A building is 'any permanent or temporary building but not any other kind of structure or erection'. When 'a building' is referred to in the regulations this includes a part of a building.

The effect of this definition is to exclude from control under the regulations such things as garden walls, fences, silos, air-supported structures and so forth.

BUILDING NOTICE – A notice in prescribed form given to the local authority under regulations 12(2)(a) and 13 informing the authority of proposed works.

BUILDING WORK – The regulations apply only to building work as defined in regulation 3(1); any work not coming within the definition is not controlled. Building work means:

- the erection or extension of a building;
- the material alteration of a building;
- the provision, extension or material alteration of services or fittings required by Schedule 1, Parts G, H, J, L or P (and called 'controlled services or fittings');
- work required by Regulation 6 – which sets out the requirements relating to 'material change of use' (see below);
- the insertion of insulating material into the cavity wall of a building;
- work involving the underpinning of a building;
- work required by regulation 4A – which sets out the requirements relating to thermal elements (see below);
- work required by regulation 4B – which sets out the requirements relating to a change of energy status (see below);
- work required by regulation 17D – consequential improvements to energy performance (applies to an existing building with a total useful floor area over 1,000 m² where the proposed building work consists of or includes an extension, the initial provision of

any fixed building services, or an increase to the installed capacity of any fixed building services).

CHANGE TO A BUILDING'S ENERGY STATUS – Any change which results in a building becoming a building to which the energy efficiency requirements of the Regulations apply, where previously it was not.

CONTROLLED SERVICE OR FITTING – This means services or fittings required by Parts G, H, J L or P, i.e. bathrooms, hot-water storage systems, sanitary conveniences, drainage and waste disposal, heat-producing appliances; and replacement doors, windows and rooflights, space heating and hot-water boilers, hot-water vessels, and electrical installations.

DAY – Any period of 24 hours commencing at midnight. It does not include weekends, Bank Holidays or public holidays.

DWELLING – This includes a dwellinghouse and a flat.

DWELLING-HOUSE excludes a flat or building containing a flat.

ELECTRICAL INSTALLATION – This means fixed electrical cables or fixed electrical equipment located on the consumer's side of the electricity supply meter.

ENERGY EFFICIENCY REQUIREMENTS – This means the requirements of regulations 4A, 17C and 17D and Part L of Schedule 1.

ENERGY RATING – A numerical indication of the energy efficiency of a dwelling calculated in accordance with a procedure approved by the Secretary of State.

EUROPEAN TECHNICAL APPROVAL ISSUING BODY – This means the issue of a favourable technical assessment of the fitness for use of a construction product for the purposes of the Construction Products Directive by an authorised body.

EXTRA-LOW VOLTAGE – voltage which is less than or equal to:

- 50 volts between conductors and earth for alternating current; or
- 120 volts between conductors for direct current.

FINAL CERTIFICATE — A certificate given by an approved inspector to a local authority under section 51 of the Building Act 1984 to indicate that a project has been successfully completed.

FIXED BUILDING SERVICES – Any part of, or any controls associated with:

- fixed internal or external lighting systems (but not including emergency escape lighting or specialist process lighting); or

● fixed systems for heating, hot-water service, air conditioning or mechanical ventilation.

FLAT – Separate and self-contained premises (including a maisonette) constructed or adapted for residential purposes and forming part of a building divided horizontally from some other part (see Fig. 2.1).

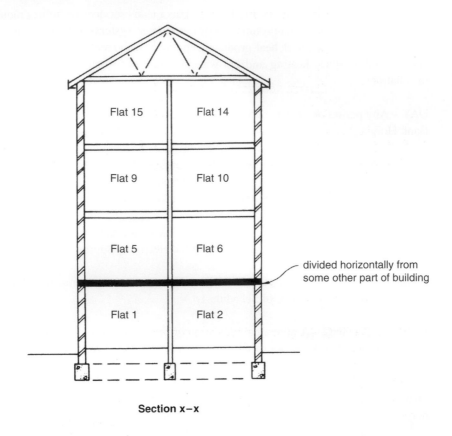

Section x–x

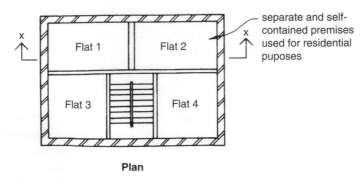

Plan

Fig. 2.1   Flat – Regulation 2.

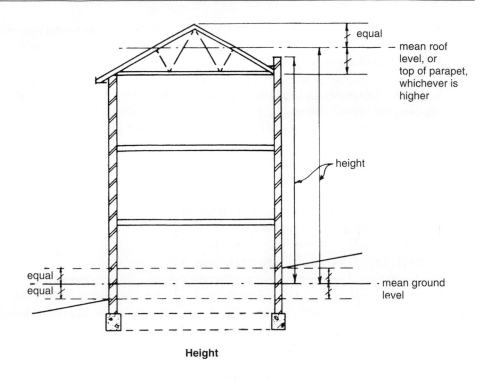

Height

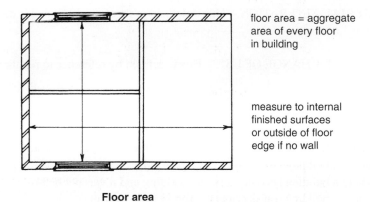

Floor area

Fig. 2.2 Floor area and height – Regulation 2.

FLOOR AREA – This means the aggregate area of every floor in a building or extension. The area is to be calculated by reference to the finished internal faces of the enclosed walls or, where there is no enclosing wall, to the outermost edge of the floor (see Fig. 2.2).

FRONTING – As section 203(3) of the Highways Act 1980 (includes being adjacent to).

FULL PLANS – Plans deposited with a local authority in accordance with regulations 12(2)(b) and 14. The Building Act 1984, section 126, gives a definition of 'plans' as including drawings of any description and specifications or other information in any form.

HEIGHT – This means the height of a building measured from the mean level of the ground adjoining the outside external walls to a level of half the vertical height of the roof, or to the top of any walls or parapet, whichever is the higher (see Fig. 2.2).

INDEPENDENT ACCESS to an extension or part of a building means access to that part which does not pass through the rest of the building;

INITIAL NOTICE – A notice given by an approved inspector to a local authority under section 47 of the Building Act 1984.

INSTITUTION – This means a hospital, home, school, etc. used as living accommodation for, or for the treatment, care, etc. of people suffering from disabilities due to illness or old age or other physical or mental disability or who are under five years old. Those concerned must sleep on the premises and so day care centres, etc. are not included.

LOW-VOLTAGE – voltage which is less than or equal to:

- 1000 volts between conductors or 600 volts between conductors and earth for alternating current; or
- 1500 volts between conductors or 900 volts between conductors and earth for direct current.

MATERIAL ALTERATION – This is defined in regulation 3(2) and is described fully in section 2.6 below.

MATERIAL CHANGE OF USE – This is defined by reference to regulation 5 and there are ten cases:

- where a building becomes a dwelling when it was not one before;
- where a building will contain a flat for the first time;
- where a building becomes a hotel or boarding house, where it previously was not;
- where a building becomes an institution, where it previously was not;
- where a building becomes a public building and it was not before;
- where a building was previously exempt from control (see Schedule 2, below), but is no longer so exempt;
- where a building containing at least one dwelling is altered so that it provides more or less dwellings than before;
- where the building contains a room for residential purposes, where previously it did not;
- where a building containing at least one room for residential purposes is altered so that it contains more or less of such rooms than it did before;
- where a building becomes a shop where it previously was not.

PRIVATE STREET – As section 203(2) of the Highways Act 1980.

PUBLIC BODY'S FINAL CERTIFICATE – This means a certificate given under paragraph 3 of Schedule 4 to the Building Act 1984.

PUBLIC BODIES NOTICE – This means a notice given under section 54 of the Building Act 1984.

PUBLIC BUILDING – This means a building which consists of or contains:

- a theatre, public library, hall or other place of public resort;
- a school or other educational establishment which is not exempt under the 1984 Act, section 4(l)(a);
- a place of public worship.

The definition is restrictive because occasional visits by the public to shops, stores, warehouses or private houses do not make the building a public building.

RENOVATION – This means the provision of a new layer in a thermal element or the replacement of an existing layer, but excludes decorative finishes.

ROOM FOR RESIDENTIAL PURPOSES – This means a room (or suite of rooms) which is not in a dwellinghouse or flat and which is used by people to live and sleep in. It includes a room in a hotel, hostel, boarding house, hall of residence or a residential home, whether or not the room is separated from or arranged in a cluster group with other rooms. By 'cluster' the regulations mean a group of rooms used for residential purposes which is:

- separated by a lockable door from the rest of the building; and
- not designed to be occupied by a single household.

An example of such a cluster may be found in student hostel accommodation where the building is designed as a series of flatlets with separate study bedrooms arranged around a common bathroom, kitchen and lounge. It does not include rooms in hospitals or other similar establishments, used for patient accommodation.

SHOP – This includes premises used by members of the public:

- for sales of food or drink for consumption on or off the premises;
- for retail sales by auction;
- as a barber's or hairdresser's business;
- for the hiring of any item;
- for the treatment or repair of goods.

THERMAL ELEMENT – This means a wall, floor or roof (but not windows, doors, roof windows or roof-lights) which separates a thermally conditioned part of the building (referred to as the 'conditioned space' in the regulations) from:

- the external environment (including the ground); or
- in the case of floors and walls, another part of the building which is:
  - unconditioned;
  - an extension falling within Class VII in Schedule 2 (i.e. conservatories, porches, covered yards or ways or a carport open on at least two sides);
  - conditioned to a different temperature.

The term covers all parts of the thermal element between the surface bounding the conditioned space and the external environment or other part of the building as appropriate. The reference above to the separation of floors and walls of a conditioned space from a space conditioned to a different temperature only applies to buildings other than dwellings, where the other part of the building is used for a purpose which is not similar or identical to the purpose for which the conditioned space is used.

2.5   Exempt buildings and work

With the exception of the work listed at the end of this section, certain buildings and extensions are granted complete exemption from control. The exempt buildings and work fall into seven classes listed in Schedule 2:

CLASS I – BUILDINGS CONTROLLED UNDER OTHER LEGISLATION
- Buildings subject to the Explosives Acts 1875 and 1923.
- Buildings (other than dwellings, offices or canteens) on a site licensed under the Nuclear Installations Act 1965.
- Buildings scheduled under section 1 of the Ancient Monuments and Archaeological Areas Act 1979.

CLASS II – BUILDINGS NOT FREQUENTED BY PEOPLE
- Detached buildings into which people do not normally go.
- Detached buildings housing fixed plant or machinery, normally visited only intermittently for the purpose of inspecting or maintaining the plant, etc. Such buildings are only exempt where they are at least one-and-a-half times their own height from the boundary of the site or any other building frequented by people.

CLASS III – GREENHOUSES AND AGRICULTURAL BUILDINGS
- A building used as a greenhouse.
  A greenhouse is not exempted if the main purpose for which it is used is retailing, packing or exhibiting, e.g. one at a garden centre.
- A building used for agriculture which is:
  - sited at a distance not less than one-and-a-half times its own height from any building containing sleeping accommodation; *and,*

○ is provided with a fire exit not more than 30 m from any point within the building. The definition of 'agriculture' includes horticulture, fruit growing, seed growing and fish farming. Agricultural buildings are not exempted if the main purpose for which they are used is retailing, packing or exhibiting.

CLASS IV – TEMPORARY BUILDINGS

- A building intended to remain where it is erected for 28 days or less, e.g. exhibition stands.

CLASS V – ANCILLARY BUILDINGS

- Buildings on a site intended to be used only in connection with the letting or sale of buildings or building plots on that estate.
- Site buildings on all construction and civil engineering sites, provided they contain no sleeping accommodation.
- Buildings, except those containing a dwelling or used as an office or showroom, erected in connection with a mine or quarry.

CLASS VI – SMALL DETACHED BUILDINGS

- Detached single storey buildings of up to 30 m² floor area, with no sleeping accommodation.

For the exemption to apply, such buildings must either be:
situated more than 1 m from the boundary of their curtilage; *or*
constructed substantially of non-combustible material.

- Detached buildings of up to 30 m² intended to shelter people from the effects of nuclear, chemical or conventional weapons and not used for any other purpose. The excavation for the building must be no closer to any exposed part of another building or structure than a distance equal to the depth of the excavation plus one metre.
- Detached buildings with a floor area not exceeding 15 m² and which do not contain sleeping accommodation, e.g. garden sheds.

CLASS VII – EXTENSIONS

- Ground-level extensions of up to 30 m² floor area which are conservatories, porches, covered yards or ways or a carport open on at least two sides.

The regulations do not apply to the erection of any building set out in Classes I to VI or to extension work in Class VII. Furthermore, they have no application at all to *any* work done to or in connection with buildings in Classes I to VII provided, of course, that the work does not involve a change of use which takes the building out of exemption, e.g. a barn conversion.

Exceptions to the general exemption granted by Schedule 2
The following work carried out to certain buildings falling within Schedule 2 must comply with the requirements indicated:

(1) A conservatory or porch which is wholly or partly glazed must satisfy the requirements of Part N (Safety in relation to impact, opening and cleaning).

(2) Any greenhouse, any small detached building falling within Class VI and any extension falling within Class VII must satisfy the requirements of Part P (Electrical safety) where such buildings receive their electricity from a source shared with or located inside a dwelling.

(3) In general, and taking account of the exceptions listed below, the energy efficiency requirements of the regulations (Part L) apply to:

(a) the erection of any building falling within Schedule 2;

(b) the extension of any such building, other than an extension falling within Class VII in Schedule 2; and

(c) the carrying out of any work to or in connection with any such building or extension.

It should be noted that the term 'building' means the building as a whole or parts of it that have been designed or altered to be used separately.

In order for these requirements to be applied the building must be of roofed construction having walls, and must use energy to condition the indoor climate.

Even with these conditions certain categories of building which might come under Schedule 2 do not have to comply with the energy efficiency requirements of the regulations. These are:

- buildings where compliance with the energy efficiency requirements would unacceptably alter their character or appearance, such as:
  ○ buildings listed in accordance with section 1 of the Planning (Listed Buildings and Conservation Areas) Act 1990;
  ○ buildings in a conservation area designated in accordance with section 69 of the Planning (Listed Buildings and Conservation Areas) Act 1990, or
    (i) buildings included in the schedule of monuments maintained under section 1 of the Ancient Monuments and Archaeological Areas Act 1979;
- buildings which are used primarily or solely as places of worship;
- temporary buildings with a planned time of use of two years or less, industrial sites, workshops, and non-residential agricultural buildings with low energy demand;
- stand-alone buildings other than dwellings with a total useful floor area of less than 50 m^2.

It should be noted that the terms 'industrial sites', 'low energy demand', 'non-residential agricultural buildings', 'places of worship', 'stand-alone', 'total useful floor area' and 'workshops', have the same meaning as in European Parliament and Council Directive 2002/91/EC on the energy performance of buildings.

2.6   Application of the Regulations

The 2000 Regulations apply only to 'building work' or to a 'material change of use', i.e. use for a different purpose. Work or a change of use not coming under these headings is not controlled.

Meaning of 'building work'

The definition of 'building work' means that the regulations apply in nine cases:

ERECTION OR EXTENSION OF A BUILDING

Subject to the exemptions set out in the preceding section, the regulations apply to the erection or extension of all buildings. No attempt is made to define what is meant by 'erection of a building', nor is any definition really necessary. There is a good deal of obscure case law under other legislation as to what amounts to 'erection of a building', but none of it is particularly helpful in the light of section 123 of the Building Act 1984.

This gives a relevant statutory definition. For the purposes of Part II of the Act and for building regulation purposes, erection will include related operations 'whether for the reconstruction of a building, [and] the roofing over of an open space between walls or buildings'.

For the purposes of Part III of the 1984 Act (other provisions about buildings) which is also relevant to building control, *certain* building operations are 'deemed to be the erection of a building'. These are:

(1) Re-erection of any building or part of a building when an outer wall has been pulled or burnt down to within ten feet (3 m) of the surface of the ground adjoining the lowest storey of the building.

It follows that the outer wall must have been demolished throughout its length to within ten feet (3 m) of ground level to constitute re-erection.

(2) The re-erection of any frame building when it has been so far pulled or burnt down that only the framework of the lowest storey remains.

(3) Roofing over any space between walls or buildings. Clearly other operations could be 'the erection of a building'.

PROVISION OR EXTENSION OF CONTROLLED SERVICES AND FITTINGS

Controlled services and fittings are those required by specified parts of Schedule 1:

- G1 – Sanitary conveniences and washing facilities.
- G2 – Bathrooms in dwellings.
- G3 – Hot-water storage systems, except space-heating systems, industrial systems, or those with a storage capacity of 15 litres or less.
- H – Drainage and waste disposal systems.
- J – Fixed heat-producing appliances burning solid or oil fuel or gas or incinerators.
- L – In non-domestic buildings, heating and hot-water systems, lighting, air-conditioning and mechanical ventilation systems.
  - In dwellings, replacement windows, doors and rooflights, space heating or hot-water service boilers and hot water vessels.
- P – Electrical Safety – design installation, inspection and testing.

MATERIAL ALTERATION OF A BUILDING OR OF A CONTROLLED SERVICE OR FITTING

The material alteration of an existing building falls within the definition of building work, and is subject to the regulation requirements. Other alterations are not controlled. There are two cases where an alteration is material, namely an alteration to a building or controlled service or fitting, or part of the work involved, which would at any stage result *either*:

- in the building or controlled service or fitting not complying with the relevant requirements of Schedule 1 where it previously did comply; *or*
- in the building, which did not comply with such requirements before work started, being made worse in relation to the requirement after the alteration.

The specified requirements (called 'relevant requirements' in the regulations) are:

- Part A (structure);
- B1 (means of warning and escape);
- B3 (internal fire spread – structure);
- B4 (external fire spread);
- B5 (access and facilities for the fire service);
- Part M (access to and use of buildings).

The work done must, of course, comply with all the requirements of Schedule 1. In general, it is not necessary to bring the existing building up to regulation standards. However, it should not be made worse when measured against the standards of the relevant requirements in Schedule 1.

WORK IN CONSEQUENCE OF A MATERIAL CHANGE OF USE

When there is a material change of use, as defined in regulation 5 (see section 2.4), work must be done to make the building comply with some of the regulations, as explained below. Such work is, of course, then subject to control, just as the material change of use is itself controlled. In practical terms, change of use is only subject to control if the change involves the provision of sleeping accommodation or use as a public building or where the building was previously exempt.

'Material change of use' requirements

Material change of use has already been defined (see section 2.4), and in the ten cases falling within that definition, specific technical requirements from Schedule 1 are made to apply in the interests of health and safety, which is the philosophy behind building control. Interestingly, there is no requirement applicable in respect of surface water drainage or stairs, nor is there any definition of 'part' of a building. The parts of the regulations applicable are set out in Table 2.1.

INSERTION OF INSULATING MATERIAL INTO A CAVITY WALL

When there is the insertion of cavity fill in an existing wall in a building, the work done must comply with certain specific regulation requirements, namely C2 and D1 (toxic substances).

UNDERPINNING OF A BUILDING

Work involving the underpinning of an existing building is 'building work' for the purposes of the regulations and so comes under control.

Table 2.1   Requirements applicable according to material change of use.

Case	Schedule 1 requirements
[1A] All cases (dwellings including conversion of single dwelling to provide greater number, flats, hotels, shops, boarding houses, institutions and public buildings, no longer exempt, rooms for residential purposes,) where there is a change of use to the whole of the building	Bl (means of warning and escape) B2 and B3 (internal fire spread) B4(2) (external fire spread – roofs) B5 (access etc. for fire services) C2(c) (interstitial and surface condensation) Fl (ventilation) Gl (sanitary conveniences & washing facilities) G2 (bathrooms) HI (foul water drainage) H6 (solid waste storage) Jl to J3 (combustion appliances) LI (conservation of fuel and power) P1 (electrical safety)
[1B] Exempt building to non-exempt, hotel, boarding house, institution, public building	As in [A] plus Al to A3 (structure)
[1C] Building more than 15 metres in height	As in [A] plus B4(l) (external fire spread -walls)
[1CC] Building used as a dwelling, hotel, boarding house or institution or containing a flat where it did not before; where more or less dwellings are provided than was originally the case; where the building contains a room for residential purposes, where previously it did not; where a building containing at least one room for residential purposes is altered so that it contains more or less of such rooms than it did before; building no longer exempt under Schedule 2, Classes I to VI, where previously it was, where the material alteration provides new residential accommodation.	As in (A) plus C1(2) resistance to contaminants
[1D] Building used as a dwelling, where previously it was not	As in [A] plus C2 (resistance to moisture)
[1E] Building used as a dwelling, hotel, boarding house or containing a flat where it did not before; where more or less dwellings are provided than was originally the case; where the building contains a room for residential purposes, where previously it did not; where a building containing at least one room for residential purposes is altered so that it contains more or less of such rooms than it did before	As in [A] plus El to E3 (resistance to passage of sound) *(Continued.)*

Table 2.1   *(Continued.).*

Case	Schedule 1 requirements
[1F] Change of use to public building consisting of or containing a school	As in **[A]** plus E4 (acoustic conditions in schools)
[1G] Change of use to hotel, boarding house, shop, institution or public building	As in **[A]** plus M1 (access to and use of buildings)
[2A], **[2B]** and **[2C]** Change of use of part only of a building in cases **[1A]**, **[1B]**, **[1D]**, **[1E]** and **[1F]**	The part itself must comply with the relevant requirements as **[1A]**, **[1B]**, **[1D]**, **[1E]** and **[1F]**. In **[1C]** the whole building must comply with B4(l)
[2D] Change of use of part only of a building in case **[1G]**	M1 (access to and use of buildings) applied to the part being changed and to any sanitary conveniences provided in or in connection with the part being changed. Whole building complies with requirement M1(a) of Schedule 1 to the extent that reasonable provision is made to provide either suitable independent access to the part being changed or suitable access through the building to that part.

WORK REQUIRED BY REGULATION 4A (REQUIREMENTS RELATING TO THERMAL ELEMENTS)

When renovating a thermal element, sufficient work must be carried out so as to ensure that the whole thermal element complies with the requirements of paragraph L1(a)(i) of Schedule 1 (i.e. to the extent of limiting gains and losses through the thermal elements and other parts of the building fabric). When replacing a thermal element, the new thermal element must also comply with the requirements of paragraph L1(a)(i) of Schedule 1.

WORK REQUIRED BY REGULATION 4B (REQUIREMENTS RELATING TO A CHANGE OF ENERGY STATUS)

When a building is changed so that it becomes a building to which the energy efficiency requirements of the Regulations apply, where previously it was not, such work, if any, must be carried out so as to ensure that the building complies with the applicable requirements of Part L of Schedule 1. In this case 'building' means the building as a whole, or parts of it that have been designed or altered so that they can be used separately.

WORK REQUIRED BY REGULATION 17D (CONSEQUENTIAL IMPROVEMENTS TO ENERGY PERFORMANCE)
Where an existing building with a total useful floor area over $1000\,m^2$:

- is extended, or
- has fixed building services installed for the first time, or
- has an increase to the installed capacity of any fixed building services.

Such work, if any, must be carried out so as to ensure that the building complies with the requirements of Part L of Schedule 1. However, the work only needs to be carried out if it is technically, functionally and economically feasible.

2.7   Regulation requirements

The Regulations impose broad general requirements on the builder. Breach of these requirements does not, of itself, involve the builder in any civil liability, although such liability may arise, quite independently, at common law.

Compliance with Schedule 1 is mandatory. All building work (except work carried out to improve energy performance, see below) must be carried out so that it complies with the requirements set out in that Schedule. The method adopted for compliance must not result in the contravention of another requirement.

The work must also be carried out so that, after completion, an existing building or controlled service or fitting to which work has been done continues to comply with the specified requirements if it previously did so comply or, if it did not so comply before in any respect, it must not be more unsatisfactory afterwards.

The special case concerning energy performance referred to above is for work carried out:

- to thermal elements;
- where there is a change in the building's energy status; and
- where there are consequential improvements to a building's energy performance.

Provided that the work does not constitute a material alteration then it only needs to comply with the applicable requirements of Part L of Schedule 1.

2.8   Schedule 1 – Technical requirements

Schedule 1 contains the technical requirements, which are discussed in Chapters 6 to 18 and which are almost all expressed functionally, e.g. Cl dealing with site preparation states that 'the ground to be covered by the building shall be reasonably free from vegetable matter'. These requirements cannot be subject to relaxation.

Which requirements apply depends on the type of building being constructed, but the majority of them are of universal application.

Materials and workmanship

Regulation 7(1) provides that any building work which is required to comply with any relevant requirement of Schedule 1:

> 'shall be carried out:
> (a) with adequate and proper materials which –
> (i)   are appropriate for the circumstances in which they are used,
> (ii) are adequately mixed and prepared, and
> (iii) are applied, used or fixed so as adequately to perform the functions for which they are designed; and
> (b) in a workmanlike manner.'

This is a general statutory obligation imposed on the builder. Guidance on how the obligation may be met is contained in Approved Document to support regulation 7 'Materials and workmanship', although that guidance is of a very general nature.

This statutory obligation is akin to a building contractor's obligation at common law when, in the absence of a contrary term in the contract, the builder's duty is to do the work in a good and workmanlike manner, to supply good and proper materials and to provide a building reasonably fit for its intended purpose: *Hancock* v. *B.W. Brazier (Anerley) Ltd* (1966) [1966] 1 WLR 1317; [1966] 2 All ER 901, CA. This threefold obligation would normally be implied in any case where a contractor was employed to both design and build, but the third limb of the duty would not arise, for example, where the client employs his own architect (*Lynch* v. *Thorne* (1956) [1956] 1 WLR 303; [1956] 1 All ER 744, CA), although the other two limbs remain.

The principal object of the regulations is to ensure that buildings meet reasonable standards of health and safety, and this is spelled out in regulation 8:

> 'Parts A to D, F to K, N and P (except for paragraphs H2 and J6) of Schedule 1 shall not require anything to be done except for the purpose of securing *reasonable standards of health and safety* for persons in or about buildings (and any others who may be affected by buildings, or matters connected with buildings).'

The obligations imposed by the regulations are not therefore absolute obligations, but rather a duty to use reasonable skill and care to secure reasonable standards of health and safety of people using the building and others who may be affected by failure to comply with the requirements of the regulations.

2.9   Relaxation of regulation requirements

Section 8 of the Building Act 1984 enables the Secretary of State to dispense with or relax any requirement of the regulations 'if he considers that the operation of [that] requirement would be unreasonable in relation to the particular case'. This power has been delegated to the local authority which may grant a relaxation if, because of special circumstances, the terms of a requirement cannot be fully met.

However, the majority of regulation requirements cannot be relaxed because they require something to be provided at an 'adequate' or 'reasonable' level, and to grant a relaxation might mean acceptance of something that was 'inadequate' or 'unreasonable'.

It should be noted that a relaxation of the requirements of the Building Regulations cannot be sought in order to thwart:

- the need for a building to achieve its target CO_2 emission rate identified under Part VA of the regulations (requirements 17A, 17B and 17C); or
- in the case of existing buildings with a total useful floor area over 1,000 m², the energy efficiency requirements of the regulations.

The application procedure is laid down in sections 9 and 10 of the 1984 Act. There is no prescribed form. Only the local authority (or the Secretary of State on appeal) can grant a relaxation; approved inspectors have no power to do so.

At least 21 days before giving a decision on an application for dispensation or relaxation of any requirement, the local authority must advertise the application in a local newspaper unless the application relates only to internal work. The notice must indicate the situation and nature of the work, and the requirement which it is sought to relax or dispense with. Objections may then be made on grounds of public health or safety. No notice need be published if the effect of the proposal is confined to adjoining premises only, but notice must then be given to the owner and occupier of those premises.

Where a local authority refuses an application it must notify the applicant of his right of appeal to the Secretary of State. This must be exercised within one month of the date of refusal. The grounds of the appeal must be set out in writing, and a copy must be sent to the local authority, which must send it to the Secretary of State with a copy of all relevant documents, and any representations it wishes to make. The applicant must be informed of the local authority's representations. There is no time limit prescribed for the Secretary of State's decision on the appeal.

Where a local authority fails to give a decision on an application within two months, it is deemed to be refused and the applicant may appeal forthwith.

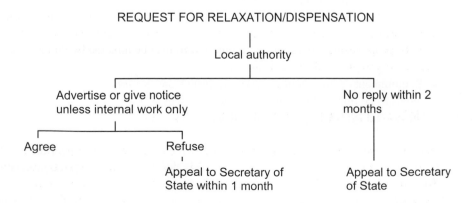

REQUEST FOR RELAXATION/DISPENSATION

Local authority

Advertise or give notice unless internal work only   No reply within 2 months

Agree   Refuse

Appeal to Secretary of State within 1 month   Appeal to Secretary of State

Neither the Secretary of State nor the local authority may give a direction for any relaxation of the regulations where, before the application is made, the local authority has become statutorily entitled to demolish, remove or alter any work to which the application relates, i.e. as a result of service of a notice under section 36 of the 1984 Act. The same prohibition applies where a court has issued an injunction requiring the work to be demolished, altered or removed.

The procedure may be summarised in tabular form:

2.10   Type relaxations

The local authority's power of dispensation and relaxation must be distinguished from that of the Secretary of State to grant a type relaxation, i.e. to dispense with a requirement of the regulations generally. A type relaxation can be made subject to conditions and can be for a limited period only. It can be issued on application to the Secretary of State, e.g. from a manufacturer, in which case a fee may be charged. The Secretary of State may also make a type relaxation of his own accord. Before granting a relaxation the Secretary of State must consult such bodies as appear to him to be representative of the interests concerned and must publish notice of any relaxation issued. No such type relaxations have been granted under the current legislation.

2.11   Continuing requirements

Under section 2 of the Building Act 1984 Building Regulations can impose continuing requirements on owners and occupiers of buildings. These requirements are of two kinds:

- Continuing requirements in respect of designated provisions of the Building Regulations, to ensure that the purpose of the provision is not frustrated.

For example, where an item is required to be provided, there could be a requirement that it should continue to be provided or kept in working order. Examples of the possible use of the power are the operation of mechanical ventilation which is necessary for health reasons or the operation of any lifts required to be provided in blocks of flats.

- Requirements with regard to services, fittings and equipment. This enables requirements to be imposed on buildings whenever they were erected and independently of the normal application of Building Regulations to a building.

A possible use of this power would be to require the maintenance and periodic inspection of lifts in flats if they are to be kept in use. This power of continuing requirements has, as yet, not been used.

2.12   Testing and sampling

Regulations 18 and 19 empower the local authority to test building work to ensure compliance with the requirements of regulation 7 and any applicable parts of Schedule 1 and to take samples of materials *to be used* in the carrying out of building work. The wording does not appear to cover materials which are already incorporated in the building, but this may prove to be of little importance if the provisions of section 33 of the Building Act 1984 are ever activated.

Under that section the local authority may test for compliance with the regulations. They will also be permitted to require a builder or developer to carry out reasonable tests or may carry out such tests themselves and also take samples for the purpose. Section 33(3) sets out the following matters with respect to which tests may be made:

- test of the soil or subsoil of the site of any building;
- tests of any material or component or combination of components;
- tests of any service, fitting or equipment.

This is not an exhaustive description of the matters which may be subjected to tests.

The cost of testing is to be borne by the builder or developer, and there will be a right to apply to a magistrates' court regarding the reasonableness of any test required or of any decision of the local authority on meeting the cost of the test. It should be noted that the local authority will have a discretionary power to bear the whole or part of the costs themselves.

In fact the power of testing is given to 'a duly authorised officer of the local authority'. 'Authorised officer' is defined in section 126 of the Building Act 1984 as:

'... an officer of the local authority authorised by them in writing, either generally or specially, to act in matters of any special kind, or in any specified matter; or ... by virtue of his appointment and for the purpose of matters within his province, a proper officer of the local authority.'

Section 95 of the 1984 Act confers upon an authorised officer appropriate powers of entry, and penalties for obstructing any person acting in the execution of the regulations are provided by section 112.

A duly authorised officer of the local authority must also be permitted to take samples of the materials used in works or fittings, to see whether they comply with the requirements of the regulations. In practice, the authorised officer may ask the builder to have the tests carried out and to submit a report to the local authority. In any event, the builder should be notified of the result of the tests.

It should be noted, however, that regulations 18 and 19 do not apply where the work is supervised by an approved inspector (see Chapter 4) or is done under a public body's notice; however, similar powers exist in the Building (Approved Inspectors, etc.) Regulations 2000 (as amended).

2.13   Unauthorised building work

Regulation 21 allows local authorities retrospectively to certify unauthorised building work carried out on or after 11 November 1985. The regulation became effective on 1 October 1994 and there are prescribed fees payable. It applies to building work which should have been subject to control, but the person who carried out the work failed to deposit plans with the authority, to give a building notice or to give an initial notice jointly with an approved inspector. The regulation enables the owner of the building (the applicant) to make a written application to the authority for a regularisation certificate.

The applicant's notice should describe the unauthorised work and, if reasonably practicable, include a plan of it as well as a plan showing any additional work needed to ensure compliance with the regulations. On receipt of the notice and the accompanying plans, the council may require the applicant to take reasonable steps to enable them to inspect the work, e.g. opening up, testing and sampling. The local authority will then notify the applicant of any work required to ensure compliance, with or without relaxation and when this has been carried out to their satisfaction they may issue a regularisation certificate. This is stated to be evidence (but not conclusive evidence) that the relevant specified requirements have been complied with.

2.14   Contravening works

Under section 36 of the Building Act 1984 where a building is erected, or work is done contrary to the regulations, the local authority may require its removal or alteration by serving notice on the owner of the building. Where work is required to be removed or altered, and the owner fails to comply with the local authority's notice within a period of 28 days, the local authority may remove the contravening work or execute the necessary work itself so as to ensure compliance with the regulations, recovering its expenses in so doing from the defaulter.

A section 36 notice may not be given after the expiration of 12 months from the date on which the work was completed. A notice cannot be served where the local authority has passed the plans and the work has been carried out in accordance with the deposited plans.

The recipient of a section 36 notice has a right of appeal to the magistrates' court. The burden of proving non-compliance with the regulations lies on the authority, but if it shows that the works do not comply with an approved document (under section 7) then the burden shifts. The appellant against the notice must then prove compliance with the regulations: *Richards* v. *Kerrier District Council* (1987) CILL 345; 4-CLD-04–26.

Section 37 provides an alternative to the ordinary appeal procedure. Under that section, the owner may notify the local authority of his intention to obtain from 'a suitably qualified person' a written report about the matter to which the section 36 notice relates. Such notices are served where the local authority considers that the technical requirements of the regulations have been infringed.

The expert's report is then submitted to the local authority. In light of it the local authority may withdraw the section 36 notice and *may* pay the owner the expenses which he has reasonably incurred in consequence of the service of the notice, including his expenses

Table 2.2   Current levels of fines for summary offences.

Level on standard scale	Amount of fine
Level 1	£200
Level 2	£500
Level 3	£1000
Level 4	£2500
Level 5	£5000

in obtaining the report. Adopting this procedure has the effect of extending the time for compliance with the notice or appeal against it from 28 to 70 days.

If the local authority rejects the report, it can then be used as evidence in any appeal under section 40 and section 40(6) provides that:

'if, on appeal ... there is produced to the court a report that has been submitted to the local authority ... the court, in making an order as to costs, may treat the expenses incurred in obtaining the report as expenses for the purposes of the appeal.'

Thus, in the normal course of events, if the appeal was successful, the owner would recover the cost of obtaining the report as well as his other costs.

The local authority – or anyone else – may also apply to the civil courts for an injunction requiring the removal or alteration of any contravening works. This power is exercisable even in respect of work which has been carried out in accordance with deposited plans, e.g. oversight or mistake on the part of the local authority. In such a case the court might well order the local authority to pay compensation to the owner. The 12 months' time limit does not apply to this procedure, which is, however, unusual and rarely invoked in practice. The Attorney-General, as guardian of public rights, may seek an injunction in similar circumstances, and in practice proceedings for an injunction must be taken in his name and with his consent.

Where a person contravenes any provision in the Building Regulations, he renders himself liable to prosecution by the local authority. The case is dealt with in the magistrates' court. Such a person is liable on summary conviction to a fine not exceeding level 5 on the 'standard scale' and to a further fine not exceeding £50 per day for a continuing offence (Building Act 1984, section 35).

The standard scale of fines for summary offences was introduced by section 37 of the Criminal Justice Act 1982 (as amended by section 17 of the Criminal Justice Act 1991). The current levels of fines are as shown in Table 2.2.

In *Torridge District Council* v. *Turner* (1991) 9-CLD-07–21, it was held, for reasons which are not entirely clear, that breach of 'do' provisions such as requirement Al which requires that a building 'shall be so constructed' as to meet the specified standards does not constitute a continuing offence, which means that the proceedings must be commenced within six months of the commission of the alleged offence. This six-month limitation period is specified by section 127(1) of the Magistrates' Courts Act 1980. The Divisional Court held that the person constructing a building commits an offence when the building works are completed in a way not complying with the regulations. He does not commit a continuing offence.

2.15 Determinations

It is sometimes the case that a local authority rejects a full plans application (or an approved inspector refuses to give a plans certificate – see Chapter 4) on the grounds that the plans show a contravention of the Building Regulations, but the applicant believes that the plans do, in fact, comply. In this case the applicant can apply to the Secretary of State for a determination as to whether or not the work complies with the regulations.

It is possible to apply for a determination at any time after the plans have been submitted to the local authority or an approved inspector has been asked for a plans certificate, but it is usually better to wait until a decision has been made and the plans have been declared unacceptable. Applications for a determination must usually be made before the commencement of work (or before commencement of that part of the work which is the subject of the determination).

The application, in the form of a letter, is made direct to the Department of Communities and Local Government if the proposal is in England, or to the National Assembly for Wales if it is in Wales, and it should include the following information:

- the names and addresses of the parties involved, including any agents;
- details of the local authority or approved inspector providing the building control service;
- the full address of where the proposed building work will be carried out;
- a statement setting out details of the building, the proposed work, and the matter in dispute;
- a statement setting out the case for compliance with the particular regulation requirement in question;
- a copy of the plans of the proposed work and any other documents which have been submitted to the local authority with the full plans application, or those submitted to the approved inspector on which he was unable to give a plans certificate;
- a copy of all relevant correspondence with the local authority/approved inspector involved, including the notice of rejection of plans if one has been issued;
- a copy of any listed building consent if required for the proposed work and any associated planning permission relevant to the listed building;
- a copy of any other documents supporting the case for compliance, including calculations;
- where appropriate, a location or block plan and photographs of the proposed work to illustrate particular points;
- the appropriate fee (i.e. half the local authority's plan charge, with a minimum of £50 and a maximum of £500 payable).

It is not necessary to obtain the permission of the local authority or approved inspector before making such an application.

3 Local authority control

3.1 Introduction

Local authorities have exercised control over buildings in England and Wales since 1189, but it was not until 1965 that uniform national Building Regulations were made applicable throughout the country generally. Inner London retained its own system based on the London Building Acts 1930 to 1978 and byelaws made thereunder until 6 January 1986. Building Regulations now apply to Inner London, although many provisions of the London Building Acts continue to apply in modified form. The Building Regulations 1985 introduced a number of substantive changes to the system of local authority control, and there have been several modifications to the system since then, culminating in the Building Regulations 2000, which came into force on 1 January 2001.

Part V of the Building Regulations 2000, as amended, contains the procedural requirements which must be observed where a person proposes to undertake building work covered by the regulations and opts for local authority control. Although a great deal of building work continues to be under local authority control and supervision, an increasing volume of work is now dealt with under an alternative system – private control and supervision by an approved inspector, as explained in Chapter 4. Until January 1997 the private system was confined to house-builders under the National House Building Council (NHBC) scheme, however, it is now possible to use an approved inspector for any class of building work.

Two main procedural options are available under the local authority system of control:

- control based on service of a building notice; and
- control based on the deposit of full plans.

There are also a number of cases – where the work relates to the installation of gas, solid fuel or oil-fired combustion appliances, drainage and plumbing works, electrical work and the installation of replacement windows, doors and rooflights – where neither notice nor deposit of plans is required. It is also possible to have an intermediate situation where plans may be passed in stages.

3.2 The local authority

The local authority for the purposes of the regulations is the district council, a London borough council, the Common Council of the City of London, the Sub-Treasurer of the

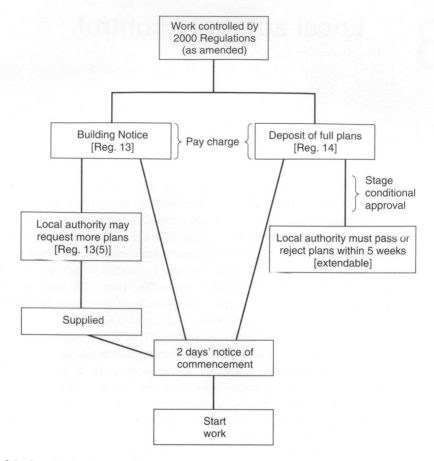

Fig. 3.1   Local authority supervision.

Inner Temple, the Under Treasurer of the Inner Temple, and the Council of the Isles of Scilly.

3.3   Building notice procedure

The major procedural innovation introduced in 1985, and now to be found in the 2000 Regulations, is based on service of a building notice. There is no approval of plans. Interestingly, this is copied from the former Inner London system.

A person intending to carry out building work, replace or renovate a thermal element in a building to which the energy efficiency requirements apply, make a change to a building's energy status or make a material change of use of a building may give a building notice to the local authority except in the following cases:

(1) where it is intended to carry out building work in relation to a building to which the Regulatory Reform (Fire Safety) Order 2005 applies, or will apply after the comple-

tion of the building work (essentially, all fixed workplaces in England and Wales, but excluding locations such as mineshafts, boreholes and open areas outside buildings in forestry and farming);

(2) where it is intended to carry out work which includes the erection of a building fronting onto a private street;

(3) where the work involves the building over of any sewers, disposal mains or drains shown on any map of sewers kept by a sewerage undertaker under section 199 of the Water Industry Act 1991 (i.e. when paragraph H4 of Schedule 1 imposes a requirement).

Normally, the building notice should be given at least two days before commencement of work. However, since there may be occasions when it is necessary to carry out emergency repairs on a building (especially with regard to building services where the work is subject to control), a building notice may be given to the local authority as soon as reasonably practicable after the work has started in these circumstances. This would not, of course, be necessary if the work was being carried out by a competent person (see section 3.4).

There is no prescribed form of building notice. The notice must be signed by the person intending to carry out the work or on his behalf, and must contain or be accompanied by the following information:

- The name and address of the person intending to carry out the work.
- A statement that it is given in accordance with regulation 12(2)(a).
- A description of the proposed building work, renovation or replacement of a thermal element, change to the building's energy status or material change of use.
- A description of the location of the building to which the proposal relates and the use or intended use of that building.
- If it relates to the erection or extension of a building it must be supported by a plan to a scale of not less than 1:1250, showing size and position of the building, its own boundaries and its relationship with adjoining boundaries, the size, position and use of every other building within its boundaries, the width and position of any streets on or within its boundaries, the number of storeys and the provisions to be made for its drainage. Where any local legislation applies, the notice must state how it will be complied with.
- Where the building notice involves cavity-wall insulation, information must be given about the insulating material to be used and whether or not it has been approved by any European Technical Approval issuing body or conforms to any national standard of a member state of the European Economic Area. The name of the body which has approved the installer must also be stated on the building notice.
- If the work includes the provision of a hot-water storage system covered by Schedule 1, G3 (e.g. an unventilated system with a storage capacity of 16 litres of more), details of the system and whether or not the system and its installer are approved.

The local authority is not required to approve or reject the building notice and, indeed, has no power to do so. However, it is entitled to ask for any plans it thinks are necessary to enable it to discharge its building control functions and may specify a time limit for their provision.

The regulations make plain that the building notice and plans shall not be 'treated as having been *deposited* in accordance with the Building Regulations'. In some ways this is an odd provision because the relevant building control sections of many of the local Acts of Parliament mentioned in Chapter 5 – and which provide for special local requirements – are triggered off by the 'deposit' of plans. At first sight, therefore, this would render such requirements inoperative, but presumably it is thought that compliance will be ensured through the requirement that the building notice must contain a statement of the steps to be taken to comply with any local enactment.

Once a building notice has been given, work can be commenced, although there is a requirement (see below) that the local authority be notified at least two days before work commences.

A building notice remains in effect for a period of three years from the date on which it was given to the local authority. If the work has not been commenced within that period, or the change to the building's energy status has not been made, or the material change of use has not been made, the building notice lapses automatically.

3.4   Exemptions from the requirement to give a building notice or deposit full plans

A person who intends to carry out building work consisting only of the work described in the first column of Schedule 2A (see Chapter 5) is not required to give a building notice or deposit full plans if the work is to be carried out by a person described in the corresponding entry in the second column of the Schedule. Additionally, this same exemption exists for work described in Schedule 2B (see Chapter 5).

3.5   Deposit of plans

This is the traditional system of building control by which full plans are deposited with the appropriate local authority in accordance with section 16 of the Building Act 1984, as supplemented by regulation 14. Section 16 imposes a duty on the building control authority to either pass or reject plans deposited for the proposed work.

In *Murphy* v. *Brentwood District Council* (1990) 20 ConLR 1, CA the Court of Appeal held that the duty is imposed on the local authority itself either to pass or reject the deposited plans, and it cannot discharge its duty by delegating performance to outside consultants. If the local authority leaves it to outside consultants to decide whether plans are passed or rejected, the local authority is vicariously responsible if the consultants are negligent, subject to proof of recoverable damage.

However, the Court of Appeal proceeded on the basis that *Anns* v. *London Borough of Merton* [1978] AC 728; [1977] 2 All ER 492, HL; 5 BLR 1 was rightly decided, and in light of the fact that *Anns* was subsequently overruled it is thought that the local authority could only be vicariously liable in these circumstances (if at all) where personal injury was suffered by the occupier or there was damage to other property. Indeed, it is probable that in the current climate of judicial opinion the local authority would be held able to discharge its section 16 duty by reliance on competent outside expertise.

If the plans submitted are not defective the authority has no alternative but to approve them unless, of course, they contravene the linked powers discussed in Chapter 1.

Where the proposed works are subject to the regulations, and it is proposed to deposit full plans, the provisions of section 16 and regulation 14 must be observed. The local authority must give notice of approval or rejection of plans within five weeks unless the period is extended by written agreement. The extended period cannot be later than two months from the deposit of plans, and any extension must be agreed before the five-week period expires. However, the five-week period does not begin to run unless the applicant submits a 'reasonable estimate' of the cost of the works (where applicable) and pays the plan charge at the same time as the plans are deposited.

The approval lapses if the work is not commenced within a period of three years from the date of the deposit of the plans, provided the local authority gives formal notice to this effect. The local authority must pass the plans of any proposed work deposited with them in accordance with the regulations unless the plans are defective, or show that the proposed work would contravene the regulations. The notice of rejection must specify the defects or non-conformity, and the applicant may then ask the Secretary of State to determine the issue. His decision is then final. The Secretary of State may refer questions of law to the High Court and must do so if the High Court so directs.

The local authority may pass plans by stages and, where it does, it must impose conditions as to the deposit of further plans. It may also impose conditions to ensure that the work does not proceed beyond the authorised stage. It has power to approve plans subject to agreed modifications, e.g. where the plans are defective in a minor respect or show a minor contravention. However, it should be noted that local authorities are not obliged to pass plans conditionally or in stages, and the applicant must agree in writing to these procedures.

The 'full plans' required under the deposit method are the same as those required under the building notice procedure, together with such other plans as are necessary to show that the work will comply with the Building Regulations.

Regulation 14 specifies that the plans must be deposited in duplicate; the local authority retains one set of plans and returns the other set to the applicant. They must be accompanied by a statement that they are deposited in accordance with regulation 12(2)(b) of the 2000 Regulations and also a statement as to whether the Regulatory Reform (Fire Safety) Order 2005 applies to the building, or will apply after the completion of the building work. Two additional copies of the plans must be submitted where Part B (Fire Safety) imposes a requirement in relation to the work and both additional plans may be retained by the local authority, although it is not necessary to provide additional copies of the plans where the proposed work relates to the erection, extension or material alteration of a dwelling-house or flat. Where it is proposed to build over a sewer and regulation H4 applies, particulars of the precautions to be taken must be provided.

Work may be commenced as soon as plans have been deposited – although the local authority must be given notice of commencement at least two days before work commences – but it is an unwise practice to commence work before notice of approval is received.

If the applicant wants the authority to issue a completion certificate (see section 3.7) in due course, a request to that effect should accompany the plans.

The advantage of the full deposit of plans method of control is that if the work is carried out exactly in conformity with the plans as passed by the local authority, they cannot take any action in respect of an alleged contravention under section 36 of the Building Act 1984.

The deposit of full plans procedure and the possible alternative solutions are shown diagrammatically in Fig. 3.2.

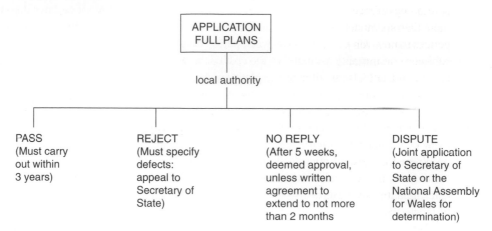

Fig. 3.2   The full plans procedure.

3.5.1   Consultation with the sewerage undertakers

Regulation 14A requires that where full plans have been deposited with the local authority and there are proposals to build over a sewer (as covered by paragraph H4 of Schedule 1 to the Building Regulations 2000), the local authority must consult the sewerage undertaker:

- as soon as practicable after the plans have been deposited; and
- before issuing any completion certificate in relation to the building work.

The local authority must give the sewerage undertaker sufficient plans to show whether the work would comply with the requirements of paragraph H4 and have regard to any views it expresses. The local authority must not pass plans or issue a completion certificate until 15 days have elapsed from the date on which the sewerage undertaker was consulted (unless, of course, the sewerage undertaker has expressed its views to the local authority before the 15 days has expired).

3.6   Notice requirements

Wherever the work is to be supervised by the local authority, in addition to the building notice or deposit of plans, the person undertaking the work must pay an inspection charge and give certain notices to the local authority. The notices must be in writing 'or by such other means as [the local authority] may agree', e.g. by telephone, but most building control authorities provide applicants who deposit plans with pre-printed postcards.

Failure to give the required notices is a criminal offence, punishable on summary conviction by a substantial fine. Under the previous regulations, the Court of Appeal has ruled that failure to deposit plans and give notices under the Building Regulations is not a continuing offence, with the result that magistrates have no power to try informations laid more than six months after the relevant period for compliance. The regulations require the notices to be given by a specified deadline and, once that deadline has passed, the offence has been committed: *Hertsmere Borough Council* v. *Alan Dunn Building Contractors Ltd* (1986) 84 LGR 214; 9-CLD-07–16.

In practice the majority of local authorities do not seek to enforce the penalty but rely on their powers to serve written notice requiring the person concerned within a reasonable time to cut into, lay open or pull down so much of the work as is necessary to enable them to check whether it complies with the regulations.

Where the person carrying out the work is advised in writing by the local authority of contravening works, and has rectified these as required by the local authority, he must give the authority written notice within a reasonable time after the completion of the further work.

'Reasonable time' is not defined in either situation; it is a question of fact in each case. The phrase has been judicially defined as being 'reasonable under ordinary circumstances': *Wright* v. *New Zealand Shipping Co.* [1878] AC 23.

Regulation 15 requires the giving of the following notices to the local authority:

- at least two days' notice of commencement before commencing the work (this does not apply in the case of emergency repair work covered by the regulations, see section 3.3).
- at least one day's notice of:
  - the covering up of any foundation excavation, foundation, damp-proof course, concrete or other material laid over a site; *or*
  - covering up of any drain or private sewer subject to the regulations.

These periods of notice commence after the day on which the notice was served. The local authority must be given notice by 'the person carrying out building work' not more than five days after the completion of:

- the laying of any drain or private-sewer, including any haunching, surrounding or trench backfilling;
- any building work covered by the regulations;
- the completion of any other work.

There is no definition of the term 'person carrying out building work' in the regulations, but in *Blaenau Gwent Borough Council* v. *Khan* (1993) 35 ConLR 65 the High Court held that the owner of a building who authorises a contractor to carry out building works on his behalf fell within the term. The court took the view that those words should not be confined so as to restrict the meaning of the phrase to the person who physically performs the work, 'but includes the owner of the premises on which the works are being performed and who had authorised the work'.

Where a building or part of a building is occupied before completion, the local authority must be given notice at least five days before occupation. This notice is additional to the required notice after completion.

Additionally, a further notice is required to be given to the local authority where a new dwelling is created either by new building work or by a material change of use. The person carrying out the building work must calculate the energy rating of the dwelling by the government-approved Standard Assessment Procedure (SAP) and must give notice of this within five days of completion of the dwelling or at least five days before occupation if this occurs before completion. A notice stating the energy rating for the dwelling must be affixed in a conspicuous place in the dwelling by the person carrying out the building work.

If a person fails to comply with the notice requirements of regulation 15(1)(2)(3), the local authority may require him by notice to cut into, lay open or pull down any work so that they may find out whether the regulations have been complied with. The person concerned must comply with the notice within a *reasonable time*, which, it is suggested, will normally be short.

Where the local authority has served notice specifying that the work contravenes the regulations, on completion of the remedial work, notice must be given to the authority within a reasonable time.

There is now a definition of *day.* It means a period of 24 hours commencing at midnight, i.e. a calendar day, but Saturdays, Sundays and Bank or public holidays are excluded.

3.7   Completion Certificate

The local authority must issue a completion certificate when they are satisfied, after having taken all reasonable steps, that the Schedule 1 requirements are met. Its issue is mandatory in respect of fire safety requirements, i.e. where the building is one to which the Regulatory Reform (Fire Safety) Order 2005 applies, or will apply after the completion of the work (see section 3.3 above), but the completion certificate need only relate to the Part B (Fire Safety) requirements. In other cases the authority need issue the certificate only where they have been requested to do so, but the completion certificate will need to relate to all applicable requirements of the regulations.

The completion certificate is evidence – but not *conclusive* evidence – that the requirements specified in the certificate have been complied with.

It should be noted that the local authority cannot be held liable for a fine if it contravenes this regulation by failing to give a completion certificate (see regulation 22).

4 Private certification

4.1 Introduction

One of the Government's aims in reforming the previous system of building control was to provide an opportunity for self-regulation by the construction industry through a scheme of private certification. This is not a complete substitute for local authority control, because local authorities will always remain responsible for taking any enforcement action which may be necessary. Indeed, in certain closely defined circumstances they may resume their control functions.

The developer is given the option of having the work supervised privately, rather than relying on the local authority control system described in the previous chapter. Essentially, the private certification scheme is based on the proposals set out in a Government White Paper *The Future of Building Control in England and Wales* published by HMSO in February 1981.

The statutory framework of the alternative system is contained in Part II of the Building Act 1984. In broad terms, this provides that the responsibility for ensuring compliance with Building Regulations may, at the option of the person intending to carry out the work, be given to an approved inspector instead of to the local authority. It also enables approved public bodies to supervise their own work. Various supplementary provisions deal with appeals, offences, and the registration of certain information.

The detailed rules and procedures relating to private certification are to be found in the Building (Approved Inspectors, etc.) Regulations 2000, as amended, which also contain prescribed forms that must be used.

It has taken some considerable time for the private certification system to become fully operational even though the first approved inspector, the National House Building Council (NHBC), was approved on 11 November 1985. Their original approval related only to dwellings of not more than four storeys but this was later extended to include residential buildings up to eight storeys and this was further extended in 1998 to include any buildings.

The approval of further corporate bodies as approved inspectors was held up by a number of factors, but was due mainly to the difficulty posed in obtaining the level of insurance cover which was required by the then Department of the Environment. After a period of consultation new proposals for insurance requirements were agreed and these were implemented on 8 July 1996. At the same time the Construction Industry Council (CIC) was designated as the body for approving non-corporate inspectors, although the Secretary of State reserved the right to approve corporate bodies.

Three further corporate bodies were approved by the Secretary of State on 13 January 1997. Others (including a number of non-corporate approved inspectors) have continued to be approved since that date, but until 31 October 2005, NHBC Building Control Services Ltd remained the only body insured to deal with speculative domestic construction (i.e. self-contained houses, flats and maisonettes built for sale to private individuals).

In this context, the DTLR issued insurance guidelines on 23 October 2001 that allowed approved inspectors to carry out their building control function on a range of different dwelling types, except so-called 'non-exempt' dwellings (see below). This definition excluded speculative dwellings constructed by house-building companies for sale to the public. During 2004 the ODPM consulted on proposals for a 'Warranty Link Rule' (see section 4.2 below) whereby Approved Inspectors requiring to perform building control duties on new dwellings built for private sale or renting could carry out this function provided that the dwelling also carried an approved housing warranty from an approved provider. This resulted, on 31 October 2005, in the opening up of the private housing market to Approved Inspectors and ended the NHBC's monopoly of this area of work. Furthermore, from 1 March 1999 the CIC became responsible also, for the approval of corporate approved inspectors.

Further information on corporate and non-corporate approved inspectors may be obtained from The Association of Consultant Approved Inspectors, c/o 14 Berkeley Street, London W1J 8DX or from their website: www.acai.org.uk.

4.2   Insurance requirements

All approved inspectors are required to carry insurance cover in accordance with a scheme approved by the Secretary of State.

Since the NHBC deals mainly with dwellings, the insurance cover required is more extensive than that needed for other types of buildings. In fact, the NHBC has to provide two different types of insurance policy:

- *Ten year no-fault insurance* against breaches of the Building Regulations relating to site preparation and resistance to moisture, structure, fire, drainage and heat-producing appliances. The limit on cover is related to the original cost of the work allowing for inflation during the ten year period up to a maximum of 12% per annum compound.
- *Insurance against the approved inspector's liabilities in negligence for* 15 years from the issue of the Final Certificate for each dwelling. The limit of cover is twice the cost of the building work (unless there is a simultaneous claim made under the no-fault policy), together with cover against claims made for personal injury (which is normally £100,000 a dwelling). This is also proof against inflation up to 12% compound per annum and is subject to a minimum of £1 million a site.

For corporate approved inspectors dealing with work other than dwellings it is necessary to provide professional indemnity insurance renewable on an annual basis. Additionally, a ten year run-off period is required where the approved inspector fails to renew his policy. Indemnity has to extend to any claim reported in writing within ten years from the acceptance of the final certificate, and claims from the owner of the work, his successors in title, or third parties must be met. Cover must be provided for claims against damage (including personal injury, illness, disease or death) resulting from the negligent performance by the Approved Inspector who issued the Initial Notice (see sections 4.9.1 and 4.9.2 below).

Unfortunately, approved inspectors covered under the above insurance scheme could not undertake building control on dwellings. Thus, although all approved inspectors were approved without any direct limitation on their approvals, NHBC Building Control Services Ltd remained the only approved inspector in a position to undertake building control on all types of buildings, including dwellings.

In an attempt to remedy this situation and to create wider consumer choice in relation to building control services for dwellings a consultation document was issued by the Department in July 1999. This resulted in new insurance schemes which were approved during 2002, allowing approved inspectors to undertake building control work on a limited range of dwellings. This situation came about because the Department, in its covering letter to the guidelines listed a range of 'exempt' dwellings (i.e. those which are covered by the guidelines and could be dealt with by Approved Inspectors).

The complete list of exempt dwellings is as follows:

- dwellings in purpose groups 2(a) or (b) (Residential (Institutional) and Residential (Other)) as defined in Appendix D to Approved Document B (2000 edition); also see Chapter 7, section 7.3;
- dwellings in purpose groups l(a), (b) or (c) (flats, maisonettes and dwelling houses):
  (1) which are being developed, for renting tenants, by a local authority, a registered social landlord, a housing association not registered with the Housing Corporation or a local housing company; or
  (2) which are being built, or created by conversion work, by or for a person on their own land and for their own occupation; or
  (3) which are flats, serving purposes that are functionally connected to one or more non-residential uses of the buildings in which they are situated, whether or not access to the flats involves passing through non-residential accommodation; or
  (4) which belong to schools, universities, hospitals or similar establishments and which are used as living accommodation for their staff, pupils or students; or
  (5) which are specifically designed for use as living accommodation for the staff, pupils or students of such establishments and which are subject to planning conditions or legal agreements restricting their use to such living accommodation.

'Non-exempt' dwellings were not covered by the guidelines and consequently approved inspectors were not able to provide a building control service for such dwellings. Non-exempt dwellings included primarily, speculative private development schemes for houses and flats, etc. for sale or private renting.

Nothing in the new guidelines affected the existing status of approved inspectors, and therefore the NHBC remained the only approved inspector permitted to carry out building control services on all types of housing developments including mixed use.

4.3   The Warranty Link Rule

In the summer of 2004 the ODPM consulted on proposals for a Warranty Link Rule (WLR) linking building control by Approved Inspectors on new dwellings built for private sale or renting to the registration of those dwellings under a designated housing warranty scheme, and for criteria and procedure for assessing warranty schemes for designation. Responses to the consultation were generally supportive of these proposals, which resulted in developers of new homes and of mixed-use buildings containing such dwellings having a choice between the local authority building control department and an Approved Inspector authorised by the CIC, as their building control provider.

On 31 October 2005 the ODPM issued a circular letter setting out details of the way in which the WLR would operate. The main points of this are as follows:

- Where Approved Inspectors intend to undertake building control on building work involving the creation by new build or conversion of any new dwellings for sale or private renting, they must check that the new dwellings are registered under a designated warranty scheme (DWS) with a designated warranty provider.
- Currently designated New Home Warranties include:
  - NHBC Buildmark, including Sections 4 and 5 cover;
  - Zurich Standard 10;
  - Premier Guarantee for New Homes.
- If the dwellings are not so registered, the Approved Inspector must not undertake the building control or issue an Initial Notice.
- Where projects are registered with NHBC, Approved Inspectors will be required to ensure the availability to them of Sections 4 and 5 of the Buildmark Warranty. Section 4 cover relates to health and safety hazards due to the home not complying with one or more of the following Building Regulations:
  - structure (Part A);
  - fire safety (Part B);
  - site preparation and resistance to moisture (Part C);
  - hygiene (Part G);
  - drainage and waste disposal (Part H);
  - combustion appliances and fuel storage systems (Part J);
  - protection from falling, collision and impact (Part K); and
  - glazing – safety in relation to impact, opening and cleaning (Part N).

  Section 5 relates to additional cover in years three to ten of the warranty for contaminated land whereby the NHBC agrees to pay the cost of treating or isolating or removing deleterious substances found in the land under the building in a controlled manner in accordance with the requirements of any Statutory Notice, or instead of paying the cost of any work arranging for that work to be carried out at its own expense. The NHBC currently makes a charge of £42 per dwelling unit for this additional Sections 4 and 5 cover.
- Since local authorities cannot refuse to undertake building control, the WLR cannot be applied to them.

An Approved Inspector may only give an Initial Notice in relation to work for which the Approved Instructor is approved, within the terms described in their re-approval from the Construction Industry Council. This is set out in Annex B of the circular letter as follows:

Limitations on dwellings

(1) Approved Inspectors are approved for the purpose of Part II of the Building Act 1984 to act for all types of building work subject only to the limitations set out below.

(2) In relation to building work consisting of the erection of a building containing any dwelling, or in connection with the change of use of a building or part of a building to create a dwelling only when the dwelling is within one of the following sub-paragraphs:

   (a) it is within purpose group 2(a) or (b) (residential (institutional) and residential (other)) in Appendix D to Approved Document B (2000 edition);

   (b) it is within purpose group 1(a), (b) or (c) (flats, maisonettes and dwelling houses) and –

   (i) is being developed for letting by a local authority (including an Arms Length Management Organisation), a registered social landlord, a local housing action trust, a housing association whether or not registered with the Housing Corporation; or

   (ii) is being developed, or created by conversion work, by or for a person on their own land and for their own occupation; or

   (iii) is a flat serving purposes that are functionally connected to one or more non-residential uses of the building in which it is situated, whether or not access to the flat involves passing through non-residential accommodation; or

   (iv) belongs to a school, university, hostel similar establishment and which is to be used as living accommodation for staff, pupils or students; or

   (v) is specifically designed for use as living accommodation for the staff, pupils or students of an establishment described in sub-paragraph (iv) and which is subject to planning conditions or legal agreements restricting its use to such living accommodation; or

   (vi) is subject to planning conditions or legal agreements restricting its use to use as holiday accommodation; or

   (vii) is registered under one of the warranty schemes as listed and detailed below in paragraph 3.

(3) Designated New Home Warranties

   (a) NHBC Buildmark, including Sections 4 and 5 cover;

   (b) Zurich Standard 10;

   (c) Premier Guarantee for New Homes.

In order to assess the current DWS the ODPM used certain criteria which are set out in Annex E of the circular letter.

Operators of designated warranty schemes and insurers need to be satisfied with the credentials of builders admitted to scheme registers. The same also applies to building control bodies. Therefore, operators of a DWS are free to carry out their own assessment of an Approved Inspector as the basis for deciding whether, and on what terms, to make the cover (i.e. including the equivalent of NHBC's Section 4) available for homes where that Approved Inspector was carrying out building control. The circular letter accepts that a fair charge may be made by scheme operators for assessments of this kind, and allows for scheme operators to make adjustments in per unit premiums, depending on which Approved Inspector is involved.

4.4 Approval of inspectors

Section 49 of the Building Act 1984 defines an 'approved inspector' as being a person approved by the Secretary of State or a body designated by him for that purpose. Part II of the Building (Approved Inspectors, etc.) Regulations 2000 (as amended) sets out the detailed arrangements and procedures for the grant and withdrawal of approval.

There are two types of Approved Inspector:

- corporate bodies, such as the NHBC or Bespoke Building Services Ltd (BBS); and
- individuals, not firms, who must be approved by a designated body.

Approval may limit the description of work in relation to which the person or company concerned is an Approved Inspector.

Approval of an inspector is not automatic. Any individual or corporate body wishing to operate as an Approved Inspector must satisfy several criteria. They must hold suitable professional qualifications, have adequate practical experience and carry suitable indemnity insurance. They must also be registered with a body designated for that purpose by the Secretary of State. At present this is the CIC. The CIC established the Construction Industry Council Approved Inspectors Register (CICAIR) to maintain and operate the Approved Inspector Register in accordance with the responsibilities entailed by CIC's appointment as a designated body on 8 July 1996.

The CICAIR route to qualification for Approved Inspectors involves the following four stages of assessment.

4.4.1 Application

Completion of an application form and a detailed knowledge base. The knowledge base addresses six key areas as follows:

- Building Regulations & Statutory Control;
- Law;
- Construction Technology & Materials;
- Fire Studies;
- Foundation & Structural Engineering;
- Building Service & Environmental Engineering.

4.4.2 Pre-qualification verification

On receipt of an application, the CICAIR Registrar will check for gaps in experience or qualification which may disqualify the application or cause delays at further stages.

4.4.3 Admissions panel

On successful completion of pre-qualification verification, the applicant becomes a *Candidate* for approved inspector. The papers are then considered by professional assessors, who decide whether the candidate has demonstrated the necessary experience and knowledge to merit a *Professional Interview.* Assessors include experts nominated from across the range of disciplines by CIC members together with qualified Approved Inspectors.

4.4.4   Professional interview

Candidates granted a professional interview will be seen by an interview panel consisting of three assessors assisted by the CICAIR Registrar. The professional interview is the final stage of assessment and is an opportunity for candidates to demonstrate their knowledge and experience, and expand upon the information provided in their application.

Successful completion of the above assessment stages will result in the candidate being invited to register as an Approved Inspector. The approval will be for a period of five years. Further terms of approval may be sought.

If an applicant/candidate is unsuccessful at any stage in the assessment, he/she will be given reasons and, on application to the Registrar, any advice CICAIR is able to give. Opportunities for appeals against decisions are provided.

The CIC can withdraw its approval – for example, if the inspector has contravened any relevant rules of conduct or shown that he or she is unfitted for the work.

More seriously, where an approved inspector is convicted of an offence under section 57 of the 1984 Act (which deals with false or misleading notices and certificates, etc.) the CIC may withdraw its approval. In this case the convicted person's name would be removed from the list for a period of five years. There is no provision for appeals or reinstatement.

The Secretary of State may himself withdraw his approval of any designated body, thus ensuring that the designated bodies act responsibly in giving approvals. Such action would not necessarily prejudice any approvals given by the designated body but the Secretary of State can, if he so desires, withdraw any approvals given by the designated body.

Provision is made for the Secretary of State to keep lists of designated bodies and inspectors approved by him, and for their supply to local authorities. He must also keep the lists up to date (if there are withdrawals or additions to the list) and must notify local authorities of these changes.

In a similar manner, designated bodies are required to maintain a list of inspectors whom they have approved. There is no express provision for these lists to be open to public inspection, although the designated body is bound to inform the appropriate local authority if it withdraws its approval from any inspector.

In approving any inspector, either the Secretary of State or a designated body may limit the description of work in relation to which the person concerned is approved. Any limitations will be noted in the official lists, as will any date of expiry of approval.

4.5   Approved persons and self-certification by competent persons

The following bodies, together with the Chartered Institution of Building Services Engineers, have been designated to approve private individuals who wish to become approved persons who can certify plans to be deposited with the local authority as complying with the energy conservation requirements:

- The Chartered Institute of Constructors
- The Faculty of Architects and Surveyors
- The Association of Building Engineers
- The Institution of Building Control Officers
- The Institution of Civil Engineers
- The Institution of Structural Engineers
- The Royal Institute of British Architects
- The Royal Institution of Chartered Surveyors

Additionally, the Institution of Civil Engineers and the Institution of Structural Engineers have been designated to approve persons to certify plans as complying with the structural requirements. Approved *Persons* under section 16(9) of the Building Act should not be confused with Approved *Inspectors* under sections 47 to 54, or *Competent Persons* operating under self-certification schemes discussed in Chapter 5. As yet however, no approved persons have been designated in England and Wales.

4.6   Self-certification schemes and the Approved Inspector

Since April 2002 a number of self-certification schemes have existed for relatively minor work where the person carrying out the work (known as a Competent Person under the Regulations) has been able to self certify to the local authority (or other registration body such as CERTASS) that the work complies with the requirements of the relevant Building Regulations (see Chapter 5).

Until the coming into force of the Building and Approved Inspectors (Amendment) Regulations 2006 this process only involved local authorities and did not apply to Approved Inspectors. A new regulation 11A introduced by the aforementioned regulations has applied this process to Approved Inspectors and has, at the same time, increased the number of such Competent Person Schemes. Approved Inspectors are now authorised to accept, as evidence that the requirements of regulations 4 and 7 of the Building Regulations 2000 have been satisfied, a certificate to that effect by a person carrying out the building work. The person must be registered under the relevant Competent Person Scheme and the details of these may be found in the Table in Schedule 2A attached to the Building and Approved Inspectors (Amendment) Regulations 2006 and reproduced in Chapter 5. The certificate must be given to the occupier of the property and the Approved Inspector not more than 30 days after completion of the work. At present the detailed procedures for giving such certificates have still to be worked out with the DCLG.

4.7   Independence of Approved Inspectors

An Approved Inspector cannot supervise work in which he or she has a professional or financial interest, unless it is 'minor work'. In this context, 'minor work' means:

(1) The material alteration or extension of a dwelling-house (not including a flat or a building containing a flat) which has two storeys or less before the work is carried

out and which afterwards has no more than three storeys. A basement is not regarded as a storey.

(2) The provision, extension or material alteration of controlled services or fittings (see section 2.4 above for definition of controlled services or fittings).

(3) Work involving the underpinning of a building.

Independence is not required of an inspector supervising minor work but the limitation on the number of storeys should be noted.

There is a broad definition of what is meant by having a professional or financial interest in the work, the effect of which is to debar the following:

- anyone who is or has been responsible for the design or construction of the work in any capacity, e.g. the architect;
- anyone who or whose nominee is a member, officer or employee of a company or other body which has a professional or financial interest in the work, e.g. a shareholder in a building company;
- anyone who is a partner or employee of someone who has a professional or financial interest in the work.

However, involvement in the work as an approved inspector on a fee basis is not a debarring interest!

4.8   Approval of public bodies

Public bodies, such as nationalised industries, are able to supervise their own building work by following a special procedure, which is detailed in the regulations.

Regulation 21 empowers the Secretary of State (or the National Assembly for Wales) to approve public bodies for this purpose, although, curiously, no criteria have been laid down as to the qualification and experience of the personnel involved. The regulation confers wide discretionary powers on the Secretary of State, but clearly approval will be limited to those bodies which may reasonably be expected to operate responsibly without detailed supervision. The regulations relating to notices, consultation with the fire authority and the sewerage undertaker, plans certificates and final certificates mirror those of Part III of the Building Act 1984 dealing with Approved Inspectors. The grounds on which the local authority may reject a public body's notice, etc. mirror those applicable to the approved inspector system, except that:

- there is no provision for cancellation of a public body's notice;
- there is no requirement that there should be an approved insurance scheme in force; and
- unlike Approved Inspectors, public bodies do not have to be independent of the work they are supervising under the regulations (i.e. they can carry out a building control service on projects which they have designed and are building themselves).

To date, only the Metropolitan Police Authority has been prescribed as a public body under the provisions of the Building Act 1984, section 5.

4.9   Private certification procedure

The procedures which operate when using an approved inspector are illustrated in Fig. 4.1. The left column shows the steps applicable to the person carrying out the building work (i.e. the client) and the right column indicates the duties and responsibilities of the approved inspector. Joint actions are indicated in the centre column.

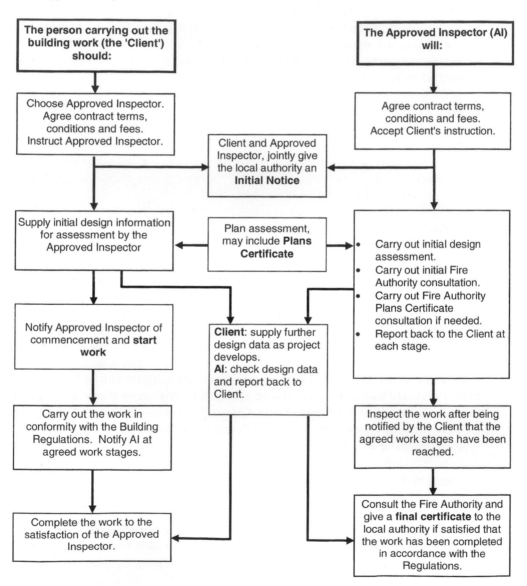

Fig. 4.1   Private certification procedure.

4.9.1   Initial notice

If the developer decides to employ an Approved Inspector, whether an individual or a corporate body, the first formal step in the process is for the applicant and the approved inspector jointly to give to the local authority in whose area the work is to be carried out an initial notice in the prescribed form. The purpose of the initial notice is to make the local authority aware that building work in their area is being properly controlled under the regulations, and to notify them of certain linked powers that they have under the Building Act 1984 and any local Acts of Parliament.

The initial notice must be signed by the Approved Inspector and the 'person intending to carry out the work'. This is the not usually the builder, but the person on whose behalf the work is being carried out (i.e. the client).

The initial notice must be in a prescribed form, and it is a contravention of the regulation to start work before the notice has been accepted. Fig. 4.2 shows a typical initial notice. It must contain:

- A description of the work.
- A declaration that an approved scheme of insurance applies to the work, which must be signed by the insurer.
- A copy of the notice of approval.
- In the case of a new building or extension, a site plan to a scale of not less than 1:1250 showing the boundaries and location of the site.
- Where the work includes the construction of a new drain or private sewer, a statement:
  - (1) as to the approximate location of the proposed connection that is to be made to the sewer; or
  - (2) where no connection is to be made to a sewer, as to the method of discharge of the proposed drain or private sewer. This could include, for example, the location of any septic tank and secondary treatment system, or any wastewater treatment system or cesspool.
- A statement of any local legislation relevant to the work and the steps to be taken to comply with it.
- A statement confirming that the work is not minor work.
- A declaration that the Approved Inspector has no financial or professional interest in the work (other than for minor work).
- An undertaking that the Approved Inspector will consult the Fire and Rescue Authority where obliged to do so.
- An undertaking that the Approved Inspector will consult the sewerage undertaker where obliged to do so (i.e. where the works involve building over or near a sewer).

It is essential that the initial notice is fully completed, because the local authority must reject it unless they are satisfied that the notice contains sufficient information. The local authority has five working days in which to consider the notice and may only reject it on prescribed grounds. These are:

A N APPROVED INSPECTORS LTD
INITIAL NOTICE

This notice is issued pursuant to section 47 of the Building Act 1984 ('the Act') and the Building (Approved Inspectors etc) Regulations 2000 ('the 2000 Regulations') as amended
FORM 1

To:(Note 1)

. . . .

. . . .

. . . .

. . . .

This notice relates to the following work:(Note 2)

Location: . . . .

. . . .

Description of work: . . . .

. . . .

Use of any building: . . . .

The person intending to carry out the work is:(Note 3)

Name: . . . .

Address: . . . .

. . . .

. . . . Postcode: . . . .

The Approved Inspector in relation to the work is: AN Approved Inspectors Ltd, 1 Any Street, Anytown, AB12 3CD

A declaration signed by the insurer for AN Approved Inspectors Ltd accompanies this notice and states that a named scheme of insurance approved by the Secretary of State applies in relation to the work described. AN Approved Inspectors Ltd was approved under section 3(1) of the 2000 regulations and a copy of the notice of approval accompanies this notice.

With this notice are the following documents, which are those relevant to the work described in this notice:(Note 4)

(a)   in the case of the erection or extension of a building, a plan to a scale of not less than 1:1250 showing the boundaries and location of the site, and where the work includes the construction of a new drain or private sewer, an indication within that plan and/or an accompanying statement:
   (i)   as to the approximate location of any proposed connection to be made to a sewer, or
   (ii)   if no connection is to be made to a sewer, as to the proposals for the discharge of the proposed drain or private sewer including the location of any septic tank and associated secondary treatment system, or any wastewater treatment system or any cesspool.(Note 5)
(b)   the following **Local Enactment** is relevant to the work (*delete if none applies*):
   . . . .
   The following steps will be taken to comply with it:
   . . . .
   . . . .

Fig. 4.2   Typical initial notice.

A N Approved Inspectors Ltd declares that:

(i)   The work is/is not* minor work[Note 6]

(ii)   It does not and will not while this notice is in force have any financial or professional interest in the work described.[Note 7]

(iii) It will/will not* be obliged to consult the fire authority by regulation 13 of the 2000 Regulations.[Note 8]

(iv)   It undertakes that before giving a plans certificate in accordance with section 50 of the Act, or a final certificate in accordance with section 51 of the Act, it will consult the fire authority in respect of any work described above.[Note 9]

(v)   It will/will not* be obliged, by regulation 13A of the AI regulations, to consult the sewerage undertaker. (*If the sewerage undertaker has to be consulted the next following declaration must be made.*)

(vi)   It undertakes that before giving a plans certificate in accordance with section 50 of the Act, or a final certificate in accordance with section 51 of the Act, it will consult the sewerage undertaker in respect of any of the work described above.[Note 9]

(vii) It is aware of the obligations laid upon it by Part II of the Act and by regulation 11 of the 2000 Regulations.

(* delete whichever statement is inapplicable)

Signed . . . .   Name . . . .

For and on behalf of A N Approved   Position . . . .
Inspectors Ltd
Approved Inspector   Date . . . .

Signed . . . .   Name . . . .

For and on behalf of the person intending to   Position . . . .
carry out the work[Note 10]
          Date . . . .

Notes

(1)   Insert the name and address of the local authority in whose area the work will be carried out. This should be addressed to the office of the building control department, not the general local authority address. Failure to insert the correct address may lead to delays in acceptance of this notice.

(2)   Location and description of the work and the use of any building to which the work relates.

(3)   Insert the name and address of the person intending to carry out the work (this will usually be the client who commissions the work, not his agents or contractors etc.).

(4)   The local authority may reject this notice only on grounds prescribed by the Secretary of State. These are set out in Schedule 3 to the 2000 Regulations. They include failure to provide the relevant documents set out in this section. The documents listed in this section of the notice relevant to the work described above should therefore be sent with this notice. Any sub-paragraph which does not apply should be deleted.

(5)   The design of any drainage system to which the requirements of Part H of Schedule 1 to the Building Regulations 2000 apply, and which therefore falls to be considered by the A N Approved Inspectors Ltd, will not necessarily be shown in full on the plans accompanying this notice. The plans will indicate the *location* of any connection to be made to a sewer, or the proposals for the *discharge* of any proposed drain or private sewer including the *location* of any septic tank and associated secondary treatment system, or any wastewater system or any cesspool.

(6)   'Minor work' has the meaning given in regulation 10(1) of the 2000 Regulations. If the work is **not** minor work, the next following declaration **must** be made.

(7)   'Professional or financial interest' has the meaning given in regulation 10 of the 2000 Regulations.

(8)   If the inspector is obliged to consult the fire authority, the next following declaration **must** be made.

(9)   Delete this statement if it does not apply.

(10) The person intending to carry out the work will usually be the client who commissions the work, not his agents or contractors etc.

Fig. 4.2   (*Contd*).

- The notice is not in the prescribed form.
- The notice has been served on the wrong local authority.
- The person who signed the notice as an Approved Inspector is not an Approved Inspector.
- The information supplied is deficient because neither the notice nor plans show the location or contain a description of the work (including the use of any building to which the work relates).
- A copy of the Approved Inspector's approval notice has not been supplied.
- Evidence of approved insurance is not supplied.
- The Approved Inspector is obliged to consult the sewerage undertaker before giving a plans certificate or final certificate and the initial notice does not contain an undertaking to do so.
- Where it is intended to erect or extend a building, the local authority considers that a proposed drain must connect to an existing sewer, but no such arrangement is indicated in the initial notice.
- The notice does not contain an undertaking to consult the fire authority (where this is appropriate).
- The notice does not contain a declaration that the Approved Inspector has no financial or professional interest in the work (this does not apply to minor work).
- Local legislative requirements will not be complied with.
- An earlier initial notice has been given for the work, which is still effective. This ground for rejection does not apply if:
  - (1) the earlier notice has ceased to be in force and the local authority have taken no positive steps to supervise the work described in it; or
  - (2) the initial notice is accompanied by an undertaking from the Approved Inspector who gave the earlier notice such that the earlier notice will be cancelled when the new notice is accepted.

If the local authority does not reject the initial notice within five working days (beginning on the day the notice is given to the local authority), it is presumed to have accepted it without imposing requirements. Therefore, an initial notice comes into force when it has been accepted by a local authority (or is deemed to have been accepted by the passing of five days). So long as the initial notice remains in force, the function of enforcing the Building Regulations, which is conferred on a local authority under the Building Act 1984, is not exercisable in relation to the work described in the initial notice.

Generally, the initial notice remains in force during the currency of the works. However, in certain circumstances, it may be cancelled, or cease to have effect after the lapse of certain defined periods of time where there has been a failure to give a final certificate to the local authority. The time periods depend on the circumstances, but the position may be summarised as follows:

- If a final certificate is rejected – four weeks from the date of rejection.
- Where there is a failure to give a final certificate:
  - (1) Eight weeks from the date of occupation for the erection, extension or material alteration of a building. This period of time is reduced to four weeks where the

building is considered to be a 'relevant building' under the terms of the Regulatory Reform (Fire Safety) Order 2005.

(2) Eight weeks after the change of use takes place where the work relates to a material change of use.

The local authority is given power to extend these time periods.

It is possible to give a final certificate for part of a building or extension if it is needed to be occupied before overall completion of the project. In these circumstances the initial notice is not cancelled but remains in force until final completion of the work.

Sometimes it may be necessary to vary work which is the subject of an initial notice (e.g. it may be decided to change the number of units being erected). In such circumstances the person who is carrying out the work and the approved inspector should give an amendment notice to the local authority.

There is a prescribed form for an amendment notice and it must contain the information which is required for an initial notice (see above) plus either:

- a statement to the effect that all plans submitted with the original notice remain unchanged; or
- amended plans are submitted with the amendment notice plus a statement that any plans not included remain unchanged.

The local authority has five working days in which to accept or reject the notice and it may only reject it on prescribed grounds. The procedure is identical to that for acceptance or rejection of an initial notice.

4.9.2   Cancellation of initial notice

In the following cases, the Approved Inspector must cancel the initial notice by issuing to the local authority a cancellation notice in a prescribed form. The grounds on which the initial notice must be cancelled are:

- The Approved Inspector has become or expects to become unable to carry out (or continue to carry out) his functions.
- The Approved Inspector believes that because of the way in which the work is being carried out he cannot adequately perform his functions.
- The Approved Inspector is of the opinion that the requirements of the regulations are being contravened and despite giving notice of contravention to the person carrying out the work that person has not complied with the notice within the three-month period allowed (Approved Inspector Regulations, regulation 19).

It is also possible for the person carrying out the work to cancel the initial notice. This arises if it becomes apparent that the approved inspector is no longer willing or able to carry out his functions (through bankruptcy, death, illness, etc.). This must be done in the prescribed form and must be served on the local authority and (where practicable), on the Approved Inspector.

Alternatively, it is possible for the person carrying out the work to give a new initial notice jointly with a new Approved Inspector, provided that the new notice is accompanied by an undertaking by the original Approved Inspector that he will cancel the earlier notice as soon as the new notice is accepted. Once the initial notice has ceased to have effect, the approved inspector will be unable to give a final certificate and the local authority's powers to enforce the Building Regulations can revive. In this case the local authority becomes responsible for enforcing the regulations and it must be provided on request with plans of the building work so far carried out. Additionally, it may require the person carrying out the work to cut into, lay open or pull down work so that it may ascertain whether any work not covered by a final certificate contravenes the regulations.

If it is intended to continue with partially completed work, the local authority must be given sufficient plans to show that the work can be completed without contravention of the Building Regulations. A fee, which is appropriate to that work, will be payable to it.

Where the work covered by the initial notice has not commenced within three years from the date on which the initial notice was accepted, the local authority may (not must) cancel the initial notice.

4.9.3   Functions of approved inspectors

The fees payable to an approved inspector are a matter for negotiation; there is no pre-scribed scale. The Approved Inspector is under an obligation to '*take such steps (which may include the making of tests of building work and the taking of samples of material) as are reasonable to enable him to be satisfied within the limits of professional care and skill*' that the Building Regulations are complied with.

The approved inspector must be satisfied that:

- The requirements relating to building work, material change of use, and materials and workmanship specified in regulations 4, 4A, 6, 7, 17C and 17D of the 2000 Regulations are complied with.
- Regulation 12 relating to energy ratings for dwellings is complied with. This means that the person carrying out the building work must supply a SAP energy rating to the approved inspector not more than five days after completion of the dwelling (see also section 3.6 above).
- Regulation 12A relating to sound insulation testing is complied with. This means that the person carrying out the building work must supply a copy of the sound insulation testing results (see Chapter 10, Sound insulation) to the approved inspector not more than five days after completion of the work to which the initial notice relates.
- A certificate of pressure testing has been given by a person registered by the British Institute of Non-Destructive Testing in respect of pressure testing for the air tightness of buildings (regulation 12B).
- A notice has been provided confirming that the fixed building services have been commissioned in accordance with a procedure approved by the Secretary of State (regulation 12C).
- A notice has been provided which specifies the target CO_2 emission rate for the building and the calculated CO_2 emission rate for the building as constructed. In these cases an Approved Inspector is also authorised to accept a certificate given by a person who is

registered by FAERO Ltd or BRE Certification Ltd in respect of the CO_2 calculation rates for the building (regulation 12D).

- (In the case of any building) the need to ensure that:
  - ○ work related to the renovation or replacement of a thermal element complies with paragraph L1(a)(i) of Schedule 1 to the Building Regulations 2000;
  - ○ where there is a change in the building's energy status the work complies with paragraph L1(a)(i) of Schedule 1 to the Building Regulations 2000;
  - ○ where a new building is erected, it does not exceed the target CO_2 emission rate for the building;
  - ○ the requirements of regulation 17D regarding consequential improvements to energy performance are met for existing buildings where the total useful floor area exceeds 1000 m².
- Where building work involves the insertion of insulating material into the cavity in a wall after that wall has been built, the Approved Inspector is not required to supervise the insertion of the material, but must state in the final certificate whether or not the material has been inserted.

The second amendment to the 2000 Regulations (SI 2002/2872) places an obligation on the Approved Inspector (in the case of dwelling-houses, flats and rooms used for residential purposes) to check that appropriate pre-completion testing (PCT) has been carried out by the developer to ensure compliance with Requirement E1 (*Protection against sound from other parts of the building and adjoining buildings*). The results of the test must be recorded in a manner approved by the Secretary of State (see paragraph 1.41 of Approved Document E) and must be given to the Approved Inspector not later than five days after completion of the works.

The requirement for PCT has been the subject of much debate, and the response to the consultation on Part E in 2002 showed that some sections of the building industry would prefer to develop robust standard details making testing of sound insulation for a sample of new homes on each site unnecessary. This was the principal reason for the deferment of pre-completion sound insulation testing of dwelling-houses or flats, initially to 1 January 2004, later further deferred to 1 July 2004. The situation now is that in such cases PCT has to be carried out to demonstrate compliance with Requirement E1 except where a robust standard detail is to be used and is shown to be an acceptable alternative. In this respect SI 2004/1466 added a new paragraph to regulation 12A (sound insulation testing) of the 2000 Regulations. This provides that any such detail has to be one approved by a company specially set up for the purpose, namely Robust Details Ltd, the Approved Inspector must be given specified information about that detail before any building work is commenced on site and, at completion, the design detail is in accordance with what was specified in the notification.

An approved inspector is liable for negligence and it is suggested that he *must* inspect the work to ensure compliance, in contrast to local authorities who have discretion as to whether or not to inspect.

In *NHBC Building Control Services Ltd* v. *Sandwell Borough Council* (1990) 50 BLR 101 the Divisional Court emphasised that regulation 11 does not require a system of individual inspection of every detail covered by the substantive requirements of the regulations. In principle, random sampling is sufficient, although in case of dispute it is for the Approved Inspector to show that adopting a system of random or selective sampling is a satisfactory way of discharging his duties.

The approach of the court to this important matter was indicated by Lord Justice Leggatt:

'Any system of inspection that is selective involves consideration not only of the importance of a risk against which the inspection is designed to guard, but of the likelihood of its occurrence. In my judgement the justices' conclusion that the [approved inspector's] system is an inadequate precaution is not one that can properly be based solely upon the fact that the risk was obvious and potentially fatal. That amounts to saying that failure in relation to an individual house to detect the absence of rockwool in the gap between the ceiling and wall of its garage could not have occurred unless the system was inadequate or the inspector had shown want of professional skill and care in operating the system. But the liability imposed is not absolute. The system has been impliedly approved by the Secretary of State. In the light of its experience the [inspector] determines the extent and closeness of the inspections to be conducted in respect of the work of any particular builder. Inherent in any selected system is the risk that some defects may escape detection. Except [for] the fact that the defect … was not spotted, there is no criticism to be made of the system. It follows that the mere fact that an important defect escaped detection in a particular instance cannot … constitute a proper basis for concluding beyond reasonable doubt that there was any failure to undertake the functions of supervision so as to render false the statement that the [inspector] had performed those functions.'

The Approved Inspector may arrange for plans or work to be inspected on his behalf by someone else (although only the Approved Inspector can give plans or final certificates), but delegation does not affect any civil or criminal liability. In particular, the 1984 Act states that:

'…an approved inspector is liable for negligence on the part of a person carrying out an inspection on his behalf in like manner as if it were negligence by a servant of his acting in the course of his employment.'

Consultation with the fire and rescue authority

Where an initial notice or an amendment notice is to be given (or has been given) in relation to the erection, extension, material alteration or change of use of a building which is, or will become a 'relevant building' covered by the terms of the Regulatory Reform (Fire Safety) Order 2005 and the Building Regulations 2000, Schedule 1, Part B (Fire safety) also applies, the Approved Inspector is required, before or as soon as practicable after giving the notice, to consult the fire and rescue authority. He must give them sufficient plans, and/or other information to show that the work described in the notice will comply with the applicable parts of the Building Regulations 2000, Schedule 1, Part B and must have regard to any views they express.

Additionally, before giving a plans certificate or final certificate to the local authority the Approved Inspector must allow the fire and rescue authority 15 working days to comment, and have regard to the views they express. Some local Acts of Parliament also impose extensive fire authority consultation requirements. The Approved Inspectors must undertake any consultation required by local legislation.

Consultations with the sewerage undertaker

Where an initial notice or amendment notice is to be given (or has been given) and it is intended to erect, extend or carry out underpinning works to a building within 3 m of the centreline of a drain, sewer or disposal main to which the Building (Amendment) Regulations 2001, Schedule 1, paragraph H4 applies, the Approved Inspector must consult the sewerage undertaker. The procedures and time periods involved parallel those described for fire authority consultations above.

4.10   Plans certificates

A plans certificate is a certificate issued by an Approved Inspector certifying that the design has been checked and that the plans comply with the 2000 Regulations. Its issue is entirely at the option of the person carrying out the work, and is issued by the Approved Inspector to the local authority and the building owner.

If the approved inspector is asked to issue a plans certificate and declines to do so on the grounds that the plans do not comply with the Building Regulations, the building owner can refer the dispute to the Secretary of State for a determination. A plans certificate can be issued at the same time as the initial notice or at a later stage, provided the work has not been carried out. There are two prescribed forms of plans certificate. There are three preconditions to its issue:

- The Approved Inspector must have inspected the plans specified in the initial notice.
- He must be satisfied that the plans are neither defective nor show any contravention of the regulation requirements.
- He must have complied with any requirements about consultation, etc.

If a plans certificate is issued and accepted and, at a later stage, the initial notice ceases to be effective, the local authority cannot take enforcement action in respect of any work described in the plans certificate if it has been done in accordance with those plans.

The local authority has five working days in which to reject the plans certificate, but may only do so on certain specified grounds:

- The plans certificate is not in the prescribed form.
- It does not describe the work to which it relates.
- It does not specify the plans to which it relates.
- Unless it is combined with an initial notice, that no initial notice is in force.
- The certificate is not signed by the approved inspector who gave the initial notice or that he is no longer an Approved Inspector.

- the required declaration of insurance is not given.
- there is no declaration that the fire authority has been consulted (if appropriate).
- the Approved Inspector was obliged to consult the sewerage undertaker before giving the certificate, but the certificate does not contain a declaration that he has done so.
- there is no declaration of independence (except for minor work).

When combined with an initial notice, the grounds for rejecting an initial notice specified in Schedule 3 (see section 4.9.1) also apply.

Plans certificates may be rescinded by a local authority if the work has not been commenced within three years from the date on which the certificate was accepted.

4.11   Final certificates

The final certificate should be issued by the Approved Inspector when the work is completed, but curiously there are no sanctions against an approved inspector who fails to issue a final certificate. The final certificate need not relate to all the work covered by the initial notice; it can, for example, be given in respect of part of a building which complies with the 2000 Regulations, or one or more of the houses on a development covered by an initial notice. Once given and accepted the initial notice ceases to apply.

It is to be issued, in a prescribed form, where an Approved Inspector is satisfied that any work specified in an initial notice given by him has been completed and certifies that 'the work described … has been completed' and that the inspector has performed the functions assigned to him by the regulations. If the local authority does not reject the final certificate within ten working days it is deemed to have accepted it. A final certificate can only be rejected on limited grounds. These are:

- The certificate is not in the prescribed form.
- It does not describe the work to which it relates.
- No initial notice relating to the work is in force.
- The certificate is not signed by the Approved Inspector who gave the notice or he is no longer an approved inspector.
- The required declaration of insurance is not provided.
- There is no declaration of independence (except for minor works).

Once the final certificate is accepted by a local authority, its powers to take proceedings against a person for contravention of Building Regulations in relation to the work referred to in the final certificate are cancelled.

4.12   Public body's notices and certificates

Part VII of the Building (Approved Inspectors, etc.) Regulations 2000 is concerned with public bodies and, read in conjunction with section 54 of the Building Act 1984, its effect is to enable designated public bodies to self-certify their own work.

Public bodies are approved by the Secretary of State, and the regulations, relating to notices, consultation with the fire authority, plans certificates and final certificates mirror those of Part III dealing with Approved Inspectors. The grounds on which the local authority may reject a public body's notice, etc., mirror those applicable to private certification, except that:

- There is no provision for cancellation of a public body's notice.
- There is no requirement that there should be an approved insurance scheme in force.

4.13   Prescribed forms

Twelve prescribed forms are set out in Schedule 2 of the Building (Approved Inspectors, etc.) Regulations 2000. Regulation 2(2) provides that where the regulations require the use of one of the numbered forms set out in Schedule 2, 'a form substantially to the like effect may be used'. Approved Inspectors, public bodies and local authorities, etc., may therefore have their own forms printed, provided they follow the precedents laid down in Schedule 2.

5 Work under the supervision of a competent person

5.1 Background

In recent years the Government has been keen to extend control over work that can affect the energy efficiency of buildings. In April 2002 new requirements came into force bringing under control replacement windows, rooflights, roof windows and doors, and hot-water vessels. At the same time the Government was conscious of the fact that the new requirements would increase the administrative burden on local authorities and Approved Inspectors for what was, in fact, fairly minor work. As a result it was thought appropriate to introduce new ways of controlling such work by means of 'self-certification' schemes and 'non-notification'. The essence of such systems is that they are self-policing. The difference between them is that self-certification schemes require the person carrying out the work to notify the local authority on completion (a certificate of compliance has also to be given to the client), whereas for non-notification no such requirement exists. One such non-notification system has been in existence for many years whereby a gas appliance could be installed by a person, or an employee of a person, approved in accordance with regulation 3 of the Gas Safety (Installation and Use) Regulations 1998. This non-notification system has now been extended to a range of building operations covered by the regulations and described more fully below.

The first self-certification scheme, the Fenestration Self-Assessment Scheme (FENSA) came into force on 1 April 2002. Subsequent changes to the regulations have resulted in a wide range of self-certification schemes being licensed by the Government and this trend is continuing. Such schemes are now more commonly referred to as 'competent person schemes' since members are deemed to possess the necessary competences to ensure that their work complies with the relevant regulations. It should be noted that the term 'person' can refer either to an individual or to a company employing individuals. All Competent Person Schemes register the company as being the legal entity and the body responsible for issuing data to local authorities and to clients; however some schemes (such as HETAS, OFTEC, NAPIT and BESCA) also require individuals working for the approved company to be separately assessed under the scheme.

In the case of both self-certification schemes and non-notification, the authorisation for the procedures is given in regulation 12 – Giving of a building notice or deposit of plans. Sub-paragraph (5) states that:

'A person who intends to carry out building work is not required to give a building notice or deposit full plans where the work consists only of work –
(a) described in column 1 of the Table in Schedule 2A if the work is to be carried out by a person described in the corresponding entry in column 2 of that Table, and

paragraphs 1 and 2 of that Schedule have effect for the purposes of the descriptions in the Table; or

(b)   described in Schedule 2B.'

Schedule 2A is reproduced as Table 5.1. Schedule 2B may be found in section 5.3.

Table 5.1   Schedule 2A – Self-certification schemes and exemptions from requirement to give building notice or deposit full plans.

Column 1	Column 2
Type of work	**Person carrying out work**
1. Installation of a heat-producing gas appliance.	A person, or an employee of a person, who is a member of a class of persons approved in accordance with regulation 3 of the Gas Safety (Installation and Use) Regulations 1998.
2. Installation of heating or hot water service system connected to a heat-producing gas appliance, or associated controls.	A person registered by CORGI Services Limited in respect of that type of work.
3. Installation of – (a) an oil-fired combustion appliance which has a rated heat output of 100 kilowatts or less and which is installed in a building with no more than three storeys (excluding any basement) or in a dwelling; (b) oil storage tanks and the pipes connecting them to combustion appliances; or (c) heating and hot water service systems connected to an oil-fired combustion appliance.	An individual registered by Oil Firing Technical Association Limited, NAPIT Certification Limited or Building Engineering Services Competence Accreditation Limited in respect of that type of work.
4. Installation of – (a) a solid fuel burning combustion appliance which has a rated heat output of 50 kilowatts or less which is installed in a building with no more than three storeys (excluding any basement); or (b) heating and hot water service systems connected to a solid fuel burning combustion appliance.	A person registered by HETAS Limited, NAPIT Certification Limited or Building Engineering Services Competence Accreditation Limited in respect of that type of work.
5. Installation of a heating or hot water service system, or associated controls, in a dwelling.	A person registered by Building Engineering Services Competence Accreditation Limited in respect of that type of work.
6. Installation of a heating, hot water service, mechanical ventilation or air conditioning system, or associated controls, in a building other than a dwelling.	A person registered by Building Engineering Services Competence Accreditation Limited in respect of that type of work.
7. Installation of an air conditioning or ventilation system in an existing dwelling, which does not involve work on systems shared with other dwellings.	A person registered by CORGI Services Limited or NAPIT Certification Limited in respect of that type of work.

Table 5.1   (*Cont*).

Column 1	Column 2
Type of work	**Person carrying out work**
8. Installation of a commercial kitchen ventilation system which does not involve work on systems shared with parts of the building occupied separately.	A person registered by CORGI Services Limited in respect of that type of work.
9. Installation of a lighting system or electric heating system, or associated electrical controls.	A person registered by The Electrical Contractors Association Limited in respect of that type of work.
10. Installation of fixed low or extra-low voltage electrical installations.	A person registered by BRE Certification Limited, British Standards Institution, ELECSA Limited, NICEIC Group Limited or NAPIT Certification Limited in respect of that type of work.
11. Installation of fixed low or extra-low voltage electrical installations as a necessary adjunct to or arising out of other work being carried out by the registered person.	A person registered by CORGI Services Limited, ELECSA Limited, NAPIT Certification Limited, NICEIC Group Limited or Oil Firing Technical Association Limited in respect of that type of electrical work.
12. Installation, as a replacement, of a window, rooflight, roof window or door (being a door which together with its frame has more than 50% of its internal face area glazed) in an existing building.	A person registered under the Fenestration Self-Assessment Scheme by Fensa Limited, or by CERTASS Limited or the British Standards Institution in respect of that type of work.
13. Installation of a sanitary convenience, washing facility or bathroom in a dwelling, which does not involve work on shared or underground drainage.	A person registered by CORGI Services Limited or NAPIT Certification Limited in respect of that type of work.

14. (1) Subject to paragraph (2), any building work, other than the provision of a masonry chimney, which is necessary to ensure that any appliance, service or fitting which is installed and which is described in the preceding entries in column 1 above, complies with the applicable requirements contained in Schedule 1.
(2) Paragraph (1) does not apply to –
(a) building work which is necessary to ensure that a heat-producing gas appliance complies with the applicable requirements contained in Schedule 1 unless the appliance –
  (i) has a rated heat output of 100 kilowatts or less; and
  (ii) is installed in a building with no more than three storeys (excluding any basement), or in a dwelling; or
(b) the provision of a masonry chimney.
	The person who installs the appliance, service or fitting to which the building work relates and who is described in the corresponding entry in column 2 above.

5.2   Self-assessment

5.2.1   Self-certification schemes (Competent Person Schemes)

Regulation 16A of the Building Regulations 2000 (as amended) allows local authorities to accept certificates given by certain persons, as evidence that the requirements of regulations 4 and 7 have been met in relation to work described in Schedule 2A and discussed below. Such persons must be suitably qualified and experienced and/or registered persons under a recognised scheme as described in Schedule 2A. Whether the local authority is prepared to accept the certificate is optional. However, if the authority does not accept it and decides to inspect the work, it may incur certain obligations and liabilities attendant upon exercising its building control function.

Under regulation 16A local authorities must either be informed within 30 days of completion of the work by the person carrying it out, or must be given the certificate referred to above. A copy of the certificate must also be given to the occupier of the building where the work is carried out.

The need to notify completion or give a certificate to the local authority and the occupier does not apply to building work described in Schedule 2B (see 5.3 below).

Interestingly, a parallel notification procedure exists where the person carrying out the building work is using an Approved Inspector instead of a local authority for building control. This is discussed in Chapter 4, section 4.6.

In general, self-certification schemes can be divided into:

- Part L Schemes featuring work to:
  - replacement windows, doors, rooflights and roof windows;
  - heating and hot water service systems and associated controls;
  - mechanical ventilation or air conditioning systems and associated controls;
  - lighting systems.
- Part J Schemes where the work is concerned with the installation (as well as replacement of) heat-producing appliances (gas, oil and solid fuel).
- Part P Schemes covering work to electrical installations.
- Part G and Part H Schemes dealing with the installation of sanitary appliances (not involving work on underground drainage).
- Building work in connection with work carried out under the above-mentioned schemes (but not the provision of a masonry chimney).

5.2.2   Schemes covering replacement windows, rooflights, roof windows and doors

To assist the effective implementation of the ongoing amendments to Part L (Conservation of Fuel and Power) three schemes for the self-certification of replacement glazing now exist:

- the CERTASS (Certification and Self Assessment) Scheme;
- the British Standards Institution (BSI) Scheme; and
- the Fenestration Self-Assessment (FENSA) Scheme.

All schemes have similar characteristics, although the actual modes of operation and the fees charged vary. Details of the CERTASS Scheme are given below.

5.2.3   Operation of the CERTASS Scheme

In order to take part in the scheme, persons (companies or individuals) must be registered with CERTASS Ltd. Currently the CERTASS Scheme permits registration to installers of replacement windows, rooflights, roof windows and doors in any building, subject to the work not having structural implications for a building.

Following completion of the replacement window installation in a building, installers who are self-certifying their work either key the necessary information to the CERTASS database or the information can be sent by other means such as fax or post. The system then consolidates all the relevant installation information and forwards this to the appropriate local authority within five days of receipt from the installer. Paper certificates are also prepared by CERTASS for forwarding to the appropriate householder. This is important, since evidence of compliance with the regulations will need to be supplied to any future purchaser of the house (this is likely to be a requirement of the Home Information Packs which will become mandatory in June 2007).

In order to operate the scheme CERTASS maintains a database of all its members and all replacement window installations that have been certified by CERTASS registered companies. The database system is used to:

- receive authenticated transaction data from installation companies;
- forward installation information to local authorities;
- receive membership data from CERTASS registered companies;
- receive and provide inspection data from/to CERTASS inspectors;
- prepare and despatch paper certificates to householders;
- permit secure interrogation for authorised persons.

Installation information is forwarded to local authorities electronically by means of a universally accepted XML Schema (all self-certification schemes have adopted the same electronic format). Additionally, CERTASS can allow access to the database so that local authorities can obtain information about individual company registrations and particular properties within their area, subject to security authentication procedures that confirm the identity of the enquirer.

When a company applies to CERTASS for registration, a number of checks are carried out to assess the creditworthiness and technical capabilities of the applicant. Additionally, independent checks of each installation company's work are carried out on an activity sampling basis by CERTASS inspectors in order to ensure that the regulation requirements are being met. The Government has laid down that 1% of a company's installations must be checked each year, with a minimum of two installations being inspected.

Where replacement windows are installed by companies or individuals not registered under a self-certification scheme, the work will be subject to the normal building control procedures whereby the person carrying out the work will need to give a building notice or submit full plans to a local authority, or give an initial notice jointly with an Approved Inspector.

5.2.4   Combustion appliances and oil storage facilities

For the following types of installation the person carrying out the work is not required to give a building notice or deposit full plans provided that they are suitably qualified and experienced and/or registered under a recognised scheme as mentioned below and shown in Schedule 2A, and they notify the local authority of completion or give a certificate as mentioned in section 5.2.1 above.

Installation of a heat-producing gas appliance

The person (or an employee of that person) carrying out the work must be approved in accordance with regulation 3 of the Gas Safety (Installation and Use) Regulations 1998. The regulations are concerned with controlling the risks which arise when using gas from either mains pipes or gas storage vessels. The work must be carried out by a competent person who is a member (or is employed by a member) of the Council for Registered Gas Installers (CORGI).

Under the CORGI Competent Person Scheme, from 1 April 2005, CORGI registered gas installers can satisfy the relevant areas of Part J (Gas appliances) and Part L (Heating services) under the Gas Registration Scheme, through Gas Work Notification.

Applicants from gas installer businesses need to be competent in order to carry out gas work and will need to be in possession of certificates of competence in the areas of gas work they intend to undertake before they can become CORGI registered. Gas operatives are required to demonstrate gas safety competence by successfully completing nationally agreed assessments, carried out at an approved assessment centre. CORGI registration is valid for a year from April through to March.

Installation of oil-fired combustion appliances and oil storage tanks

The self-certification option applies only to:

- oil-fired combustion appliances with a rated heat output of 100 kW or less, where the appliance is installed in a building of three storeys or less (not including basements) or in a dwelling;
- oil storage tanks (including the pipes which connect the tank to the combustion appliance).

The installation must be carried out by a person registered for that type of work by:

- Oil Firing Technical Association Ltd (OFTEC);
- NAPIT Certification Ltd; or
- Building Engineering Services Competence Accreditation Ltd (BESCA).

Each scheme has its own competency requirements; however, for OFTEC, the operatives concerned must be registered as:

- installation technicians and domestic/light commercial commissioning and service technicians for pressure jet and for vaporising burners; and
- commercial industrial commissioning and service technicians; and
- oil storage tank installation technicians.

Although the self-certification option is limited to registered individuals, others who are not registered often assist installation. In these cases, supervision by the registered person is the same as installation by that person.

Installation of solid-fuel burning combustion appliances

The self-certification option applies only to solid-fuel burning combustion appliances:

- with a rated heat output of 50 kW or less;
- installed in a building of three storeys or less (not including basements).

The installation must be carried out by a person registered for that type of work by:

- HETAS Ltd,
- NAPIT Certification Ltd, or
- Building Engineering Services Competence Accreditation Ltd (BESCA).

Although the non-notification option is limited to registered individuals, others who are not registered often assist installation. In these cases, supervision by the registered person is the same as installation by that person.

An example of one Competent Person Scheme in the area of solid-fuel burning appliances is that operated by HETAS Ltd. In conjunction with the Solid Fuel Association, HETAS Ltd has introduced a Registration Scheme for Engineers and Companies working in the domestic solid fuel market.

The objective of the scheme is to ensure that any consumer wishing to have a new appliance or their existing appliance serviced can contact a contractor employing competent engineers.

Registered Companies must employ Registered Engineers and must have the appropriate level of insurance cover in respect of third party liability.

Individual engineers have to be registered with HETAS Ltd for which they are required to declare their Technical Qualifications, training and work experience in connection with domestic solid-fuel appliances. They must be employed by a Registered Company and their registration is only valid when working for that company.

Companies can seek Registration as being competent to carry out work in the following areas:

- full systems – the installation of solid-fuel fired central heating systems;
- dry systems – the installation of open fires and solid-fuel fired appliances not connected to a hot-water system;
- service and maintenance of domestic solid-fuel fired appliances;

- chimney relining: (a) twin wall flexible, (b) rigid sectional – metal and pumice, (c) other systems meeting requirements of Approved Document J;
- specialist company: e.g. technical consultant, installer of mechanically fed or other specialist appliances;
- prefabricated chimney systems.

Installation and commissioning

Whereas for gas and solid fuel the installer also commissions or brings into use the appliance, this is not always the case for oil-fired combustion appliances. For this reason individuals are required to register separately as installers and commissioning technicians. They may be the same person, or two different individuals; however, both aspects of the work, initial installation and commissioning, must have been completed before the installation can be regarded as complete.

5.2.5   Heating and hot-water service systems, mechanical ventilation or air conditioning systems

The following systems can be installed without giving a building notice or depositing full plans if carried out by the persons indicated:

- The installation of a heating or hot-water service system connected to:
  - a heat-producing gas appliance (including associated controls) – a person registered by CORGI Services Ltd for that type of work;
  - an oil-fired combustion appliance – a person registered by Oil Firing Technical Association Ltd (OFTEC), NAPIT Certification Ltd or Building Engineering Services Competence Accreditation Ltd (BESCA), for that type of work;
  - a solid-fuel burning combustion appliance – a person registered by HETAS Ltd, NAPIT Certification Ltd or Building Engineering Services Competence Accreditation Ltd (BESCA), for that type of work.
- The installation of a heating or hot-water service system, or associated controls, in a dwelling – a person registered by Building Engineering Services Competence Accreditation Ltd (BESCA), for that type of work.
- The installation of a heating, hot-water service, mechanical ventilation or air conditioning system, or associated controls, in a building other than a dwelling – a person registered by Building Engineering Services Competence Accreditation Ltd (BESCA), for that type of work.
- The installation of an air conditioning or ventilation system in an existing dwelling, which does not involve work on systems shared with other dwellings – a person registered by CORGI Services Ltd or NAPIT Certification Ltd for that type of work.
- The installation of a commercial kitchen ventilation system which does not involve work on systems shared with parts of the building occupied separately – a person registered by CORGI Services Ltd for that type of work.

5.2.6   Electric lighting and heating systems and other electrical work

The following types of electrical work can be installed without giving a building notice or depositing full plans if carried out by the persons indicated:

- The installation of a lighting system or electric heating system, or associated electrical controls – a person registered by the Electrical Contractors Association Ltd for that type of work.
- The installation of fixed low or extra-low voltage electrical installations – a person registered by BRE Certification Ltd, British Standards Institution, ELECSA Ltd, NICEIC Group Ltd or NAPIT Certification Ltd for that type of work.
- The installation of fixed low or extra-low voltage electrical installations as a necessary adjunct to or arising out of other work being carried out by the registered person (such as electrical work needed for the installation of boiler controls) – a person registered by CORGI Services Ltd, ELECSA Ltd, NAPIT Certification Ltd, NICEIC Group Ltd or Oil Firing Technical Association Ltd for that type of electrical work.

An example of one Competent Person Scheme in the area of electrical work is the NICEIC Domestic Installer Scheme. The NICEIC launched its new Domestic Installer Scheme in January 2005 to satisfy the requirements for Part P. The Scheme enables competent installers to register for Full Scope or Defined Competence for domestic electrical installation work. The scheme covers design, installation and inspection and testing of electrical installation work intended to operate at low or extra low voltage.

Full Scope – covers the full range of domestic electrical installation work and is intended for businesses that carry out electrical installation work as their main activity.
Defined Scope – limited to defined electrical installation work in connection with, or ancillary to, some other non-electrical activity.

There are several key requirements which must be satisfied before a contractor can become registered onto the Domestic Installer Scheme. They must:

- have £2 million public liability insurance;
- compile certification as detailed in the models in BS 7671: 2001 (incorporating Amendments No 1: 2002 and No 2: 2004). *Requirements for Electrical Installations (IEE Wiring Regulations 16th Edition)* published by The Institution of Electrical Engineers (see Chapter 19);
- nominate two sites where competence can be demonstrated;
- provide details of two sites where domestic electrical installation work has been carried out by their company in the last six months;
- nominate a Qualified Supervisor/responsible person, who must:
  - ○ have 16th Edition IEE Wiring Regulations (BS 7671) qualifications;
  - ○ have an adequate knowledge and understanding of inspection and testing verification and certification procedures for the range of electrical work conducted;
  - ○ is a full-time principal or employee of the company.

The Government requires all registered installers to offer an insurance-backed warranty to their customers and, as a result, NICEIC contractors provide a warranty on all electrical installation work carried out to their customers. The warranty lasts for six years from the date of completion of the work.

NICEIC lists the following benefits of using a contractor registered with the NICEIC Domestic Installer Scheme:

- All NICEIC Contractors are covered under the NICEIC's complaints procedure.
- A sample of the contractors work is assessed annually to ensure that it continues to meet the requirements of Part P.
- NICEIC is the brand leader in the market place.
- Customers can be assured that the business is reputable, reliable and has satisfied the criteria in order to be deemed competent to self-certify their work under Part P of the Building Regulations.

Further information about the NICEIC Domestic Installer Scheme can be obtained by visiting www.niceic.com or by contacting NICEIC on 01582 531000.

5.2.7   Sanitary conveniences, washing facilities and bathrooms in dwellings

Provided that the installation does not involve work on shared or underground drainage, a person registered by CORGI Services Ltd or NAPIT Certification Ltd may install a sanitary convenience, washing facility or bathroom in a dwelling without giving a building notice or depositing full plans.

5.2.8   Other building work

The self-certification option extends to any building work which is necessary to ensure that any appliance, service or fitting which is installed and which is described above, complies with the applicable requirements contained in Schedule 1 of the Building Regulations 2000. The person carrying out the building work would be the person who installed the appliance, service or fitting referred to in the sections above.

For example, where a combustion appliance was being installed, this could mean the provision of flue pipes, and work required to ensure compliance with Part L. In the case of the installation of cast in situ flue linings, the material and installation procedures would need to be independently certified. Even if he does not actually carry out the building work himself, the installer of the combustion appliance must supervise the work and, through that supervision, take responsibility for it.

The building work referred to in this section, which is necessary to ensure that a gas combustion appliance complies with the Schedule 1 of the Building Regulations, can only be carried out by the person referred to in section 5.2.4 if the appliance has a net rated heat input of 100 kW or less and is installed in a building of three storeys or less (not counting basements).

The self-certification option does not extend, however, to the provision of a masonry chimney.

5.3   Non-notification

5.3.1   Installation of fixed low or extra-low voltage electrical installations

Part P of the Building Regulations 2000 (as amended) applies to electrical installations that are intended to operate at low or extra-low voltage (see Chapters 2 and 19 for definitions). Such installations are generally associated with dwellings (including common parts in flats, sheds and greenhouses, and outdoor lighting, etc.) and Part P is restricted to such uses. In general, the self-certification option is available as described above for installations covered by Part P. However, this is further extended to a 'non-notification' system of control for the following types of work described in Schedule 2B to the Building Regulations 2000 (as amended):

(1) Work consisting of:
   (a) replacing any fixed electrical equipment which does not include the provision of –
      (i)   any new fixed cabling; or
      (ii) a consumer unit;
   (b) replacing a damaged cable for a single circuit only;
   (c) re-fixing or replacing enclosures or existing installation components, where the circuit protective measures are unaffected;
   (d) providing mechanical protection to an existing fixed installation, where the circuit protective measures and current carrying capacity of the conductors are unaffected by the increased thermal insulation;
   (e) installing or upgrading main or supplementary equipotential bonding;
   (f) in heating or cooling systems –
      (i)   replacing control devices that utilise existing fixed control wiring or pneumatic pipes;
      (ii) replacing a distribution system output device;
      (iii) providing a valve or a pump;
      (iv) providing a damper or a fan;
   (g) in hot-water service systems, providing a valve or a pump;
   (h) replacing an external door (where the door together with its frame has not more than 50% of its internal face area glazed);
   (i) in existing buildings other than dwellings, providing fixed internal lighting where no more than 100 m^2 of the floor area of the building is to be served by the lighting.
(2) Work which:
   (a) is not in a kitchen, or a special location;
   (b) does not involve work on a special installation; and
   (c) consists of –
      (i)   adding light fittings and switches to an existing circuit; or
      (ii) adding socket outlets and fused spurs to an existing ring or radial circuit.

(3) Work on.
   (a) telephone wiring or extra-low voltage wiring for the purposes of communications, information technology, signalling, control and similar purposes, where the wiring is not in a special location;
   (b) equipment associated with the wiring referred to in sub-paragraph (a);
   (c) pre-fabricated equipment sets and associated flexible leads with integral plug and socket connections.
(4) For the purposes of Schedule 2B:
   • 'kitchen' means a room or part of a room which contains a sink and food preparation facilities;
   • 'special installation' means an electric floor or ceiling heating system, an outdoor lighting or electric power installation, an electricity generator, or an extra-low voltage lighting system which is not a pre-assembled lighting set bearing the CE marking referred to in regulation 9 of the Electrical Equipment (Safety) Regulations 1994; and
   • 'special location' means a location within the limits of the relevant zones specified for a bath, a shower, a swimming or paddling pool or a hot air sauna in the *Wiring Regulations*, 16th edition, published by the Institution of Electrical Engineers and the British Standards Institution as BS 7671: 2001 and incorporating amendments 1 and 2.

Therefore, all work to electrical installations in, or associated with, dwellings must now comply with Part P of the Building Regulations 2000 (as amended).

II | Technical

6 Structural stability (Part A)

6.1 Introduction

Part A of Schedule 1 to the 2000 Regulations is concerned with the strength, stability and resistance to deformation of the building and its parts. The loads to be allowed for in the design calculations are specified, and recommendations as to construction are given in Approved Document A.

In line with the Government's intention to remove from the regulations those matters which are not directly concerned with public health and safety or the conservation of fuel and power, the previous requirements regarding the ability of a building structure or foundation to resist *damage* due to settlement, etc. have been omitted.

Additionally, control of deflection or deformation of the building structure under normal loading conditions is only relevant if it would impair the stability of another building.

It is conceivable, therefore, that a building constructed under the regulations could be safe and stable but could settle and deflect to such an extent that it would be unusable. In that event, of course, the owner would probably have redress against the designer and/or builder under the general law by way of an action for damages. Insurance cover might be somewhat hard to obtain for such a building.

The section of the regulations dealing with disproportionate collapse was simplified by the revocation (in the 1994 amendment) of paragraph A4 of Schedule 1. This was concerned with maintaining structural stability in public buildings and shops in the event of roof failures, where roof spans exceeded nine metres. The original requirements concerning the failure of long span roof structures were introduced in response to a number of roof collapses which occurred in the 1970s and which led to the banning of high alumina cement in structural work. As this ban still exists, the problem of such failures seems largely to have been solved without additional regulatory safeguard.

Structural safety depends on the successful interrelationship between design and completed construction, particularly with regard to:

- the design – involving identification of the hazards to which the structure is likely to be subjected and assessment of the likely risks, and the selection of relevant critical situations reflecting the conditions that can reasonably be foreseen during future use;
- degree of loading – dead, imposed and wind loads should be assessed in accordance with the current Codes of Practice referred to in Section 1 of AD 1/2 and referred to in section 6.4 below;
- the properties of the materials chosen;
- the detailed design and structural assembly;

- safety factors; and
- standards of workmanship.

It is essential that the numeric values of the safety factors which are used are derived from a consideration of the above factors, since a change in any one of these could disturb the safety of the structure as a whole.

Additionally, loads used in calculations should take account of possible dynamic, concentrated and peak loads which may arise. For example, grandstands (and other similar structures erected in places of public assembly) may need to carry the synchronous or rhythmical movement caused by large numbers of people. These factors should be taken into account in the design of the structure so that it will not be impaired or cause alarm to the people using it. For interim guidance on the design of grandstands see *Dynamic performance requirements for permanent grandstands subject to crowd action, Interim Guidance on assessment and design* published by the Institution of Structural Engineers, November 2001. Additionally, the Institution of Structural Engineers has published (in June 2002) supplementary advice on the dynamic testing of grandstands and seating decking.

Approved Document A is arranged in five sections and gives guidance that may be adopted, if relevant, at the discretion of the designer. Where precise guidance is not given, due regard should be paid to the factors listed above.

- Section 1 lists various codes and standards for structural design and construction and is relevant to all types of buildings. Further information sources are included for the first time regarding landslip and structural appraisal of existing buildings subject to a change of use.
- Section 2 allows the sizes of certain structural members to be assessed in certain small residential buildings and other small buildings of traditional construction.
- Section 3 gives guidance on the fixing and support of external wall cladding.
- Section 4 makes it clear that certain roof re-covering operations may constitute a material alteration to the building and gives guidance to that effect.
- Section 5 gives guidance on reducing the sensitivity of a building to disproportionate collapse in the event of an accident.

6.2   Loading

Paragraph A1 of Schedule 1 to the Building Regulations 2000 requires buildings to be constructed so that all dead, imposed and wind loads are sustained and transmitted to the ground:

- safely; and,
- without causing such settlement of the ground, or such deflection or deformation of the building, as will impair the stability of any part of another building.

The imposed and wind loads referred to above are those to which the building is likely to be subjected in the normal course of its use and for the purpose for which it is intended.

6.3   Ground movement

In addition to the provisions of paragraph Al above regarding loading, there are requirements in paragraph A2 of Schedule 1 that the building shall be so constructed that movement of the ground caused by:

- swelling, shrinking or freezing of the subsoil; or
- landslip or subsidence (other than subsidence arising from shrinkage),

will not impair the stability of any part of the building.

It should be noted that the requirement as to landslip and subsidence applies to the extent
that the risk can be reasonably foreseen.

6.4   Guidance on structural design in buildings of all types

6.4.1   Foundations and ground movement

Foundations should be designed in accordance with BS 8002: 1994 *Code of practice
for earth retaining structures* or BS 8004: 1986 *Code of practice for foundations* as
appropriate.

Paragraph A2 of Schedule 1 to the 2000 Regulations requires that ground movement caused by landslip or subsidence must not impair the stability of any part of the
building.

When developing a site it is always essential that a site investigation is carried out.
This is discussed fully in Chapter 8 (sections 8.5.2 and 8.5.3). The desk study and site
reconnaissance and walkover survey should reveal conditions of ground instability (e.g.
arising from landslides, disused mines or unstable strata), which, if ignored, can very
seriously affect the safety of a building and its environs. The design of the building and
its foundations should take such conditions into account. Guidance on the broad planning and technical issues relating to development on unstable land can be found in DOE
Planning Policy Guidance Note 14 *Development on unstable land*, which may be obtained
from HM Stationery Office.

For guidance on determining the scale and nature of problems arising from mining
instability, natural underground cavities and adverse foundation conditions, the DOE has
sponsored a series of reviews. These contain databases of both subsidence incidents and
subsidence potential and may be obtained from the following licence holders:

- British Geological Survey, Sir Kingsley Dunham Centre, Keyworth, Nottingham
  NG12 5GG;
- Landmark, 7 Abbey Court, Eagle Way, Exeter, Devon EX2 7HY;
- Peter Brett Associates, 16 Westcote Road, Reading, Berkshire RG20 2DE;
- Catalytic Data Ltd, The Spinney, 19 Woodlands Road, Bickley, Kent BR1 2AD.

The reports from these reviews, which include 1:250,000 scale maps showing the distribution of the physical constraints are available from the following organisations:

- Arup Geotechnics, 1991. *Review of mining instability in Great Britain*. Obtainable from Arup Geotechnics, Bede House, All Saints, Newcastle-upon-Tyne NE1 2EB;
- Applied Geology Ltd, 1994. *Review of instability due to natural underground cavities in Great Britain*. Obtainable from Kennedy & Donkin Ltd, 14 Calthorpe Road, Edgbaston, Birmingham Bl5 1TH;
- Wimpey Environmental Ltd, and National House Building Council, 1995. *Foundation conditions in Great Britain, a guide for planners and developers*. Obtainable from ESNR International Ltd, 16 Frogmore Road, Hemel Hempstead, Hertfordshire HP3 9RW.

6.4.2   Loading

Dead and imposed loads may be assessed by reference to BS 6399 *Loading for buildings*: Part 1: 1996 *Code of practice for dead and imposed loads*. Similarly, wind loads may be assessed by using the same code (Part 2: 1997) or by using BRE Digest 436 Parts 1, 2 and 3. Imposed roof loads are also covered in BS 6399 (see Part 3: 1988 *Code of practice for imposed roof loads*).

6.4.3   Structure above foundations

Structural work of reinforced, prestressed or plain concrete should comply with:

- BS 8110 *Structural use of concrete:*
  - (a) Part 1: 1997 *Code of practice for design and construction;*
  - (b) Part 2: 1985 *Code of practice for special circumstances;* and
  - (c) Part 3: 1995 *Design charts for singly reinforced beams, doubly reinforced beams and rectangular columns.*
- BS 8103 *Structural design of low rise buildings*: Part 4: 1995 *Code of practice for suspended concrete floors for housing.*

Structural work of aluminium should comply with:

- BS 8118 *Structural use of aluminium*:
  - (a) Part 1: 1991 *Code of practice for design*; and
  - (b) Part 2: 1991 *Specification for materials, workmanship and protection.*

Structural work of masonry should comply with:

- BS 5628 *Code of practice for use of masonry*:
  - (a) Part 1: 1992 *Structural use of unreinforced masonry*;
  - (b) Part 2: 2000 *Structural use of reinforced and prestressed masonry*; and
  - (c) Part 3: 2001 *Materials and components, design and workmanship.*
- BS 8103 *Structural design of low-rise buildings*:

(a) Part 1: 1995 *Code of practice for stability, site investigation, foundations and ground floor slabs for housing*; and
(b) Part 2: 1996 *Code of practice for masonry walls for housing.*

Structural work of timber should comply with:

- BS 5268 *Structural use of timber*:
  (a) Part 2: 2002 *Code of practice for permissible stress design, materials and workmanship*; and
  (b) Part 3: 1998 *Code of practice for trussed rafter roofs.*
- BS 8103 *Structural design of low-rise buildings*:
  (a) Part 3: 1996 *Code of practice for timber floors and roofs for housing.*

Structural work of steel should comply with:

- BS 5950 *Structural use of steelwork in building*:
  (a) Part 1: 2000 *Code of practice for design in simple and continuous construction: hot rolled and welded sections*;
  (b) Part 2: 2001 *Specification for materials, fabrication and erection: hot rolled sections*;
  (c) Part 3: *Design in composite construction, section* 3.1: 1990 *Code of practice for design of simple and continuous composite beams*;
  (d) Part 4: 1994 *Code of practice for design of floors with profiled steel sheeting*; and
  (e) Part 5: 1998 *Code of practice for design of cold formed thin gauge sections.*
- BRE Digest 437 – *Industrial platform floors: mezzanine and raised storage.*

6.5   Structural requirements in existing buildings subject to change of use

In certain circumstances, the structural requirements of Part A apply to buildings subject to a material change of use (see section 2.4, regulations 5 and 6). In these cases it is necessary to carry out structural appraisals of the existing buildings to see if they are capable of coping with the changed loading conditions necessitated by the change of use. Guidance concerning this may be found in the following documents:

- BRE Digest 366: *Structural Appraisal of Existing Buildings for Change of Use*;
- The Institution of Structural Engineers Report, Appraisal of Existing Structures, 1996.

The Institution of Structural Engineers' Report contains an item on design checks where a choice of various partial factors should be made to suit the individual circumstances of each case.

6.6   Design of structural members in houses and other small buildings

6.6.1   Definitions

The following definitions apply throughout section 1 of AD Al/2.

BUTTRESSING WALL – A wall which provides lateral support, from base to top, to another wall perpendicular to it.

CAVITY WIDTH – The horizontal distance between the leaves in a cavity wall.

COMPARTMENT WALL – See Chapter 7, Fire.

DEAD LOAD – The load due to the weight of all roofs, floors, walls, services, finishes and partitions, i.e. all the permanent construction.

IMPOSED LOAD – The load assumed to be produced by the intended occupancy or use, including moveable partitions, distributed, concentrated, impact, inertia and snow loads, but *excluding* wind loads.

PIER – An integral part of a wall which consists of a thickened section occurring at intervals along a wall to which it is bonded or securely tied so as to afford lateral support.

SEPARATING WALL – A wall which is common to two adjoining buildings (see Chapter 7, Fire).

SPACING – The centre to centre distance between two adjacent timbers measured in a plane parallel to the plane of the structure of which they form part.

SPAN – The distance measured along the centreline of a member between centres of adjacent bearings. (However, it should be noted that the spans given in the tables in the document *Span tables for solid timber members in floors, ceilings and roofs (excluding trussed rafter roofs) for dwellings*, published by TRADA, are *clear spans*, i.e. measured between the faces of supports.)

SUPPORTED WALL – A wall which is supported by buttressing walls, piers or chimneys, or floor or roof lateral support arrangements.

WIND LOAD – Any load due to the effect of wind pressure or suction.

6.6.2   Structural stability

The basic stability of a small house of traditional masonry construction is largely dependent on the provision of intermediate floors and a braced roof structure which are adequately anchored to walls restrained laterally by buttressing walls, piers or chimneys. In effect, the floors and roof act as horizontal diaphragms, which are capable of transmitting forces from the wind to the buttressing elements of the building. If this can be achieved, then it should not be necessary to take additional precautions against wind loading.

To achieve the necessary transfer of wind loading the following conditions should be met:

- the internal and external walls must be adequately connected to each other using masonry bonding or mechanical connections;
- the layout of the walls must be such that a robust three-dimensional box structure is created with limitations placed on the maximum size of rooms;
- the overall size and proportions of the building are limited.

A traditionally constructed cut timber roof generally provides in-built resistance to instability from features such as hipped ends, tiling battens, rigid sarking boards, etc. However, where this is not provided then extra wind bracing may be required in a similar manner to that provided for trussed rafter roofs (see below) and especially in the case of single-hipped or non-hipped roofs to detached houses with pitches greater than 40°.

Trussed rafter roofs have, in the past, been susceptible to collapse during high winds. If this form of construction is used it should be braced in accordance with BS 5268 *Structural use of timber*: Part 3: 1998 *Code of practice for trussed rafter roofs* or Annex H of BS 8103 *Structural design of low-rise buildings*: Part 3: 1996 *Code of practice for timber floors and roofs for housing*.

6.6.3   Structural work of timber in single occupancy dwellings

Section 2B of AD Al/2 provides that if the work concerned is in a floor, ceiling or roof of a single occupancy dwelling of not more than three storeys, that work will be satisfactory if the grades and dimensions of the timbers used are at least equal to those given in Tables 1 to 45 of the TRADA publication *Span tables for solid timber members in floors, ceilings and roofs (excluding trussed rafter roofs) for dwellings* (obtainable from TRADA Technology Ltd, Chiltern House, Stocking Lane, Hughenden Valley, High Wycombe, Buckinghamshire HP14 4ND).

Alternatively the guidance in BS 5268 *Structural use of timber*: Part 2: 2002 *Code of Practice for permissible stress design, materials and workmanship*, BS 5268 Part 3: 1998 *Code of Practice for trussed rafter roofs* and BS 8103 *Structural design of low-rise buildings*: Part 3: 1996 *Code of Practice for timber floors and roofs for dwellings* may also be used.

The span tables in the TRADA publication replace Tables A1 to A24, which were included in previous editions of AD A. The advantage of using such span tables is that it is not necessary to calculate the size of joists, rafters, purlins, etc.; one merely selects the appropriate sizes from the tables.

The decision by the Government to outsource the preparation of these tables means that it is now necessary for designers, etc. to purchase another guidance document (in addition to AD A) with resulting increases in costs to the consumer. Since one of the original purposes of the Approved Documents was to provide relatively simple solutions to building regulation compliance issues (especially for dwellings) in single documents, it is to be regretted that the Government now deems it necessary to reduce the content and worth of the Approved Documents by placing ever increasing reliance on guidance documents produced by other, non-governmental bodies.

In order to assist our readers by providing, where possible, as complete a guide to the regulations and Approved Documents as we can in one place, we have reproduced the original tables from the previous edition of AD A at the end of this chapter. Unfortunately, for copyright reasons we are unable to reproduce the Tables from the TRADA document; however, a comparison between the previous Approved Document tables and those in the TRADA publication reveals that, in all cases, use of the previous tables from AD A will result in compliance with requirement A1. The main difference is that the TRADA tables often provide a more economical solution when choosing a particular timber member.

For example, TRADA Table 15 compares with Table A6 of the previous edition of AD A. These tables both relate to the design of purlins supporting rafters with pitches between 15° and 22.5°. Considering a 75 × 225 purlin carrying a dead load between 0.75 and 1.00 kN/m² at a spacing of 3000 mm – Table A6 gives a maximum permissible span of 2.38 m and TRADA Table 15 gives a maximum permissible span of 2.48 m (i.e. an improvement in efficiency of about 5%). In many cases no improvement in efficiency will be gained by using the TRADA tables since the figures for maximum span are the same in both documents.

There are, however, differences between the content of the two documents some of which are shown below:

- TRADA gives span tables for timber strength classes C16 and C24. AD A gives span tables for timber strength classes SC3 and SC4. For all practical purposes C16 equates to SC3 and C24 equates to SC4.
- In some cases timber sizes given in the AD A tables (notably 50 mm wide members) are not included in the TRADA tables. Similarly, some timber sizes shown in the TRADA tables are not included in the AD A tables (e.g. TRADA Table 40 has a flat-roof joist size of 38 × 235, which is not included in the equivalent Table A21 of AD A).
- There are also very slight variations in a few of the timber sizes given between the two sets of tables which in no way affect the ability of the chosen roof design to achieve compliance with requirement A1 and are only of academic interest.

The following notes refer to the use of Tables A1 to A24 of the 1992 edition (amended 1994) of AD A and these tables are reproduced below. In many cases the TRADA document gives the same design guidance and we have indicated the relevant points of interest in brackets in the text.

(1) Tables Al to A24 apply to all floor, ceiling and roof timbers in a single occupancy house of three storeys or less (TRADA Tables 1 to 45 also cover this).
(2) The timber used for any binder, beam, joist, purlin or rafter must be of a species, origin and grade specified in Table 1 to AD A1/2 (see below) or as given in the more comprehensive Tables of BS 5268: Part 2: 2000. (Table 2 of the TRADA document is similar in content; however the strength classes referred to are C16, C18, C22 and C24. C16 to C22 are equivalent to SC3 and C24 is equivalent to SC4.)

(3) When using Tables Al to A24 the following points should also be taken into account:

The imposed load to be sustained by the floor, ceiling or roof of which the member forms part should not exceed:

(a) In the case of a floor: 1.5 kN/m^2 – see Tables Al and A2. (TRADA includes a concentrated load limitation of 1.4 kN – see Tables 6 and 7.)

(b) In the case of a ceiling: 0.25 kN/m^2 and a concentrated load of 0.9 kN acting with the imposed load – see Tables A3 and A4. (TRADA – see Tables 8 to 11.)

(c) In the case of a flat roof with access not limited to the purposes of maintenance or repair: 1.5 kN/m^2 or a concentrated load of 0.9 kN – see Tables A21 and A22. (TRADA – the maximum concentrated load is allowed to be 1.8 kN – see Tables 40 and 41.)

(d) In the case of a roof (flat or pitched up to 45°) with access only for maintenance: 0.75 kN/m^2 or 1.00 kN/m^2, measured on plan (depending on the location, see below), or a concentrated load of 0.9 kN – see Tables A5 to A20 inclusive. (TRADA – see Tables 12 to 39 inclusive.)

(e) In the case of a roof supporting sheeting or decking pitched at between 10° and 35°: 0.75 kN/m^2 or 1.00 kN/m^2 measured on plan (depending on the location, see below), or a concentrated load of 0.9 kN – see Tables A23 and A24. (TRADA – see Tables 42 to 45.)

The loading variations on the roofs mentioned in (d) and (e) above are due to the different imposed snow loadings which vary with altitude and location in England and Wales.

Diagram 2 from ADA 1/2 is reproduced below and shows how the values of 0.75 kN/m^2 and 1.00 kN/m^2 are chosen. It is important to note that Diagram 2 has been substantially revised by the Building Research Establishment and the new version is shown in the TRADA document as Figure 1. For safety, the principal changes which should be taken into account when using Diagram 2 are as follows:

- For all altitudes up to 100 m above sea level a figure for snow roof loads of 0.75 kN/m^2 may be adopted.
- Within the hatched area at altitudes between 100 m and 200 m use BS 6399: Part 3 to determine the imposed snow roof load.

(4) Information on floorboarding can be found in BS 8103: Part 3 1996 *Code of practice for timber floors and roofs for housing*. Table 6.1 is based on Table 2 from BS 8103 and shows typical thicknesses and spans of tongued and grooved softwood floorboards for use in dwellings.

(5) As stated in the footnotes to Tables Al to A24 in AD A1/2, the cross-sectional dimensions given are applicable to either basic sawn or regularised sizes from

Table 6.1   Softwood floorboards (tongued and grooved).

Finished thickness of board (mm)	Maximum span of board (centre to centre of joists) (mm) up to
16	505
19	600

BS 4471: 1987 *Specification for sizes of sawn and processed softwood*. For North American timber (CLS/ALS) the tables apply to surface sizes only unless the timber has been resawn to BS 4471 requirements. (TRADA updates the references above to the tolerance classes specified in BS EN 336: 2003 Tables NA2, NA3 and NA4 respectively.)

(6) Notches and holes in floor and roof joists should comply with Fig. 6.1. However, no notches or holes should be cut in rafters except for birdsmouths at supports. The rafter may be birdsmouthed to a depth of up to one third the rafter depth. Notches and holes should not be cut in purlins or binders unless checked by a competent person. (TRADA restates these recommendations but adds that 'notches or holes should not be cut in rafters, purlins or binders unless approved by the building designer'.)

(7) Bearing areas and workmanship should be in accordance with BS 5268: Part 2: 2000 and the following minimum bearing lengths should be provided unless different figures can be justified by specialist calculation:
   (a) floor joists and flat roof joists – 35mm (now amended to 40mm in the TRADA document);
   (b) ceiling joists, rafters and binders – 35 mm;
   (c) purlins – 50 mm.

(8) If the spans of purlins or rafters are unequal, the section sizes chosen should relate to the longer span.

(9) On floor joists, no allowances have been made for additional loadings due to baths or partitions. It is recommended that all joists under baths should be doubled up but no advice is given regarding partition loads in AD A. (TRADA recommends that for lightweight non-loadbearing partitions weighing not more than 0.8 kN per metre run placed parallel to floor joists, one or two extra joists can be placed immediately below them depending on the dead load of the floor.) Similarly, when choosing ceiling joist sizes no account has been taken of trimming or of other additional loads such as water tanks.

(10) Purlins are assumed to be placed perpendicular to the roof slope and adequate connections between the various roof members should be provided as appropriate.

(11) Tables A1 to A24 (TRADA Tables 1 to 45) are not applicable to trussed rafter roofs.

(12) Example applications of Tables A1 to A24 are given in Fig. 6.2. It should be remembered that all spans, except for floorboards, are measured as the clear dimension between supports, and all spacings are the dimensions between longitudinal centres of members.

(13) Strutting of joists – where floor joists span more than 2.4 m (amended to 2.5 m in TRADA) they should be strutted with one or more rows of:
   •   solid timber at least 38 mm wide and 0.75 times the joist depth; or
   •   herringbone strutting in 38 mm × 38 mm timber except where the distance between the joists is greater than three times the joist depth.
   In this latter case the alternatives to timber herringbone strutting are not specified and it is not clear if solid timber strutting would be recommended or whether proprietary steel herringbone strutting, for example, could be used. (TRADA adds that proprietary herringbone strutting systems are permitted if used in accordance with manufacturer's instructions.)

AD Al/2 (1992 edition revised 1994)

Table 1 Common species/grade combinations which satisfy the requirements for the strength classes to which tables A1–A24 in Appendix A relate.

Species	Origin	Grading Rules	Grades to satisfy strength class SC3			SC4	
Redwood or whitewood	Imported	BS 4978	GS	MGS	M50	SS	MSS
Douglas Fir	UK	BS 4978	M50	SS	MSS	—	—
Larch	UK	BS 4978	GS	MGS	M50	SS	MSS
Scotch Pine	UK	BS 4978	GS	MGS	M50	SS	MSS
Corsican Pine	UK	BS 4978		M50		SS	MSS
European Spruce	UK	BS 4978		M75			
Sitka Spruce	UK	BS 4978		M75			
Douglas Fir-Larch Hem-Fir Spruce-Pine-Fir	CANADA	BS 4978	GS	MGS	M50	SS	MSS
Douglas Fir-Larch Hem-Fir Spruce-Pine-Fir	CANADA	NLGA	Joist & Plank Struct. L.F.	No.1 & No.2 No.1 & No.2		Joist & Plank Struct. L.F.	Select Select
Douglas Fir-Larch Hem-Fir Spruce-Pine-Fir	CANADA	MSR		Machine Stress-Rated 1450f-1.3E			Machine Stress-Rated 1650f-1.5E
Douglas Fir-Larch	USA	BS 4978	GS	MGS		SS	MSS
Hem-Fir	USA	BS 4978	GS	MGS	M50	SS	MSS
Western Whitewoods	USA	BS 4978	SS	MSS		—	—
Southern Pine	USA	BS 4978	GS	MGS		SS	MSS
Douglas Fir-Larch	USA	NGRDL	Joist & Plank Struct. L.F.	No.1 & No.2 No.1 & No.2		Joist & Plank Struct. L.F.	Select Select
Hem-Fir	USA	NGRDL	Joist & Plank Struct. L.F.	No.1 & No.2 No.1 & No.2		Joist & Plank Struct. L.F.	Select Select
Western Whitewoods	USA	NGRDL	Joist & Plank Struct. L.F.	Select Select		—	
Southern Pine	USA	NGRDL	Joist & Plank	No.3 Stud grade		Joist & Plank	Select
Douglas Fir-Larch Hem-Fir Southern Pine	USA	MSR		Machine Stress-Rated 1450f-1.3E			Machine Stress-Rated 1650f-1.5E

Notes: The common species/grade combinations given in this table are for particular use with the other tables in Appendix A and for cross section sizes given in those tables.

Definitive and more comprehensive tables for assigning species/grade combinations to strength classes are given in BS 5268: Part 2: 2000 . The grading rules for American and Canadian Lumber are those approved by the American Lumber Standards (ALS)

Board of Review and the Canadian Lumber Standards (CLS) Accreditation Board respectively (see BS 5268: Part 2: 2000).

NGLA denotes the National Lumber Grading Association.

NGRDL denotes the National Grading Rules for Dimension Lumber.

MSR denotes the North American Export Standard for Machine Stress-Rated Lumber.

AD Al/2 (1992 edition revised 1994)

Diagram 2   Imposed snow roof loading

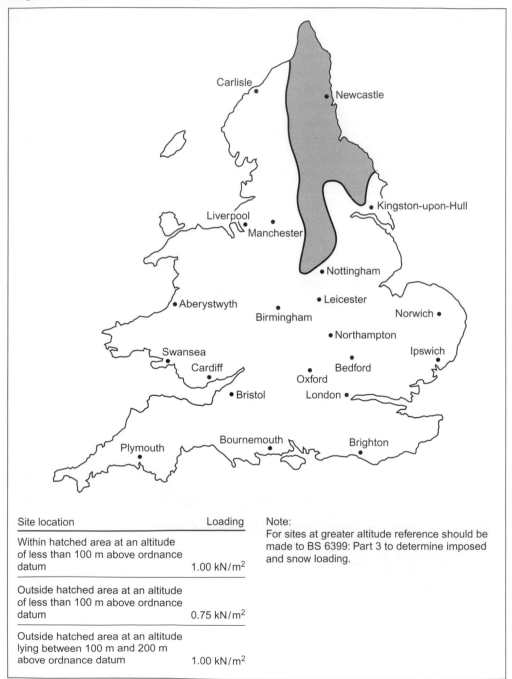

Site location	Loading
Within hatched area at an altitude of less than 100 m above ordnance datum	1.00 kN/m²
Outside hatched area at an altitude of less than 100 m above ordnance datum	0.75 kN/m²
Outside hatched area at an altitude lying between 100 m and 200 m above ordnance datum	1.00 kN/m²

Note:
For sites at greater altitude reference should be made to BS 6399: Part 3 to determine imposed and snow loading.

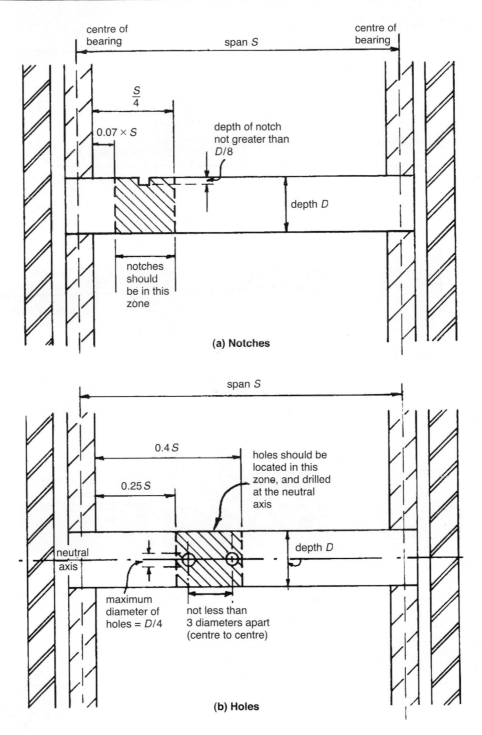

Fig. 6.1   Notches and holes in floor and roof joists.

(a) Floor joists, small house

dead load not more than 0.25 kN/m^2 clear span 4 m, centres 400 mm, timber of strength class SC3 (TRADA Strength Classes C16, C18 and C22)

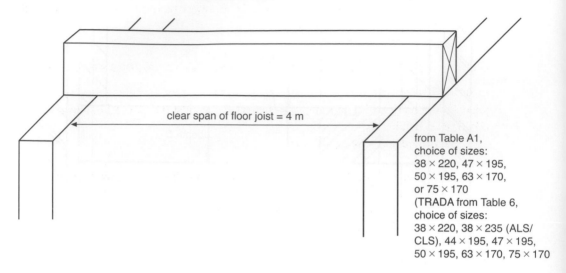

clear span of floor joist = 4 m

from Table A1,
choice of sizes:
38 × 220, 47 × 195,
50 × 195, 63 × 170,
or 75 × 170
(TRADA from Table 6,
choice of sizes:
38 × 220, 38 × 235 (ALS/
CLS), 44 × 195, 47 × 195,
50 × 195, 63 × 170, 75 × 170

(b) Rafter, small house

pitch 20°, dead load not more than 0.50 kN/m^2 clear span 2.90 m, centres 400 mm, timber of strength class SC3 (TRADA Strength Class C16, C18 and C22), imposed loading 0.75 kN/m^2

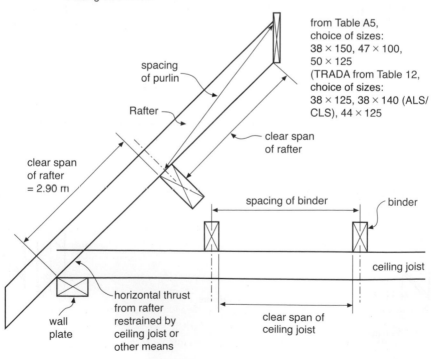

from Table A5,
choice of sizes:
38 × 150, 47 × 100,
50 × 125
(TRADA from Table 12,
choice of sizes:
38 × 125, 38 × 140 (ALS/
CLS), 44 × 125

spacing of purlin

Rafter

clear span of rafter

clear span of rafter = 2.90 m

spacing of binder

binder

ceiling joist

horizontal thrust from rafter restrained by ceiling joist or other means

wall plate

clear span of ceiling joist

Fig. 6.2 Example of application of Tables A1 to A24.

One row of strutting at mid-span is recommended for joist spans between 2.5 m and 4.5 m. Above 4.5 m, two rows of strutting at the one third positions would be required.

TRADA adds that the outer joist should be solidly blocked to the perimeter wall at the end of each row of strutting.

6.6.4   Special treatment against House Longhorn Beetle infestation

In specified areas in the south of England all softwood roof timbers, including ceiling joists should be treated with a suitable preservative against the House Longhorn Beetle (Hylotrupes bajulus L).

The specified areas are as follows:

- in the Borough of Bracknell Forest the parishes of Sandhurst and Crowthorne;
- the Borough of Elmbridge;
- in the District of Hart, the parishes of Hawley and Yateley;
- the District of Runnymede;
- the Borough of Spelthorne;
- the Borough of Surrey Heath;
- in the Borough of Rushmoor, the area of the former district of Farnborough, the Borough of Woking.

No specific forms of treatment are recommended; however, guidance on suitable preservative treatment may be found in the *BWPDA Manual* (2000 revision) published by the British Wood Preserving and Damp-Proofing Association (available from 1 Gleneagles House, Vernongate, South Street, Derby DE1 1UP).

6.6.5   Structural work of bricks, blocks and plain concrete

If a wall of these materials comes within the scope of section 2C of AD A1/2, it is not necessary to calculate loads or wall thicknesses, provided the wall is built with the thicknesses required by section 2C and complies with the rules therein.

Section 2C may be applied to any wall which is:

- an external wall, compartment wall, internal load-bearing wall or separating wall of a residential building of not more than three storeys; *and*
- an external wall or internal load-bearing wall of a small single-storey non-residential building or small annexe to a residential building (such as a garage or outbuilding) provided that:
  - (a) the building design complies with the requirements of paragraphs 2C5 to 2C13 of section 2C; and
  - (b) the wall construction details comply with the relevant requirements of BS 5628 *Code of practice for use of masonry:* Part 3: 2001 *Materials and components, design and workmanship* except as regards the conditions given in paragraphs 2C4 and 2C14 to 2C38 of section 2C.

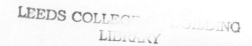
LEEDS COLLEGE ... ...DING
LIBRARY

It should be noted that when the guidance for section 2C was formulated the worst combination of circumstances likely to arise was taken into account. This may therefore result in a somewhat conservatively designed structure. AD A 1/2 makes it clear that where a particular requirement is considered to be too onerous it may be appropriate to consider a minor departure from the recommendations of section 2C on the basis of judgement and experience. Alternatively, calculations may be carried out to show adequacy in respect of the aspect of the wall which is subject to the departure rather than for the entire wall.

The guidance given is based on the compressive strengths of bricks and blocks shown in section 6.6.7 below.

Where the suitability for use of masonry units of other compressive strengths is being considered, design strengths for walls may be found in BS 5628 Part 1: 1992 *Structural use of unreinforced masonry*.

6.6.6 Building design requirements (section 2C, paragraphs 2C4, 2C14 to 2C16 and 2C38)

These are concerned with the maximum allowable height of the building (which is related to the maximum wind load), the maximum imposed load, the building proportions and the plan area of each storey or sub-division.

MAXIMUM ALLOWABLE HEIGHT OF THE BUILDING. AD A 1/2 gives a simplified way of determining the maximum allowable height of a building from a map (Diagram 6 of AD A 1/2 reproduced below) and a series of tables (tables a, b and c in Diagram 7 (also reproduced below). This method is based on BS 6399: *Loading for buildings:* Part 2: 1997 *Code of practice for wind loads.*

In order to determine the maximum height of a building use the following procedure:

- Use Diagram 6 Figure (a) (Map showing wind speeds in m/s for maximum height of buildings) to determine the wind speed (V) relative to the location of the building.
- Find the topographic zone for the site from Diagram 6 Figure (b) and use this in Table a of Diagram 7 to obtain Factor T (it should be noted that Table a has been wrongly labelled as Table b and Factor T is referred to as Factor A in the heading to the table in the 2004 edition of AD A).
- Use the site altitude in metres to determine Factor A in Table b of Diagram 7.
- Calculate Factor S from the formula $S = V \times T \times A$.
- Using the calculated Factor S look up the maximum allowable building height permitted in Table c (Maximum allowable building height).

The height obtained should not be exceeded in the building design.

IMPOSED LOADS. These should not exceed:
- on any floor, 2.0 kN/m² distributed;
- on any ceiling, 0.25 kN/m² distributed and 0.9 kN concentrated; *and*
- on any roof, 1.00 kN/m² for spans not exceeding 12 m, or 1.5kN/m² for spans not exceeding 6 m.

These recommendations are illustrated in Fig. 6.3.

AD A1/2

Diagram 6   Map showing wind speeds in m/s for maximum height of buildings

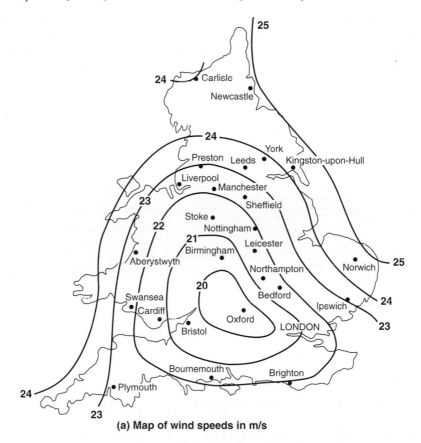

(a) Map of wind speeds in m/s

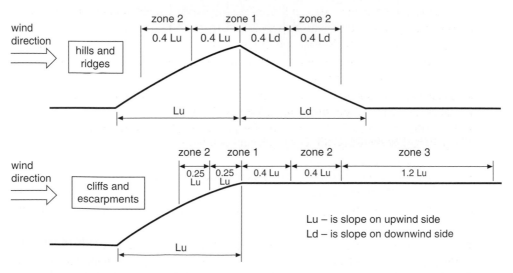

Lu – is slope on upwind side
Ld – is slope on downwind side

(b) Topographic zones for Factor T

AD A1/2

Diagram 7   Maximum height of buildings

Table a Factor T			
Topographic category and average slope of whole hillside, ridge, cliff or escarpment	Factor T		
	Zone 1	Zone 2	Zone 3
Category 1: Nominally flat terrain, average slope < 1/20	1.0	1.0	1.0
Category 2: Moderately steep terrain, average slope < 1/5	1.24	1.13	1.10
Category 3: Steep terrain, average slope > 1/5	1.36	1.20	1.15

Note: outside these zones factor T = 10

Table b Factor A	
Site altitude (m)	Factor A
0	1.00
50	1.05
100	1.10
150	1.15
200	1.20
300	1.30
400	1.40

Table c Maximum allowable building height (m)						
Factor	Country sites			Town sites*		
S	Distance to the coast			Distance to the coast		
	< 10 km	10–50 km	> 50 km	< 10 km	10–50 km	> 50 km
24	15	15	15	15	15	15
25	11.5	14.5	15	15	15	15
26	8	10.5	13	15	15	15
27	6	8.5	10	15	15	15
28	4.5	6.5	8	13.5	15	15
29	3.5	5	6	11	13	14.5
30	3	4	5	9	11	12.5
31		3.5	4	8	9.5	10.5
32		3	3.5	7	8.5	9.5
33			3	6	7.5	8.5
34				5	7	8
35				4	6	7
36				3	5.5	6
37					4.5	5.5
38					4	5
39					3	4
40						3

*For sites on the outskirts of towns not sheltered by other buildings use the values for country sites

BUILDING PROPORTIONS. For residential buildings of not more than three storeys:

- The maximum height of the building should not exceed 15 m (and may need to be less than this as it will be subject to the calculation shown above under 'maximum allowable height of building'). The height is measured from the lowest finished ground level adjacent to the building to the highest point of any wall or roof.
- The width of the building should not be less than at least half the height of the building.
- Any wing of the building which projects more than twice its own width from the remainder of the building should have a width at least equal to half its height.

These recommendations are illustrated in Fig. 6.3.

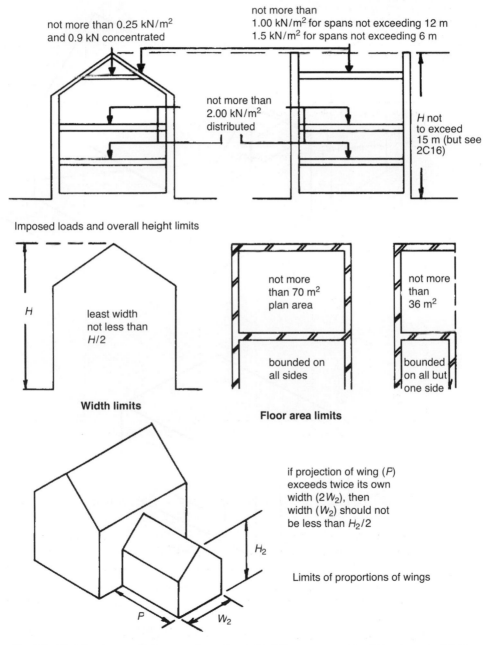

Fig. 6.3   Building design requirements for residential buildings not exceeding three storeys in height.

For small single-storey non-residential buildings:

- The height of the building should not exceed the dimensions shown in Fig. 6.4(a).
- The greatest length or width of the building should not exceed 9 m.

For annexes attached to residential buildings the height of any part should not exceed the dimensions shown in Fig. 6.4(b).

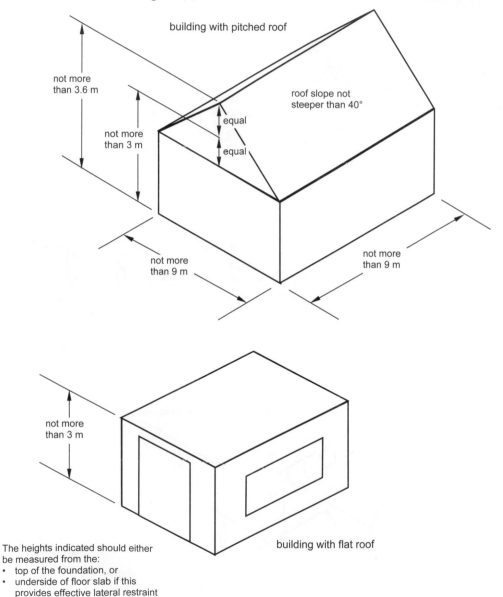

The heights indicated should either be measured from the:
- top of the foundation, or
- underside of floor slab if this provides effective lateral restraint

(a) Small single-storey non-residential buildings

Fig. 6.4   Building design requirements for non-residential buildings and annexes.

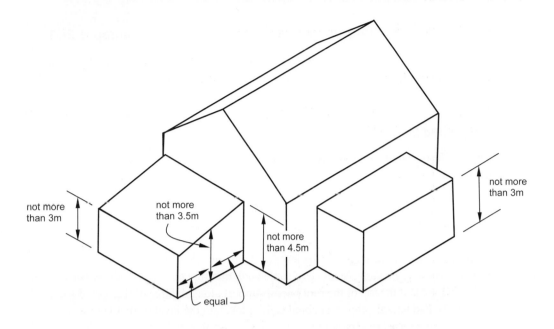

The heights indicated should either
be measured from the:
- top of the foundation, or
- underside of floor slab if this
  provides effective lateral restraint

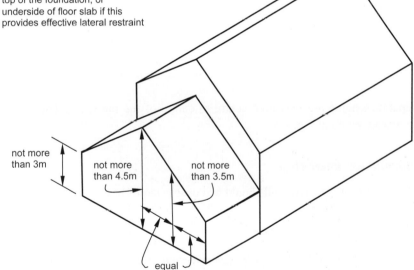

(b) Annexes attached to residential buildings

Fig. 6.4   *(Contd).*

PLAN AREA OF STOREY. The plan area of each storey which is completely bounded
by structural walls on all sides should not be more than 70 m². However, if the storey is
bounded in this way on all sides but one, the limiting area is 36 m². See Fig. 6.3.

6.6.7   Wall construction requirements (section 2C, paragraphs 2C17 to 2C38)

These are concerned with height and length, materials and workmanship, buttressing, loading conditions, openings and recesses, and lateral support.

Height and length

The height or length of a wall should not be more than 12 m and together with storey heights should be measured in accordance with the following rules:

- The height of the ground storey of a building is measured from the base of the wall to the underside of the next floor above.
- The height of an upper storey is measured from the level of the underside of the floor of that storey, in each case to the level of the underside of the next floor above.
- For a top storey which comprises a gable wall, measure to a level midway between the gable base and the top of the roof lateral support along the line of the roof slope, but if there is also lateral support at about ceiling level, to the level of that lateral support.
- Where an internal or separating wall comprises a gable and is built up to the underside of the roof, measure the height from its base to the base of the gable.
- Any external wall which includes a gable should be measured from its base to half the height of the gable.
- Any external wall which is not a gable wall should be measured from its base to its highest part, excluding any parapet not exceeding 1.2 m in height.
- Walls are regarded as being divided into separate lengths by securely tied buttressing walls, piers or chimneys for the purposes of measuring their length.

These separate lengths are measured centre to centre of the piers, etc. These special requirements are noted in Fig. 6.5.

Materials and workmanship

BRICKS AND BLOCKS. The wall should be constructed of bricks or blocks, properly bonded and solidly put together with mortar. The materials should comply with the following standards:

- clay bricks or blocks to BS 3921: 1985 *Specification for clay bricks* or BS 6649: 1985 *Specification for clay and calcium silicate modular bricks* or BS EN 771 *Specification for Masonry Units*: Part 1: 2003 *Specification for masonry units*;
- calcium silicate bricks to BS 187: 1978 *Specification for calcium silicate (sandlime and flintlime) bricks* or BS 6649: 1985 or BS EN 771 Part 2:2001 *Calcium silicate masonry units*;
- concrete bricks or blocks to BS 6073: *Precast concrete masonry units:* Part 1: 1981 *Specification for precast concrete masonry units* or BS EN Part 3: *Aggregate concrete masonry units* or Part 4: 2001 *Autoclaved aerated concrete masonry units*;

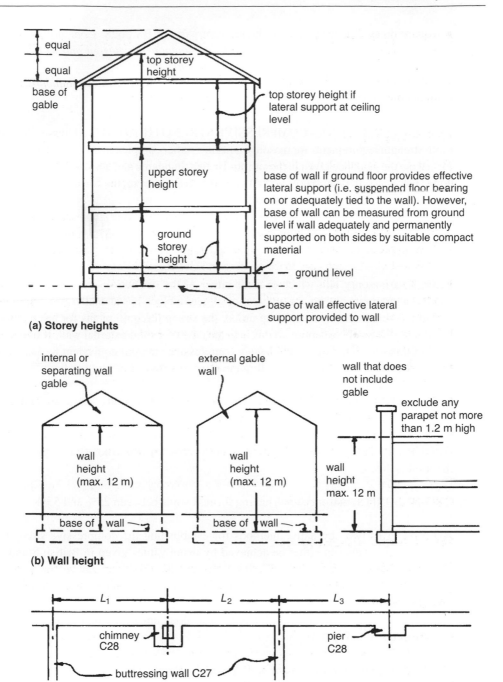

(a) Storey heights

(b) Wall height

division of wall into separate effective lengths on plan: L_1, L_2, L_3 – each not more than 12 m

(c) Wall length

Fig. 6.5   Rules for measurement, 1C18, 1C19.

- square dressed natural stone to the relevant parts of BS 5628: *Code of practice for use of masonry:* Part 3: 2001 *Materials and components, design and workmanship* or BS EN 771 Part 6: 2001 *Natural stone masonry units*;
- manufactured stone to BS 6457: 1984 and BS EN 771 Part 5: *Manufactured stone masonry units.*

BRICKS AND BLOCKS – COMPRESSIVE STRENGTH. AD A1/2 addresses the minimum strength requirements for masonry units in a rather complex manner when it attempts to analyse the guidance given in the various British Standards and BS EN Standards which cover this issue. It is probably best to consider the requirements for bricks and blocks separately and these are defined as follows:

BRICK: a masonry unit having work sizes not greater than 337.5 mm in length and 112.5 mm in height.

BLOCK: a masonry unit which exceeds either of the two work sizes given for bricks and has a minimum bed height of 190 mm. Where a block has a smaller bed height than 190 mm (excluding cuts or make up units), the strength requirements for bricks should be adopted. The only exception to this is in the case of a solid external wall. If blocks are used in this situation they should have a compressive strength equivalent to that shown for a block inner leaf of a cavity wall in the same position.

Reference is also made in AD A1/2 to Group 1 and Group 2 units. These are defined as follows:

GROUP 1 UNITS: masonry units having up to 25% formed voids (for frogged bricks the limit is 20%).

GROUP 2 UNITS: masonry units having formed voids between 25% and 55%.

AD A1/2 gives two alternative methods for establishing the compressive strength of masonry units. This can either be achieved by using values given in British Standards BS 3921, BS 6073 Part 1, BS 187, BS 5390 and BS 6649 (see above for details of standards and Fig 6.6 for details of compressive strengths) or can be achieved by selecting masonry units with declared compressive strengths of at least the values given in Table 6.2 below for brick units or Table 6.3 for block units for Conditions A, B and C as appropriate (see Fig 6.6).

MORTAR. The mortar used in any wall to which section 2C of AD A1/2 applies should be:

- to the proportions given in BS 5628 *Code of practice for use of masonry*: Part 3: 2001 *Materials and components, design and workmanship for mortar designation* (iii);
- strength class M4 from BS EN 998 *Specification for mortar for masonry:* Part 2: 2002 *Masonry mortar*;

Table 6.2 Declared compressive strength* of masonry units (bricks) complying with BS EN 771: Parts 1 to 5 (N/mm²).

Masonry unit	Clay masonry units to BS EN 771: Part 1		Calcium silicate masonry units to BS EN 771: Part 2		Aggregate concrete masonry units to BS EN 771: Part 3	Autoclaved aerated concrete masonry units to BS EN 771: Part 4	Manufactured masonry units to BS EN 771: Part 5
Condition A (See diagram)							Any unit complying with BS EN 771: Part 5 will be acceptable for conditions A, B and C
Brick	Group 1 6.0	Group 2 9.0	Group 1 6.0	Group 2 9.0	6.0	—	
Condition B (See diagram)							
Brick	Group 1 9.0	Group 2 13.0	Group 1 9.0	Group 2 13.0	9.0	—	
Condition C (See diagram)							
Brick	Group 1 18.0	Group 2 25.0	Group 1 18.0	Group 2 25.0	18.0	—	

Note: the values of declared compressive strengths given above are mean values.

Table 6.3 Normalised compressive strength of block masonry units (blocks) complying with BS EN 771: Parts 1 to 5 (N/mm²).

Masonry unit	Clay masonry units to BS EN 771: Part 1		Calcium silicate masonry units to BS EN 771: Part 2		Aggregate concrete masonry units to BS EN 771: Part 3	Autoclaved aerated concrete masonry units to BS EN 771: Part 4	Manufactured masonry units to BS EN 771: Part 5
Condition A (See diagram)							Any unit complying with BS EN 771: Part 5 will be acceptable for conditions A, B and C
Block	Group 1 5.0	Group 2 8.0	Group 1 5.0	Group 2 8.0	3.1*	3.1	
Condition B (See diagram)							
Block	Group 1 7.5	Group 2 11.0	Group 1 7.5	Group 2 11.0	7.7*	7.7	
Condition C (See diagram)							
Block	Group 1 15.0	Group 2 21.0	Group 1 15.0	Group 2 21.0	7.7*	7.7	

* These values are dry ground strengths to BS EN 771: Part 1

Notes:

This table applies to masonry units where the work size is greater than 337.5 mm in length or 112.5 mm in height

Values in this Table are normalised compressive strengths (N/mm²). Compressive strengths of masonry units should be derived according to BS EN 772: Part 1.

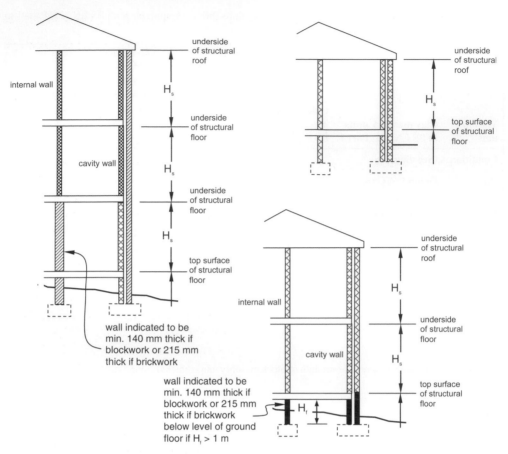

Table to Fig 6.6 Compressive strengths of masonry units				
Key	**Compressive strength of masonry units as given in British Standards**		**Declared compressive strength of masonry units (see Tables 6.1 and 6.2)**	
	Brick 5 N/mm² Block 2.8 N/mm²		Condition A	
	Brick 7 N/mm² Block 7 N/mm²		Condition B	
	Brick 15 N/mm² Block 7 N/mm²		Condition C	
	$H_f \leq 1$ m Brick 5 N/mm² Block 2.8 N/mm²	$H_f > 1$ m Brick 7 N/mm² Block 7 N/mm²	$H_f \leq 1$ m Condition A	$H_f > 1$ m Condition B
	If $H_s < 2.7$m compressive strength of masonry to be as shown above	If $H_s > 2.7$m compressive strength of masonry to be as shown above or 7N/m² whichever is greater	If $H_s < 2.7$m compressive strength of masonry to be as shown above	If $H_s > 2.7$m compressive strength of masonry to be as shown above or at least condition B whichever is greater

Notes: 1. For external walls of solid construction, strength of masonry to be at least as shown for the internal leaf of cavity wall in the same position.
2. The above guidance to compressive strengths of bricks or blocks for two and three storey buildings is only applicable where the roof structure is of timber construction.

Fig. 6.6   Compressive strengths of masonry units.

- at least equal in strength to a 1:1:5 or 6 CEM1/lime/fine aggregate mortar measured by volume of dry materials; or
- of equivalent or greater strength and durability than any of the specifications previously referred to.

The statement that the mortar should be compatible with the masonry units and the position of use contained in the previous edition of AD A1/2, has been omitted from the current edition even though it is sound advice.

WALL TIES. These should comply with:

- BS 1243: 1978 *Specification for metal ties for cavity wall construction*, DD 140: Part 2: 1987 *Recommendations for design of wall ties*; or
- BS EN 845 *Specification for ancillary components for masonry:* Part 1: 2001 *Ties, tension straps, hangers and brackets.*

Additionally, they should be material references 1 or 3 in BS EN 845 Table A1 austenitic stainless steel and should be selected in accordance with Table 5 from AD A1/2 which is reproduced below. The effect of this recommendation is to make it necessary to use stainless steel wall ties in all houses irrespective of where those houses are located. Previously, stainless steel wall ties were only required if a house was sited where conditions of severe exposure occurred.

AD A1/2
Table 5   Cavity walls

Normal cavity width mm (Note 1)	Permissible type of tie		
	Tie length mm (Note 2)	Tie shape in accordance with BS 1243*	BS EN 845-1 tie (Note 4)
50 to 75	200	Butterfly, double triangle or vertical twist	Types 1, 2, 3 or 4 to DD 140-2* and selected on the basis of the design loading and design cavity width
76 to 90	225	Double triangle or vertical twist	
91 to 100	225	Double triangle (Note 3) or vertical twist	
101 to 125	250	Vertical twist	
126 to 150	275	Vertical twist	* Although BS 1243 and DD 140: Part 2 were withdrawn on 1 February 2005, the tile user classes (types given in Tables 1 and 3 of the latter document) can continue to be used
151 to 175	300	Vertical twist	
176 to 300	(See Note 2)	Vertical twist style	

Notes

1   Where face insulated blocks are used the cavity width should be measured from the face of the masonry unit.

2   The embedment depth of the tie should not be less than 50mm in both leaves. For cavities wider than 180mm calculate the length as the structural cavity width plus 125mm and select the nearest stock length.

3   Double triangle ties of this shape having a strength to satisfy Type 2 of DD 140-2*, are manufactured. Specialist tie manufacturers should be consulted if 225mm long double triangle format ties are needed for 91 to 100mm cavities.

4   Where BS EN 845-1 ties are used reference needs to be additionally made to DD 140-2* for the selection of the type (i.e. types 1, 2, 3 or 4) relevant to the performance levels given in DD 140-2.

6.7   Buttressing walls, piers and chimneys

6.7.1   Introduction

Every wall should be bonded or securely tied at each end to a buttressing wall, pier or chimney. These supporting elements should be of such dimensions as to provide effective lateral support over the full wall height from its base to its top.

If, additionally, such supporting elements are bonded or securely tied to the supporting wall at intermediate points in the length of the wall within each storey, then the wall may be regarded as being divided into separate distinct lengths by these buttressing walls, piers or chimneys. Each of the distinct lengths may then be regarded as a supported wall, and the length of any wall is the distance between adjacent supporting elements. The intermediate buttressing walls, piers or chimneys should provide lateral restraint for the full height of the supported wall, but they may be staggered within each storey.

BUTTRESSING WALLS should have:

- one end bonded or securely tied to the supported wall;
- the other end bonded or securely tied to another buttressing wall, pier or chimney;
- no opening or recess greater than 0.1 m^2 in area within a horizontal distance of 550 mm from the junction with the supported wall, and openings and recesses generally disposed so as not to impair the supporting effect of the buttressing wall;
- a length of not less than one sixth of the height of the supported wall;
- the minimum thickness required by the appropriate rule, according to whether the buttressing wall is actually an external, compartment, separating or internal load-bearing wall or a wall of a small building or annexe; but if the buttressing wall is none of these and is not itself a supported wall, then a thickness, t (see Fig. 6.7), of not less than:
  - ○ half the thickness required of an external or separating wall of similar height and length as the buttressing wall, less 5 mm; or
  - ○ if the buttressing wall is part of a dwelling-house and is not more than 6 m high and 10 m in length, 75 mm; or
  - ○ in any other case, 90 mm (see Fig. 6.7).

PIERS may project on either or both sides of the supported wall and should:

- run from the base to the top of the supported wall;
- have a thickness, measured at right angles to the length of the supported wall and including the thickness of that wall, of at least three times the thickness required of the supported wall; and
- measure at least 190 mm in width (the measurement being parallel to the length of the supported wall).

CHIMNEYS should have:

- a horizontal cross-section area, excluding any fireplace opening or flue, of not less than the area required of a pier in the same wall; and
- a thickness overall of at least twice the thickness required of the supported wall.

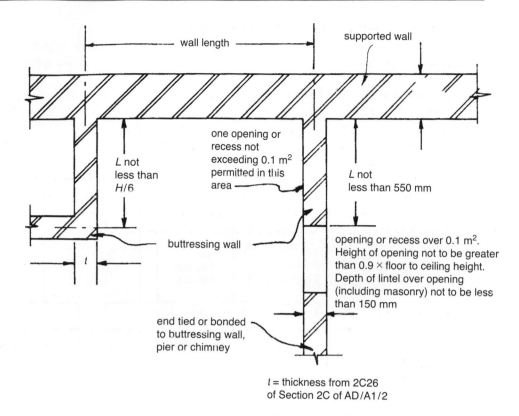

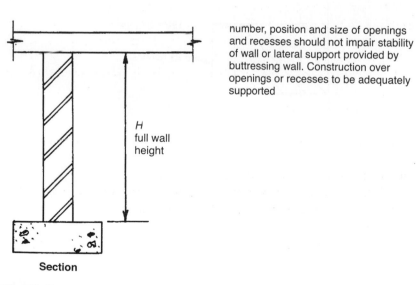

Section

Fig. 6.7 Buttressing walls.

The requirements for piers and chimneys are shown in Fig. 6.8.

It should be noted that requirements in respect of plan dimensions of piers do not apply to piers in walls of small buildings and annexes, for which there are special rules (see Fig. 6.19).

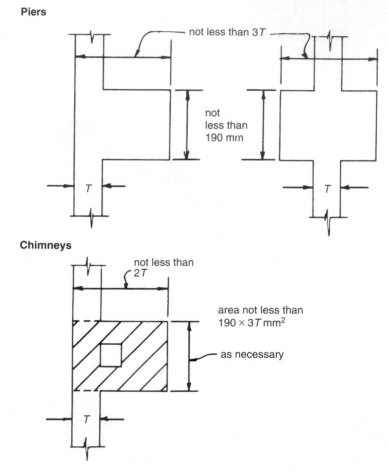

Fig. 6.8 Piers and chimneys .

6.7.2   Loading conditions

FLOOR SPANS. The wall should not support any floor members with a span of more than 6 m. (Span is measured centre to centre of bearings.)

LATERAL THRUST. Where the levels of the ground or oversite concrete on either side of a wall differ, the thickness of the wall as measured at the higher level should not be less than one quarter of the difference in level or 1 m whichever is the least. In the case of a cavity wall, the thickness is taken as the sum of the leaf thicknesses. However, if the cavity is filled with fine concrete, the overall thickness may be taken.

The recommendations described above and illustrated in Fig 6.9(b) apply equally where the retained soil is internal (i.e. the soil under the building is at a higher level than the outside ground) or external (i.e. the soil under the building is at a lower level than the outside ground). In the latter case (e.g. on sloping sites) the ground level adjacent to the building on the higher side should be maintained for a distance of not less than $1.25 \times H$ (H is defined in Fig 6.9).

The lateral thrust occasioned in these circumstances is the only one which a wall must be expected to sustain, apart from that due to direct wind load and the transmission of wind load.

VERTICAL LOADING. The total dead and imposed load transmitted by a wall at its base should not exceed 70 kN/m. All vertical loads carried by a wall should be properly distributed. This may be assumed for pre-cast concrete floors, concrete floor slabs and timber floors complying with section 2B of AD Al/2. Distributed loading may also be assumed for lintels with a bearing length of 150 mm or more. Where the clear span of the lintel is 1200 mm or less the bearing length may be reduced to 100 mm. These recommendations are summarised in Fig. 6.9.

6.7.3   Openings and recesses

Openings or recesses in a wall should not be placed in such a manner as to impair the stability of any part of it or to adversely affect the lateral restraint offered to the wall by a buttressing wall. Adequate support for the superstructure should be provided over every opening and recess.

As a general rule, any opening or recess in a wall should be flanked on each side by a length of wall equal to at least one sixth of the width of the opening or recess, in order to provide the required stability. Accordingly, the minimum length of wall between two openings or recesses should not be less than one sixth of the *combined* width of the two openings or recesses.

However, where long span roofs or floors bear onto a wall containing openings or recesses it may be necessary to increase the width of the flanking portions of wall. Table 8 of section 2C of AD A1/2 (see below) contains factors that enable this to be done.

Where several openings and/or recesses are formed in a wall, their total width should, at any level, be not more than two thirds of the length of the wall at that level and should not in any case exceed 3 m in total.

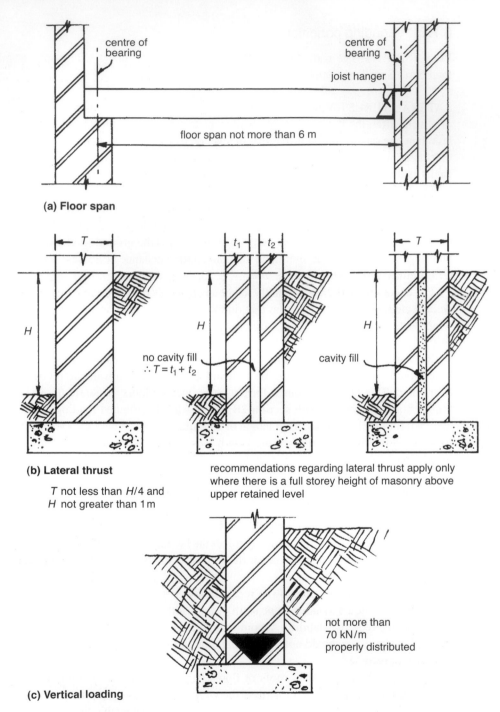

(a) Floor span

(b) Lateral thrust

T not less than $H/4$ and
H not greater than 1 m

no cavity fill
$\therefore T = t_1 + t_2$

cavity fill

recommendations regarding lateral thrust apply only
where there is a full storey height of masonry above
upper retained level

(c) Vertical loading

not more than
70 kN/m
properly distributed

Fig. 6.9   Loading requirements.

AD A1/2, section 2C

Table 8   Value of factor 'X'

Nature of roof span	Maximum roof span [m]	Minimum thickness of wall inner leaf [mm]	Span of floor parallel to wall	Span of timber floor into wall		Span of concrete floor into wall	
				max 4.5 m	max 6.0 m	max 4.5 m	max 6.0 m
				Value of factor 'X'			
roof span parallel to wall	not applicable	100	6	6	6	6	6
		90	6	6	6	6	5
timber roof span into wall	9	100	6	6	5	4	3
		90	6	4	4	3	3

The only openings permitted in walls below ground floor level are small holes for services and ventilation, etc. These should not exceed 0.1 m^2 in area and should be at least 2 m apart.

These requirements are illustrated in Fig. 6.10.

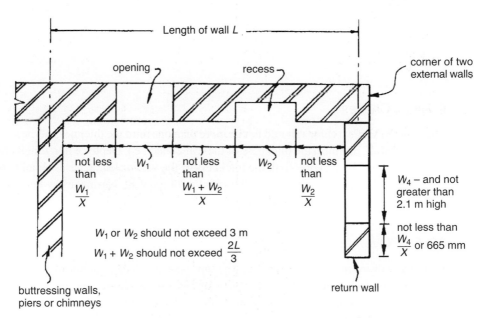

Note: value of X comes from Table 8 of section 1C of AD A1/2 which is reproduced above. OR it may be given the value 6 provided the compressive strength of the blocks or bricks (or cavity wall loaded leaf) is not less than 7 N/mm^2

Fig. 6.10   Openings and recesses.

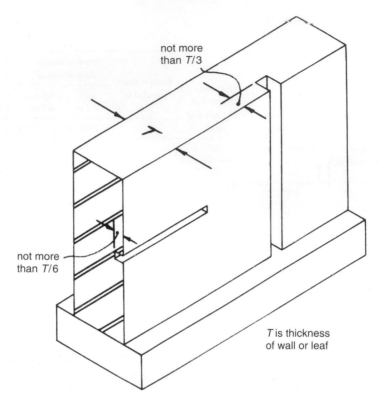

not more
than *T*/3

not more
than *T*/6

T is thickness
of wall or leaf

Fig. 6.11   Chases.

6.7.4   Chases

The depth of vertical chases should not be more than one third the thickness of the wall, or in a cavity wall, one third the thickness of the leaf concerned. Depth of horizontal chases should be not more than one sixth the thickness of the wall or leaf. Chases should not be placed in such a manner as to impair the stability of the wall, particularly where hollow blocks are used (see Fig. 6.11).

6.7.5   Overhanging

Where a wall overhangs a supporting structure beneath it, the amount of the overhang should not be such as to impair the stability of the wall. No limits are specified, but this would generally be interpreted as allowing an overhang of one third the thickness of the wall (see Fig. 6.12).

6.7.6   Lateral support

Floor or roof lateral support is horizontal support or stiffening, intended to stabilise or stiffen a wall by restraining its movement in a direction at right angles to the wall length. At least one wall in each storey of a building should extend to the full height of that storey

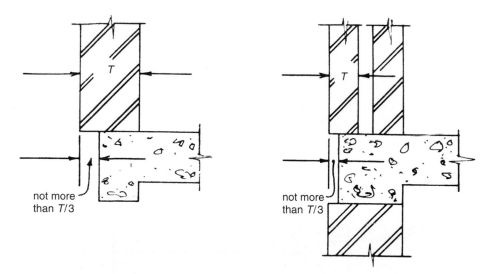

Fig. 6.12 Overhanging.

and this should be provided with restraint or support by connecting a floor or roof to the wall in such a way that the floor or roof acts as a stiffening frame or diaphragm transferring the lateral forces to walls, buttressing walls, piers or chimneys.

ROOF LATERAL SUPPORT. This should be provided for all external, compartment, separating and internal load-bearing walls irrespective of their length, at the point of junction between the roof and supported wall (i.e. at eaves level and along the verges).

Walls should be strapped to roofs at not exceeding 2 m centres using galvanised mild steel or other durable metal straps (referred to as tension straps conforming to BS EN 845 *Specification for ancillary components for masonry:* Part 1: 2001 *Ties, tension straps, hangers and brackets* in AD A1/2), with a minimum cross section of 30 mm × 5 mm, and a minimum length of 1 m for eaves strapping. Durable metal straps are defined (for corrosion resistance purposes) as being of material reference 14 or 16.1 or 16.2 (galvanised steel) or other more resistant specifications including material references 1 or 3 (austenitic stainless steel) with a declared tensile strength of at least 8 kN.

Eaves strapping need not be provided for a roof which:

- has a pitch of 15° or more;
- is tiled or slated;
- is of a type known by local experience as being resistant to damage by wind gusts;
- has main timber members spanning onto the supported wall at intervals of not more than 1.2 m.

Figure 6.13 shows methods of providing satisfactory lateral support at separating or gabled end wall positions.

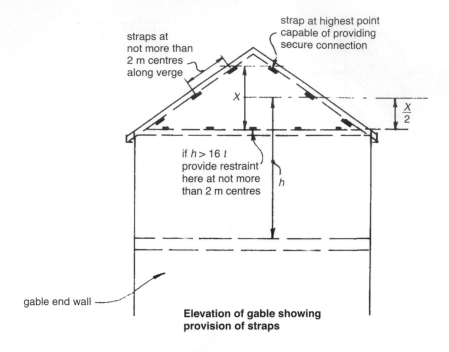

straps at
not more than
2 m centres
along verge

strap at highest point
capable of providing
secure connection

X

$\dfrac{X}{2}$

if $h > 16\,t$
provide restraint
here at not more
than 2 m centres

h

gable end wall

**Elevation of gable showing
provision of straps**

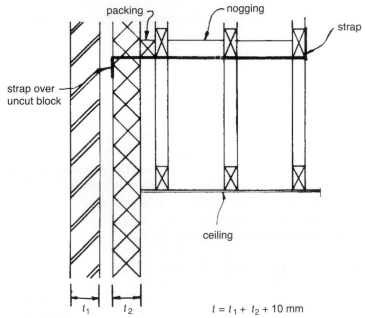

packing

nogging

strap

strap over
uncut block

ceiling

t_1 t_2 $t = t_1 + t_2 + 10$ mm

(a) Section through gable at roof level showing method of strapping

Fig. 6.13 Lateral support for roof.

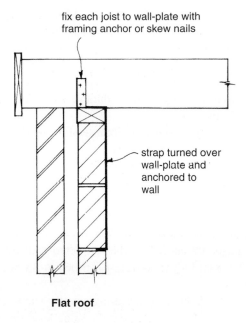

fix each joist to wall-plate with
framing anchor or skew nails

strap turned over
wall-plate and
anchored to
wall

Flat roof

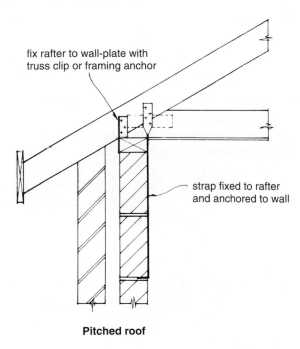

fix rafter to wall-plate with
truss clip or framing anchor

strap fixed to rafter
and anchored to wall

Pitched roof

(b) Section at eaves level showing method of strapping

Fig. 6.13   (*Contd*).

FLOOR LATERAL SUPPORT. This should be provided for any external, compartment or separating wall which exceeds 3 m in length.

It should also be provided for any internal load-bearing wall (which is not compartment or separating wall) at the top of each storey, irrespective of length.

Walls should be strapped to floors above ground level at not exceeding 2 m centres using galvanised mild steel or other durable metal straps, with a minimum cross section of 30 mm × 5 mm.

There are certain cases where, because of the nature of the floor construction, it is not necessary to provide restraint straps:

- Where a floor forms part of a house having not more than two storeys and:
  - (a) has timber members spanning so as to penetrate into the supported wall at intervals of not more than 1.2 m with at least 90 mm bearing directly on the walls or 75 mm bearing onto a timber wall plate; or
  - (b) the joists are carried on the supported wall by *restraint* type joist hangers to BS EN 845: Part 1: 2001, as described in BS 5628: Part 1, at not more than 2 m centres.
- Where a concrete floor has a bearing onto the supported wall of at least 90 mm.
- Where two floors are at or about the same level on either side of a supported wall, contact between floors and wall may be continuous or intermittent. If intermittent, the points of contact should be at or about the same positions on plan at intervals not exceeding 2 m. Figure 6.14 summarises these provisions.

6.7.7 Interruption of lateral support

It is clear that in certain circumstances it may be necessary to interrupt the continuity of lateral support for a wall. This occurs chiefly where a stairway or similar structure adjoins a supported wall and necessitates the formation of an opening in a floor or roof. This is permitted provided certain precautions are taken:

- The opening extends for a distance not exceeding 3 m measured parallel to the supported wall.
- If the connection between wall and floor or roof is provided by means of mild steel anchors, these should be spaced closer than 2 m on either side of the opening so as to result in the same number of anchors being used as if there were no opening.
- Other forms of connection (i.e. than mild steel anchors) should be provided throughout the length of each part of the wall on either side of the opening.
- There should be no other interruption of lateral support (see Fig. 6.15).

6.7.8 Thickness of walls

Provided the building design and wall construction requirements discussed above are satisfied, it is permissible to determine the thickness of a wall without calculation.

The minimum thicknesses required depend upon the wall height and length, and the rules applying to walls of bricks or blocks are set out in Table 3 of section 2C of AD A1/2 (see below) and illustrated in Fig. 6.16.

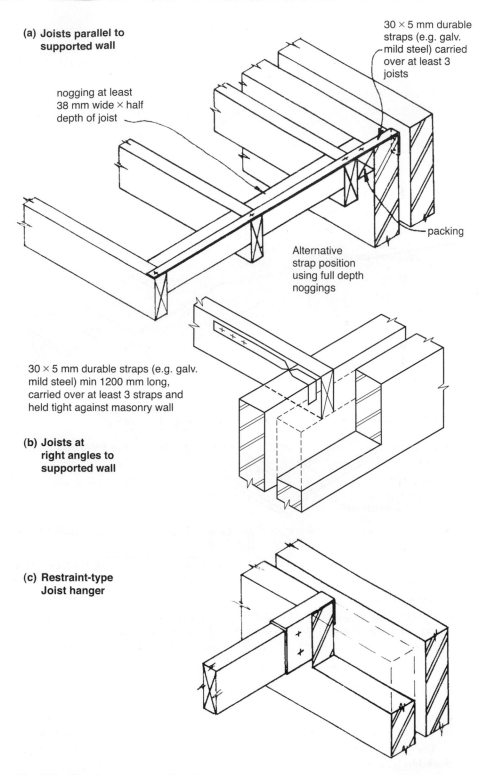

(a) Joists parallel to supported wall

30 × 5 mm durable straps (e.g. galv. mild steel) carried over at least 3 joists

nogging at least 38 mm wide × half depth of joist

packing

Alternative strap position using full depth noggings

30 × 5 mm durable straps (e.g. galv. mild steel) min 1200 mm long, carried over at least 3 straps and held tight against masonry wall

(b) Joists at right angles to supported wall

(c) Restraint-type Joist hanger

Fig. 6.14   Floor lateral support. (*Contd*).

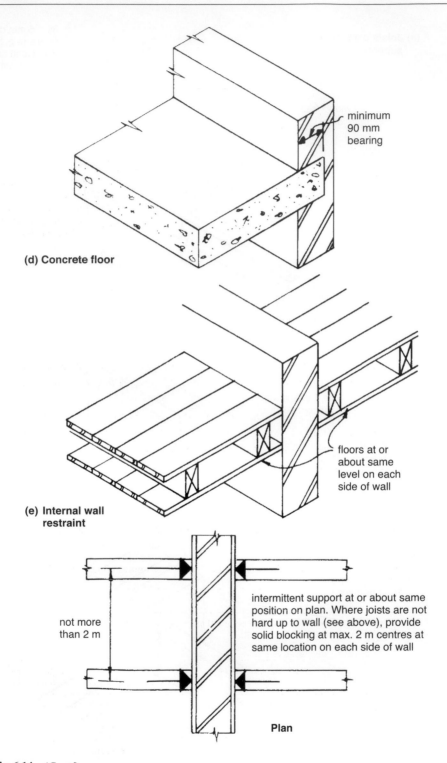

(d) Concrete floor

minimum
90 mm
bearing

floors at or
about same
level on each
side of wall

**(e) Internal wall
restraint**

not more
than 2 m

intermittent support at or about same
position on plan. Where joists are not
hard up to wall (see above), provide
solid blocking at max. 2 m centres at
same location on each side of wall

Plan

Fig. 6.14   (*Contd*).

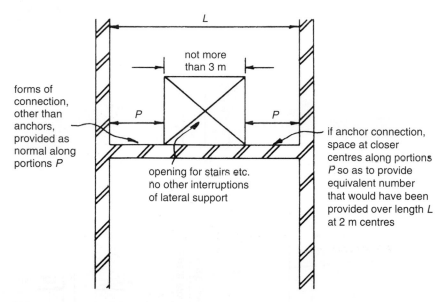

forms of connection, other than anchors, provided as normal along portions *P*

if anchor connection, space at closer centres along portions *P* so as to provide equivalent number that would have been provided over length *L* at 2 m centres

opening for stairs etc. no other interruptions of lateral support

Fig. 6.15   Interruption of lateral support.

These thicknesses do not apply to parapet walls, for which there are special rules (see below) or to bays, and gables over bay windows above the level of the lowest window sill.

As a general rule, the thickness of any storey of a brick or block wall should not be less than one sixteenth of the height of that storey. However, walls of uncoursed stone, flints, clunches of bricks or other burnt or vitrified material should have a thickness of at least $1^{1}/_{3}$ times the thickness required of brick or block walls.

AD A1/2, section 2C

Table 3   Minimum thickness of certain external walls, compartment walls and separating walls.

(1) Height of wall	(2) Length of wall	(3) Minimum thickness of wall
Not exceeding 3.5 m	Not exceeding 12 m	190 mm for the whole of its height
Exceeding 3.5 m but not exceeding 9 m	Not exceeding 9 m	190 mm for the whole of its height
	Exceeding 9 m	290 mm from the base for the height of one storey, and 190 mm for the rest of its height
Exceeding 9 m but not exceeding 12 m	Not exceeding 9 m	290 mm from the base for the height of one storey, and 190 mm for the rest of its height
	Exceeding 9 m but not exceeding 12 m	290 mm from the base for the height of two storeys, and 190 mm for the rest of its height

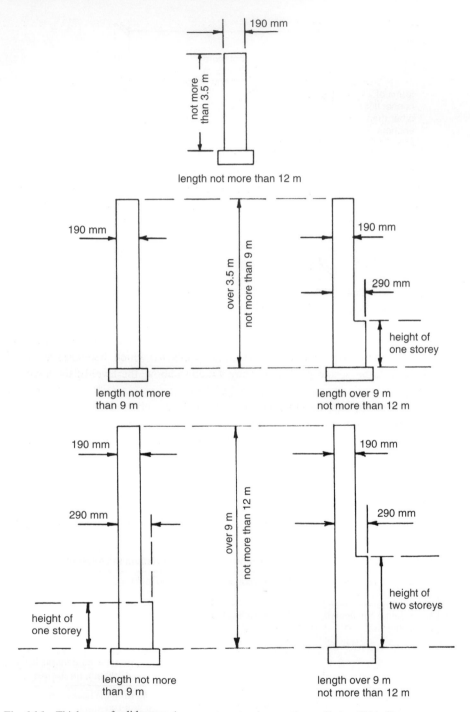

Fig. 6.16   Thickness of solid external, compartment and separating walls (see Table 3).

Irrespective of the materials used in construction, no part of a wall should be thinner than any other part of the wall that it supports.

6.7.9 Solid internal load-bearing walls which are not compartment or separating walls

For these walls the sum of the wall thickness, plus 5 mm, should be equal to at least half the thickness that would be required by Table 3 for an external wall, compartment wall or separating wall of the same height and length.

Where a wall forms the lowest storey of a three-storey building, and it carries loading from both upper storeys, its thickness should not be less than the thickness calculated above or 140 mm, *whichever is greater.* Thus there is an absolute minimum thickness of 140 mm for such walls.

6.7.10 Cavity walls

Any external, compartment or separating wall which is built as a cavity wall should consist of two leaves, each leaf built of bricks or blocks.

The leaves of these walls should be properly tied together with wall ties in compliance with Table 5 of section 2C of AD A1/2 (see section 6.6.7 above). Ties should be placed at centres 900 mm horizontally and 450 mm vertically (i.e. 2.5 ties per m²), and at any opening, movement joint or roof verge at least one tie should be provided for each 300 mm of height within 225 mm of the opening. The cavity should be at least 50 mm wide and each leaf should be at least 90 mm thick at any level.

The sum of the thicknesses of the two leaves, plus 10 mm, should not be less than the thickness required for a solid wall of the same height and length by Table 3. (See also Fig. 6.17.)

6.7.11 Parapets

The minimum thicknesses for both solid and cavity parapet walls are related to their heights as is shown in Fig. 6.18.

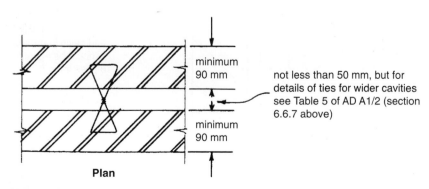

Plan

Fig. 6.17 Cavity walls.

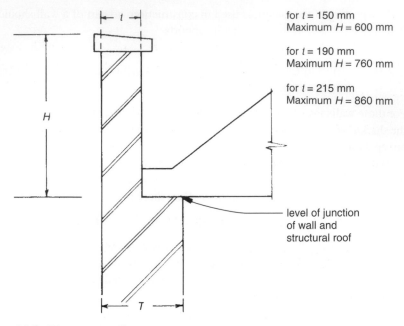

for t = 150 mm
Maximum H = 600 mm

for t = 190 mm
Maximum H = 760 mm

for t = 215 mm
Maximum H = 860 mm

level of junction
of wall and
structural roof

(a) Solid parapet walls

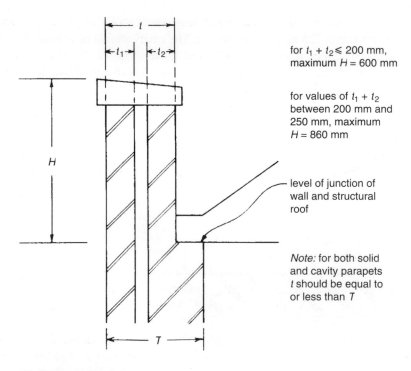

for $t_1 + t_2 \leqslant$ 200 mm,
maximum H = 600 mm

for values of $t_1 + t_2$
between 200 mm and
250 mm, maximum
H = 860 mm

level of junction of
wall and structural
roof

Note: for both solid
and cavity parapets
t should be equal to
or less than T

(b) Cavity parapet walls

Fig. 6.18   Height of parapet walls.

6.7.12   Block and brick dimensions

The wall thicknesses specified in section 2C relate to the *work size* of the materials used. This means the size specified in the relevant British Standard as the size to which the brick or block must conform, account being taken of any permissible deviations or tolerances specified in the British Standard.

Some walls may be constructed of bricks or blocks having modular dimensions derived from BS 6649: 1985 *Specification for clay and calcium silicate modular bricks*. In these cases, the wall thicknesses prescribed in section 2C may be reduced by an amount not exceeding that allowed in a British Standard for equivalent sized bricks or blocks of the same material.

6.7.13   External walls of small buildings and annexes

The external walls of small single-storey non-buildings and of annexes have to comply with special rules. The external walls of such buildings may be not less than 90 mm thick if:

- piers are provided at intervals and sizes as shown in Fig. 6.20 below, tied (using flat stainless steel min 20 mm x 30 mm in cross section, placed in pairs minimum 300 mm vertical centres) or bonded to walls if length of wall exceeds 2.5 m;
- the enclosed floor area does not exceed 36 m²;
- the walls are solidly constructed of bricks and blocks using materials described in section 6.6.7 above;
- for buildings or annexes with floor areas greater than 10 m² the walls have a mass of at least 130 kg/m² (it should be noted that this surface mass limitation does not apply where the floor area is less than 10 m²);
- roof access is for the purposes of maintenance and repair only;
- the only lateral loads are wind loads (e.g. the roof does not transmit a lateral load to the walls);
- the building or annexe has a maximum length or width which does not exceed 9 m;
- the height of the building or annexe does not exceed the lower value derived from Figs. 6.4(a) and (b) in section 6.6.6 above;
- Bracing is provided to the roof:
  - at rafter level,
  - horizontally at eaves level, and
  - at the base of any gable
  by rigid sarking, timber roof decking or diagonal bracing, etc., in accordance with BS 5268 *Structural use of timber*: Part 3: 1998 *Code of practice for trussed rafter roofs*;
- the walls are tied to the roof structure vertically and horizontally in accordance with section 6.7.6 above;
- horizontal lateral restraint is provided at roof level as is shown in Fig. 6.19;
- for an annexe, the roof structure is fixed to the structure of the main building at both rafter and eaves level;
- the size and location of any openings is restricted to those shown in Fig. 6.20 below.

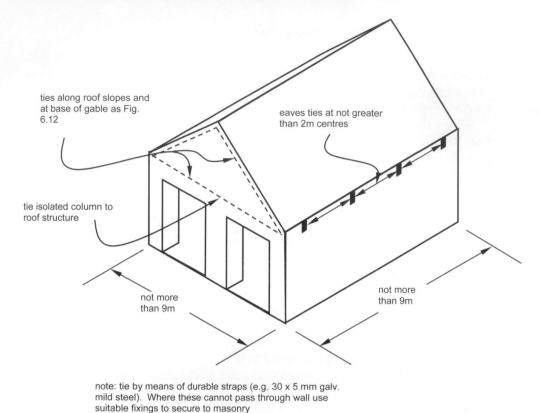

ties along roof slopes and at base of gable as Fig. 6.12

eaves ties at not greater than 2m centres

tie isolated column to roof structure

not more than 9m

not more than 9m

note: tie by means of durable straps (e.g. 30 x 5 mm galv. mild steel). Where these cannot pass through wall use suitable fixings to secure to masonry

Fig. 6.19   Small buildings and annexes – lateral restraint at roof level.

6.7.14   Dimensions of chimneys

The wholly external part of a chimney, constructed of masonry and not supported by adequate ties or otherwise stabilised will be deemed satisfactory if the width of the chimney, at the level of the highest point in the line of junction with the roof and at any higher level, is such that its height as measured from that level to the top of the external part of the chimney is not more than $4^1/_2$ times that width. That height includes any pot or flue terminal on a chimney. Additionally, the masonry should have a density greater than 1500 kg/m³.

The width of chimney at any level is taken as the smallest width which can be shown on an elevation of the chimney from any direction. This is illustrated in Fig. 6.21.

6.7.15   Foundation recommendations

Section 2E of AD A1/2 provides rules for the construction of strip foundations and trench fill foundations of plain concrete. It should be remembered that section 2 applies to certain residential buildings of not more than three storeys, small single-storey non-residential buildings and annexes. Strictly speaking, the guidance given in section 2 of AD A1/2 should not be used for any other building types.

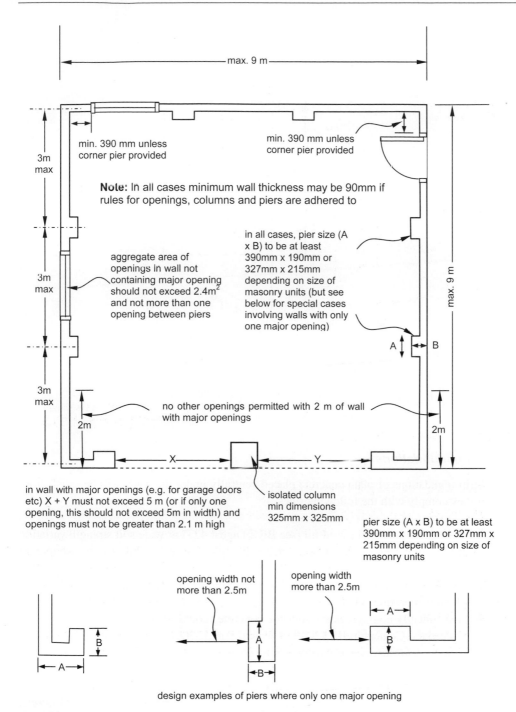

Fig. 6.20   Small buildings and annexes – design of openings, piers and columns.

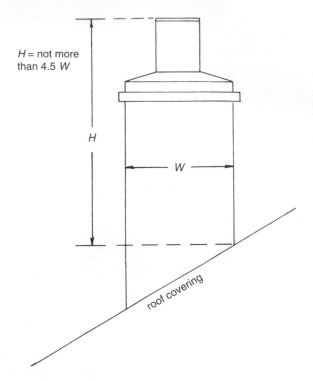

H = not more
than 4.5 W

H

W

roof covering

Fig. 6.21   External part of chimneys, 1D1.

Strip foundations

Strip foundations of plain concrete placed centrally under the walls will be satisfactory if they comply with the following rules:

- There is no non-engineered fill (see BRE Digest 427) or wide soil strength variation in the loaded area or weak soil patches likely to cause instability in the supported structure.
- The width of foundation strip is in accordance with Table 10 to section 2E of AD A1/2 which is reproduced below.
- In chemically non-aggressive soils the concrete should be composed of:
  - Portland Cement to BS EN 197 *Cement: Part 1: 2000 Composition, specifications and conformity criteria for common elements* and Part 2:2000 *Conformity evaluation*; and
  - coarse and fine aggregate to BS EN 12620 2002 *Aggregates for concrete* .
- For foundations in chemically aggressive soils, the guidance in BS 8500 *Concrete. Complementary British Standard* to BS EN 206–1 Part 1:2002 *Method of specifying and guidance for the specifier* should be followed.
- The concrete mix is:
  - in the proportion 50 kg of cement: 0.1 m³ fine aggregate: 0.2 m³ coarse aggregate, i.e. 1:3:6 or better; *or*

    ○ Grade ST2 concrete or Grade GEN 1 concrete to BS 8500: Part 2: 2002 *Specification for constituent materials and concrete.*

- The concrete strip thickness is equal to or greater than the projection from the wall face, and never less than 150 mm.
- The upper level of a stepped foundation overlaps the lower level by twice the height of the step, by the thickness of the foundation or 300 mm, whichever is the greater.
- The height of a step is not greater than the thickness of the foundation.
- The foundation strip projects beyond the faces of any pier, buttress or chimney forming part of a wall by at least as much as it projects beyond the face of the wall proper.

Strip foundations should be laid to the following minimum depths:

- On rock – no minimum depth.
- On other soils except shrinkable clays – 450 mm to avoid the action of frost. However, in areas subject to long periods of frost or in order to transfer the loading onto satisfactory ground, this depth will commonly need to be increased.
- On clay soils subject to volume change on drying (commonly known as 'shrinkable clays', i.e. clays having a plasticity index greater than or equal to 10%) – 750 mm. However, this depth will very often need to be increased in order to transfer the loading onto satisfactory ground. The actual depth decided upon will be based on an assessment of all site conditions including the influence of vegetation and trees, in order that ground movements can be anticipated and will not impair the stability of any part of the building.

Table 10 to section E2 of AD A1/2 specifies seven subsoil types, and the minimum strip widths to use vary according to the calculated load per metre run of the wall at foundation level. The Table is reproduced below.

    Where a wall load exceeds 70 kN per metre run the foundation will be outside the scope of section E2 and must be properly designed on structural principles.

Trench fill foundations

Trench fill foundations are permitted as an alternative to strip foundations. Where used, the overlap at a step should be the greater of twice the height of the step or 1 m.

    These recommendations are illustrated in Fig. 6.22.

6.8   External wall cladding

6.8.1   Introduction

In recent years a number of accidents have occurred involving heavy concrete cladding panels. Failure of the fixings and deterioration of the concrete has resulted in parts and, in some cases, whole panels becoming detached with the resultant danger to people in the street below. There have also been instances of glazing failures endangering the public.

AD A1/2, section 2E

Table 10 Minimum width of strip footings

Type of Ground (including engineered fill)	Condition of ground	Field test applicable	Total load of load-bearing walling not more than (kN/linear metre)					
			20	30	40	50	60	70
			Minimum width of strip foundation (mm)					
I Rock	Not inferior to sandstone, limestone or firm chalk	Requires at least a pneumatic or other mechanically operated pick for excavation	In each case equal to the width of wall					
II Gravel or Sand	Medium dense	Requires pick for excavation. Wooden peg 50 mm square in cross section hard to drive beyond 150 mm	250	300	400	500	600	650
III Clay Sandy Clay	Stiff Stiff	Can be indented slightly by thumb	250	300	400	500	600	650
IV Clay Sandy Clay	Firm Firm	Thumb makes impression easily	300	350	450	600	750	850
V Sand Silty Sand Clayey Sand	Loose Loose Loose	Can be excavated with a spade. Wooden peg 50 mm square in cross section can be easily driven	400	600	Note Foundations on soil types V and V1 do not fall within the provisions of this section if the total load exceeds 30 kN/m.			
VI Silt Clay Sandy Clay Clay or silt	Soft Soft Soft Soft	Finger pushed in up to 10 mm	450	650				
VII Silt Clay Sandy Clay Clay or silt	Very soft Very soft Very soft Very soft	Finger easily pushed in up to 25 mm	Refer to specialist advice					

The table is applicable only within the strict terms of the criteria described within it.

Guidance is provided in section 3 of AD A1/2 which relates to all forms of cladding including curtain walling and glass façades.

Weather-resistance of wall cladding is not covered by the guidance in AD A1/2; Approved Document C (Site preparation and resistance to contaminants and moisture) should be consulted for this.

Examples: cavity wall 60 kN/m run in different soil types, to rules of Table 10

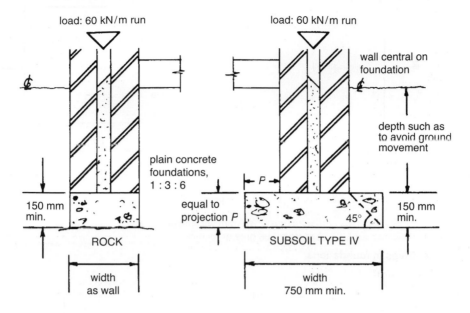

no made ground, no weak patches, no strength variation

(a) Plain strip foundation

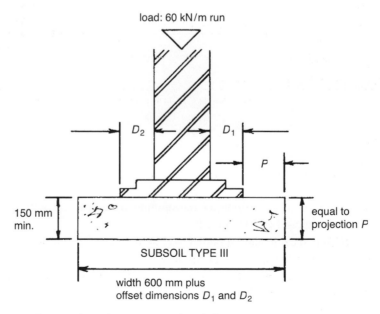

no made ground, no weak patches, no strength variation

(b) Strip foundation with footing

Fig. 6.22   Strip foundations of plain concrete. (*Contd*).

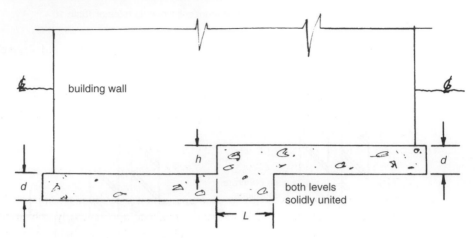

$L = 2h$ or d or 300 mm whichever is greater,
h must not be greater than d

(c) Steps in foundations

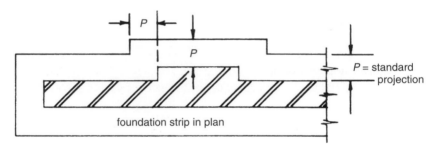

(d) Projections

Fig. 6.22   (*Contd*).

6.8.2   Performance

Wall cladding should be:

- capable of safely carrying and transmitting to the structure of the building the combined dead, imposed and wind loads;
- securely fixed to and supported by the structure of the building, the fixing comprising both vertical support and lateral restraint;
- capable of accommodating differential movement between the cladding and the building support structure;
- manufactured of durable materials (including any fixings and associated support components which should also have an anticipated life at least equal to that of the cladding).

Any fixings should resist corrosion and should be of a material type appropriate for the local environment.

6.8.3   Loading

Apart from the dead load of the cladding itself the following loads should also be taken into account:

- Wind loading – see BS 6399 *Loading for buildings*: Part 2: 1997 *Code of practice for wind loads*. Due consideration should be given to the funnelling effect of the wind through gaps between buildings since this can give rise to local increases in wind suction. A useful source of information on this phenomenon is BRE Digest 436 *Wind loading on buildings – Brief guidance for using BS 6399 Part 2: 1997* which is available from the BRE.
- An assessment of the imposed forces from maintenance equipment such as ladders or access cradles which should be based on the actual equipment likely to be used.
- Loading from fixtures such as handrails, and fittings (e.g. antennae or signboards) supported by the cladding.
- Lateral loads where the cladding is required to act as pedestrian guarding to stairs, ramps and open wells, or as a vehicle barrier. Refer to Approved Document K (Protection from falling, collision and impact) for loading requirements (see Chapter 15).
- Lateral pressures from crowds where the wall cladding is required to act as a barrier. Appropriate design loadings are given in BS 6399 *Loading for buildings*: Part 1: 1996 *Code of practice for dead and imposed loads* and in the publication entitled *Guide to Safety at Sports Grounds* (4th edition, 1997) where the wall cladding is required to act as spectator barriers at sports stadia requiring a safety certificate.

6.8.4   Design and testing of fixings and anchors

The guidance given in section 3 of AD A1/2 regarding the design and testing of fixings and anchors is only of a very general nature. Certain principles are stated and reference is made to a large number of 'further guidance' documents. This reflects the specialist nature of the products, since most cladding systems are or have been the subject of a great deal of research and development involving extensive in-situ testing before they are deemed safe to use.

When a fixing or anchor is selected for a particular cladding support application it is necessary to consider not only the proven performance of the fixing but also the risks associated with the particular application. Thus the required reliability of the fixing or anchor will depend on whether the application is considered as being redundant (i.e. where failure or excessive movement of one fixing results in load sharing by adjacent fixings) or non-redundant (i.e. where failure of a single fixing could lead to detachment of the cladding). In this regard it is possible to obtain fixings and anchors with a European Technical Approval (ETA) gained in accordance with the requirements of ETAG 001: 1997 *Guideline for European Technical Approvals of Metal Anchors for use in Concrete* Parts 1–6, (covering both redundant and non-redundant applications). ETAG 001 Part 6 contains, in an annex, the UK definition of 'Multiple use'. The way in which this definition is framed allows all applications to be validated as to whether or not they conform to this category without calculation. Copies of all ETAG parts may be downloaded in English from www.eota.be.

The strength of a fixing is a function of the fixing itself and the material into which it is fixed. Therefore, its strength should be derived from tests using materials which are representative of the true in-situ condition. In this way inherent weaknesses in the support structure, such as shrinkage or flexure cracks in concrete and voids in masonry, will be highlighted and may be taken into account in the final design of the fixing. Design loads can usually be obtained from manufacturer's test data either determined from a European Technical Approval (ETA – see definition in Chapter 8) or from a British Standard.

It should be noted that European Technical Approvals are available which cover use in:

- cracked and non-cracked concrete (higher loads being allowed for non-cracked concrete); or
- non-cracked concrete only.

In determining whether a particular concrete section may be regarded as cracked or non-cracked, reference should be made to the publication *Use of anchors with European Technical Approvals. UK Guidance – Distinction between cracked and non-cracked concrete*. This allows the distinction to be made without reverting to stress calculations. The publication may be obtained from the BBA website www.bbacerts.co.uk by clicking the 'ETA' tab.

6.8.5   Wall claddings – further guidance

Large glass panels – special consideration needs to be given to the use of large panels of glass in the cladding of walls and roofs (i.e. where the cladding is not divided into small areas by load bearing framing). Guidance is given in the following:

- *Structural use of glass in buildings* (1999 edition) published by The Institution of Structural Engineers, available from 11 Upper Belgrave Street, London SW1X 8BH;
- *Nickel sulfide in toughened glass* (2000 edition) published by the Centre for Window Cladding and Technology.

Cladding – Further guidance on cladding is given in the following:

- *Aspects of Cladding* (1995 edition) published by The Institution of Structural Engineers;
- *Guide to the structural use of adhesives* (1999 edition) published by The Institution of Structural Engineers;
- BS 8297: 2000 *Code of practice for the design and installation of non-loadbearing precast concrete cladding*;
- BS 8298: 1994 *Code of practice for the design and installation of natural stone cladding and lining*.

Fixings – Additional guidance on fixings is given in the following:

- ETAG No. 001 1997 *Guideline for European Technical Approvals of Metal Anchors for use in Concrete*, European Organisation for Technical Approvals (EOTA) (see section 6.8.4 for further details). An English version of this document, published by the British Board of Agreement (BBA), may be obtained from BBA, PO Box 195, Bucknalls Lane, Garston, Watford, Hertfordshire WD25 9BA. ETAG 001 contains the following parts:
  - ○ Part 1 Anchors in general
  - ○ Part 2 Torque controlled anchors
  - ○ Part 3 Undercut anchors
  - ○ Part 4 Deformation controlled anchors
  - ○ Part 5 Bonded anchors
  - ○ Part 6 Metal anchors for redundant use in concrete for lightweight systems.

  It should be noted that all EOTA parts may be downloaded in English from www.eota.be.
- BS 5080 *Structural fixings in concrete and masonry:* Part 1: 1993 *Method of test for tensile loading.* This standard describes a method for testing fixings, such as expanding anchors installed in solid materials either on site or for comparative purposes in a standard material.
- CIRIA Report RP 566 *Cladding Fixings: Good practice guidance*, available from 6 Storey's Gate, London SW1P 3AU.
- CIRIA Reports C579 and C589 *Retention of masonry facades – Best practice guide*.
- The following Guidance Notes are published by the Construction Fixings Association, c/o Institute of Spring Technology, Henry Street, Sheffield, South Yorks S3 7EQ:
  - ○ *Procedure for Site Testing Construction Fixings* (1994)
  - ○ *European Technical Approvals for Construction Fixings* (1998)
  - ○ *Anchor Selection* (1995)
  - ○ *Fixings and Fire* (1998)
  - ○ *Anchor Installation* (1996)
  - ○ *Bonded Anchors* (1999)
  - ○ *Heavy Duty Expansion Anchors* (1997)
  - ○ *Fixings for Brickwork and Blockwork* (1997)
  - ○ *Undercut Anchors* (1998)
  - ○ *Fixings and Corrosion* (2002).

6.9   Replacement of roof coverings

6.9.1   Introduction

It is possible that the re-roofing of a building may result in the existing roof structure having to carry substantially more or less load than it did before the works were carried out. This may be due to the inclusion of underdrawing, additional insulation, or lighter or heavier roof-covering materials (such as the replacement of slate with plain tiles). Section 4 of AD A1/2 indicates that, where the work involves a significant change in the applied loading, the replacement works (including any necessary strengthening of the existing support structure) would constitute a material alteration under the provisions of regulation

3(2). A significant change is defined as *'when the loading upon the roof is increased by more than 15%'*. This is curious since it would appear that where there is a substantial decrease in roof loading then the work would not be regarded as a material alteration and the requirements of the regulations would not apply to it, even though AD A1/2 recommends that the roof structure and its anchorage to the supporting structure should be checked to ensure that an adequate factor of safety is maintained against roof uplift under imposed wind loading. In any event, all the materials used to cover a roof must be capable of safely withstanding the concentrated imposed that are imposed on the roof as specified in BS 6399 *Loading for buildings:* Part 3: 1988 *Code of practice for imposed roof loads*. It should be noted that this requirement applies to transparent or translucent materials, but does not include windows of glass in residential buildings where the roof pitch is greater than or equal to 15°.

6.9.2   Checking the roof structure

As is indicated in section 6.9.1, a significant change in the applied loading may constitute a material alteration under the regulations. This will mean that the roof structure and its supporting structure will need to be checked to ensure that the completed work will be no less compliant with requirement A1 than the original roof was before the work was carried out. If such a check reveals that the existing construction is unable to sustain the new loading without the need for strengthening of the roof structure (or replacement of some roof members, etc.) then this will constitute a material alteration and an application under the regulations will have to be made. It is possible that a check on the roof structure may reveal that some existing roof members are carrying greater loads than they were before (with resultant increase in stresses). This does not necessarily mean that the roof structure is less compliant than it was originally if it can be shown that an adequate factor of safety is still being maintained.

AD A1/2 no longer contains guidance on how the existing roof structure may be assessed to see if it is capable of coping with the changed loading conditions. The previous edition of Approved Document A did contain such guidance and this is repeated below since it gives an indication of the assessment stages involved in the checking procedure.

There are three stages to the assessment procedure:

Stage 1 Compare the proposed and original roof loadings.

This should include an allowance for the increase in loading due to water absorption which may be only 0.3% for oven dry slates but up to 10.5% for plain clay or concrete tiles. These figures are based on the dry mass per unit area of roof coverings.

Stage 2 Carry out a structural inspection on the original roof.

The roof structure must be checked to see if:

- it is capable of sustaining the increased load; *or*
- it contains sufficient vertical restraints to cope with the wind uplift forces as a result of the lighter roof covering or addition of underlay.

Stage 3 Carry out appropriate strengthening measures.

These may include:

- replacement of defective parts of the roof, such as structural members, nails or other fixings and vertical restraints;
- provision of additional structural members as necessary to take the increased loads, such as rafters, purlins, binders or trusses, etc.;
- provision of additional restraint straps, ties or fixings to walls as necessary to resist wind uplift forces.

6.10   Disproportionate collapse

6.10.1   Introduction

In May 1968 a gas explosion on the eighteenth floor of a block of flats in London, known as Ronan Point, caused a large portion of the corner of the block to collapse. Following on from the subsequent tribunal and public inquiry into the disaster, new Building Regulations were formulated and introduced in 1970 with the express purpose of preventing further similar occurrences of this kind where the extent of the collapse is disproportionate to its cause.

These regulations have been updated and revised in line with current experience and knowledge, the main requirement being stated in paragraph A3 of Schedule 1 to the 2000 Regulations.

Buildings are required to be constructed so that in the event of an accident they will not suffer collapse to an extent disproportionate to the cause of that collapse.

Approved Document A3 contains guidance on measures designed to reduce the sensitivity of a building to disproportionate collapse in the event of an accident.

The approach now adopted in section 5 of Approved Document A is to divide buildings into three classes categorised by building type and occupancy, as defined in Table 11 (reproduced below). A building falling within a particular class can then be assessed as shown in sections 6.10.3 to 6.10.7 below, to determine whether it is sufficiently robust to sustain damage or failure to a limited extent, without collapse.

6.10.2   Definitions

The following definitions apply in section 5 of Approved Document A:

NOMINAL LENGTH OF LOAD-BEARING WALL – this should be taken as follows, where H is the storey height in metres:

- reinforced concrete wall – the distance between lateral supports but with a maximum length of $2.25H$;
- external masonry wall, timber stud wall, steel stud wall – the length measured between vertical lateral supports;

- internal masonry wall, internal timber stud wall, internal steel stud wall – a maximum length of 2.25*H*.

KEY ELEMENTS – these should be capable of sustaining an accidental design loading of 34 kN/m^2. It is assumed that this loading will be applied to the member and any attached components (e.g. cladding, etc.) in the horizontal and vertical directions taking one direction at a time. Regard must be paid to the ultimate strength of such components and their connections. Such accidental design loading is assumed to act simultaneously with one third of all normal characteristic loading such as wind and imposed loading.

LOAD-BEARING WALL CONSTRUCTION – this term includes walls consisting of close centred timber or lightweight steel section studs and masonry cross-wall construction.

6.10.3   Class 1 buildings

For buildings that have been designed and constructed in accordance with the guidance rules for meeting compliance with requirements A1 and A2 in normal use given in this Chapter (including the other guidance referred to in sections 6.4 and 6.5), additional measures are unlikely to be necessary.

6.10.4   Class 2A buildings

Provide effective horizontal ties, or effective anchorage of suspended floors to walls, as described in the following Codes and Standards for framed and load-bearing wall construction (see definition in section 6.10.2):

- BS 8110 *Structural use of concrete*: Part 1: 1997 *Code of practice for design and construction* and Part 2: 1985 *Code of practice for special circumstances, for structural work of reinforced, pre-stressed or plain concrete*;
- BS 5950 *Structural use of steelwork in building*: Part 1: 2000 *Code of practice for design. Rolled and welded sections, for structural work of steel*;
- BS 5628 *Code of practice for use of masonry*: Part 1: 1992 *Structural use of unreinforced masonry, for structural use of masonry*.

It should be noted that in addition to details of effective horizontal and vertical ties, these Codes and Standards also contain details of the design approaches that can be adopted for checking the integrity of a building following the notional removal of vertical members and the design of key elements as described in sections 6.10.5 and 6.10.6.

6.10.5   Class 2B buildings

For framed and load-bearing wall construction, provide effective horizontal ties as described in the Codes and Standards listed in section 6.10.4 above, together with one of the following alternatives:

- effective vertical ties, as defined in the same Codes and Standards, in all supporting columns and walls; or
- check that upon the notional removal of the following structural elements (i.e. one element at a time in each storey of the building):
  - (a) each supporting column and each beam supporting one or more columns, or
  - (b) any nominal length of load-bearing wall (see definition in section 6.10.2),
  the building remains stable and that the area of floor at any storey at risk of collapse does not exceed 15% of the floor area of that storey or 70 m^2, whichever is smaller, and does not extend further than the immediate adjacent storeys.

  If the area and storeys put at risk cannot be limited as described above when a structural element is notionally removed, then the structural element should be designed as a 'key element' (see definition in section 6.10.2).

6.10.6   Class 3 buildings

For Class 3 buildings it will be necessary to undertake a systematic risk assessment of the building. This should take into account all the normal hazards that may reasonably be foreseen, together with any abnormal hazards.

This will entail selecting critical situations for design that reflect the conditions that can reasonably be foreseen as possible during the life of the building. After this the structural form and concept of the building can be chosen, including any protective measures. The detailed design of the structure and its elements can then be undertaken in accordance with the recommendations given in the Codes and Standards listed in section 6.10.4.

6.10.7   Alternative approach for other buildings

Clearly, not all building types will fall into the classes listed under Table 11, or there may be some buildings for which the consequences of collapse may warrant particular examination of the risks involved. In these cases the performance may be met by the recommendations given in the following Reports:

- 'Guidance on Robustness and Provision against Accidental Actions' dated July1999, together with the accompanying BRE Report No. 200682;
- 'Calibration of Proposed Revised Guidance on meeting Compliance with the Requirements of Building Regulation Part A3'.

Both of the above documents are available on the DCLG website at www.communities.gov.uk.

AD A3, section 5

Table 11   Building Classes

Class	Building Type and Occupancy
1	Houses not exceeding 4 storeys Agricultural buildings Buildings into which people rarely go, provided no part of the building is closer to another building, or area where people do go, than a distance of 1.5 times the building height
2A	5 storey single occupancy house Hotels not exceeding 4 storeys Flats, apartments and other residential buildings not exceeding 4 storeys Offices not exceeding 4 storeys Industrial buildings not exceeding 3 storeys Retailing premises not exceeding 3 storeys or less than 2000 m² floor area in each storey Single storey educational buildings All buildings not exceeding 2 storeys to which members of the public are admitted and which contain floor areas not 2000 m² floor area in each storey
2B	Hotels, flats, apartments and other residential buildings greater than 4 storeys but not exceeding 15 storeys Educational buildings greater than 1 storey but not exceeding 15 storeys Retailing premises greater than 3 storeys but not exceeding 15 storeys Hospitals not exceeding 3 storeys Offices greater than 4 storeys but not exceeding 15 storeys All buildings to which members of the public are admitted which contain floor areas exceeding 2000 m² floor area but less than 5000 m² at each storey Car parking not exceeding 6 storeys
3	All buildings defined above as Class 2A and 2B that exceed the limits on area and/or number of storeys Grandstands accommodating more than 5000 spectators Buildings containing hazardous substances or processes

Note 1: For buildings intended for more than one type of use the Class should be that pertaining to the most onerous type.

Note 2: In determining the number of storeys in a building, basement storeys may be excluded provided such basement storeys fulfil the robustness requirements of Class 2B buildings.

AD A1/2, Appendix A

Table A1 Floor joists

Maximum clear span of joist (m) Timber of strength class SC3 (see Table 1)

Size of joist (mm × mm)	Dead Load [kN/m²] excluding the self weight of the joist								
	Not more than 0.25			More than 0.25 but not more than 0.50			More than 0.50 but not more than 1.25		
	Spacing of joists (mm)								
	400	450	600	400	450	600	400	450	600
38 × 97	1.83	1.69	1.30	1.72	1.56	1.21	1.42	1.30	1.04
38 × 122	2.48	2.39	1.93	2.37	2.22	1.76	1.95	1.79	1.45
38 × 147	2.98	2.87	2.51	2.85	2.71	2.33	2.45	2.29	1.87
38 × 170	3.44	3.31	2.87	3.28	3.10	2.69	2.81	2.65	2.27
38 × 195	3.94	3.75	3.26	3.72	3.52	3.06	3.19	3.01	2.61
38 × 220	4.43	4.19	3.65	4.16	3.93	3.42	3.57	3.37	2.92
47 × 97	2.02	1.91	1.58	1.92	1.82	1.46	1.67	1.53	1.23
47 × 122	2.66	2.56	2.30	2.55	2.45	2.09	2.26	2.08	1.70
47 × 147	3.20	3.08	2.79	3.06	2.95	2.61	2.72	2.57	2.17
47 × 170	3.69	3.55	3.19	3.53	3.40	2.99	3.12	2.94	2.55
47 × 195	4.22	4.06	3.62	4.04	3.89	3.39	3.54	3.34	2.90
47 × 220	4.72	4.57	4.04	4.55	4.35	3.79	3.95	3.74	3.24
50 × 97	2.08	1.97	1.67	1.98	1.87	1.54	1.74	1.60	1.29
50 × 122	2.72	2.62	2.37	2.60	2.50	2.19	2.33	2.17	1.77
50 × 147	3.27	3.14	2.86	3.13	3.01	2.69	2.81	2.65	2.27
50 × 170	3.77	3.62	3.29	3.61	3.47	3.08	3.21	3.03	2.63
50 × 195	4.31	4.15	3.73	4.13	3.97	3.50	3.65	3.44	2.99
50 × 220	4.79	4.66	4.17	4.64	4.47	3.91	4.07	3.85	3.35
63 × 97	2.32	2.20	1.92	2.19	2.08	1.82	1.93	1.84	1.53
63 × 122	2.93	2.82	2.57	2.81	2.70	2.45	2.53	2.43	2.09
63 × 147	3.52	3.39	3.08	3.37	3.24	2.95	3.04	2.92	2.58
63 × 170	4.06	3.91	3.56	3.89	3.74	3.40	3.50	3.37	2.95
63 × 195	4.63	4.47	4.07	4.44	4.28	3.90	4.01	3.85	3.35
63 × 220	5.06	4.92	4.58	4.91	4.77	4.37	4.51	4.30	3.75
75 × 122	3.10	2.99	2.72	2.97	2.86	2.60	2.68	2.58	2.33
75 × 147	3.72	3.58	3.27	3.56	3.43	3.13	3.22	3.09	2.81
75 × 170	4.28	4.13	3.77	4.11	3.96	3.61	3.71	3.57	3.21
75 × 195	4.83	4.70	4.31	4.68	4.52	4.13	4.24	4.08	3.65
75 × 220	5.27	5.13	4.79	5.11	4.97	4.64	4.74	4.60	4.07
38 × 140	2.84	2.73	2.40	2.72	2.59	2.17	2.33	2.15	1.75
38 × 184	3.72	3.56	3.09	3.53	3.33	2.90	3.02	2.85	2.47
38 × 235	4.71	4.46	3.89	4.43	4.18	3.64	3.80	3.59	3.11

AD A1/2, Appendix A

Table A2 Floor joists

Maximum clear span of joist (m) Timber of strength class SC4 (see Table 1)

Size of joist (mm × mm)	Dead Load [kN/m²] excluding the self weight of the joist								
	Not more than 0.25			More than 0.25 but not more than 0.50			More than 0.50 but not more than 1.25		
	Spacing of joists (mm)								
	400	450	600	400	450	600	400	450	600
38 × 97	1.94	1.83	1.59	1.84	1.74	1.51	1.64	1.55	1.36
38 × 122	2.58	2.48	2.20	2.47	2.37	2.08	2.18	2.07	1.83
38 × 147	3.10	2.98	2.71	2.97	2.85	2.59	2.67	2.56	2.31
38 × 170	3.58	3.44	3.13	3.43	3.29	2.99	3.08	2.96	2.68
38 × 195	4.10	3.94	3.58	3.92	3.77	3.42	3.53	3.39	3.07
38 × 220	4.61	4.44	4.03	4.41	4.25	3.86	3.97	3.82	3.46
47 × 97	2.14	2.03	1.76	2.03	1.92	1.68	1.80	1.71	1.50
47 × 122	2.77	2.66	2.42	2.65	2.55	2.29	2.38	2.27	2.01
47 × 147	3.33	3.20	2.91	3.19	3.06	2.78	2.87	2.75	2.50
47 × 170	3.84	3.69	3.36	3.67	3.54	3.21	3.31	3.18	2.88
47 × 195	4.39	4.22	3.85	4.20	4.05	3.68	3.79	3.64	3.30
47 × 220	4.86	4.73	4.33	4.71	4.55	4.14	4.26	4.10	3.72
50 × 97	2.20	2.09	1.82	2.08	1.98	1.73	1.84	1.75	1.54
50 × 122	2.83	2.72	2.47	2.71	2.60	2.36	2.43	2.33	2.06
50 × 147	3.39	3.27	2.97	3.25	3.13	2.84	2.93	2.81	2.55
50 × 170	3.91	3.77	3.43	3.75	3.61	3.28	3.38	3.25	2.94
50 × 195	4.47	4.31	3.92	4.29	4.13	3.75	3.86	3.72	3.37
50 × 220	4.93	4.80	4.42	4.78	4.64	4.25	4.35	4.18	3.80
63 × 97	2.43	2.32	2.03	2.31	2.19	1.93	2.03	1.93	1.71
63 × 122	3.05	2.93	2.67	2.92	2.81	2.55	2.63	2.53	2.27
63 × 147	3.67	3.52	3.21	3.50	3.37	3.07	3.16	3.04	2.76
63 × 170	4.21	4.06	3.70	4.04	3.89	3.54	3.64	3.51	3.19
63 × 195	4.77	4.64	4.23	4.61	4.45	4.05	4.17	4.01	3.65
63 × 220	5.20	5.06	4.73	5.05	4.91	4.56	4.68	4.51	4.11
75 × 122	3.22	3.10	2.83	3.09	2.97	2.71	2.78	2.68	2.43
75 × 147	3.86	3.72	3.39	3.70	3.57	3.25	3.34	3.22	2.93
75 × 170	4.45	4.29	3.91	4.27	4.11	3.75	3.86	3.71	3.38
75 × 195	4.97	4.83	4.47	4.82	4.69	4.29	4.41	4.25	3.86
75 × 220	5.42	5.27	4.93	5.25	5.11	4.78	4.88	4.74	4.35
38 × 140	2.96	2.84	2.58	2.83	2.72	2.47	2.54	2.44	2.17
38 × 184	3.87	3.72	3.38	3.70	3.56	3.23	3.33	3.20	2.90
38 × 235	4.85	4.71	4.31	4.70	4.54	4.12	4.24	4.08	3.70

AD A1/2, Appendix A

Table A3   Ceiling joists

Maximum clear span of joist (m) Timber of strength class SC3 and SC4 (see Table 1)

Size of joist (mm × mm)	Dead Load [kN/m²] excluding the self weight of the joist — SC3						Dead Load [kN/m²] excluding the self weight of the joist — SC4					
	Not more than 0.25			More than 0.25 but not more than 0.50			Not more than 0.25			More than 0.25 but not more than 0.50		
	Spacing of joists (mm)											
	400	450	600	400	450	600	400	450	600	400	450	600
38 × 72	1.15	1.14	1.11	1.11	1.10	1.06	1.21	1.20	1.17	1.17	1.16	1.12
38 × 97	1.74	1.72	1.67	1.67	1.64	1.58	1.84	1.82	1.76	1.76	1.73	1.66
38 × 122	2.37	2.34	2.25	2.25	2.21	2.11	2.50	2.46	2.37	2.37	2.33	2.22
38 × 147	3.02	2.97	2.85	2.85	2.80	2.66	3.18	3.13	3.00	3.00	2.94	2.79
38 × 170	3.63	3.57	3.41	3.41	3.34	3.16	3.81	3.75	3.58	3.58	3.51	3.32
38 × 195	4.30	4.23	4.02	4.02	3.94	3.72	4.51	4.43	4.22	4.22	4.13	3.89
38 × 220	4.98	4.88	4.64	4.64	4.54	4.27	5.21	5.11	4.86	4.86	4.75	4.47
47 × 72	1.27	1.26	1.23	1.23	1.21	1.17	1.35	1.33	1.30	1.30	1.28	1.24
47 × 97	1.92	1.90	1.84	1.84	1.81	1.73	2.03	2.00	1.93	1.93	1.90	1.83
47 × 122	2.60	2.57	2.47	2.47	2.42	2.31	2.74	2.70	2.60	2.60	2.55	2.43
47 × 147	3.30	3.25	3.11	3.11	3.05	2.90	3.47	3.42	3.27	3.27	3.21	3.04
47 × 170	3.96	3.89	3.72	3.72	3.64	3.44	4.15	4.08	3.89	3.89	3.81	3.61
47 × 195	4.68	4.59	4.37	4.37	4.28	4.04	4.90	4.81	4.57	4.57	4.47	4.22
47 × 220	5.39	5.29	5.03	5.03	4.91	4.63	5.64	5.53	5.25	5.25	5.14	4.84
50 × 72	1.31	1.30	1.27	1.27	1.25	1.21	1.39	1.37	1.34	1.34	1.32	1.28
50 × 97	1.97	1.95	1.89	1.89	1.86	1.78	2.08	2.06	1.99	1.99	1.96	1.88
50 × 122	2.67	2.63	2.53	2.53	2.49	2.37	2.81	2.77	2.66	2.66	2.62	2.49
50 × 147	3.39	3.34	3.19	3.19	3.13	2.97	3.56	3.50	3.35	3.35	3.29	3.12
50 × 170	4.06	3.99	3.81	3.81	3.73	3.53	4.25	4.18	3.99	3.93	3.91	3.69
50 × 195	4.79	4.70	4.48	4.48	4.38	4.13	5.01	4.92	4.68	4.68	4.58	4.32
50 × 220	5.52	5.41	5.14	5.14	5.03	4.73	5.77	5.66	5.37	5.37	5.25	4.95
38 × 89	1.54	1.53	1.48	1.48	1.46	1.41	1.63	1.62	1.57	1.57	1.55	1.49
38 × 140	2.84	2.79	2.68	2.68	2.63	2.50	2.99	2.94	2.82	2.82	2.77	2.63
38 × 184	4.01	3.94	3.75	3.75	3.68	3.47	4.20	4.13	3.94	3.94	3.85	3.64

AD A1/2, Appendix A

Table A4 Binders supporting ceiling joists

Maximum clear span of binder (m) Timber of strength class SC3 and SC4 (see Table 1)

Dead Load [kN/m²] excluding the self weight of the binder

Spacing of binders (mm)

	Not more than 0.25						More than 0.25 but not more than 0.50					
Size of binder (mm × mm)	1200	1500	1800	2100	2400	2700	1200	1500	1800	2100	2400	2700
SC3												
47 × 150	2.17	2.05	1.96	1.88	1.81		1.99	1.87				
47 × 175	2.59	2.45	2.33	2.24	2.15	2.08	2.37	2.23	2.11	2.02	1.94	1.87
50 × 150	2.22	2.11	2.01	1.93	1.86		2.04	1.92	1.83			
50 × 175	2.65	2.51	2.39	2.29	2.21	2.13	2.42	2.28	2.16	2.07	1.99	1.91
50 × 200	3.08	2.91	2.77	2.65	2.55	2.47	2.81	2.64	2.50	2.39	2.29	2.21
63 × 125	1.97	1.87					1.82					
63 × 150	2.44	2.31	2.20	2.12	2.04	1.97	2.23	2.11	2.00	1.91	1.84	
63 × 175	2.90	2.74	2.61	2.51	2.41	2.33	2.65	2.49	2.37	2.26	2.17	2.10
63 × 200	3.37	3.18	3.03	2.90	2.79	2.69	3.07	2.88	2.74	2.61	2.51	2.42
63 × 225	3.83	3.61	3.44	3.29	3.16	3.05	3.49	3.27	3.10	2.96	2.84	2.74
75 × 125	2.12	2.01	1.92	1.85			1.95	1.84				
75 × 150	2.61	2.47	2.36	2.26	2.18	2.11	2.39	2.25	2.14	2.05	1.97	1.90
75 × 175	3.10	2.93	2.79	2.68	2.58	2.49	2.83	2.66	2.53	2.42	2.32	2.24
75 × 200	3.59	3.39	3.23	3.09	2.98	2.88	3.27	3.08	2.92	2.79	2.68	2.58
75 × 225	4.08	3.85	3.66	3.51	3.37	3.26	3.71	3.50	3.31	3.16	3.03	2.92
SC4												
47 × 150	2.28	2.16	2.06	1.98	1.90	1.84	2.09	1.97				
47 × 175	2.72	2.57	2.45	2.34	2.26	2.18	2.48	2.34	2.22	2.12	2.03	1.96
50 × 150	2.33	2.21	2.11	2.02	1.95	1.89	2.14	2.02	1.92	1.83		
50 × 175	2.78	2.63	2.51	2.40	2.31	2.23	2.54	2.39	2.27	2.17	2.08	2.01
50 × 200	3.23	3.05	2.90	2.78	2.67	2.58	2.95	2.77	2.62	2.51	2.40	2.32
63 × 125	2.07	1.97	1.88	1.81			1.91	1.80				
63 × 150	2.56	2.42	2.31	2.22	2.14	2.07	2.34	2.21	2.10	2.01	1.93	1.86
63 × 175	3.04	2.87	2.74	2.62	2.53	2.44	2.78	2.61	2.48	2.37	2.28	2.20
63 × 200	3.52	3.32	3.16	3.03	2.92	2.82	3.21	3.02	2.86	2.73	2.63	2.53
63 × 225	4.00	3.77	3.59	3.44	3.31	3.19	3.65	3.42	3.24	3.10	2.97	2.86
75 × 125	2.22	2.11	2.01	1.94	1.87	1.81	2.04	1.93	1.84			
75 × 150	2.73	2.59	2.47	2.37	2.28	2.21	2.50	2.36	2.24	2.15	2.06	1.99
75 × 175	3.24	3.07	2.92	2.80	2.70	2.61	2.96	2.79	2.65	2.53	2.43	2.35
75 × 200	3.75	3.54	3.37	3.23	3.11	3.00	3.42	3.22	3.05	2.92	2.80	2.70
75 × 225	4.26	4.02	3.82	3.66	3.52	3.40	3.88	3.65	3.46	3.30	3.17	3.06

AD A1/2, Appendix A

Table A5 Common or jack rafters for roofs having a pitch more than 15° but not more than 22.5° with access only for purpose of maintenance or repair. Imposed loading 0.75 kN/m² (see Diagram 2).

Maximum clear span of rafter (m) Timber of strength class SC3 and SC4 (see Table 1)

	Dead Load [kN/m²] excluding the self weight of the rafter								
	Not more than 0.50			More than 0.50 but not more than 0.75			More than 0.75 but not more than 1.00		
	Spacing of rafters (mm)								
Size of rafter (mm × mm)	400	450	600	400	450	600	400	450	600
SC3									
38 × 100	2.10	2.05	1.93	1.93	1.88	1.75	1.80	1.75	1.61
38 × 125	2.89	2.79	2.53	2.63	2.55	2.34	2.44	2.35	2.15
38 × 150	3.47	3.34	3.03	3.26	3.14	2.78	3.08	2.96	2.57
47 × 100	2.46	2.40	2.18	2.25	2.19	2.03	2.10	2.03	1.87
47 × 125	3.10	2.99	2.72	2.92	2.81	2.56	2.78	2.67	2.41
47 × 150	3.71	3.57	3.25	3.50	3.36	3.06	3.32	3.20	2.86
50 × 100	2.54	2.45	2.23	2.35	2.29	2.09	2.19	2.12	1.95
50 × 125	3.17	3.05	2.78	2.98	2.87	2.61	2.83	2.73	2.48
50 × 150	3.78	3.64	3.32	3.57	3.43	3.12	3.39	3.26	2.94
38 × 89	1.76	1.72	1.63	1.63	1.59	1.49	1.53	1.49	1.38
38 × 140	3.24	3.12	2.83	3.05	2.93	2.61	2.82	2.72	2.41
SC4									
38 × 100	2.42	2.33	2.11	2.28	2.19	1.99	2.16	2.08	1.88
38 × 125	3.01	2.90	2.64	2.83	2.73	2.48	2.69	2.59	2.35
38 × 150	3.60	3.47	3.16	3.39	3.26	2.97	3.22	3.10	2.82
47 × 100	2.59	2.49	2.27	2.44	2.35	2.13	2.32	2.23	2.02
47 × 125	3.22	3.11	2.83	3.04	2.92	2.66	2.89	2.78	2.53
47 × 150	3.85	3.71	3.38	3.63	3.50	3.18	3.45	3.32	3.02
50 × 100	2.64	2.54	2.32	2.49	2.40	2.18	2.37	2.28	2.07
50 × 125	3.29	3.17	2.89	3.10	2.98	2.72	2.95	2.83	2.58
50 × 150	3.93	3.78	3.45	3.70	3.57	3.25	3.52	3.39	3.09
38 × 89	2.16	2.07	1.88	2.03	1.95	1.77	1.92	1.85	1.68
38 × 140	3.37	3.24	2.95	3.17	3.05	2.77	3.01	2.90	2.63

AD A1/2, Appendix A

Table A6 Purlins supporting rafters to which Table A5 refers (Imposed loading 0.75 kN/m²)

Maximum clear span of purlin (m) Timber of strength class SC3 and SC4 (see Table 1)

	Dead Load [kN/m²] excluding the self weight of the purlin																	
	Not more than 0.50						More than 0.50 but not more than 0.75						More than 0.75 but not more than 1.00					
	Spacing of purlins (mm)																	
Size of purlin (mm × mm)	1500	1800	2100	2400	2700	3000	1500	1800	2100	2400	2700	3000	1500	1800	2100	2400	2700	3000
SC3																		
50 × 150	1.90																	
50 × 175	2.22	2.08	1.96	1.87			2.08	1.95	1.84				1.97	1.84				
50 × 200	2.53	2.37	2.24	2.13	2.02	1.92	2.38	2.22	2.10	1.97	1.85		2.25	2.10	1.95	1.82		
50 × 225	2.84	2.66	2.52	2.40	2.26	2.14	2.67	2.50	2.35	2.20	2.07	1.96	2.53	2.36	2.18	2.03	1.91	1.81
63 × 150	2.06	1.94	1.83				1.94	1.82					1.84					
63 × 175	2.41	2.26	2.13	2.03	1.95	1.87	2.26	2.12	2.00	1.91	1.82		2.14	2.01	1.90	1.80		
63 × 200	2.75	2.58	2.44	2.32	2.22	2.14	2.58	2.42	2.29	2.18	2.08	1.97	2.45	2.29	2.16	2.05	1.93	1.83
63 × 225	3.09	2.89	2.74	2.61	2.50	2.40	2.90	2.72	2.57	2.45	2.33	2.20	2.75	2.58	2.43	2.29	2.16	2.04
75 × 125	1.83																	
75 × 150	2.19	2.06	1.95	1.86			2.06	1.94	1.83				1.96	1.83				
75 × 175	2.56	2.40	2.27	2.17	2.08	2.00	2.41	2.26	2.13	2.03	1.95	1.87	2.28	2.14	2.02	1.92	1.84	
75 × 200	2.92	2.74	2.59	2.47	2.37	2.28	2.75	2.58	2.44	2.32	2.22	2.14	2.61	2.44	2.31	2.20	2.10	2.00
75 × 225	3.28	3.08	2.91	2.78	2.66	2.56	3.09	2.89	2.74	2.61	2.50	2.40	2.93	2.74	2.60	2.47	2.36	2.23
SC4																		
50 × 150	1.99	1.86					1.87											
50 × 175	2.32	2.17	2.05	1.95	1.87		2.18	2.04	1.92	1.83			2.06	1.93	1.82			
50 × 200	2.64	2.48	2.34	2.23	2.14	2.05	2.49	2.33	2.20	2.09	2.00	1.92	2.36	2.20	2.08	1.98	1.89	
50 × 225	2.97	2.78	2.63	2.51	2.40	2.31	2.79	2.62	2.47	2.35	2.25	2.16	2.65	2.48	2.34	2.22	2.12	2.04
63 × 150	2.16	2.02	1.91	1.82			2.03	1.90					1.92	1.80				
63 × 175	2.51	2.36	2.23	2.13	2.04	1.96	2.36	2.22	2.10	2.00	1.91	1.84	2.24	2.10	1.99	1.89	1.81	
63 × 200	2.87	2.69	2.55	2.43	2.33	2.24	2.70	2.53	2.39	2.28	2.18	2.10	2.56	2.40	2.27	2.16	2.06	1.98
63 × 225	3.22	3.02	2.86	2.73	2.61	2.52	3.03	2.84	2.69	2.56	2.45	2.36	2.88	2.70	2.55	2.43	2.32	2.23
75 × 125	1.91																	
75 × 150	2.29	2.15	2.04	1.94	1.86		2.16	2.02	1.91	1.82			2.05	1.92	1.82			
75 × 175	2.67	2.51	2.37	2.26	2.17	2.09	2.51	2.36	2.23	2.13	2.04	1.96	2.39	2.24	2.12	2.02	1.93	1.85
75 × 200	3.05	2.86	2.71	2.58	2.48	2.39	2.87	2.69	2.55	2.43	2.33	2.24	2.72	2.55	2.42	2.30	2.20	2.12
75 × 225	3.42	3.21	3.04	2.90	2.78	2.68	3.22	3.02	2.86	2.73	2.62	2.52	3.06	2.87	2.72	2.59	2.48	2.38

AD A1/2, Appendix A

Table A7   Common or jack rafters for roofs having a pitch more than 15° but not more than 22.5° with access only for purposes of maintenance or repair. Imposed loading 1.00 kN/m² (see Diagram 2).

Maximum clear span of rafter (m) Timber of strength class SC3 and SC4 (see Table 1)

	Size of rafter (mm × mm)	Dead Load [kN/m²] excluding the self weight of the rafter								
		Not more than 0.50			More than 0.50 but not more than 0.75			More than 0.75 but not more than 1.00		
		Spacing of rafters (mm)								
		400	450	600	400	450	600	400	450	600
SC3	38 × 100	2.10	2.05	1.90	1.93	1.88	1.75	1.80	1.75	1.61
	38 × 125	2.73	2.63	2.35	2.59	2.49	2.17	2.44	2.34	2.03
	38 × 150	3.27	3.14	2.79	3.10	2.97	2.58	2.94	2.78	2.41
	47 × 100	2.35	2.26	2.05	2.23	2.15	1.95	2.10	2.03	1.83
	47 × 125	2.93	2.82	2.56	2.78	2.68	2.41	2.66	2.56	2.26
	47 × 150	3.50	3.37	3.07	3.33	3.20	2.86	3.18	3.06	2.68
	50 × 100	2.40	2.31	2.10	2.28	2.19	1.99	2.18	2.09	1.88
	50 × 125	2.99	2.88	2.62	2.84	2.73	2.48	2.71	2.61	2.33
	50 × 150	3.57	3.44	3.13	3.40	3.27	2.95	3.25	3.12	2.76
	38 × 89	1.76	1.72	1.63	1.63	1.59	1.49	1.53	1.49	1.38
	38 × 140	3.05	2.94	2.61	2.90	2.78	2.42	2.76	2.61	2.26
SC4	38 × 100	2.28	2.19	1.99	2.16	2.08	1.89	2.07	1.99	1.80
	38 × 125	2.84	2.73	2.48	2.70	2.59	2.35	2.58	2.48	2.25
	38 × 150	3.40	3.27	2.97	3.23	3.10	2.82	3.09	2.97	2.69
	47 × 100	2.44	2.35	2.14	2.32	2.23	2.03	2.22	2.13	1.94
	47 × 125	3.04	2.93	2.67	2.89	2.78	2.53	2.77	2.66	2.42
	47 × 150	3.64	3.50	3.19	3.46	3.33	3.03	3.31	3.18	2.89
	50 × 100	2.49	2.40	2.18	2.37	2.28	2.07	2.27	2.18	1.98
	50 × 125	3.10	2.99	2.72	2.95	2.84	2.58	2.82	2.72	2.47
	50 × 150	3.71	3.57	3.26	3.46	3.40	3.09	3.38	3.25	2.95
	38 × 89	2.03	1.95	1.77	1.93	1.85	1.68	1.84	1.77	1.60
	38 × 140	3.18	3.06	2.78	3.02	2.90	2.63	2.88	2.77	2.52

AD A1/2, Appendix A

Table A8 Purlins supporting rafters to which Table A7 refers (Imposed loading 1.00 kN/m²)

Maximum clear span of purlin (m) Timber of strength class SC3 and SC4 (see Table 1)

	Dead Load [kN/m²] excluding the self weight of the purlin																	
	Not more than 0.50						More than 0.50 but not more than 0.75						More than 0.75 but not more than 1.00					
	Spacing of purlins (mm)																	
Size of purlin (mm × mm)	1500	1800	2100	2400	2700	3000	1500	1800	2100	2400	2700	3000	1500	1800	2100	2400	2700	3000
SC3																		
50 × 175	2.09	1.95	1.84	—	—	—	1.97	1.85	—	—	—	—	1.88	—	—	—	—	—
50 × 200	2.38	2.23	2.10	1.97	1.85	—	2.26	2.11	1.96	1.82	—	—	2.15	1.98	1.83	—	—	—
50 × 225	2.68	2.50	2.36	2.20	2.07	1.96	2.54	2.36	2.18	2.04	1.92	1.81	2.42	2.21	2.04	1.90	—	—
63 × 150	1.94	1.82	—	—	—	—	1.84	—	—	—	—	—	—	—	—	—	—	—
63 × 175	2.27	2.12	2.01	1.91	1.83	—	2.15	2.01	1.90	1.81	—	—	2.05	1.92	1.81	—	—	—
63 × 200	2.59	2.42	2.29	2.18	2.09	1.98	2.45	2.30	2.17	2.06	1.94	1.83	2.30	2.19	2.06	1.92	1.81	—
63 × 225	2.91	2.72	2.58	2.45	2.33	2.21	2.76	2.58	2.44	2.30	2.16	2.05	2.63	2.46	2.30	2.15	2.02	1.91
75 × 150	2.07	1.94	1.83	—	—	—	1.96	1.84	—	—	—	—	1.87	—	—	—	—	—
75 × 175	2.41	2.26	2.14	2.04	1.95	1.88	2.29	2.14	2.03	1.93	1.85	—	2.18	2.04	1.93	1.84	—	—
75 × 200	2.75	2.58	2.44	2.33	2.23	2.14	2.61	2.45	2.31	2.20	2.11	2.01	2.49	2.33	2.20	2.10	1.98	1.88
75 × 225	3.09	2.90	2.74	2.61	2.50	2.41	2.92	2.75	2.60	2.48	2.36	2.24	2.80	2.62	2.48	2.35	2.21	2.09
SC4																		
50 × 150	1.87	—	—	—	—	—	—	—	—	—	—	—	—	—	—	—	—	—
50 × 175	2.18	2.04	1.93	1.83	—	—	2.07	1.93	1.82	—	—	—	1.97	1.84	—	—	—	—
50 × 200	2.49	2.33	2.20	2.10	2.00	1.92	2.36	2.21	2.08	1.98	1.89	—	2.25	2.10	1.98	1.88	—	—
50 × 225	2.80	2.62	2.48	2.36	2.25	2.16	2.65	2.48	2.34	2.23	2.13	1.95	2.53	2.36	2.23	2.12	2.02	—
63 × 150	2.03	1.90	1.80	—	—	—	1.93	1.81	—	—	—	—	1.84	—	—	—	—	—
63 × 175	2.37	2.22	2.10	2.00	1.91	1.84	2.25	2.10	1.99	1.89	1.81	—	2.14	2.01	1.90	1.81	—	—
63 × 200	2.70	2.53	2.40	2.28	2.19	2.10	2.57	2.40	2.27	2.16	2.07	1.99	2.45	2.29	2.16	2.06	1.96	1.86
63 × 225	3.04	2.85	2.70	2.57	2.46	2.36	2.88	2.70	2.55	2.43	2.32	2.23	2.75	2.58	2.44	2.31	2.20	2.09
75 × 125	1.80	—	—	—	—	—	—	—	—	—	—	—	—	—	—	—	—	—
75 × 150	2.16	2.03	1.92	1.83	—	—	2.05	1.92	1.82	—	—	—	1.96	1.83	—	—	—	—
75 × 175	2.52	2.36	2.24	2.13	2.04	1.96	2.39	2.24	2.12	2.02	1.93	1.86	2.28	2.14	2.02	1.92	1.84	—
75 × 200	2.87	2.70	2.55	2.43	2.33	2.24	2.73	2.56	2.42	2.31	2.21	2.12	2.61	2.44	2.31	2.20	2.10	2.02
75 × 225	3.23	3.03	2.87	2.74	2.62	2.52	3.07	2.88	2.72	2.59	2.48	2.39	2.93	2.75	2.60	2.47	2.36	2.27

AD A1/2, Appendix A

Table A9 Common or jack rafters for roofs having a pitch more than 22.5° but not more than 30° with access only for purposes of maintenance or repair. Imposed loading 0.75 kN/m² (see Diagram 2).

Maximum clear span of rafter (m) Timber of strength class SC3 and SC4 (see Table 1)

Size of rafter (mm × mm)	Dead Load [kN/m²] excluding the self weight of the rafter								
	Not more than 0.50			More than 0.50 but not more than 0.75			More than 0.75 but not more than 1.00		
	Spacing of rafters (mm)								
	400	450	600	400	450	600	400	450	600
SC3									
38 × 100	2.18	2.13	2.01	2.01	1.96	1.82	1.88	1.82	1.68
38 × 125	2.97	2.86	2.60	2.74	2.66	2.44	2.54	2.46	2.25
38 × 150	3.55	3.42	3.11	3.34	3.21	2.92	3.17	3.04	2.72
47 × 100	2.55	2.46	2.23	2.35	2.28	2.10	2.18	2.12	1.95
47 × 125	3.18	3.06	2.79	2.99	2.88	2.62	2.84	2.73	2.48
47 × 150	3.80	3.66	3.33	3.57	3.44	3.13	3.39	3.27	2.97
50 × 100	2.60	2.51	2.28	2.45	2.36	2.14	2.28	2.21	2.03
50 × 125	3.24	3.12	2.84	3.05	2.53	2.67	2.89	2.79	2.53
50 × 150	3.87	3.73	3.40	3.65	3.51	3.20	3.46	3.33	3.03
38 × 89	1.82	1.79	1.69	1.69	1.65	1.55	1.59	1.55	1.44
38 × 140	3.32	3.19	2.90	3.12	3.00	2.72	2.94	2.84	2.55
SC4									
38 × 100	2.48	2.38	2.17	2.33	2.24	2.03	2.21	2.12	1.93
38 × 125	3.08	2.97	2.70	2.90	2.75	2.53	2.75	2.65	2.40
38 × 150	3.69	3.55	3.23	3.47	3.34	3.04	3.29	3.17	2.88
47 × 100	2.65	2.55	2.32	2.49	2.40	2.18	2.37	2.28	2.07
47 × 125	3.30	3.18	2.90	3.11	2.99	2.72	2.95	2.84	2.58
47 × 150	3.94	3.80	3.46	3.71	3.58	3.26	3.53	3.40	3.09
50 × 100	2.71	2.61	2.37	2.55	2.45	2.23	2.42	2.32	2.11
50 × 125	3.37	3.24	2.96	3.17	3.05	2.78	3.01	2.90	2.63
50 × 150	4.02	3.87	3.53	3.79	3.65	3.32	3.60	3.46	3.15
38 × 89	2.21	2.12	1.93	2.07	1.99	1.81	1.97	1.89	1.72
38 × 140	3.45	3.32	3.02	3.24	3.12	2.84	3.08	2.96	2.69

AD A1/2, Appendix A

Table A10 Purlins supporting rafters to which Table A9 refers (Imposed loading 0.75 kN/m²)

Maximum clear span of purlin (m) Timber of strength class SC3 and SC4 (see Table 1)

Dead Load [kN/m²] excluding the self weight of the purlin — Spacing of purlins (mm)

Size of purlin (mm × mm)	Not more than 0.50						More than 0.50 but not more than 0.75						More than 0.75 but not more than 1.00					
	1500	1800	2100	2400	2700	3000	1500	1800	2100	2400	2700	3000	1500	1800	2100	2400	2700	3000
SC3																		
50 × 150	1.95	1.83					1.83											
50 × 175	2.27	2.12	2.01	1.92	1.83		2.13	1.99	1.88				2.02	1.89				
50 × 200	2.59	2.43	2.30	2.19	2.09	1.99	2.43	2.28	2.15	2.03	1.91	1.81	2.30	2.15	2.01	1.88		
50 × 225	2.92	2.73	2.58	2.46	2.34	2.22	2.74	2.56	2.42	2.27	2.14	2.02	2.59	2.42	2.25	2.10	1.98	1.87
63 × 150	2.12	1.98	1.88				1.99	1.86					1.88					
63 × 175	2.47	2.31	2.19	2.09	2.00	1.92	2.32	2.17	2.05	1.95	1.87		2.19	2.05	1.94	1.85		
63 × 200	2.81	2.64	2.50	2.38	2.28	2.19	2.64	2.48	2.34	2.23	2.13	2.04	2.50	2.35	2.22	2.11	1.99	1.89
63 × 225	3.16	2.97	2.81	2.68	2.56	2.47	2.97	2.78	2.63	2.51	2.40	2.28	2.82	2.64	2.49	2.37	2.23	2.11
75 × 125	1.88																	
75 × 150	2.25	2.11	2.00	1.91	1.83		2.11	1.98	1.87				2.00	1.88				
75 × 175	2.62	2.46	2.33	2.22	2.13	2.05	2.46	2.31	2.19	2.08	1.99	1.92	2.33	2.19	2.07	1.97	1.89	1.81
75 × 200	2.99	2.81	2.66	2.54	2.43	2.34	2.81	2.64	2.50	2.38	2.28	2.19	2.67	2.50	2.36	2.25	2.15	2.07
75 × 225	3.36	3.15	2.99	2.85	2.73	2.63	3.16	2.96	2.80	2.67	2.56	2.46	3.00	2.81	2.66	2.53	2.42	2.31
SC4																		
50 × 150	2.04	1.91	1.81				1.91						1.31					
50 × 175	2.37	2.22	2.10	2.00	1.92	1.84	2.23	2.09	1.97	1.88			2.11	1.97	1.86			
50 × 200	2.71	2.54	2.40	2.29	2.19	2.11	2.54	2.38	2.25	2.14	2.05	1.97	2.41	2.26	2.13	2.02	1.94	1.84
50 × 225	3.05	2.86	2.70	2.57	2.46	2.37	2.86	2.68	2.53	2.41	2.30	2.21	2.71	2.54	2.39	2.28	2.18	2.07
63 × 125	1.84																	
63 × 150	2.21	2.07	1.96	1.87			2.08	1.95	1.84				1.96	1.84				
63 × 175	2.57	2.42	2.29	2.18	2.09	2.01	2.42	2.27	2.15	2.04	1.96	1.88	2.29	2.15	2.03	1.93	1.85	
63 × 200	2.94	2.76	2.61	2.49	2.39	2.30	2.76	2.59	2.45	2.33	2.24	2.15	2.62	2.45	2.32	2.21	2.11	2.03
63 × 225	3.30	3.10	2.93	2.80	2.68	2.58	3.10	2.91	2.75	2.62	2.51	2.42	2.94	2.76	2.61	2.48	2.38	2.28
75 × 125	1.96	1.84					1.84											
75 × 150	2.35	2.20	2.09	1.99	1.91	1.84	2.21	2.07	1.96	1.87			2.09	1.96	1.86			
75 × 175	2.73	2.57	2.43	2.32	2.22	2.14	2.57	2.41	2.28	2.18	2.09	2.01	2.44	2.29	2.16	2.06	1.97	1.90
75 × 200	3.12	2.93	2.78	2.65	2.54	2.45	2.93	2.75	2.61	2.49	2.38	2.29	2.79	2.61	2.47	2.35	2.26	2.17
75 × 225	3.50	3.29	3.12	2.98	2.86	2.75	3.30	3.10	2.93	2.80	2.68	2.58	3.13	2.94	2.78	2.65	2.54	2.44

AD A1/2, Appendix A

Table A11 Common or jack rafters for roofs having a pitch more than 22.5° but not more than 30° with access only for purposes of maintenance or repair. Imposed loading 1.00 kN/m² (see Diagram 2).

Maximum clear span of rafter (m) Timber of strength class SC3 and SC4 (see Table 1)

	Size of rafter (mm × mm)	Dead Load [kN/m²] excluding the self weight of the rafter								
		Not more than 0.50			More than 0.50 but not more than 0.75			More than 0.75 but not more than 1.00		
		Spacing of rafters (mm)								
		400	450	600	400	450	600	400	450	600
SC3	38 × 100	2.18	2.13	1.96	2.01	1.95	1.82	1.88	1.82	1.68
	38 × 125	2.80	2.69	2.45	2.65	2.55	2.30	2.53	2.44	2.15
	38 × 150	3.35	3.22	2.93	3.18	3.06	2.73	3.03	2.92	2.55
	47 × 100	2.41	2.32	2.11	2.28	2.20	2.00	2.18	2.10	1.90
	47 × 125	3.00	2.89	2.63	2.85	2.74	2.49	2.72	2.62	2.37
	47 × 150	3.59	3.46	3.14	3.41	3.28	2.98	3.25	3.13	2.83
	50 × 100	2.46	2.37	2.15	2.33	2.24	2.04	2.23	2.14	1.94
	50 × 125	3.06	2.95	2.68	2.91	2.80	2.54	2.78	2.67	2.43
	50 × 150	3.66	3.52	3.21	3.48	3.34	3.04	3.32	3.20	2.90
	38 × 89	1.82	1.79	1.69	1.69	1.65	1.55	1.59	1.55	1.44
	38 × 140	3.13	3.01	2.74	2.97	2.85	2.56	2.83	2.72	2.29
SC4	38 × 100	2.34	2.25	2.04	2.21	2.13	1.93	2.11	2.03	1.84
	38 × 125	2.91	2.80	2.55	2.76	2.66	2.41	2.64	2.53	2.30
	38 × 150	3.48	3.35	3.05	3.30	3.18	2.89	3.16	3.04	2.76
	47 × 100	2.51	2.41	2.19	2.38	2.29	2.08	2.27	2.18	1.98
	47 × 125	3.12	3.00	2.73	2.96	2.85	2.59	2.83	2.72	2.47
	47 × 150	3.73	3.59	3.27	3.54	3.41	3.10	3.38	3.26	2.96
	50 × 100	2.56	2.46	2.24	2.42	2.33	2.12	2.32	2.23	2.02
	50 × 125	3.18	3.06	2.79	3.02	2.91	2.64	2.89	2.78	2.52
	50 × 150	3.80	3.66	3.34	3.61	3.48	3.16	3.45	3.32	3.02
	38 × 89	2.08	2.00	1.82	1.97	1.90	1.72	1.88	1.81	1.64
	38 × 140	3.25	3.13	2.85	3.09	2.97	2.70	2.95	2.84	2.57

AD A1/2, Appendix A

Table A12   Purlins supporting rafters to which Table A11 refers (Imposed loading $1.00\ \text{kN/m}^2$)

Maximum clear span of purlin (m) Timber of strength class SC3 and SC4 (see Table 1)

Size of purlin (mm × mm)	Dead Load [kN/m²] excluding the self weight of the purlin																	
	Not more than 0.50						More than 0.50 but not more than 0.75						More than 0.75 but not more than 1.00					
	Spacing of purlins (mm)																	
	1500	1800	2100	2400	2700	3000	1500	1800	2100	2400	2700	3000	1500	1800	2100	2400	2700	3000
SC3																		
50 × 150	1.84																	
50 × 175	2.14	2.00	1.89				2.03	1.89					1.93	1.80				
50 × 200	2.45	2.29	2.16	2.05	1.93	1.82	2.31	2.16	2.03	1.89			2.20	2.05	1.89			
50 × 225	2.75	2.57	2.43	2.29	2.15	2.04	2.60	2.43	2.26	2.11	1.99	1.88	2.48	2.29	2.11	1.97	1.85	
63 × 150	2.00	1.87	1.85				1.89						1.88					
63 × 175	2.33	2.18	2.06	1.96	1.88	1.80	2.20	2.06	1.95	1.85			2.10	1.96	1.85			
63 × 200	2.66	2.49	2.35	2.24	2.14	2.05	2.51	2.35	2.22	2.12	2.00	1.90	2.40	2.24	2.12	1.99	1.87	
63 × 225	2.98	2.80	2.65	2.52	2.41	2.29	2.83	2.65	2.50	2.38	2.24	2.12	2.69	2.52	2.38	2.22	2.09	1.98
75 × 150	2.12	1.99	1.88				2.01	1.88					1.92					
75 × 175	2.47	2.32	2.20	2.09	2.00	1.93	2.34	2.20	2.08	1.98	1.89	1.82	2.24	2.09	1.98	1.88	1.80	
75 × 200	2.82	2.65	2.51	2.39	2.29	2.20	2.68	2.51	2.37	2.26	2.16	2.08	2.55	2.39	2.26	2.15	2.05	1.94
75 × 225	3.17	2.98	2.82	2.68	2.57	2.47	3.01	2.82	2.67	2.54	2.43	2.32	2.87	2.69	2.54	2.42	2.29	2.17
SC4																		
50 × 150	1.92						1.82											
50 × 175	2.24	2.10	1.98	1.89	1.80		2.12	1.98	1.87				2.02	1.89				
50 × 200	2.56	2.39	2.26	2.15	2.06	1.98	2.42	2.26	2.14	2.03	1.94	1.86	2.31	2.16	2.03	1.93	1.81	
50 × 225	2.87	2.69	2.54	2.42	2.32	2.22	2.72	2.55	2.40	2.29	2.18	2.09	2.59	2.42	2.29	2.17	2.04	1.83
63 × 150	2.09	1.95	1.85				1.98	1.85					1.88					
63 × 175	2.43	2.28	2.16	2.05	1.97	1.89	2.30	2.16	2.04	1.94	1.86		2.20	2.06	1.94	1.85		
63 × 200	2.77	2.60	2.46	2.35	2.25	2.16	2.63	2.46	2.33	2.22	2.12	2.04	2.51	2.35	2.22	2.11	2.02	1.94
63 × 225	3.12	2.92	2.77	2.64	2.52	2.43	2.95	2.77	2.62	2.49	2.39	2.29	2.82	2.64	2.49	2.37	2.27	2.18
75 × 125	1.85																	
75 × 150	2.22	2.08	1.97	1.88			2.10	1.97	1.86				2.01	1.88				
75 × 175	2.58	2.42	2.29	2.19	2.10	2.02	2.45	2.30	2.17	2.07	1.98	1.91	2.34	2.19	2.07	1.97	1.89	1.81
75 × 200	2.95	2.77	2.62	2.50	2.39	2.30	2.80	2.62	2.48	2.36	2.26	2.18	2.67	2.50	2.37	2.25	2.16	2.07
75 × 225	3.31	3.11	2.94	2.81	2.70	2.59	3.14	2.95	2.79	2.66	2.55	2.45	3.00	2.81	2.66	2.53	2.42	2.33

AD A1/2, Appendix A

Table A13 Common or jack rafters for roofs having a pitch more than 30° but not more than 45° with access only for purposes of maintenance or repair. Imposed loading 0.75 kN/m² (see Diagram 2).

Maximum clear span of rafter (m) Timber of strength class SC3 and SC4 (see Table 1)

	Size of rafter (mm × mm)	Dead Load [kN/m²] excluding the self weight of the rafter								
		Not more than 0.50			More than 0.50 but not more than 0.75			More than 0.75 but not more than 1.25		
		Spacing of rafters (mm)								
		400	450	600	400	450	600	400	450	600
SC3	38 × 100	2.28	2.23	2.10	2.10	2.05	1.91	1.96	1.91	1.76
	38 × 125	3.07	2.95	2.69	2.87	2.77	2.52	2.65	2.56	2.35
	38 × 150	3.67	3.53	3.22	3.44	3.31	3.01	3.26	3.14	2.85
	47 × 100	2.64	2.54	2.31	2.45	2.38	2.17	2.28	2.21	2.04
	47 × 125	3.29	3.17	2.88	3.09	2.97	2.70	2.92	2.81	2.56
	47 × 150	3.93	3.78	3.45	3.69	3.55	3.23	3.50	3.37	3.06
	50 × 100	2.69	2.59	2.36	2.53	2.43	2.21	2.38	2.30	2.09
	50 × 125	3.35	3.23	2.94	3.15	3.03	2.76	2.98	2.87	2.61
	50 × 150	4.00	3.86	3.52	3.76	3.62	3.30	3.57	3.44	3.13
	38 × 89	1.91	1.87	1.77	1.77	1.73	1.62	1.67	1.62	1.50
	38 × 140	3.43	3.30	3.01	3.22	3.10	2.82	3.05	2.93	2.66
SC4	38 × 100	2.56	2.47	2.24	2.40	2.31	2.10	2.28	2.19	1.99
	38 × 125	3.19	3.07	2.80	2.99	2.88	2.62	2.84	2.73	2.48
	38 × 150	3.81	3.67	3.35	3.58	3.45	3.14	3.39	3.27	2.97
	47 × 100	2.74	2.64	2.41	2.58	2.48	2.25	2.44	2.35	2.13
	47 × 125	3.41	3.29	3.00	3.21	3.09	2.81	3.04	2.93	2.66
	47 × 150	4.08	3.93	3.59	3.83	3.69	3.36	3.64	3.50	3.19
	50 × 100	2.80	2.70	2.45	2.63	2.53	2.30	2.49	2.40	2.18
	50 × 125	3.48	3.35	3.06	3.27	3.15	2.87	3.10	2.99	2.72
	50 × 150	4.16	4.01	3.66	3.91	3.77	3.43	3.71	3.57	3.25
	38 × 89	2.28	2.20	2.00	2.14	2.06	1.87	2.03	1.95	1.77
	38 × 140	3.56	3.43	3.13	3.35	3.22	2.93	3.17	3.05	2.77

AD A1/2, Appendix A

Table A14 Purlins supporting rafters to which Table A13 refers. (Imposed loading 0.75 kN/m²)

Maximum clear span of purlin (m) Timber of strength class SC3 and SC4 (see Table 1)

Dead Load [kN/m²] excluding the self weight of the purlin

Size of purlin (mm × mm)	Not more than 0.50						More than 0.50 but not more than 0.75						More than 0.75 but not more than 1.00					
Spacing of purlins (mm)	1500	1800	2100	2400	2700	3000	1500	1800	2100	2400	2700	3000	1500	1800	2100	2400	2700	3000
SC3																		
50 × 150	2.02	1.89					1.89											
50 × 175	2.36	2.21	2.09	1.99	1.90	1.83	2.21	2.06	1.95	1.86			2.08	1.95	1.84			
50 × 200	2.69	2.52	2.38	2.27	2.17	2.09	2.52	2.36	2.23	2.12	2.01	1.90	2.38	2.23	2.10	1.97	1.85	
50 × 225	3.02	2.83	2.68	2.55	2.44	2.34	2.83	2.65	2.50	2.38	2.24	2.12	2.68	2.50	2.36	2.20	2.07	1.96
63 × 125	1.83																	
63 × 150	2.19	2.06	1.95	1.85			2.05	1.93	1.82				1.94	1.82				
63 × 175	2.55	2.40	2.27	2.16	2.07	1.99	2.39	2.24	2.12	2.02	1.94	1.86	2.27	2.12	2.01	1.91	1.83	
63 × 200	2.91	2.74	2.59	2.47	2.37	2.28	2.73	2.56	2.42	2.31	2.21	2.13	2.59	2.42	2.29	2.18	2.09	1.98
63 × 225	3.28	3.07	2.91	2.78	2.66	2.56	3.07	2.88	2.73	2.60	2.49	2.39	2.91	2.72	2.58	2.45	2.33	2.21
75 × 125	1.94	1.82					1.82											
75 × 150	2.33	2.19	2.07	1.97	1.89	1.82	2.18	2.05	1.94	1.85			2.07	1.94	1.83			
75 × 175	2.71	2.55	2.41	2.30	2.21	2.12	2.55	2.39	2.26	2.15	2.06	1.99	2.41	2.26	2.14	2.04	1.95	1.87
75 × 200	3.10	2.91	2.75	2.63	2.52	2.43	2.91	2.73	2.58	2.46	2.36	2.27	2.75	2.58	2.44	2.33	2.23	2.14
75 × 225	3.48	3.27	3.10	2.95	2.83	2.73	3.26	3.06	2.90	2.77	2.65	2.55	3.09	2.90	2.74	2.61	2.50	2.41
SC4																		
50 × 150	2.11	1.98					1.98						1.87					
50 × 175	2.46	2.31	2.18	2.08	1.99	1.91	2.31	2.16	2.04	1.94	1.86		2.18	2.04	1.93	1.83		
50 × 200	2.81	2.63	2.49	2.37	2.27	2.19	2.63	2.47	2.33	2.22	2.12	2.04	2.49	2.33	2.20	2.09	2.00	1.92
50 × 225	3.16	2.96	2.80	2.67	2.56	2.46	2.96	2.77	2.62	2.50	2.39	2.30	2.80	2.62	2.48	2.36	2.25	2.16
63 × 125	1.91	1.87																
63 × 150	2.29	2.15	2.03	1.94	1.86		2.15	2.01	1.90	1.81			2.03	1.90	1.80			
63 × 175	2.67	2.50	2.37	2.26	2.17	2.08	2.50	2.35	2.22	2.12	2.03	1.95	2.37	2.22	2.10	2.00	1.91	1.84
63 × 200	3.04	2.86	2.71	2.58	2.47	2.38	2.86	2.68	2.54	2.42	2.31	2.23	2.70	2.53	2.40	2.28	2.19	2.10
63 × 225	3.42	3.21	3.04	2.90	2.78	2.68	3.21	3.01	2.85	2.72	2.60	2.50	3.04	2.85	2.70	2.57	2.47	2.36
75 × 125	2.03	1.90	1.80				1.90						1.80					
75 × 150	2.43	2.28	2.16	2.06	1.98	1.91	2.28	2.14	2.03	1.93	1.85		2.16	2.03	1.92	1.83		
75 × 175	2.83	2.66	2.52	2.40	2.31	2.22	2.66	2.49	2.36	2.25	2.16	2.08	2.52	2.36	2.24	2.13	2.04	1.96
75 × 200	3.23	3.03	2.88	2.74	2.63	2.54	3.03	2.85	2.70	2.57	2.47	2.37	2.88	2.70	2.55	2.43	2.33	2.24
75 × 225	3.63	3.41	3.23	3.08	2.96	2.85	3.41	3.20	3.03	2.89	2.77	2.67	3.23	3.03	2.87	2.74	2.62	2.52

AD A1/2, Appendix A

Table A15 Common or jack rafters for roofs having a pitch more than 30° but not more than 45° with access only for purposes of maintenance or repair. Imposed loading 1.00 kN/m^2 (see Diagram 2).

Maximum clear span of rafter (m) Timber of strength class SC3 and SC4 (see Table 1)

	Size of rafter (mm × mm)	Dead Load [kN/m^2] excluding the self weight of the rafter								
		Not more than 0.50			More than 0.50 but not more than 0.75			More than 0.75 but not more than 1.00		
		Spacing of rafters (mm)								
		400	450	600	400	450	600	400	450	600
SC3	38 × 100	2.28	2.23	2.03	2.10	2.05	1.91	1.96	1.91	1.76
	38 × 125	2.90	2.79	2.54	2.75	2.64	2.40	2.62	2.52	2.26
	38 × 150	3.47	3.34	3.04	3.29	3.16	2.87	3.13	3.01	2.69
	47 × 100	2.50	2.40	2.18	2.36	2.27	2.06	2.25	2.17	1.97
	47 × 125	3.11	2.99	2.72	2.94	2.83	2.58	2.81	2.70	2.45
	47 × 150	3.72	3.58	3.26	3.52	3.39	3.08	3.36	3.23	2.94
	50 × 100	2.55	2.45	2.23	2.41	2.32	2.11	2.30	2.21	2.01
	50 × 125	3.17	3.05	2.78	3.00	2.89	2.63	2.87	2.76	2.51
	50 × 150	3.79	3.65	3.33	3.59	3.46	3.15	3.43	3.30	3.00
	38 × 89	1.91	1.87	1.77	1.77	1.73	1.62	1.67	1.62	1.50
	38 × 140	3.24	3.12	2.84	3.07	2.95	2.68	2.93	2.82	2.52
SC4	38 × 100	2.42	2.33	2.12	2.29	2.20	2.00	2.18	2.10	1.90
	38 × 125	3.02	2.90	2.64	2.86	2.75	2.50	2.72	2.62	2.38
	38 × 150	3.61	3.47	3.16	3.42	3.29	2.99	3.26	3.14	2.85
	47 × 100	2.60	2.50	2.27	2.46	2.36	2.15	2.34	2.25	2.05
	47 × 125	3.23	3.11	2.83	3.06	2.95	2.68	2.92	2.81	2.55
	47 × 150	3.86	3.72	3.39	3.66	3.52	3.21	3.49	3.36	3.06
	50 × 100	2.65	2.55	2.32	2.51	2.41	2.19	2.39	2.30	2.09
	50 × 125	3.30	3.17	2.89	3.12	3.01	2.73	2.98	2.87	2.61
	50 × 150	3.94	3.79	3.46	3.73	3.60	3.27	3.57	3.43	3.12
	38 × 89	2.16	2.08	1.89	2.04	1.96	1.78	1.95	1.87	1.70
	38 × 140	3.37	3.25	2.95	3.19	3.07	2.79	3.05	2.93	2.66

AD A1/2, Appendix A

Table A16 Purlins supporting rafters to which Table A15 refers. (Imposed loading $1.00\,\text{kN/m}^2$)

Maximum clear span of purlin (m) Timber of strength class SC3 and SC4 (see Table 1)

Size of purlin (mm × mm)	Dead Load [kN/m²] excluding the self weight of the purlin																	
	Not more than 0.50						More than 0.50 but not more than 0.75						More than 0.75 but not more than 1.00					
	Spacing of purlins (mm)																	
	1500	1800	2100	2400	2700	3000	1500	1800	2100	2400	2700	3000	1500	1800	2100	2400	2700	3000
SC3																		
50 × 150	1.91	—	—	—	—	—	1.80	—	—	—	—	—	—	—	—	—	—	—
50 × 175	2.22	2.08	1.97	1.87	—	—	2.10	1.96	1.85	—	—	—	2.00	1.87	—	—	—	—
50 × 200	2.54	2.38	2.25	2.14	2.03	1.92	2.40	2.24	2.12	1.99	1.87	—	2.28	2.13	1.99	1.85	—	—
50 × 225	2.85	2.67	2.53	2.40	2.27	2.15	2.70	2.52	2.38	2.22	2.09	1.98	2.56	2.40	2.22	2.07	1.95	1.84
63 × 150	2.07	1.94	1.84	—	—	—	1.96	1.83	—	—	—	—	1.86	—	—	—	—	—
63 × 175	2.41	2.26	2.14	2.04	1.95	1.88	2.28	2.14	2.02	1.92	1.84	—	2.17	2.03	1.92	1.83	—	—
63 × 200	2.76	2.58	2.44	2.33	2.23	2.14	2.61	2.44	2.31	2.20	2.10	2.00	2.48	2.32	2.19	2.09	1.97	1.86
63 × 225	3.10	2.90	2.75	2.62	2.51	2.41	2.93	2.74	2.59	2.47	2.36	2.23	2.79	2.61	2.47	2.33	2.20	2.08
75 × 125	1.84	—	—	—	—	—	1.84	—	—	—	—	—	—	—	—	—	—	—
75 × 150	2.20	2.07	1.95	1.86	—	—	2.08	1.95	1.85	—	—	—	1.98	1.86	—	—	—	—
75 × 175	2.57	2.41	2.28	2.17	2.08	2.00	2.43	2.28	2.15	2.05	1.96	1.89	2.31	2.17	2.05	1.95	1.87	—
75 × 200	2.93	2.75	2.60	2.48	2.38	2.29	2.77	2.60	2.46	2.34	2.24	2.16	2.64	2.47	2.34	2.23	2.13	2.04
75 × 225	3.29	3.09	2.92	2.79	2.67	2.57	3.12	2.92	2.76	2.63	2.52	2.43	2.97	2.78	2.63	2.50	2.40	2.28
SC4																		
50 × 150	1.99	1.87	—	—	—	—	1.88	—	—	—	—	—	—	—	—	—	—	—
50 × 175	2.32	2.18	2.06	1.96	1.88	1.80	2.20	2.06	1.94	1.85	—	—	2.09	1.96	1.85	—	—	—
50 × 200	2.65	2.49	2.35	2.24	2.14	2.06	2.51	2.35	2.22	2.11	2.02	1.94	2.39	2.23	2.11	2.00	1.91	—
50 × 225	2.98	2.79	2.64	2.52	2.41	2.31	2.82	2.64	2.49	2.37	2.27	2.18	2.68	2.51	2.37	2.25	2.15	2.01
63 × 125	1.81	—	—	—	—	—	—	—	—	—	—	—	—	—	—	—	—	—
63 × 150	2.16	2.03	1.92	1.83	—	—	2.05	1.92	1.81	—	—	—	1.95	1.83	—	—	—	—
63 × 175	2.52	2.36	2.24	2.13	2.04	1.97	2.39	2.24	2.11	2.01	1.93	1.85	2.27	2.13	2.01	1.91	1.83	—
63 × 200	2.88	2.70	2.56	2.44	2.33	2.24	2.72	2.55	2.41	2.30	2.20	2.12	2.59	2.43	2.30	2.19	2.09	2.01
63 × 225	3.23	3.03	2.87	2.74	2.62	2.52	3.06	2.87	2.71	2.59	2.48	2.38	2.92	2.73	2.58	2.46	2.35	2.26
75 × 125	1.92	—	—	—	—	—	1.82	—	—	—	—	—	—	—	—	—	—	—
75 × 150	2.30	2.16	2.04	1.95	1.87	—	2.18	2.04	1.93	1.84	—	—	2.07	1.94	1.84	—	—	—
75 × 175	2.68	2.51	2.38	2.27	2.18	2.10	2.54	2.38	2.25	2.15	2.06	1.98	2.42	2.27	2.14	2.04	1.96	1.88
75 × 200	3.06	2.87	2.72	2.59	2.49	2.39	2.89	2.72	2.57	2.45	2.35	2.26	2.76	2.59	2.45	2.33	2.23	2.15

AD A1/2, Appendix A

Table A17 Joists for flat roofs with access only for purposes of maintenance or repair. Imposed loading $0.75\,\text{kN/m}^2$ (see Diagram 2).

Maximum clear span of joist (m) Timber of strength class SC3 (see Table 1)

Size of joist (mm × mm)	Dead Load [kN/m²] excluding the self weight of the joist								
	Not more than 0.50			More than 0.50 but not more than 0.75			More than 0.75 but not more than 1.00		
	Spacing of joists (mm)								
	400	450	600	400	450	600	400	450	600
38 × 97	1.74	1.72	1.67	1.67	1.64	1.58	1.61	1.58	1.51
38 × 122	2.37	2.34	2.25	2.25	2.21	2.11	2.16	2.11	2.01
38 × 147	3.02	2.97	2.85	2.85	2.80	2.66	2.72	2.66	2.51
38 × 170	3.63	3.57	3.37	3.41	3.34	3.17	3.24	3.17	2.98
38 × 195	4.30	4.23	3.86	4.03	3.94	3.63	3.81	3.72	3.45
38 × 220	4.94	4.76	4.34	4.64	4.49	4.09	4.38	4.27	3.88
47 × 97	1.92	1.90	1.84	1.84	1.81	1.74	1.77	1.74	1.65
47 × 122	2.60	2.57	2.47	2.47	2.43	2.31	2.36	2.31	2.19
47 × 147	3.30	3.25	3.12	3.12	3.06	2.90	2.96	2.90	2.74
47 × 170	3.96	3.89	3.61	3.72	3.64	3.40	3.53	3.44	3.23
47 × 195	4.68	4.53	4.13	4.37	4.28	3.89	4.14	4.04	3.70
47 × 220	5.28	5.09	4.65	4.99	4.81	4.38	4.75	4.58	4.17
50 × 97	1.97	1.95	1.89	1.89	1.86	1.78	1.81	1.78	1.70
50 × 122	2.67	2.64	2.53	2.53	2.49	2.37	2.42	2.37	2.25
50 × 147	3.39	3.34	3.19	3.19	3.13	2.97	3.04	2.97	2.80
50 × 170	4.06	3.99	3.69	3.81	3.73	3.47	3.61	3.53	3.30
50 × 195	4.79	4.62	4.22	4.48	4.36	3.97	4.23	4.13	3.78
50 × 220	5.38	5.19	4.74	5.09	4.90	4.47	4.85	4.67	4.25
63 × 97	2.19	2.16	2.09	2.09	2.06	1.97	2.01	1.97	1.87
63 × 122	2.95	2.91	2.79	2.79	2.74	2.61	2.66	2.61	2.47
63 × 147	3.72	3.66	3.44	3.50	3.43	3.25	3.33	3.26	3.07
63 × 170	4.44	4.35	3.97	4.16	4.07	3.74	3.95	3.85	3.56
63 × 195	5.14	4.96	4.54	4.86	4.69	4.28	4.61	4.47	4.07
63 × 220	5.77	5.57	5.10	5.46	5.27	4.82	5.21	5.02	4.59
75 × 122	3.17	3.12	3.00	3.00	2.94	2.80	2.86	2.80	2.65
75 × 147	3.98	3.92	3.64	3.75	3.67	3.44	3.56	3.48	3.27
75 × 170	4.74	4.58	4.19	4.44	4.35	3.96	4.21	4.11	3.77
75 × 195	5.42	5.23	4.79	5.13	4.95	4.53	4.89	4.72	4.31
75 × 220	6.07	5.87	5.38	5.76	5.56	5.09	5.50	5.30	4.85
38 × 140	2.84	2.79	2.68	2.68	2.63	2.51	2.56	2.51	2.37
38 × 184	4.01	3.94	3.64	3.76	3.68	3.43	3.56	3.48	3.25

AD A1/2, Appendix A

Table A18   Joists for flat roofs with access only for purposes of maintenance or repair. Imposed loading 0.75 kN/m² (see Diagram 2).

Maximum clear span of joist (m) Timber of strength class SC4 (see Table 1)

Size of joist (mm × mm)	Dead Load [kN/m²] excluding the self weight of the joist								
	Not more than 0.50			More than 0.50 but not more than 0.75			More than 0.75 but not more than 1.00		
	Spacing of joists (mm)								
	400	450	600	400	450	600	400	450	600
38 × 97	1.84	1.82	1.76	1.76	1.73	1.66	1.69	1.66	1.59
38 × 122	2.50	2.46	2.37	2.37	2.33	2.22	2.27	2.22	2.11
38 × 147	3.18	3.13	3.00	3.00	2.94	2.79	2.85	2.79	2.64
38 × 170	3.81	3.75	3.50	3.58	3.51	3.30	3.40	3.32	3.12
38 × 195	4.51	4.40	4.01	4.22	4.13	3.78	3.99	3.90	3.59
38 × 220	5.13	4.95	4.51	4.85	4.67	4.25	4.59	4.44	4.04
47 × 97	2.03	2.00	1.94	1.94	1.91	1.83	1.86	1.83	1.74
47 × 122	2.74	2.70	2.60	2.60	2.55	2.43	2.48	2.43	2.30
47 × 147	3.47	3.42	3.26	3.27	3.21	3.04	3.11	3.04	2.87
47 × 170	4.15	4.08	3.76	3.89	3.81	3.54	3.69	3.61	3.36
47 × 195	4.88	4.70	4.29	4.58	4.44	4.05	4.33	4.22	3.85
47 × 220	5.48	5.29	4.83	5.18	5.00	4.56	4.94	4.76	4.33
50 × 97	2.08	2.06	1.99	1.99	1.96	1.88	1.91	1.88	1.79
50 × 122	2.81	2.77	2.66	2.66	2.62	2.49	2.54	2.49	2.36
50 × 147	3.56	3.50	3.32	3.35	3.29	3.12	3.19	3.12	2.94
50 × 170	4.26	4.18	3.83	3.99	3.91	3.61	3.78	3.69	3.43
50 × 195	4.97	4.80	4.38	4.68	4.53	4.13	4.43	4.31	3.93
50 × 220	5.59	5.39	4.93	5.28	5.09	4.65	5.04	4.85	4.42
63 × 97	2.31	2.28	2.20	2.20	2.16	2.07	2.11	2.07	1.97
63 × 122	3.10	3.05	2.93	2.93	2.88	2.74	2.80	2.74	2.59
63 × 147	3.90	3.84	3.58	3.67	3.60	3.38	3.49	3.41	3.21
63 × 170	4.65	4.51	4.12	4.35	4.26	3.89	4.13	4.03	3.70
63 × 195	5.33	5.15	4.71	5.05	4.87	4.45	4.82	4.64	4.24
63 × 220	5.98	5.78	5.30	5.57	5.47	5.00	5.41	5.22	4.76
75 × 122	3.33	3.27	3.14	3.14	3.08	2.93	2.99	2.93	2.77
75 × 147	4.17	4.10	3.78	3.92	3.84	3.57	3.73	3.64	3.40
75 × 170	4.92	4.75	4.35	4.64	4.50	4.11	4.40	4.29	3.92
75 × 195	5.61	5.42	4.97	5.32	5.14	4.70	5.08	4.90	4.48
75 × 220	6.29	6.08	5.59	5.97	5.77	5.28	5.70	5.50	5.04
38 × 140	2.99	2.94	2.82	2.82	2.75	2.63	2.69	2.63	2.49
38 × 184	4.21	4.13	3.79	3.94	3.85	3.57	3.73	3.64	3.39

AD A1/2, Appendix A

Table A19 Joists for flat roofs with access only for purposes of maintenance or repair. Imposed loading 1.0 kN/m² (see Diagram 2).

Maximum clear span of joist (m) Timber of strength class SC3 (see Table 1)

Size of joist (mm × mm)	Dead Load [kN/m²] excluding the self weight of the joist								
	Not more than 0.50			More than 0.50 but not more than 0.75			More than 0.75 but not more than 1.00		
	Spacing of joists (mm)								
	400	450	600	400	450	600	400	450	600
38 × 97	1.74	1.72	1.67	1.67	1.64	1.58	1.61	1.58	1.51
38 × 122	2.37	2.34	2.25	2.25	2.21	2.11	2.16	2.11	2.01
38 × 147	3.02	2.97	2.75	2.85	2.80	2.61	2.72	2.66	2.49
38 × 170	3.62	3.49	3.17	3.41	3.31	3.01	3.24	3.17	2.88
38 × 195	4.15	3.99	3.63	3.94	3.79	3.45	3.77	3.63	3.29
38 × 220	4.67	4.49	4.09	4.44	4.27	3.88	4.25	4.09	3.71
47 × 97	1.92	1.90	1.84	1.84	1.81	1.74	1.77	1.74	1.65
47 × 122	2.60	2.57	2.45	2.47	2.43	2.31	2.36	2.31	2.19
47 × 147	3.30	3.24	2.95	3.12	3.06	2.80	2.96	2.90	2.68
47 × 170	3.88	3.74	3.40	3.69	3.56	3.23	3.53	3.40	3.09
47 × 195	4.44	4.27	3.89	4.23	4.07	3.70	4.05	3.89	3.54
47 × 220	4.99	4.81	4.38	4.75	4.58	4.17	4.55	4.38	3.99
50 × 97	1.97	1.95	1.89	1.89	1.86	1.78	1.81	1.78	1.70
50 × 122	2.67	2.64	2.50	2.53	2.49	2.37	2.42	2.37	2.25
50 × 147	3.39	3.31	3.01	3.19	3.13	2.86	3.04	2.97	2.73
50 × 170	3.96	3.81	3.47	3.77	3.63	3.30	3.61	3.47	3.16
50 × 195	4.53	4.36	3.97	4.31	4.15	3.78	4.13	3.97	3.61
50 × 220	5.09	4.90	4.47	4.85	4.67	4.25	4.65	4.47	4.07
63 × 97	2.19	2.16	2.09	2.09	2.06	1.97	2.01	1.97	1.87
63 × 122	2.95	2.91	2.70	2.79	2.74	2.57	2.66	2.61	2.46
63 × 147	3.70	3.56	3.25	3.50	3.39	3.09	3.33	3.25	2.95
63 × 170	4.26	4.10	3.74	4.06	3.91	3.56	3.89	3.74	3.41
63 × 195	4.86	4.69	4.28	4.64	4.47	4.07	4.45	4.28	3.90
63 × 220	5.46	5.27	4.82	5.21	5.02	4.59	5.00	4.82	4.39
75 × 122	3.17	3.12	2.86	3.00	2.94	2.72	2.86	2.80	2.60
75 × 147	3.90	3.76	3.44	3.72	3.59	3.27	3.56	3.44	3.13
75 × 170	4.49	4.33	3.96	4.29	4.13	3.77	4.11	3.96	3.61
75 × 195	5.13	4.95	4.53	4.89	4.72	4.31	4.70	4.53	4.13
75 × 220	5.76	5.56	5.09	5.50	5.30	4.85	5.28	5.09	4.65
38 × 140	2.84	2.79	2.62	2.68	2.63	2.48	2.56	2.51	2.37
38 × 184	3.92	3.77	3.43	3.73	3.58	3.25	3.56	3.43	3.11

AD A1/2, Appendix A

Table A20   Joists for flat roofs with access only for purposes of maintenance or repair. Imposed loading 1.0 kN/m² (see Diagram 2).

Maximum clear span of joist (m) Timber of strength class SC4 (see Table 1)

Size of joist (mm × mm)	Dead Load [kN/m²] excluding the self weight of the joist								
	Not more than 0.50			More than 0.50 but not more than 0.75			More than 0.75 but not more than 1.00		
	Spacing of joists (mm)								
	400	450	600	400	450	600	400	450	600
38 × 97	1.84	1.82	1.76	1.76	1.73	1.66	1.69	1.66	1.59
38 × 122	2.50	2.46	2.37	2.37	2.33	2.22	2.27	2.22	2.11
38 × 147	3.18	3.13	2.86	3.00	2.94	2.71	2.85	2.79	2.59
38 × 170	3.77	3.63	3.30	3.58	3.45	3.13	3.40	3.30	2.99
38 × 195	4.31	4.15	3.78	4.10	3.95	3.59	3.93	3.78	3.43
38 × 220	4.85	4.67	4.25	4.61	4.44	4.04	4.42	4.25	3.86
47 × 97	2.03	2.00	1.94	1.94	1.91	1.83	1.86	1.83	1.74
47 × 122	2.74	2.70	2.55	2.60	2.55	2.42	2.48	2.43	2.30
47 × 147	3.47	3.37	3.07	3.27	3.21	2.91	3.11	3.04	2.79
47 × 170	4.03	3.88	3.54	3.84	3.70	3.36	3.68	3.54	3.22
47 × 195	4.61	4.44	4.05	4.39	4.23	3.85	4.21	4.05	3.68
47 × 220	5.18	5.00	4.56	4.94	4.76	4.33	4.73	4.56	4.15
50 × 97	2.08	2.06	1.99	1.99	1.96	1.88	1.91	1.88	1.79
50 × 122	2.81	2.77	2.60	2.66	2.62	2.47	2.54	2.49	2.36
50 × 147	3.56	3.44	3.13	3.35	3.27	2.97	3.19	3.12	2.85
50 × 170	4.11	3.96	3.61	3.92	3.77	3.43	3.75	3.61	3.28
50 × 195	4.70	4.53	4.13	4.48	4.31	3.93	4.29	4.13	3.76
50 × 220	5.28	5.09	4.65	5.04	4.85	4.42	4.83	4.65	4.23
63 × 97	2.31	2.28	2.20	2.20	2.16	2.07	2.11	2.07	1.97
63 × 122	3.10	3.05	2.81	2.93	2.88	2.67	2.80	2.74	2.56
63 × 147	3.84	3.70	3.38	3.66	3.52	3.21	3.49	3.38	3.07
63 × 170	4.42	4.26	3.89	4.21	4.06	3.70	4.04	3.89	3.54
63 × 195	5.05	4.87	4.45	4.81	4.64	4.24	4.62	4.45	4.06
63 × 220	5.67	5.47	5.00	5.41	5.22	4.76	5.19	5.00	4.56
75 × 122	3.33	3.26	2.97	3.14	3.08	2.83	2.99	2.93	2.71
75 × 147	4.05	3.91	3.57	3.86	3.72	3.40	3.71	3.57	3.25
75 × 170	4.66	4.50	4.11	4.45	4.29	3.92	4.27	4.11	3.75
75 × 195	5.32	5.14	4.70	5.08	4.90	4.48	4.88	4.70	4.29
75 × 220	5.97	5.77	5.28	5.70	5.50	5.04	5.48	5.28	4.83
38 × 140	2.99	2.94	2.72	2.82	2.77	2.59	2.69	2.63	2.47
38 × 184	4.07	3.92	3.57	3.87	3.73	3.39	3.71	3.57	3.24

AD A1/2, Appendix A

Table A21 Joists for flat roofs with access not limited to the purposes of maintenance or repair. Imposed loading 1.50 kN/m².

Maximum clear span of joist (m) Timber of strength class SC3 (see Table 1)

Size of joist (mm × mm)	Dead Load [kN/m²] excluding the self weight of the joist								
	Not more than 0.50			More than 0.50 but not more than 0.75			More than 0.75 but not more than 1.00		
	Spacing of joists (mm)								
	400	450	600	400	450	600	400	450	600
38 × 122	1.80	1.79	1.74	1.74	1.71	1.65	1.68	1.65	1.57
38 × 147	2.35	2.33	2.27	2.27	2.25	2.18	2.21	2.18	2.09
38 × 170	2.88	2.85	2.77	2.77	2.74	2.64	2.68	2.64	2.53
38 × 195	3.47	3.43	3.29	3.33	3.28	3.16	3.21	3.16	3.02
38 × 220	4.08	4.03	3.71	3.90	3.84	3.56	3.75	3.68	3.43
47 × 122	2.00	1.99	1.94	1.94	1.93	1.87	1.89	1.87	1.81
47 × 147	2.60	2.58	2.51	2.51	2.48	2.40	2.44	2.40	2.31
47 × 170	3.18	3.14	3.06	3.06	3.02	2.91	2.95	2.91	2.78
47 × 195	3.82	3.78	3.54	3.66	3.61	3.40	3.52	3.46	3.28
47 × 220	4.48	4.38	3.99	4.27	4.20	3.83	4.10	4.03	3.70
50 × 122	2.06	2.05	2.00	2.00	1.98	1.93	1.95	1.93	1.86
50 × 147	2.68	2.65	2.59	2.59	2.56	2.47	2.51	2.47	2.38
50 × 170	3.27	3.23	3.14	3.14	3.10	2.99	3.04	2.99	2.86
50 × 195	3.93	3.88	3.61	3.76	3.70	3.47	3.62	3.56	3.35
50 × 220	4.60	4.47	4.07	4.38	4.30	3.91	4.21	4.13	3.78
63 × 97	1.67	1.66	1.63	1.63	1.61	1.57	1.59	1.57	1.53
63 × 122	2.31	2.29	2.24	2.24	2.21	2.15	2.17	2.15	2.07
63 × 147	2.98	2.95	2.87	2.87	2.84	2.74	2.78	2.74	2.63
63 × 170	3.62	3.59	3.41	3.48	3.43	3.28	3.36	3.30	3.16
63 × 195	4.34	4.29	3.90	4.15	4.08	3.75	3.99	3.92	3.62
63 × 220	5.00	4.82	4.39	4.82	4.64	4.22	4.62	4.48	4.08
75 × 122	2.50	2.48	2.42	2.42	2.40	2.32	2.35	2.32	2.24
75 × 147	3.23	3.19	3.11	3.11	3.07	2.96	3.00	2.96	2.84
75 × 170	3.91	3.87	3.61	3.75	3.69	3.47	3.61	3.55	3.35
75 × 195	4.66	4.53	4.13	4.45	4.36	3.97	4.28	4.20	3.84
75 × 220	5.28	5.09	4.65	5.09	4.90	4.47	4.92	4.74	4.32
38 × 140	2.19	2.17	2.12	2.12	2.10	2.04	2.07	2.04	1.94
38 × 184	3.21	3.17	3.08	3.08	3.04	2.93	2.98	2.93	2.80

AD A1/2, Appendix A

Table A22   Joists for flat roofs with access not limited to the purposes of maintenance or repair. Imposed loading 1.50 kN/m².

Maximum clear span of joist (m) Timber of strength class SC4 (see Table 1)

Size of joist (mm × mm)	Not more than 0.50			More than 0.50 but not more than 0.75			More than 0.75 but not more than 1.00		
	400	450	600	400	450	600	400	450	600
38 × 122	1.91	1.90	1.86	1.86	1.84	1.79	1.81	1.79	1.73
38 × 147	2.49	2.46	2.40	2.40	2.38	2.30	2.33	2.30	2.21
38 × 170	3.04	3.01	2.93	2.93	2.89	2.79	2.83	2.79	2.67
38 × 195	3.66	3.62	3.43	3.51	3.46	3.29	3.38	3.33	3.18
38 × 220	4.30	4.25	3.86	4.10	4.04	3.71	3.94	3.87	3.58
47 × 122	2.12	2.10	2.06	2.06	2.04	1.98	2.00	1.98	1.91
47 × 147	2.75	2.73	2.66	2.66	2.62	2.54	2.57	2.54	2.44
47 × 170	3.35	3.32	3.22	3.22	3.18	3.06	3.11	3.06	2.93
47 × 195	4.03	3.98	3.68	3.85	3.80	3.54	3.71	3.64	3.42
47 × 220	4.71	4.56	4.15	4.49	4.39	3.99	4.31	4.23	3.85
50 × 122	2.19	2.17	2.12	2.12	2.10	2.04	2.06	2.04	1.97
50 × 147	2.83	2.81	2.73	2.73	2.70	2.61	2.65	2.61	2.51
50 × 170	3.45	3.41	3.28	3.31	3.27	3.15	3.20	3.15	3.01
50 × 195	4.14	4.09	3.76	3.96	3.90	3.61	3.81	3.74	3.49
50 × 220	4.83	4.65	4.23	4.61	4.47	4.07	4.42	4.32	3.93
63 × 97	1.77	1.75	1.72	1.72	1.71	1.66	1.68	1.66	1.61
63 × 122	2.44	2.42	2.36	2.36	2.34	2.27	2.30	2.27	2.18
63 × 147	3.15	3.12	3.03	3.03	2.99	2.89	2.93	2.89	2.77
63 × 170	3.82	3.78	3.54	3.66	3.61	3.41	3.53	3.47	3.29
63 × 195	4.56	4.45	4.06	4.36	4.29	3.90	4.19	4.11	3.77
63 × 220	5.19	5.00	4.56	5.00	4.82	4.39	4.84	4.66	4.24
75 × 122	2.64	2.62	2.56	2.56	2.53	2.45	2.48	2.45	2.36
75 × 147	3.40	3.36	3.25	3.27	3.23	3.11	3.16	3.11	2.98
75 × 170	4.11	4.07	3.75	3.94	3.88	3.61	3.79	3.73	3.49
75 × 195	4.79	4.70	4.29	4.67	4.53	4.13	4.49	4.38	3.99
75 × 220	5.48	5.28	4.83	5.28	5.09	4.65	5.11	4.93	4.49
38 × 140	2.32	2.30	2.25	2.25	2.22	2.16	2.19	2.16	2.08
38 × 184	3.39	3.35	3.24	3.25	3.21	3.09	3.14	3.09	2.95

Dead Load [kN/m²] excluding the self weight of the joist

Spacing of joists (mm)

AD A1/2, Appendix A

Table A23 Purlins supporting sheeting or decking for roofs having a pitch more than 10° but not more than 35°. Imposed loading 0.75 kN/m².

Maximum clear span of purlin (m) Timber of strength class SC3 and SC4 (see Table 1)

	Dead Load [kN/m²] excluding the self weight of the purlin																	
	Not more than 0.25						More than 0.25 but not more than 0.50						More than 0.50 but not more than 0.75					
	Spacing of purlins (mm)																	
Size of purlin (mm × mm)	900	1200	1500	1800	2100	2400	900	1200	1500	1800	2100	2400	900	1200	1500	1800	2100	2400
SC3																		
50 × 100	1.68	1.63	1.51	1.42	1.34	1.28	1.55	1.48	1.40	1.31	1.24	1.18	1.45	1.37	1.31	1.22	1.16	1.10
50 × 125	2.24	2.03	1.88	1.77	1.67	1.60	2.06	1.88	1.74	1.63	1.54	1.47	1.91	1.77	1.63	1.53	1.44	1.37
50 × 150	2.68	2.44	2.26	2.12	2.01	1.91	2.49	2.26	2.09	1.96	1.85	1.76	2.34	2.12	1.96	1.83	1.73	1.65
50 × 175	3.12	2.84	2.63	2.47	2.34	2.23	2.90	2.63	2.43	2.28	2.16	2.06	2.72	2.47	2.28	2.13	2.02	1.92
50 × 200	3.56	3.24	3.00	2.82	2.67	2.55	3.31	3.00	2.78	2.60	2.46	2.35	3.11	2.81	2.60	2.44	2.30	2.19
50 × 225	4.00	3.63	3.37	3.17	3.00	2.86	3.71	3.37	3.12	2.93	2.77	2.64	3.49	3.16	2.92	2.74	2.59	2.47
63 × 100	1.87	1.77	1.64	1.54	1.46	1.39	1.72	1.64	1.51	1.42	1.34	1.28	1.60	1.52	1.42	1.33	1.26	1.20
63 × 125	2.42	2.20	2.04	1.92	1.82	1.73	2.25	2.04	1.89	1.77	1.68	1.60	2.10	1.91	1.77	1.66	1.57	1.50
63 × 150	2.90	2.63	2.44	2.30	2.18	2.08	2.69	2.44	2.26	2.12	2.01	1.92	2.53	2.29	2.12	2.00	1.88	1.79
63 × 175	3.37	3.07	2.85	2.67	2.54	2.42	3.13	2.84	2.63	2.47	2.34	2.23	2.94	2.67	2.47	2.32	2.19	2.09
63 × 200	3.84	3.50	3.25	3.05	2.89	2.76	3.57	3.24	3.01	2.82	2.67	2.55	3.36	3.05	2.82	2.65	2.51	2.39
63 × 225	4.31	3.92	3.64	3.43	3.25	3.10	4.01	3.64	3.38	3.17	3.01	2.87	3.77	3.42	3.17	2.97	2.82	2.68
SC4																		
50 × 100	1.79	1.71	1.58	1.48	1.40	1.34	1.64	1.57	1.46	1.37	1.30	1.23	1.53	1.45	1.37	1.28	1.21	1.15
50 × 125	2.34	2.13	1.97	1.85	1.75	1.67	2.17	1.97	1.82	1.71	1.62	1.54	2.02	1.85	1.71	1.60	1.51	1.44
50 × 150	2.80	2.55	2.36	2.22	2.10	2.00	2.60	2.36	2.18	2.05	1.94	1.85	2.44	2.21	2.05	1.92	1.81	1.73
50 × 175	3.26	2.97	2.75	2.58	2.45	2.34	3.03	2.75	2.54	2.39	2.26	2.15	2.85	2.58	2.39	2.24	2.12	2.01
50 × 200	3.72	3.38	3.14	2.95	2.79	2.67	3.45	3.13	2.90	2.73	2.58	2.46	3.25	2.94	2.72	2.55	2.42	2.30
50 × 225	4.17	3.80	3.52	3.31	3.14	3.00	3.88	3.52	3.26	3.06	2.90	2.77	3.65	3.31	3.06	2.87	2.72	2.59
63 × 100	1.99	1.84	1.71	1.61	1.52	1.45	1.81	1.71	1.58	1.49	1.41	1.34	1.69	1.60	1.48	1.39	1.32	1.26
63 × 125	2.53	2.30	2.13	2.00	1.90	1.81	2.35	2.13	1.97	1.85	1.76	1.68	2.21	2.00	1.85	1.74	1.65	1.57
63 × 150	3.02	2.75	2.55	2.40	2.28	2.17	2.81	2.55	2.37	2.22	2.10	2.01	2.64	2.40	2.22	2.08	1.97	1.88
63 × 175	3.52	3.20	2.97	2.80	2.65	2.53	3.27	2.97	2.75	2.59	2.45	2.34	3.08	2.79	2.59	2.43	2.30	2.19
63 × 200	4.01	3.65	3.39	3.19	3.03	2.89	3.73	3.39	3.14	2.95	2.80	2.67	3.51	3.19	2.95	2.77	2.62	2.50
63 × 225	4.49	4.10	3.81	3.58	3.40	3.25	4.18	3.80	3.55	3.32	3.15	3.00	3.94	3.58	3.32	3.11	2.95	2.81

AD A1/2, Appendix A

Table A24 Purlins supporting sheeting or decking for roofs having a pitch more than 10° but not more than 35°. Imposed loading $1.0\,\text{kN/m}^2$.
Maximum clear span of purlin (m) Timber of strength class SC3 and SC4 (see Table 1)

	Size of purlin (mm × mm)	Dead Load [kN/m²] excluding the self weight of the purlin																	
		Not more than 0.25						More than 0.25 but not more than 0.50						More than 0.50 but not more than 0.75					
		Spacing of purlins (mm)																	
		900	1200	1500	1800	2100	2400	900	1200	1500	1800	2100	2400	900	1200	1500	1800	2100	2400
SC3	50 × 100	1.67	1.51	1.40	1.31	1.24	1.18	1.55	1.42	1.31	1.22	1.16	1.10	1.45	1.34	1.24	1.16	1.09	1.04
	50 × 125	2.08	1.88	1.74	1.64	1.55	1.47	1.95	1.77	1.63	1.53	1.45	1.38	1.85	1.67	1.54	1.44	1.36	1.30
	50 × 150	2.49	2.26	2.09	1.96	1.85	1.77	2.34	2.12	1.96	1.83	1.73	1.65	2.22	2.00	1.85	1.73	1.64	1.56
	50 × 175	2.90	2.63	2.43	2.28	2.16	2.06	2.73	2.47	2.28	2.14	2.02	1.92	2.58	2.34	2.16	2.02	1.91	1.81
	50 × 200	3.31	3.00	2.78	2.61	2.47	2.35	3.11	2.82	2.60	2.44	2.31	2.20	2.95	2.67	2.46	2.31	2.18	2.07
	50 × 225	3.72	3.37	3.12	2.93	2.77	2.64	3.49	3.16	2.93	2.74	2.59	2.47	3.31	3.00	2.77	2.59	2.45	2.31
	63 × 100	1.80	1.64	1.51	1.42	1.35	1.28	1.69	1.54	1.42	1.33	1.26	1.20	1.60	1.45	1.34	1.26	1.19	1.13
	63 × 125	2.25	2.04	1.89	1.77	1.68	1.60	2.11	1.92	1.77	1.66	1.57	1.50	2.00	1.81	1.68	1.57	1.49	1.41
	63 × 150	2.69	2.44	2.26	2.13	2.01	1.92	2.53	2.29	2.12	1.99	1.88	1.80	2.40	2.17	2.01	1.88	1.78	1.70
	63 × 175	3.13	2.85	2.64	2.48	2.35	2.24	2.95	2.67	2.47	2.32	2.20	2.09	2.80	2.53	2.34	2.20	2.08	1.98
	63 × 200	3.57	3.25	3.01	2.83	2.68	2.55	3.36	3.05	2.82	2.65	2.51	2.39	3.19	2.89	2.67	2.51	2.37	2.26
	63 × 225	4.01	3.65	3.33	3.18	3.01	2.87	3.77	3.43	3.17	2.98	2.82	2.69	3.58	3.25	3.01	2.82	2.67	2.54
SC4	50 × 100	1.74	1.58	1.46	1.37	1.30	1.24	1.64	1.48	1.37	1.28	1.21	1.16	1.53	1.40	1.30	1.21	1.15	1.09
	50 × 125	2.17	1.97	1.82	1.71	1.62	1.54	2.04	1.85	1.71	1.60	1.52	1.44	1.94	1.75	1.62	1.51	1.43	1.36
	50 × 150	2.60	2.36	2.19	2.05	1.94	1.85	2.45	2.22	2.05	1.92	1.82	1.73	2.32	2.10	1.94	1.82	1.72	1.63
	50 × 175	3.03	2.75	2.55	2.39	2.26	2.16	2.85	2.58	2.39	2.24	2.12	2.02	2.70	2.45	2.26	2.12	2.00	1.90
	50 × 200	3.46	3.14	2.91	2.73	2.58	2.46	3.25	2.95	2.73	2.56	2.42	2.30	3.08	2.79	2.58	2.42	2.28	2.17
	50 × 225	3.88	3.52	3.27	3.07	2.90	2.77	3.65	3.31	3.06	2.87	2.72	2.59	3.46	3.14	2.90	2.72	2.57	2.44
	63 × 100	1.89	1.71	1.58	1.49	1.41	1.34	1.77	1.61	1.49	1.39	1.32	1.26	1.68	1.52	1.41	1.32	1.25	1.19
	63 × 125	2.35	2.13	1.98	1.86	1.76	1.68	2.21	2.00	1.85	1.74	1.65	1.57	2.10	1.90	1.76	1.65	1.56	1.48
	63 × 150	2.81	2.55	2.37	2.22	2.11	2.01	2.65	2.40	2.22	2.08	1.97	1.88	2.51	2.27	2.10	1.97	1.87	1.78
	63 × 175	3.27	2.97	2.76	2.59	2.46	2.34	3.08	2.79	2.59	2.43	2.30	2.19	2.92	2.65	2.45	2.30	2.18	2.07
	63 × 200	3.73	3.39	3.15	2.96	2.80	2.67	3.51	3.19	2.95	2.77	2.63	2.50	3.33	3.02	2.80	2.63	2.48	2.37
	63 × 225	4.18	3.81	3.53	3.32	3.15	3.01	3.94	3.58	3.32	3.12	2.95	2.81	3.74	3.30	3.15	2.95	2.79	2.66

7 Fire (Part B)

7.1 Introduction

Part B of Schedule 1 to the Building Regulations 2000 is concerned with means of escape from buildings, fire spread within and between buildings, and access for the fire services to fight fires. Since the regulations are made in the interests of public health and safety, they do not attempt to achieve non-combustible buildings. The aims are to ensure the safety of the occupants and others who may be affected by the building, and to provide assistance for fire fighters. The protection of property, including the building itself, is the province of insurers who may require additional measures to provide higher standards before accepting the insurance risk. Guidance on the protection of property in the event of fire is given in the Loss Prevention Council's *Design guide for the fire protection of buildings*. For the protection of assets in the Civil and Defence Estates reference may be made to the *Crown Fire Standards* published by the Property Advisers to the Civil Estate (PACE).

Buildings must therefore be constructed so that, in the event of a fire:

- the occupants are given sufficient warning and are able to reach a place of safety
- they will resist collapse for a sufficient period of time to allow evacuation of the occupants and prevent further rapid fire spread
- the spread of fire within and between buildings is kept to a minimum
- there is satisfactory access for fire appliances and facilities are provided to assist firefighters in the saving of lives.

The first requirement is met by providing a means of warning and an adequate number of exits and protected escape routes. The second is met by setting reasonable standards of fire-resistance for the structural elements – the floors, roofs, load-bearing walls and frames. The third is met by:

- dividing large buildings into *compartments* and requiring higher standards of fire-resistance of the walls and floors bounding a compartment
- setting standards of fire-resistance for external walls
- controlling the surface linings of walls and ceilings to inhibit flame spread
- sealing and sub-dividing concealed spaces in the structure or fabric of a building to prevent the spread of unseen fire and smoke and
- setting standards of resistance to fire penetration and flame spread for roof coverings.

The fourth is met by:

- providing access for fire appliances and firefighting personnel;
- providing fire mains within the building; and
- making sure that heat and smoke may be vented from basement areas.

In some large and complex buildings (and in those containing different uses, such as airport terminals) the provisions contained in Approved Document (AD) B may prove inadequate or difficult to apply. In such buildings, the only viable way to achieve a satisfactory standard of fire safety may be to adopt a fire safety engineering approach which takes into account the total fire safety package. This approach may also be appropriate for solving a building design problem which otherwise follows the provisions of AD B. For a framework and guidance on the design and assessment of fire safety measures in buildings it may be useful to follow the discipline of British Standard Draft for Development (DD) 240 *Fire safety engineering in buildings*. This should enable designers and building control bodies to:

- be aware of the relevant issues
- be aware of the need to consider complete fire-safety systems; and
- follow a disciplined analytical framework.

Some difficulty may also be encountered when trying to apply the provisions of AD B to existing buildings, particularly when they are of special historic or architectural importance. In buildings of this type it may be appropriate to carry out an assessment of the potential fire hazard or risk to life, and then incorporate in the design a sufficient number of fire safety features to alleviate the danger. The risk assessment should take account of:

- the likelihood of a fire occurring
- the anticipated severity of the fire
- how well the structure of the building is able to resist the spread of smoke and flames
- the consequential danger to persons in or near the building.

Fire safety measures which can be incorporated in the design include:

- an assessment of the adequacy of the means to prevent fire
- the installation of automatic fire detection and warning systems
- the provision of adequate means of escape
- the provision of smoke control
- design features aimed at controlling the rate of growth of a fire if one does occur
- an assessment of the ability of a structure to resist the effects of fire
- the extent of fire containment offered by the building
- the fire separation from other buildings or parts of the same building
- the standard of firefighting equipment in the building
- the ease with which the fire service may gain access to fight a potential fire

- the existence of legislative controls to require staff training in fire safety and fire routines (e.g. the Fire Precautions Act 1971, the Fire Precautions (Workplace) Regulations 1997 or licensing and registration controls)
- the existence of continuing control so that fire safety systems can be seen to be maintained (as in the certification procedure under the Fire Precautions Act)
- the installation of suitable fire management procedures.

Some factors in the measures listed above may be assessed using quantitative techniques and given numerical values in some circumstances. Quantitative techniques may also be used to evaluate risk and hazard although the assumptions made need to be carefully assessed.

An example of an overall approach to fire safety may be found in BS 5588 *Fire precautions in the design, construction and use of buildings*, Part 10: 1991 *Code of practice for shopping complexes*. (See section 7.29 below for further reference to enclosed shopping complexes.)

Other examples where the guidance in AD B may be inappropriate include:

- Where a building contains an atrium which passes through compartment floors. Special fire safety measures may need to be incorporated in the design. BS 5588 *Fire precautions in the design, construction and use of buildings*, Part 7: 1997 *Code of practice for the incorporation of atria in buildings* contains guidance on suitable fire safety measures.
- Hospitals. The design of fire safety in hospitals is dealt with in Health Technical Memorandum (HTM) 81 *Fire precautions in new hospitals* (1996 revision). Part B of the Building Regulations 2000 will be satisfied where HTM 81 is used.

Fire safety in buildings is a complex matter and in recent years a number of reviews and enquiries have been set up to consider different aspects of fire safety law. Additionally, matters have been further complicated by the need to comply with various European Community directives.

Following the 2001 General Election, there was a reorganisation of Government Departmental responsibilities for fire safety. It brought the Home Office's Fire Service Inspectorate and the Building Regulations Division together, under the Department for Transport, Local Government and the Regions (DTLR) for the first time. The DTLR has since been broken up and responsibility for Building Regulations now rests with the Office of the Deputy Prime Minister (ODPM). There may therefore be a unified approach to fire safety in the future.

7.2   Terminology

Certain terms which apply generally throughout this chapter are defined here. Other terms are defined in the specific section to which they apply.

BASEMENT STOREY – A storey which has some part of the perimeter of its floor more than 1200 mm below the highest level of the ground adjoining that part of the

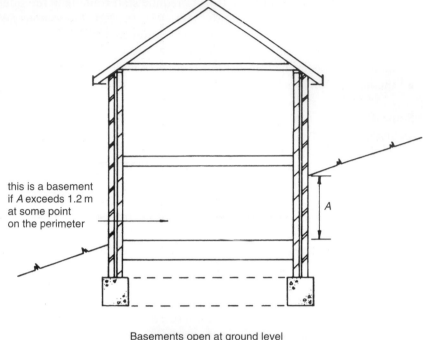

this is a basement
if *A* exceeds 1.2 m
at some point
on the perimeter

A

Basements open at ground level
on one side allow smoke venting
or access for fire fighting. Therefore
treat as above ground structure for
fire resistance

Fig. 7.1 Basement storey.

floor. A basement storey may be treated as the above ground structure for fire-resistance purposes, if one side is open at ground level for smoke venting or fire-fighting. (See Fig. 7.1.)

BOUNDARY – When referring to any side of a building or compartment (including any external wall or part), means the usual legal boundary adjacent to that side, being taken up to the centre of any abutting railway, street, canal or river.

CEILING – Includes any soffit, rooflight or other part of a building which encloses and is exposed overhead in a room, circulation space or protected shaft (but not including the surface of the frame of any rooflight, the upstand of which is considered as part of the wall).

CIRCULATION SPACE – A space (including a protected stairway) used mainly as a means of access between a room and an exit from the compartment or building.

COMPARTMENT – Any part of a building (including rooms, spaces or storeys) which is constructed to stop fire spreading to or from another part of the same building, or an adjacent building. If any part of the top storey of a building comes

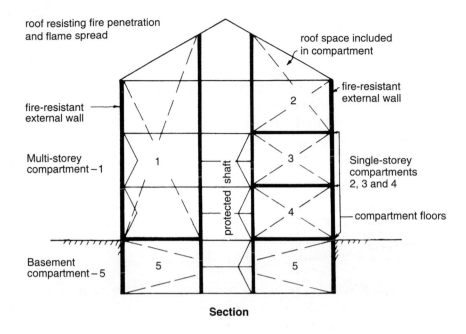

roof resisting fire penetration
and flame spread

roof space included
in compartment

fire-resistant
external wall

fire-resistant
external wall

Multi-storey
compartment – 1

Single-storey
compartments
2, 3 and 4

protected shaft

compartment floors

Basement
compartment – 5

Section

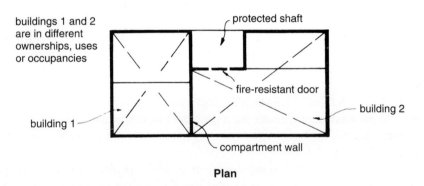

buildings 1 and 2
are in different
ownerships, uses
or occupancies

protected shaft

fire-resistant door

building 1

building 2

compartment wall

Plan

Fig. 7.2 Division of buildings into compartments.

within a compartment, that compartment is taken to include any roof space above that part of the top storey (see Fig. 7.2). See also 'Separated part' below.

COMPARTMENT WALL/FLOOR – Fire-resisting construction provided to separate one fire compartment from another for the purpose of preventing fire spread.

CONCEALED SPACE (CAVITY) – A space which is concealed by the elements of a building (such as a roof space or the space above a suspended ceiling) or contained within an element (such as the cavity in a wall). This definition does *not* include a room, cupboard, circulation space, protected shaft or space within a flue, chute, duct, pipe or conduit.

ELEMENTS OF STRUCTURE –

- Any member forming part of the structural frame of a building or any other beam or column. (This does not normally include members which form part of a roof structure only unless the roof performs the function of a floor, or the roof structure provides stability for fire-resisting walls.)
- A floor (but not the lowest floor in a building, or a platform floor).
- An external wall.
- A compartment wall (including a wall which is common to two or more buildings).
- A load-bearing wall or the load-bearing part of a wall.
- A gallery (but not a loading gallery, fly gallery, lighting bridge, stage grid or any gallery provided for similar purposes or for maintenance and repair).

These elements are illustrated in Fig. 7.3.

EXTERNAL WALL – Includes a portion of a roof sloping at 70° or more to the horizontal if it adjoins a space within the building to which persons have access, other than for occasional maintenance and repair (see Fig. 7.4).

FIRE DOOR – Includes any shutter, cover or other form of protection to an opening in any fire-resisting wall or floor of a building or in the structure surrounding a protected shaft. The fire door should be able to resist the passage of fire and/or gaseous products of combustion in accordance with specified criteria (see Fire doors in section 7.22.3 below). A fire door may have one or more leaves.

FIRE SEPARATING ELEMENT – Includes a compartment wall or floor, a cavity barrier and construction enclosing a protected escape route and/or place of special fire hazard.

GALLERY – A floor (or raised storage area) which projects into another space but has less than half the floor area of that space.

HABITABLE ROOM – A room used for dwelling purposes, including a kitchen (in Part B) but not a bathroom.

NOTIONAL BOUNDARY – A boundary which is assumed to exist between two buildings in the residential, and assembly and recreation purpose groups, on the same site where there is no actual boundary. The notional boundary line should be so placed that neither building contravenes any of the requirements of AD B relevant to the external walls facing each other (see Fig. 7.5).

OCCUPANCY TYPE – Part of a purpose group (see section 7.3 below).

PLACES OF SPECIAL FIRE HAZARD – Oil-filled switchgear and transformer rooms, boiler rooms, stores for fuel or other highly flammable substances, and

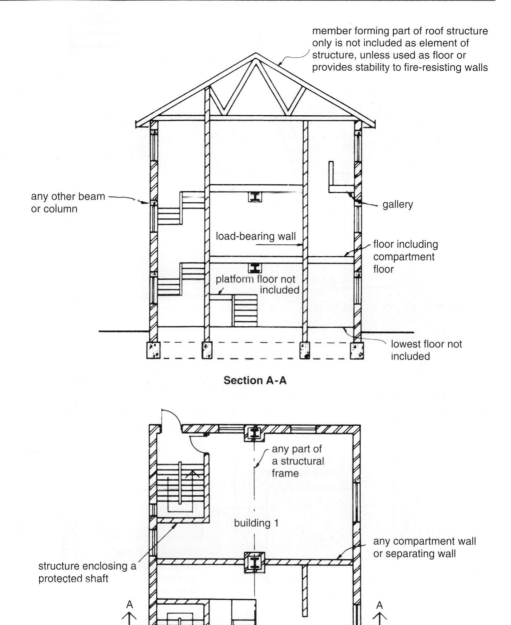

member forming part of roof structure only is not included as element of structure, unless used as floor or provides stability to fire-resisting walls

any other beam or column

gallery

load-bearing wall

floor including compartment floor

platform floor not included

lowest floor not included

Section A-A

any part of a structural frame

building 1

any compartment wall or separating wall

structure enclosing a protected shaft

building 2

any external wall

Plan

Fig. 7.3 Elements of structure.

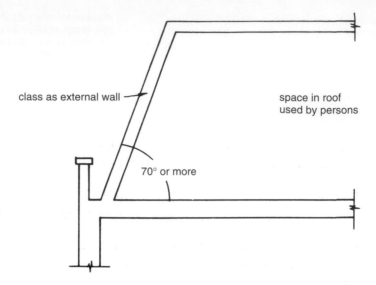

Fig. 7.4   Steeply pitched roofs.

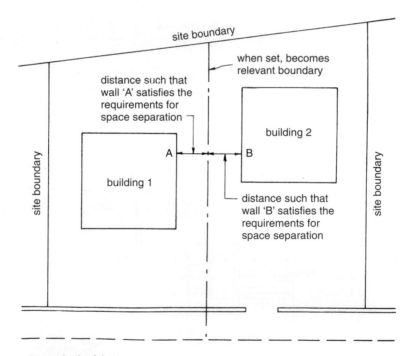

One or both of the buildings new.
One or other of the buildings of residential or assembly and recreation use.
Existing building treated as identical new building but with existing unprotected area and fire resistance in external wall.

Fig. 7.5   Notional boundary.

rooms containing fixed internal combustion engines. Additionally, in schools – laboratories, technology rooms with open heat sources, kitchens and stores for PE mats or chemicals.

PLATFORM FLOOR (sometimes called an access or raised floor) – A floor over a concealed space intended to house services, which is supported by a structural floor.

PROTECTED CORRIDOR/LOBBY – A corridor or lobby protected from fire in adjacent accommodation by fire-resisting construction.

PROTECTED SHAFT – A shaft enclosed with fire-resisting construction which enables persons, things or air to pass between different compartments.

PROTECTED STAIRWAY – A stair adequately protected with fire-resisting construction discharging to place of safety through final exit (includes passage from foot of stair to final exit).

RELEVANT BOUNDARY – For a boundary to be considered relevant it should:

- be coincident with, or
- be parallel to, or
- not make an angle of more than 80° with the external wall (see Fig. 7.6).

In certain circumstances a 'notional' boundary, as defined above, will be the relevant boundary. A wall may have more than one relevant boundary.

ROOFLIGHT – Includes any domelight, lanternlight, skylight, ridge light, glazed barrel vault or other element which is intended to admit daylight.

ROOM – An enclosed space in a building, but not one used solely as a circulation space. This term would also include cupboards that were not fittings and large rooms such as warehouses and auditoria. Excluded are voids such as ducts, roof spaces and ceiling voids.

SEPARATED PART (OF A BUILDING) – Where a compartment wall completely divides a building from top to bottom and is in one plane, the divided sections of the building are referred to as separated parts. The height of each separated part may then be treated individually (see Fig. 7.7 and section 7.4 Rules for measurement below).

SINGLE-STOREY BUILDING – A building which consists of a ground storey only. (A separated part consisting of a ground storey only and with a roof which is accessible only for the purposes of maintenance and repair may be treated as part of a single-storey building.) Basements are not included when counting the number of storeys in a building.

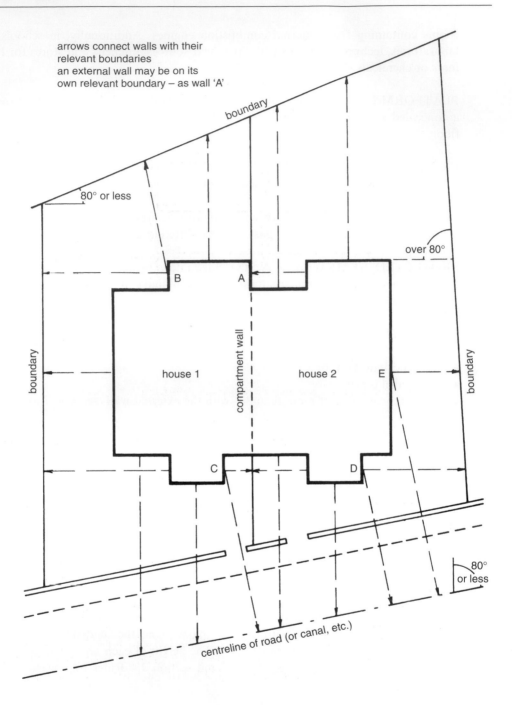

arrows connect walls with their
relevant boundaries
an external wall may be on its
own relevant boundary – as wall 'A'

Fig. 7.6   Relevant boundaries.

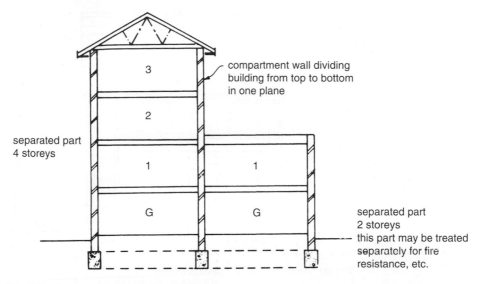

Fig. 7.7   Separated part.

SITE (OF A BUILDING) – The land occupied by the building up to the boundaries with land in other ownership.

STOREY – Included in this definition are the following:

- any gallery in an assembly building (PG5); and
- a gallery in any other building if its area exceeds half that of the space into which it projects; and
- a roof, unless used only for maintenance or repair.

UNPROTECTED AREA – In relation to an external wall or side of a building, means:

- a window, door or other opening (although windows designed and glazed to provide the necessary standard of fire resistance and which are not openable, and recessed car parking areas are not regarded as unprotected areas)
- any part of an external wall of fire-resistance less than that required by Section 13 of AD B4 (see section 7.27)
- any part of an external wall with external facing attached or applied, whether as cladding or not, the facing being of combustible material more than 1mm thick (combustible in this context means any material which does not have a Class 0 rating) (see Fig. 7.8).

7.3   Purpose Groups

The fire hazard presented by a building will, to a large extent, depend on the use to which the building is put. Many of the provisions concerning means of warning and

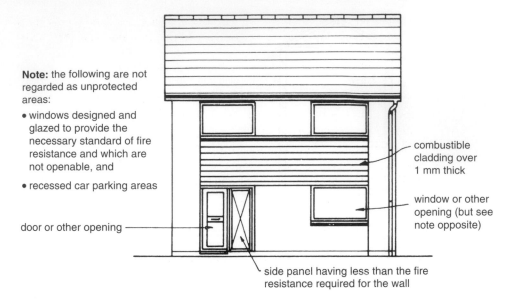

Fig. 7.8   Unprotected areas.

escape, fire-resistance, compartmentation, etc. are directly related to these use classifications, which in AD B are termed purpose groups (PG).

The seven PGs are set out in Table D1 (Classification of purpose groups) of Appendix D of AD B. They are as follows:

1   **Residential (dwellings)** – This includes parts of a dwelling used by the occupant in a professional or business capacity (such as a surgery, consulting room, office or other accommodation), not exceeding $50 \, m^2$ in total. This group is further sub-divided into:

1(a)   flat or maisonette

1(b)   dwellinghouse (with a habitable storey more than 4.5 m above ground level)

1(c)   dwellinghouse (no habitable storey above 4.5 m from ground level). 1(c) also includes any detached garage or open carport not exceeding $40 \, m^2$ in area, or a detached building consisting of a garage and open carport, neither of which exceeds $40 \, m^2$ in area, irrespective of whether or not they are associated with a dwelling.

2(a)   **Residential (institutional)** – Includes a hospital, home, school or other similar establishment. The premises will be used as living (and sleeping) accommodation for, or for the treatment, care or maintenance of:

• persons suffering from disabilities due to illness or old age or other physical or mental incapacity

• children under the age of five years

• persons in a place of lawful detention.

2(b)   Residential (other) – Includes a hotel, boarding house, residential college, hall of residence, hostel, and any other residential purpose not described above.

3   Office – Includes offices or premises used for the purpose of:
- administration
- clerical work (including writing, book keeping, sorting papers, filing, typing, duplicating, machine calculating, drawing and the editorial preparation of matter for publication, police and fire service work)
- handling money (including banking and building society work)
- communications (including postal, telegraph and radio communications) radio, television, film, audio or video recording, or performance (not open to the public) and their control.

4   Shop and Commercial – Shops or premises used for a retail trade or business including:
- the sale to members of the public of food or drink for immediate consumption
- retail by auction, self-selection and over-the-counter wholesale trading
- the business of lending books or periodicals for gain
- the business of a barber or hairdresser
- premises to which the public is invited to deliver or collect goods in connection with their hire, repair or other treatment, or (except in the case of repair of motor vehicles) where they themselves may carry out such repairs or other treatments.

5   Assembly and Recreation – Places of assembly, entertainment or recreation including:
- broadcasting, recording and film studios open to the public
- bingo halls, casinos, dance halls
- entertainment, conference, exhibition and leisure centres
- funfairs and amusement arcades
- museums and art galleries
- non-residential clubs
- theatres, cinemas and concert halls
- educational establishments
- dancing schools, gymnasia, swimming pool buildings, riding schools, skating rinks, sports pavilions, sports stadia
- law courts
- churches and other buildings of worship
- crematoria
- libraries open to the public
- non-residential day centres, clinics, health centres and surgeries
- passenger stations and termini for air, rail, road or sea travel
- public toilets
- zoos and menageries.

6   Industrial – Factories and other premises used for:

- manufacturing, altering, repairing, cleaning, washing, breaking-up, adapting or processing any article
- generating power
- slaughtering livestock.

7 **Storage and other non-residential** – This group is further sub-divided into:

    7(a) Place for the storage or deposit of goods or materials [other than described under 7(b)] and any building not within any of the purpose groups 1 to 6, and

    7(b) Car parks designed to admit and accommodate only cars, motorcycles and passenger or light goods vehicles weighing no more than 2500 kg gross.

Normally the PG is applied to the whole building, or (where a building is compartmented) to a compartment in the building, by reference to the main use of the building or compartment. Parts of the building put to different uses can be treated as ancillary to the main use, and therefore as though they are in the same use. However, a different use in the same building is not regarded as ancillary, and is therefore treated as belonging to a PG in its own right:

- where the ancillary use is a flat or maisonette
- where the building or compartment exceeds $280\,m^2$ in area and the ancillary use exceeds one-fifth of the total floor area of the building or compartment
- where the building is a shop or commercial building or compartment of Purpose Group (4) and contains a storage area which exceeds one third of the total floor area of the building or compartment and the building or compartment is more than $280m^2$ in area.

Where a building contains different main uses which are not ancillary to one another, each use should be considered as belonging to a purpose group in its own right.

    Some large buildings, such as shopping complexes, may involve complicated mixes of purpose groups. In these cases special precautions may need to be taken to reduce any additional risks caused by the interaction of the different purpose groups.

7.4 Rules for measurement

Many of the requirements concerning compartmentation and fire resistance, etc. in AD B are based on the height, area and cubic capacity of the building, compartment or separated part. For consistency, it is necessary to have a standard way of measuring these proportions. Appendix C of AD B indicates diagrammatically how the various forms of measurement should be made. These rules can be summarised as follows:

- HEIGHT – The height of the building or part is measured from the mean level of the ground adjoining the outside of the building's external walls to the level of half the vertical height of the roof, or to the top of the walls or parapet, whichever is the higher. This rule applies to double pitch, mono-pitch, flat and mansard type roofs (see Fig. 7.9(a)).

- AREA – The area of any storey of a building, compartment or separated part should be calculated as the total area in that storey within the finished inner surfaces of the enclosing walls. If there is no enclosing wall, the area is measured to the outermost edge of the floor on that side. The area should include any internal walls or partitions.

  The area of a room, garage, conservatory or outbuilding is calculated by measuring to the inner surface of the enclosing walls.

  The area of any part of a roof should be calculated as the actual visible area of that part, as measured on a plane parallel to the roof slope. For a lean-to roof, the measurement should be taken from the wall face to the outer edge of the roof slope. For a hipped roof, the measurement should be to the outer point of the roof as a base area (see Fig. 7.9(a)).

- CUBIC CAPACITY – The cubic capacity of a building, compartment or separated part should be calculated as the volume of the space between the finished surfaces of the enclosing walls, the upper surface of its lowest floor and the under surface of the roof or ceiling surface as appropriate. If there is no enclosing wall, the measurement should be taken to the outermost edge of the floor on that side. The cubic capacity should again include space occupied by other walls, shafts, ducts or structures within the measured space (see Fig. 7.9(b)).

- NUMBER OF STOREYS – The number of storeys in a building or separated part should be calculated at the position which gives the maximum number. Basement storeys are not counted. In most purpose group buildings galleries are also not counted as storeys. However, in assembly buildings, a gallery is included as a storey unless it is a fly gallery, loading gallery, stage grid, lighting bridge or other similar gallery, or is for maintenance and repair purposes. (The common factor here is that these excluded galleries are not generally accessible to the public.) (See Fig. 7.9(b).)

- HEIGHT OF TOP STOREY – This is measured from ground level on the lowest side of the building to the upper surface of the top floor. Roof-top plant areas are excluded from this measurement as are any top storeys consisting exclusively of plant rooms.

7.5   Means of warning and escape in case of fire

7.5.1   The mandatory requirement

Buildings must be designed and constructed so that there are appropriate provisions for the early warning of fire, and appropriate means of escape in case of fire capable

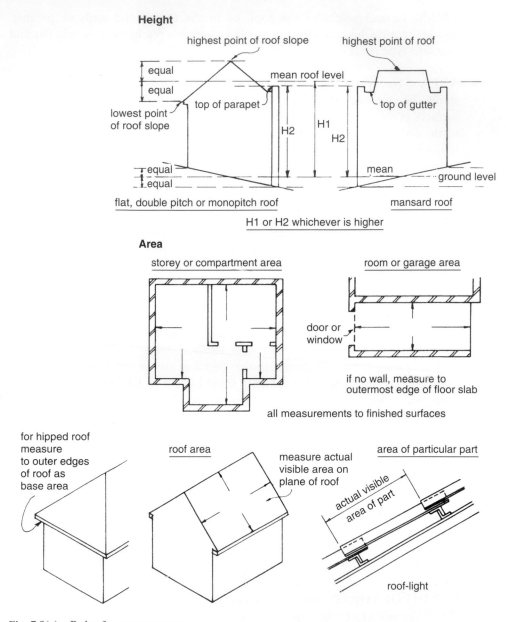

Height

highest point of roof slope   highest point of roof

equal

mean roof level

equal

top of parapet   top of gutter

lowest point
of roof slope

H2   H1

H2

equal   mean

equal   ground level

flat, double pitch or monopitch roof   mansard roof

H1 or H2 whichever is higher

Area

storey or compartment area   room or garage area

door or
window

if no wall, measure to
outermost edge of floor slab

all measurements to finished surfaces

for hipped roof
measure
to outer edges   roof area   measure actual   area of particular part
of roof as   visible area on
base area   plane of roof

actual visible
area of part

roof-light

Fig. 7.9(a)   Rules for measurement.

of being used safely and effectively at all material times. (No definition is given of material times.) The mandatory requirement goes on to state that the means of escape must be to a place of safety outside the building; however, it will be seen below that for certain classes of buildings the place of safety may be within the building itself. This requirement applies to all buildings except prisons provided under section 33 of the Prisons Act 1952.

Cubic capacity

cubic capacity of compartment or building

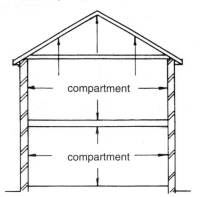 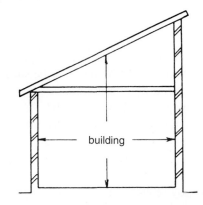

Number of storeys

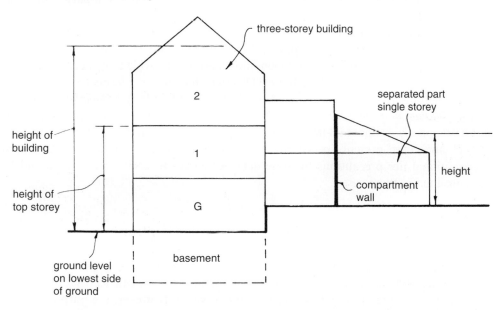

Fig. 7.9(b) Rules for measurement.

7.5.2 Standard of performance to meet the mandatory requirement

The mandatory requirement may be met by:

- providing escape routes which are suitably located and of sufficient number and size to enable the occupants to escape to a place of safety in the event of fire
- making sure that escape routes are adequately lit and suitably signed

- providing appropriate facilities to limit the ingress of smoke to the escape route or to restrict the fire and remove smoke
- enclosing escape routes where necessary, so that they are sufficiently protected from the effects of fire

to an extent necessary depending on the size, height and use of the building.

Additionally, there must be sufficient means for giving early warning of fire for the building's occupants.

Approved Document B1, Means of warning and escape, contains details of practical ways to satisfy the mandatory requirement and the performance standard referred to above. It does this in two ways:

- by giving actual recommendations in the text (described in detail below); and
- by reference to other sources of guidance.

Generally, BS 5588: *Fire precautions in the design, construction and use of buildings*, Part 0: 1996 *Guide to fire safety codes of practice for particular premises/applications* contains references to a great many codes and guides dealing with means of escape provision. If it is decided to use one of these codes or guides, the parts which are relevant to means of escape should be followed rather than a mixture of the specific publication and the relevant sections of AD B1. Having said this, it may still be necessary to supplement the Approved Document information with guidance from another publication where insufficient information is provided, or specific reference is made. For example, section 18 of AD B (which deals with access to buildings for fire-fighting personnel) recommends the use of BS 5588: Part 5 *Code of practice for firefighting stairs and lifts* since AD B contains no other guidance on the subject. Additionally, some buildings which house particular industrial or commercial activities may present special fire hazards (filling stations, etc.) and may need additional fire precautions to those detailed in AD B. BS5588: Part 0 should be consulted in these cases.

7.6   Means of escape – links with other legislation

In addition to building regulations there is a large amount of other legislation covering means of escape in case of fire. Reference to some other provisions will be found in Chapter 1 (Building control: an overview). However, the purpose of this book is not to review all the legislation on this subject, but to show how it may interact with and affect building regulations. For a comprehensive guide to means of escape legislation see *Means of escape from fire* by M.J. Billington, Anthony Ferguson and A.G. Copping (Blackwell Science 2002). Some of the most important provisions are mentioned below.

7.6.1   Regulatory Reform (Fire Safety) Order 2005

Up until 1 April 2006, most existing buildings, other than dwellings, were controlled when in use by two major pieces of legislation:

- The Fire Precautions Act 1971; and
- Fire Precautions (Workplace) Regulations 1997.

Together, these probably represented the most significant pieces of legislation which were linked with the Building Regulations.

Whilst these pieces of legislation have been repealed by the Regulatory Reform (Fire Safety) Order 2005, their main provisions (including the 1999 amendments) continue to be enacted in a new form as set out by the Order.

The centrepiece of the new provisions is an obligation placed on a 'responsible person' to carry out a risk assessment of the premises under their control along the same lines as originally contained in the Fire Precautions (Workplace) Regulations. This is in direct contrast to the procedure under the Fire Precautions Act 1971 where the fire authority was placed under an obligation to inspect premises and to determine appropriate measures.

In conjunction with the process of making the new Regulatory Reform (Fire Safety) Order the Government is continuing to implement a modernisation of the fire and rescue services principally by means of the powers contained in the Fire and Rescue Services Act 2004. This legislation subsumes the Fire Services Act 1947 and radically redefines the powers, duties and responsibilities of the former Fire Authorities, now recast as Fire and Rescue Authorities.

The Regulatory Reform (Fire Safety) Order applies to most non-domestic premises used or operated by employers, the self-employed and the voluntary sector. It does not apply to offshore installations, mines and boreholes, fields, woods and agricultural land since these have their own separate fire regimes.

7.6.2   Health and Safety at Work etc. Act 1974

In the case of some highly specialised industrial or storage premises where hazardous processes are carried out or materials are kept the Health and Safety Executive may have certification responsibilities similar to those of the fire authority.

7.6.3   Housing Act 1985

Local authorities are obliged to require means of escape from houses in multiple occupation, i.e. houses which are occupied by persons who do not form part of a single household. (For guidance on the interpretation of this definition see DOE/ Home Office/Welsh Office Circular *Memorandum on overcrowding and houses in multiple occupation*.) Many local authorities use the services of the fire authority in deciding on the adequacy of an existing means of escape in such buildings. If new dwellinghouses, flats and maisonettes are designed in accordance with AD B1, they should be acceptable for use as 'houses' in multiple occupation.

7.7   Interpretation of AD B1

A large number of terms are used in Approved Document Bl which relate specifically to means of escape:

ACCOMMODATION STAIR – A stair which is provided for the convenience of the occupants of a building and is additional to those required for means of escape. (When calculating the number of stairs required in a building, accommodation stairs are ignored.)

ALTERNATIVE ESCAPE ROUTES – Routes which are sufficiently separated from one another by fire-resisting construction or space and direction so that one route will still be available even if the other is affected by fire. In most buildings this will mean alternative protected corridors and stairs; however, in dwellings an alternative escape route could be via a second stair, balcony or flat roof if it enabled a person to reach a place free from danger of fire.

ALTERNATIVE EXIT – One of two or more exits, each of which is separate from the other. (There are rules in AD Bl for determining whether exits are sufficiently far enough apart to be considered as alternatives and these are illustrated in section 7.15.2 below (Fig. 7.24).)

ATRIUM – A space in a building (not necessarily vertically aligned) which passes through one or more structural floors. (Enclosed escalator wells, enclosed lift wells, building services ducts or stairways are not classified as atria.)

COMMON BALCONY – An escape route from more than one flat or maisonette which is formed by a walkway open to the air on one or more sides.

COMMON STAIR – An escape stair which serves more than one maisonette or flat.

CORRIDOR ACCESS – A common horizontal internal access or circulation space which serves each dwelling in a building containing flats. It may include a common entrance hall.

DEAD END – An area from which it is only possible to escape in one direction.

DIRECT DISTANCE – The shortest distance which can be measured from within the floor area of the inside of the building, to the storey exit. All internal walls, partitions and fittings are ignored when measuring the direct distance, except the walls enclosing the protected stairway.

DWELLING – A unit of residential accommodation which is occupied by:

- a single person or family; or
- not more than six residents living together as a single household, including a household where the residents receive care.

The dwelling need not be the sole or main residence of the occupants. (This is an important definition since it shows the difference, for the purposes of means of escape, between a dwelling and a house in multiple occupation. It also appears to indicate that certain buildings, where people receive care, may be regarded as dwellings and not institutional buildings, again, for the purposes of means of escape.)

EMERGENCY LIGHTING – Lighting which is provided to be used when the normal lighting fails.

ESCAPE LIGHTING – Part of the emergency lighting which is provided specifically to light escape routes.

ESCAPE ROUTE – That part of the means of escape from any point in the building to the final exit.

EVACUATION LIFT – A lift used to evacuate disabled people in the event of fire.

FINAL EXIT – The termination of an escape route from a building sited so that people may be able rapidly to get clear of any danger from smoke or fire in the vicinity of the building. It should give direct access to a street, passageway, walkway or open space. (It should be noted that windows are not acceptable as final exits.)

INNER ROOM – A room contained within another room (termed the access room). Escape is only possible by passing through the access room.

MAISONETTE – A 'flat' on more than one level.

MEANS OF ESCAPE – Structural means whereby a safe route or routes is or are provided for persons to travel to a place of safety from any point in the building, in the event of fire.

OPEN SPATIAL PLANNING – The internal arrangements of a building whereby a number of floors are contained within one undivided space, e.g. split-level floors. AD B1 makes a distinction between an atrium space and open spatial planning.

PROTECTED CIRCUIT – An electrical circuit which is protected against fire.

PROTECTED ENTRANCE HALL/LANDING – A hall or circulation space in a dwelling which is protected by fire-resisting construction (other than any part of the wall which is external).

SMOKE ALARM – A device for detecting smoke which gives an audible alarm. All the components will be contained within one housing (except possibly the energy source).

STOREY EXIT – A doorway giving direct access to:

- a protected stairway (defined in section 7.2)
- a firefighting lobby (defined in section 7.30.2)
- an external escape route.

Also the following:

- a final exit (see above for definition)
- a door in a compartment wall in an institutional building if the building is planned for progressive horizontal evacuation (see section 7.17.1).

TRAVEL DISTANCE – The actual distance travelled by a person from any point in the floor area to the nearest storey exit. In this case the layout of the floor in terms of walls, partitions and fittings *is* taken into account. (Cf. Direct distance above.)

7.8   General requirements for means of warning and escape

There are certain basic principles which govern the design of means of warning and escape in buildings and which apply to all building types. In general the design should be based on an assessment of the risk to the occupants should a fire occur and should take account of:

- the use of the building (and the activities of the users)
- the nature of the building structure
- the processes undertaken and/or the materials stored in the building
- the potential fire sources
- the potential for fire spread throughout the building
- the standard of fire safety management to be installed.

In assessing the above, judgements regarding the likely level of provision may have to be made when the exact details are unknown.

The following assumptions must be made in order that a safe and economical design may be achieved:

- In general, when a fire occurs, the occupants should be able to escape safely, without external assistance or rescue from the fire service or anyone else. Obviously, there are some institutional buildings where it is not practical to expect the occupants to escape unaided and special arrangements are necessary in these cases. Similar considerations apply to disabled people. Aided escape is also permitted in certain low-rise dwellings.
- Fires do not normally break out in two different parts of a building at the same time.
- Fires are most likely to occur in the furnishings and fittings of a building or in other items which are not controlled by the Building Regulations.
- Fires are less likely to originate in the building structure and accidental fires in circulation spaces, corridors and stairways are unlikely due to the restriction on the use of combustible materials in these areas.
- When a fire breaks out the initial hazard is to the immediate area in which it occurs and it is unlikely that a large area will be affected at this stage. When fire spread does occur it is usually along circulation routes.
- The primary danger in the early stages of a fire is the production of smoke and noxious gases. These obscure the way to escape routes and exits and are responsible for the most casualties. Therefore, limiting the spread of smoke and fumes is vital in the design of a safe means of escape.
- Buildings covered are assumed to be properly managed. Where there is a failure of management responsibility, the building owner or occupier may be prosecuted under the Fire Precautions Act or the Health and Safety at Work etc. Act, which may result in prohibition of the use of the building.

7.8.1 Alternative escape routes

When a fire occurs it should be possible for people to turn their backs on it and travel away from it to either a final exit or a protected escape route leading to a place of safety. This means that alternative escape routes should be provided in most situations.

The basic criteria governing the design of means of escape are as follows.

- The first part of the escape route will be within the accommodation or circulation areas and will usually be unprotected. It should be of limited length so that people are not exposed to fire and smoke for any length of time. Where the horizontal escape route is protected it should still be of limited length since there is always the risk of premature failure.
- The second part of the escape route will usually be in a protected stairway designed to be virtually 'fire sterile'. Once inside it should be possible to proceed direct to a place of safety without rushing. Therefore, flames, smoke and gases must be excluded from these routes by fire-resisting construction or adequate smoke control measures or by both these methods. This does not preclude the use of unprotected stairs for normal everyday use; however their relative vulnerability to fire situations means that they can only be of limited use for escape purposes.

- The protected stairway should lead direct to a place of safety or it may do this via a protected corridor. The ultimate place of safety is open air clear of the effects of fire; however in certain large and complex buildings reasonable safety may be provided within the building if suitable planning and protection measures can be included in the design.

7.8.2 Dead ends

Ideally, alternative escape routes should be provided from all points in a building, since there is always the possibility that the path of a single escape route may become impassable due to the presence of fire, smoke or fumes. Escape in one direction only (a dead end) is acceptable under certain conditions depending on:

- the use of the building
- its associated fire risk
- its size and height
- the length of the dead end
- the number of people accommodated in the dead end.

7.8.3 Unacceptable means of escape

Certain paths of travel are not acceptable as means of escape, including the following.

- Lifts, unless designed and installed as evacuation lifts for disabled people in the event of fire.
- Portable or throw-out ladders.
- Manipulative apparatus and appliances such as fold-down ladders and chutes.
- Escalators. These should not be counted as additional escape routes due to the uneven nature of the top and bottom steps; however it is likely that people would use them in the event of a fire. Mechanised walkways could be acceptable if they were properly assessed for capacity as a walking route in the static mode.

7.8.4 Security

It is possible that security measures intended to prevent unauthorised access to a building may hinder the entry of the fire services when they need access to fight a fire or rescue trapped occupants. Advice may be sought from architectural liaison officers attached to most police forces so that possible conflicts between security and access may be solved at the design stage.

AD B1 does not deem it appropriate to control, under the Building Regulations, the type of lock used on the front doors to dwellings. Guidance on door security in buildings other than dwellings is given in section 7.16.2 below.

7.9   Rules for measurement

In addition to the rules for measurement described in section 7.4 above, which apply to all parts of AD B, there are certain rules which relate only to AD B1. These are concerned with assessing the length and capacity of escape routes which may, in turn, have a bearing on the quantity of routes and the number of stairways that it is necessary to provide.

7.9.1   Occupant capacity

In order to design a safe means of escape, it is necessary to assess the number of people who are likely to be present in the different parts of the building (i.e. the occupant capacity). Occupant capacity is the total number of people that a building, storey or room is designed to contain and it depends partly on the use (or purpose group) of the building and partly on the use of individual rooms in the building. It has a direct effect on the numbers and widths of:

- the exits from any room, storey or tier; and
- escape stairs and final exits from the building.

For some building uses (such as theatres or restaurants) occupant capacity can be calculated by totalling the number of seats and then adding an allowance for staff.

In other buildings (such as speculative office developments, shops and super-markets), the designer will not be able to do this since he will never be sure how intensively the building will be used. In these circumstances, Table 1 of Approved Document B1 provides floor space factors (expressed in m^2 per person) which will give a value for the occupant capacity when divided into the relevant floor area.

The table is based on the type of accommodation contained in the building (i.e. the use of individual rooms) and therefore the following procedure could be adopted to calculate the occupant capacity for the building:

- choose the top floor and inspect each room to determine its use
- for each room calculate its floor area and divide it by the relevant floor space factor to determine the occupant capacity for the room
- add together the individual room capacities to obtain the occupant capacity for the floor
- repeat this process for each floor
- total the occupants of all floors to obtain the occupant capacity for the building.

Floor space factors based on Table 1 of AD B1 are given in Table 7.1 below. The following should also be noted when using the floor space factors:

- It is not necessary to calculate occupancy capacities for stair enclosures, lifts or sanitary accommodation and fixed parts of the building structure may also be excluded from the calculation (although the area taken up by counters and display units etc. should not be excluded).

Table 7.1 Floor space factors.

Floor space factor (m² per person)	Type of accommodation/use of room	Notes
0.3	Standing spectator areas. Bars and other similar refreshment areas, without seating	
0.5	Amusement arcade, assembly hall (including a general purpose place of assembly), bingo hall, crush hall, dance floor or hall, venue for pop concert or similar events	
0.7	Concourse, queuing area or shopping mall	See also section 4 of BS 5588: Part 10: 1991 Code of practice for shopping complexes for detailed guidance on the calculation of occupancy in common public areas in shopping complexes.
1.0	Committee room, common room, conference room, dining room, licensed betting office (public area), lounge or bar (other than above), meeting room, reading room, restaurant, staff room or waiting room	In many of these uses the occupants will normally be seated. In such cases occupant capacity may be taken as the number of *fixed* seats provided
1.5	Exhibition hall or studio (film, radio, television, recording)	
2.0	Skating rink, shop sales area (1)	(1) Shops, such as supermarkets and the main sales areas of department stores; shops for personal services (e.g. hairdressers); shops for the delivery or collection of goods for cleaning, repair or other treatment either by the company or by the public themselves. For other shops see (2) below
5.0	Art gallery, dormitory, factory production area, museum or workshop	
6.0	Office	
7.0	Kitchen or library, shop sales area (2)	(2) Shops trading in large items such as furniture, floor coverings, cycles, prams, large domestic appliances or other bulky goods, or cash and carry wholesalers. For other shops see (1) above
8.0	Bedroom or study-bedroom	
10.0	Bed-sitting room, billiards or snooker room or hall	
30.0	Storage and warehousing Car park	Occupant capacity based on two persons per parking space

- Where the descriptions do not completely cover the accommodation it is acceptable to use a value based on a similar use.
- Where any part of the building is likely to have multiple uses (for example, a multi-purpose hall might be used for a low density activity like gymnastics or a high density use such as a disco) the most onerous floor space factor should be used.

As an example, take an exhibition hall with a floor area (excluding stair enclosures, lifts and sanitary accommodation) of 2250 m^2. From Table 7.1, the floor space factor is 1.5 m^2 per person. Therefore, the occupant capacity is 2250 divided by 1.5 i.e. 1500 people.

Alternatively, some designers or developers may be able to obtain actual data relating to occupancy numbers, taken from existing premises which are similar to those being designed. Where this data is available it should reflect the average occupant density at a peak trading time of year.

7.9.2 Travel distance

This is measured along the shortest, most direct route. Where there is fixed seating or other fixed obstructions, the travel distance is measured along the centreline of the seating or gangways. If a stair (such as an accommodation stair) is included in the escape route, the travel distance is measured along the pitch line on the centre line of travel.

7.9.3 Width

Usually, the narrowest part of an escape route will be at the door openings which form the room or storey exits. These are measured as shown in Fig. 7.10 below.

Escape route widths are measured at a height of 1500 mm above floor level where the route is defined by walls. Elsewhere, the width will be measured between any fixed obstructions.

Stair widths are measured clear between walls and balustrades. Strings may be

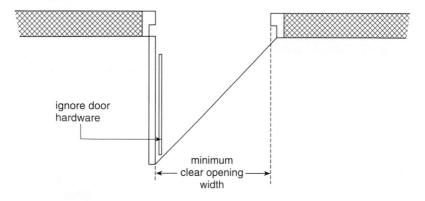

Fig. 7.10 Door width measurement.

ignored as may handrails which project less than 100 mm. Additionally, where a stairlift is installed, the guide rail may be ignored; however, the chair or carriage must be capable of being parked where it does not obstruct either the stair or the landing.

7.10   Fire alarm and fire detection systems

7.10.1   Fire detection and alarm in dwellings

Dwellinghouses

In all dwellinghouses (including bungalows), safety levels can be significantly increased by installing systems which automatically give early warning of fire. Approved Document B1 gives the following range of options for dwellinghouses:

- an automatic fire detection and alarm system in accordance with the relevant recommendations of BS 5839 *Fire detection and alarm systems for buildings*, Part 1:1988 *Code of practice for system design, installation and servicing* to at least the L3 standard specified in the code; or
- an automatic fire detection and alarm system in accordance with the relevant recommendations of BS 5839 *Fire detection and alarm systems for buildings*, Part 6:1995 *Code of practice for the design and installation of fire detection and alarm systems in dwellings* to at least a Grade E type LD3 standard; or
- a suitable number of smoke alarms provided as indicated below.

Large dwellinghouses

A large dwellinghouse is defined in AD B1 as having one or more floors greater than 200 m^2 in area. Where the house has more than three storeys (including basements) the following recommendations are given in AD B1.

- It may be fitted with an L2 system as described in BS 5839 Part 1 but the provisions in clause 16.5 regarding duration of the standby supply may be disregarded.
- Where the system is unsupervised, the standby supply should be able to automatically maintain the system in normal operation for 72 hours (but with audible and visible indication of mains failure). At the end of the 72 hours, sufficient capacity should remain to supply the maximum alarm for a minimum of 15 minutes.

Where a large house has no more than three storeys (including basements) it may be fitted with an automatic fire detection and alarm system of Grade B type LD3 in accordance with BS 5839: Part 6 instead of the L2 system referred to above.

Loft conversions

Where it is proposed to convert a loft space to habitable accommodation in a one or two storey dwellinghouse, an automatic smoke detection and alarm system based on linked smoke alarms should be installed throughout the dwellinghouse (i.e. not just in the extended part). The installation should follow the guidance described above and below for the size of dwelling which will be created by the extension (i.e. either a two storey or three storey house). Further details of the means of escape provisions for loft conversions are given in section 7.12.10 below.

Sheltered housing

Sheltered housing usually consists of a block or group or dwellings designed specifically for persons who might require assistance (such as elderly people) where some form of assistance will be available at all times (although not necessarily on the premises). Each unit of accommodation will have its own cooking and sanitary facilities and amenities common to all occupiers may also be provided such as communal lounges.

BS 5839 Part 6 recommends the installation of a Grade C, Type LD3 system. The detection equipment should have a connection to a central monitoring point (or central relay station) so that the warden or supervisor is able to identify the dwelling in which the fire has occurred. It is not intended that the provisions in AD B1 or BS 5839 Part 6 be applied to the common parts of sheltered accommodation and they do not apply to sheltered accommodation in Purpose Groups 2(a) Residential (Institutional) or 2(b) Other Residential (see section 7.3 above).

Flats and maisonettes

The principles for the provision of fire alarm and detection systems in flats and maisonettes are the same as those which apply to dwellinghouses described above, with the following additions.

- There is no need to apply the provisions to the common parts of blocks of flats.
- The systems in individual flats do not need to be interconnected.
- A maisonette (i.e. a 'flat' in which the accommodation is contained on more than one level and which may be entered from either the higher or lower level) should be treated in the same way as a house with more than one storey.

Student residential accommodation

Traditional halls of residence consisting of individual study bedrooms and shared dining and washing facilities should be designed for general evacuation in the same way as non-residential buildings (see section 7.15 below).

Where the accommodation is arranged as in a block of flats, with a group of students sharing an individual flat with its own entrance door, it is appropriate for an automatic detection system to be provided in each flat. In such cases the flats will be compartmented from each other and the automatic detection system need only give warning in the flat of fire origin.

Smoke alarms in dwellings

Smoke alarms should fulfil these criteria.

- Mains-operated, and designed and manufactured in accordance with BS 5446 *Components of automatic fire alarm systems for residential premises*, Part 1: 1990 *Specification for self-contained smoke alarms and point-type detectors.* They may have a secondary power supply (such as a capacitor or rechargeable or replaceable battery, see clause 13 of BS 5839 Part 6). BS 5446: Part 1 deals with smoke alarms based on ionization chamber smoke detectors and optical (photo-electric) smoke detectors. Each type of detector responds differently to smouldering and fast flaming fires. Therefore, optical smoke alarms should be installed in circulation spaces (e.g. hallways and landings) and ionization chamber based smoke alarms may be more appropriate if placed where a fast burning fire presents the greater danger to occupants (such as in living or dining rooms). Additionally, optical detectors are less affected by low levels of 'invisible' smoke that often cause false alarms.
- Located in circulation areas between sleeping places and places where fires are likely to start (kitchens and living rooms) and within 7.5 m of the door to every habitable room.
- Fixed to the ceiling and at least 300 mm from any walls and light fittings (unless there is test evidence to prove that the detector will not be adversely affected by the proximity of a light fitting). Units specially designed for wall mounting are acceptable provided that they are mounted above the level of doorways into the space and are fixed in accordance with manufacturers' instructions.
- Sited so that the sensor, for ceiling mounted devices, is between 25 mm and 600 mm below the ceiling (between 25 mm and 150 mm for heat detectors), assuming that the ceiling is predominantly flat and level.
- Fixed in positions that allow for routine maintenance, testing and cleaning (i.e. not over a stairwell or other floor opening).
- Sited away from areas where steam, condensation or fumes could give false alarms (this would include heaters, air-conditioning outlets, bathrooms, showers, cooking areas or garages, etc.).
- Sited away from areas that get very hot (e.g. boiler rooms) or very cold (e.g. an unheated porch). They should not be fitted to surfaces which are either much hotter or much colder than the rest of the room since air currents might be created which would carry smoke away from the unit.

The number of alarms which are installed will depend on the size and complexity of the layout of the dwelling and should be based on an analysis of the risk to life from fire. The following minimum provisions should be observed.

- At least one alarm should be installed in each storey of the dwelling.
- Where more than one smoke alarm is installed they should be interconnected so that the detection of smoke in any unit will activate the alarm signal in all of them. Provided that the lifetime or duration of any standby power supply is not

reduced, smoke alarms may be interconnected using radio-links. Manufacturers' instructions should be adhered to regarding the maximum number of units that can be interconnected.

- In open plan designs where the kitchen is not separated from the stairway or circulation space by a door, the kitchen should contain a compatible interlinked heat detector. This should be in addition to the normal provision of smoke detector(s) in the circulation space(s).

It should be noted that maintenance of the system in use is of utmost importance. Since this cannot be made a condition of the passing of plans for Building Regulation purposes it is important to ensure that developers and builders provide occupiers with full details of the use of the equipment and its maintenance (or guidance on suitable maintenance contractors). BS 5839: Parts 1 and 6 also recommend that occupiers receive manufacturers' operating and maintenance instructions.

Power supplies to smoke alarms

Power supplies for smoke alarm systems should be derived from the dwelling's mains electricity supply and connected to the smoke alarms through a separately fused circuit at the distribution board (consumer unit).

The power supply options described here are all based on using the mains supply at the normal 240 volts. Other effective (though possibly more expensive) options exist, such as, reducing the mains supply to extra low voltage in a control unit incorporating a standby trickle-charged battery, before distributing the power to the smoke alarms at that voltage.

The smoke alarm system can include a stand-by power supply which will operate during mains failure. This can allow the system to obtain its power by connection to a regularly used local lighting circuit with the advantage that the circuit will be unlikely to be disconnected for any prolonged period. Where the system does not include a stand-by power supply, no other electrical equipment should be connected to the smoke alarm circuit (except for a mains failure monitoring device, see below).

The mains supply to the smoke alarm system can be monitored by a device which will give warning in the event of failure of the supply. The warning of failure may be visible or audible (in which case it should be possible to silence it) and should be sited so that it is readily apparent to occupants. The circuit for the mains failure monitor should be designed so that any significant reduction in the reliability of the mains supply is avoided.

Ideally, the smoke alarm circuit should not be protected by any residual current device (rcd) such as a miniature circuit breaker or earth leakage trip. However, sometimes it is necessary for reasons of electrical safety that such devices be used and in these cases, either:

- the rcd should serve only the circuit supplying the smoke alarms, or
- the rcd protection of a fire alarm circuit should operate independently of any rcd protection for circuits supplying socket outlets or portable equipment.

Since it does not need any fire survival properties, the mains supply to smoke alarms, and any interconnecting wiring, may comprise any cable which is suitable for ordinary domestic mains wiring. Cables used for interconnections should be readily identifiable from those supplying power (e.g. by colour coding).

7.11   Fire detection and alarm in buildings other than dwellings

Introduction

It is extremely important to realise that there is a causal connection between the design of the means of warning and escape in a building and the eventual management of that means of warning and escape. Therefore, the escape strategy to be adopted in a particular building will be based on one of the following.

- Simultaneous evacuation; all the occupants are expected to leave the building at the same time.
- Phased evacuation; only the storeys most affected by the fire (e.g. the floor of origin and the floor above it) are evacuated immediately. Subsequently, two floors at a time are evacuated if the need arises.
- Progressive horizontal evacuation; the concept which is usually adopted in the in-patient parts of hospitals and similar healthcare premises where total evacuation of the building is inappropriate. In-patients are evacuated, in the event of fire, to adjoining compartments or sub-divisions of compartments, the object being to provide a place of relative safety within a short distance. If necessary, further evacuation can be made from these safe places but under less pressure of time.

Therefore, in buildings other than dwellings, selection of the appropriate fire detection and alarm system will depend on the means of escape strategy adopted. For example, in residential accommodation (where the occupants sleep on the premises) the threat posed by a fire will be much greater than in premises where the occupants are fully alert. In these circumstances an escape strategy based on simultaneous evacuation will mean that all fire sounders will operate almost instantaneously once a manual call point or fire detector has been activated. If however, the escape strategy is based on phased evacuation, a staged alarm system may be more appropriate. Two or more stages of alarm may be given within a particular area corresponding to 'alert' and 'evacuate' signals.

Fire alarm systems

All buildings should have arrangements for detecting fire and in most buildings this will be done directly by people through observation or smell. In many small buildings where there is no sleeping risk this may be all that is needed. Similarly, the means of raising the alarm may be simple in such buildings and where the occupants are in sight and hearing of each other a shouted warning may be sufficient. Clearly, it is necessary to assess the risk in each set of circumstances and decide standards on a case by case basis.

The risk analysis will consider the likelihood of a fire occurring and the degree to which the alarm can be heard by all the occupants. Therefore any of the following may need to be incorporated in the building.

- Manually operated sounders (e.g. rotary gongs or handbells)
- Simple manual callpoints combined with bell, battery and charger.
- Electrically operated fire warning system with manual callpoints sited adjacent to exit doors, combined with sufficient sounders to ensure that the alarm can be heard throughout the building. The system should comply with BS 5839: Part 1 (see below) and the call points with BS 5839: *Fire detection and alarm systems for buildings:* Part 2: 1983 *Specification for manual call points* or Type A of BS EN 54 *Fire detection and fire alarm systems – Part 11: Manual call points.* BS EN 54-11 covers two types of call points:
- Type A (direct operation) – the change to the alarm condition is automatic (i.e. without the need for further manual action) when the frangible element is broken or displaced, and
- Type B (indirect operation) – the change to the alarm condition requires a separate manual operation of the operating element by the user after the frangible element is broken or displaced. (Type B call points should only be used with the approval of the Building Control Body).

Four types of fire alarm and detection system are specified in BS 5839: Part 1 as follows:

- Type L – for the protection of life
- Type M – manual alarm systems
- Type P – for the protection of property
- Type X – for multi-occupancy buildings.

Type L systems are further subdivided into:

- L1 – systems installed throughout the protected building
- L2 – systems installed only in defined parts of the protected building (but always providing the coverage required of a Type L3 system)
- L3 – systems installed only for the protection of escape routes.

Type P systems are further subdivided into:

- P1 – systems installed throughout the protected building
- P2 – systems installed only in defined parts of the building.

In certain premises where large numbers of the public are present (e.g. large shops and places of assembly) it may be undesirable for an initial general alarm to be sounded since this may cause unnecessary confusion. Therefore it is essential in these

circumstances that staff are trained to effect pre-planned procedures for safe evacuation. Usually, actuation of the fire alarm system will alert staff by means of
discreet sounders or personal pagers first. This will enable them to be prepared and
in position should it be necessary for a general evacuation to be initiated by means of
sounders or an announcement over the public address system. In all other respects
any staff system should conform to BS 5839: Part 1.

Voice alarm systems are useful in circumstances where it is considered that people
might not respond quickly to a fire warning, or where they are unfamiliar with fire
warning arrangements. An audible fire warning signal can be given via a public
address system, provided that it is distinct from other signals which are in general
use and is accompanied by clear verbal instructions. Voice alarms should comply
with BS 5839: *Fire detection and alarm systems for buildings:* Part 8 1998 *Code of
practice for the design, installation and servicing of voice alarm systems.*

Automatic fire detection and alarm systems

Automatic fire detection systems involve a sensor network plus associated control
and indicating equipment. Sensors may detect heat, smoke or radiation and it is
usual for the control and indicating equipment to operate a fire alarm system. It may
also perform other signalling or control functions, such as, the operation of an
automatic sprinkler system.

Automatic fire detection and alarm systems should be installed in Institutional
(Purpose Group 2(a)) buildings and in Other Residential (Purpose Group 2(b))
occupancies.

Automatic fire detection systems are not normally needed in non-residential
occupancies; however it may be desirable to install a fire detection system in the
following circumstances:

- to compensate for the fact that it has not been possible to follow all the guidance
  in Approved Document B
- where it is necessary as part of a fire protection operating system, such as a
  pressurised staircase or automatic door release mechanism
- where a fire could occur unseen in an unoccupied or rarely visited part of a
  building and prejudice the means of escape from the occupied parts.

Where a building is designed for phased evacuation (see above) it should be fitted
with an appropriate fire warning system conforming to at least the L3 standard
given in BS 5839: Part 1. Additionally, an internal speech communication system
(telephone, intercom etc.) should be provided so that conversation is possible
between a fire warden at every floor level and a control point at the fire service access
level.

Where a building contains an atrium and is designed in accordance with BS 5588:
Fire precautions in the design, construction and use of buildings, Part 7: *Code of
practice for the incorporation of atria in buildings,* then the relevant recommendations of that code should be followed for the installation of fire alarm and/or fire
detection systems.

Further guidance on the standard of automatic fire detection that may need to be provided in a building can be found in:

- Home office guides that support the Fire Precautions Act 1971 and the Fire Precautions (Workplace) Regulations 1997.
- The NHS Estates 'Firecode' documents for buildings in the Institutional Purpose Group 2(a).

Design, maintenance and installation of systems

Fire warning and detection systems must be properly designed, installed and maintained. Installation and commissioning certificates should be obtained wherever a fire alarm system is installed. Additionally, third party certification schemes for fire protection products and related services offer an effective means for providing the fullest possible assurances that the required level of performance will be achieved, and that the products actually supplied are provided to the same specification or design as those that have been tested or assessed.

7.12   Means of escape in dwellinghouses

7.12.1   Introduction

Approved Document B1 deals with means of escape from dwellinghouses according to the height of the top storey above ground level (i.e. the ground level on the lowest side of the building).

This is probably a sensible approach since storey heights can vary and means of escape through upper windows become more hazardous with increasing height. Thus, the divisions chosen are:

- houses with all floors not more than 4.5 m above ground (i.e. ground and first floor only)
- houses with one floor more than 4.5 m above ground (i.e. ground floor, first floor and second floor)
- houses with two or more floors more than 4.5 m above ground (i.e. ground floor and three or more upper floors).

Therefore, as the height of the top floor increases above ground level, the means of escape provisions become more complex and these are dealt with under separate sections below. Certain recommendations however, are common to all dwellings and these include:

- the provision of an automatic fire detection and alarm system (see above)
- special provisions to deal with basements and inner rooms
- windows and external doors used for escape purposes
- balconies and flat roofs.

The guidance contained in this section is also applicable to houses in multiple occupation (HMOs) provided that there are no more than six residents. An HMO is defined in section 345 of the Housing Act 1985 as 'a house which is occupied by persons who do not form part of a single household'. For HMOs containing greater numbers of occupants, technical guidance may be sought in DOE Circular 12/92 (see below) or Welsh Office Circular 25/92.

It may also be possible to treat an unsupervised group home for mentally ill or mentally impaired people with up to six residents as an ordinary dwelling but this will depend on the nature of the occupants and their management. Such premises have to be registered, therefore the registration authority should be consulted to establish if there are any additional fire safety measures needed by the authority.

7.12.2   Basements and inner rooms

With certain dwelling designs (such as open-plan layouts and the provision of sleeping galleries) it is possible that that a situation will be created whereby the innermost room (termed the inner room) will be put at risk by a fire occurring in the room that gives access to it (termed the access room), since escape is only possible by passing through that access room. Therefore, an inner room should only be used as:

- a kitchen, laundry or utility room
- a dressing room
- a bathroom, shower room or WC
- any other room which has a suitable escape window or door (see next section for details), provided the room is in the basement, or is on the ground or first floor
- a sleeping gallery (see section 7.12.5 below for details).

Escape from a basement fire may be particularly hazardous if an internal stair has to be used, since it will be necessary to pass through a layer of smoke and hot gases. Therefore, any habitable room in a basement should have either:

- an alternative escape route via a suitable door or window, or
- a protected stairway leading from the basement to a final exit.

7.12.3   Windows and external doors used for escape purposes

To be suitable for escape purposes, windows and external doors should conform to the dimensions given in Fig. 7.11. Dormer windows and roof windows situated above the ground storey should comply with Fig. 7.16. Escape should be to a place of safety free from the effects of fire. Where this is to an enclosed back garden or yard from which escape may be made only by passing through other buildings, its length should be at least equivalent to the height of the dwelling (see Fig. 7.12).

7.12.4   Balconies and flat roofs

If used as an escape route, a flat roof should:

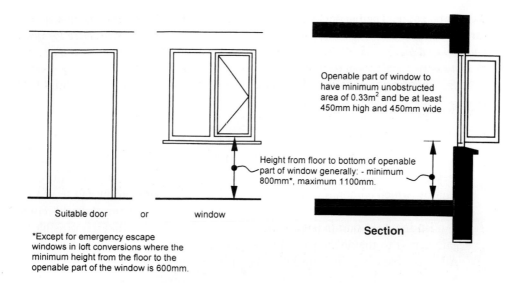

Fig. 7.11   Windows and doors for escape purposes.

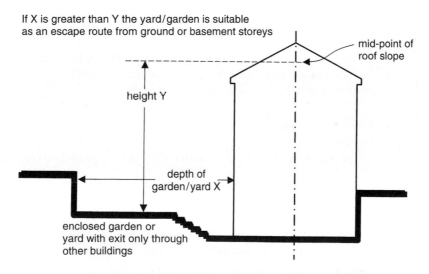

Fig. 7.12   Enclosed yard or garden suitable for escape purposes – dwellinghouses.

- be part of the same building from which escape is being made
- lead to a storey exit or external escape route
- be provided with 30 minutes fire resistance. This applies only to the part of the flat roof forming the escape route, its supporting structure and any opening within 3 m of the escape route.

Balconies and flat roofs provided for escape purposes should be guarded in accordance with the provisions of Approved Document K. (This relates to the provision of barriers at least 1100 mm high designed to prevent people falling from the escape route. The barriers should be capable of resisting at least the horizontal force given in BS 6399 *Loading for buildings,* Part 1: *1984 Code of practice for dead and imposed loads.*

7.12.5   Dwellinghouses with floors not more than 4.5 m above ground

For dwellings with floors not more than 4.5 m above ground, the means of escape measures outlined above need only to be augmented by the following provisions:

(1)   All habitable rooms (except kitchens) in the upper storey(s) of a dwellinghouse served by only one stairway, should be provided with an external door or window which complies with Fig. 7.11. A single door or window can serve two rooms provided that they each have their own access to the stairs. A communicating door should be provided between the rooms so that it is possible to gain access to the door or window without entering the stair enclosure.
(2)   All habitable rooms (except kitchens) in the ground storey should either:
   - open directly onto a hall which leads to an entrance or other suitable exit, or
   - be provided with a door or window which complies with Fig. 7.11.
(3)   Where the dwelling contains a sleeping gallery (i.e. a floor which is used as a bedroom but which is open on at least one side to some other part of the dwelling) which is not more than 4.5 m above ground level:
   - The distance between the foot of the access stair to the gallery and the door to the room which contains the gallery should not be more than 3 m.
   - An alternative exit from the gallery complying with Fig. 7.11 should be provided, if the distance from the head of the access stair to any point in the gallery exceeds 7.5 m.
   - Unless they are enclosed with fire-resisting construction, any cooking facilities within a room containing a gallery should be remote from the stair to the gallery and positioned so that they do not prejudice the escape route from the gallery.

7.12.6   Houses with one floor more than 4.5 m above ground

Houses with one floor more than 4.5 m above ground are likely to consist of ground, first and second floors. Two alternative solutions are recommended in AD B1.

(1)   Provide the first and second floors with a protected stairway which should either:
   - terminate directly in a final exit, or
   - give access to a minimum of two escape routes at ground level, separated from each other by self-closing fire doors and fire-resisting construction and each leading to a final exit (these alternatives are illustrated in Fig. 7.13) or
(2)   Separate the top floor (and the main stairs leading to it) from the other floors by fire-resisting construction and provide it with an alternative means of escape leading to its own final exit. (A variation on this solution can be used where it is intended to convert the roof space of a two-storey dwelling, effectively making it three-storey. Further details of this are described below.)

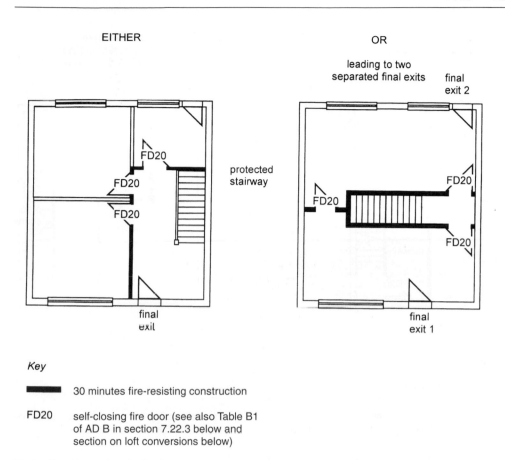

EITHER

OR

leading to two
separated final exits   final
exit 2

protected
stairway

FD20

FD20

FD20

FD20

FD20

FD20

final
exit

final
exit 1

Key

▬▬▬▬   30 minutes fire-resisting construction

FD20   self-closing fire door (see also Table B1
of AD B in section 7.22.3 below and
section on loft conversions below)

Fig. 7.13   Alternative final exit arrangements.

7.12.7   Houses with two or more floors more than 4.5 m above ground

Normally, this would include houses with three or more storeys above the ground floor. In most circumstances it will be necessary to provide an alternative means of escape for all floors which are 7.5 m or more above ground.

Where it is intended that the alternative escape route be accessed via the protected stairway (i.e. to an upper storey, see Fig. 7.14(a)) or via a landing within its enclosure (i.e. on the same storey, see Fig. 7.14(b)) it is possible a fire in the lower part of the dwellinghouse might make access to the alternative route impassable. Therefore, the protected stairway at or about 7.5 m above ground level should be separated from the lower storeys by fire-resisting construction.

7.12.8   Air circulation systems in houses with any floor more than 4.5 m above ground

Modern houses are often fitted with air circulation systems designed for heating, energy conservation or condensation control. With such systems, there is the possibility that fire or smoke may spread (possibly by forced convection, natural

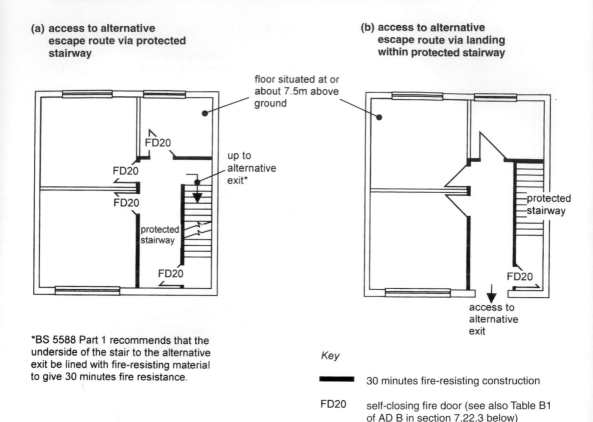

(a) access to alternative escape route via protected stairway

FD20

FD20

FD20

protected stairway

FD20

floor situated at or about 7.5m above ground

up to alternative exit*

(b) access to alternative escape route via landing within protected stairway

protected stairway

FD20

access to alternative exit

*BS 5588 Part 1 recommends that the underside of the stair to the alternative exit be lined with fire-resisting material to give 30 minutes fire resistance.

Key

▬▬▬ 30 minutes fire-resisting construction

FD20   self-closing fire door (see also Table B1 of AD B in section 7.22.3 below)

Fig. 7.14   Alternative exit arrangements in houses with more than one floor over 4.5 m above ground.

convection or fire-induced convection) from the room of origin of the fire to the protected stairway. Clearly, the risk to life is greater in dwellings where there are floors at high levels so AD B1 and BS 5588 Part 1 make the following recommendations for such buildings:

- The walls enclosing a protected stairway should not be fitted with transfer grilles.
- Where ductwork passes through the enclosure to a protected stairway, it should be fitted so that all joints between the ductwork and the enclosure are fire-stopped.
- Ductwork used to convey air into the protected stairway through its enclosure should be ducted back to the plant.
- Grilles or registers which supply or return air should be positioned not more than 450 mm above floor level.
- Any room thermostat for a ducted warm air heating system should be mounted in the living room at a height between 1370 mm and 1830 mm and its maximum setting should not be more than 27°C.
- Mechanical ventilation systems which recirculate air should comply with BS 5588 *Fire precautions in the design, construction and use of buildings*, Part 9: 1989 *Code of practice for ventilation and air conditioning ductwork*.

7.12.9   Passenger lifts in dwellinghouses

If a passenger lift is installed in a dwellinghouse which serves any floor situated more than 4.5 m above ground level, it should either be contained in a fire-resisting lift shaft or be located in the enclosure to a protected stairway.

7.12.10   Conversions to provide rooms in a roof space

The following provisions apply when it is proposed to convert the roof space of a two-storey house to provide living accommodation. The additional floor provided is most likely to be more than 4.5 m above ground and will therefore require additional protection to the means of escape or an alternative escape route. The measures described do not apply if:

- the new floor exceeds 50 m^2 in area; or
- it is proposed to provide more than two habitable rooms in the new storey.

In these cases, the loft conversion will have to meet the full provisions of the Building Regulations which apply to dwellings of three or more storeys.

The recommendations for means of escape are illustrated in Fig. 7.15(a) and 7.15(b), the general principles being:

- Every doorway within the existing stairway enclosure should be fitted with a door.
- Doors to habitable rooms in the existing stair enclosure should be self-closing to prevent the movement of smoke into the means of escape. For this purpose rising butt hinges are acceptable as self-closing devices in dwellinghouses.
- New doors to habitable rooms should be fire-resisting and self-closing, but existing doors will only need to be fitted with rising butt hinges.
- The stairway in the ground and first floors should be adequately protected with fire-resisting construction and should terminate as shown in Fig. 7.13 above.
- Glazing (whether new or existing) in the existing stair enclosure should be fire-resisting and retained by a suitable glazing system and beads compatible with the type of glass. This includes glazing in all doors (whether or not they need to be fire doors) except those to bathrooms or WCs. Fire-resisting glazing will usually consist of traditional annealed wired glass based on soda-lime-silica which is able to satisfy the integrity requirements of BS 476 Part 22. There are also some unwired glass products capable of satisfying the requirements for integrity. There is no limit on the area of such glass which can be incorporated in walls and doors above 100 mm from floor level.
- The new storey should be served by a stair which is located as shown in Fig. 7.15(a) and 7.15(b) and complies with Approved Document K. This could be an alternating tread stair.
- The new storey should be separated from the remainder of the dwelling by fire-resisting construction. Any openings should be protected by self-closing fire doors situated as shown in 7.15(a) and 7.15(b).

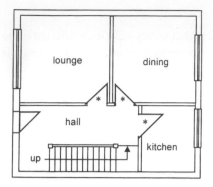

Existing ground floor common to all examples

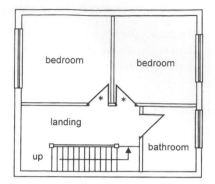

Proposed first floor - example A

Example A - new stairway rising in same staircase enclosure over existing stairway

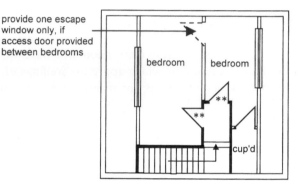

Proposed second floor - example A

Key
*   self-closing door
**   FD20 self-closing fire door (20 minutes integrity)
▬   30 minutes fire-resisting construction

Fig. 7.15(a)   Loft conversion to existing two-storey dwellinghouse.

- Windows and doors provided for emergency escape in the basement, ground and first floor of the existing dwelling provide a means of self-rescue. At higher levels, it is acceptable in loft conversions for escape to depend on a ladder being set up to a suitable escape window, or to a roof terrace, which is accessible by a door from the loft conversion. Any escape window should be large enough and suitably positioned to allow escape by means of a ladder to the ground. See Fig. 7.16.

- There should be suitable means of pedestrian access to the place at which the ladder would be set, to allow fire service personnel to carry a ladder from their vehicle (although it should not be assumed that only the fire service will effect a rescue). A ladder fixed to the slope of the roof is not recommended due to the danger inherent in using such ladders.

- Where the loft conversion consists of two rooms a single window can serve both rooms provided that they each have their own access to the stairs. A communicating door should be provided between the rooms so that it is possible to gain access to the window without entering the stair enclosure.

Example B - new
stairway rising from
existing staircase
through existing room

Example C - new
stairway separated
from existing staircase
at first floor level

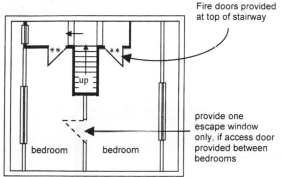

Fire doors provided
at top of stairway

provide one
escape window
only, if access door
provided between
bedrooms

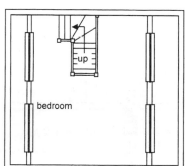

Proposed second floor - example B

Proposed second floor - example C

30 minutes fire-resisting
construction including underside of
staircase, if exposed in bedroom

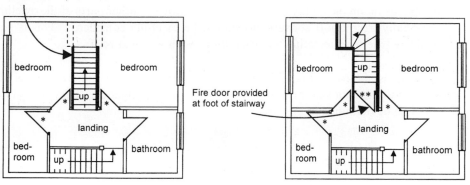

Fire door provided
at foot of stairway

Proposed first floor - example B

Proposed first floor - example C

Key

*   self-closing door

**   FD20 self-closing fire door
(20 minutes integrity)

■■   30 minutes fire-resisting
construction

Fig. 7.15(b)   Loft conversion to existing two-storey dwellinghouse.

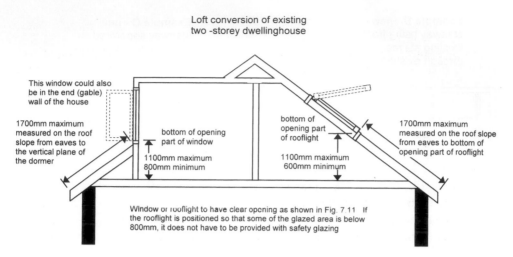

Loft conversion of existing
two -storey dwellinghouse

This window could also
be in the end (gable)
wall of the house

1700mm maximum
measured on the roof
slope from eaves to
the vertical plane of
the dormer

bottom of opening
part of window

1100mm maximum
800mm minimum

bottom of
opening part
of rooflight

1100mm maximum
600mm minimum

1700mm maximum
measured on the roof slope
from eaves to bottom of
opening part of rooflight

Window or rooflight to have clear opening as shown in Fig. 7.11   If
the rooflight is positioned so that some of the glazed area is below
800mm, it does not have to be provided with safety glazing

Fig. 7.16   Window positions for emergency escape dormers and rooflights in loft conversions.

- Escape across the roof of a ground storey extension is acceptable provided that the roof complies with the recommendations listed in section 7.12.4 above. Additionally, when it is proposed to erect an extension to a dwelling, the effect of the extension on the ability to escape from windows in other parts of the dwelling (especially from a loft conversion) should be carefully considered. This is particularly true when erecting a conservatory with a glazed roof.

The examples shown in Fig. 7.15(a) and 7.15(b) are typical illustrations of the principles outlined above for one particular form of dwellinghouse. They should not be taken to be definitive in any sense.

7.13   Flats and maisonettes

7.13.1   Introduction

Approved Document B1 deals with a limited range of common designs for flats and maisonettes. Where less common arrangements are desired (for example, where flats are entered above or below accommodation level, or flats contain sleeping galleries) the principles described in this chapter may still be applied or reference can be made to BS 5588: Part 1:1990, clauses 9 and 10.

The means of warning and escape recommendations listed above for dwelling-houses which consist of basement, ground and first floors only, apply equally to flats and maisonettes situated at these levels. At higher levels, escape through upper windows becomes more hazardous and more complex provisions are necessary, especially in maisonettes where internal stairs will need protection.

In addition to the general assumptions stated in section 7.8 above for all building types, the following assumptions are made when considering the means of warning and escape from flats and maisonettes.

- Fires generally originate in the dwelling.
- Rescue by ladders is not considered suitable.
- Fire spread beyond the dwelling of origin is unlikely due to the compartmentation recommendations in Approved Document B3, therefore simultaneous evacuation of the building should be unnecessary.
- Fires which occur in common areas are unlikely to spread beyond the immediate vicinity of the outbreak due to the materials and construction used there.

The guidance contained in this section is also applicable to flats and maisonettes when they are considered to be houses in multiple occupation. These are defined in section 7.12.1 above where references will be found to further guidance. Additionally, much of the following guidance will be applicable to flats used as sheltered housing although the nature of the occupancy may necessitate some additional fire protection measures. Guidance on these may be found in clause 17 of BS 5588: Part 1: 1990.

7.14   Means of escape in flats and maisonettes

There are two main components to the means of escape from flats and maisonettes:

- escape from within the dwelling itself; and
- escape from the dwelling to the final exit from the building (usually along a common escape route).

The following sections consider these two components.

7.14.1   Means of escape from within flats and maisonettes

The provisions relating to inner rooms, basements, balconies and flat roofs in dwellings (see sections 7.12.2 and 7.12.4 above) also apply to the inner parts of flats and maisonettes. However, where a balcony is provided as an alternative exit to a dwelling situated more than 4.5 m above ground it should be designed as a common balcony and should meet the conditions given in section 7.14.2 below for alternative exits.

Flats and maisonettes with floors not more than 4.5 m above ground

Generally, where the floor of a flat or maisonette is not more than 4.5 m above ground it should be planned so that any habitable room in a ground or upper storey is provided with a suitable escape window or door, as shown in Fig. 7.11 above.

  A single door or window in an upper storey can serve two rooms provided that they each have their own access to the stair enclosure or entrance hall. A communicating door should be provided between the rooms so that it is possible to gain access to the door or window without entering the stair enclosure or entrance hall.

Alternatively, upper floors can be designed to follow the remainder of the guidance described below, in which case escape windows in upper storeys will be unnecessary.

Flats and maisonettes with floors more than 4.5 m above ground

Provided that the restrictions on inner rooms are observed, three possible solutions to the internal planning of flats are given in AD B1.

- All the habitable rooms in the flat are arranged to have direct access to a protected entrance hall. The maximum distance from the entrance door to the door of a habitable room should not exceed 9 m. (See Fig. 7.17(a)).
- In flats where a protected entrance hall is not provided, the 9 m distance referred to in (a) should be taken as the furthest distance from any point in the flat to the entrance door. Cooking facilities should be remote from the entrance door and positioned so that they do not prejudice the escape route from any point in the flat. (See Fig. 7.17(b)).
- Provide an alternative exit from the flat. In this case, the internal planning will be more flexible. An example of a typical flat plan where an alternative exit is provided, but not all the habitable rooms have direct access to the entrance hall, is shown in Fig. 7.17(c).

Where flats and maisonettes with floors more than 4.5m above ground are fitted with air circulation systems designed for heating, energy conservation or condensation control there is the possibility that fire or smoke may spread from the room of origin of the fire to the protected entrance hall or landing. Clearly, the risk to life is greater where there are floors at high levels so AD B1 and BS 5588 Part 1 make the following recommendations for such buildings:

- Any wall, door, floor or ceiling enclosing a protected entrance hall of a dwelling or protected stairway and landing of a maisonette should not be fitted with transfer grilles.
- Where ductwork passes through the enclosure to a protected entrance hall or protected stairway and landing, it should be fitted so that all joints between the ductwork and the enclosure are fire-stopped.
- Ductwork used to convey air into the protected entrance hall of a dwelling or protected stairway and landing of a maisonette through the enclosure of the protected hall or stairway, should be ducted back to the plant.
- Grilles or registers which supply or return air should be positioned not more than 450 mm above floor level.
- Any room thermostat for a ducted warm air heating system should be mounted in an area from which air is drawn directly to the heating unit at a height between 1370 mm and 1830 mm and its maximum setting should not be more than 27°C.
- Mechanical ventilation systems which recirculate air should comply with BS 5588 Part 9.

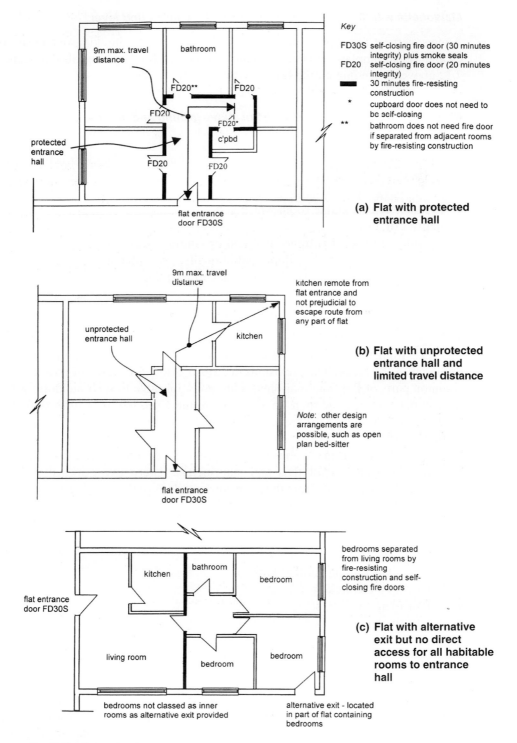

Key

FD30S   self-closing fire door (30 minutes integrity) plus smoke seals

FD20   self-closing fire door (20 minutes integrity)

█   30 minutes fire-resisting construction

*   cupboard door does not need to be self-closing

**   bathroom does not need fire door if separated from adjacent rooms by fire-resisting construction

9m max. travel distance

bathroom

FD20**

FD20

FD20

FD20*

c'pbd

protected entrance hall

FD20

FD20

flat entrance door FD30S

(a) Flat with protected entrance hall

9m max. travel distance

kitchen remote from flat entrance and not prejudicial to escape route from any part of flat

unprotected entrance hall

kitchen

Note: other design arrangements are possible, such as open plan bed-sitter

flat entrance door FD30S

(b) Flat with unprotected entrance hall and limited travel distance

bedrooms separated from living rooms by fire-resisting construction and self-closing fire doors

kitchen

bathroom

bedroom

flat entrance door FD30S

(c) Flat with alternative exit but no direct access for all habitable rooms to entrance hall

living room

bedroom

bedroom

bedrooms not classed as inner rooms as alternative exit provided

alternative exit - located in part of flat containing bedrooms

Fig. 7.17   Examples of alternative internal layouts to flats.

Maisonette with independent external entrance at ground level

A maisonette of this type is similar to a dwellinghouse and should have a means of escape which complies with the recommendations for dwellings described in section 7.12 above, depending on the height of the top storey above ground.

Maisonette with floor more than 4.5 m above ground and no external entrance at ground level

Two internal planning arrangements are described in AD B1 for maisonettes of this type. These are illustrated in Fig. 7.18(a) and 7.18(b), as follows:

- Provide each habitable room which is not on the entrance floor with an alternative exit, or
- Provide a protected entrance hall and/or landing entered directly from all the habitable rooms on that floor. Additionally, one alternative exit should be provided on each floor which is not the entrance floor.

7.14.2   Alternative exits from flats and maisonettes

That part of the means of escape from the entrance door of a flat or maisonette to a final exit from the building is often by way of a route which is common to all dwellings in the block. The provisions described below for means of escape in the common parts of flats and maisonettes are not applicable to such buildings where the top floor is not more than 4.5 m above ground level. However, they should be read in conjunction with the section below on 'Provision and protection of stairs' in section 7.15.8.

Reference has been made above, to the provision of alternative exits in certain planning arrangements for flats and maisonettes. Alternative exits will only be effective if they are remote from the main entrance door to the dwelling and lead to a common stair or final exit by means of:

- a door to an access corridor, access lobby or common balcony, or
- an internal private stairway leading to an access corridor, access lobby or common balcony on another level, or
- a door onto a common stair, or
- a door to an external stair, or
- a door to an escape route over a flat roof.

7.14.3   Means of escape in the common parts of flats and maisonettes

In general, flats and maisonettes should have access to an alternative means of escape. In this way, it will be possible to escape from a fire in a neighbouring flat by walking away from it. It is not always possible to provide alternative escape routes (which means providing two or more staircases) in all buildings containing flats and maisonettes, therefore single staircase buildings are permissible in certain circumstances. Typical examples of single and multi-stair buildings are described below.

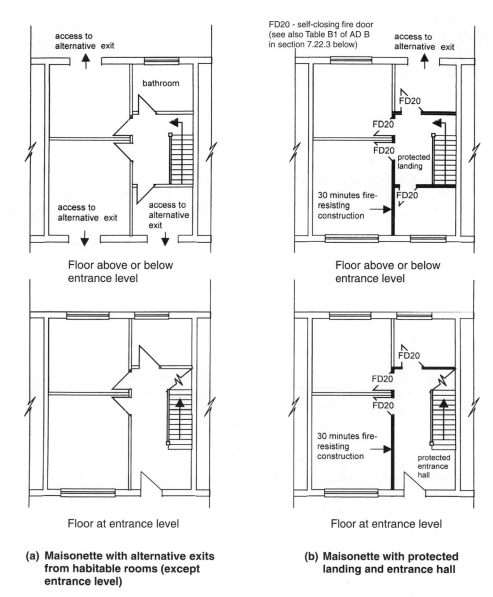

(a) Maisonette with alternative exits from habitable rooms (except entrance level)

(b) Maisonette with protected landing and entrance hall

Fig. 7.18 Maisonette with no independent access at ground level and at least one storey more than 4.5 m above ground.

Flats and maisonettes with single common stairs

In larger buildings, it will be necessary to separate the entrance to each dwelling from the common stair by a protected lobby or common corridor. The maximum distance from any entrance door to the common stair or protected lobby should not exceed 7.5 m. (See Fig. 7.19(a) and 7.19(b).) These recommendations may be modified for smaller buildings in the following cases:

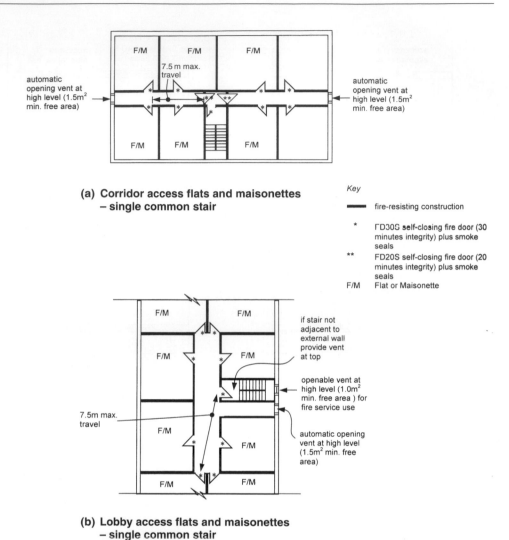

**(a) Corridor access flats and maisonettes
– single common stair**

Key

—— fire-resisting construction

*   FD30S self-closing fire door (30
      minutes integrity) plus smoke
      seals

**   FD20S self-closing fire door (20
      minutes integrity) plus smoke
      seals

F/M   Flat or Maisonette

**(b) Lobby access flats and maisonettes
– single common stair**

Fig. 7.19   Flats and maisonettes with single common stairs.

- The building consists of a ground storey and no more than three other storeys above the ground storey.
- The top floor does not exceed 11m above ground level.
- The stair does not connect to a covered car park unless it is open-sided.
- The stair does not also serve ancillary accommodation (see section 7.3 above), although this does not apply to ancillary accommodation:
  (a)   in any storey which does not contain dwellings, and
  (b)   which is separated from the stair by a ventilated protected lobby or ventilated protected corridor (i.e. provide permanent ventilation of not less than $0.4\,m^2$ or a mechanical smoke control system to prevent the ingress of smoke).

The modified recommendations are illustrated in Fig. 7.20(a). The maximum distance from the dwelling entrance door to the stair entrance should be reduced to 4.5 m. If the intervening lobby is provided with an automatic opening vent this distance may be increased to 7.5 m. Where the building contains only two flats per floor further simplifications as shown in Fig. 7.20(b) and (c) are possible.

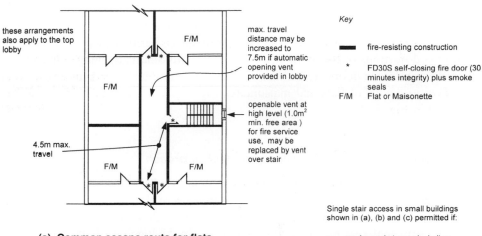

Key

━━ fire-resisting construction

* FD30S self-closing fire door (30 minutes integrity) plus smoke seals

F/M Flat or Maisonette

these arrangements also apply to the top lobby

max. travel distance may be increased to 7.5m if automatic opening vent provided in lobby

openable vent at high level (1.0m² min. free area) for fire service use, may be replaced by vent over stair

4.5m max. travel

(a) Common escape route for flats and maisonettes in small single stair building

Single stair access in small buildings shown in (a), (b) and (c) permitted if:

- maximum 4 storeys including ground storey.
- top floor not greater than 11m above ground level.
- stair does not connect to a covered car park unless it is open-sided.
- stair does not also serve certain types of ancillary accommodation (see section 7.3)

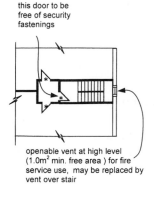

this door to be free of security fastenings

openable vent at high level (1.0m² min. free area) for fire service use, may be replaced by vent over stair

(b) Small single stair building – maximum two dwellings per floor

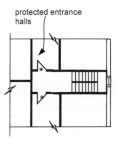

protected entrance halls

in (b) the lobby may be omitted if the dwellings have protected entrance halls

(c) Small single stair building – maximum two dwellings per floor with protected entrance halls

Fig. 7.20 Flats and maisonettes – small single stair buildings.

Flats and maisonettes with more than one common stair

Where escape is possible in two directions from the dwelling entrance door, the maximum escape distance to a storey exit may be increased to 30 m.

Furthermore, if all the dwellings on a storey have independent alternative means of escape, the maximum travel distance of 30 m does not apply. (There is still, however, the need to comply with the fire service access recommendations in Approved Document B5, where vehicle access for a pump appliance should be within 45 m of every dwelling entrance door.)

In buildings of this type it is possible to have a dead end situation where the stairs are not located at the extremities of each storey. This is permissible provided that the dead end portions of the corridor are fitted with automatic opening vents and the dwelling entrance doors are within 7.5 m of the common stair entrance.

Typical details of flats and maisonettes with more than one common stair are shown in Fig. 7.21.

Flats and maisonettes with balcony or deck approach

This is a fairly common arrangement whereby all dwellings are accessed by a continuous open balcony or deck on one or both sides of the block. AD B1 refers the reader to clause 13 of BS 5588 Part 1 on which the following notes are based.

The principal risk in such arrangements is smoke-logging of the balconies or decks.

This is less likely to occur when the balconies are relatively narrow, therefore the only considerations necessary are these.

- That vehicle access for a fire service pump appliance is within 45 m of every dwelling entrance door, and all parts of the building are within 60 m of a fire main.
- In the case of single stair buildings, that persons wishing to escape past the dwelling on fire can do so safely. This is usually achieved by ensuring that the external part of each dwelling facing the balcony is protected by 30 minutes fire-resisting construction for a distance of 1100 mm from the balcony floor level.

Smoke-logging is more likely to occur with the adoption of wider balconies or a deck approach. The provision of downstands from the soffit above a deck or balcony at right angles to the face of the building can reduce the possibility of smoke from any dwelling on fire spreading laterally along the deck. This would also reduce the chances of smoke logging on the decks above. Therefore, where the soffit above a deck or a balcony has a width of 2 m or more:

- it should be designed with downstands placed at 90° to the face of the building (on the line of separation between individual dwellings), and
- the down-stand should project 300 mm to 600 mm below any other beam or downstand parallel to the face of the building.

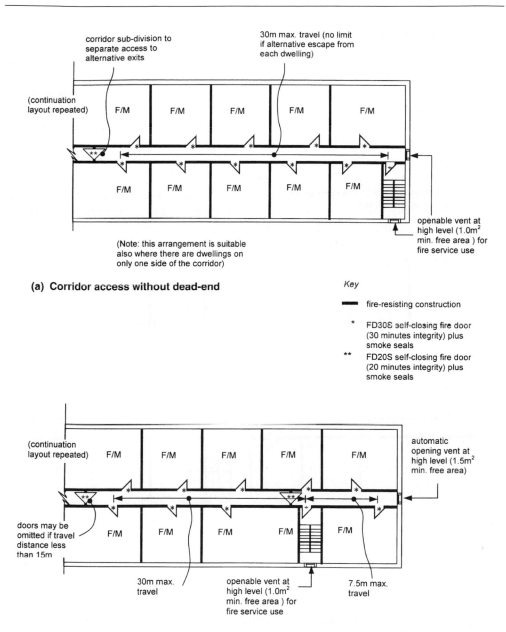

(a) Corridor access without dead-end

(Note: this arrangement is suitable also where there are dwellings on only one side of the corridor)

Key

━━ fire-resisting construction

* FD30S self-closing fire door (30 minutes integrity) plus smoke seals

** FD20S self-closing fire door (20 minutes integrity) plus smoke seals

(b) Corridor access with dead-end

Fig. 7.21 Flats and maisonettes with more than one common stair.

There is a risk that occupants of dwellings with wider balconies or deck approach will use this opportunity of greater depth to erect 'external' stores or other fire risks. Therefore, no store or other fire risk should be erected externally on the balcony or deck.

Examples of common escape routes for dwellings with balcony or deck access are shown on Fig. 7.22 below.

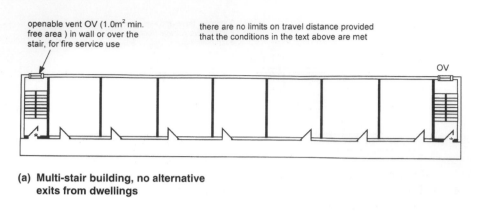

openable vent OV (1.0m² min. free area) in wall or over the stair, for fire service use

there are no limits on travel distance provided that the conditions in the text above are met

(a) Multi-stair building, no alternative exits from dwellings

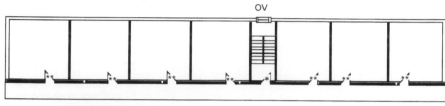

(b) Single stair building, no alternative exits from dwellings

either front or back wall to be fire resisting (not both)

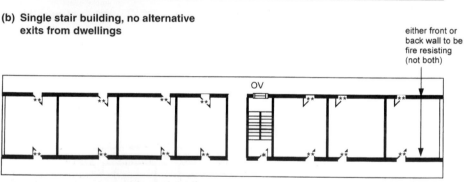

(c) Single stair building with alternative exit from each dwelling

Key

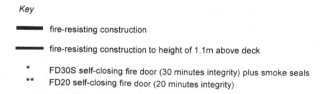

━━━ fire-resisting construction

── fire-resisting construction to height of 1.1m above deck

*   FD30S self-closing fire door (30 minutes integrity) plus smoke seals
**   FD20 self-closing fire door (20 minutes integrity)

Fig. 7.22 Flats and maisonettes with balcony/deck approaches.

Additional provisions for common escape routes

Common escape routes should be planned so that they are not put at risk by a fire in any of the dwellings or in any stores or ancillary accommodation. The following recommendations are designed to provide additional protection to these routes.

- It should not be necessary to pass through one stairway enclosure to reach another. Where this is unavoidable a protected lobby should be provided to the stairway. This lobby may be passed through in order to reach the other stair.
- Common corridors should be designed as protected corridors and should be constructed to be fire-resisting.
- The wall between each dwelling and the common corridor should be a compartment wall.
- A common corridor connecting two or more storey exits should be sub-divided with a self-closing fire door and/or fire-resistant screen, positioned so that smoke will not affect access to more than one storey exit.
- A dead end section of a common corridor should be separated in a similar manner from the rest of the corridor.
- Protected lobbies and corridors should not contain any stores, refuse chutes, refuse storage areas or other ancillary accommodation. See section 7.16.10 for for further details of the provision of refuse chutes and stores.

Ventilation of common escape routes

Although precautions can be taken to prevent the ingress of smoke onto common corridors and lobbies, it is almost inevitable that there will be some leakage since a flat entrance door must be opened in order that the occupants can escape. Provisions for ventilation of the common areas (which also provide some protection for common stairs) are therefore vital and may be summarised as follows.

- Subject to the variations shown in Fig. 7.19(a) and 7.19(b), common corridors or lobbies in larger, single-stair buildings should be provided with automatic opening ventilators, triggered by automatic smoke detection located in the space to be ventilated. These should be positioned as shown in the figure, should have a free area of at least 1.5 m^2 and be fitted with a manual override.
- Small single-stair buildings should conform to the guidance shown in Fig. 7.20(a), 7.20(b) and 7.20(c).
- Common corridors in multi-stair buildings should extend at both ends to the external face of the building where openable ventilators, or automatic opening ventilators, should be fitted for fire service use. They should have a free area of 1.0 m^2 at each end of the corridor (see Fig. 7.21).
- It is possible to protect escape stairways, corridors and lobbies by means of smoke control systems employing pressurisation. These systems should comply with BS 5588: *Fire precautions in the design, construction and use of buildings,* Part 4:1998 *Code of practice for smoke control using pressure differentials.* Where these are provided the cross corridor fire doors and the openable and automatic opening ventilators referred to above should be omitted.

Escape routes across flat roofs

Where more than one escape route exists from a storey or part of a building, one of those routes may be across a flat roof if the following conditions are observed.

- The flat roof should be part of the same building.
- The escape route over the flat roof should lead to a storey exit or external escape route.
- The roof and its structure forming the escape route should be fire-resisting.
- Any opening within 3 m of the route should be fire-resisting.
- The route should be adequately defined and guarded in accordance with Approved Document K. (This relates to the provision of barriers at least 1100 mm high designed to prevent people falling from the escape route. The barriers should be capable of resisting at least the horizontal force given in BS 6399: Part 1).

Provision of common stairs in flats and maisonettes

Stairs which are used for escape purposes should provide a reasonable degree of safety during evacuation of a building. Since they may also form a potential route for fire spread from floor to floor there are recommendations contained in Approved Document B3 which are designed to prevent this. Stairs may also be used for firefighting purposes. In this case reference should be made to the recommendations contained in Approved Document B5 (see section 7.30.2). The following recommendations are specifically for means of escape purposes.

- Each common stair should be situated in a fire-resisting enclosure with the appropriate level of fire resistance taken from Tables A1 and A2 of Appendix A of Approved Document B.
- Each protected stair should discharge either:
  (a) direct to a final exit, or
  (b) by means of a protected exit passageway to a final exit.
- If two protected stairways or protected exit passageways leading to different final exits are adjacent, they should be separated by an imperforate enclosure.
- A protected stairway should not be used for anything else apart from a lift well or electricity meters.
- Openings in the external walls of protected stairways should be protected from fire in other parts of the building if they are situated where they might be at risk. (See section 7.14.3 'Additional provisions for common escape routes' above for details of protection measures.)
- A stair of acceptable width for everyday use will also be sufficient for escape purposes (BS 5588: Part 1 recommends a minimum width of 1 m); however if the stair is also a firefighting stair this should be increased to 1.1 m.
- Basement stairs will need to comply with special measures (see section 7.15.9 below).
- Gas service pipes and meters should only be installed in protected stairways if the installation complies with the requirements for installation and connection set out in the Pipelines Safety Regulations 1996, SI 1996/825 and the Gas Safety (Installation and Use) Regulations 1998 SI 1998/2451.
- A common stair which forms part of the only escape route from a flat or maisonette should not also serve any fire risk area such as a covered car park, boiler room, fuel storage space or other similar ancillary accommodation on the

same storey as that dwelling (but see the exceptions to this in section 7.14.3 'Flats and maisonettes with single common stairs' above).

- Where, in addition to the common stair, an alternative escape route is provided from a dwelling, it is permitted to serve ancillary accommodation from the common stair, provided that it is separated from that accommodation by a protected lobby or protected corridor.
- Where any stair serves an enclosed car park or place of special fire hazard (see section 7.2 above) it should be separated from that accommodation by a ventilated lobby or ventilated corridor (i.e. provide permanent ventilation of not less than $0.4\,m^2$ or a mechanical smoke control system to prevent the ingress of smoke).

7.14.4   External access and escape stairs

Where the building (or any part of it) is permitted to be served by a single access stair, that stair may be placed externally if it serves a floor which is not more than 6 m above ground level and it complies with the provisions listed at (2) to (6) below (see Fig. 7.23(a)).

Where there is more than one escape route available from a storey or part of a building some of the escape routes may be by way of an external escape stair if there is at least one internal escape stair serving every part of each storey (excluding plant areas) if the following provisions can be met.

(1)   The stair should not serve any floors which are more than 6 m above the ground or a roof or podium. The roof or podium should itself be served by an independent protected stair.

(2)   If it is more than 6 m in vertical extent, it is sufficiently protected from adverse weather. This does not necessarily mean that full enclosure will be necessary. The stair may be located so that protection may be obtained from the building itself. In deciding on the degree of protection it is necessary to consider the height of the stair, the familiarity of the occupants with the building, and the likelihood of the stair becoming impassable as a consequence of adverse weather conditions.

(3)   Any part of the building (including windows and doors etc.) which is within 1.8 m of the escape route from the stair to a place of safety should be protected with fire-resisting construction. This does not apply if there is a choice of routes from the foot of the stair thereby enabling the people escaping to avoid the effects of fire in the adjoining building. Additionally, any part of an external wall which is within 1.8 m of an external escape route (other than a stair) should be of fire-resisting construction up to a height of 1.1 m from the paving level of the route.

(4)   All the doors which lead onto the stair should be fire-resisting and self-closing. This does not apply to the only exit door to the landing at the head of a stair which leads downward.

(5)   Any part of the external envelope of the building which is within 1.8 m of (and 9 m vertically below) the flights and landings of the stair, should be of fire-

LEEDS COLLEGE OF BUILDING
LIBRARY

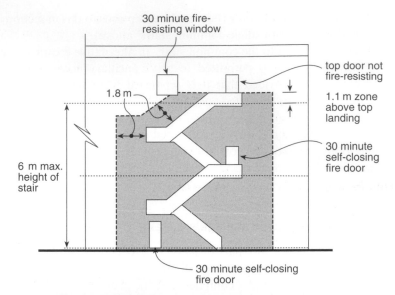

(a) Flats and maisonettes with external single access stair

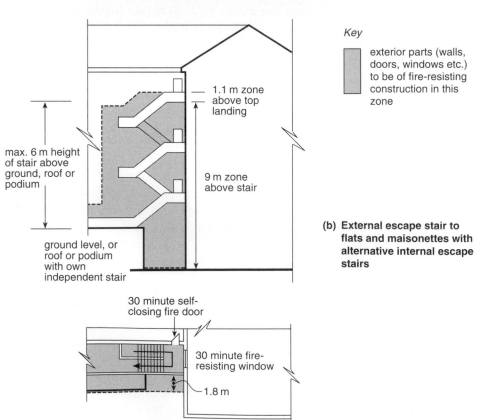

Fig. 7.23   External escape stairs to flats and maisonettes.

resisting construction. This 1.8 m dimension may be reduced to 1.1 m above the top landing level provided that this is not the top of a stair up from basement level to ground.

(6) Any glazing which is contained within the fire-resisting areas mentioned above should also be fire-resisting in terms of maintaining its integrity in a fire, and be fixed shut. (For example, Georgian wired glass is adequate; it does not also have to meet the requirements for insulation.)

These provisions are illustrated in Fig. 7.23.

Stairs to dwellings in mixed use buildings

Many buildings consist of a mix of dwellings (i.e. flats and maisonettes) and other uses. Sometimes the dwellings are ancillary to the main use (such as a caretaker's flat in an office block), and sometimes they form a distinct separate use (as in the case of shops with flats over). Clearly, the degree of separation of the uses for means of escape purposes will depend on the height of the building and the extent to which the uses are interdependent.

Where a building has no more than three storeys above the ground storey, the stairs may serve both non-residential and dwelling uses, with the proviso that each occupancy is separated from the stairs by protected lobbies at all levels.

In larger buildings where there are more than three storeys above the ground storey, stairs may serve both the dwellings and the other occupancies if:

- the dwelling is ancillary to the main use of the building and is provided with an independent alternative escape route;
- the stair is separated from other occupancies in the building at lower storey levels by protected lobbies at those levels and has the same standard of fire resistance as that required by Approved Document B for the rest of the building (including any additional provisions if it is a firefighting stair);
- any automatic fire detection and alarm system fitted in the main part of the building is extended to the flat; and
- any security measures (usually installed for the benefit of the non-dwelling use) do not prevent escape at all material times.

Where fuels, such as petrol and liquid petroleum gas are stored, additional measures (such as an increase in the fire resistance period for the structure between the storage area and the dwelling) may be necessary.

7.15 Means of escape from buildings other than dwellinghouses, flats and maisonettes

7.15.1 Introduction

So far we have discussed the provisions contained in AD B1 for means of escape in buildings of Purpose Groups 1(a), (b) and (c) i.e. dwellinghouses, flats and maisonettes. Although only a few of the more common arrangements for flats and maisonettes are covered in the Approved Document, the recommendations which are given are quite detailed and should form the basis for sound design guidance.

All other building types are grouped together in AD Bl and design recommendations are given for horizontal escape in Section 4, and vertical escape in Section 5. Of necessity, the guidance given is general in nature and is aimed at smaller, simpler types of buildings. For more complex or specialised buildings designers may well find that it is better to use other relevant design documents (such as the BS5588 series of codes) where more comprehensive guidance may be given.

7.15.2   Horizontal escape routes in buildings other than dwellings

Section 4 of AD B 1 deals with the provision of means of escape from any point in the floor of a building to the storey exit of that floor. It covers all buildings apart from dwellinghouses, flats and maisonettes. Whilst most of the guidance given in section 4 is related to general issues of design, the layouts of certain institutional buildings may warrant special provisions and some guidance on this is given in section 4.

The main decision that needs to be taken when designing the means of escape from a building is the number of escape routes and exits that are required. This will depend on:

- the maximum travel distance which is permitted to the nearest exit; and
- the number of occupants in the room, tier or storey under consideration.

Maximum travel distances and alternative escape routes in buildings other than dwellings

Ideally, there should be alternative escape routes provided from every part of the building. This is especially important in multi-storey buildings and in buildings where a mixture of Purpose Groups are present. In fact, if a mixed use building also contains Residential, or Assembly and Recreation purpose groups, these should be served by their own independent means of escape. (But see the exceptions to this for flats, described under 'Stairs to dwellings . . .' above.)

Where alternative escape routes are provided, escape will be possible in more than one direction. AD B1 places limits on the travel distance from any part of a room, tier or storey to a storey exit and these are shown in Table 3 from AD B1. The substance of Table 3 is summarised in Table 7.2 below. It should be read in conjunction with the following comments.

- The Table dimensions are actual travel distances and are measured along the shortest route taken by a person escaping in the event of a fire.
- Where there is fixed seating or there are other fixed obstructions, the travel distance is measured along the centre line of the seatways or gangways.
- Where the route of travel includes a stair it is measured along the pitch line on the centre line of travel.

- Where the layout of a room or storey is not known at the design stage, the direct distance measured in a straight line should be taken. Direct distances should be taken as two-thirds of the travel distance.

Once it has been established that *at least one exit* is within the distance limitations given in Table 7.2 the other exits may be further away than the distances given.

It will be observed from Table 7.2 that where escape is possible in one direction only the travel distances are much reduced. However, where a storey exit can be reached within these one-directional travel distances, it is not necessary to provide an alternative route except in the case of a room or storey that:

- has an occupant capacity exceeding 60 in the case of places of assembly or bars; or
- has an occupant capacity exceeding 30 if the building is in Purpose Group 2(a) Residential (institutional);
- is used for in-patient care in hospitals.

Similarly, it is often the case that there will not be alternative escape routes, especially at the beginning of an escape route. A room may have only one exit onto a corridor from where it may be possible to escape in two directions. This is permissible provided that:

- the overall distance from the furthest point in the room to the storey exit complies with the multi-directional travel distance from Table 7.2; and
- the single direction part of the route (in this case, in the room) complies with the 'one direction' travel distance specified in Table 7.2.

Although a choice of escape routes may be provided from a room or storey, it is possible that they may be so located, relative to one another, that a fire might disable them both. In order to consider them as true alternatives they should be positioned as shown in Fig. 7.24, i.e. the angle which is formed between the exits and any point in the space should be at least 45°. Where this angle cannot be achieved:

- the maximum travel distance for escape in one direction will apply; or
- the alternative escape routes should be separated from each other by fire-resisting construction.

Figure 7.24 illustrates these rules covering alternative escape routes.

Special rules apply where a dead-end situation exists in an open storey layout as shown in Fig. 7.25 as follows:

- XY should be within the one direction travel distance from Table 7.2
- whichever is the least distance of WXY and ZXY should be within the multi-direction travel distance from Table 7.2
- angle WXZ should be at least 45° plus 2.5° for each metre travelled in a single direction from Y to X.

Table 7.2   Travel distance limitations.

Purpose Group	Maximum Travel Distance (m) in:		Notes
	One direction	Multi-direction	
2(a) Institutional	9	18	In hospitals or other healthcare premises where the means of escape is being designed using the Department of Health's '*Firecode*' documents, the relevant travel distances recommended in those documents should be used
2(b) Other residential (i) in bedrooms	9	18	This is the maximum part of the travel distance within the bedroom but includes any associated dressing room, bathroom or sitting room etc. It is measured to the door onto the protected corridor serving the bedroom or suite
(ii) in bedroom corridors	9	35	This is the distance from the door onto the protected corridor serving the bedroom or suite to the storey exit
(iii) elsewhere	18	35	
3 Office	18	45	
4 Shop and commercial	18	34	For shopping malls see BS 5588: Part 10. This document applies more restrictive provisions to units with only one exit in covered shopping complexes. See also BRE Report (BR 368) *Design methodologies for smoke and heat exhaust ventilation* for guidance on associated smoke control measures
5 Assembly and recreation (i) buildings mainly for disabled people (not schools)	9	18	
(ii) schools	18	45	
(iii) areas with seating in rows	15	32	
(iv) elsewhere	18	45	

6	**Industrial and**		
7	**Storage & other non-residential**		
(i)	'normal' fire risk	45	25	For 'normal' fire risk as defined in Home Office *Guide to fire precautions in existing places of work that require a fire certificate: Factories, Offices, Shops and Railway Premises*
(ii)	'high' fire risk	25	12	For 'high' fire risk as defined in Home Office *Guide to fire precautions in existing places of work that require a fire certificate: Factories, Offices, Shops and Railway Premises*
2–7	**Places of special fire hazard**	18	9	This is the maximum part of the travel distance within the room or area. The travel distance outside such room or area should comply with the limits for the purpose group as shown above. Places of special fire hazard are: oil-filled transformer and switch gear rooms, boiler rooms, storage space for fuel or other highly flammable substances, and rooms housing fixed internal combustion engine. Plus, in schools: laboratories, technology rooms with open heat sources, kitchens and stores for PE mats or chemicals
2–7	**Plant room or rooftop plant:**			
(i)	distance within plant room	35	9	
(ii)	escape route not in open air	45	18	Overall travel distance
(iii)	escape route in open air	100	60	Overall travel distance

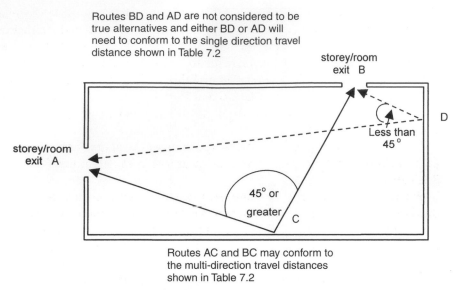

Fig. 7.24   Alternative escape routes.

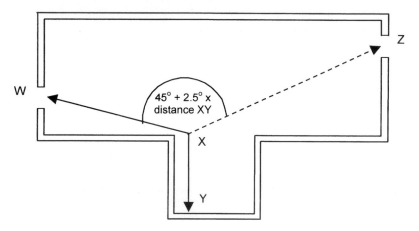

Fig. 7.25   Dead-end situation – single-storey layout.

Number and widths of exits related to the number of occupants in buildings other than dwellings

The number of occupants in a room, tier or storey influences the numbers and widths of the exits and escape routes that need to be provided from that area. Table 4 of AD B1 lists the minimum number(s) of escape routes or exits which should be provided relative to the maximum numbers of occupants as follows:

- up to 60 persons – 1 exit
- 61 to 600 persons – 2 exits
- over 600 persons – 3 exits

Realistically, the figures given above will only serve to define the absolute minimum number of exits for a means of escape. In practical terms the actual number of exits will be determined by travel distances and exit widths. The width of an escape route or exit may be determined by reference to Table 5 of AD B1. The information contained in this Table is restructured in Table 7.3 below to include relevant data from Approved Document M (Access and facilities for disabled people) since the minimum widths shown in Table 5 of AD B1 may not be adequate for disabled access.

Usually, the narrowest part of an escape route will be at the door openings which form the room or storey exits. These are measured as shown in Fig. 7.10 in section 7.9.3 above.

Where a storey has two or more exits it is assumed that one of them will be disabled by a fire. Therefore the remaining exits should have sufficient width to take the occupants safely and quickly. This means that the widest exit should be discounted and the remainder should be designed to take the occupants of the storey. Since stairs need to be as wide as the exit leading onto them, this recommendation for exit width may influence the width of the stairways. (Stairways may also need to be discounted and this is discussed in section 7.15.7 below.)

Except in doorways, all escape routes should have clear headroom of at least 2 m.

7.15.3   Horizontal escape routes – factors affecting internal planning

The efficacy of an escape route in a storey may be affected by a number of internal planning considerations, such as:

- the need for inner rooms
- the relationship between circulation routes and stairways
- the need for different occupancies in a building to use the same escape route
- the design and layout of means of escape corridors.

These are considered in more detail in the following paragraphs.

Provision of inner rooms

The rules governing the provision of inner rooms are more stringent than those for dwellings. Inner rooms are only acceptable under the following conditions.

- The occupant capacity of the inner room should not exceed 60 (or 30 for institutional buildings in Purpose Group 2(a)).
- The inner room should not be a bedroom.
- Only one access room should be passed through when escaping from the inner room.
- The maximum travel distance from the furthest point in the inner room to the exit from the access room should not exceed the appropriate limit given in Table 7.2 above.
- The access room should be in the control of the same occupier as the inner room.

Table 7.3 Widths of exits and escape routes.

Maximum number of persons	Min. width of exit*	Min. width of escape route*	Notes
Up to 50	750	750[1]	[1] Does not apply to: • schools where minimum width in corridors is 1050 (and 1600 in dead ends) • areas accessible to disabled people where minimum width in corridors is 1200 (or 1000 where lift access is not provided to the corridor or it is situated in an extension approached through an existing building) • gangways between fixed storage racking in Purpose Group 4 (Shop and Commercial) where minimum width may be 530 mm (but not in public areas) Widths of escape routes and exits less than 1050 should not be interpolated
51 to 110	850	850[2]	[2] Does not apply to: • schools where minimum width in corridors is 1050 (and 1600 in dead ends) • areas accessible to disabled people where minimum width in corridors is 1200 (or 1000 where lift access is not provided to the corridor or it is situated in an extension approached through an existing building) Widths of escape routes and exits less than 1050 should not be interpolated
111 to 220	1050	1050[3]	[3] Does not apply to: • schools where minimum width in corridor dead ends is 1600 mm • areas accessible to disabled people where minimum width in corridors is 1200
over 220	5 mm/person	5 mm/person	This method of calculation should not be used for any opening serving less than 220 persons [e.g. three exits each 850 mm wide will accommodate $3 \times 110 = 330$ people, not the 510 (i.e. $2550 \div 5$) people that $3 \times 850 = 2550$ mm would accommodate]

* For method of measuring widths of escape routes and exits see section 7.9 above.

- The access room should not be a place of special fire hazard (e.g. a boiler room).

Where these conditions are met the inner room should be designed to conform to one of the following arrangements:

- the walls or partitions of the inner room should stop at least 500 mm from the ceiling, or
- a vision panel, which need not be more than 0.1 m^2 in area, should be situated in the walls or door of the inner room (this is to enable the occupiers to see if a fire has started in the access room), or
- a suitable automatic fire detection and alarm system should be fitted in the access room which will give warning of fire in that room to the occupiers of the inner room.

Horizontal escape routes and stairways

Care must be taken in the design of horizontal escape routes since they also form part of the normal circulation in a building and may jeopardise access to stairways unless the following points are considered.

- In any storey which has more than one escape stair, it should not be necessary to pass through one stairway to reach another. Where this is unavoidable a protected lobby should be provided to the stairway. This lobby may be passed through in order to reach the other stair.
- As part of the normal circulation in a building, it should not be necessary to pass through a stairway enclosure in order to reach another part of the building on the same level. (Such circulation patterns should be avoided since familiarity breeds contempt, and fire doors may become ineffective due to excessive use or misuse.) Such an arrangement is permissible if the doors to the protected stairway and any associated exit passageway are fitted with an automatic release mechanism (see below section 7.22.3).
- Where buildings are planned with more than one exit round a central core, these exits should be remote from each other and no two exits should be approached from the same lift hall, common lobby or undivided corridor, or linked together by any of these (see Fig. 7.26).

Use of common escape routes by different occupancies

It is common in mixed-use or multi-tenanted buildings for common escape routes to be used by all the occupants. There are restrictions on this and these have been referred to above in section 7.15.2. Where common escape routes are permitted the following rules should be observed.

- The means of escape from one occupancy should not pass through another.
- Common corridors or circulation spaces should either be fire-protected or fitted

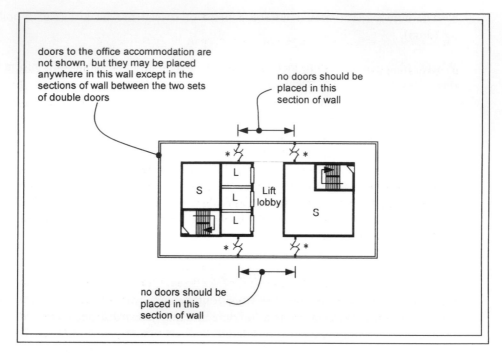

Key

*   FD20 self-closing fire door (20 minutes integrity)

L   Lift

S   services, toilets etc.

Fig. 7.26   Corridor layout: building with central core.

with an automatic fire detection and alarm system which extends throughout the storey.

Storeys containing areas for consumption of food and/or drink by customers

In some buildings (such as department stores and shops) it may be desirable to provide an area for the consumption of food and/or drink by customers where this is ancillary to the main use of the building. Such an arrangement is permissible if the following conditions are met.

- At least two escape routes should be provided from each area (inner rooms which follow the guidance contained in section 7.15.3 'Provision of inner rooms' above are exempt from this); and
- Each escape route should lead directly to a storey exit without having to pass through a kitchen or similar area of high fire hazard.

Means of escape corridors – design factors

The following means of escape corridors should be fire-protected.

- Every dead-end corridor (although small recesses and extensions less than 2 m long and referred to in figures 10 and 11 of BS 5588: *Fire precautions in the design, construction and use of buildings:* Part 11: 1997 *Code of practice for shops, offices, industrial, storage and other similar buildings*, may be ignored).
- Every corridor serving bedrooms.
- Every corridor or circulation space common to two or more 'different occupancies' (i.e. where the premises are split into separate ownerships or tenancies of different organisations). In this case the need for fire protection may be omitted where an automatic fire detection and alarm system is installed throughout the storey. Even so, the means of escape from one occupancy should not pass through any other occupancy.

The way in which a storey layout is planned can have an effect on its means of escape characteristics. For example, whilst it is perfectly acceptable to have open plan floor areas, they offer no impediment to smoke spread but do have the advantage that occupants can become aware of a fire more quickly. On the other hand, the provision of a cellular layout, where the means of escape is enclosed by partitions, means that some defence is provided against smoke spread in the early stages of a fire even though the partitions may have no fire resistance rating.

To maintain the effectiveness of the partitions, they should be carried up to ceiling level (i.e. either the soffit of the structural floor above or to a suspended ceiling) and room openings should be fitted with doors (which do not need to be fire resisting).

Corridors which give access to alternative escape routes may become blocked by smoke before all the occupants of a building have escaped and may make both routes impassable. Additionally, the means of escape from any permitted dead-end corridors may be blocked. Therefore, corridors connecting two or more storey exits should be sub-divided by means of self-closing fire doors (and screens, if necessary) if they exceed 12 m in length. The doors (and screens) should be positioned so that:

- they are approximately mid-way between the two storey exits; and
- the route is protected from smoke, having regard to any adjacent fire risks and the layout of the corridor.

Unless the escape stairway and its associated corridors are protected by a pressurisation system complying with BS 5588: *Fire precautions in the design, construction and use of buildings:* Part 4: 1998 *Code of practice for smoke control using pressure differentials*, dead-end corridors exceeding 4.5 m in length giving access to a point from which alternative escape routes are available, should also be provided with fire doors so positioned that the dead-end is separated from any corridor which:

- provides two directions of escape; or
- continues past one storey exit to another.

These provisions for means of escape corridors are summarised in Fig. 7.27.

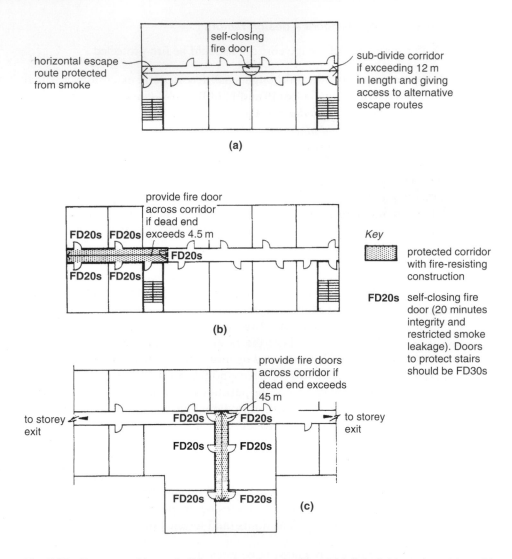

Fig. 7.27   Escape corridors – buildings other than dwellings. (a) Sub-division of corridors. (b) Dead end separation from continuation corridor. (c) Dead end separation from alternative escape routes.

7.15.4   Escape routes across flat roofs

The recommendations given above for escape over flat roofs from flats and maisonettes (section 7.14.3) also apply to all other building types except where the route serves an institutional building or part of a route used by members of the public.

7.15.5   Vertical escape routes in buildings other than dwellings

Section 5 of AD B1 deals with the provisions for vertical escape, by means of a sufficient number of adequately sized and protected escape stairs, for all buildings apart from dwellinghouses, flats and maisonettes.

The main decision to be taken when designing the vertical means of escape in a building is the number of stairways that need to be provided. It has already been shown above that alternative means of escape are required where horizontal constraints are imposed by travel distances, exit widths and numbers of occupants. Additionally, in buildings where there is a mix of different uses it may be the case that a fire in an unattended shop or office might have serious consequences for a residential or hotel use in the same building. Therefore, it is important to analyse the risks involved and to consider whether completely separate escape routes should be provided from each different use or whether other effective means of protecting common escape routes can be provided. (See above section 7.14 for examples of the use of common stairs in buildings which contain both dwellings and other uses).

Section 5 of AD B1 provides additional recommendations for assessing the number of stairways that are needed with regard to:

- the acceptability of single stairs for means of escape in a building; and
- the influence that adequate width of stairs may have on their provision whilst allowing for the fact that a stair may have to be discounted due to the effects of fire or smoke.

One further influence on the provision of stairs is the necessity to provide fire-fighting stairs in larger buildings. This is covered by AD B5 (see section 7.30) and may mean that extra stairways are required beyond those needed merely for means of escape purposes.

7.15.6   The provision of single stairs in buildings other than dwellings

Assuming that the building is not excluded from having a single escape route by virtue of the recommendations listed above, it may be served by a single escape stair in the following circumstances.

- Where it serves a basement which is allowed to have a single horizontal escape route (i.e. storey occupancy not exceeding 60, maximum travel distance within the limits for travel in one direction).
- Where it serves what are termed '*small premises*' and the recommendations of clause 10 of BS 5588 Part 11: 1997 *Code of practice for shops, offices, industrial, storage and other similar buildings* are followed (see section 7.15.8 below).
- Where it serves a building which has no floor more than 11m above ground and in which every floor is allowed to have a single horizontal escape route.

It should be noted that in schools where single stairs are provided the storeys above first floor level should only be occupied by adults. Additionally, where a two-storey school building (or part of a building) is provided with a single stair the following conditions apply.

- There should be no more than 120 pupils plus supervisors on the first floor.
- The first floor should not contain a place of special fire hazard.
- Classrooms and stores should not open onto the stairway.

7.15.7   Escape stair design – widths

Clearly, the width of escape stairs is related to the number of people that they can carry in an evacuation situation. AD B1 contains a number of provisions which enable the width of stairs to be calculated by reference to:

- the number of people who will use them,
- whether or not it will be necessary to discount any of the stairs, and
- their mode of use (i.e. simultaneous or phased evacuation).

Escape stairs should be at least as wide as any exits giving access to them and should not reduce in width as they approach the final exit. Additionally, if the exit route from a stair also picks up occupants of the ground and basement storeys, it may need to be increased in width accordingly. Although stairs need to be sufficiently wide for escape purposes, research has shown that people prefer to stay close to a handrail when making a long descent. Therefore the centre of a very wide stairway would be little used and might, in fact, be hazardous. For this reason, AD B1 puts a maximum limit of 1400 mm on the width of a stairway where its vertical extent exceeds 30 m unless it is centrally divided with a handrail. Where the design of the building calls for a stairway that is wider than 1400 mm, it should be at least 1800 mm wide and contain a central handrail. In this case, the stair width on either side of the central handrail will need to be considered separately when assessing stair capacity.

Minimum stair widths can, in the first instance, be assessed using Table 7.4 below. This is based on Table 6 from AD B1 and is suitable for most simple building designs where the maximum number of people served by the stair(s) does not exceed 220.

Where two or more stairways are provided, it is possible that one of the stairs may be inaccessible due to fire or smoke unless special precautions are taken. Therefore, it may necessary to discount each stair in turn in order to check that the remaining stairways are capable of coping with the demand. Discounting is unnecessary if:

- the escape stairs are approached through a protected lobby at each floor level (although a lobby is not needed for the top floor for the exception still to apply); or
- the stairs are protected by a pressurisation smoke control system designed in accordance with BS 5588: Part 4.

As Table 7.4 suggests, in multi-storey buildings where the number of occupants exceeds 220 it may be necessary to consider the mode of evacuation and use other methods to calculate stair widths.

Where it is assumed that all the occupants would be evacuated, this is termed 'simultaneous evacuation' and this should be the design approach for:

- all stairs which serve basements
- all stairs which serve buildings with open spatial planning (i.e. where the building is arranged internally so that two or more floors are contained within one undivided volume)
- all stairs which serve Assembly and Recreation buildings (PG 5) or Other Residential buildings (PG 2(b)).

Table 7.4 Minimum widths of escape stairs.

Description of stair	Numbers of people assessed as using stair in emergency[1]	Minimum width of stair (mm)
Escape stairs in any building (but see footnotes for exceptions)	Up to 50	800[2]
	51 to 150	1000
	151 to 220	1100
	Over 220	See note 3

Notes
(1) For methods of assessing occupancy see section 7.9.1 above.
(2) This minimum stair width does not apply:
    (a) in an Institutional buildings unless the stair will only be used by staff
    (b) in an assembly building unless the area served is less than 100 m² and/or is not for assembly purposes (e.g. office)
    (c) to any areas which are accessible to disabled people.
(3) See AD B1 Table 7 (and Formula 7.1) for simultaneous evacuation and AD B1 Table 8 (and Formula 7.2) for phased evacuation.

Using this approach the escape stairs should be wide enough to allow all the floors to be evacuated simultaneously. The calculations take into account the number of people temporarily housed in the stairways during evacuation.

A simple way of assessing the escape stair width is to use Table 7 from AD B1. This covers the capacity of stairs with widths from 1000 mm to 1800 mm for buildings up to ten storeys high (although the capacity of stairs serving more than ten storeys can be obtained from Table 7 by using linear extrapolation).

In fact, the capacities given in the table for stair widths of 1100 mm and greater are derived from the formula:

$$P = 200w + 50(w - 0.3)(n - 1) \dots\dots\dots\dots\dots\dots\text{(Formula 7.1)}$$

where:
P = the number of people that can be served by the stair
w = the width of the stair in metres
n = the number of storeys in the building.

Formula 7.1 can be used for any size of building with no limit being placed on the occupant capacity or number of floors and it is probably advisable to use it for buildings which are larger than those covered by Table 7 from AD B1. It should be noted that separate calculations should be made for stairs serving basements and for those serving upper storeys.

The formula is particularly useful where the occupants of a building are not evenly distributed – either within a storey or between storeys. However, it cannot be used for stairs which are narrower than 1100 mm, so for stairs which are allowed to be 1000 mm wide, in buildings up to ten storeys high, the values have been extracted from Table 7 and are presented in Fig. 7.28.

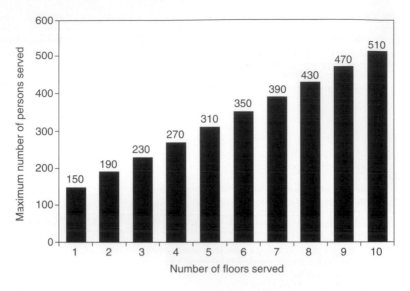

Fig. 7.28   Stair capacity for simultaneous evacuation – 1000 mm wide stairs.

In certain buildings it may be more advantageous to design stairs on the basis of 'phased evacuation'. Indeed, in high buildings it may be impractical or unnecessary to evacuate the building totally, especially if the recommendations regarding fire resistance, compartmentation and the installation of supporting facilities such as sprinklers and fire alarms are adhered to.

In phased evacuation, people with reduced mobility and those most immediately affected by the fire (i.e. those people on the floor of fire origin and the one above it) are evacuated first. After that, if the need arises, floors can be evacuated two at a time. Phased evacuation allows narrower stairs to be used and has the added advantage that it causes less disruption in large buildings than total evacuation.

Phased evacuation may be used for any buildings unless they are of the types listed above as needing simultaneous evacuation.

Where a building is designed for phased evacuation the following conditions should be met.

- The stairs should be approached through a protected lobby or protected corridor at each floor level (this does not apply to a top storey).
- The lifts should be approached through a protected lobby at each floor level.
- Each floor should be a compartment floor.
- If the building has a floor which is more than 30 m above ground, it should be protected throughout by an automatic sprinkler system which complies with the relevant requirements of BS 5306 *Fire extinguishing installations and equipment on premises,* Part 2:1990 *Specification for sprinkler systems* (i.e. the sections dealing with the relevant occupancy rating and the additional requirements for life safety). This provision does not apply to flats of PG 1(a) in a mixed use building.
- An appropriate fire warning system should be fitted which complies with

BS 5839 *Fire detection and alarm systems for buildings,* Part 1: 1988 *Code of practice for system design, installation and servicing,* to at least the L3 standard.
- An internal speech communication system (such as a telephone, intercom system or similar) should be provided so that conversation is possible between a fire warden at every floor level and a control point at the fire service access level.
- Where it is deemed appropriate to install a voice alarm, the recommendations regarding phased evacuation in BS 5839: Part 1 should be followed and the voice alarm system itself should conform to BS 5839: Part 8 1998 *Code of practice for the design, installation and servicing of voice alarm systems.*

When phased evacuation is used as the basis for design, the minimum stair width needed may be taken from Table 8 of AD B1 assuming phased evacuation of not more than two floors at a time. The data from Table 8, which has been reconfigured in Fig 7.29 below, is derived from the formula:

$$w = [(P \times 10) - 100] \text{ mm} \dots\dots\dots\dots\dots\dots\dots\dots\dots\dots\dots\dots\dots\dots\dots(\text{Formula } 7.2)$$

where:
w = the minimum width of stair (w must not be less than 1000 mm)
P = the number of people on the most heavily occupied storey.

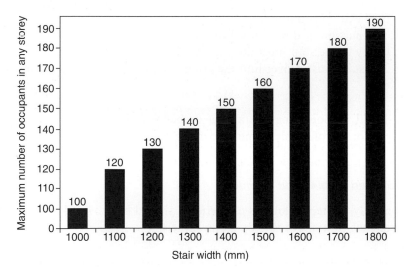

Fig. 7.29 Minimum width of stairs for phased evacuation.

7.15.8 Provision and protection of stairs

To be effective as an area of relative safety during a fire, escape stairs need to have an adequate standard of fire protection. This relates not only to the presence of fire-resisting enclosures but also to the provision of protected lobbies, corridors and final exits.

Fire-resisting enclosures

Each internal escape stair should be a protected stair situated in a fire-resisting enclosure. Additional measures may also be necessary for a stairway which is also a protected shaft (penetrating one or more compartment floors, see AD B section 9) or a firefighting shaft (see AD B Section 18). However, this does not preclude the provision of an accommodation stair (i.e. a stair which is provided for the convenience of occupants and is additional to those required for means of escape) if the design of the building so warrants it.

Exceptionally, an unprotected stair can form part of the internal escape route to a storey or final exit in low risk buildings if the number of people and the travel distance involved are very limited.

For example, BS 5588 Part 11 contains details in clause 10 of the use of an unprotected stair for means of escape in what are termed 'small premises'. Typically, the premises will be used as a small shop and will have two-storeys consisting of:

- ground and first floor; or
- basement and ground floor.

The following conditions will also need to be complied with.

- The maximum floor area in any storey must not exceed 90 m^2.
- The maximum direct travel distance from any point in the ground storey to the final exit must not exceed 18 m.
- The maximum direct travel distance from any point in the basement or first storey to the stair must not exceed 12 m.
- The stair must deliver into the ground storey not more than 3 m from the final exit.

It should be noted that restaurant or bar premises and those used for the sale, storage or use of highly flammable materials are barred from this arrangement.

Protected lobbies and corridors

Generally, protected lobbies or corridors should be provided at all levels including basements (but not at the top storey) where:

- the building has a single stair and there is more than one floor above or below the ground storey (except for small premises, see above)
- the building has a floor which is more than 18 m above ground
- the building is designed for phased evacuation
- the option has been taken to not discount one stairway when calculating stair widths (see section 7.15.7 above).

In these cases an alternative to a protected lobby or corridor is the use of a smoke control system designed in accordance with BS 5588: Part 4.

Protected lobbies are also needed where:

- the stairway is a firefighting stair (see AD B5)
- the stairway serves a place of special fire hazard (i.e. oil-filled transformer and switch gear rooms, boiler rooms, storage space for fuel or other highly flammable substances, rooms housing a fixed internal combustion engine, and in schools – laboratories, technology rooms with open heat sources, kitchens and stores for PE mats or chemicals). In this case, the lobby should be ventilated by permanent vents with an area of at least 0.4 m^2, or should be protected by a mechanical smoke control system.

Final exits from protected stairs

Ideally, every protected stairway should discharge directly to a final exit, i.e. it should be possible to leave the staircase enclosure and immediately reach a place of safety outside the building and away from the effects of fire.

Obviously, it is not always possible to achieve this ideal, especially where the building design calls for stairs to be remote from external walls. Therefore it is permissible for a protected stairway to discharge into a protected exit passageway, which in turn leads to a final exit from the building. Such a passageway can contain doors (e.g. to allow people on the ground floor to use the escape route) but they will need to be fire doors in order to maintain fire integrity and may need to be lobbied if the stairway needs to be served by lobbies. Therefore, if the exit route from a stair also picks up occupants of the ground and/or basement storeys, it may need to be increased in width accordingly. Thus, the width of the protected exit passageway will need to be designed in accordance with Table 7.3, for the estimated numbers of people that will use it in an emergency.

Sometimes the design of the building will call for two protected stairways or protected exit passageways to be adjacent to each other. Where this happens they should be separated by an imperforate enclosure.

Restrictions on the use of space in protected stairways

Since a protected stairway is considered to be a place of relative safety, it should be free of potential sources of fire. Therefore, the facilities that may be included in protected stairways are restricted to the following.

- Washrooms or sanitary accommodation provided that the accommodation is not used as a cloakroom. The only gas appliances that may be installed are water heaters or sanitary towel incinerators.
- A lift well, on condition that the stairway is not a firefighting stair.
- An enquiry office or reception desk area at ground or access level of not more than 10 m^2, provided that there is more than one stair serving the building.
- Fire-protected cupboards, provided that there is more than one stair serving the building.
- Gas service pipes and meters, but only if the gas installation is in accordance with

the requirements for installation and connection set out in the *Pipelines Safety Regulations 1996*, SI 1996/825 and the *Gas Safety (Installation and Use) Regulations 1998* SI 1998/2451.

Protection of external walls of protected stairways

If a protected stairway is situated on the external wall of a building it is not necessary for the external part of the enclosure to be fire-protected and in many cases it may be fully glazed. This is because fires are unlikely to start in protected stairways. Therefore these areas will not contribute to the radiant heat from a building fire which might put at risk another building.

In some building designs the stairway may be situated at an internal angle in the building façade (see Fig. 7.30) and may be jeopardised by smoke and flames coming from windows in the facing walls. This may also be the case if the stair projects from the face of the building (see Fig. 7.31) or is recessed into it. In these cases any windows or other unprotected areas in the face of the building and in the stairway should be separated by at least 1800 mm of fire-resisting construction. This provision also applies to flats and maisonettes.

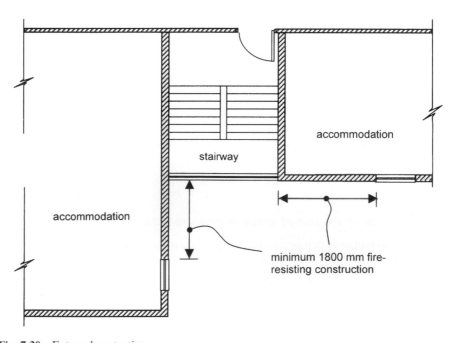

Fig. 7.30 External protection.

7.15.9 Basement stairs

Basement fires are particularly serious since combustion products tend to rise and find their way into stairways unless other smoke venting measures are taken. (See AD B5, section 7.30.13 below). Therefore, it is necessary to take additional pre-

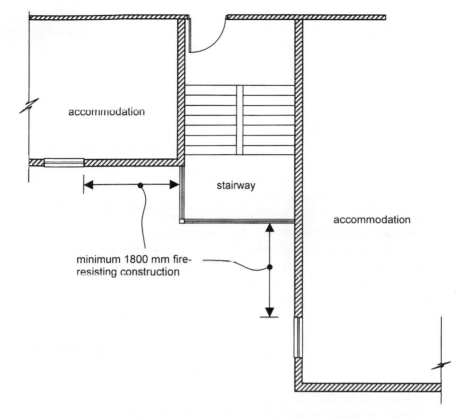

Fig. 7.31 External protection – protected stairway projecting beyond face of building.

cautions to prevent a basement fire endangering upper storeys in a building as follows.

- In most buildings with only one escape stair serving the upper storeys, this stair should not continue down to the basement, i.e. the basement should be served by a separate stair.
- In buildings containing more than one escape stair, at least one of the stairs should terminate at ground level and not continue down to the basement. The other stairs may terminate at basement level on condition that the basement accommodation is separated from the stair(s) by a protected lobby or corridor at basement level.

These provisions apply to all buildings, including flats and maisonettes.

7.15.10   External escape stairs

External escape stairs have long been used to provide additional vertical means of escape where parts of the building would otherwise be contained in long dead-ends or would be too distant from internal stairways. It is uncommon for such stairways to be fully protected from the elements therefore, they are not normally used for everyday access and egress around the building. Thus it can be argued that external escape stairs are not subject to the requirements of Part K of Schedule 1 to the Building Regulations 2000 because they do not form part of a building.

In buildings other than dwellings, an external escape stair may be used as an alternative means of escape provided that there is more than one escape route available from a storey, or part of a building, and the following conditions are met.

- There is at least one internal escape stair available from every part of each storey (plant areas excluded).
- It is not intended for use by members of the public if installed in assembly and recreation buildings (PG 5).
- It serves only office or residential staff accommodation if installed in an institutional building (PG 2(a)).
- If it is more than 6 m in vertical extent it is sufficiently protected from adverse weather. This does not necessarily mean that full enclosure will be necessary. The stair may be located so that protection may be obtained from the building itself. In deciding on the degree of protection it is necessary to consider the height of the stair, the familiarity of the occupants with the building, and the likelihood of the stair becoming impassable as a consequence of adverse weather conditions.
- Any part of the building (including windows and doors etc) which is within 1.8 m of the escape route from the stair to a place of safety should be protected with fire-resisting construction. This does not apply if there is a choice of routes from the foot of the stair thereby enabling the people escaping to avoid the effects of fire in the adjoining building. Additionally, any part of an external wall which is within 1.8 m of an external escape route (other than a stair) should be of fire-resisting construction up to a height of 1.1 m from the paving level of the route.
- All the doors which lead onto the stair should be fire-resisting and self-closing. This does not apply to the only exit door to the landing at the head of a stair which leads downward.
- Any part of the external envelope of the building which is within 1.8 m of (and 9 m vertically below) the flights and landings of the stair, should be of fire-resisting construction. This 1.8 m dimension may be reduced to 1.1 m above the top landing level provided that this is not the top of a stair up from basement level to ground.
- Any glazing which is contained within the fire-resisting areas mentioned above should also be fire-resisting in terms of maintaining its integrity in a fire, and be fixed shut. (For example, Georgian wired glass is adequate; it does not also have to meet the requirements for insulation).

7.16 General recommendations common to all buildings except dwellinghouses

Section 6 of AD B1 gives general guidance on a number of features of escape routes which apply to all buildings, except dwellinghouses, concerning:

- the standard of protection necessary for the elements enclosing the means of escape
- the provision of doors
- the construction of escape stairs
- the position and design of final exits
- lighting and signing
- mechanical services including lift installations
- protected circuits for the operation of equipment in the event of fire
- refuse chutes and storage
- the provision of fire safety signs.

These recommendations should be read in conjunction with the provisions described above for flats and maisonettes, and other buildings except dwellinghouses.

7.16.1 Protection of escape routes – standards and general constructional provisions

Those parts of a means of escape which are required by Part B1 to be fire-resisting should comply with the recommendations given in AD B3 (Internal fire spread – structure) or AD B5 (Access and facilities for the fire service) in addition to AD B1. In most cases 30 minutes fire protection is sufficient for the protection of a means of escape. The exceptions to this are when the element also performs a fire-separating function, or separates areas of different fire risk, such as:

- a compartment floor
- a compartment wall
- an external wall
- a protected shaft, or
- a firefighting shaft.

In these cases the element should achieve the standard of fire resistance given in Table A2 of Appendix A to Approved Document B. This may require considerably more than 30 minutes fire resistance (the fire resistance periods in Table A2 range from 30 minutes to 120 minutes).
   The following general constructional provisions should also be met.

- Fully enclosed walk-in store rooms in shops should be separated from retail areas with 30 minute fire-resisting construction (see Table A1 of Appendix A of AD B) if they are sited so as to prejudice the means of escape. This does not apply if the store room is fitted with an automatic fire detection and alarm system or sprinklers.
- Glazed elements in fire-resisting enclosures and doors, which are only able to meet the requirements for integrity in the event of a fire, will be limited in area to the amounts shown in Table A4 of Appendix A of AD B (see section 7.22.3).
- There are no limitations on the use of glazed elements that can meet both the integrity and insulation performance recommendations of AD B1. However,

there may be some restrictions on the use of glass in firefighting stairs and lobbies in BS 5588: *Fire precautions in the design, construction and use of buildings* Part 5:1991 *Code of practice for firefighting stairways and lift* under the recommendations for robust construction. This is referred to in AD B5 and is mentioned below.

- Glazed elements may also need to comply with AD N (Chapter 18).
- All escape routes should have a minimum headroom of 2 m. The only projections allowed below this are for door frames.
- The floors of escape routes, including the surfaces of steps and ramps, should be chosen so that they are not unduly slippery when wet.
- Sloping floors or tiers should not have a pitch greater than 35° to the horizontal.
- Further guidance on the provision of ramps, stairs, aisles and gangways may be found in AD K (Chapter 15) and AD M (Chapter 17).

7.16.2   The provision of doors on escape routes

The time taken to pass through a closed door can be critical when escaping from a building in a fire situation. Doors on escape routes should be readily openable if undue delay in escaping from a building is to be avoided. They should also comply with the following general provisions.

- Doors on escape routes often need to be fire-resisting. This means that certain test criteria and performance standards as set out in Appendix B of Approved Document B (Table B1) will need to be met (see section 7.22.3 below).
- In general, escape doors should open in the direction of the means of escape where it is reasonably practicable to do so and should always do so if more than 60 people are likely to use the door in an emergency. However, for some industrial activities where there is a very high risk with potential for rapid fire growth it may be necessary for escape doors to open in the direction of escape for lower occupant numbers than 60. The exact figure will depend on the individual circumstances of the case and there is no specific guidance laid down in AD B.
- Ideally, doors on escape routes (whether or not they are fire doors) should not be fitted with fastenings unless these are simple to use and can be operated from the side of the door which is approached by people escaping. Any fastenings should be able to be operated without a key and without having to operate more than one mechanism, however this does not prevent doors being fitted with ironmongery which allows them to be locked when the rooms are empty. For example, this would permit a hotel bedroom to be fitted with a lock which could be operated from the outside with a key and from the inside by a knob or lever.
- Where security of final exit doors is important, as in Assembly and Recreation (PG 5) and Shop and Commercial (PG 4) buildings, panic bolts may be used. Additionally, it is accepted that in non-residential buildings it is appropriate for final exit doors to be locked when the building is empty. Clearly, a good deal of responsibility must be placed on management procedures for the safe use of these locks.

- Recommendations for self-closers and hold-open devices for fire doors are contained in Appendix B of AD B (see section 7.22.3 below).
- Doors on escape routes should swing through at least 90° to open, and should not reduce the effective width of any escape route across a landing. The swing should be clear of any changes in floor level, although a single step or threshold on the line of a door opening is permitted.
- Any door that opens towards a corridor or stairway should be recessed so that it does not encroach on or reduce the effective width of the corridor or stairway.
- Doors on escape routes which subdivide corridors, or are hung to swing in two directions, should contain vision panels. (See also Approved Document M and Chapter 17 of this book for vision panels in doors across accessible corridors).
- If revolving or automatic doors, or turnstiles, are placed across an escape route it is possible that they might obstruct the passage of people escaping. Therefore, they should not be placed across an escape route unless:
  - (a) in the case of automatic doors which are the correct width for the design of the route they:
    - (i) will fail safely to become outward opening from any position of opening, or
    - (ii) are provided with a monitored failsafe system for opening the doors in the event of mains power failure, or
    - (iii) fail safely in the open position in the event of mains power failure, or
  - (b) they have non-automatic swing doors of the required width adjacent to them which can provide an alternative exit.

7.16.3 The construction of escape stairs – conventional stairs

Escape stairs and their associated landings in certain high risk situations or buildings require the extra safeguard of being constructed in materials of limited combustibility. These are composite materials (such as plasterboard) which include combustible materials in their composition so that they cannot be classed as totally non-combustible. When exposed as linings to walls or ceilings they must achieve certain low flame spread ratings.

This recommendation applies in the following cases:

- where a building has only one stair serving it (this does not apply to two and three-storey flats and maisonettes)
- where a stair is located in a basement storey (except if it is a private stair in a maisonette)
- to any stair serving a storey in a building which is more than 18 m above ground or access level
- to any external stair (except where it connects the ground floor or paving level to a floor or flat roof which is not more than 6 m above ground)
- if the stair is a firefighting stair.

In all the above, except for the firefighting stair, it is permissible to add combustible materials to the upper surface of the stair.

Where possible, single steps should be avoided on escape routes unless prominently marked, since they can cause falls. It is permissible though, to have a single step on the line of a doorway.

7.16.4   The construction of escape stairs – special stairs and ladders

Although spiral and helical stairs, and fixed ladders are not as inherently safe as conventional stairs they may be used as part of a means of escape if the following restrictions are observed.

- Spiral and helical stairs should be designed in accordance with BS 5395 *Stairs, ladders and walkways,* Part 2:1984 *Code of practice for the design of helical and spiral stairs.* They are not suitable for use by pupils in schools, and if used by members of the public they should be type E (public) stair from the above standard.
- Fixed ladders are not suitable as a means of escape for members of the public. They should only be used to access areas which are not normally occupied, such as plant rooms, where it is not practical to provide a conventional stair. They should be constructed of non-combustible materials.

7.16.5   The position and design of final exits

Final exits should not be narrower than the escape routes they serve and should be positioned to facilitate evacuation of people out of and away from the building. This means that they should be:

- positioned so that rapid dispersal of people is facilitated to a street, passageway, walkway or open space clear of the effects of fire and smoke. The route from the building should be well defined and guarded if necessary;
- clearly apparent to users. This is very important where stairs continue up or down past the final exit level in a building; and
- sited so that they are clear of the effects of fire from risk areas in buildings such as basements (e.g. outlets for basement smoke vents), and openings to transformer chambers, refuse chambers, boiler rooms and other similar risk areas.

7.16.6   Lighting and signing

All escape routes should have adequate artificial lighting. In certain cases escape lighting which illuminates the route if the mains supply fails, should also be provided. These are listed in Table 7.5 below which is based on Table 9 to AD B1.

The lighting to escape stairs will also need to be on a separate circuit from that which supplies any other part of the escape route.

Standards for installation of escape lighting systems are given in BS 5266 *Emergency lighting,* Part 1:1988 *Code of practice for the emergency lighting of premises other than cinemas and certain other specified premises used for entertainment,* or CP 1007: 1955 *Maintained lighting for cinemas.*

Table 7.5 Provision of escape lighting.

Purpose Group	Description of building or part	Areas where escape lighting is required	Areas where escape lighting is *not* required
1(a) 2(a) 2(b)	Flat or maisonette Institutional Other Residential	All common escape routes (including external routes)	Common escape routes in two-storey flats. Dwellinghouses in PG 1(b) & 1(c)
3 4 6 7(a)	Office Shop and Commercial[1] Industrial Storage and other non-residential	(i) Underground or windowless accommodation (ii) Stairways in a central core or serving storey(s) over 18 m from ground level (iii) Internal corridors more than 30 m long (iv) Open-plan areas exceeding 60 m²	
4 7(b)	Shop and Commercial[2] Car parks which admit the public	All escape routes (including external routes)	Escape routes in shops[3] of 3 or less storeys (with no sales floor exceeding 280 m²)
5	Assembly and Recreation	All escape routes (including external routes), and accommodation	(i) Accommodation open on one side to view sport or entertainment during normal daylight hours (ii) Parts of school buildings with natural light and used only during normal school hours
All	All	(i) Windowless toilet accommodation with floor area not exceeding 8 m² (ii) All toilet accommodation with floor area exceeding 8 m² (iii) Electricity and generator rooms (iv) Switch room/battery room for emergency lighting system (v) Emergency control room	Dwellinghouses in PG 1(b) & 1(c)

Notes:
1. Those parts of the premises where the public are not admitted.
2. Those parts of the premises where the public are admitted.
3. Any 'shop' (see definition section 7.3 above) which is a restaurant or bar will require escape lighting as indicated in column 3.

Except in dwellinghouses, flats and maisonettes, emergency exit signs should be provided to every escape route. It is not necessary to sign exits which are in ordinary, daily use. The exit should be distinctively and conspicuously marked by a sign with letters of adequate size complying with the *Health and Safety (Safety signs and signals) Regulations 1996 (SI 1996/2341)*. In general, these regulations may be satisfied by signs containing symbols or pictograms which are in accordance with

BS 5499 *Fire safety signs, notices and graphic symbols,* Part 1: 1990 *Specification for fire safety signs.*

In some buildings other legislation may require additional signs.

7.16.7   Lift installations

Lifts are not normally used for means of escape since there is always the danger that they may become immobilised due to power failure and may trap the occupants. It is possible to provide lifts as part of a management plan for evacuating disabled people if the lift installation is appropriately sited and protected. It should also contain sufficient safety devices to ensure that it remains usable during a fire. Further details may be found in BS 5588: *Fire precautions in the design, construction and use of buildings* Part 8: 1988 *Code of practice for means of escape for disabled people.*

A further problem with lifts is that they connect floors and may act as a vertical conduit for smoke or flames thus prejudicing escape routes. This may be prevented if the following recommendations are observed.

- Lift wells should be enclosed throughout their height with fire-resisting construction if their siting would prejudice an escape route. Alternatively, they should be contained within the enclosure of a protected stairway.
- Any lift well which connects different compartments in a building should be constructed as a protected shaft.
- In buildings where escape is based on the principles of phased or progressive horizontal evacuation, if the lift well is not within the enclosure of a protected stairway, its entrance should be separated from the floor area on each storey by a protected lobby.
- Similarly, unless the lift is in a protected stairway enclosure, it should be approached through a protected lobby or corridor:
  - (a)   if it is situated in a basement or enclosed car park, or
  - (b)   where the lift delivers directly into corridors serving sleeping accommodation if any of the storeys also contain high fire risk areas such as kitchens, lounges or stores.
- A lift should not continue down to serve a basement if there is only one escape stairway in the building (since smoke from a basement fire might be able to prejudice the escape routes in the upper storeys) or if it is in an enclosure to a stairway which terminates at ground level.
- Lift machine rooms should be located over the lift shaft wherever possible. Where a lift is within the only protected stairway serving a building and the machine room cannot be located over the lift shaft, then it should be sited outside the protected stairway. This is to prevent smoke from a fire in the machine room from blocking the stair.
- Wall-climber and feature lifts often figure in large volume spaces such as open malls and atria. Such lifts do not have a conventional well and may place their occupants at risk if they pass through a smoke reservoir. Care will be needed in the design in order to maintain the integrity of the reservoir and protect the occupants of the lift.

7.16.8   Mechanical ventilation and air-conditioning services

Mechanical ventilation systems should be designed so that in a fire:

- air is drawn away from protected escape routes and exits, or
- the system (or the appropriate part of it) is closed down.

Systems which recirculate air should comply with BS 5588: *Fire precautions in the design, construction and use of buildings* Part 9:1989 *Code of practice for ventilation and air conditioning ductwork* for operation under fire conditions.

Recommendations for the use of mechanical ventilation in a place of assembly are given in BS 5588: *Fire precautions in the design, construction and use of buildings* Part 6:1991 *Code of practice for assembly buildings*.

Guidance on the design and installation of mechanical ventilation and air-conditioning plant is given in BS 5720: 1979 *Code of practice for mechanical ventilation and air conditioning in buildings*.

Where a pressure differential system is installed in a building (in order to keep smoke from entering the means of escape) it should be compatible with any ventilation or air-conditioning systems in the building, when operating under fire conditions.

7.16.9   Protected circuits for the operation of equipment in the event of fire

Protected power circuits are provided in situations where it is critical that the circuit should continue to function during a fire. For example, this will apply where the circuits provide power to:

- fire extinguishing systems
- smoke control systems
- sprinkler systems
- firefighting shaft systems (such as firefighting lifts)
- motorised fire shutters
- CCTV systems installed for monitoring means of escape
- data communications systems that link fire safety systems.

The cable used in a protected power circuit for operation of equipment in the event of fire should:

- meet the requirements for classification as CWZ in accordance with BS 6387: 1994 *Specification for performance requirements for cables required to maintain circuit integrity under fire conditions*;
- follow a route which passes through parts of the building in which there is negligible fire risk; and
- be separate from circuits which are provided for other purposes.

7.16.10   Refuse chutes and storage

Fires in refuse chute installations are extremely common and they are required to be built of non-combustible materials in Approved Document B3. So that escape routes are not jeopardised, refuse chutes and rooms for refuse storage should:

- be separated, by fire-resisting construction, from the rest of the building; and
- not be located in protected lobbies or stairways.

Rooms which store refuse or contain refuse chutes should:

- be approached directly from the open air; or
- be approached via a protected lobby provided with at least 0.2 m^2 of permanent ventilation.

Refuse storage chamber access points should be sited away from escape routes, final exits and windows to dwellings.

Refuse storage chambers, chutes and hoppers should be sited and constructed in accordance with BS 5906:1980 *Code of practice for storage and on-site treatment of solid waste from buildings.*

7.17   Alternative approach to the provision of means of escape in selected premises

Reference has been made throughout the text above to the use of design guides, other than AD B1, in the provision of means of escape. There are certain specialised types of premises where it is recommended that this other guidance be used in preference to the more general guidance in AD B1. Additionally, whilst AD M covers access and facilities for disabled people, there are no specific recommendations for means of escape for disabled people in AD B1. It may not be necessary to provide special structural measures to aid the escape of disabled people other than suitable management arrangements to cater for emergencies. Where it is felt that special provisions for means of escape are desirable, reference should be made to BS 5588: Part 8: 1988 *Code of practice for means of escape for disabled people.* Advice is given in the Code on refuges, evacuation lifts and the need for efficient management of escape.

7.17.1   Means of escape in health care premises and hospitals

Health care premises such as hospitals, nursing homes and homes for the elderly differ from other premises in that they contain people who are bed-ridden or who have very restricted mobility. In such buildings it is unrealistic to expect that the patients will be able to leave without assistance, or that total evacuation of the building is feasible.

Hence, the approach to the design of means of escape in these premises will demand a very different approach to that embodied in much of AD B1, and NHS Estates has prepared a set of guidance documents under the general title of *Firecode* for use in health care buildings. These documents are also applicable to non-National Health Service premises and are as follows.

- Means of escape in new hospitals – *Firecode. Health Technical Memorandum (HTM)* 81, *Fire precautions in new hospitals.* NHS Estates, HMSO, 1996.
- Work that affects the means of escape in existing hospitals – *Firecode. HTM 85 Fire precautions in existing hospitals.*
- Existing residential care premises – *Draft guide to fire precautions in existing residential care premises.* Home Office/Scottish Home and Health Department, 1983. This document is under review.

If an existing house of not more than two storeys is converted for use as an unsupervised Group Home for not more than seven mentally impaired or mentally ill people, it may be regarded as a dwellinghouse (PG 1(c)) if it has means of escape designed in accordance with *Firecode. HTM 88 Guide to fire precautions in NHS housing in the community for mentally handicapped (or mentally ill) people.* If the building is new, it might be better to regard it as being in PG 2(b) Residential (Other).

It should be noted that the *Firecode* documents contain managerial and other fire safety provisions which are outside the scope of the Building Regulations.

Progressive horizontal evacuation

Since total evacuation of health care premises is inappropriate in most cases, the *Firecode* documents contain guidance on progressive horizontal evacuation of premises. In-patients are evacuated, in the event of fire, to adjoining compartments or sub-divisions of compartments, the object being to provide a place of relative safety within a short distance. If necessary, further evacuation can be made from these safe places but under less pressure of time.

Section 4 of AD B1 gives a limited amount of guidance on progressive horizontal evacuation in some other residential buildings to which the *Firecode* documents do not apply, such as residential rest homes. When storeys are being planned for progressive horizontal evacuation, the following conditions should be considered when they are being divided into compartments.

- The compartment into which the evacuation is to take place should be large enough to take the occupants from the adjoining compartment and its own occupants. The design occupancy figures should be used to assess the total number of people involved.
- Each compartment should have an alternative escape route which is independent of the adjoining compartment. This may be through another compartment which also has an independent escape route.

7.17.2   Sheltered housing

Sheltered housing schemes which consist of specially adapted groups of houses, bungalows or two-storey flats with warden assistance and few communal facilities need not be treated differently from other one or two-storey houses or flats.

Sheltered accommodation in the institutional or other residential purpose groups would need to comply with the provisions for other buildings listed above. Additional guidance may be found in BS 5588: Part 1: 1990, clause 17.

7.17.3   Assembly buildings

A principal problem with assembly buildings is the difficulty in escaping from fixed seating. This occurs in theatres, concert halls, conference centres and at sports events. Specific guidance on this may be obtained from:

- sections 3 and 5 of BS 5588: Part 6: 1991 *Code of practice for places of assembly*, where guidance on the spacing of fixed seating and other aspects of means of escape may be found;
- *Guide to fire precautions in existing places of entertainment and like premises*. Home Office/Scottish Home and Health Department, HMSO, 1990; and
- *Guide to safety at sports grounds*, Department of National Heritage/Scottish Office. HMSO, 1997, for sports stadia, etc.

7.17.4   Schools and other educational buildings

As a result of the coming into force of the *Education (Schools and Further and Higher education)(Amendment) Regulations 2001* on 1 April 2001, schools in England are no longer exempt from the Building Regulations. Therefore, the fire safety objectives of the Department for Education and Skills may be met by following the guidance contained in **AD B** outlined in this chapter. It should be noted however that the Department for Education and Skills has issued guidance on constructional standards for schools (see *Guidance on the Constructional Standards for Schools*: DfES 142/2001-09-11), indicating where it is necessary to supplement the Approved Documents to take account of the specific requirements of schools. It is proposed that in due course the Approved Documents will be amended to take account of these. Further information on this can be obtained from the School Premises Team, Department for Education and Skills, Caxton House, Room 762, 6–12 Tothill Street, London SW1H 9NA, Tel 020 7273 6023, E-mail: premises.schools@dfes.gsi.gov.uk.

7.17.5   Shops and shopping complexes

British Standard BS 5588: Part 11: 1997 *Code of practice for shops, offices, industrial, storage and other similar buildings* may be used instead of AD B1.

Shopping complexes are not covered adequately by AD B1 and should be designed in accordance with section 4 of BS 5588: Part 10: 1991 *Code of practice for shopping complexes*. It should be noted that BS 5588: Part 10 applies more restrictive provisions to units with only one exit in covered shopping complexes than may be found in BS 5588: Part 11.

7.18   Internal fire spread (linings)

7.18.1   Introduction

Although the linings of walls and ceilings are unlikely to be the materials first ignited in a fire (this is more likely to occur in furniture and fittings), they can significantly affect the spread of fire and its rate of growth. This is especially true in circulation areas where rapid fire spread may prevent occupants from escaping. Part B2 of Schedule 1 to the 2000 Regulations seeks to control these surface linings.

7.18.2   Control of wall and ceiling linings

The spread of fire within a building may be inhibited by paying attention to the lining materials used on walls, ceilings, partitions and other internal structures. These linings must:

- offer adequate resistance to spread of flame over their surfaces; and
- if ignited, have a rate of heat release or fire growth which is reasonable in the circumstances.

It should be noted that floors and stairs are not covered by the requirements since they are not usually involved in a fire until it is well established. Consequently, they will not contribute to fire spread in the early stages of a fire.

7.18.3   Methods of test

In order to meet the requirements of AD B2 it is necessary for materials or products to meet certain levels of performance in appropriate tests.

Under National classifications the surface spread of flame characteristics of a material may be determined by testing it in accordance with the method specified in BS 476 *Fire tests on building materials and structures*: Part 7

- 1971 *Surface spread of flame tests for material; or*
- 1987 *Method for classification of the surface spread of flame of product; or*
- 1997 *Method of test to determine the classification of surface spread of flame of products.*

A strip of the material under test is placed with one end resting against a furnace and the rate at which flames spread along the material is measured.

Materials or products are thus placed in Classes 1, 2, 3 or 4, Class 1 representing a surface of very low flame spread. Class 4 (a surface of rapid flame spread) is not acceptable under the provisions of the approved document.

Under the European classifications, lining systems are classified in accordance with BS EN 1350: *Fire classification of construction products and building elements,* Part 1: 2002: *Classification using data from reaction to fire tests.* Materials or products are classified as A1, A2, B, C, D, E or F, with A1 being the highest. When a classification includes 's3, d2', it means that there is no limit set for smoke production and/or flaming droplets/particles. The relevant European test methods are specified as follows:

- BS EN ISO 1182:2002 *Reaction to fire tests for building products – Non combustibility test*;
- BS EN ISO 1716:2002 *Reaction to fire tests for building products – Determination of the gross calorific value*;
- BS EN 13823:2002 *Reaction to fire tests for building products – Building products excluding floorings exposed to the thermal attack by a single burning item*;
- BS EN ISO 11925 *Reaction to fire tests for building products*, Part: 2002 – *Ignitability when subjected to direct impingement of flame*; and
- BS EN 13238: 2001: *Reaction to fire tests for building products – conditioning procedures and general rules for selection of substrates*.

In the event of fire, some materials have a higher rate of heat release or ignite more easily than others. They are therefore more hazardous and this may mean a reduced time to flashover. In the National test, the way in which the rate of heat release may be assessed is contained in BS 476: Part 6: 1981 or 1989 *Method of test for fire propagation of products*.

The material or product is tested for a certain period of time in a furnace and is given two numerical indices related to its performance. The sub-index (i_1) is derived from the first three minutes of the test whilst the overall test performance is denoted by the index of performance (I).

7.18.4   Class 0 materials

In order to establish a high product performance classification for lining materials in high risk areas (such as circulation spaces), AD B2 recommends that materials in these areas should conform to either the Class 0 (National) standard or European Class B-s3, d2 or better. Class 0 is not a classification found in any British Standard test as such; however it is evident that it draws on BS 476 test results as the following definition shows.

The Class 0 standard will be achieved by any material or the surface of a composite product which:

(a)   is composed of materials of limited combustibility (see below) throughout; or
(b)   is a material of Class 1 which has an index of performance (I) of not more than 12 and a sub-index (i_1) of not more than 6.

7.18.5   Interpretation

The following terms are common to all parts of AD B but occur most frequently in AD B2, 3 and 4:

NON-COMBUSTIBLE MATERIAL – A material which has the highest level of reaction to fire performance when tested as follows:

- (National classes) to BS 476: Part 11; the material does not flame and there is no rise in temperature on either the centre (specimen) or furnace thermocouples; or
- (European classes) when classified as class A1 in accordance with BS EN13501 *Fire classification of construction products and building elements,* Part 1: 2002: *Classification using data from reaction to fire tests*:
    (a) BS EN ISO 1182: 2002: *Reaction to fire tests for building products – Non-combustibility test*; and
    (b) BS EN ISO 1716:2002 *Reaction to fire tests for building products – Determination of the gross calorific value.*

Table A6 from Appendix A of AD B (which is reproduced below) lists some examples of non-combustible materials and gives details of where they should be used.

MATERIALS OF LIMITED COMBUSTIBILITY – Materials in this group, whilst not regarded as non-combustible, would contribute little heat energy to a fire. Therefore, they can be used in situations where control of fire spread is essential, such as stairs to basements. They are defined in AD B, Table A7 (see below) when tested as follows:

- (National classes) to BS 476: Part 11, the material does not flame and there is no rise in temperature on either the centre (specimen) or furnace thermocouples; or
- (European classes) when classified as class A2-s3, d2 in accordance with BS EN13501 *Fire classification of construction products and building elements,* Part 1: 2002: *Classification using data from reaction to fire* test when tested to:
    (a) BS EN ISO 1182: 2002: *Reaction to fire tests for building products – Non-combustibility test,* or
    (b) BS EN ISO 1716: 2002 *Reaction to fire tests for building products – Determination of the gross calorific value,* and BS EN 13823: 2002, *Reaction to fire tests for building products – Building products excluding floorings exposed to the thermal attack by a single burning item.*

It should be noted that certain insulating materials in group (d) of the table are of a lower standard than the group (a), (b) or (c) materials and may only be used in the situations shown in items 9 and 10 of the table. It is of course permissible to use non-combustible materials whenever a recommendation for materials of limited combustibility is specified.

7.18.6   Materials for surface linings

As a guide to the materials which may be used for wall and ceiling linings, AD B lists in Table A8 (see below) the typical performance ratings for some generic materials and products. Test results for proprietary materials may be obtained from manufacturers and trade associations. However, small differences in detail (e.g. thickness, substrate, colour, form, fixings, adhesives, etc.) can significantly affect the rating. Therefore, the reference used to substantiate the spread of flame rating should be carefully checked to ensure that it is suitable, adequate and applicable to the construction to be used.

AD B, Appendix A

Table A6   Use and definitions of non-combustible materials

References in AD B guidance to situations where such materials should be used	Definitions of non-combustible materials	
	National class	**European class**
1. ladders referred to in the guidance to B1, paragraph 6.22 (see section 7.16.4) 2. refuse chutes meeting the provisions in the guidance to B3, paragraph 9.35c (see section 7.22.2) 3. suspended ceilings and their supports where there is provision in the guidance to B3, paragraph 10.13, for them to be constructed of non-combustible materials (see section 7.23.2) 4. pipes meeting the provisions in the guidance to B3, Table 15 (see section 7.24.1) 5. flue walls meeting the provisions in the guidance to B3, Diagram 39 (see section 7.24.3) 6. construction forming car parks referred to in the guidance to B3, paragraph 12.3 (see section 7.28.1)	a. Any material which when tested to BS 476: Part 11 does not flame nor cause any rise in temperature on either the centre (specimen) or furnace thermocouples b. Totally inorganic materials such as concrete, fired clay, ceramics, metals, plaster and masonry containing not more than 1% by weight or volume of organic material. (Use in buildings of combustible metals such as magnesium/aluminium alloys should be assessed in each individual case c. Concrete bricks or blocks meeting BS 6073: Part 1 d. Products classified as non-combustible under BS 476: Part 4	a. Any material classified as class A1 in accordance with BS EN 13501-1: 2002, Fire classification of construction products and building elements, Part 1 – Classification using data from reaction to fire tests b. Products made from one or more of the materials considered as Class A1 without the need for testing, as defined in Commission Decision 96/603/EC of 4th October 1996 establishing the list of products belonging to Class A1 'No contribution to fire' provided for in the Decision 94/611/EC implementing Article 20 of the Council Directive 89/106/EEC on construction products. None of the materials shall contain more than 1.0% by weight or volume (whichever is the lower) of homogeneously distributed organic material
		Note: The National classifications do not automatically equate with the equivalent classifications in the European column, therefore products cannot typically assume a European class unless they have been tested accordingly

7.18.7   Thermoplastic materials

A thermoplastic material is a synthetic polymer which has a softening point below 200°C when tested in accordance with BS 2782 Methods of testing plastics, Part 1 Thermal properties: Method 120A: 1990 Determination of the Vicat softening temperature of thermoplastics. If the thickness of the product to be tested is less than 2.5 mm then specimens for the test may be fabricated from the original polymer.

When used in isolation as a wall or ceiling lining, a thermoplastic material cannot be assumed to protect a substrate. The surface rating of both the substrate and the lining would need to meet the required classification. However, where the thermoplastic material is fully bonded to a non-thermoplastic substrate then only the surface rating of the composite would need to comply.

AD B, Appendix A

Table A7   Use and definitions of materials of limited combustibility.

References in AD.B guidance to situations where such materials should be used	Definitions of non-combustible materials	
	National class	**European class**
1. stairs where there is provision in the guidance to B1 for them to be constructed of materials of limited combustibility (see 6.19 and section 7.16.3) 2. materials above a suspended ceiling meeting the provision in the guidance to B3, paragraph 10.13 (see section 7.23.2) 3. reinforcement/support for fire-stopping referred to in the guidance to B3, see 11.13 (see section 7.24.4) 4. roof coverings meeting provisions: a. in the guidance to B3, paragraph 10.11 or (see section 7.23.2) b. in the guidance to B4, Table 17 or c. in the guidance to B4, Diagram 47 (see section 7.18.9) 5. roof deck meeting the provisions of the guidance to B3, Diagram 28a (see section 7.22.1) 6. class 0 materials meeting the provisions in Appendix A, paragraph 13(a) (see section 7.18.4) 7. ceiling tiles or panels of any fire protecting suspended ceiling (Type Z) in Table A3 (see section 7.19) 8. compartment walls and compartment floors in hospitals referred to in paragraph 9.32	a. Any non-combustible material listed in Table A6 b. Any material of density 300/ kg/m^2 or more, which when tested to BS 476: Part 11, does not flame and the rise in temperature on the furnace thermocouple is not more than 20°C c. Any material with a non-combustible core at least 8 mm thick having combustible facings (on one or both sides) not more than 0.5 mm thick. (Where a flame spread rating is specified, these materials must also meet the appropriate test requirements)	a. Any material listed in Table A6 b. Any material/product classified as Class A2-s3, d2 or better in accordance with BS EN 13501-1: 2002, *Fire classification of construction products and building elements, Part 1 – Classification using data from reaction to fire tests*
9. insulation material in external wall construction referred to in paragraph 13.7 (see section 7.27) 10. Insulation above any fire-protecting suspended ceiling (Type Z) in Table A3 (see section 7.19)	Any of the materials (a), (b) or (c) above, or: d. Any material of density less than 300 kg/m^3, which when tested to BS 476: Part 11, does not flame for more than 10 seconds and the rise in temperature on the centre (specimen) thermocouple is not more than 35°C and on the furnace thermocouple is not more than 25°C	Any of the materials/products (a) or (b) above
		Notes: 1. The National classifications do not automatically equate with the equivalent classifications in the European column, therefore products cannot typically assume a European class unless they have been tested accordingly 2. When a classification includes 's3, d2', this means that there is no limit set for smoke production and/or flaming droplets/particles

AD B, Appendix A

Table A8 Typical performance ratings of some generic materials and products.

Rating	Material or product
Class 0 (National)	1. any non-combustible material or material of limited combustibility. (Composite products listed in Table A7 must meet test requirements given in Appendix A, paragraph 13(b)) 2. brickwork, blockwork, concrete and ceramic tiles 3. plasterboard (painted or not with a PVC facing not more than 0.5 mm thick) with or without an air gap or fibrous or cellular insulating material behind 4. woodwool cement slabs 5. mineral fibre tiles or sheets with cement or resin binding
Class 3 (National)	6. timber or plywood with a density more than $400\,kg/m^3$, painted or unpainted 7. wood particle board or hardboard, either untreated or painted 8. standard glass reinforced polyesters
Class A1 (European)	9. any material that achieves this class and is defined as 'classified without further test' in a published Commission Decision
Class A2-s3, d2 (European)	10. any material that achieves this class and is defined as 'classified without further test' in a published Commission Decision
Class B-s3, d2 (European)	11. any material that achieves this class and is defined as 'classified without further test' in a published Commission Decision
Class C-s3, d2 (European)	12. any material that achieves this class and is defined as 'classified without further test' in a published Commission Decision
Class D-s3, d2 (European)	13. any material that achieves this class and is defined as 'classified without further test' in a published Commission Decision

Notes (National):
1. Materials and products listed under Class 0 also meet Class 1.
2. Timber products listed under Class 3 can be brought up to Class 1 with appropriate proprietary treatments.
3. The following materials and products may achieve the ratings listed below. However, as the properties of different products with the same generic description vary, the ratings of these materials/products should be substantiated by test evidence.
   Class 0 – aluminium faced fibre insulating board, flame retardant decorative laminates on a calcium silicate board, thick polycarbonate sheet, phenolic sheet and UPVC;
   Class 1 – phenolic or melamine laminates on a calcium silicate substrate and flame retardant decorative laminates on a combustible substrate.

Notes (European):
For the purposes of the Building Regulations
1. Materials and products listed under Class A1 also meet Classes A2-s3, d2, B-s3, d2, C-s3, d2 and D-s3, d2.
2. Materials and products listed under Class A2-s3, d2 also meet Classes B-s3, d2, C-s3, d2 and D-s3, d2.
3. Materials and products listed under Class B-s3, d2 also meet Classes C-s3, d2 and D-s3, d2.
4. Materials and products listed under Class C-s3, d2 also meet Class D-s3, d2.
5. The performance of timber products listed under Class D-s3, d2 can be improved with appropriate proprietary treatments.
6. Materials covered by the CWFT process (classification without further testing) can be found by accessing the European Commission's website via the link on the ODPM's web site www.odpm.gov.uk/bregs/cpd/index.htm.
7. The National classifications do not automatically equate with the equivalent classifications in the European column, therefore products cannot typically assume a European class unless they have been tested accordingly.
8. When a classification includes 's3, d2', this means that there is no limit set for smoke production and/or flaming droplets/particles.

Some thermoplastic materials can be tested under BS 476: Parts 6 and 7, and can be used in accordance with their ratings as described above. Alternatively, they may be classified as TP(a) rigid, TP(a) flexible, or TP(b), as described below, but their use would be restricted to rooflights, lighting diffusers, suspended ceiling panels and external window glazing (except in circulation areas). These uses are described more fully below.

TP(a) rigid means:

- rigid solid PVC sheet
- solid (i.e. not double- or multi-skin) polycarbonate sheet at least 3 mm thick
- multi-skinned rigid sheet made from uPVC or polycarbonate with a BS 476: Part 7 rating of Class 1
- any other rigid thermoplastic product which, when tested in accordance with BS 2782: Part 5: 1970 (1974): Method 508A, performs so that the flame extinguishes before reaching the first mark, and the duration of the flame or afterglow after removal of the burner does not exceed five seconds.

TP(a) flexible means:

- flexible products not more than 1 mm thick complying with the Type C requirements of BS 5867 *Specification for fabrics for curtains and drapes,* Part 2: 1980 *Flammability requirements,* when tested to BS 5438: Test 2, 1989. In the BS 5438 test, the flame should be applied to the specimens for 5,15,20 and 30 seconds respectively, although it is not necessary to include the cleansing procedure described in the British Standard.

TP(b) means:

- rigid solid polycarbonate sheet products less than 3 mm thick, or multi-skin polycarbonate sheet products which do not qualify as TP(a) by test
- any other product, a specimen of which between 1.5 mm and 3 mm thick, when tested in accordance with BS 2782: Part 5: 1970 (1974): Method 508A, has a rate of burning not exceeding 50 mm/minute.

If it is not possible to cut or machine a 3 mm thick test specimen from the product, then it is permissible to mould one from the original material used for the product.

Currently, no new guidance is possible on the assessment or classification of thermoplastic materials under the European system since there is no generally accepted European test procedure and supporting comparative data.

7.18.8   Specific recommendations for wall and ceiling linings

Table 10 from Section 6 of AD B2 gives the recommended flame spread classifications for the surfaces of walls and ceilings in any room or circulation space, for all building types.

Different standards are set for 'small rooms', which are totally enclosed rooms with floor area of not more than 4 m^2 in residential buildings or 30 m^2 in non-residential buildings, and for other rooms and circulation spaces.

When considering the performance of wall linings, window glazing and ceilings which slope at more than 70° to the horizontal are treated as a wall. Certain vertical surfaces are excluded from this definition, such as:

- doors, door frames and glazing in doors
- window frames and other frames containing glazing
- narrow members, such as architraves, cover moulds, picture rails and skirtings
- fireplace surrounds, mantle shelves and fitted furniture.

Similarly, ceiling surfaces include glazing and walls which slope at 70° or less to the horizontal, but exclude:

- trap doors and frames
- window frames, rooflight frames and other frames in which glazing is fitted
- narrow members, such as architraves, cover moulds, exposed beams and picture rails.

Therefore, bearing in mind the above definitions and exclusions the linings of walls and ceilings should conform to the following classifications:

- in circulation spaces in buildings other than dwellings (but including the common areas of flats and maisonettes) – not lower than Class 0 (or European Class B-s3, d2);
- in circulation spaces within dwellings – not lower than Class 1 (or European Class C-s3, d2);
- in any other rooms in any building (other than in small rooms as defined above) – not lower than Class 1 (or European Class C-s3, d2);
- in small rooms (see definition above) – not lower than Class 3 (or European Class D-s3, d2); and
- in domestic garages of area not exceeding 40 m^2 – not lower than Class 3. Above this floor area – not lower than Class 1 (or European Class D-s3, d2).

(Note: the National classifications do not automatically equate with the equivalent European classifications; therefore products cannot typically assume a European class, unless they have been tested accordingly. When a classification includes 's3, d2', this means that there is no limit set for smoke production and/or flaming droplets/particles).

Approved Document B2 allows certain variations from the strict lining classifications shown above provided that no lining is lower than Class 3 (National Class) or D-s3, d2 (European Class), as follows.

- Wall linings in rooms may be lower than Class 1 if their area does not exceed half the floor area of the room subject to the following maxima:

(a) in residential buildings – 20 m^2

(b) in non-residential buildings – 60 m^2.

- Plastic rooflights, and lighting diffusers fitted in suspended ceilings, may be lower than Class 0 or Class 1 (but not lower than Class 3) if they comply with recommendations for thermoplastic rooflights or diffusers shown below. Rooflights of other materials should comply with the lining classifications shown above.
- External windows to rooms may be glazed with a TP(a) rigid thermoplastic product, but not in circulation areas. However, internal glazing should comply with the lining classifications shown above.
- Suspended or stretched-skin ceilings made from thermoplastic material with a TP(a) flexible classification are permitted provided they do not form part of a fire-resisting ceiling. Each panel should be supported on all its sides and should not exceed 5 m^2 in area.

7.18.9   Rooflights and lighting diffusers

Rooflights, and lighting diffusers which form an integral part of a ceiling (i.e. not attached to the soffit or suspended beneath the ceiling), may:

- comply with the classification shown above; or
- consist of thermoplastic materials with a TP(a) rigid or TP(b) classification or at least a Class 3 rating if used under the conditions described in Table 11 from AD B2, and limited in extent and layout as illustrated in Diagram 24 from AD B2. (It should be noted that no guidance is currently possible on the performance requirements in the European fire tests as there is no generally accepted test and classification procedure.) These conditions and limitations are combined in Fig. 7.32 below. Rooflights made from these materials must not be used in a protected stairway.

The space in the ceiling above a lighting diffuser should comply with the lining classifications shown above for flame spread, according to the type of space below the ceiling. Lighting diffusers should only be used in fire-resisting or fire-protecting ceilings if they have been satisfactorily tested as part of the ceiling system being used to provide the appropriate fire protection.

   The external surfaces of rooflights may also need to follow the recommendations for roofs contained in AD B4. Tables 18 and 19 of AD B4 give details of the limitations on the use of plastic rooflights in roofs. The main recommendations of the tables are combined in Table 7.6 and are summarised below.

- TP(a) rigid thermoplastic material may be used for a rooflight over any space except a protected stairway. Unless located in any of the buildings listed in (a) and (b) immediately below, the rooflight should be at least 6 m from any point on a boundary,
- Rooflights serving:

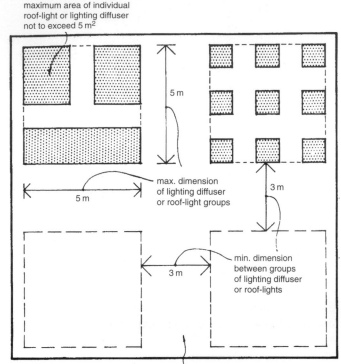

maximum area of individual
roof-light or lighting diffuser
not to exceed 5 m²

5 m

5 m

max. dimension
of lighting diffuser
or roof-light groups

3 m

3 m

min. dimension
between groups
of lighting diffuser
or roof-lights

Notes:

1. For Class 3 or TP(b), total area of lighting diffusers or roof-lights should not exceed:

● 50% of floor area if located in rooms; or

● 15% of floor area if located in circulation spaces

2. Plastics roof-lights and lighting diffusers should not be used in protected stairways

3. There are no restrictions on the use of Class 3 materials in small rooms

4. There are no restrictions on the use of TP (a) materials except for note 2 above

5. Class 3 rooflights to rooms in industrial and other non-residential purpose groups may be spaced 1800mm apart provided the rooflights are evenly distributed and do not exceed 20% of the area of the room

6. The minimum 3m separation between each 5m² must be maintained. Therefore, in some cases it may not be possible to use the maximum percentage quoted.

Upper and lower surface of suspended ceiling between
roof-lights or diffusers to comply with AD B2, paragraph 7.1
regarding lining classification

Ceiling plan

Fig. 7.32   Limitations on use of Class 3 plastics rooflights, TP(b) rooflights and TP(b) lighting diffusers in suspended ceilings.

(a)   balconies, verandas, carports, covered ways or loading bays with one longer side permanently open, or detached swimming pools; or

(b)   garages, conservatories or outbuildings with a floor area not exceeding 40 m²;

may consist of plastics materials with a lower surface of not less than Class 3 surface spread of flame or TP(b). The rooflight should be 6 m from any point on the boundary if the external surface is designated AD, BD, CA, CB, CC, CD or TP(b). (For details of designatory letters see section 7.27.9.) If the external surface is designated DA, DB, DC or DD the rooflight should be 20 m from any point on the boundary.

●   The internal and external surface limits specified in (b) immediately above also apply to roof-lights in all other types of buildings. However, there are additional limitations on the area and spacing of the rooflights.

Table 7.6 Plastics rooflights: limitations on use and boundary distance.

| Space which rooflight can serve | Min. classification on lower surface[1] | Minimum distance from any point on relevant boundary to rooflight with | | | |
| | | an external designation[2] of: | | an external surface classification[1] of: | |
		AB BD CA CB CC CD	DA DB DC DD	TP(a)	TP(b)
1. Any space except a protected stairway	TP(a) rigid	N/A	N/A	6 m	N/A
2. Balcony, veranda, carport, covered way or loading bay, which has at least one longer side wholly or permanently open	TP(b)	N/A	N/A	N/A	6 m
3. Detached swimming pool	Class 3	6 m	20 m	N/A	N/A
4. Conservatory, garage or outbuilding, with a maximum floor area of 40 m²					
5. Circulation space[3] (except a protected stairway)	TP(b)	N/A	N/A	N/A	6 m[4]
6. Room[3]	Class 3	6 m	20 m	N/A	N/A

Table References:

(1) See also the guidance to AD B2 (section 7.18.3).
(2) The designation of external roof surfaces is explained in Appendix A of AD B (see section 7.27.9).
(3) Single skin rooflight only, in the case of non-thermoplastic material.
(4) The rooflight should also meet the provision of Diagram 47 of AD B4 (see Fig. 7.32).

Notes

(a) N/A Not applicable.
(b) Polycarbonate and PVC rooflights which achieve a Class 1 rating by test, see paragraph 15.7 of AD B4 (section 7.27.9), may be regarded as having an AA rating.
(c) None of the above designations are suitable for protected stairways.
(d) Products may have upper and lower surfaces with different properties if they have double skins or are laminates of different materials. In which case the more onerous distance applies.
(e) Where the roof covering continues over a compartment wall, rooflights should be at least 1.5 m from the compartment wall (see section 7.22.1).

- Individual rooflights serving circulation spaces or rooms should not exceed 5 m² in area. They should be separated by roof covering materials of limited combustibility at least 3 m wide. If the rooflight is not thermoplastic it should consist of a single-skin material.

These recommendations for plastic rooflights and lighting diffusers are illustrated in Fig. 7.33.

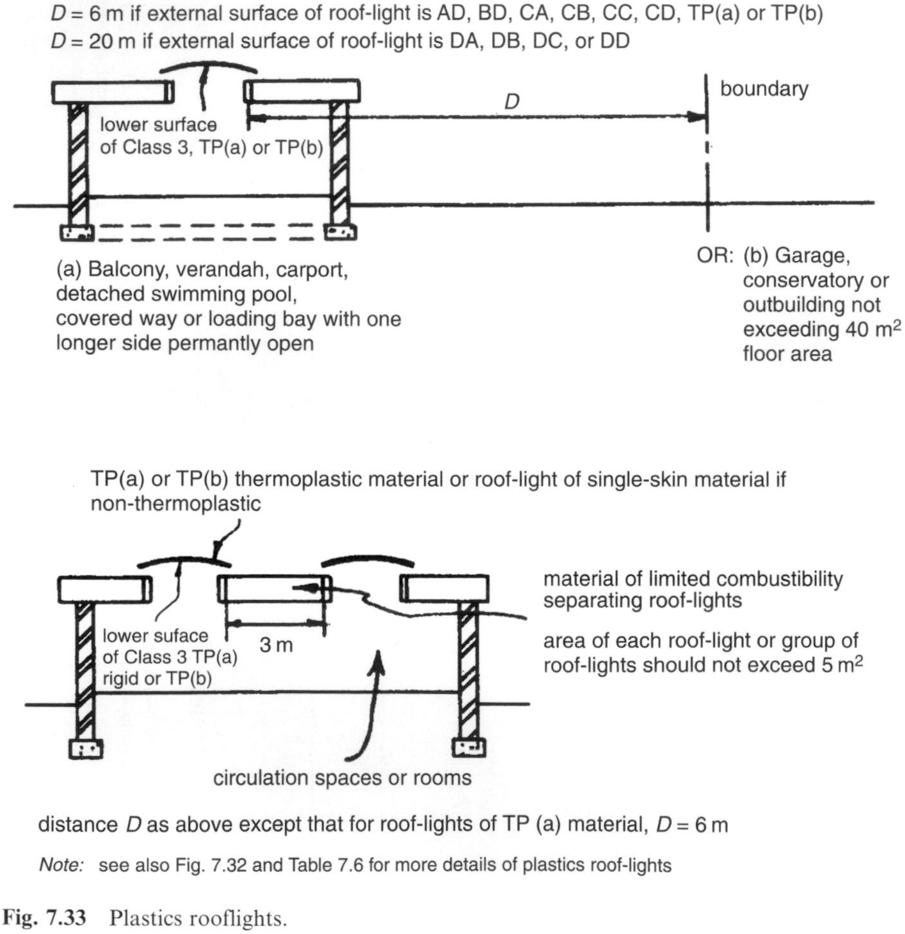

D = 6 m if external surface of roof-light is AD, BD, CA, CB, CC, CD, TP(a) or TP(b)
D = 20 m if external surface of roof-light is DA, DB, DC, or DD

boundary

lower surface
of Class 3, TP(a) or TP(b)

D

(a) Balcony, verandah, carport, detached swimming pool, covered way or loading bay with one longer side permanently open

OR: (b) Garage, conservatory or outbuilding not exceeding 40 m² floor area

TP(a) or TP(b) thermoplastic material or roof-light of single-skin material if non-thermoplastic

lower suface of Class 3 TP(a) rigid or TP(b)

3 m

material of limited combustibility separating roof-lights

area of each roof-light or group of roof-lights should not exceed 5 m²

circulation spaces or rooms

distance D as above except that for roof-lights of TP (a) material, D = 6 m

Note: see also Fig. 7.32 and Table 7.6 for more details of plastics roof-lights

Fig. 7.33 Plastics rooflights.

7.18.10   Flexible membranes and air supported structures

In recent years there has been a move towards the use of flexible membranes to provide the external envelope of a structure, often where this is to be of a temporary nature. The most famous example of this was the Millennium Dome which was supported by masts and cables. It is equally possible for the structure to be air supported. Since such materials are unlikely to satisfy the strict criteria laid down for the linings of walls and ceilings in AD B2, the Approved Document gives a number of alternative sources where guidance may be sought as follows.

- For any flexible membrane covering a structure (except an air supported structure) see Appendix A of BS 7157: 1989 *Method of test for ignitability of fabrics used in the construction of large tented structures.*
- Guidance on the use of PTFE-based materials for tension-membrane roof structures may be found in BRE Report 274: 1994 *Fire safety of PTFE-based materials used in buildings.*
- Air supported structures should follow the guidance given in BS 6661: 1986 *Guide for the design, construction and maintenance of single-skin air supported structures.*

7.19 Further requirements for ceilings

Fire-protecting suspended ceilings may be used to contribute towards the fire-resistance of a floor if they meet certain criteria. These are contained in Table A3 of Appendix A of AD B which is reproduced below. The highest grade of suspended ceiling is type Z. This (together with any insulating material) should be constructed of materials of limited combustibility (National) or Class A2-s3, d2 or better

AD B, Appendix A

Table A3 Limitations on fire-protecting suspended ceilings (see Table A1, Note 4).

Height of building or separated part (m)	Type of floor	Provision for fire resistance or floor (minutes)	Description of suspended ceiling
less than 18	not compartment	60 or less	Type W, X, Y or Z
	compartment	less than 60	
		60	Type X, Y or Z
18 or more	any	60 or less	Type Y or Z
no limit	any	more than 60	Type Z

Notes:
1. Ceiling type and description (the change from Types A–D to Types W–Z is to avoid confusion with Classes A–D (European)):
   W. Surface of ceiling exposed to the cavity should be Class 0 or Class 1 (National) or Class C-s3, d2 or better (European).
   X. Surface of ceiling exposed to the cavity should be Class 0 (National) or Class B-s3, d2 or better (European).
   Y. Surface of ceiling exposed to the cavity should be Class 0 (National) or Class B-s3, d2 or better (European). Ceiling should not contain easily openable access panels.
   Z. Ceiling should be a material of limited combustibility (National) or of Class A2-s3, d2 or better (European) and not contain easily openable access panels. Any insulation above the ceiling should be of a material of limited combustibility (National) or Class A2-s3, d2 or better (European).
2. Any access panels provided in fire protecting suspended ceilings of type Y or Z should be secured in position by releasing devices or screw fixings, and they should be shown to have been tested in the ceiling assembly in which they are incorporated.
3. European classifications.
The National classifications do not automatically equate with the equivalent European classifications, therefore products cannot typically assume a European class unless they have been tested accordingly.
When a classification includes 's3, d2', this means that there is no limit set for smoke production and/or flaming droplets/particles.

(European) and should not contain easily openable access panels. Type Z ceilings may be used anywhere under any conditions and should be used where the fire-resistance of the total floor/ceiling assembly exceeds 60 minutes. All the ceiling types need to comply with certain surface spread of flame requirements for the upper surface facing into the cavity, in addition to the Table 10 recommendations of AD B2 for the lower surface.

Similarly, fire-resisting ceilings may be used to reduce the need for cavity barriers in concealed void spaces in some floors and roofs. These are discussed in section 7.23.2 below, in the section on cavity barriers (see Fig. 7.41).

7.20 Internal fire spread (structure)

Paragraph B3 of Schedule 1 to the Building Regulations lists a number of factors which must be considered in order to reduce the effects of fire spread throughout the structure of a building as follows.

- The building must be so designed and constructed that its stability will be maintained for a reasonable period during a fire.
- Walls which are common to two or more buildings must be designed and constructed so that they resist the spread of fire between those buildings. Semi-detached and terraced houses are treated as separate buildings for the purposes of this requirement.
- The building must be subdivided by fire-resisting construction, depending on its size and intended use, where this is necessary to inhibit the spread of fire. (This requirement does not apply to material alterations to prisons provided under section 33 of the Prisons Act 1952.)
- Fire and smoke may spread unseen through concealed spaces in the structure and fabric of a building. The building must be designed and constructed so that this fire and smoke spread is inhibited.

7.21 Fire resistance and structural stability

If the structural elements of a building can be satisfactorily protected against the effects of fire for a reasonable period, it will be possible for the occupants to be evacuated safely and also the spread of fire throughout the building will be kept to a minimum. The risk to firefighters (who may have to search for or rescue people who are trapped) will be reduced and there will be less risk to people in the vicinity of the building from falling debris or as a result of an impact with an adjacent building from the collapsing structure.

One way to measure the standard of protection to be provided is by reference to the fire resistance of the elements under consideration. A number of factors which have a bearing on fire resistance are considered in Appendix A of AD B including the following.

- Fire severity – estimated from the purpose group (and therefore, the use) of the building. This assumes that the contents (which constitute the fire load) are the same for buildings of similar usage and that the contents of some building types will be more hazardous than others.
- Height of the top floor above ground – affects ease of escape, firefighting and the consequences of a large-scale collapse.
- Building occupancy – influences the speed of evacuation.
- The presence of basements – lack of venting may increase heat build-up and the duration of a fire, and hinder firefighting.
- The number of floors – escape from single-storey buildings is easier and a structural failure is unlikely to happen before evacuation has taken place.

It can be seen from the foregoing that an assessment of the standard of fire resistance in a building is a complicated matter. It is further complicated by the fact that there can be little control exercised over a building's future fire load unless a material change of use occurs. If a fire engineering approach is adopted for the assessment of fire severity based on fire load for a particular use, then future changes in use should also be borne in mind.

The method of assessment of fire resistance contained in AD B is based on the performance of an element of structure, door or other part of a building by reference to standard tests contained in:

- BS 476: Parts 20-24: 1987 (or to BS 476: Part 8: 1972, for items tested prior to 1 January 1988); or
- (European tests) Commission Decision 2000/367/EC of 3 May 2000 implementing Council Directive 89/106/EEC as regards the classification of the resistance to fire performance of construction products, construction works and parts thereof.

All products are classified in accordance with:

- BS EN 13501: *Fire classification of construction products and building elements, Part 2: xxxx: Classification using data from fire resistance tests (excluding products for use in ventilation systems)*
- BS EN 13501: *Fire classification of construction products and building elements, Part 3: xxxx: Classification using data from fire resistance tests on components of normal building service installations (other than smoke control systems)*
- BS EN 13501: *Fire classification of construction products and building elements, Part 4: xxxx: Classification using data from fire resistance tests on smoke control systems.*

(It should be noted that the designation of xxxx above is used for the year reference for standards that are not yet published. It is permissible to use the latest version of any standard provided that it continues to address the relevant requirements of the Regulations.)

The tests relate to the ability of the element:

- to resist a fire without collapse (loadbearing capacity), denoted R in the European classification of the resistance to fire performance;
- to resist fire penetration (integrity), denoted E in the European classification of the resistance to fire performance; and
- to resist excessive heat penetration so that fire is not spread by radiation or conduction (insulation), denoted I in the European classification of the resistance to fire performance.

Clearly the criteria of resistance to fire and heat penetration are applicable only to fire-separating elements, such as walls and floors. The criterion of resistance to collapse is applicable to all load-bearing elements, such as columns and beams, in addition to floors and load-bearing walls; however, it does not apply to external curtain walling or other claddings which transmit only self-weight and wind loads.

Table A1 to Appendix A of AD B (see below) shows the method of exposure required for the various elements of structure and other forms of construction, together with the BS 476 requirements which should be satisfied in terms of load-bearing capacity, integrity and insulation, and the minimum provisions when tested to the relevant European standard. For some items the table indicates the actual period of fire resistance recommended under each heading, but for others Table A2 of Appendix A gives the detailed recommendations in respect of fire resistance periods. The performance standards for doors are contained in Table B1 of AD B which is reproduced in the section on fire doors below.

In addition to the elements of structure defined in section 7.2 and illustrated in Fig. 7.3 above, there are requirements for some other elements of the building to be of fire-resisting construction. Included in this category are some doors, pipe casings and cavity barriers. These are considered later under the actual element references.

7.21.1   Minimum period of fire resistance

In order to establish the minimum period of fire resistance for the elements of structure of a building, it is necessary, first, to determine the building's use or Purpose Group. The fire resistance period will then depend on the height of the top storey of the building above ground or the depth of the lowest basement storey below ground.

It will be seen that the fire resistance recommendations for basements are generally more onerous than for ground floors in the same building. This reflects the greater difficulty experienced in dealing with a basement fire. However, it is sometimes the case that due to the slope of the ground, at least one side of a basement is accessible at ground level. This gives opportunities for smoke venting and firefighting and in these circumstances it may be reasonable to adopt the less onerous fire resistance provisions of the upper elements of the construction for the elements of structure in the basement.

The minimum periods of fire resistance recommended for the elements of structure in the basements, ground or upper storeys of a building are given in Table A2 from Appendix A of AD B which is reproduced below.

AD B, Appendix A

Table A1   Specific provisions of test for fire resistance of elements of structure etc.

Part of building	Minimum provisions when tested to the relevant part of BS 476(1) (minutes)			Minimum provisions when tested to the relevant European standard (minutes) (12)	Method of exposure
	Loadbearing capacity (2)	**Integrity**	**Insulation**		
1. **Structural** frame, beam or column	see Table A2	not applicable	not applicable	R see Table A2	exposed faces
2. **Loadbearing wall** (which is not also a wall described in any of the following items)	see Table A2	not applicable	not applicable	R see Table A2	each side separately
3. **Floors (3)** a. in upper storey of two-storey dwelling house (but not over garage or basement)	30	15	15	REI 30 (9)	
b. between a shop and flat above	60 or see Table A2 (whichever is greater)	60 or see Table A2 (whichever is greater)	60 or see Table A2 (whichever is greater)	REI 60 or see Table A2 (whichever is greater)	from underside (4)
c. any other floor, including compartment floors	see Table A2	see Table A2	see Table A2	REI see Table A2	
4. **Roofs** a. any part forming an escape route	30	30	30	REI 30	from underside (4)
b. any roof that performs the function of a floor	see Table A2	see Table A2	see Table A2	REI see Table A2	
5. **External walls** a. any part less than 1000 mm from any point on the relevant boundary	see Table A2	see Table A2	see Table A2	REI see Table A2	each side separately
b. any part 1000 mm or more from the relevant boundary (5)	see Table A2	see Table A2	15	REI see Table A2 (10)	from inside the building
c. any part adjacent to an external escape route (see Section 6, Diagram 22)	30	30	no provision (6)(7)	RE 30	from inside the building
6. **Compartment wall** Separating occupancies (see 9.20f)	60 or see Table A2 (whichever is less)	60 or see Table A2 (whichever is less)	60 or see Table A2 (whichever is less)	REI 60 or see Table A2 (whichever is less)	each side separately

(Contd)

Table A1 (*Contd*).

Part of building	Minimum provisions when tested to the relevant part of BS 476(1) (minutes)			Minimum provisions when tested to the relevant European standard (minutes) (12)	Method of exposure
	Loadbearing capacity (2)	Integrity	Insulation		
7. **Compartment walls** (other than in item 6)	see Table A2	see Table A2	see Table A2	REI see Table A2	each side separately
8. **Protected shafts**, excluding any firefighting shaft a. any glazing described in Section 9, Diagram 30	not applicable	30	no provision (7)	E 30	
b. any other part between the shaft and a protected lobby/corridor described in Diagram 30 above	30	30	30	REI 30	each side separately
c. any part not described in (a) or (b) above	see Table A2	see Table A2	see Table A2	REI see Table A2	
9. **Enclosure** (which does not form part of a compartment wall or a protected shaft) to a: a. protected stairway	30	30	30 (8)	REI 30 (8)	each side separately
b. lift shaft	30	30	30	REI 30	
10. **Firefighting shafts** a. construction separating firefighting shaft from rest of building;	120	120	120	REI 120	from side remote from shaft
	60	60	60	REI 60	from shaft side
b. construction separating firefighting stair, firefighting lift shaft and firefighting lobby	60	60	60	REI 60	each side separately
11. **Enclosure** (which is not a compartment wall or described in item 8) to a: a. protected lobby	30	30	30 (8)	REI 30 (8)	each side separately
b. protected corridor	30	30	30 (8)	REI 30 (8)	

(*Contd*)

Table A1 (*Contd*).

Part of building	Minimum provisions when tested to the relevant part of BS 476(1) (minutes)			Minimum provisions when tested to the relevant European standard (minutes) (12)	Method of exposure
	Loadbearing capacity (2)	Integrity	Insulation		
12. **Sub-division of a corridor**	30	30	30 (8)	REI 30 (8)	each side separately
13. **Wall separating** an attached or integral garage from a dwellinghouse	30	30	30 (8)	REI 30 (8)	from garage side
14. **Enclosure** in a flat or maisonette to a protected entrance hall, or to a protected landing	30	30	30 (8)	REI 30 (8)	each side separately
15. **Fire-resisting construction** a. in dwellings not described elsewhere	30	30	30 (8)	REI 30 (8)	each side separately
b. enclosing places of special fire hazard (see 9.12)	30	30	30	REI 30	
c. between store rooms and sales area in shops (see 6.54)	30	30	30	REI 30	
d. fire-resisting subdivision described in Section 10, Diagram 34(b)	30	30	30	REI 30	
16. **Cavity barrier**	not applicable	30	15	EI 30 (11)	each side separately
17. **Ceiling** described in Section 10, Diagram 33 or Diagram 35	not applicable	30	30	EI 30	from underside
18. **Duct** described in paragraph 10.14e	not applicable	30	no provision	E 30	from outside
19. **Casing** around a drainage system described in Section 11, Diagram 38	not applicable	30	no provision	E 30	from outside

(*Contd*)

Table A1   (*Contd*).

Part of building	Minimum provisions when tested to the relevant part of BS 476(1) (minutes)			Minimum provisions when tested to the relevant European standard (minutes) (12)	Method of exposure
	Loadbearing capacity (2)	Integrity	Insulation		
20. **Flue walls** described in Section 11, Diagram 39	not applicable	half the period specified in Table A2 for the compartment wall/floor	half the period specified in Table A2 for the compartment wall/floor	EI half the period specified in Table A2 for the compartment wall/floor	from outside
21. **Construction** described in Note (a) to paragraph 15.9	not applicable	30	30	EI 30	from underside
22. **Fire doors**		see Table B1		see Table B1	

Notes:

1. Part 21 for loadbearing elements, Part 22 for non-loadbearing elements, Part 23 for fire-protecting suspended ceilings, and Part 24 for ventilation ducts. BS 476: Part 8 results are acceptable for items tested or assessed before 1st January 1988.
2. Applies to loadbearing elements only (see B3.ii and Appendix E).
3. Guidance on increasing the fire resistance of existing timber floors is given in BRE Digest 208 increasing the fire resistance of existing timber floors (BRE 1988).
4. A suspended ceiling should only be relied on to contribute to the fire resistance of the floor if the ceiling meets the appropriate provisions given in Table A3.
5. The guidance in Section 14 allows such walls to contain areas which need not be fire-resisting (unprotected areas).
6. Unless needed as part of a wall in item 5a or 5b.
7. Except for any limitations on glazed elements given in Table A4.
8. See Table A4 for permitted extent of uninsulated glazed elements.
9. For the purposes of meeting the Building Regulations floors under item 3a will be deemed to have satisfied the provisions above, provided that they achieve loadbearing capacity of at least 30 minutes and integrity and insulation requirements of at least 15 minutes when tested in accordance with the relevant European test.
10. For the purposes of meeting the Building Regulations external walls under item 5b will be deemed to have satisfied the provisions above, provided that they achieve the loadbearing capacity and integrity requirements as defined in Table A2 and an insulation requirement of at least 15 minutes.
11. For the purposes of meeting the Building Regulations cavity barriers will be deemed to have satisfied the provisions above, provided that they achieve an integrity requirement of at least 30 minutes and an insulation requirement of at least 15 minutes.
12. The National classifications do not automatically equate with the equivalent classifications in the European column, therefore products cannot typically assume a European class unless they have been tested accordingly.
'R' is the European classification of the resistance to fire performance in respect of loadbearing capacity;
'E' is the European classification of the resistance to fire performance in respect of integrity; and
'I' is the European classification of the resistance to fire performance in respect of insulation.

The following points should also be taken into account when using Table A2.

- Any element of structure should have fire resistance at least equal to the fire resistance of any element which it carries, supports or to which it gives stability. This principle may be varied where:
  (a) the supporting structure is in the open air and would be unlikely to be affected by a fire in the building, or

AD B, Appendix A

Table A2 Minimum periods of fire resistance.

Purpose group of building	Minimum periods (minutes) for elements of structure in a:					
	Basement storey ($) including floor over		**Ground or upper storey**			
of a lowest basement	**Depth (m) of a lowest basement**		**Height (m) of top floor above ground, in a building or separated part of a building**			
	more than 10	**not more than 10**	**not more than 5**	**not more than 18**	**not more than 30**	**more than 30**

1. Residential (domestic):
  a. flats and maisonettes
  b. and c. dwellinghouses
	90
	not relevant
		60
		30*
			30*
			30*
				60**†
				60@
					90**
					not relevant
						120**
						not relevant

2. Residential:
  a. institutional œ
  b. other residential
	90
	90
		60
		60
			30*
			30*
				60
				60
					90
					90
						120#
						120#

3. Office:
  – not sprinklered
 	
  – sprinklered (2)
	90
	
	60
		60
		
		60
			30*
			
			30*
				60
				
				30*
					90
					
					60
						not permitted
						120#

4. Shop and commercial:
  – not sprinklered
 	
  – sprinklered (2)
	90
	
	60
		60
		
		60
			60
			
			30*
				60
				
				60
					90
					
					60
						not permitted
						120#

5. Assembly and recreation:
  – not sprinklered
 	
  – sprinklered (2)
	90
	
	60
		60
		
		60
			60
			
			30*
				60
				
				60
					90
					
					60
						not permitted
						120#

6. Industrial:
  – not sprinklered
 	
  – sprinklered (2)
	120
	
	90
		90
		
		60
			60
			
			30*
				90
				
				60
					120
					
					90
						not permitted
						120#

7. Storage and other non-residential:
  a. any building or part not described elsewhere:
  – not sprinklered
 	
  – sprinklered (2)
	120
	
	90
		90
		
		60
			60
			
			30*
				90
				
				60
					120
					
					90
						not permitted
						120#

b. car park for light vehicles:
  i. open sided car park (3)
  ii. any other car park
	not applicable
	90
		not applicable
		60
			15* + (4)
			30*
				15* + (4)
				60
					15* + (4)
					90
						60
						120#

(Contd)

Table A2   (*Contd*).

Single storey buildings are subject to the periods under the heading 'not more than 5'. If they have basements, the basement storeys are subject to the period appropriate to their depth.

Modifications referred to in Table A2: (for application of the table notes in text above)

$ The floor over a basement (or if there is more than 1 basement, the floor over the topmost basement) should meet the provisions for the ground and upper storeys if that period is higher.

*   Increased to a minimum of 60 minutes for compartment walls separating buildings.

**   Reduced to 30 minutes for any floor within a maisonette, but not if the floor contributes to the support of the building.

œ   Multi-storey hospitals designed in accordance with the NHS Firecode document should have a minimum 60 minutes standard.

Reduced to 90 minutes for elements not forming part of the structural frame.

+   Increased to 30 minutes for elements protecting the means of escape.

†   Refer to p. 7.149 below regarding the acceptability of 30 minutes in flat conversions.

@   30 minutes in the case of three storey dwellinghouses, increased to 60 minutes minimum for compartment walls separating buildings.

Notes:
1. Refer to Table A1 for the specific provisions of test.
2. 'Sprinklered' means that the building is fitted throughout with an automatic sprinkler system meeting the relevant recommendations of BS 5306 Fire extinguishing installations and equipment on premises. Part 2 Specification for sprinkler systems; ie the relevant occupancy rating together with the additional requirements for life safety.
3. The car park should comply with the relevant provisions in the guidance on requirement B3, Section 12.
4. For the purposes of meeting the Building Regulations, the following types of steel elements are deemed to have satisfied the minimum period of fire resistance of 15 minutes when tested to the European test method;
    (i)   Beams supporting concrete floors, maximum Hp/A = 230m-1 operating under full design load.
    (ii) Free standing columns, maximum Hp/A = 180m-1 operating under full design load.
    (iii) Wind bracing and struts, maximum HP/A = 210m-1 operating under full design load.
Guidance is also available in BS 5950 Structural use of steelwork in building. Part 8 Code of practice for fire resistant design.

    (b)   where a roof top plant room needs a higher standard of fire resistance than the structure supporting it, or

    (c)   the supporting and supported structures are in different compartments (the separating element between the compartments would have to have the higher standard of fire resistance; see next item).

- If an element of structure forms part of more than one building or compartment, and is thus subject to two or more different fire resistances, it is the greater of these which applies.

- A structural frame, beam, column or load-bearing wall of a *single-storey building* (or which is part of the ground storey of a building that consists of a ground storey and one or more basement storeys) is generally not required to have fire resistance. (This reflects the view that, given satisfactory means of escape, and the restricted use of combustible materials as wall and ceiling linings, fire resistance in the elements of structure in the ground storey will contribute little to the safety of the occupants.) However, the above concession will only apply if the element of structure:

    (a)   which is part of or supports an external wall, is sufficiently far from its relevant boundary to be regarded as a totally unprotected area;

    (b)   is not part of and does not support a compartment wall or a wall which is common to two or more buildings;

(c)   is not a wall between a house and an attached or integral garage; or

(d)   does not support a gallery.

- Single-storey buildings should comply with the fire resistance periods under the heading 'not more than 5'. Where they have basements, these should, of course, comply with the recommendations for basement storeys depending on their depth below ground level.
- Further fire resistance provisions relating to the following elements may be found in the relevant sections below:

   (a)   compartment walls, external walls and the wall between a dwelling-house and a domestic garage;

   (b)   walls which enclose a firefighting shaft or protect a means of escape;

   (c)   compartment floors.

7.21.2   Meeting the performance recommendations

Reference has been made above to the BS 476 tests, where the fire resistance period of a material, product, structure or system may be assessed. It should be realised that the aim of the standard fire tests is to measure or assess the response of the sample to one or more aspects of fire behaviour under standardised conditions. The tests cannot normally measure fire hazard and represent only one aspect of the total fire safety package.

In a real fire the conditions will not be standard and there is always the possibility of premature failure of a particular component due to faulty design or workmanship. Therefore, the periods stated in Table A2 should be used for guidance and not taken as 'cast in tablets of stone', there being little correlation between the period derived from the standard test and the performance in a real fire. They do however, enable different systems to be compared under similar circumstances and are useful in this sense.

Therefore, in order that a material, product or structure may be used in a building it should:

- be in accordance with a specification or design which has been shown by test to be capable of meeting a performance standard referred to in a relevant British or European Standard or European Technical Approval (British Standards may continue to be used for products or materials where European standards or approvals are not yet available). For this purpose, laboratories accredited by the United Kingdom Accreditation Service (UKAS) for conducting the relevant tests would be expected to have the necessary expertise; or
- be assessed from test evidence against appropriate standards, or by using relevant design guides, as meeting the relevant performance (suitably qualified fire safety engineers and laboratories accredited by UKAS, might be expected to have the necessary expertise to carry out the assessment. Additionally, any body notified to the UK Government by the Government of another member state of the European Union as capable of assessing materials and products against the relevant British Standards, may also be expected to have the necessary expertise); or

- comply with an appropriate specification given in relevant tables of notional performance included in AD B. (Over the years since the 1976 Regulations, there has been a tendency to reduce the practical guidance given on forms of construction which will satisfy the fire resistance requirements. The 1992 edition of AD B contains no such examples, the reader is merely referred to the publication listed in the next point); or
- for fire-resisting elements,
  - (a) conform with an appropriate specification from Part II of the BRE Report *Guidelines for the construction of fire-resisting structural elements* (BR 128, BRE, 1988)
  - (b) be designed in accordance with a relevant British Standard or Eurocode.

Where test evidence is used to substantiate a fire resistance rating of a construction, care should be taken to check that it demonstrates compliance which is adequate and applicable to the intended use. For example, small differences in detail (e.g. fixing method, joint details, dimensional variations etc.) may significantly affect the rating.

In order to provide some assistance to readers, Table 7.7 gives notional periods of fire resistance for some common floor and wall constructions. It is based on a selection of constructions from Table A3 of DOE's 1985 Approved Document B2/3/4, which in turn is based on the BRE Report. The BRE Report mentioned above also contains much information on fire protection to structural frameworks of beams and columns. Additionally, there are available various mineral based insulating boards, sprayed coatings and intumescent paint systems which are capable of providing differing degrees of fire protection depending on their thickness and method of fixing. Reference should be made to individual manufacturers or their trade associations for further details.

Information on tests on fire-resisting elements is also given in such publications as:

- Association for Specialist Fire Protection/Steel Construction Institute/Fire Test Study Group *Fire protection for structural steel in buildings*, second edition (revised, 1992 and available from the ASFP, Association House, 99 West Street, Farnham, Surrey GU9 7EN and the Steel Construction Institute, Silwood Park, Ascot, Berks SL5 7QN);
- PD 6520 *Guide to fire test methods for building materials and elements of construction* (available from the British Standards Institution);
- BS 6336 *Guide to development and presentation of fire tests and their use in hazard assessment.*

For the first two items in Table 7.7 it should be noted that the upper floor of a two-storey dwelling is regarded as a special case. Such a floor, when tested for fire resistance from the underside is required only to provide

- stability for 30 minutes
- integrity for 15 minutes
- insulation for 15 minutes.

Table 7.7   Notional periods of fire resistance of some common constructions.

These constructions are a selection from Table A3 of DOE's 1985 Approved Document B2/3/4.
  A large number of constructions other than those shown are capable of providing the fire resistance looked for. For example, various mineral based insulating boards can be used. Because their performance varies and it dependent on their thickness, it is not possible to give specific thicknesses in this table. However, manufacturers will normally be able to say what thickness would be needed to achieve the particular performance.

Floors: timber joist

Modified 30 minutes (stability 30 minutes) (integrity 15 minutes) (insulation 15 minutes)	**1**	any structurally suitable flooring: floor joists at least 37 mm wide ceiling: (a) 12.5 mm plasterboard[a] with joints taped and filled and backed by timber, or (b) 9.5 mm plasterboard[a] with 10 mm lightweight gypsum plaster finish
	2	at least 15 mm t&g boarding or sheets of plywood or wood chipboard, floor joists at least 37 mm wide ceiling: (a) 12.5 mm plasterboard[a] with joints taped and filled, or (b) 9.5 mm plasterboard[a] with at least 5 mm neat gypsum plaster finish
30 minutes	**3**	at least 15 mm t&g boarding or sheets of plywood or wood chipboard, floor joists at least 37 mm wide ceiling: 12.5 mm plasterboard[a] with at least 5 mm neat gypsum plaster finish
	4	at least 21 mm t&g boarding or sheets of plywood or wood chipboard, floor joists at least 37 mm wide ceiling: 12.5 mm plasterboard[a] with joints taped and filled
60 minutes	**5**	at least 15 mm t&g plywood or wood chipboard, floor joists at least 50 mm wide ceiling: not less than 30 mm plasterboard[a] with joints staggered and exposed joints taped and filled

Floors: concrete

60 minutes	**6**	reinforced concrete floor not less than 95 mm thick, with not less than 20 mm cover on the lowest reinforcement

(Contd)

Table 7.7 (*Contd*).

Walls: internal			
30 minutes load-bearing	7		framing members at least 44 mm wide[b] and spaced at not more than 600 mm apart, with lining (both sides) of 12.5 mm plasterboard[a] with all joints taped and filled
	8		100 mm reinforced concrete wall[c] with minimum cover to reinforcement of 25 mm
60 minutes load-bearing	9		framing members at least 44 mm wide[b] and spaced at not more than 600 mm apart, with lining (both sides) at least 25 mm plasterboard[a] in 2 layers with joints staggered and exposed joints taped and filled
	10		solid masonry wall (with or without plaster finish) at least 90 mm thick (75 mm if non-load-bearing) *Note:* for masonry cavity walls, the fire resistance may be taken as that for a single wall of the same construction, whichever leaf is exposed to fire
	11		120 mm reinforced concrete wall[c] with at least 25 mm cover to the reinforcement

Walls: external			
Modified 30 minutes (stability 30 minutes) (integrity 30 minutes) (insulation 15 minutes) load-bearing wall 1 m or more from relevant boundary	12		any external weathering system with at least 8 mm plywood sheathing, framing members at least 37 mm wide and spaced not more than 600 mm apart internal lining: 12.5 mm plasterboard[a] with at least 10 mm lightweight gypsum plaster finish
30 minutes load-bearing wall less than 1 m from the relevant boundary	13		100 mm brickwork or blockwork external face (with, or without, a plywood backing); framing members at least 37 mm wide and spaced not more than 600 mm apart internal lining: 12.5 mm plasterboard[a] with at least 10 mm lightweight gypsum plaster finish

(Contd)

Table 7.7 (*Contd*).

60 minutes load-bearing wall less than 1 m from the relevant boundary	**14**	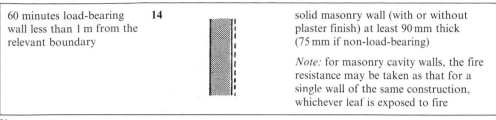	solid masonry wall (with or without plaster finish) at least 90 mm thick (75 mm if non-load-bearing) *Note:* for masonry cavity walls, the fire resistance may be taken as that for a single wall of the same construction, whichever leaf is exposed to fire

Notes

[a] Whatever the lining material, it is important to use a method of fixing that the manufacturer says would be needed to achieve the particular performance. For example, if the lining is plasterboard, the fixings should be at 150 mm centres as follows (where two layers are being used each should be fixed separately):
9.5 mm thickness, use 30 mm galvanised nails
12.5 mm thickness, use 40 mm galvaniscd nails
19 mm–25 mm thickness, use 60 mm galvanised nails.
[b] Thinner framing members, such as 37 mm, may be suitable depending on the loading conditions.
[c] A thinner wall may be suitable depending on the density of the concrete and the amount of reinforcement. (See *Guidelines for the construction of fire-resisting structural elements* (BRE, 1988).)

This is termed 'modified 30 minutes' fire resistance in the BRE Report (see Table A1, Appendix A, item 3, 'Floors').

7.21.3   Compartmentation

In order to prevent the rapid spread of fire within buildings (which could trap the occupants) and to restrict the size of any fires which do occur, AD B3 contains provisions for subdividing a building into compartments separated from one another by fire-resisting walls and floors.

The extent of the subdivisions will depend on the same factors which were considered for fire resistance, namely:

- the severity of the fire
- the height of the top floor above ground
- the building occupancy
- the presence of basements
- the number of floors.

These items are considered more fully in section 7.21 above. Additionally, the provision of a sprinkler system can affect the growth rate of a fire and may suppress it altogether.

The subdivision is achieved by means of compartment walls and compartment floors and since these come under the definition of elements of structure they should be fire resisting (but not, of course, the lowest floor in the building).

Most multi-storey buildings should be compartmented because of the increased risk to life should a fire occur. However, in the case of a two storey building in Purpose Group 4 – Shop and Commercial, or Purpose Group 6 – Industrial, the ground storey may be treated as a single storey building for the purposes of fire compartmentation if the use of the upper storey is ancillary to the use of the ground storey. This concession is dependent on the following conditions:

- the upper storey floor area should not exceed the lesser of 20% of the ground storey area or 500 m^2;
- the upper storey should be compartmented from the ground storey; and
- there should be an independent means of escape from the upper storey which is separated from the ground storey escape routes.

In single storey buildings, where the risk to life is obviously less than in multi-storey buildings, compartmentation is recommended only for:

- single-storey hospitals with a floor area limit of 3000 m^2;
- schools, where the floor area limit is 800 m^2; and
- unsprinklered shops, limited to a maximum floor area of 2000 m^2.

Where atria (see section 7.7 above for definition of an atrium) are incorporated into the design of a building there are obvious implications for the integrity of any compartmentation in the building. Detailed advice on all issues relating to atria is given in BS 5588: Part 7: 1997 *Code of practice for the incorporation of atria in buildings*; however, the standard is only relevant where compartmentation is breached by the atria.

Generally, two main principles are adopted when deciding how to compartment a building:

(1) For non-residential buildings, the compartment sizes should be restricted to the maximum dimensions shown in Table 12 of AD B3, which is reproduced below, and the following points should be noted.
   - Limits are given by reference to maximum floor areas for differing top storey heights (for multi-storey buildings) for assembly and recreation, shop and commercial, and industrial buildings.
   - For storage and other non-residential buildings, the limits are according to maximum compartment volumes for differing top storey heights.
   - The installation of sprinklers allows the compartment size limits to be doubled.
   - If a building has a storey which is more than 30 m above ground or a basement more than 10 m below ground, all floors should be constructed as compartment floors, including any ground floor over a basement (except for small premises described in section 7.15.8 above).
(2) For all buildings, the following walls and floors which separate buildings into different ownerships, uses, or occupancies (or protect particular risk areas) should be constructed as compartment walls and floors:
   - Walls which are common to two or more buildings (including walls between semi-detached and terraced houses). These walls should run the full height of the building in a continuous vertical plane, the adjoining buildings being separated only by walls, not floors.
   - Walls and floors dividing parts of buildings used for different purposes. This does not apply where one use is ancillary to another (see section 7.3).

AD B3, Section 9

Table 12 Maximum dimensions of building or compartment (non-residential buildings).

Purpose Group of building or part	Height of floor of top storey above ground level (m)	Floor area of any one storey in the building or any one storey in a compartment (m²)	
		in multi-storey buildings	in single storey buildings
Office	no limit	no limit	no limit
Assembly & Recreation Shop & Commercial: a. schools	no limit	800	800
b. shops – not sprinklered	no limit	2000	2000
shops – sprinklered (1)	no limit	4000	no limit
c. elsewhere – not sprinklered	no limit	2000	no limit
elsewhere – sprinklered (1)	no limit	4000	no limit
Industrial (2) not sprinklered	not more than 18 more than 18	7000 2000 (3)	no limit no limit
sprinklered (1)	not more than 18 more than 18	14000 4000 (3)	no limit no limit
	Height of floor of top storey above ground level (m)	**Maximum compartment volume (m³)**	
		in multi-storey buildings	in single storey buildings
Storage (2) & other non-residential: a. car park for light vehicles	no limit	no limit	no limit
b. any other building or part: not sprinklered	not more than 18 more than 18	20000 4000 (3)	no limit no limit
sprinklered (1)	not more than 18 more than 18	40000 8000 (3)	no limit no limit

Notes:
1. 'Sprinklered' means that the building is fitted throughout with an automatic sprinkler system meeting the relevant recommendations of BS 5306: Part 2, ie the relevant occupancy rating together with the additional requirements for life safety.
2. There may be additional limitations on floor area and/or sprinkler provisions in certain industrial and storage uses under other legislation, for example in respect of storage of LPG and certain chemicals.
3. This reduced limit applies only to storeys that are more than 18 m above ground level. Below this height the higher limit applies.

- Walls dividing buildings into separated parts. These walls should also run the full height of the building in a continuous vertical plane, in a similar manner to common walls above (see section 7.10) and the two separated parts can have different standards of fire resistance.
- Walls and floors bounding a protected shaft. (Protected shafts are considered in more detail below.)
- Construction enclosing places of special fire hazard (see section 7.2).

Curiously, any walls or floors enclosing such places are not considered to be compartment walls and floors for the purposes of AD B3.

Finally, the following walls and floors in particular Purpose Groups should be constructed as compartment walls and floors:

- Any floor or wall separating an attached or integral garage from a dwellinghouse should be constructed as shown in Fig. 7.34.
- Any floor in an institutional or other residential building.
- Any floor in flats and maisonettes, except an internal floor in an individual dwelling.
- Any wall separating a flat or maisonette from any other part of the same building.
- Any wall enclosing a refuse storage chamber in flats and maisonettes.

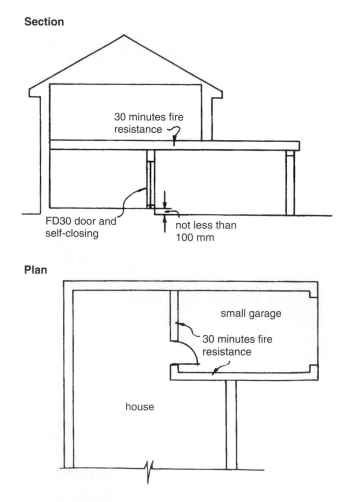

Fig. 7.34 Attached small garages.

- Any wall or floor in a shopping complex referred to in Section 5 of BS 5588: Part 10 as needing to be constructed to the compartmentation standard.
- Any wall or floor provided to divide a building into separate occupancies (i.e. parts of the building used by different organisations irrespective of whether or not they fall within the same Purpose Group) in Shop and Commercial, Industrial or Storage premises.
- Any wall between compartments in the upper storeys of health-care premises used for in-patients. The upper storeys of such buildings should be divided into at least two compartments (so that no compartments exceed $2000\,\text{m}^2$) so as to permit progressive horizontal evacuation of each compartment (see section 7.17.1 above).

7.22   Compartment walls and floors – construction details

Since the purpose of compartment walls and floors is to form a complete barrier to the passage of fire between the compartments which they separate, it follows that they should have the appropriate standards of fire resistance indicated in Tables A1 and A2 above.

Additionally, compartment walls and floors in hospitals designed on the basis of *Firecode* (see section 7.17.1 above) should be constructed of materials of limited combustibility if they have a fire resistance of 60 minutes or more. This does not apply if the building is fitted throughout with a suitable sprinkler system as detailed in *Firecode*.

Any points of weakness in compartment walls and floors should be adequately protected. These points of weakness occur:

- at junctions with other compartment walls, external walls and roofs
- where timber beams, joists, purlins and rafters are built into or pass through a compartment wall
- at openings for doors, pipes and ducts of various kinds, refuse chutes and protected shafts.

7.22.1   Junction details

Where a compartment wall and roof meet, the wall may be carried at least 375 mm above the roof covering surface, measured at right angles to the roof slope, or the wall may be taken up to meet the underside of the roof covering or deck, and the junction fire-stopped. Acceptable design solutions are illustrated in Fig. 7.35 for buildings in different purpose groups.

Generally, the covering in a 1.5 m wide zone on either side of the wall should be designated AA, AB or AC (see below for designations) and it should be carried on a substrate or deck consisting of a material of limited combustibility. The roof covering and deck could be of a composite structure such as profiled steel cladding. Where double skinned insulated roof sheeting is specified it should incorporate a band of material of limited combustibility at least 300 mm wide centred over the wall.

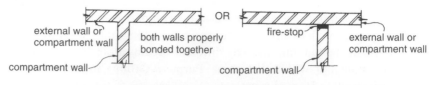

junction of external and compartment walls

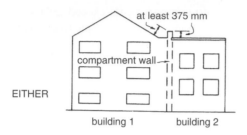

EITHER

OR   **(a) Building or compartment of any use or height**

- roof covering to be AA, AB or AC on a substrate or deck of material of limited combustibility
- roof covering and deck could be of a composite structure such as profiled steel cladding
- double skinned insulated roof sheeting should incorporate a band of material of limited combustibility at least 300 mm wide centred over the wall.

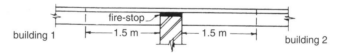

(b) Dwellinghouse, office, assembly and recreation, or residential (not institutional) nor more than 15 m high

roof covering: AA, AB or AC

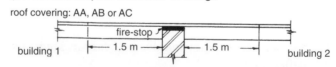

see below for materials which may be carried over or through compartment wall

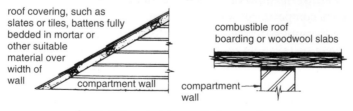

Combustible materials permitted over wall

Fig. 7.35   Junction details – compartment walls.

Any roof support members which pass through the wall may need to have fire protection on their undersides for at least 1.5 m on either side of the wall to delay distortion at the junction.

Some exceptions to the need for the substrate or deck to be in materials of limited combustibility are permitted in the case of certain roof constructions in buildings not more than 15 m high in dwellinghouses, offices, assembly and recreation buildings, and residential buildings (but not institutional). In the roof shown at (b) in Fig. 7.35, the following combustible materials may be carried over the top of the compartment wall:

- roof boarding serving as a base for roof covering; or
- woodwool slabs; or
- timber slating or tiling battens
- sarking felt.

The materials should be fully bedded in mortar or similar material.

Where a compartment wall or floor meets another compartment wall or an external wall, the junction should maintain the fire resistance of the compartmentation. This will normally mean that the various structures should be bonded together or the junction fire-stopped (see Fig. 7.35). The compartment wall should also be continued across any eaves cavity.

It is permissible for timber beams, joists, purlins and rafters to be built into or carried through a concrete or masonry compartment wall provided that the openings for them are kept as small as is practicable and are then fire-stopped.

If trussed rafters bridge a compartment wall they should be designed so that failure of any part of the truss caused by fire in one compartment does not lead to failure of any part of the truss in another compartment.

7.22.2 Openings in compartment walls and floors

Generally, the only openings permitted in compartment walls and floors are one or more of the following.

- An opening fitted with a door which has the appropriate fire resistance given in Table B1 of Appendix B to AD B and is fitted in accordance with Appendix B of AD B (see Fire doors below).
- An opening for a protected shaft (see Protected shafts in section 7.22.4).
- An opening for a refuse chute of non-combustible construction.
- An opening for a pipe, ventilation duct, chimney, appliance ventilation duct or duct encasing one or more flue pipes, provided it complies with the relevant parts of Section 11 of AD B (see Pipes, ventilation ducts and flues in section 7.24).
- Atria designed in accordance with BS 5588: Part 7.

In the case of compartment walls which are common to two or more buildings or which separate different occupancies in the same building, it would not be sensible or necessary to allow all of the above openings to exist. Therefore such walls should be imperforate except for:

- an opening for a door which is needed as a means of escape in case of fire. The door should have the same fire resistance as the wall and should be fixed in accordance with the provisions of Appendix B to AD B (see below); or
- an opening for a pipe complying with the provisions of Section 11 of AD B (see section 7.24).

7.22.3   Fire doors

All fire doors should be fitted with an automatic self-closing device unless they are to cupboards or service ducts which are normally kept locked shut. Rising butt hinges are not considered as automatic self-closing devices unless the door is:

- to or within a dwelling
- in a cavity barrier
- between a dwellinghouse and a garage.

As a general rule, no device should be provided to hold a door open. However, in some cases a self-closing device may be considered a hindrance to normal use. In such cases a fire-resisting door may be held open by:

- a fusible link device, provided that the door is not fitted in an opening used as a means of escape (this does not apply where two doors are provided in the opening as mentioned below);
- a door closure delay device; or
- an automatic release mechanism, actuated by an automatic fire detection and alarm system.

In this context an automatic release mechanism is one which automatically closes a door in the event of each or any one of:

- smoke detection by appropriate apparatus suitable in nature, quality and location
- manual operation by a suitably located switch
- failure of the electricity supply to the device, smoke apparatus, or switch
- operation of a fire alarm system, if fitted.

All fire-resisting doors should have the appropriate fire performance described in Table Bl of Appendix B (see below), either:

- in accordance with a rating given in terms of their performance under test to BS 476: Part 22. This rating relates to the ability of the door to maintain its integrity for a specified period in minutes, e.g. FD30. Doors should be tested from each side separately; however, since lift doors are only at risk from one side in the event of a fire, it is only necessary to test these from the landing side. The rating for some doors has the suffix S added where restricted smoke leakage is needed at ambient temperatures. The leakage rate should not exceed $3\,m^3/m/hour$ (head and jambs only) when tested at 25 Pa to BS 476: Section 31.1, unless pressurisation techniques complying with BS 5588: Part 4 are used; or

- as determined with reference to Commission Decision 2000/367/EC of 3 May 2000 implementing Council Directive 89/106/EEC as regards the classification of the resistance to fire performance of construction products, construction works and parts thereof. All such fire doors should be classified in accordance with BS EN 13501 *Fire classification of construction products and building elements,* Part 2: xxxx: *Classification using data from fire resistance tests (excluding products for use in ventilation systems).* They are tested to the relevant European method from the following:
  - (a) BS EN 1634 *Fire resistance tests for door and shutter assemblies:* Part 1: 2000 *Fire doors and shutters;*
  - (b) BS EN 1634 *Fire resistance tests for door and shutter assemblies:* Part 2: xxxx *Fire door hardware;*
  - (c) BS EN 1634 *Fire resistance tests for door and shutter assemblies:* Part 3: xxxx *Smoke control doors.*

The performance requirement is in terms of integrity (E) for a period of minutes. An additional classification of Sa is used for all doors where restricted smoke leakage at ambient temperatures is needed. The designation of xxxx is used for standards that are not yet published. The latest version of any standard may be used provided that it continues to address the relevant requirements of the Regulations. Until such time that the relevant harmonised product standards are published, for the purposes of meeting the Building Regulations, products tested in accordance with BS EN 1634: Part 1 (with or without pre-fire test mechanical conditioning) will be deemed to have satisfied the provisions provided that they achieve the minimum fire resistance in terms of integrity, as detailed in Table B1.

No fire door should be hung on hinges made of a material with a melting point less than 800°C, unless the hinges can be shown to be satisfactory when tested as part of a fire door assembly.

Although each fire door should have the appropriate period of fire resistance defined in Table Bl, it is permissible for two fire doors to be fitted in an opening if each door is capable of closing the opening and the required level of fire resistance can be achieved by the two doors together. However, if these two doors are fitted in an opening provided for a means of escape, both doors should be self-closing. One of them may be held open by a fusible link and be fitted with an automatic self-closing device if the other is easily openable by hand and has at least 30 minutes fire resistance.

Doors which are to be kept closed or locked when not in use, or which are held open by an automatic release mechanism, should be marked with the appropriate fire safety sign in accordance with BS 5499 *Fire safety signs, notices and graphic symbols,* Part 1: 1990 *Specification for fire safety signs.* The signs should be on both

AD B, Appendix B

Table B1 Provisions for fire doors.

Position of door	Minimum fire resistance of door in terms of integrity (minutes) when tested to BS 476 part 22(1)	Minimum fire resistance of door in terms of integrity (minutes) when tested to the relevant European standard
1. In a compartment wall separating buildings	As for the wall in which the door is fitted, but a minimum of 60	As for the wall in which the door is fitted, but a minimum of 60
2. In a compartment wall: a. if it separates a flat or maisonette from a space in common use;	FD 30S (2)	E30 S. (3)
b. enclosing a protected shaft forming a stairway situated wholly or partly above the adjoining ground in a building used for Flats, Other Residential, Assembly and Recreation, or Office purposes;	FD 30S (2)	E30 S. (3)
c. enclosing a protected shaft forming a stairway not described in (b) above;	Half the period of fire resistance of the wall in which it is fitted, but 30 minimum and with suffix S (2)	Half the period of fire resistance of the wall in which it is fitted, but 30 minimum and with suffix S. (3)
d. enclosing a protected shaft forming a lift or service shaft;	Half the period of fire resistance of the wall in which it is fitted, but 30 minimum	Half the period of fire resistance of the wall in which it is fitted, but 30 minimum
e. not described in (a), (b), (c) or (d) above	As for the wall it is fitted in, but add S (2) if the door is used for progressive horizontal evacuation under the guidance to B1	As for the wall it is fitted in, but add S. (3) if the door is used for progressive horizontal evacuation under the guidance to B1
3. In a compartment floor	As for the floor in which it is fitted	As for the floor in which it is fitted
4. Forming part of the enclosures of: a. a protected stairway (except where described in item 9); or	FD 30S (2)	E20 S. (3)
b. a lift shaft (see paragraph 6.42b); which does not form a protected shaft in 2(b), (c) or (d) above	FD 30	E30
5. Forming part of the enclosure of: a. a protected lobby approach (or protected corridor) to a stairway;	FD 30S (2)	E30 S. (3)
b. any other protected corridor; or	FD 20 (S)	E20 S. (3)
c. a protected lobby approach to a lift shaft (see paragraph 6.42)	FD 30S (2)	E30 S. (3)
6. Affording access to an external escape route	FD 30	E30

(Contd)

Table B1   (*Contd*).

Position of door	Minimum fire resistance of door in terms of integrity (minutes) when tested to BS 476 part 22(1)	Minimum fire resistance of door in terms of integrity (minutes) when tested to the relevant European standard
7. Sub-dividing: a. corridors connecting alternative exits;	FD 20S (2)	E20 S. (3)
b. dead-end portions of corridors from the remainder of the corridor	FD 20S (2)	E20 S. (3)
8. Any door: a. within a cavity barrier;	FD 30	E30
b. between a dwellinghouse and a garage	FD 30	E20
9. Any door: a. forming part of the enclosures to a protected stairway in a single family dwellinghouse;	FD 20	E20
b. forming part of the enclosure to a protected entrance hall or protected landing in a flat or maisonette;	FD 20	E20
c. within any other fire-resisting construction in a dwelling not described elsewhere in this table	FD 20	E20

Notes:
1. To BS 476: Part 22 (for BS 476: Part 8 subject to paragraph 5 in Appendix A).
2. Unless pressurization techniques complying with BS 5588: Part 4 *Fire precautions in the design, construction and use of buildings, Code of practice for smoke control using pressure differentials* are used, these doors should also either:
   (a) have a leakage rate not exceeding $3\,m^3/m/hour$ (head and jambs only) when tested at 25 Pa under BS 476 *Fire tests on building materials and structures*, Section 31.1 *Methods for measuring smoke penetration through doorsets and shutter assemblies, Method of measurement under ambient temperature conditions*; or
   (b) meet the additional classification requirement of S. when tested to BS EN 1634-3:xxxx, *Fire resistance tests for door and shutter assemblies, Part 3 – Smoke control doors*.
3. The National classifications do not automatically equate with the equivalent classifications in the European column, therefore products cannot typically assume a European class unless they have been tested accordingly.

sides of the fire doors, except for cupboards and service ducts where it is only necessary to mark the doors on the outside. This recommendation does not apply to:

- fire doors within dwellinghouses
- fire doors to and within flats or maisonettes
- bedroom doors in other residential buildings (PG 2(b))
- lift entrance/landing doors.

Normal fire doors do not provide any significant amount of insulation. Therefore it is necessary to limit the proportion of doorway openings in compartment walls to 25% of the length of the wall, unless the doors provide both integrity and insulation (see Appendix A, Table A2).

Further sources of guidance

The following additional sources of guidance to fire doors and ironmongery are referred to in AD B and may be found useful.

- Recommendations for the specification, design, construction, installation and maintenance of fire doors constructed with non-metallic leaves may be found in BS 8214: *Code of practice for fire door assemblies with non-metallic leaves*, 1990.
- Guidance on timber fire-resisting doorsets, in relation to the new European test standard, may be found in *Timber Fire-Resisting Doorsets: maintaining performance under the new European test standard* published by TRADA.
- Guidance for metal doors is given in *Code of practice for fire-resisting metal doorsets* published by the DSMA (Door and Shutter Manufacturers' Association) in 1999.
- Ironmongery used on fire doors can significantly affect their performance in a fire. Further guidance is available in *Hardware for timber and escape doors* published by the Builders Hardware Industry Federation, 2000.

Rolling shutters

Rolling shutters are sometimes used to protect compartments and means of escape. They are usually held open by automatic release mechanisms (see above). Where rolling shutters are provided across a means of escape, they should only be released by a heat sensor (such as a fusible link or electric heat detector) situated in the immediate vicinity of the door. Closure initiated by smoke detectors or a fire alarm system should not be considered unless it is also the intention that the shutter will descend partially to form a boundary to a smoke reservoir.

All rolling shutters should be capable of manual operation for firefighting purposes.

Glazing to fire doors

It is often desirable to provide glazed vision panels in fire doors. Where the glazing can satisfy the relevant insulation criteria from Table A1 of Appendix A, there are no limitations on its use in fire doors. Where this is not the case, the uninsulated glazing should comply with the recommendations of Table A4 of Appendix A, which is reproduced below. Table A4 also contains details of the use of uninsulated glazing in protected stairways, lobbies and corridors. This is described more fully under 'Protected shafts' below.

7.22.4   Protected shafts

Protected shafts are needed when it is necessary to pass persons, things or air between compartments. Therefore, they should only be used to accommodate:

AD B, Appendix A

Table A4   Limitations on the use of uninsulated glazed elements on escape routes. (These limitations do not apply to glazed elements which satisfy the relevant insulation criterion, see Table A1) (See BS 5588: Part 7 for glazing to atria; see BS 5588: Part 8 for glazing to refuges)

Position of glazed element	Maximum total glazed area in parts of a building with access to:			
	a single stairway		more than one stairway	
	walls	door leaf	walls	door leaf
Single family dwellinghouses 1. a. Within the enclosures of: i. a protected stairway, or within fire-resisting separation shown in Fig. 7.13; or	fixed fanlights only	unlimited	fixed fanlights only	unlimited
ii. an existing stair (see section 7.12.10)	unlimited	unlimited	unlimited	unlimited

b. Within fire-resisting separation:
    i. shown in Fig. 7.14; or
    ii. described in section 7.12.6
	unlimited above 100 mm from floor	unlimited above 100 mm from floor	unlimited above 100 mm from floor	unlimited above 100 mm from floor

c. Existing window between an attached/integral garage and the house	unlimited	not applicable	unlimited	not applicable
Flats and maisonettes 2. Within the enclosures of a protected entrance hall or protected landing or within fire-resisting separation shown in Fig. 7.17	fixed fanlights only	unlimited above 1100 mm from floor	fixed fanlights only	unlimited above 1100 mm from floor
General (except dwellinghouses) 3. Between residential/sleeping accommodation and a common escape route (corridor, lobby or stair)	nil	nil	nil	nil

4. Between a protected stairway (1) and:
   a. the accommodation; or
   b. a corridor which is not a protected corridor. Other than in item 3 above
	nil	25% of door area	unlimited above 1100 mm (2)	50% of door area

5. Between:
   a. a protected stairway (1) and a protected lobby or protected corridor; or
   b. accommodation and a protected lobby. Other than in item 3 above
	unlimited above 1100 mm from floor	unlimited above 100 mm from floor	unlimited above 100 mm from floor	unlimited above 100 mm from floor

(Contd)

Table A4   (*Contd*).

Position of glazed element	Maximum total glazed area in parts of a building with access to:			
	a single stairway		more than one stairway	
	walls	door leaf	walls	door leaf
6. Between the accommodation and a protected corridor forming a dead end. Other than in item 3 above	unlimited above 1100 mm from floor	unlimited above 100 mm from floor	unlimited above 1100 mm from floor	unlimited above 100 mm from floor
7. Between accommodation and any other corridor; or subdividing corridors. Other than in item 3 above	not applicable	not applicable	unlimited above 100 mm from floor	unlimited above 100 mm from floor
8. Adjacent an external escape route described in section 7.15.10	unlimited above 1100 mm from paving	unlimited above 1100 mm from paving	unlimited above 1100 mm from paving	unlimited above 1100 mm from paving
9. Adjacent an external escape stair (see section 7.15.10 and Fig. 7.23) or roof escape (see section 7.15.4)	unlimited	unlimited	unlimited	unlimited

Notes:
1. If the protected stairway is also a protected shaft (see section 7.22.4) or a firefighting stair (see section 7.30.2) there may be further restrictions on the uses of glazed elements.
2. Measured vertically from the landing floor level or the stair pitch line.
3. The 100 mm limit is intended to reduce the risk of fire spread from a floor covering.
4. Items 1c, 3 and 6 apply also to single storey buildings.

- stairs, lifts and escalators
- pipes, ducts or chutes
- sanitary accommodation and/or washrooms.

They should form a complete barrier between the different compartments which they connect and, except for glazed screens which should meet the recommendations referred to below, should have fire resistance as specified in Table A1 of Appendix A (see above).

A protected shaft containing a stairway is often approached by way of a corridor or lobby. It is sometimes desirable to glaze the wall between the shaft and the corridor or lobby in order to allow light and visibility in both directions.

This glazing is permitted provided that it has at least 30 minutes fire resistance in terms of integrity and the following conditions are met:

- the stair enclosure is not required to have more than 60 minutes fire resistance
- the corridor or lobby has at least 30 minutes fire separation from the rest of the floor (including doors)
- the protected shaft is not a firefighting shaft.

These recommendations are illustrated in Fig. 7.36. Where these provisions cannot be met, the guidance shown in Table A4 of Appendix A relating to the limits on areas of uninsulated glazing will apply. There should be no oil pipe or ventilating

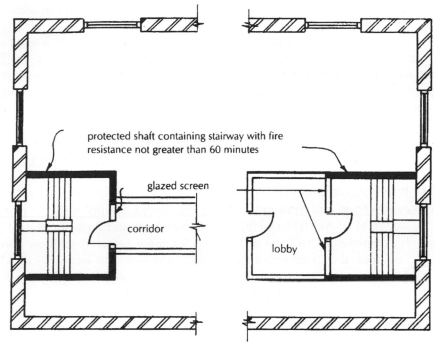

glazed screen, walls to corridor or lobby and doors to have at least 30 minutes fire
resistance

Fig. 7.36 Glazed screen separating protected shaft from lobby or corridor.

duct within any protected shaft which contains any stairway and/or lift (although
pipes which convey oil for hydraulic lift mechanisms and ducts used in pressurisation systems aimed at keeping stairs smoke-free are permitted).

Where a protected shaft contains a pipe carrying natural gas or LPG, the pipe
should be of screwed steel or of all welded steel construction in accordance with the
Pipelines Safety Regulations 1996, S.I. 1996 No. 825, and Gas Safety (Installation &
Use) Regulations 1998, SI 1998 No 2451. Where a pipe is completely separated from
a protected shaft by fire-resisting construction it is not considered to be contained
within a protected shaft. The shaft should be adequately ventilated direct to external
air by ventilation openings at high and low level in the shaft and any extension of the
floor of the storey into the shaft should not compromise the free movement of air
over the entire length of the shaft. Guidance on shafts, such as those which convey
piped flammable gas, may be found in BS 8313: 1989 *Code of practice for accommodation of building services in ducts.*

Ideally, protected shafts should be imperforate except for certain openings
mentioned below. The number of openings permitted will depend, to a great extent,
on the function of the wall surrounding the protected shaft.

Generally, external walls to protected shafts do not need to be fire-resisting and
hence there are no restrictions on the number of openings in such walls. This would
not be so if the external wall was part of a firefighting shaft; in this case, reference
should be made to BS 5588: Part 5: 1991 *Code of practice for firefighting stairs and*

lifts. See also *Protection of external walls of protected stairways* in section 7.15.8 above.

Where part of the enclosure to a protected shaft consists of a wall which is common to two or more buildings, the only openings permitted are those referred to in section 7.22.2 above.

Any other walls which make up the enclosure should have no openings other than those referred to below.

- A protected shaft containing one or more lifts may have openings to allow lift cables to pass to the lift motor room. If this is at the bottom of the shaft, the openings should be kept as small as possible.
- Where a protected shaft contains or is itself a ventilating duct, any inlets to, outlets from and openings for the duct should comply with the guidance in section 11 of AD B3 (see Pipes, ventilation ducts and flues in section 7.24).
- It is permissible to form an opening for pipes (other than those specifically forbidden above), provided that the pipe complies with section 11 of AD B3.
- Any opening, other than those detailed above, should be fitted with a fire door complying with Table B1 of Appendix B of AD B.

Figure 7.37 illustrates the principles of protected shafts.

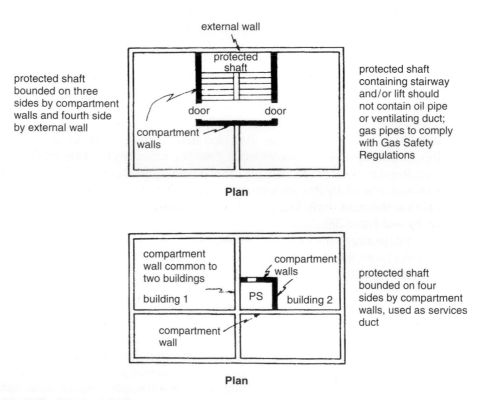

Fig. 7.37   Protected shafts.

7.23 Concealed spaces

Many buildings constructed today contain large hidden void spaces within floors, walls and roofs. This is particularly true of system-built housing, schools and other local authority buildings such as old people's homes.

These buildings may also contain combustible wall panels, frames and insulation, thereby increasing the risk of unseen smoke and flame spread through these concealed spaces.

Therefore, despite compartmentation and the use of fire-resistant construction, many buildings have been destroyed as a result of fire spreading through cavities formed by, or in, constructional elements, and by-passing compartment walls/ floors, etc.

Section 10 of AD B contains provisions designed to reduce the chance of hidden fire spread by making sure that the edges of openings are closed and that cavities are:

- interrupted if there is a chance that the cavity could form a route around a barrier to fire (such as a compartment wall or floor)
- subdivided if they are very large.

This interruption or sub-division of concealed spaces is achieved by using *cavity barriers*. These are defined in Appendix E of AD B as any form of construction (other than a smoke curtain) which is intended to close a cavity (concealed space) and prevent the penetration of smoke or flame, or is fitted inside a cavity in order to restrict the movement of smoke or flame within the cavity. Therefore, provided that it meets the requirements for cavity barriers, a form of construction designed for some other use (such as a compartment wall) may be acceptable as a cavity barrier.

7.23.1 Interruption of cavities

Where an element which is required to form a barrier to fire abuts another element containing a cavity, there is a risk that smoke and flame could by-pass the fire barrier via the cavity. Cavity barriers should, therefore, be provided to interrupt the cavity at the point of contact between the fire barrier and the element containing the cavity (see Fig. 7.38).

The degree to which cavity barriers should be provided will depend on the use (or Purpose Group) of the building concerned and is set out in Table 13 of AD B3. These provisions are summarised below.

Cavities in some elements are excluded from the above provisions regarding interruption, mainly on the grounds that they present little risk to unseen fire spread. These include any cavity:

- in an external wall of masonry construction which complies with the guidance shown in Fig. 7.39 below; and
- in a wall which requires fire resistance only because it is load-bearing (therefore it does not form a barrier to fire).

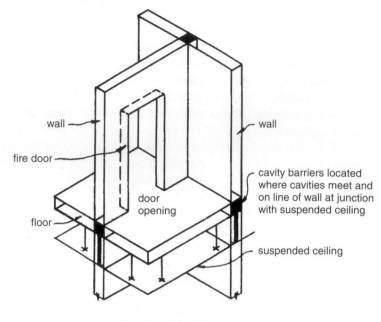

Closing of cavities

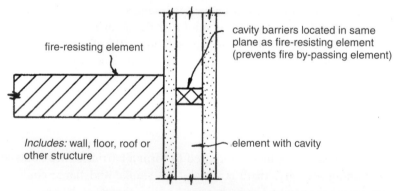

Excludes: wall required to have fire resistance only because it is load-bearing

Interrupting cavities

Fig. 7.38   Cavity barriers.

7.23.2   Cavity subdivision

Any cavity, including a roof space, should generally be subdivided into separate sections by cavity barriers placed across the cavity at intervals not greater than the distances specified in Table 14 to section 10 of AD B3, which is reproduced below. These distances are measured along the members bounding the cavity and depend on the cavity location and the class of surface exposed within it (see Fig. 7.40).

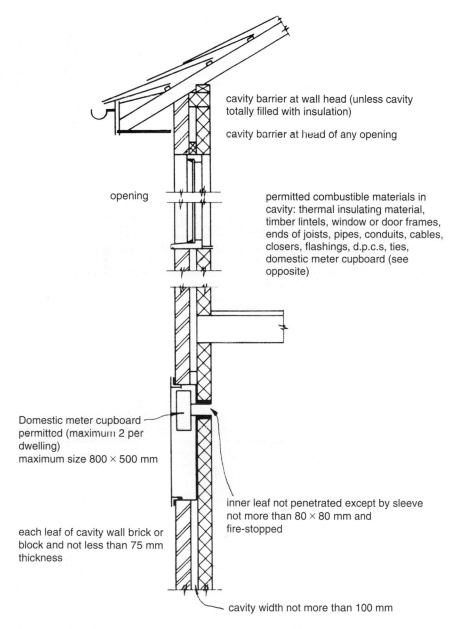

cavity barrier at wall head (unless cavity totally filled with insulation)

cavity barrier at head of any opening

opening

permitted combustible materials in cavity: thermal insulating material, timber lintels, window or door frames, ends of joists, pipes, conduits, cables, closers, flashings, d.p.c.s, ties, domestic meter cupboard (see opposite)

Domestic meter cupboard permitted (maximum 2 per dwelling) maximum size 800 × 500 mm

inner leaf not penetrated except by sleeve not more than 80 × 80 mm and fire-stopped

each leaf of cavity wall brick or block and not less than 75 mm thickness

cavity width not more than 100 mm

Fig. 7.39   Wall cavity exempt from section 10 of AD B3, Table 13.

Section 10 of AD B3 permits certain variations to the dimensions given in Table 14 as follows.

- If a room under a ceiling cavity exceeds the dimensions given in Table 14, then cavity barriers need only be placed on the line of the enclosing walls or partitions of that room, subject to a maximum cavity barrier spacing of 40 m, provided that

AD B3, Section 10

Table 14   Maximum dimensions of cavities in non-domestic buildings (Purpose Groups 2–7).

Locations of cavity	Class of surface/product exposed in cavity (excluding the surface of any pipe, cable or conduit, or any insulation to any pipe)		Maximum dimensions in any direction (m)
	National class	European class	
Between roof and a ceiling	Any	Any	20
Any other cavity	Class 0 or Class 1	Class A1 or Class A2-s3, d2 or Class B-s3, d2 or Class C-s3, d1	20
	Not Class 0 or Class 1	Not any of the above classes	10

Notes:
1. Exceptions to these provisions are given in paragraphs 10.11–10.13 of AD B3 and are summarised in section 7.23.2.
2. The National classifications do not automatically equate with the equivalent classifications in the European column, therefore products cannot typically assume a European class unless they have been tested accordingly.
3. When a classification includes 's3, d2', this means that there is no limit set for smoke production and/or flaming droplets/particles.

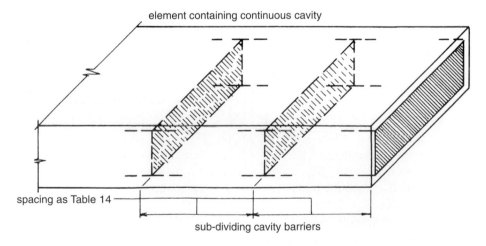

This requirement does not apply to:

- certain cavity walls in masonry construction (see Fig. 7.39)
- floors next to ground or oversite concrete if no access or not more than 1 m high space and no possibility of rubbish accumulating below floor
- cavity between non-combustible roof sheeting under certain conditions
- cavity in an underfloor service void

Fig. 7.40   Subdivision of cavities.

the surfaces exposed in the cavity are Class 0 or Class 1 (National class) or Class C-s3, d2 or better (European class).

- Cavities over undivided areas which exceed the 40 m limit mentioned above in both directions need not be divided with cavity barriers if the following conditions can be met:
  (a) both room and cavity are compartmented from the rest of the building;
  (b) an automatic fire detection and alarm system to BS 5839: Part 1: 1988 is fitted in the building (although detectors are not required in the cavity);
  (c) if the cavity is used as a plenum for ventilating and air-conditioning ductwork the recommendations about recirculating air distribution systems in BS 5588: Part 9, should be followed;
  (d) the ceiling surface exposed in the cavity is Class 0 (National class) or Class C-s3, d2 or better (European class) and any supports or fixings for the ceiling are non-combustible;
  (e) any pipe insulation system should have a Class 1 flame spread rating;
  (f) electrical wiring in the void should be laid in metal trays or metal conduit;
  (g) any other materials in the cavity should be of limited combustibility.

Additionally, the following low-risk cavities are excluded from the provisions of Table 14 (as well as those specified above for cavity interruption):

- under a floor next to the ground or oversite concrete, provided that either:
  (a) the height of the cavity is not more than 1 m, or
  (b) there is no access to the cavity for persons;

This exclusion does not apply if it is possible for combustible material to accumulate in the cavity through openings in the floor (such as happened at the Bradford City Football Club fire). In this case the cavity should be accessible for cleaning and should contain cavity barriers as described in Table 14.

- between double-skinned corrugated or profiled insulated roof sheeting consisting of materials of limited combustibility, provided that the sheets are separated by insulating material having surfaces of Class 0 or Class 1(National class) or Class C-s3, d2 or better (European class), and that insulating material is in contact with both the inner and outer liner sheets. It should be noted that when a classification includes 's3, d2', this means that there is no limit set for smoke production and/or flaming droplets/particles;
- within an underfloor service void;
- within a floor or roof, or enclosed by a roof, provided the lower side of that cavity is enclosed by a ceiling which:
  (a) extends throughout the whole building or compartment;
  (b) is not designed to be demountable;
  (c) has at least 30 minutes fire resistance;
  (d) is imperforate except for any openings permitted in a cavity barrier (see below);

(e)   has an upper surface facing the cavity of at least Class 1 surface spread of flame;

(f)   has a lower surface of Class 0 (National class) or Class C-s3, d2 or better (European class).

Cavities above such fire-resisting ceilings are subject to an overall limit of 30 m in extent (see Fig. 7.41).

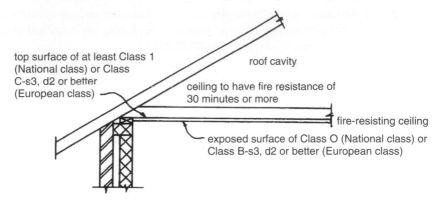

top surface of at least Class 1 (National class) or Class C-s3, d2 or better (European class)

roof cavity

ceiling to have fire resistance of 30 minutes or more

fire-resisting ceiling

exposed surface of Class O (National class) or Class B-s3, d2 or better (European class)

(i)   ceiling should extend throughout building or compartment
(ii) ceiling should be imperforate except for any allowable openings (see text)
(iii) ceiling should not be demountable

Fig. 7.41   Fire-resisting ceilings.

7.23.3   Provision of cavity barriers

Bearing in mind the exclusions already mentioned above, cavity barriers should be provided as indicated below:

In all buildings

(a)   At the junction between a compartment wall which separates buildings and an external cavity wall. The top of the external wall in this case should also be closed with cavity barriers.

(b)   At the edges of cavities (including around openings). Such cavity barriers may be formed by a window or door frame if this is appropriate.

In a dwellinghouse with a storey which is more than 4.5 m above ground level

● Above the enclosure to a protected stairway unless a fire-resisting ceiling is provided as shown in Figs. 7.41 and 7.42.

In flats and maisonettes

● At the junction between every compartment wall and compartment floor and an external cavity wall.

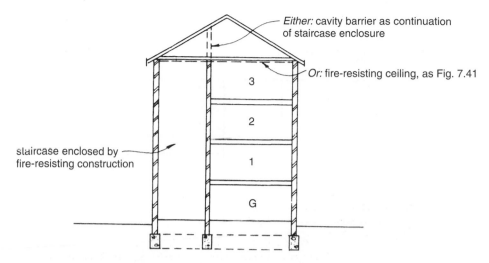

Fig. 7.42   Provision of cavity barriers in dwellinghouses of three or more storeys.

- At the junction between a cavity wall and every compartment wall, compartment floor or other wall or door assembly which forms a fire-resisting barrier.
- Above fire-resisting construction in a protected escape route if this is not carried up to the full storey height (or to the underside of the roof, if it is in the top storey).

In office, shop and commercial, assembly and recreation, industrial, storage and other non-residential buildings

- All as recommended for flats and maisonettes above.
- Above corridor enclosures where the corridor is subdivided to prevent smoke or fire affecting two escape routes simultaneously. The cavity barriers are not needed if the corridor enclosure is carried up to the full storey height (or to the underside of the roof, if it is in the top storey) or if the cavity is enclosed on the lower side by a fire-resisting ceiling complying with Fig. 7.41. (See also Fig.7.43.)
- To subdivide extensive cavities in floors, roofs, walls, etc. (but not in underfloor service voids) so that the distance between cavity barriers does not exceed the dimensions given in Table 14 (see above).

In other residential and institutional buildings:

- All as recommended for Office, etc., buildings above.
- Above any bedroom partitions if these are not carried up to the full storey height (or to the underside of the roof, if they are in the top storey).

Rain screen cladding

Rain screen cladding is a form of construction often used in refurbishment works to improve the weather resistance and thermal performance of the external walls of

To prevent both storey exits from becoming blocked by smoke:

EITHER

(a) sub-divide storey with fire-resisting construction

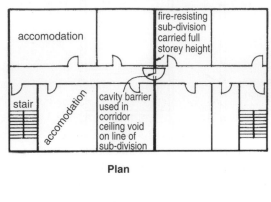

Plan

OR

(b) provide cavity barriers above corridor enclosure if enclosures are not carried up to full storey height

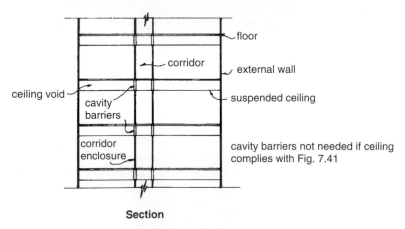

Section

Fig. 7.43   Alternative forms of corridor enclosure.

buildings. It has been used regularly to upgrade the performance of blocks of high-rise flats (often built in the 1960s), where an external weatherproof cladding is lined internally with combustible insulation and this is fixed to the outer surface of the external walls. There have been occasions when fire has occurred within the void space containing the insulation and fire has spread upwards within the cladding, eventually affecting the interior of the building. Table 13 of AD B3 recommends that cavity barriers be provided within the void behind the external face of rain screen cladding at every floor level, and on the line of compartment walls abutting the external wall, where the building has any floor which is more than 18 m above ground level. This applies to Flats, Maisonettes, Other residential buildings and Institutional buildings. The spacing recommendations of Table 14 do not apply to such rain screen cladding, or to the over-cladding of an external masonry (or concrete) external wall, or to an over-clad concrete roof where the cavity does not contain combustible insulation and the cavity barriers are positioned as described above.

7.23.4   Construction of cavity barriers

A cavity barrier may be formed by construction provided for another purpose provided that it meets the recommendations for cavity barriers. However, where compartment walls are provided these should be taken up the full storey height to a compartment floor or roof and not completed by cavity barriers above them. This is because the fire resistance standards for compartment walls are higher than those for cavity barriers, and compartment walls should therefore be continued through the cavity to maintain the fire resistance standard.

Table A1 of Appendix A recommends that all cavity barriers should have a minimum standard of fire resistance of 30 minutes with regard to integrity and 15 minutes with regard to insulation. The only exception to this is in the case of a cavity barrier in a stud wall or partition which is permitted to be formed of:

- steel at least 0.5 mm thick
- timber at least 38 mm thick
- polythene sleeved mineral wool, or mineral wool slabs, in either case under compression when installed in the cavity
- calcium silicate, cement based or gypsum based boards at least 12 m thick.

Cavity barriers should be tightly fitted against rigid construction and mechanically fixed in position where possible. Where they abut against slates, tiles, corrugated sheeting and similar non-rigid construction, the junctions should be fire-stopped as described below.

Cavity barriers should also be fixed in such a way that their performance is unlikely to be affected by:

- building movements due to shrinkage, subsidence, thermal change, or the movement of the external envelope due to wind; or
- collapse in a fire of any services which penetrate them; or
- failure of their fixings, or any construction or material which they abut, due to fire. However, where cavity barriers are provided in roof spaces, the roof members to which they are fitted are not expected to have fire resistance.

7.23.5   Openings in cavity barriers

Cavity barriers should be imperforate except for one or more of the following openings:

- for a pipe which complies with Section 11 of AD B3
- for a cable, or for a conduit containing one or more cables

- if fitted with a suitably mounted fire damper
- for a duct (unless it is fire-resisting), fitted with an automatic fire damper where it passes through the barrier
- if fitted with a door which complies with Appendix B of AD B, and having at least 30 minutes fire resistance.

The above provisions do not apply to any cavity barrier provided above any bedroom partitions which are not carried up to the full storey height (or to the underside of the roof, if they are in the top storey).

7.24   Pipes, ventilation ducts and flues

It is impossible to construct a building without passing some pipes or ducts through the walls and floors, and such penetration of the fire separating elements of structure is a potential source of flame and smoke spread. Section 11 of AD B3 therefore attempts to control the specifications of such pipes and ducts and of their associated enclosing structures. The measures in section 11 are primarily designed to delay the passage of fire. They may also have the added benefit of retarding smoke spread, but the integrity test specified in Appendix A of AD B does not cover criteria for the passage of smoke as such.

7.24.1   Pipes

For the purposes of section 11, the term 'pipe' includes a ventilating pipe for an above-ground drainage system, but does not include any flue pipe or other form of ventilating pipe.

As is usual, the expression 'pipe' here may be read as 'pipeline', and should be taken to include all pipe fittings and accessories.

Requirements

Where a pipe as defined passes through an opening in:

- a compartment wall or compartment floor (unless the pipe is wholly enclosed within a protected shaft); or
- a cavity barrier;

then either a proprietary sealing system should be used which will maintain the fire resistance of the floor, wall or cavity barrier (and has been shown by test to do so) or, the nominal internal diameter of the pipe should not exceed the relevant dimension listed in Table 15 of section 11 of AD B3 (see below). The opening should be as small as practicable and fire-stopped around the pipe.

Where a pipe of specification (b) of Table 15 penetrates a structure, it is permissible to pass it through or connect it to a pipe or sleeve of specification (a) of Table 15 provided the pipe or sleeve of specification (a) extends on both sides of the

AD B3, Section 11

Table 15 Maximum nominal internal diameter of pipes passing through a compartment wall/floor (see para 11.5 et seq)

Situation	Pipe material and maximum nominal internal diameter (mm)		
	(a) Non-combustible material (1)	(b) Lead, aluminium, aluminium alloy, uPVC (2), fibre cement	(c) Any other material
1. Structure (but not a wall separating buildings) enclosing a protected shaft which is not a stairway or a lift shaft	160	110	40
2. Wall separating dwelling houses, or compartment wall or compartment floor between flats	160	160 (stack pipe) (3) 110 (branch pipe) (3)	40
3. Any other situation	160	40	40

Notes:
1. Any non-combustible material (such as cast iron, copper or steel) which if exposed to a temperature of 800°C, will not soften or fracture to the extent that flame or hot gas will pass through the wall of the pipe.
2. uPVC pipes complying with BS 4514 and uPVC pipes complying with BS 5255.
3. These diameters are only in relation to pipes forming part of an above-ground drainage system and enclosed as shown in Diagram 38. In other cases the maximum diameters against situation 3 apply.

structure for a minimum distance of 1m. The sleeve should be in contact with the pipe. (See Fig. 7.44.)

The following above-ground drainage system pipes complying with specification (b) of Table 15 may be passed through openings in a wall which separates houses, or through openings in a compartment wall or compartment floor in flats.

- A stack pipe of not more than 160 mm nominal internal diameter, provided it is contained within an enclosure in each storey; and
- a branch pipe of not more than 110 mm nominal internal diameter, provided it discharges into a stack pipe which is contained in an enclosure, the enclosure being partly formed by the wall penetrated by the branch pipe.

The enclosures referred to in both cases immediately above should comply with the following requirements.

- In any storey, the enclosure should extend from floor to ceiling or from floor to floor if the ceiling is suspended.
- Each side of the enclosure should be formed by a compartment wall or floor, external wall, intermediate floor or casing.
- The internal surface of the enclosure should meet the requirements of Class 0 (National class) or Class C-s3, d2 or better (European class), except for any supporting members.

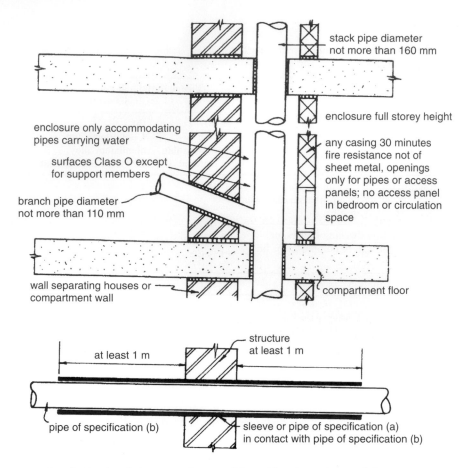

stack pipe diameter
not more than 160 mm

enclosure full storey height

enclosure only accommodating
pipes carrying water

any casing 30 minutes
fire resistance not of
sheet metal, openings
only for pipes or access
panels; no access panel
in bedroom or circulation
space

surfaces Class O except
for support members

branch pipe diameter
not more than 110 mm

wall separating houses or
compartment wall

compartment floor

structure
at least 1 m

at least 1 m

pipe of specification (b)

sleeve or pipe of specification (a)
in contact with pipe of specification (b)

openings in structure to be as small as possible and fire-stopped around pipes – see
Table 15 of AD B3 for pipe material specifications

Fig. 7.44 Penetration of structure by pipes.

- No access panel to the enclosure should be fitted in any bedroom or circulation space.
- The enclosure should not be used for any purpose except to accommodate drainage or water supply pipes.

The 'casing' referred to in section 11, Diagram 34 and the second requirement immediately above, should provide at least 30 minutes fire resistance, including any access panel, and it should not be formed of sheet metal. The only openings permitted in a casing are openings for the passage of a pipe, or openings fitted with an access panel. The pipe opening, whether it be in the structure or the casing, should be as small as is practicable and fire-stopped around the pipe (see Fig. 7.44). A casing to the drainage or water supply pipes should always be provided if a wall which separates houses is penetrated by a branch pipe in the top storey.

7.24.2   Ventilating ducts

Ventilating ducts, normally forming part of an air-conditioning system, convey air to various parts of a building. It is, therefore, inevitable that they will need to pass through compartment walls and floors at some stage and it is important that the integrity of these fire separating elements is maintained. This can be achieved by following the guidance in Part 9 of BS 5588 where alternative ways of protecting compartmentation are described when air handling ducts pass from one compartment to another.

7.24.3   Flues

Where any flue, appliance ventilation duct or duct containing one or more flues:

- passes through a compartment wall or floor, or
- is built into a compartment wall,

the flue or duct walls should be separated from the compartment wall or floor by non-combustible construction of fire resistance equal to at least half that required for the compartment wall or floor (see Fig. 7.45).

   For the purposes of the above, an appliance ventilation duct is a duct provided to convey combustion air to a gas appliance.

7.24.4   Fire stops

A fire stop is defined in Appendix E of AD B as a seal provided to close an imperfection of fit or design tolerance between elements or components, to restrict the passage of fire and smoke.

   Therefore, fire stops should be provided:

- at junctions or joints between elements which are required to act as a barrier to fire; and
- where pipes, ducts, conduits or cables pass through openings in cavity barriers (see section 7.23.5 above), or elements which serve as a barrier to fire.

In the second case above, the openings should be kept as small and as few in number as possible, and the fire-stopping should not restrict the thermal movement of pipes or ducts.

   Fire-stopping materials should be reinforced with or supported by materials of limited combustibility to prevent displacement if:

- the unsupported span exceeds 100 mm; and
- in any other case, non-rigid materials have been used (unless these have been shown by test to be satisfactory).

Suitable fire-stopping materials include:

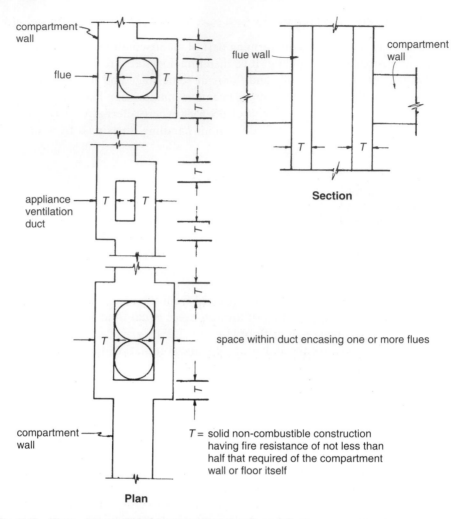

Section

space within duct encasing one or more flues

T = solid non-combustible construction
having fire resistance of not less than
half that required of the compartment
wall or floor itself

Plan

Fig. 7.45 Flues, etc., contained in compartment walls and floors.

- cement mortar
- gypsum based plaster
- cement or gypsum based vermiculite/perlite mixes
- glassfibre, crushed rock, blast furnace slag or ceramic based products (with or without resin binders)
- intumescent mastics
- any proprietary fire-stopping or sealing systems (including those designed for penetration by services) capable of maintaining the fire resistance of the element concerned (test results would be necessary to prove acceptability).

These materials should be used in appropriate situations, i.e. they may not all be suitable in every situation.

7.25   Variations to the provisions of Approved Document B3

As has been mentioned in the introduction to this chapter (see section 7.1 above), some difficulty may be encountered when trying to apply the provisions of AD B to existing buildings. Accordingly, AD B3 contains a number of specific recommendations related to raised storage areas, floors in domestic loft conversions and to the conversion of buildings to flats, where the 'normal' provisions are somewhat reduced.

7.25.1   Varying the provisions – raised storage areas

Sometimes, raised free standing floors (usually called mezzanine floors) are erected (often in single-storey industrial buildings) for the purposes of storage. They may be regarded merely as galleries or may be large enough to be considered as a floor forming an additional storey. In such cases, the normal recommendations regarding the fire resistance of elements of structure may prove to be unduly onerous if applied to raised storage areas.

It may be possible to reduce the level of fire resistance or even allow unprotected steelwork if the following precautions are taken.

(1) The structure should be used for storage purposes only and should contain only one tier.
(2) The number of people using the floor at any time should be limited. Members of the public should not be admitted.
(3) The floor is open both above and below to the space in which it is located.
(4) Means of escape is provided from the floor which meets the recommendations of Sections 4, 5 and 6 of AD B1.
(5) The floor should comply with the following parameters regarding its size:
   - it should not be more than 10 m in length and/or width and should not be greater than half the floor area of the space in which it is located;
   - the floor size may be increased to not more than 20 m in length and width where an automatic fire detection and alarm system is provided on the lower level, which complies with BS 5839: Part 1;
   - there are no limits on the size of the floor if the building is fitted throughout with an automatic sprinkler system which meets the relevant recommendations of BS 5306: Part2. (This relates to the relevant occupancy rating together with the additional requirements for life safety).

7.25.2   Varying the provisions – dwellinghouses

Under the recommendations for means of escape in case of fire (AD B1) certain special provisions apply where it is proposed to construct one or two rooms in the roof space of a two-storey dwelling thereby creating a three-storey dwelling (see section 7.12.10 above).

Floors in an existing two-storey dwelling may only be capable of achieving a modified 30 minutes fire resistance (see section 7.21.2 (Table 7.7) above). However,

AD B3 recommends that floors in a dwelling of three or more storeys should have a full 30 minutes fire resistance.

It is considered reasonable to relax the recommendation for the *existing* floor (thereby allowing the modified 30 minutes standard) provided the following provisions are complied with by way of compensation.

- Only one storey is being added with a floor area not exceeding 50 m^2.
- No more than two habitable rooms are provided in the new storey.
- The existing floor should only separate rooms (not circulation spaces).
- The means of escape provisions relevant to loft conversions in AD Bl (see section 7.12.10 above) should be complied with.

The relaxed recommendation will only apply to any floor which separates rooms (i.e. not circulation spaces). Therefore, the full 30 minute standard will need to be provided where the floor forms part of the enclosures to the circulation space between the loft conversion and the final exit.

It is sometimes the case that a floor is only capable of achieving a modified 30 minutes standard of fire resistance because it is constructed with plain edged boarding on the upper surface. The addition of a 3.2 mm thickness of standard hardboard nailed to the floor boards can usually upgrade the floor to the full 30 minutes standard (see Fig. 7.46).

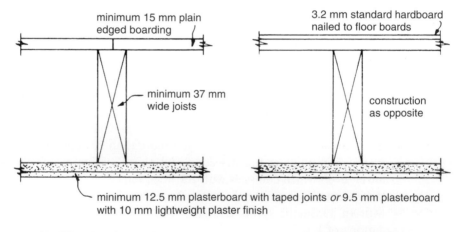

Modified 30 minutes floor   **Full 30 minutes floor**

Fig. 7.46   Upgrading of existing floors in dwellings.

Other methods of upgrading existing timber floors can be found in *BRE Digest* 208.

7.25.3   Varying the provisions – conversion to flats

If it is proposed to convert a building into flats, Approved Document B of Schedule 1 of the 2000 Regulations will apply due to the material change of use.

It is often the case that the existing floors are of timber construction and have insufficient fire resistance for the proposed change of use.

The provision of an adequate, fully protected means of escape which complies with the recommendations of Section 3 of AD B1 will allow a 30 minute standard of fire resistance in the elements of structure in a building of not more than three storeys.

The full standard of fire resistance given in Table A2 of Appendix A would normally be required if the converted building contained four or more storeys.

7.26   External fire spread

The external walls of a building are required to adequately resist the spread of fire over their surfaces and from one building to another. In assessing the adequacy of resistance to fire spread, regard must be given to the height, use and position of the building.

The roof of a building must also offer adequate resistance to the spread of fire across its surface and from one building to another, having regard to the use and position of the building.

7.27   External walls

External walls serve to restrict the outward spread of fire to a building beyond the property boundary and also help resist fire from outside the building. This is achieved by ensuring that the walls have adequate fire resistance and external surfaces with restricted fire spread and low rates of heat release. Fire spread between buildings usually occurs by radiation through openings in the external walls (termed 'unprotected areas'). The risk of fire spread and its consequences are related to:

- the severity of the fire
- the fire resistance offered by the facing external walls including the number and disposition of the unprotected areas
- the distance between the buildings
- the risk presented to people in the opposite building.

In general, the severity of a fire will be related to the amount of combustible material contained in the building per unit of floor area (termed the 'fire load density'). Certain types of buildings, such as shops, industrial buildings and warehouses, may contain large quantities of combustible materials and are usually required to be sited further from their boundaries than other types of buildings.

7.27.1   External walls – general constructional recommendations

External walls are elements of structure and therefore they should have the relevant period of fire resistance specified in Appendix A of AD B. However, the provisions for space separation mentioned below allow increasingly large areas of the external walls of a building to be unprotected as the distance to the relevant boundary increases. A point will eventually be reached where the whole of a wall may be unprotected. In such a case only the load-bearing parts of the wall would need fire resistance. Similarly, where a wall is 1 m or more from the relevant boundary it only needs to resist fire from the inside and the insulation criteria of fire resistance are, in most cases, not applied.

The combustibility of the external envelope of a building is also controlled in certain circumstances. The limiting factors are the height of the building, its use and its distance from the relevant boundary. Table 7.8 sets out the recommendations for the external surfaces of walls and it is generally the case that buildings which are less than 1 m from the relevant boundary should have external surfaces of Class 0 (National class) or class B-s3, d2 or better (European class). For buildings which are 18 m or more in height, there are restrictions on the external surface materials irrespective of the distance to the boundary.

After the disastrous fire at the Summerland Leisure complex on the Isle of Man, special recommendations were introduced to prevent other assembly and recreation buildings from suffering a similar fate. The Summerland centre was constructed largely of plastics materials which extended to ground level. A fire was deliberately started adjacent to the building which, because of its rapid surface spread of flame characteristics, quickly became engulfed in flames. Therefore, any Assembly and Recreation Purpose Group building which has more than one storey (galleries counted, but not basements) should have only those external surfaces indicated in Fig. 7.47 below. This also applies in mixed use buildings which include Assembly and Recreation Purpose Group accommodation.

The provisions described above for the combustibility of external wall surfaces may, of course, be affected by the recommendations for space separation and the limits on unprotected areas contained in Section 14 of AD B and described below.

Mention has already been made of the risks involved with the use of combustible insulation in rain screen cladding of buildings (see section 7.23.3). In such a system the surface of the outer cladding which faces the cavity should comply with the provisions of Table 7.8 and Fig. 7.47. Furthermore, any insulation used in ventilated cavities in the external walls of a building over 18 m in height should be composed of materials of limited combustibility, although this restriction does not apply to insulation in masonry cavity walls which comply with Fig. 7.39 above. (Reference should also be made to the BRE Report *Fire performance of external thermal insulation for walls of multi-storey buildings*, BR 135, 1988.)

7.27.2   External walls and steel portal frames

Steel portal frames are commonly used in single-storey industrial and commercial buildings. Structurally, the portal frame acts as a single member. Therefore, where the column sections are built into the external walls, collapse of the roof sections may result in destruction of the walls.

If the building is so situated that the external walls cannot be totally unprotected,

Table 7.8 Limitations on external wall surfaces (all buildings).

Maximum height of building (m)	Distance of external surface from any point on the relevant boundary[a]		
	Less than 1 m	**1 m or more**	
Up to 18	Class 0[1]	No provision	
Over 18	Class 0[1]	Any surface less than 18 m above the ground	Timber at least 9 mm thick; or any material with an index of performance (*I*) not more than 20
		Any surface 18 m or more above the ground	Class 0

Notes:
1. Either class 0 (National class) or class B-s3, d2 or better (European class).
    For meaning of class 0 and index of performance (*I*) see section 7.18.3 above.

Note: the provisions described in Table 7.8 and illustrated in Fig. 7.47 might also be met by following the guidance in BRE Fire Note 9 *Assessing the fire performance of external cladding systems: a test method* (BRE, 1999).

Assembly and recreation buildings of more than one storey

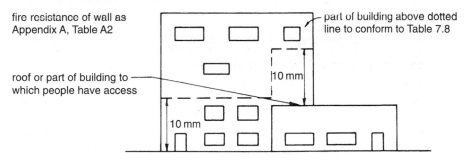

fire resistance of wall as Appendix A, Table A2

part of building above dotted line to conform to Table 7.8

roof or part of building to which people have access

10 mm

10 mm

Part of building below dotted lines:
Surfaces
(a) Class O (National Class) or class B-s3, d2 or better (European class) if less than 1 m to boundary
(b) index of performance (*I*) not more than 20, or timber cladding at least 9 mm thick, if more than 1 m from the boundary

Fig. 7.47 External walls – special provisions for assembly and recreation buildings.

the provisions of AD B may recommend that both rafter and column sections be fire protected. This would result in an uneconomic building which would defeat the object of using a portal frame.

Investigations have been carried out into the behaviour of steel portal frames in fire. Provided that the connection between the frame and its foundation can be made sufficiently rigid to transfer the over-turning moment caused by collapse, in a fire, of the rafters, purlins and some of the roof cladding, it may be possible to remove the fire protection to the rafters and purlins while still allowing the external wall to perform its structural function.

Additional measures may be necessary in certain circumstances to ensure the stability of the external walls. Full details of the design method may be found in the publication *Fire and Steel Construction: The Behaviour of Steel Portal Frames in Boundary Conditions*, 1990 (2nd edition) which is available from the Steel Construction Institute, Silwood Park, Ascot, Berks, SL5 7QN. The publication also contains guidance on many aspects of portal frames including multi-storey types. The recommendations of this publication for designing the foundation to resist overturning need not be followed if the building is fitted with a sprinkler system that follows the relevant provisions in BS 5306: Part 2 (i.e. the relevant occupancy rating together with the additional requirements for life safety).

Normally, reinforced concrete portal frames can support external walls without specific measures at the base to resist overturning.

The following design guidance (which some existing buildings may already comply with) is also acceptable.

- To resist overturning, the column members should be rigidly fixed to a base of suitable size and depth.
- Brick, block or concrete protection should be provided up to a protected ring beam giving lateral support.
- Some form of roof venting should be provided to give early heat release (e.g. pvc roof lights covering at least 10% of the floor area evenly spaced out).

7.27.3   Space separation – permitted limits of unprotected areas

Unprotected areas in the external walls of a building are those areas which have less fire resistance than that recommended by Table A2 of Appendix A of AD B. Areas such as doors, windows, ventilators or combustible cladding are permitted in the external walls but their extent is limited depending on the use of the building and its distance from the relevant boundary. In order that a reasonable standard of space separation may be specified for buildings, the following basic assumptions are made in AD B, Section 14.

- A fire in a compartmented building will be restricted to that compartment and will not spread to adjoining compartments.
- The intensity of the fire is related to the purpose group of the building and it can be moderated by a sprinkler system.
- Residential, and Assembly and Recreation Purpose Groups represent a greater risk to life than other uses.
- Where buildings are on the same site, the spread of fire between them represents a low risk to life and can be discounted. This does not apply to buildings in the Residential, and Assembly and Recreation Purpose Groups.
- There is a building on the far side of the boundary situated an equal distance away with an identical elevation to the building in question.

- The amount of thermal radiation that passes through an external wall which has fire resistance may be discounted.
- A roof which is pitched at less than 70° to the horizontal does not need to comply with the recommendations in section 14 of AD B4. (See also definition of external wall in section 7.2 above).
- Vertical parts of a pitched roof (such as dormer windows) generally do not need to comply with the recommendations in section 14 of AD B4, unless they are part of a roof which is pitched at greater than 70° to the horizontal. However, a continuous run of dormer windows occupying most of a steeply pitched roof might need to be treated as a wall rather than a roof. This will be a matter of individual judgement.

It follows from the above that reduced separation distances (or increased amounts of unprotected areas) may be obtained by dividing a building into compartments.

7.27.4 Boundaries

It is clear from AD B4 that the separation distances referred to are those to the relevant boundaries of the site of the building in question. (Relevant boundary is defined in section 7.2 above and is illustrated in Fig. 7.6). Where the site boundary adjoins an area which is unlikely to be developed, such as a street, canal, railway or river, then the relevant boundary is usually taken as the centreline of that area.

Where buildings share the same site, the separation distance between them is usually discounted. However, if either or both of the buildings are in the Residential, or Assembly and Recreation Purpose Groups, then a notional boundary is assumed to exist between them such that they both comply with the space separation recommendations of AD B4. (Notional boundary is defined in section 7.2 and is illustrated in Fig. 7.5.)

7.27.5 Unprotected areas that can be discounted

Certain openings, etc. in walls have little effect on fire protection. Accordingly, AD B4 provides that four areas may be discounted when calculating the permitted limits of unprotected areas in the external walls of a building.

- Any unprotected area of not more than 0.1 m² which is at least 1.5 m away from any other unprotected area in the same side of the building or compartment, except an area of external wall forming part of a protected shaft.
- One or more unprotected areas, with a total area of not more than 1 m², which is at least 4 m away from any other unprotected area in the same side of the building or compartment, except a small area of not more than 0.1 m² as described above.
- Any unprotected area in an external wall of a stairway in a protected shaft. (But see Fig. 7.31 above for further provisions affecting stairways.)
- An unprotected area in the side of an uncompartmented building, if the area is at least 30 m above the ground adjoining the building.

Where part of an external wall is regarded as an unprotected area merely because of combustible cladding more than 1mm thick, the unprotected area presented by that cladding is to be calculated as only half the actual cladding area (see Fig. 7.48). Any cladding with a Class 0 (National class) or class B-s3, d2 or better (European class) surface spread of flame rating need not be counted as an unprotected area.

Therefore any wall which is situated within 1 m of the relevant boundary should contain only those unprotected areas listed above, and illustrated in Fig. 7.48. The

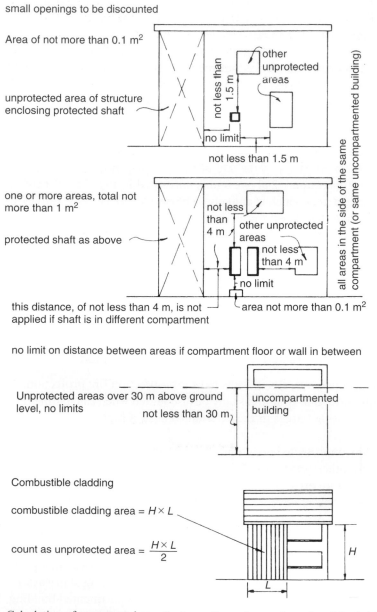

small openings to be discounted

Area of not more than 0.1 m²

unprotected area of structure enclosing protected shaft

other unprotected areas

not less than 1.5 m

no limit

not less than 1.5 m

one or more areas, total not more than 1 m²

protected shaft as above

not less than 4 m

other unprotected areas

not less than 4 m

no limit

this distance, of not less than 4 m, is not applied if shaft is in different compartment

area not more than 0.1 m²

all areas in the side of the same compartment (or same uncompartmented building)

no limit on distance between areas if compartment floor or wall in between

Unprotected areas over 30 m above ground level, no limits

not less than 30 m

uncompartmented building

Combustible cladding

combustible cladding area = $H \times L$

count as unprotected area = $\dfrac{H \times L}{2}$

H

L

Fig. 7.48 Calculation of unprotected area limit, small openings and combustible cladding.

rest of the wall will need to meet the fire resistance requirements contained in Table A2 of Appendix A of AD B.

7.27.6   Unprotected areas – methods of calculation

Where a wall is situated 1m or more from the relevant boundary, the permitted limit of unprotected areas may be determined by either of two methods described in full in AD B4. The Approved Document also permits other methods, described in a BRE Report *External fire spread: Building separation and boundary distances*, BR 187, BRE, 1991. Part 1 of this report covers the 'Enclosing Rectangle' and 'Aggregate Notional Area' methods which were originally contained in the 1985 edition of AD B2/3/4 and which are also described below. An applicant may use whichever of these methods gives the most favourable result for his own building. Again, the rest of the wall should meet the fire resistance recommendations of Table 2 of Appendix A.

The basis of the two methods described in AD B4 is contained in the BRE Report mentioned above. The building should be separated from its boundary by at least half the distance at which the total thermal radiation intensity received from all unprotected areas in the wall would be 12.6 kW/m^2 in still air, assuming that the radiation intensity at each unprotected area is:

- 84kW/m^2 for buildings in the Residential, Office or Assembly and Recreation Purpose Groups, and
- 168 kW/m^2 for buildings in the Shop and Commercial, Industrial, Storage or Other non-residential Purpose Groups.

AD B, Section 14

Diagram 46   Permitted unprotected areas in small residential buildings.

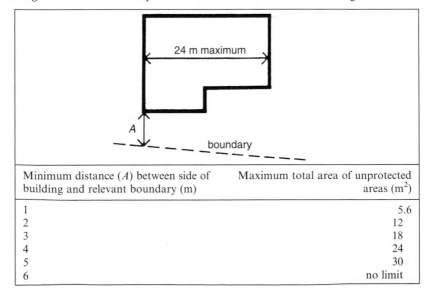

Minimum distance (A) between side of building and relevant boundary (m)	Maximum total area of unprotected areas (m^2)
1	5.6
2	12
3	18
4	24
5	30
6	no limit

This clearly illustrates the different fire load densities assumed for the two groups of buildings.

Where a sprinkler system complying with BS 5306: Part 2 is installed throughout a building, it is reasonable to assume that the extent and intensity of a fire will be reduced. In these circumstances the permitted boundary distances may be halved subject to a minimum distance of 1 m. Alternatively, if the boundary distance is kept the same, the amount of unprotected area can be doubled.

Method 1 – small residential buildings

Method 1 applies only to dwellinghouses, flats, maisonettes or other residential buildings (not institutional buildings) which:

- are not less than 1m from the relevant boundary
- are not more than three storeys high (basements not counted)
- have no side which exceeds 24 m in length.

The permitted limit of unprotected area in an external wall of any of these buildings is given in Diagram 46 of AD B4 which is reproduced above. It varies according to the size of the building and the distance of the side from the relevant boundary. Any parts of the side in excess of the maximum unprotected area should have the recommended fire resistance. The small areas referred to above may be discounted.

Method 2 – all buildings or compartments

This method applies to buildings and compartments in any purpose group which:

- are not less than 1m from the relevant boundary
- are not more than 10 m high (except open-sided car parks in Purpose Group 7(b)).

The permitted limits of unprotected areas given by this method are contained in Table 16 of AD B4 which is reproduced below. It should be noted that actual areas are not given. Column 3 of Table 16 expresses the permitted unprotected areas in percentage terms but fails to state what the percentage relates to. It is assumed that the percentages given refer to the total area of the side of the building in question. Thus, if a shop is at least 2 m from its relevant boundary then it is permitted to have 8% of its external wall area on that side as unprotected. Any other areas (except the small permitted areas) would need the requisite fire resistance.

For buildings or compartments that exceed 10m in height the methods set out in BRE Report, BR 187 can be applied. These are explained in the following paragraphs.

Additional methods – enclosing rectangles

This method of calculating the permitted limit of unprotected areas is based on the smallest rectangle of a height and width taken from Table 7.9 (which follows and is

AD B, Section 14

Table 16 Permitted unprotected areas in small buildings or compartments.

Minimum distance between side of building and relevant boundary (m)		Maximum total percentage of unprotected area %
Purpose groups		
Residential, Office, Assembly and Recreation (1)	Shop & Commercial, Industrial, Storage & other Non-residential (2)	(3)
n.a.	1	4
1	2	8
2.5	5	20
5	10	40
7.5	15	60
10	20	80
12.5	25	100

Notes:
n.a. = not applicable.
a. Intermediate values may be obtained by interpolation.
b. For buildings which are fitted throughout with an automatic sprinker system, see section 7.27.6.
c. In the case of open-sided car parks in Purpose Group 7(b), the distances set out in column (1) may be used instead of those in column (2).

based on Table J2 of the 1985 edition of AD B2/3/4), which would totally enclose all the relevant unprotected areas in the side of a building or compartment. This is referred to as the *enclosing rectangle* and is usually larger than the actual rectangle that would enclose these areas (see Fig. 7.49).

The unprotected areas are projected at right angles onto a *plane of reference* and Table 7.9 then gives the distance that the relevant boundary must be from the plane of reference according to the *unprotected percentage*, the height and width of the enclosing rectangle and the Purpose Group of the building. The figures in Table 7.9 relate to Shop and Commercial, Industrial and Storage and other non-residential buildings, whilst those in brackets relate to Residential, Office or Assembly and Recreation buildings.

The plane of reference is a vertical plane which touches some part of the outer surface of a building or compartment. It should not pass through any part of the building (except projections such as balconies or copings) and it should not cross the relevant boundary. It can be at any angle to the side of the building and in any position which is most favourable to the building designer, although it is usually best if roughly parallel to the relevant boundary. This method can be used to determine the maximum permitted unprotected areas for a given boundary position (Fig. 7.50) *or* how close to the boundary a particular design of building may be (Fig. 7.51).

It is permissible to calculate the enclosing rectangle separately for each compartment in a building. Therefore the provision of compartment walls and floors in a building can effectively reduce the enclosing rectangle thereby decreasing the

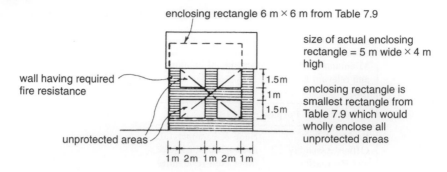

enclosing rectangle 6 m × 6 m from Table 7.9

size of actual enclosing rectangle = 5 m wide × 4 m high

wall having required fire resistance

enclosing rectangle is smallest rectangle from Table 7.9 which would wholly enclose all unprotected areas

1.5m
1m
1.5m

unprotected areas

1m 2m 1m 2m 1m

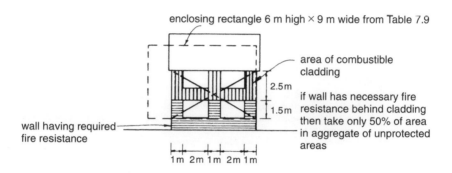

enclosing rectangle 6 m high × 9 m wide from Table 7.9

area of combustible cladding

2.5m
1.5m

wall having required fire resistance

if wall has necessary fire resistance behind cladding then take only 50% of area in aggregate of unprotected areas

1m 2m 1m 2m 1m

Note: plane of reference is taken to coincide with surface of wall

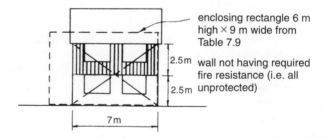

enclosing rectangle 6 m high × 9 m wide from Table 7.9

2.5m

wall not having required fire resistance (i.e. all unprotected)

2.5m

7m

Fig. 7.49   Enclosing rectangles.

distance to the boundary without affecting the amount of unprotected areas pro-vided. This technique is demonstrated in Fig. 7.52 which also shows how the enclosing rectangle method is applied in practice.

   This method is quick and easy to use in practice but in certain circumstances it may give an uneconomical result with regard to the permitted distance from the boundary and it may unduly restrict the designer's freedom in choice of window areas, etc. It takes no account of the true distance from the boundary of unprotected areas in deeply indented buildings since all unprotected areas must be projected onto

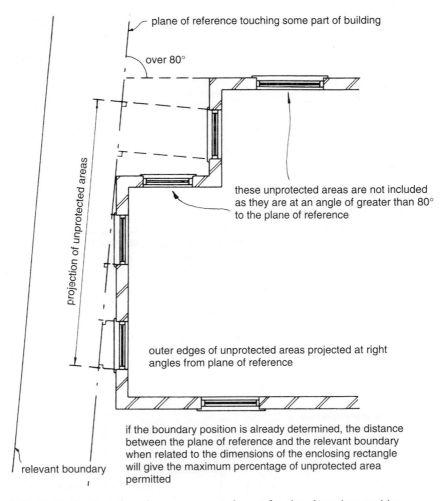

plane of reference touching some part of building

over 80°

projection of unprotected areas

these unprotected areas are not included as they are at an angle of greater than 80° to the plane of reference

outer edges of unprotected areas projected at right angles from plane of reference

if the boundary position is already determined, the distance between the plane of reference and the relevant boundary when related to the dimensions of the enclosing rectangle will give the maximum percentage of unprotected area permitted

relevant boundary

Fig. 7.50 Determination of maximum unprotected areas for given boundary position.

a single plane of reference. It also assumes that the effects of a fire will be equally felt at all points on the boundary from all unprotected areas despite the fact that some windows, for example, may be shielded from certain parts of the boundary by fire resistant walls. For these reasons it may be preferable to use the following method – the *aggregate notional areas* technique – as this method is usually more accurate in practice.

Additional methods – aggregate notional areas

The basis of the method of aggregate notional areas is to assess the effect of a building fire at a series of points 3 m apart on the relevant boundary.

A *vertical datum* of unlimited height is set at any position on the relevant boundary (see point P on Fig. 7.53). A *datum line* is drawn from this point to the nearest point on the building or compartment. A base line is then constructed at 90°

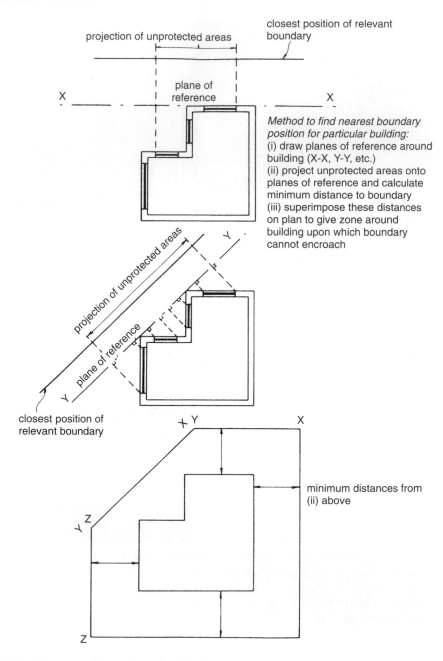

Fig. 7.51 Nearest position to boundary for given building design.

to the datum line and an arc of 50 m radius is drawn centred on the vertical datum to meet the base line.

Using this method it is possible to exclude certain unprotected areas that would have to be considered under the enclosing rectangles method (see Fig. 7.53).

For those unprotected areas which remain (that is, those that cannot be

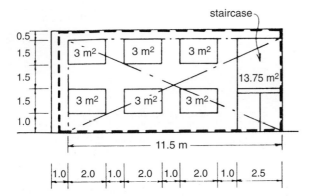

Uncompartmented building

(i) minimum size of rectangle enclosing unprotected areas = 11.5 m wide × 5.5 m high
(ii) from Table 7.9 enclosing rectangle = 12 m wide × 6 m high (take next highest values)
(iii) calculate aggregate of unprotected areas = 18 + 13.75 m² = 31.75 m²
(iv) calculate unprotected percentage (aggregate of unprotected areas as percentage of enclosing
       rectangle) = $\dfrac{31.75}{12 \times 6} \times 100 = 44\%$ ∴ use 50% column in Table 7.9
(v) select distance from Table 7.9 (second part of table, fourth column, fourth row, figure in brackets)
       permitted distance to boundary = 3.5 m

Note:
(a) minimum size of rectangle indicated by diagonal lines
(b) enclosing rectangle indicated by dotted lines
(c) relevant boundary is parallel with wall face, plane of reference coincides with wall face (this need
       not be the case)

The situation can be improved if the staircase is enclosed in a protecting structure and a compartment
floor is provided:

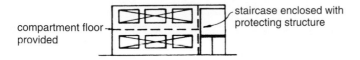

(i) each compartment now considered separately, the minimum rectangle being shown by diagonal
       lines above = 8 m × 1.5 m
(ii) enclosing rectangle from Table 7.9 = 9 m wide × 3 m high
(iii) aggregate of unprotected areas = 9 m²
(iv) unprotected percentage = $\dfrac{9}{9 \times 3} \times 100 = 33\frac{1}{4}\%$ i.e. 40% in Table 7.9
(v) distance from Table 7.9 = 1.5 m

Compartment has therefore reduced the permitted distance to the boundary from 3.5 m to 1.5 m

Fig. 7.52   Enclosing rectangles – effects of compartmentation.

excluded), it is necessary to measure the distance of each from the vertical datum.
Table J3 from the 1985 edition of AD B2/3/4 is reproduced below as Table 7.10.
This contains a series of multiplication factors which are related to the distance
from the vertical datum. The Table is based on the fact that the amount of heat
caused by a fire issuing from an unprotected area will decrease in proportion to

Table 7.9 Permitted unprotected percentages in relation to enclosing rectangles.

Width of enclosing rectangle (m)	Distance from relevant boundary for unprotected percentage not exceeding								
	20%	30%	40%	50%	60%	70%	80%	90%	100%
	Minimum boundary distance (m); figures in brackets are for residential, office or assembly								
Enclosing rectangle 3 m high									
3	1.0 (1.0)	1.5 (1.0)	2.0 (1.0)	2.0 (1.5)	2.5 (1.5)	2.5 (1.5)	2.5 (2.0)	3.0 (2.0)	3.0 (2.0)
6	1.5 (1.0)	2.0 (1.0)	2.5 (1.5)	3.0 (2.0)	3.0 (2.0)	3.5 (2.0)	3.5 (2.5)	4.0 (2.5)	4.0 (3.0)
9	1.5 (1.0)	2.5 (1.0)	3.0 (1.5)	3.5 (2.0)	4.0 (2.5)	4.0 (2.5)	4.5 (3.0)	5.0 (3.0)	5.0 (3.5)
12	2.0 (1.0)	2.5 (1.5)	3.0 (2.0)	3.5 (2.0)	4.0 (2.5)	4.5 (3.0)	5.0 (3.0)	5.5 (3.5)	5.5 (3.5)
15	2.0 (1.0)	2.5 (1.5)	3.5 (2.0)	4.0 (2.5)	4.5 (2.5)	5.0 (3.0)	5.5 (3.5)	6.0 (3.5)	6.0 (4.0)
18	2.0 (1.0)	2.5 (1.5)	3.5 (2.0)	4.0 (2.5)	5.0 (2.5)	5.0 (3.0)	6.0 (3.5)	6.5 (4.0)	6.5 (4.0)
21	2.0 (1.0)	3.0 (1.5)	3.5 (2.0)	4.5 (2.5)	5.0 (3.0)	5.5 (3.0)	6.0 (3.5)	6.5 (4.0)	7.0 (4.5)
24	2.0 (1.0)	3.0 (1.5)	3.5 (2.0)	4.5 (2.5)	5.0 (3.0)	5.5 (3.5)	6.0 (3.5)	7.0 (4.0)	7.5 (4.5)
27	2.0 (1.0)	3.0 (1.5)	4.0 (2.0)	4.5 (2.5)	5.5 (3.0)	6.0 (3.5)	6.5 (4.0)	7.0 (4.0)	7.5 (4.5)
30	2.0 (1.0)	3.0 (1.5)	4.0 (2.0)	4.5 (2.5)	5.5 (3.0)	6.0 (3.5)	6.5 (4.0)	7.5 (4.0)	8.0 (4.5)
40	2.0 (1.0)	3.0 (1.5)	4.0 (2.0)	5.0 (2.5)	5.5 (3.0)	6.5 (3.5)	7.0 (4.0)	8.0 (4.0)	8.5 (5.0)
50	2.0 (1.0)	3.0 (1.5)	4.0 (2.0)	5.0 (2.5)	6.0 (3.0)	6.5 (3.5)	7.5 (4.0)	8.0 (4.0)	9.0 (5.0)
60	2.0 (1.0)	3.0 (1.5)	4.0 (2.0)	5.0 (2.5)	6.0 (3.0)	7.0 (3.5)	7.5 (4.0)	8.5 (4.0)	9.5 (5.0)
80	2.0 (1.0)	3.0 (1.5)	4.0 (2.0)	5.0 (2.5)	6.0 (3.0)	7.0 (3.5)	8.0 (4.0)	9.0 (4.0)	9.5 (5.0)
no limit	2.0 (1.0)	3.0 (1.5)	4.0 (2.0)	5.0 (2.5)	6.0 (3.0)	7.0 (3.5)	8.0 (4.0)	9.0 (4.0)	10.0 (5.0)
Enclosing rectangle 6 m high									
3	1.5 (1.0)	2.0 (1.0)	2.5 (1.5)	3.0 (2.0)	3.0 (2.0)	3.5 (2.0)	3.5 (2.5)	4.0 (2.5)	4.0 (3.0)
6	2.0 (1.0)	3.0 (1.5)	3.5 (2.0)	4.0 (2.5)	4.5 (3.0)	5.0 (3.0)	5.5 (3.5)	5.5 (4.0)	6.0 (4.0)
9	2.5 (1.0)	3.5 (2.0)	4.5 (2.5)	5.0 (3.0)	5.5 (3.5)	6.0 (4.0)	6.0 (4.5)	7.0 (4.5)	7.0 (5.0)
12	3.0 (1.5)	4.0 (2.5)	5.0 (3.0)	5.5 (3.5)	6.5 (4.0)	7.0 (4.5)	7.5 (5.0)	8.0 (5.0)	8.5 (5.5)
15	3.0 (1.5)	4.5 (2.5)	5.5 (3.0)	6.0 (4.0)	7.0 (4.5)	7.5 (5.0)	8.0 (5.5)	9.0 (5.5)	9.0 (6.0)
18	3.5 (1.5)	4.5 (2.5)	5.5 (3.5)	6.5 (4.0)	7.5 (4.5)	8.0 (5.0)	9.0 (5.5)	9.5 (6.0)	10.0 (6.5)
21	3.5 (1.5)	5.0 (2.5)	6.0 (3.5)	7.0 (4.0)	8.0 (5.0)	9.0 (5.5)	9.5 (6.0)	10.0 (6.5)	10.5 (7.0)
24	3.5 (1.5)	5.0 (2.5)	6.0 (3.5)	7.0 (4.5)	8.5 (5.0)	9.5 (5.5)	10.0 (6.0)	10.5 (7.0)	11.0 (7.0)
27	3.5 (1.5)	5.0 (2.5)	6.5 (3.5)	7.5 (4.5)	8.5 (5.0)	9.5 (6.0)	10.5 (7.6)	11.0 (7.0)	12.0 (7.5)

30	3.5 (1.5)	5.0 (2.5)	6.5 (3.5)	8.0 (4.5)	9.0 (5.0)	10.0 (6.0)	11.0 (6.5)	12.0 (7.0)	12.5 (8.0)
40	3.5 (1.5)	5.5 (2.5)	7.0 (3.5)	8.5 (4.5)	10.0 (5.5)	11.0 (6.5)	12.0 (7.0)	13.0 (8.0)	14.0 (8.5)
50	3.5 (1.5)	5.5 (2.5)	7.5 (3.5)	9.0 (4.5)	10.5 (5.5)	11.5 (6.5)	13.0 (7.5)	14.0 (8.0)	15.0 (9.0)
60	3.5 (1.5)	5.5 (2.5)	7.5 (3.5)	9.5 (5.0)	11.0 (5.5)	12.0 (6.5)	13.5 (7.5)	15.0 (8.5)	16.0 (9.5)
80	3.5 (1.5)	6.0 (2.5)	7.5 (3.5)	9.5 (5.0)	11.5 (6.0)	13.0 (7.0)	14.5 (7.5)	16.0 (8.5)	17.5 (9.5)
100	3.5 (1.5)	6.0 (2.5)	8.0 (3.5)	10.0 (5.0)	12.0 (6.0)	13.5 (7.0)	15.0 (8.0)	16.5 (8.5)	18.0 (10.0)
120	3.5 (1.5)	6.0 (2.5)	8.0 (3.5)	10.0 (5.0)	12.0 (6.0)	14.0 (7.0)	15.5 (8.0)	17.0 (8.5)	19.0 (10.0)
no limit	3.5 (1.5)	6.0 (2.5)	8.0 (3.5)	10.0 (5.0)	12.0 (6.0)	14.0 (7.0)	16.0 (8.0)	18.0 (8.5)	19.0 (10.0)

Enclosing rectangle 9 m high

3	1.5 (1.0)	2.5 (1.0)	3.0 (1.5)	3.5 (1.5)	4.0 (2.5)	4.0 (2.5)	4.5 (3.0)	5.0 (3.0)	5.0 (3.5)
6	2.5 (1.0)	3.5 (2.0)	4.5 (2.5)	5.0 (3.0)	5.5 (3.5)	6.0 (4.0)	6.5 (4.5)	7.0 (4.5)	7.0 (5.0)
9	3.5 (1.5)	4.5 (2.5)	5.5 (3.5)	6.0 (4.0)	6.5 (4.5)	7.5 (5.0)	8.0 (5.5)	8.5 (5.5)	9.0 (6.0)
12	3.5 (1.5)	5.0 (3.0)	6.0 (3.5)	7.0 (4.5)	7.5 (5.0)	8.5 (5.5)	9.0 (6.0)	9.5 (6.5)	10.5 (7.0)
15	4.0 (2.0)	5.5 (3.0)	6.5 (4.0)	7.5 (5.0)	8.5 (5.5)	9.5 (6.0)	10.0 (6.5)	11.0 (7.0)	11.5 (7.5)
18	4.5 (2.0)	6.0 (3.5)	7.0 (4.5)	8.5 (5.0)	9.5 (6.0)	10.0 (6.5)	11.0 (7.0)	12.0 (8.0)	12.5 (8.5)
21	4.5 (2.0)	6.5 (3.5)	7.5 (4.5)	9.0 (5.5)	10.0 (6.5)	11.0 (7.0)	12.0 (7.5)	13.0 (8.5)	13.5 (9.0)
24	5.0 (2.0)	6.5 (3.5)	8.0 (5.0)	9.5 (5.5)	11.0 (6.5)	12.0 (7.5)	13.0 (8.0)	13.5 (9.0)	14.5 (9.5)
27	5.0 (2.0)	7.0 (3.5)	8.5 (5.0)	10.0 (6.0)	11.5 (7.0)	12.5 (7.5)	13.5 (8.5)	14.5 (9.5)	15.0 (10.0)
30	5.0 (2.0)	7.0 (3.5)	9.0 (5.0)	10.5 (6.0)	12.0 (7.0)	13.0 (8.0)	14.0 (9.0)	15.0 (9.5)	16.0 (10.5)
40	5.5 (2.0)	7.5 (3.5)	9.5 (5.5)	11.5 (6.5)	13.0 (7.5)	14.5 (8.5)	15.5 (9.5)	17.0 (10.5)	17.5 (11.5)
50	5.5 (2.0)	8.0 (4.0)	10.0 (5.5)	12.5 (6.5)	14.0 (8.0)	15.5 (9.0)	17.0 (10.0)	18.5 (11.5)	19.5 (12.5)
60	5.5 (2.0)	8.0 (4.0)	11.0 (5.5)	13.0 (7.0)	15.0 (8.0)	16.5 (9.5)	18.0 (11.0)	19.5 (11.5)	21.0 (13.0)
80	5.5 (2.0)	8.5 (4.0)	11.5 (5.5)	13.5 (7.0)	16.0 (8.5)	17.5 (10.0)	19.5 (11.5)	21.5 (12.5)	23.0 (13.5)
100	5.5 (2.0)	8.5 (4.0)	11.5 (5.5)	14.5 (7.0)	16.5 (8.5)	18.5 (10.0)	21.0 (11.5)	22.5 (12.5)	24.5 (14.5)
120	5.5 (2.0)	8.5 (4.0)	11.5 (5.5)	14.5 (7.0)	17.0 (8.5)	19.5 (10.0)	21.5 (11.5)	23.5 (12.5)	26.0 (14.5)
no limit	5.5 (2.0)	8.5 (4.0)	11.5 (5.5)	15.0 (7.0)	17.5 (8.5)	20.0 (10.5)	22.5 (12.0)	24.5 (12.5)	27.0 (15.0)

(Contd)

Table 7.9 (*Contd*).

Width of enclosing rectangle (m)	Distance from relevant boundary for unprotected percentage not exceeding								
	20%	30%	40%	50%	60%	70%	80%	90%	100%
	Minimum boundary distance (m); figures in brackets are for residential, office or assembly								
Enclosing rectangle 12 m high									
3	2.0 (1.0)	2.5 (1.5)	3.0 (2.0)	3.5 (2.0)	4.0 (2.5)	4.5 (3.0)	5.0 (3.0)	5.5 (3.5)	5.5 (3.5)
6	3.0 (1.5)	4.0 (2.5)	5.0 (3.0)	5.5 (3.5)	6.5 (4.0)	7.0 (4.5)	7.5 (5.0)	8.0 (5.0)	8.5 (5.5)
9	3.5 (1.5)	5.0 (3.0)	6.0 (3.5)	7.0 (4.5)	7.5 (5.0)	8.5 (5.5)	9.0 (6.0)	9.5 (6.5)	10.5 (7.0)
12	4.5 (1.5)	6.0 (3.5)	7.0 (4.5)	8.0 (5.0)	9.0 (6.0)	9.5 (6.0)	11.0 (7.0)	11.5 (7.5)	12.0 (8.0)
15	5.0 (2.0)	6.5 (3.5)	8.0 (5.0)	9.0 (5.5)	10.0 (6.5)	11.0 (7.0)	12.0 (8.0)	13.0 (8.5)	13.5 (9.0)
18	5.0 (2.5)	7.0 (4.0)	8.5 (5.0)	10.0 (6.0)	11.0 (7.0)	12.0 (7.5)	13.0 (8.5)	14.0 (9.0)	14.5 (10.0)
21	5.5 (2.5)	7.5 (4.0)	9.0 (5.5)	10.5 (6.5)	12.0 (7.5)	13.0 (8.5)	14.0 (9.0)	15.0 (10.0)	16.0 (10.5)
24	6.0 (2.5)	8.0 (4.5)	9.5 (6.0)	11.5 (7.0)	12.5 (8.0)	14.0 (8.5)	15.0 (9.5)	16.0 (10.5)	16.5 (11.5)
27	6.0 (2.5)	8.0 (4.5)	10.5 (6.0)	12.0 (7.0)	13.5 (8.0)	14.5 (9.0)	16.0 (10.5)	17.0 (11.0)	17.5 (12.0)
30	6.5 (2.5)	8.5 (4.5)	10.5 (6.5)	12.5 (7.5)	14.0 (8.5)	15.0 (9.5)	16.5 (10.5)	17.5 (11.5)	18.5 (12.5)
40	6.5 (2.5)	9.5 (5.0)	12.0 (6.5)	14.0 (8.0)	15.5 (9.5)	17.5 (10.5)	18.5 (12.0)	20.0 (13.0)	21.0 (14.0)
50	7.0 (2.5)	10.0 (5.0)	13.0 (7.0)	15.0 (8.5)	17.0 (10.0)	19.0 (11.0)	20.5 (13.0)	23.0 (14.0)	23.0 (15.0)
60	7.0 (2.5)	10.5 (5.0)	13.5 (7.0)	16.0 (9.0)	18.0 (10.5)	20.0 (12.0)	21.5 (13.5)	23.5 (14.5)	25.0 (16.0)
80	7.0 (2.5)	11.0 (5.0)	14.5 (7.0)	17.0 (9.0)	19.5 (11.0)	21.5 (13.0)	23.5 (14.5)	26.0 (16.0)	27.5 (17.0)
100	7.5 (2.5)	11.5 (5.0)	15.0 (7.5)	18.0 (9.5)	21.0 (11.5)	23.0 (13.5)	25.5 (15.0)	28.0 (16.5)	30.0 (18.0)
120	7.5 (2.5)	11.5 (5.0)	15.0 (7.5)	18.5 (9.5)	22.0 (11.5)	24.0 (13.5)	27.0 (15.0)	29.5 (17.0)	31.5 (18.5)
no limit	7.5 (2.5)	12.0 (5.0)	15.5 (7.5)	19.0 (9.5)	22.5 (12.0)	22.5 (14.0)	28.0 (15.5)	30.5 (17.0)	34.0 (19.0)
Enclosing rectangle 15 m high									
3	2.0 (1.0)	2.5 (1.5)	3.5 (2.0)	4.0 (2.5)	4.5 (2.5)	5.0 (3.0)	5.5 (3.5)	6.0 (3.5)	6.0 (4.0)
6	3.0 (1.5)	4.5 (2.5)	5.5 (3.0)	6.0 (4.0)	7.0 (4.5)	7.5 (5.0)	8.0 (5.5)	9.0 (5.5)	9.0 (6.0)
9	4.0 (2.0)	5.5 (3.0)	6.5 (4.0)	7.5 (5.0)	8.5 (5.5)	9.5 (6.0)	10.0 (6.5)	11.0 (7.0)	11.5 (7.5)
12	5.0 (2.0)	6.5 (3.5)	8.0 (5.0)	9.0 (5.5)	10.0 (6.5)	11.0 (7.0)	12.0 (8.0)	13.0 (8.5)	13.5 (9.0)
15	5.5 (2.0)	7.0 (4.0)	9.0 (5.5)	10.0 (6.5)	11.5 (7.0)	12.5 (8.0)	13.5 (9.0)	14.5 (9.5)	15.0 (10.0)
18	6.0 (2.5)	8.0 (4.5)	9.5 (6.0)	11.0 (7.0)	12.5 (8.0)	13.5 (8.5)	14.5 (9.5)	15.5 (10.5)	16.5 (11.0)

21	6.5 (2.5)	8.5 (5.0)	10.5 (6.5)	12.0 (7.5)	13.5 (8.5)	14.5 (9.5)	16.0 (10.5)	16.5 (11.0)	17.5 (12.0)
24	6.5 (3.0)	9.0 (5.0)	11.0 (6.5)	13.0 (8.0)	14.5 (9.0)	15.5 (10.0)	17.0 (11.0)	18.0 (12.0)	19.0 (13.0)
27	7.0 (3.0)	9.5 (5.5)	11.5 (7.0)	13.5 (8.5)	15.0 (9.5)	16.5 (10.5)	18.0 (11.5)	19.0 (12.5)	20.0 (13.5)
30	7.5 (3.0)	10.0 (5.5)	12.0 (7.5)	14.0 (8.5)	16.0 (10.0)	17.0 (11.0)	18.5 (12.0)	20.0 (13.5)	21.0 (14.0)
40	8.0 (3.0)	11.0 (6.0)	13.5 (8.0)	16.0 (9.5)	18.0 (11.0)	19.5 (12.5)	21.0 (13.5)	22.5 (15.0)	23.5 (16.0)
50	8.5 (3.5)	12.0 (6.0)	15.0 (8.5)	17.5 (10.0)	19.5 (12.0)	21.5 (13.5)	23.0 (15.0)	25.0 (16.5)	26.0 (17.5)
60	8.5 (3.5)	12.5 (6.5)	15.5 (8.5)	18.0 (10.5)	21.0 (17.5)	23.5 (14.0)	25.0 (15.5)	27.0 (17.0)	28.0 (18.0)
80	9.0 (3.5)	13.5 (6.5)	17.0 (9.0)	20.0 (11.0)	23.0 (13.5)	25.5 (15.0)	28.0 (17.0)	30.0 (18.5)	31.5 (20.0)
100	9.0 (3.5)	14.0 (6.5)	18.0 (9.0)	21.5 (11.5)	24.5 (14.0)	27.5 (16.0)	30.0 (18.0)	32.5 (19.5)	34.5 (21.5)
120	9.0 (3.5)	14.0 (6.5)	18.5 (9.0)	22.5 (11.5)	25.5 (14.0)	28.5 (16.5)	31.5 (18.5)	34.5 (20.5)	37.0 (22.5)
no limit	9.0 (3.5)	14.5 (6.5)	19.0 (9.0)	23.0 (12.0)	27.0 (14.5)	30.0 (17.0)	34.0 (19.0)	36.0 (21.0)	39.0 (23.0)

Enclosing rectangle 18 m high

3	2.0 (1.0)	2.5 (1.5)	3.5 (2.0)	4.0 (2.5)	5.0 (2.5)	5.0 (3.0)	6.0 (3.5)	6.5 (4.0)	6.5 (4.0)
6	3.5 (1.5)	4.5 (2.5)	5.5 (3.5)	6.5 (4.0)	7.5 (4.5)	8.0 (5.0)	9.0 (5.5)	9.5 (6.0)	10.0 (6.5)
9	4.5 (2.0)	6.0 (3.5)	7.0 (4.5)	8.5 (5.0)	9.5 (6.0)	10.0 (6.5)	11.0 (7.0)	12.0 (8.0)	12.5 (8.5)
12	5.0 (2.5)	7.0 (4.0)	8.5 (5.0)	10.0 (6.0)	11.0 (7.0)	12.0 (7.5)	13.0 (8.5)	14.0 (9.0)	14.5 (10.0)
15	6.0 (2.5)	8.0 (4.5)	9.5 (6.0)	11.0 (7.0)	12.5 (8.0)	13.5 (8.5)	14.5 (9.5)	15.5 (10.5)	16.5 (11.0)
18	6.5 (2.5)	8.5 (5.0)	11.0 (6.5)	12.0 (7.5)	13.5 (8.5)	14.5 (9.5)	16.0 (11.0)	17.0 (11.5)	18.0 (13.0)
21	7.0 (3.0)	9.5 (5.5)	11.5 (7.0)	13.0 (8.0)	14.5 (9.5)	16.0 (10.5)	17.0 (11.5)	18.0 (12.5)	19.5 (13.0)
24	7.5 (3.0)	10.0 (5.5)	12.0 (7.5)	14.0 (8.5)	15.5 (10.0)	16.5 (11.0)	18.5 (12.0)	19.5 (13.0)	20.5 (14.0)
27	8.0 (3.5)	10.5 (6.0)	12.5 (8.0)	14.5 (9.0)	16.5 (10.5)	17.5 (11.5)	19.5 (12.5)	20.5 (13.5)	21.5 (14.5)
30	8.0 (3.5)	11.0 (6.5)	13.5 (8.0)	15.5 (9.5)	17.0 (11.0)	18.5 (12.0)	20.5 (13.5)	21.5 (14.5)	22.5 (15.5)
40	9.0 (4.0)	12.0 (7.0)	15.0 (9.0)	17.5 (11.0)	19.5 (12.0)	21.5 (13.5)	23.5 (15.0)	25.0 (16.5)	26.0 (17.5)
50	9.5 (4.0)	13.0 (7.0)	16.5 (9.5)	19.0 (11.5)	21.5 (13.0)	23.5 (15.0)	26.0 (16.5)	27.5 (18.0)	29.0 (19.0)
60	10.0 (4.0)	14.0 (7.5)	17.5 (10.0)	20.5 (12.0)	23.0 (14.0)	26.0 (16.0)	27.5 (17.5)	29.5 (19.5)	31.0 (20.5)
80	10.0 (4.0)	15.0 (7.5)	19.0 (10.0)	22.5 (13.0)	26.0 (15.0)	28.5 (17.0)	31.0 (19.0)	33.5 (21.0)	35.0 (22.5)
100	10.0 (4.0)	16.0 (7.5)	20.5 (10.0)	24.0 (13.5)	28.0 (16.0)	31.0 (18.0)	33.5 (20.5)	36.0 (22.5)	38.5 (24.0)
120	10.0 (4.0)	16.5 (7.5)	21.0 (10.0)	25.5 (14.0)	29.5 (16.5)	32.5 (19.0)	35.5 (21.0)	39.0 (23.5)	41.5 (25.5)
no limit	10.0 (4.0)	17.0 (8.0)	22.0 (10.0)	26.5 (14.0)	30.5 (17.0)	34.0 (19.5)	37.0 (22.0)	41.0 (24.0)	43.5 (26.5)

(Contd)

Table 7.9 *(Contd.)*.

Width of enclosing rectangle (m)	Distance from relevant boundary for unprotected percentage not exceeding								
	20%	30%	40%	50%	60%	70%	80%	90%	100%
	Minimum boundary distance (m); figures in brackets are for residential, office or assembly								
Enclosing rectangle 21 m high									
3	2.0 *(1.0)*	3.0 *(1.5)*	3.5 *(2.0)*	4.5 *(2.5)*	5.0 *(3.0)*	5.5 *(3.0)*	5.0 *(3.5)*	6.5 *(4.0)*	7.0 *(4.5)*
6	3.5 *(1.5)*	5.0 *(2.5)*	6.0 *(3.5)*	7.0 *(4.0)*	8.0 *(5.0)*	9.0 *(5.5)*	9.5 *(6.0)*	10.0 *(6.5)*	10.5 *(7.0)*
9	4.5 *(2.0)*	6.5 *(3.5)*	7.5 *(4.5)*	9.0 *(5.5)*	10.0 *(6.5)*	11.0 *(7.0)*	12.0 *(7.5)*	13.0 *(8.5)*	13.5 *(9.0)*
12	5.5 *(2.5)*	7.5 *(4.0)*	9.0 *(5.5)*	10.5 *(6.5)*	12.0 *(7.5)*	13.0 *(8.5)*	14.0 *(9.0)*	15.0 *(10.0)*	16.0 *(10.5)*
15	6.5 *(2.5)*	8.5 *(5.0)*	10.5 *(6.5)*	12.0 *(7.5)*	13.5 *(8.5)*	14.5 *(9.5)*	16.0 *(10.5)*	16.5 *(11.0)*	17.5 *(12.0)*
18	7.0 *(3.0)*	9.5 *(5.5)*	11.5 *(7.0)*	13.0 *(8.0)*	14.5 *(9.5)*	16.0 *(10.5)*	17.0 *(11.5)*	18.0 *(12.5)*	19.5 *(13.0)*
21	7.5 *(3.0)*	10.0 *(6.0)*	12.5 *(7.5)*	14.0 *(9.0)*	15.5 *(10.0)*	17.0 *(11.0)*	18.5 *(12.5)*	20.0 *(13.5)*	21.0 *(14.0)*
24	8.0 *(3.5)*	10.5 *(6.0)*	13.0 *(8.0)*	15.0 *(9.5)*	16.5 *(10.5)*	18.0 *(12.0)*	20.0 *(13.0)*	21.0 *(14.0)*	22.0 *(15.0)*
27	8.5 *(3.5)*	11.5 *(6.5)*	14.0 *(8.5)*	16.0 *(10.0)*	18.0 *(11.5)*	19.0 *(13.0)*	21.0 *(14.0)*	22.5 *(15.0)*	23.5 *(16.0)*
30	9.0 *(4.0)*	12.0 *(7.0)*	14.5 *(9.0)*	16.5 *(10.5)*	18.5 *(12.0)*	20.5 *(13.0)*	22.0 *(14.5)*	23.5 *(16.0)*	25.0 *(16.5)*
40	10.0 *(4.5)*	13.5 *(7.5)*	16.5 *(10.0)*	19.0 *(12.0)*	21.5 *(13.5)*	23.0 *(15.0)*	25.5 *(16.5)*	27.0 *(18.0)*	28.5 *(19.0)*
50	11.0 *(4.5)*	14.5 *(8.0)*	18.0 *(11.0)*	21.0 *(13.0)*	23.5 *(14.5)*	25.5 *(16.5)*	28.0 *(18.0)*	30.0 *(20.0)*	31.5 *(21.0)*
60	11.5 *(4.5)*	15.5 *(8.5)*	19.5 *(11.5)*	22.5 *(13.5)*	25.5 *(15.5)*	28.0 *(17.5)*	30.5 *(19.5)*	32.5 *(21.0)*	33.5 *(22.5)*
80	12.0 *(4.5)*	17.0 *(8.5)*	21.0 *(12.0)*	25.0 *(14.5)*	28.5 *(17.0)*	31.5 *(19.0)*	34.0 *(21.0)*	36.5 *(23.5)*	38.5 *(25.0)*
100	12.0 *(4.5)*	18.0 *(9.0)*	22.5 *(12.0)*	27.0 *(15.5)*	31.0 *(18.0)*	34.5 *(20.5)*	37.0 *(22.5)*	40.0 *(25.0)*	42.0 *(27.0)*
120	12.0 *(4.5)*	18.5 *(9.0)*	23.5 *(12.0)*	28.5 *(16.0)*	32.5 *(18.5)*	36.5 *(21.5)*	39.5 *(23.5)*	43.0 *(26.5)*	45.5 *(28.5)*
no limit	12.0 *(4.5)*	19.0 *(9.0)*	25.0 *(12.0)*	29.5 *(16.0)*	34.5 *(19.0)*	38.0 *(22.0)*	41.5 *(25.0)*	45.5 *(26.5)*	48.0 *(29.5)*
Enclosing rectangle 24 m high									
3	2.0 *(1.0)*	3.0 *(1.5)*	3.5 *(2.0)*	4.5 *(2.5)*	5.0 *(3.0)*	5.5 *(3.5)*	6.0 *(3.5)*	7.0 *(4.0)*	7.5 *(4.5)*
6	3.5 *(1.5)*	5.0 *(2.5)*	6.0 *(3.5)*	7.0 *(4.5)*	8.5 *(5.0)*	9.5 *(5.5)*	10.0 *(6.0)*	10.5 *(7.0)*	11.0 *(7.0)*
9	5.0 *(2.0)*	6.5 *(3.5)*	8.0 *(5.0)*	9.5 *(5.5)*	11.0 *(6.5)*	12.0 *(7.5)*	13.0 *(8.0)*	13.5 *(9.0)*	14.5 *(9.5)*
12	6.0 *(2.5)*	8.0 *(4.5)*	9.5 *(6.0)*	11.5 *(7.0)*	12.5 *(8.0)*	14.0 *(8.5)*	15.0 *(9.5)*	16.0 *(10.5)*	16.5 *(11.5)*
15	6.5 *(3.0)*	9.0 *(5.0)*	11.0 *(6.5)*	13.0 *(8.0)*	14.5 *(9.0)*	15.5 *(10.0)*	17.0 *(11.0)*	18.0 *(12.0)*	19.0 *(13.0)*
18	7.5 *(3.0)*	10.0 *(5.5)*	12.0 *(7.5)*	14.0 *(8.5)*	15.5 *(10.0)*	16.5 *(11.0)*	18.5 *(12.0)*	19.5 *(13.0)*	20.5 *(14.0)*

21	8.0 (3.5)	10.5 (6.0)	13.0 (8.0)	15.0 (9.5)	16.5 (10.5)	18.0 (12.0)	20.0 (13.0)	21.0 (14.0)	22.0 (15.0)
24	8.5 (3.5)	11.5 (6.5)	14.0 (8.5)	16.0 (10.0)	18.0 (11.5)	19.5 (12.5)	21.0 (14.0)	22.5 (15.0)	24.0 (16.0)
27	9.0 (4.0)	12.5 (7.0)	15.0 (9.0)	17.0 (11.0)	19.0 (12.5)	20.5 (13.5)	22.5 (15.0)	24.0 (16.0)	25.5 (17.0)
30	9.5 (4.0)	13.0 (7.5)	15.5 (9.5)	18.0 (11.5)	20.0 (13.0)	21.5 (14.0)	23.5 (15.5)	25.0 (17.0)	26.5 (18.0)
40	11.0 (4.5)	14.5 (8.5)	18.0 (11.0)	20.5 (13.0)	23.0 (14.5)	25.0 (16.0)	27.5 (18.0)	29.0 (19.0)	30.5 (20.5)
50	12.0 (5.0)	16.0 (9.0)	19.5 (12.0)	22.5 (14.0)	25.5 (16.0)	27.5 (17.5)	30.0 (19.5)	32.0 (21.0)	33.5 (22.5)
60	12.5 (5.0)	17.0 (9.5)	21.0 (12.5)	24.5 (15.0)	27.5 (17.0)	30.0 (19.0)	32.5 (21.0)	35.0 (23.0)	36.5 (24.5)
80	13.5 (5.0)	18.5 (10.0)	23.5 (13.5)	27.5 (16.5)	31.0 (18.5)	34.5 (21.0)	37.0 (23.5)	39.5 (25.5)	41.5 (27.5)
100	13.5 (5.0)	20.0 (10.0)	25.0 (13.5)	29.5 (17.0)	33.5 (20.0)	37.0 (20.0)	40.0 (25.0)	43.0 (27.5)	45.5 (29.5)
120	13.5 (5.5)	20.5 (10.0)	26.5 (13.5)	31.0 (17.5)	36.0 (20.5)	39.5 (23.5)	43.0 (26.5)	46.5 (29.0)	49.0 (31.0)
no limit	13.5 (5.5)	21.0 (10.0)	27.5 (13.5)	32.5 (18.0)	37.5 (21.0)	42.0 (24.0)	45.5 (27.5)	49.5 (30.0)	52.0 (32.5)

Enclosing rectangle 27 m high

3	2.0 (1.0)	3.0 (1.5)	4.0 (2.0)	4.5 (2.5)	5.5 (3.0)	6.0 (3.5)	6.5 (4.0)	7.0 (4.0)	7.5 (4.5)
6	3.5 (1.5)	5.0 (2.5)	6.5 (3.5)	7.5 (4.5)	8.5 (5.0)	9.5 (6.0)	10.5 (6.5)	11.0 (7.0)	12.0 (7.5)
9	5.0 (2.0)	7.0 (3.5)	8.5 (5.0)	10.0 (6.0)	11.5 (7.0)	12.5 (7.5)	13.5 (8.5)	14.5 (9.5)	15.0 (10.0)
12	6.0 (2.5)	8.0 (4.5)	10.5 (6.0)	12.0 (7.0)	13.5 (8.0)	14.5 (9.0)	16.0 (10.5)	17.0 (11.0)	17.5 (12.0)
15	7.0 (3.0)	9.5 (5.5)	11.5 (7.0)	13.5 (8.5)	15.0 (9.5)	16.5 (10.5)	18.0 (11.5)	19.0 (12.5)	20.0 (13.5)
18	8.0 (3.5)	10.5 (6.0)	12.5 (8.0)	14.5 (9.0)	16.5 (10.5)	17.5 (11.5)	19.5 (12.5)	20.5 (13.5)	21.5 (14.5)
21	8.5 (3.5)	11.5 (6.5)	14.0 (8.5)	16.0 (10.0)	18.0 (11.5)	19.0 (13.0)	21.0 (14.0)	22.5 (15.0)	23.5 (16.0)
24	9.0 (3.5)	12.5 (7.0)	15.0 (9.0)	17.0 (11.0)	19.0 (12.5)	20.5 (13.5)	22.5 (15.0)	24.0 (16.0)	25.5 (17.0)
27	10.0 (4.0)	13.0 (7.5)	16.0 (10.0)	18.0 (11.5)	20.0 (13.0)	22.0 (14.0)	24.0 (16.0)	25.5 (17.0)	27.0 (18.0)
30	10.0 (4.0)	13.5 (8.0)	17.0 (10.0)	19.0 (12.0)	21.0 (13.5)	23.0 (15.0)	25.0 (17.0)	26.5 (18.0)	28.0 (19.0)
40	11.5 (5.0)	15.5 (9.0)	19.0 (11.5)	22.0 (14.0)	24.5 (15.5)	26.5 (17.5)	29.0 (19.0)	30.5 (20.5)	32.5 (22.0)
50	12.5 (5.5)	17.0 (9.5)	21.0 (12.5)	24.0 (15.0)	27.0 (17.0)	29.5 (19.0)	32.0 (21.0)	34.5 (22.5)	36.0 (24.0)
60	13.5 (5.5)	18.5 (10.5)	22.5 (13.5)	26.5 (16.0)	29.5 (18.5)	32.0 (20.5)	35.0 (22.5)	37.0 (24.5)	39.0 (26.5)
80	14.5 (6.0)	20.5 (11.0)	25.0 (14.5)	29.5 (17.5)	33.0 (20.5)	36.5 (22.5)	39.5 (25.0)	42.0 (27.5)	44.0 (29.5)
100	15.5 (6.0)	21.5 (11.0)	27.0 (15.5)	32.0 (19.0)	36.5 (21.5)	40.5 (24.5)	43.0 (27.0)	46.5 (30.0)	48.5 (32.0)
120	15.5 (6.0)	22.5 (11.5)	28.5 (15.5)	34.0 (19.5)	39.0 (22.5)	43.0 (26.0)	46.5 (28.5)	50.5 (32.0)	53.0 (34.0)
no limit	15.5 (6.0)	23.5 (11.5)	29.5 (15.5)	35.0 (20.0)	40.5 (23.5)	44.5 (27.0)	48.5 (29.5)	52.0 (33.0)	55.5 (35.0)

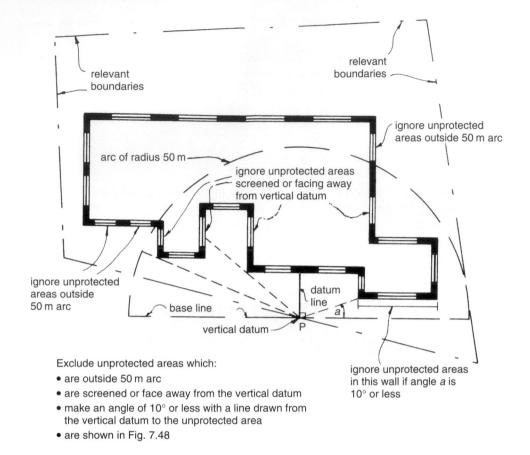

Exclude unprotected areas which:
• are outside 50 m arc
• are screened or face away from the vertical datum
• make an angle of 10° or less with a line drawn from
  the vertical datum to the unprotected area
• are shown in Fig. 7.48

Fig. 7.53   Aggregate notional areas.

Table 7.10   Multiplication factors for aggregate notional area.

Distance of unprotected area from vertical datum (m)		Multiplication factor
Not less than	**Less than**	
1.0	1.2	80.0
1.2	1.8	40.0
1.8	2.7	20.0
2.7	4.3	10.0
4.3	6.0	4.0
6.0	8.5	2.0
8.5	12.0	1.0
12.0	18.5	0.5
18.5	27.5	0.25
27.5	50.0	0.1
50.0	no limit	0.0

its distance from the boundary (it does, in fact, correspond to an inverse square law of the type $y = 1/x^2$).

Therefore, each unprotected area is multiplied by its factor (which depends on its distance from the vertical datum) and these areas are then totalled to give the *aggregate notional area* for that particular vertical datum. The aggregate notional area thus achieved should not exceed:

- $210 \, m^2$ for Residential, Assembly and Recreation or Office buildings; or
- $90 \, m^2$ for Shop and Commercial, Industrial or Other non-residential buildings.

In order to confirm that the unprotected areas in the building comply at other points on the boundary, it is necessary to repeat the above calculations at a series of points 3 m apart starting from the original vertical datum. In practice it is usually possible, by observation, to place the first vertical datum at the worst position thereby obviating the need for further calculations.

The series of measurements and calculations mentioned above may be simplified if a number of protractors are made corresponding to different scales (i.e. 1:50, 1:100 and 1:200). A typical example is illustrated in Fig. 7.54.

7.27.7   Canopy structures and space separation

Since Building Regulations apply to the erection of a building, a canopy would need to comply with the provisions concerning space separation (referred to above) unless it falls into one of the exempt classes (see Class VI and Class VII in section 2.5 above). For free-standing canopies, this might prove unduly onerous (for example, canopies over petrol pumps), since the open sides would be regarded as unprotected areas. Paragraph 14.11 of AD B4 allows space separation recommendations to be disregarded for free-standing canopies which are more than 1m from the relevant boundary in view of the high degree of ventilation and heat dissipation achieved by the open sided construction. For canopies attached to the side of a building, provided that the edge of the canopy is at least 2 m from the boundary, the separation distance may be judged from the side of the building rather than the edge of the canopy. This exception would not apply if the canopy had side walls (such as in an enclosed loading bay).

7.27.8   Atrium buildings

An atrium is defined in section 7.7 as a space in a building (not necessarily vertically aligned) which passes through one or more structural floors. Clearly, the atrium effectively joins up all the relevant floors in the building to the extent that they can no longer be regarded as being compartmented from one another. The effect that this can have on space separation is explained above. In such buildings, AD B4 recommends the use of clause 28.2 of BS 5588: Part 7: 1997. This explains that, if the atrium building is fitted with a sprinkler system, the area affected in a fire will be sufficiently reduced so that the potential for fire spread to adjacent buildings will be comparable to that of an equivalent non-atrium building that is compartmented at each level and protected by a sprinkler system.

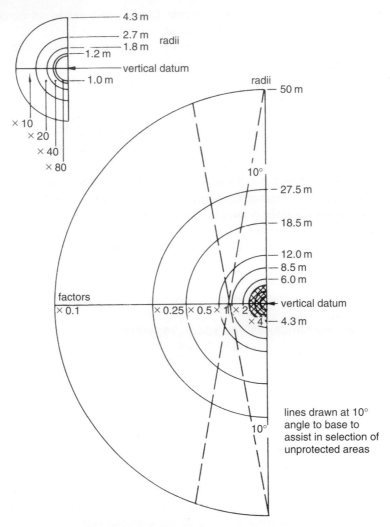

radii
- 4.3 m
- 2.7 m
- 1.8 m
- 1.2 m
- vertical datum
- 1.0 m

× 10
× 20
× 40
× 80

radii
- 50 m

10°
- 27.5 m
- 18.5 m
- 12.0 m
- 8.5 m
- 6.0 m

factors
× 0.1   × 0.25 × 0.5 × 1 × 2

vertical datum
× 4 4.3 m

lines drawn at 10°
angle to base to
assist in selection of
unprotected areas

10°

Note: the figure shown above is an enlargement to show radii and factors applicable near to the vertical datum (covering the portion shown hatched)

Fig. 7.54 Aggregate notional areas protractor.

Conversely, if the atrium building is unsprinklered, space separation needs to be calculated on the basis that all storeys not separated from the atrium by fire-resisting construction may be involved in the fire.

7.27.9 Roofs

Roofs are not required to provide fire resistance from the inside of a building but should resist fire penetration from outside and the spread of flame over their surfaces. The term roof covering means constructions which may contain one or more layers of material but it does not refer to the roof structure as a whole.

In addition to the recommendations for roof coverings contained in AD B4, reference should also be made to AD B 1 (Roofs as part of a means of escape), AD B2 (Internal surfaces of roof-lights) and AD B3 (Roofs used as part of a floor and roofs passing over the top of compartment walls).

The type of construction permitted for a roof depends on the purpose group and size of the building and its distance from the boundary.

Types of construction are specified by the two-letter designations from BS 476: Part 3: 1958 *External fire exposure roof test.* (It should be noted that this is not the most recent version of the standard but it is the one referred to in the latest edition of AD B. The current European standard for roofs is DD ENV 1187: 2002: *Test methods for external fire exposure to roofs.* Unfortunately, no guidance is possible on the performance in terms of the resistance of roofs to external fire exposure as determined by the methods specified in this document, since there is no accompanying classification procedure and no comparative supporting data.)

The first letter in the BS 476 designation method refers to flame penetration:

A – not penetrated within one hour
B – penetrated in not less than half an hour
C – penetrated in less than half an hour
D – penetrated in preliminary flame test.

The second letter in the BS 476 designation method refers to the surface spread of flame test:

A – no spread of flame
B – not more than 21 inches (533.4 mm) spread
C – more than 21 inches (533.4 mm) spread
D – those continuing to burn for five minutes after withdrawal of the test flame, or with a spread of more than 15 inches (381 mm) across the region of burning in the preliminary test.

Example

Roof surface classified AA. This means no fire penetration within one hour and no spread of flame.

Table A5 to AD B (reproduced below) gives a series of roof constructions together with their two-letter notional designations. In the example shown above, a roof constructed in accordance with Part 1 of Table A5 would satisfy the AA rating if it was of natural slates, fibre reinforced cement slates, clay tiles or concrete tiles and it was supported as shown in the table.

Table 17 of AD B4 is reproduced below and gives the notional two-letter designations for roofs in different buildings according to the distance of the roof from the relevant (or notional) boundary. Once the two-letter designation has been established a form of construction may be chosen from Table A5. Where it has been decided to use a different form of roof construction, the manufacturer's details should be consulted to confirm that the necessary designation will be achieved. It

AD B, Section 14

Table A5   Notional designations of roof coverings.

Part i: Pitched roofs covered with slates or tiles		
Covering material	**Supporting structure**	**Designation**
1. Natural slates 2. Fibre reinforced cement slates 3. Clay tiles 4. Concrete tiles	timber rafters with or without underfelt, sarking, boarding, woodwool slabs, compressed straw slabs, plywood, wood chipboard, or fibre insulating board	AA

Note: Although the Table does not include guidance for roofs covered with bitumen felt, it should be noted that there is a wide range of materials on the market and information on specific products is readily available from manufacturers.

Part ii: Pitched roofs covered with self-supporting sheet			
Roof covered material	**Construction**	**Supporting structure**	**Designation**
1. Profiled sheet of galvanised steel, aluminium, fibre reinforced cement, or pre-painted (coil coated) steel or aluminium with a pvc or pvf2 coating	single skin without underlay, or with underlay or plasterboard, fibre insulating board, or woodwool slab	structure of timber, steel or concrete	AA
2. Profiled sheet of galvanised steel, aluminium, fibre reinforced cement, or pre-painted (coil coated) steel or aluminium with a pvc or pvf2 coating	double skin without interlayer, or with interlayer of resin bonded glass fibre, mineral wool slab, polystyrene, or polyurethane	structure of timber, steel or concrete	AA

Part iii: Flat roofs covered with bitumen felt

A flat roof comprising of bitumen felt should (irrespective of the felt specification) be deemed to be of designation AA if the felt is laid on a deck constructed of 6 mm plywood, 12.5 mm wood chipboard, 16 mm (finished) plain edged timber boarding, compressed straw, slab, screeded wood wool slab, profiled fibre reinforced cement or steel deck (single or double skin) with or without fibre insulating board overlay, profiled aluminium deck (single or double skin) with or without fibre insulating board overlay, or concrete or clay pot slab (insitu or pre cast), and has a surface finish of:

a. bitumen-bedded stone chippings covering the whole surface to a depth of at least 12.5 mm;
b. bitumen-bedded tiles of a non-combustible material;
c. sand and cement screed; or
d. macadam.

Part iv. Pitched or flat roofs covered with fully supported material		
Covering material	**Supporting structure**	**Designation**
1. Aluminium sheet 2. Copper sheet 3. Zinc sheet 4. Lead sheet 5. Mastic asphalt 6. Vitreous enamelled steel 7. Lead/tin alloy coated steel sheet 8. Zinc/aluminium alloy coated steel sheet 9. Pre-painted (coil coated) steel sheet including liquid-applied pvc coatings	timber joists and: tongued and grooved boarding, or plain edged boarding	AA*
	steel or timber joists with deck of: woodwool slabs, compressed straw slab, wood chipboard, fibre insulating board, or 9.5 mm plywood	AA
	concrete or clay pot slab (insitu or pre-cast) or non-combustible deck of steel, aluminium, or fibre cement (with or without insulation)	AA

Notes:
* Lead sheet supported by timber joists and plain edged boarding should be regarded as having a BA designation.

AD B4

Table 17 Limitations on roof coverings*

Designation† of covering of roof or part of roof	Minimum distance from any point on relevant boundary			
	Less than 6 m	At least 6 m	At least 12 m	At least 20 m
AA, AB or AC	●	●	●	●
BA, BB or BC	○	●	●	●
CA, CB or CC	○	● (1)(2)	● (1)	●
AD, BD or CD (1)	○	● (2)	●	●
DA, DB, DC or DD (1)	○	○	○	● (2)

Notes:

* See paragraph 15.8 for limitations on glass; paragraph 15.9 for limitations on thatch and wood shingles; and paragraphs 15.6 and 15.7 and Tables 18 and 19 for limitations on plastics rooflights.

† The designation of external roof surfaces is explained in Appendix A. (See Table A5, for notional designations of roof coverings.)

Separation distances do not apply to the boundary between roofs of a pair of semi-detached houses (see 15.5) and to enclosed/covered walkways. However, see Diagram 28 if the roof passes over the top of a compartment wall. Polycarbonate and PVC rooflights which achieve a Class 1 rating by test, see paragraph 15.7 may be regarded as having an AA designation.

● Acceptable.
○ Not acceptable.

1. Not acceptable on any of the following buildings:
   a. Houses in terraces of three or more houses;
   b. Industrial, Storage or Other non-residential purpose group buildings of any size;
   c. Any other buildings with a cubic capacity of more than 1500 m³.

2. Acceptable on buildings not listed in Note 1, if part of the roof is no more than 3 m² in area and is at least 1500 mm from any similar part, with the roof between the parts covered with a material or limited combustibility.

should be noted that there are no restrictions on the use of roof coverings which are designated AA, AB or AC. Also, the boundary formed by the wall separating two semi-detached houses may be disregarded for the purposes of roof designations.

Where plastics rooflights form part of a roof structure they should comply with the provisions of AD B2, paragraph 7.13 and Tables 18 and 19 of AD B4 (see section 7.18.9 above) with regard to their separation, area and disposition. The following rooflight materials may be regarded as having an AA designation:

- rigid thermoplastic sheet made from polycarbonate or unplasticised PVC, which achieves a Class 1 surface spread of flame rating when tested to BS476: Part 7: 1971 or 1987 or 1997
- unwired glass at least 4 mm thick.

7.27.10 Thatch and wood shingles

Thatch and wood shingles that cannot achieve the performance specified in BS 476: Part 3: 1958 should be regarded as having a designation of AD/BD/CD in Table 17. However, it may be possible to locate thatch-roofed buildings closer to the boundary than the distances permitted by Table 17 if the following precautions (taken from *Thatched buildings. New properties and extensions* [the 'Dorset Model'], obtainable on www.dorset-technical-committee.org.uk) are incorporated into the design:

- the rafters are overdrawn with construction having at least 30 minutes fire resistance;
- the guidance in Approved Document J *Combustion appliances and fuel storage systems* is followed; and
- the smoke alarm system recommended in AD B1 is extended to the roof space.

7.28 Special provisions relating to shopping complexes and buildings used as car parks

Section 12 of AD B3 describes additional considerations which apply to the design and construction of buildings used as car parks, and to shopping complexes. Although these provisions are nominally placed in AD B3 (Internal fire spread – structure), most parts of AD B do have a bearing on these structures. Accordingly, the recommendations are dealt with in this separate section.

7.28.1 Car parks

A considerable amount of research has been carried out into the behaviour of fire in buildings used as parking for cars and light vans, with the following results.

- The fire load is not particularly high and is well defined.
- If the car park is well ventilated, there is a low risk of fire spread from one storey to another.

Because of the above, car parks are not normally sprinklered.

The best natural ventilation is achieved in open-sided car parks. Where this cannot be attained, heat and smoke will not be as readily dissipated and fewer concessions will apply.

Whatever standard of ventilation is achieved, certain provisions are common to all car parks as follows.

- The relevant provisions for means of escape in case of fire in AD B1 will apply.
- The recommendations of AD B5 regarding access and facilities for the fire service will apply.
- All materials used in the construction of the car park building should be non-combustible except for:
  (a)   any floor or roof surface finish,
  (b)   any fire door,
  (c)   any attendant's kiosk not greater than 15 m^2 in area,
  (d)   any shop mobility facility.
- Surface finishes in buildings, compartments or separated parts which are within the structure enclosing the car park do not need to be constructed from non-combustible materials but they should comply with the relevant provisions of AD B2 and AD B4.

7.28.2   Open-sided car parks

To be regarded as open-sided, the car park should comply with the following provisions.

- It should comply with the four recommendations immediately above.
- It should contain no basement storeys.
- Natural ventilation should be provided to each storey at each car parking level by permanent openings having an aggregate area of at least 5% of the floor area at that level, and at least half of the ventilation area (i.e. 2.5% of the floor area) should be equally provided in opposing walls.
- Where the building containing the car park is also used for other purposes, the part which contains the car park should be a separated part (for definition of 'separated part' see section 7.2 above).

If the above provisions can be met, the car park may be regarded as a small building or compartment for the purposes of space separation in Table 16 of AD B above (see section 7.27.6) and column (1) of that table may be used. Effectively, this halves the required distance to the relevant boundary (or doubles the permitted limit of unprotected areas). Additionally, the fire resistance recommendations of section 7b(i) of Table A2 of Appendix A (see section 7.21.1) will apply, reducing the fire resistance period to only 15 minutes in many cases.

7.28.3   Car parks not regarded as open-sided

If the ventilation recommendations mentioned above cannot be achieved, the car park cannot be regarded as open-sided and the fire resistance recommendations of section 7b(ii) of Table A2 of Appendix A (see section 7.21.1) will apply without any concessions. All car parks require some ventilation, therefore the provisions in the following section apply whatever standard of ventilation is provided.

7.28.4   Ventilation provisions

Ventilation may be provided by natural or mechanical means as follows.

- Natural ventilation to car parks that are not open-sided; provide either:
  - (a)   permanent openings to each storey at each car parking level with an aggregate area of at least 2.5% of the floor area at that level. At least half of the ventilation should be equally provided between two opposing walls; or
  - (b)   smoke vents at ceiling level with an aggregate area of permanent opening of at least 2.5% of the floor area, arranged to give a through draught.

(Reference should also be made to Approved Document F *Ventilation*, for additional guidance on normal ventilation of car parks.)

- Mechanical ventilation systems for basements and enclosed car parks should be:

(a) independent of any other ventilating system (other than any system pro-
     viding normal ventilation to the car park);
(b) designed to operate at ten air changes per hour in a fire situation.
(c) designed to run in two parts, each capable of extracting half of the amount
     which would be extracted at the rates set out in (b) above and designed so
     that each part may operate singly or simultaneously;
(d) provided with an independent power supply for each part of the system,
     capable of operating in the event of failure of the main supply;
(e) provided with extract points arranged with half the points at high level and
     half the points at low level;
(f) provided with fans rated at 300°C for a minimum of 60 minutes;
(g) provided with ductwork and fixings constructed with materials having a
     melting point of at least 800°C.

(Further information on equipment for removing hot smoke may be obtained from
BS 7346: Part 2 *Components for smoke and heat control systems, Specification for
powered smoke and heat exhaust ventilators*. For an alternative method of providing
smoke ventilation from enclosed car parks see BRE Report *Design methodologies for
smoke and heat exhaust ventilation* (BR 368, 1999).)

7.29 Shopping complexes

Individual shops contained in single separate buildings should generally be capable
of conforming to the recommendations of AD B. However, where a shop unit forms
part of a covered shopping complex, certain difficulties may arise. Such complexes
often include covered malls providing access to a number of shops and shared
servicing areas. Clearly, it is not practical to compartment a shop from a mall serving it and provisions dealing with maximum compartment sizes may be difficult to
meet. Certain other problems may arise concerning fire resistance, walls separating
shop units, surfaces of walls and ceilings, and distances to boundaries.

In order to achieve a satisfactory standard of fire safety certain alternative
arrangements and compensatory features to those set out in AD B may be appropriate. Reference should be made to sections 5 and 6 of BS 5588: Part 10: 1991 *Code
of practice for enclosed shopping complexes* and the relevant recommendations of
those sections should be followed.

7.30 Access and facilities for the fire service

7.30.1 Introduction

Part B5 of Schedule 1 to the 1991 Regulations introduced totally new requirements
for dealing with access and facilities for the fire service. It reflected guidance produced by the Home Office Fire Department for the Fire Service which had been in
use for many years, and replaced goodwill recommendations with statutory

requirements. Many local authorities had, through the medium of local Acts of Parliament, applied means of access regulations in their own districts or boroughs for a considerable number of years. These regulations tended to vary from authority to authority so the new requirements brought consistency to this important area of control. In order to avoid duplication of control, most provisions in local Acts of Parliament contain a statutory bar which gives precedence to Building Regulations.

7.30.2 Interpretation

The following terms occur throughout AD B5:

FIRE SERVICE VEHICLE ACCESS LEVEL – The level at which the fire service gain access to a building. This may not always be at ground level, e.g. in a podium design the access may be above ground level.

FIREFIGHTING LIFT – A lift which is designed to have additional fire protection in which the controls may be overriden by the fire service so that they may control it directly for use in fighting fires (see fig. 7.55).

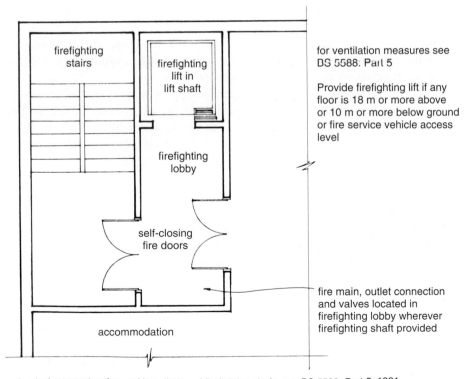

for design, construction and installation of firefighting shafts see BS 5588: Part 5: 1991

Fig. 7.55 Firefighting shafts and components.

FIREFIGHTING LOBBY – A protected lobby usually situated between a fire-fighting stair and the accommodation of a building which may also give access to any associated firefighting lift (see Fig. 7.55).

FIREFIGHTING SHAFT – A fire protected enclosure which contains a fire-fighting stair and firefighting lobbies. It may also contain a firefighting lift, if this is included in the building, together with its machine room. (See Fig. 7.55.)

FIREFIGHTING STAIR – A fire protected stair which is separated from the accommodation of the building by a firefighting lobby (see Fig. 7.55).

PERIMETER (of building) – The maximum aggregate plan perimeter. This is found by vertical projection of the building onto a horizontal plane and is illustrated in Fig. 7.56.

7.30.3 Access and facilities for the fire service – the statutory requirements

Buildings must be designed and constructed so as to provide reasonable facilities to assist firefighters in the protection of life. There must also be reasonable provision made within the site of the building to enable fire appliances to gain access to the building.

These requirements are interesting in that they apply provisions to the site of the building and not just to the building itself. This approach may be compared to that for access for disabled people in Chapter 17 of this book where site recommenda-tions are also made. It should be noted that there is no definition of 'site' contained in the regulations or approved documents.

The main factor that determines the facilities which are needed to assist the fire service is the size of the building, since it is the philosophy of firefighting in the United Kingdom that this be carried out inside the building if it is to be effective. This philosophy ensures that the water used for firefighting actually reaches the fire and means that effective search and rescue can be carried out as close as possible to the source of the fire since it is at this point that trapped people will be in most peril.

Therefore, in order to meet these statutory requirements it is necessary to provide:

(1) in most buildings:
   - sufficient means of vehicular access across the site of the building to enable fire appliances to be brought near to the building for effective use; and
   - sufficient means of access for firefighting personnel into and within the building so that they may effect rescue and fight fire;
(2) in large buildings and/or buildings with basements:
   - sufficient fire mains and other facilities, such as fire fighting shafts, to assist firefighters; and
   - adequate means of venting heat and smoke from basement fires.

It should be noted that these arrangements for access and firefighting facilities are required in order to secure reasonable standards of health and safety for people

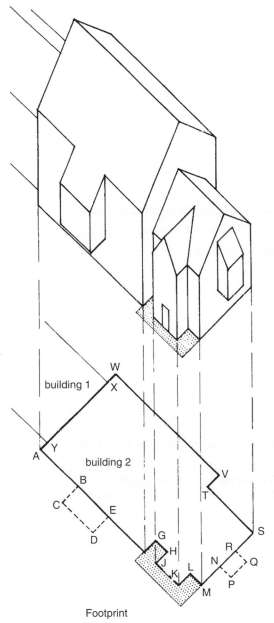

To calculate length of perimeter for purposes of Table 20 of AD B5, add together individual lengths (i.e. AB + BC + CD etc.). Do not include XY, as this is common to adjacent building and cannot be accessed anyway.

To find maximum aggregate plan perimeter (or footprint), project overhanging storeys onto ground floor plan, then footprint is outline denoted by letters opposite.

Example calculation
1. Total floor area of building = 600 m^2
2. Height of top storey above ground = 11 m
3. Perimeter length (less XY) = 60 m
4. From Table 20 provide access to 15% of perimeter i.e. 11 m shown by hatching on plan
5. Provide suitable door in this part of perimeter, min. 750 mm wide

Fig. 7.56   Calculation of perimeter.

(including firefighters) in or about buildings and for the purposes of protecting life by assisting the fire service.

7.30.4   Access facilities for fire appliances

Vehicle access to the exterior of a building is needed:

- to enable high-reach appliances (i.e. turntable ladders and hydraulic platforms) to be used; and
- to enable pumping appliances to supply water and equipment for rescue activities and firefighting.

Clearly, the requirements for access to buildings increase with building size and height. In large buildings it may be necessary to provide firefighting shafts and fire mains. Fire mains are provided in buildings to enable firefighters to connect their hoses to a convenient water supply at the floor level of the fire. Therefore, where these are fitted, it will be necessary for pumping appliances to gain access to the perimeter of the building at points near to the mains. This is especially so in the case of dry mains since these will need to be connected by hose to the pumping appliance. The provision of fire mains and firefighting shafts is described more fully below.

7.30.5   Buildings not fitted with fire mains

Fire mains need only be provided where there is a necessity to provide a firefighting shaft (see section 7.30.9 below). Therefore, in buildings not fitted with fire mains, access for fire service vehicles should be provided in accordance with Table 20 of AD B5 which is reproduced below. It should be noted that Table 20 does not apply to buildings with fire mains, or to blocks of flats and maisonettes, because every dwelling entrance door in such buildings should be within 45 m of a pump appliance.

   Buildings with a total aggregate floor area of up to 2000 m^2 and a top storey less

AD B5

Table 20   Fire service vehicle access to buildings (excluding blocks of flats) not fitted with fire mains.

Total floor area (1) of building m^2	Height of floor of top storey above ground (2)	Provide vehicle access (3)(4) to:	Type of appliance
up to 2000	up to 11 over 11	see paragraph 17.2 15% of perimeter (5)	pump high reach
2000–8000	up to 11 over 11	15% of perimeter (5) 50% of perimeter (5)	pump high reach
8000–16,000	up to 11 over 11	50% of perimeter (5) 50% of perimeter (5)	pump high reach
16,000–24,000	up to 11 over 11	75% of perimeter (5) 75% of perimeter (5)	pump high reach
over 24,000	up to 11 over 11	100% of perimeter (5) 100% of perimeter (5)	pump high reach

Notes:
1. The total floor area is the aggregate of all floors in the building (excluding basements).
2. In the case of Purpose Group 7(a) (storage) buildings, height should be measured to mean roof level, see Methods of Measurement in Appendix C of AD B
3. An access door is required to each such elevation (see paragraph 17.5).
4. See paragraph 17.9 for meaning of access.
5. Perimeter is described in Fig. 7.56.

than 11 m above ground level are referred to as 'small buildings' in AD B5 (see the first row of Table 20). There should be vehicle access to 15% of, or to within 45 m of every point on, the maximum aggregate plan perimeter (or footprint, see Fig. 7.56) of such buildings, whichever figure is the less onerous. For single family dwelling houses, the 45 m may be measured to a door to the dwelling.

In Table 20, the key figure to remember is 11 m above ground level for the top storey of the building. Buildings above this height not fitted with fire mains will need access for high-reach appliances as well as pumping appliances. There should be a suitable door giving access to the interior of the building, at least 750 mm wide, situated in any elevations which are required to be accessed by virtue of Table 20. A typical example of this is shown in Fig. 7.56.

7.30.6 Buildings fitted with fire mains

As mentioned above, buildings provided with firefighting shafts should also have fire mains.

Where dry fire mains are fitted:

- access should be provided for a pumping appliance to within 18 m of each fire main inlet connection point
- the inlet should be visible from the appliance.

Where wet fire mains are fitted:

- access should be provided for a pumping appliance to within 18 m of a suitable entrance giving access to the main
- the entrance should be visible from the appliance
- the inlet for the emergency replenishment of the suction tank for the main should be visible from the appliance.

Sometimes, fire mains are fitted in buildings even though there is no recommendation in AD B5 that they should have firefighting shafts. In such cases the provisions for access listed above may be used instead of Table 20.

7.30.7 Access routes and hardstandings

In order to provide access for fire service vehicles across the site of a building, it is necessary to design a road or other route which is wide enough and has sufficient load-carrying capacity (including manhole covers, etc.) to take the necessary vehicles. Unfortunately, fire appliances are not standardised and so it is a wise precaution to check with the local fire service in a particular area in order to establish their weight and size requirements. Some design guidance is given in Table 21 of AD B5 (reproduced below), where typical vehicle access route specifications are shown. It should be noted that the typical minimum carrying capacity for a high-reach appliance is 17 tonnes. A roadbase designed to take 12.5 tonnes should be satisfactory for high-reach vehicles since the use would be infrequent and the weight of the vehicle is distributed over a number of axles. However, structures such as bridges should still be designed to take the full 17 tonnes.

In general, where access is provided to an elevation in accordance with Table 20, the access routes and hardstandings should be designed in accordance with the following provisions:

- for buildings up to 11 m high, access should be provided for a pump appliance adjacent to the building, for the percentage of the total perimeter specified in Table 20 (this does not apply where a pump appliance can get to within 45 m of every point on the perimeter of a building or to within 45 m of every dwelling entrance door in blocks of flats or maisonettes)
- for buildings over 11 m high, a zone should be established in accordance with Fig. 7.57, which should be kept clear of overhead obstructions, since it is possible that overhead obstructions such as cables and branches might interfere with the setting of ladders or the swing of high-reach appliances.

Any dead-end access route which is more than 20 m long should be provided with a turning point, such as a hammerhead or turning circle, designed in accordance with dimensions given in Table 21 of AD B5 (see below). This is to ensure that fire service vehicles do not have to reverse more than 20 m from the end of an access road.

In all the above considerations for site access it should be borne in mind that requirements cannot be made under the Building Regulations for work to be done outside the site of the works shown on the submitted plans, building notice or initial

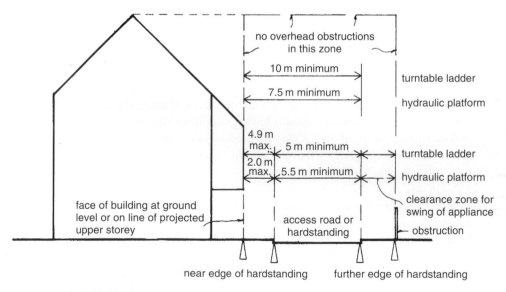

Provide hardstanding:
- as level as possible (no gradient steeper than 1 in 12);
- capable of withstanding a point load of 8.3 kg/cm^2 to accommodate jacks

Fig. 7.57   Overhead obstructions – access dimensions for high-reach appliances.

AD B5

Table 21 Typical vehicle access route specification.

Appliance type	Minimum width of road between kerbs (m)	Minimum width of gateways (m)	Minimum turning circle between kerbs (m)	Minimum turning circle between walls (m)	Minimum clearance height (m)	Minimum carrying capacity (tonnes)
Pump	3.7	3.1	16.8	19.?	3.7	12.5
High-reach	3.7	3.1	26.0	29.0	4.0	17.0

notice. Therefore, it may not always be reasonable to upgrade an existing route across a site to a small building, such as a single dwelling house. The reasonableness of any proposals to carry out upgrading of certain features (such as the removal of a sharp bend) should be considered by the local authority or approved inspector in consultation with the fire service.

7.30.8 Access for firefighting personnel to buildings

As has been mentioned above, it is important that fire service personnel are able to gain access to buildings in order to reach the seat of the fire and to carry out effective search and rescue.

This is especially true in tall buildings and in those with deep basements. In such buildings firefighters will need additional facilities contained within a protected firefighting shaft, such as firefighting lifts, firefighting stairs and firefighting lobbies, equipped with fire mains. Additionally, fire appliances will need access to entry points near the fire mains. These facilities are necessary in order to avoid delay in tackling the fire and to provide a safe working base from which effective action may be taken.

In other buildings, the normal means of escape will offer personnel access and this, when combined with the ability to work from ladders and appliances on the perimeter, will generally mean that special internal arrangements are unnecessary. Vehicle access will usually be needed to some or all of the perimeter, but the extent of this will depend on the size of the building.

Dwellings and other small buildings should be sufficiently close to as point accessible to fire brigade vehicles – no other provisions are necessary.

In taller blocks of flats, the high degree of compartmentation means that the access facilities may be simpler than in other types of tall buildings, however, fire brigade personnel access facilities will still be needed within the building.

In basement fires, products of combustion tend to escape via stairways. This can make it difficult for fire service personnel to gain access to the seat of the fire. The problem can be reduced by providing smoke vents, which improve visibility, reduce temperatures and make search, rescue and firefighting less difficult.

7.30.9 Firefighting shafts

Firefighting shafts should be provided to serve the storeys indicated in the following list.

(1) Buildings which contain two or more basement storeys each exceeding 900 m^2 in area.

(2) Buildings in Purpose Groups 4, 6 and 7(a) (Shop and Commercial, Industrial, and Storage and other non-residential) with any storey of 900 m^2 or more in area situated more than 7.5 m above ground or fire service access level.

(3) Buildings with any floor more than 18 m above ground or fire service vehicle access level.

(4) Buildings with any floor more than 10 m below ground or fire service vehicle access level.

(5) Shopping complexes, in accordance with the recommendations of Section 3 of BS 5588: Part 10: 1991 *Code of practice for enclosed shopping complexes*.

Firefighting shafts should conform to the following recommendations.

- Those provided in (3) and (4) above should also contain firefighting lifts.
- Where provided to serve a basement under (1) or (4) above, there is no need to serve the upper floors also unless they qualify in their own right because of the size or height of the building.
- Similarly, where a shaft is provided to serve upper floors in (2) or (3) above, it need not also serve a basement which is not large or deep enough to qualify on its own.
- Where they are provided, firefighting shafts and lifts should serve all intermediate floors between the lowest and the highest in the building.

7.30.10   Standard of provision of firefighting shafts

The number of firefighting shafts which needs to be provided in a building may be obtained from Table 7.11 below. It can be seen that if the building is fitted with a sprinkler system meeting the relevant recommendations of BS 5306: Part 2: 1990 *Specification for sprinkler systems*, then it is possible to reduce the number of firefighting shafts that are provided.

Firefighting shafts should be located so that every part of each storey is within 60 m of the entrance to a firefighting lobby measured along a route which is suitable for laying fire hoses. This figure is reduced to 40 m where the internal layout of the building is not known at the design stage, the 40 m being measured in a straight line from every point in the storey to the entrance of the firefighting lobby. These distance recommendations do not apply to accommodation situated at fire service access level.

7.30.11   Layout and construction of firefighting shafts

Firefighting shafts should be designed and constructed to encompass the following recommendations.

(1) Except in blocks of flats and maisonettes, access to the accommodation in a building from a firefighting lift or stair should be through a firefighting lobby.

Table 7.11   Provision of firefighting shafts in buildings.

Area of largest qualifying floor (m²)ᵃ	Number of firefighting shafts to be provided
A. With sprinklers fitted in the building (except basements)	
Under 900	1
900 to 2000	2
Over 2000	2 (plus 1 shaft for every extra 1500 m² or part thereof)
B. Without sprinklers and in any qualifying basement	
Up to 900	1
Over 900	2 (plus 1 shaft for every extra 900 m² or part thereof)

Notes:
ᵃ This is the largest floor area which is situated:
(a)   Over 18 m above fire service vehicle access level, or
(b)   Over 7.5 m or more above fire service vehicle access levels and 900 m² in area, or
(c)   In any qualifying basement (see section 7.30.9).

(2)   Every firefighting shaft should be equipped with a fire main which should have outlet connections and valves situated in fire-fighting lobbies.

(3)   Firefighting shafts should comply with the following parts of BS 5588: Part 5: 1991 *Code of practice for fire fighting stairs and lifts*:
- section 2: *Planning and construction*
- section 3: *Firefighting lift installation*
- section 4: *Electrical services*.

The various components of a firefighting shaft are illustrated in Fig. 7.55 above.

7.30.12   Provision of fire mains

Fire mains are provided to enable the fire service to fight fires inside the building. They are equipped with valves which permit direct connection of fire hoses. This assists firefighters by making it unnecessary to take hoses up stairways from the pumping appliance at ground or access level, thus saving time and avoiding blockage of the escape route.

Fire mains which serve floors above ground or access level are commonly known as rising mains and those which serve floors below ground or access level (such as basements) are usually referred to as falling mains.

Where it is necessary to provide firefighting shafts in a building (see section 7.30.10 above), each shaft should contain a fire main.

There are two types of fire main.

- Wet mains (often called 'wet risers') are usually kept full of water by header tanks and pumps in the building. Since there is the danger that the water supply may run out in a serious fire, there should be a facility to replenish the wet main from the pumping appliance in an emergency. Wet risers should be provided in any building which has a floor situated more than 60 m above ground or access level. They may, of course, serve lower floors if so desired.

- Dry mains ('dry risers') are normally kept empty and are charged with water from a fire service pumping appliance in the event of a fire. Where provided, they may serve any floor which is less than 60 m above ground or access level.

The outlets from fire mains at each floor level in the building should be situated in fire fighting lobbies giving access to the accommodation from a firefighting shaft.

Further guidance on the design and construction of fire mains may be obtained from sections 2 and 3 of BS 5306 *Fire extinguishing installations and equipment on premises,* Part 1: 1976 (1988) *Hydrant systems, hose reels and foam inlets.*

7.30.13   Smoke and heat venting of basements

A basement fire differs from a fire in another part of a building in that it is difficult for heat and smoke to be adequately vented. Products of combustion from basement fires tend to escape via stairways making it difficult for fire service personnel to gain access to the fire. Consequently, visibility will be reduced and temperatures will tend to be higher in a basement fire making search, rescue and firefighting more difficult. If smoke outlets (or smoke vents) are installed they can provide a route for heat and smoke, enabling it to escape direct to outside air from the basement and permitting the ingress of cooler air. Two typical designs for smoke outlet shafts are shown in Fig. 7.58.

7.30.14   Standard of provision for smoke outlets

Basement storeys which exceed 200 m^2 in area or are more than 3 m below ground level should be provided with smoke outlets connected directly to outside air. Some basements are excepted from this rule as follows:

- any basement in a single family dwellinghouse (PG l(b) or l(c))
- any strong room.

If possible each basement room or space should have one or more smoke outlets. In some basements the plan may be too deep, or there may be insufficient external wall areas to permit this. An acceptable solution might be to vent perimeter spaces directly and to allow internal spaces to be vented indirectly by the fire service by means of connecting doors. This solution is not acceptable if the basement is compartmented since each compartment should have direct access to venting without the use of intervening doors between compartments.

7.30.15   Means of venting

Smoke venting may be by natural or mechanical means.

Natural smoke venting may be achieved by providing smoke outlets which conform to the following recommendations.

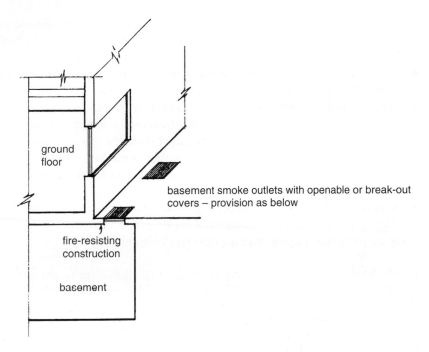

basement smoke outlets with openable or break-out covers – provision as below

outlets evenly distributed around perimeter and not placed where they would jeopardise escape routes

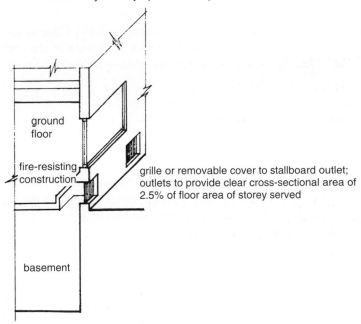

grille or removable cover to stallboard outlet; outlets to provide clear cross-sectional area of 2.5% of floor area of storey served

Fig. 7.58 Construction of smoke outlets.

- The total clear cross-sectional area of all the smoke outlets should be at least 2.5% of the floor area of the storey they serve.
- Places of special fire risk should be provided with separate outlets.
- Outlets should be positioned so that they do not compromise escape routes from the building.
- Outlets which terminate in readily accessible positions may be covered by pavement lights, stallboards or panels which can be broken out or opened. They should be suitably marked to indicate their position.
- Outlets which terminate in less accessible positions should be kept unobstructed and should only be covered by a louvre or grille which is non-combustible.
- Smoke outlets should be sited at high level (i.e. in the ceiling or wall of the space they serve) and should be distributed evenly around the perimeter of the building so as to discharge into open air outside the building.

Mechanical smoke extraction may be used as an alternative to natural venting if a sprinkler system conforming to BS 5306: Part 2 is installed in the basement. Unless needed for other reasons, it is not necessary to install sprinklers on the other storeys in the building merely because they are provided to allow mechanical smoke extraction in the basement to be used.

Any mechanical smoke extraction system should:

- achieve at least ten air changes per hour
- be capable of handling gas temperatures of 400°C for at least one hour
- come into operation automatically on activation of the sprinkler system, or be activated by an automatic fire detection system conforming to BS 5839: Part 1: 1988 (at least L3 standard).

7.30.16   Construction of smoke vent outlet ducts and shafts

Outlet ducts and shafts for smoke and heat venting should be enclosed in non-combustible fire-resisting construction. This applies equally to any bulkheads over the ducts or shafts as indicated in Fig. 7.58.

Natural smoke outlets from different compartments in the same basement storey or from different basement storeys should also be separated from each other by non-combustible fire-resisting construction.

7.30.17   Ventilation of basement car parks

The provisions contained in section 12 of AD B (see section 7.28.3 above) regarding the ventilation of basement car parks, may be regarded as satisfying the requirements contained in AD B5 for the smoke venting of any basement which is used as a car park.

7.31   Firefighting and the use of insulating core panels for internal structures

7.31.1   Introduction

Appendix F of Approved Document B provides a limited amount of guidance on the fire behaviour of insulating core panels when used for internal structures, where particular fire spread problems have been known to occur.

Internal insulating core panels are commonly used to provide chilled or subzero environment enclosures for the production, preservation, storage and distribution of perishable foodstuffs. They are also used where it is necessary to provide a hygienic environment.

A typical panel might consist of an inner insulating core sandwiched between, and bonded to, facing sheets of galvanised steel (sometimes bonded with a PVC facing where a hygienic finish is needed). The panels can be formed into a structure which can be free standing or can be attached to the building by lightweight fixings and hangers.

Common materials for the insulating core include:

- expanded or extruded polystyrene
- polyurethane
- mineral fibre
- polyisocyanurate
- modified phenolic.

7.31.2   The behaviour in fire of the core materials and fixing systems

When exposed to radiated or conducted heat from a fire, polymeric materials can be expected to degrade (and will produce large quantities of smoke). Fires involving mineral fibre cores produce less potential problems than those with polymeric cores.

It is also known that panels will tend to delaminate (the core separating from the facings because of expansion of the membrane and softening of the bond line) when exposed to the high temperatures of a developed fire, irrespective of the type of core material used.

Once it is involved in a fire, the panel will lose most of its structural integrity and the stability of the system will then depend on the residual structural strength of the non-exposed facing, the joint between panels and the fixing system. Because most panel jointing or fixing systems have an extremely limited structural integrity performance in fire conditions there is a real chance of total and unexpected collapse of the panel system (together with any associated equipment) if the fire starts to heat up the support fixings or structure to which they are attached. Fire can also spread behind the panels, where it may be hidden from the occupants of the building and this can prove to be a particular problem to fire fighters as, due to the insulating properties of the cores, it may not be possible to track the spread of fire, even using infra red detection equipment.

7.31.3   Problems for firefighters

When encountering a fire involving insulating core panel systems, firefighters may be confronted by the following problems:

* fire spread hidden within the panels;
* large quantities of black toxic smoke being produced; and
* rapid fire spread leading to flashover.

These characteristics are common to both polyurethane and polystyrene cored panels (although the rate of fire spread in polyurethane cores is significantly less than that of polystyrene cores) especially when any external heat source is removed.

Additionally, all systems are susceptible to the following problems, irrespective of the type of panel core used:

* delamination of the steel facing;
* collapse of the system; and
* hidden fire spread behind the system.

7.31.4   Design solutions

Risk assessment techniques should be used to identify the appropriate solution. The following design strategies can be adopted once the potential fire risks within the panel system enclosures have been identified:

* remove the risk
* provide an appropriate separation distance between the risk and the panels
* provide a fire suppression system for the risk
* provide a fire suppression system for the enclosure
* provide panels with suitable fire resistance
* specify appropriate materials, fixing and jointing systems.

Overall, the fire performance of the panel system should not be seen in isolation but should be considered in relation to the fire performance of the building as a whole, including the insulating envelope, the superstructure, the substructure etc.

7.31.5   Specification of panel core materials

Panels should be specified with core materials, which are appropriate to the application under consideration. This will help to ensure an acceptable level of performance for panel systems, under fire conditions. Table 7.12 below gives examples of core materials and appropriate applications.

7.31.6   Specification of materials/fixing and jointing systems

The aim of the specification should be to improve the stability of the panel system in the event of fire. Similarly, the aim of construction detailing should be to prevent the

Table 7.12 Insulating core panels – specification of core materials by application.

Core material	Applications[1]
Mineral fibre cores	Cooking areas, bakeries and other hot areas
	Fire breaks in combustible panels, fire stop panels and general fire protection
All core materials	Chill stores, cold stores and blast freezers
	Food factories
	Clean rooms

Notes: (1) Core materials may be used in other circumstances where a risk assessment has been made and other appropriate fire precautions have been put in place.

core materials from becoming exposed to the fire and contributing to the fire load. Therefore, it might be appropriate to consider the following.

- Design insulating envelopes, support systems, and supporting structure to maintain structural stability following failure of the bond line between insulant core and facing materials, by using alternative methods, such as catenary action. In a typical case this could require positive attachment of the lower faces of the insulant panels to supports.
- Where a supplementary support method is provided to support the panels, this should remain stable under fire conditions for an appropriate time. Therefore, consider fire protecting the building superstructure, together with any elements providing support to the insulating envelope, to prevent early collapse of the structure or the envelope.
- Light gauge steel members such as purlins and sheeting rails, which provide stability to building superstructures, may be compromised at an early stage of a fire. Although it is not practical to fire protect such members it may be possible to provide supplementary fire protected heavier gauge steelwork members at wider intervals than purlins to provide restraint in the event of a fire.
- In designated high risk areas, fire propagation through the insulant can be prevented by incorporating barriers consisting of, for example, non-combustible insulant cored panels into wall and ceiling construction at intervals, or strips of non-combustible material into specified wall and ceiling panels.
- The combustible insulant should be fully encapsulated by non-combustible facing materials that remain in place during a fire.
- Service penetration should be catered for by incorporating pre-finished and sealed areas in the panels.
- Do not allow panels or panel systems to support machinery or other permanent loads.
- Any cavity created by the arrangement of panels, their supporting structure or other building elements should be provided with suitable cavity barriers.

7.31.7 Further guidance on insulating core panels

Reference should be made to *Design, construction, specification and fire management of insulated envelopes for temperature controlled environments* published by the International Association of Cold Storage Contractors (European Division). This document contains examples of possible solutions and general guidance on insulating core panels construction. Chapter 8 of the document is of particular relevance since it gives guidance on the design, construction and management of insulated structures and is considered to be appropriate for most insulating core panel applications.

Important note

See January 2007 revision to Approved Document B summarised on page xiv.

8 Materials, workmanship, site preparation and moisture exclusion (Part C)

8.1 Materials and workmanship

8.1.1 Introduction

Regulation 7 of the 2000 Regulations is concerned with the fitness and use of the materials necessary for carrying out building work. It is supported by its own Approved Document entitled, rather aptly, *Approved Document to support Regulation 7* (AD Regulation 7). Apart from dealing generally with the standards of materials and workmanship needed for building work, AD Regulation 7 is also concerned with:

- the use of materials which are susceptible to changes in their properties;
- resistance to moisture and deleterious substances in the subsoil; and
- short-lived materials.

The use of materials which are unsuitable for permanent buildings is covered by section 19 of the Building Act 1984. Local authorities are enabled to reject plans for the construction of buildings of short-lived or otherwise unsuitable materials, or to impose a limit on their period of use. The Secretary of State may, by Building Regulations, prescribe materials which are considered unfit for particular purposes. Tables 1 and 2 of the 1976 Regulations listed materials which were considered unfit for the weather-resisting part of any external wall or roof. Neither the 2000 Regulations nor AD Regulation 7 prescribe any materials as unfit for particular purposes as yet; however the AD does lay down some general criteria against which materials may be judged (see sections 8.4.5 and 8.4.6). Bearing this in mind, it is unlikely that section 19 can be used by local authorities to proscribe certain materials. Since it is now possible to use materials and methods of workmanship which comply with European Standards or Technical Approvals, all references to Agrément Certificates in the 1992 edition of AD Regulation 7 were replaced in the 1999 edition by references to national or European certificates issued by a European Technical Approvals issuing body.

8.1.2 Interpretation

A large number of terms and abbreviations appear in AD Regulation 7 as follows:

BS – British Standard, issued by the British Standards Institution (BSI). To achieve British Standard status a draft document is prepared by relevant experts and is submitted for public consultation. Comments received are considered and consensus is reached before the proposed document is issued. More information on British Standards may be obtained from the BSI at 389 Chiswick High Road, London W4 4AL (www.bsi.org.uk).

BUILDING CONTROL BODY – Includes both local authorities and approved inspectors.

CE MARKING – Materials bearing the CE mark are presumed to comply with the minimum legal requirements as set out in the Construction Products Regulations 1991. This is described more fully in section 8.4.2.

CEN – Comité Européen de Normalisation. This is the body recognised by the European Commission (see below) to prepare harmonised standards to support the Construction Products Directive (CPD) (see section 8.3). The committee comprises representatives of the standards bodies of participating members of the EU and EFTA (European Free Trade Association).

CONSTRUCTION PRODUCTS REGULATIONS 1991 – These regulations (as amended) apply to any construction products which are produced for incorporation in a permanent manner in construction works, and which were supplied after 27 December 1991. The regulations are designed to ensure that when products are used in construction work, the work itself will satisfy any of the six relevant 'essential requirements' of the Construction Products Directive (see section 8.3)

CPD – Construction Products Directive (see section 8.3).

EEA (EUROPEAN ECONOMIC AREA) – Those states which signed the Agreement at Oporto on 2 May 1992 plus the Protocol adjusting that Agreement signed in Brussels on 17 March 1993. The states are Austria, Belgium, Denmark, Finland, France, Germany, Greece, Iceland, Ireland, Italy, Luxembourg, Liechtenstein, Netherlands, Norway, Portugal, Spain, Sweden and the United Kingdom.

EOTA – European Organisation for Technical Approvals. This is the umbrella organisation for bodies which issue European Technical Approvals for individual products. Whilst EOTA operates over the same area as CEN, it complements their work by producing guidelines for innovative and other products for which standards do not exist.

EN – These letters indicate that a European standard has been implemented in a particular Member State. Thus, in the United Kingdom the designation will be BS EN, followed by the relevant standard number. A British Standard shown in this way will be identical to the standards of other Member States, but will also include additional guidance regarding its use and its relationship with other standards in the same group. An EN does not have a separate existence as a formally published document.

EUROPEAN COMMISSION – Based in Brussels, this is the executive organisation of the EU. The Commission ensures that Community rules are implemented and observed and it alone has power to propose legislation based on the Treaties. It also executes decisions taken by the Council of Ministers.

EUROPEAN TECHNICAL APPROVAL – A technical assessment which specifies that a construction product is fit for its intended use. It is issued for the purposes of the Construction Products Directive (CPD) and the issuing body must be authorised by a Member State and be notified to the European Commission under section 10 of the CPD. Details of the approval issuing bodies are published in the 'C' series of the Official Journal of the European Communities. For the United Kingdom, in addition to the British Board of Agrément the current listing also includes BRE Certification Ltd, Garston, Watford, Herts WD25 9XX. A current list can be found on the Building Regulations pages of the DCLG website at: www.communities.gov.uk

EU – The 15 countries of the European Union, *viz.* Austria, Belgium, Denmark, Finland, France, Germany, Greece, Ireland, Italy, Luxembourg, Netherlands, Portugal, Spain, Sweden and the United Kingdom.

ISO – International Organisation for Standardisation. This is the worldwide standards organisation and it is likely that some ISO standards may be adapted for use with the CPD. Such standards are identified by 'ISO' and a number and in the UK they may appear as BS ISO or, if they are adopted as European standards, as BS EN ISO. Unlike ENs, ISOs are published separately.

TECHNICAL SPECIFICATION – A standard or European Technical Approval Guide. A document against which compliance can be shown (e.g. for standards) and against which an assessment is made in order to deliver the European Technical Approval.

UKAS – United Kingdom Accreditation Service. An accreditation body for quality assurance and management schemes. Further details may be obtained from UKAS, 21–47 High Street, Feltham, Middlesex TW3 4UN.

8.2   The influence of European standards

In order to understand the changes which have come about with the advent of the Single European Market, the following brief summary sets out the context of these changes and their influence on the building control system in England and Wales.

A main goal of the European Community is to allow free movement among the Member States of goods, services, people and capital. Free movement of goods may be hampered by physical, technical or fiscal barriers.

Significant technical barriers arise from the use of different technical requirements or regulations in Member States. This results in the necessity to produce slightly different versions of the same product to satisfy different markets and different methods of test for suitability must be used, resulting in undue expense and waste of resources by manufacturers and suppliers.

To overcome these difficulties, early directives were issued for which it was necessary to resolve technical issues with unanimous agreement by all Member States. This turned out to be a slow and cumbersome process.

8.2.1   The New Approach

The Single European Act in 1986 declared an agreement to establish the Single European Market by the end of 1992. Progress to the single market has been aided by a radical change in the approach to the writing of directives and European standards – the so-called New Approach. This recognises that EU legislation should only apply to areas already subject to existing national laws or regulations and also allows qualified majority voting into the decision-making process.

The New Approach Directives express requirements in broad terms, called the Essential Requirements. Member States presume conformity with these requirements where a product satisfies a harmonised European technical specification or, as an interim measure, a national standard accepted by the Commission. The advent of European standards will prevent Member States from using their own standards to protect their own markets.

A number of New Approach product directives have been adopted which are relevant to the construction industry, the most significant being the Construction Products Directive 89/106 EEC.

8.3   The Construction Products Directive (CPD)

The full title of this directive reflects its objectives:

> 'Council Directive of 21 December 1988 on the approximation of laws, regulations and the administrative provisions of the Member States relating to construction products.'

The CPD was implemented in the UK on 27 December 1991 by the Construction Products Regulations 1991 (see also DOE Circular 13/91).

Construction products are defined as:

> 'Those produced for incorporation in a permanent manner in construction works, in so far as the Essential Requirements (ERs) relate to them.'

There are six Essential Requirements as follows:

(1) Mechanical resistance and stability.
(2) Safety in case of fire.
(3) Hygiene, health and the environment.
(4) Safety in use.
(5) Protection against noise.
(6) Energy economy and heat retention.

The link between the Essential Requirements and the product on the market is made in the Interpretative Documents (IDs).

8.4   Interpretative Documents

There is one ID for each Essential Requirement. These interpret the ERs more fully and indicate:

- appropriate product characteristics;
- appropriate topics for harmonised technical specifications;
- the need for different levels or classes of performance to allow for different regulation requirements in different Member States.

The UK is represented by experts from the DCLG and BRE on the technical committees and drafting panels for the IDs. The IDs are intended mainly for standards' writers and enforcement authorities and will be of less use to manufacturers and suppliers.

8.4.1   Technical specifications

Sometimes referred to as harmonised technical specifications, these are deemed to comply with the Essential Requirements and so products meeting the specifications will immediately demonstrate their fitness to be placed on the market.

Three types exist:

(1) *Harmonised Standards* – ideally the best route for demonstrating compliance for a product. These standards are developed mainly by CEN on the basis of standardisation requests (called mandates) from the Commission. Only those parts of standards which relate to the ERs are mandated and these are the parts which support fixing of the CE mark to products.

(2) *European Technical Approvals* – these have replaced the Agrément certificate form of product assessment and are carried out by bodies designated by individual Member States which are then notified by that Member State to the European Commission. All such bodies are members of the European Organisation for Technical Approvals (EOTA), and they operate under a common set of rules.

(3) *National Specifications* – although a possibility, the recognition of national specifications at community level is likely to be rare. The procedures have to be initiated by the Commission and their use may be expected only in situations where a barrier to trade has been demonstrated and the production of a Technical Specification is some way off.

8.4.2   The CE mark and attestation of conformity

The purpose of attestation of conformity with technical specifications is to assure purchasers and regulators that products placed on the market comply with the ERs. Such products may carry the CE mark.

Each Member State is required to maintain a register of designated or notified bodies which identifies:

- Test laboratories – for testing samples;
- Inspection bodies – for inspecting factories and processes; and
- Certification authorities – to interpret results,

so as to allow the manufacturer to affix the CE mark to his product. The marking may be placed on the product itself, a label, the packaging or on the accompanying commercial information. It will be accompanied by a reference to the technical specification to which it conforms and, where appropriate, by an indication of its characteristics. In the UK the register of designated bodies is maintained by the DCLG.

It is important to appreciate that the CE mark is not a quality mark. It signifies only that the product satisfies the requirements. The CE mark is therefore primarily intended for enforcement officers and only states that the product may legally be placed on the market.

8.4.3   Interim procedures

It will be apparent that in order to harmonise the standards of all the countries in the European Union it is necessary for there to be a transitional period during which existing standards will have to co-exist with new standards.

With specific reference to British Standards, nearly all those which are related to construction products will be revised to become the British 'transposition' of the new European Standards (ENs) which are currently being drafted. In the past it has been the practice to adopt the old British Standard number when transposing an EN which is based on substantially the same material. Current BSI numbering policy is to adopt the CEN numbering, prefaced with BS.

Although British Standards are normally withdrawn when their equivalent European Standards are published the following circumstances may require a deferred withdrawal of the British Standard:

- when it is necessary for the BS to remain available for work which has already commenced;
- when a BS is called up in an Approved Document which has yet to be revised.

In practice it may be necessary for some BSs to remain valid for a number of years, fully maintained alongside the new transposed standards of European origin. Therefore, it will be necessary for controllers, designers, etc., to check the applicability of the standard in each context. Where an old standard is retained it is reasonable to presume that it will satisfy the requirements of regulation 7, and where a new standard is introduced it must be checked for applicability during the transitional period and if found suitable, compliance may reasonably also be presumed.

Standards of European origin have clauses specifically identified which relate to the 'harmonised' requirements relevant to the Building Regulations (i.e. those dealing with health and safety matters) and 'non-harmonised' requirements which contain additional material relating to trading requirements of the construction industry. The non-harmonised requirements are not the concern of regulation 7 and are not covered by the AD Regulation 7 recommendations.

Interestingly, it is possible for a product to be tested and certified as complying with a British Standard by an approved body in another Member State of the European Community under the procedure covered by article 16 of the CPD. In this case it should normally be accepted by the building control body as complying with that standard. If there are reasons to doubt its acceptance then the burden of proof is on the controlling body which is obliged to notify the Trading Standards Officer so that the UK Government can notify the Commission.

With regard to CE marking the UK Construction Product Regulations state only that products must be 'fit for intended use', without reference to how this is demonstrated. The Regulations also apply only to products supplied after 27 December 1991. Therefore, any product which was legally usable before that date can continue to be sold.

Where a manufacturer identifies an existing barrier to trade with another Member State, a procedure exists under article 16 of the CPD to overcome this. This involves the testing of products against the requirements of the importing country by a notified body in the exporting country. This procedure will not lead to a CE mark.

Not all materials will necessarily be CE marked under the CPD and, from a practical viewpoint, it will not be possible for all products to be CE marked until all the relevant technical specifications are available. However, it should be noted that for some products CE marking is compulsory under other Directives (such as gas boilers).

8.4.4   Materials and workmanship generally

Building work must be carried out:

- with proper and adequate materials which are:
  (1) appropriate for the circumstances in which they are used,
  (2) adequately mixed and prepared, and
  (3) applied, used or fixed so as adequately to perform the functions for which they are designed; and
- in a workmanlike manner.

Guidance on the choice and use of materials, and on ways of establishing the adequacy of workmanship, is given in AD Regulation 7. It should be noted however, that materials and workmanship are controlled only to the extent of:

- securing reasonable standards of health and safety for persons in or about buildings for Parts A to D, F to K, N and P of Schedule 1 (except for paragraphs H2 and J6);
- conserving fuel and power in Part L; and
- providing access and facilities for people in Part M.

Therefore, although it may be desirable for reasons of consumer satisfaction or protection to require higher standards, this cannot be required under building regulations.

In order to achieve a satisfactory standard of performance **materials** should be:

LEEDS COLLEGE OF BUILDING
LIBRARY

- suitable in nature and quality in relation to the purposes for which, and the conditions in which, they are used.

Additionally, **workmanship** should be such that, where relevant, materials are:

- adequately mixed and prepared (for example, in concrete mixes, the correct proportions must be used, there should be an appropriate water/cement ratio, mixing should be thorough, etc.); and
- applied, used or fixed so as to adequately perform their intended functions (for example, for reinforced concrete this would give control over the actual placing of the concrete, positioning of reinforcement, curing, etc.).

The definition of materials is quite broad and covers products, components, fittings, naturally occurring materials (such as timber, stone and thatch), items of equipment, and materials used in the backfilling of excavations in connection with building work. It should be noted that building regulations do not seek to control the use of materials after completion of the building work.

In order to reduce the environmental impact of building work careful thought should be given to the choice of materials, and where appropriate, recycled or recyclable materials should be considered. Obviously, the use of such materials must not have an adverse effect on the health and safety standards of the building work.

8.4.5   Fitness of materials

A number of ways of establishing the fitness of materials are dealt with in AD Regulation 7 and whilst this is mostly by reference to British Standards or certificates issued by European Technical Approvals issuing bodies other materials or products may be suitable in the particular circumstances. The following aids to establishing the fitness of materials are given in the Approved Document.

- A material may conform to the relevant provisions of an appropriate British Standard (but see the notes under interim procedures in section 8.4.3).
- A material may conform to the national technical specifications of other Member States which are contracting parties to the European Economic Area. It should be noted that where a person intends to use a product which complies with a national technical specification of another Member State, the onus is on that person to show that the product is equivalent to the relevant British Standard (and it would be necessary to provide a translation).
- A material may be covered by a national or European certificate issued by a European Technical Approvals issuing body. The conditions of use must be in accordance with the terms of the certificate and again, it will be up to the person intending to use the product to demonstrate equivalence and provide a translation.
- A material may bear a CE marking confirming that it conforms with a harmonised European standard or European Technical Approval together with the appropriate attestation procedure. Materials bearing the CE marking must be accepted if they are appropriate for the circumstances in which they are used. AD Regulation 7 qualifies

this by saying that a CE marked product can only be rejected by a building control body on the basis that:

o its performance is not in accordance with its technical specification, *or*

o where a particular declared value or class of performance is stated for a product, the resultant value does not meet Building Regulation requirements.

The burden of proof is on the controlling body which is obliged to notify the Trading Standards Officer so that the UK Government can notify the Commission.

- Independent certification schemes (e.g. the kitemark scheme operated by the British Standards Institution) may also serve to show that a material is suitable for its purpose. However, some materials which are not so certified may still conform to a relevant standard. In the UK, many certification bodies which approve such schemes are accredited by UKAS.

- A material may be shown to be capable of performing its function by the use of tests, calculations or other means. It is important to ensure that tests are carried out in accordance with recognised criteria. UKAS run an accreditation scheme for testing laboratories, and together with similar schemes run by equivalent certification bodies (including accreditation schemes operated by other Member States of the EU), this ensures that standards of testing are maintained.

- In some cases past experience of a material in use in a building may be relied upon to ensure that it is capable of adequately performing its function.

- Local authorities are entitled under regulation 19 to take and test samples of materials in order to confirm compliance. Approved inspectors have similar powers under regulation 11 of the Building (Approved Inspectors etc.) Regulations 2000 (as amended).

8.4.6   Short-lived materials

Only general guidance is given on the use of short-lived materials. These are materials which may be considered unsuitable due to their rapid deterioration when compared to the life of the building.

The main criteria to be considered are:

- accessibility for inspection, maintenance and replacement;
- the effects of failure on public health and safety.

Clearly, if a material or component is inaccessible and its failure would create a serious health risk, it is unlikely that the material or component would be suitable. (See also the reference to section 19 of the Building Act 1984 in section 8.1.1.)

8.4.7   Materials subject to changes in their properties

Under certain environmental conditions, some materials may undergo a change in their properties which may affect their performance over time. A notable example of this occurred during the 1970s to structures constructed using high alumina cement (HAC). The subsequent deterioration of the concrete led to the collapse of a number of long span roof structures and the use of HAC was banned for all work except when the material was used as a heat-resisting material. It is known that a number of other materials (such as certain stainless steels, structural silicone sealants and intumescent paints) may also be susceptible to changes in their properties under certain environmental conditions. In order to use such materials it will be necessary to estimate their final residual properties (including their structural properties) at the time the materials are incorporated into the work. It should then be shown that these residual properties will be adequate for the building to perform its intended function for its expected life.

8.4.8   Resistance to moisture and soil contaminants

Materials which are likely to suffer from the adverse effects of condensation, ground water or rain and snow may be satisfactory if:

- the construction of the building is such as to prevent moisture from reaching the materials; or
- the materials are suitably treated or otherwise protected from moisture.

Similarly, materials which are in contact with the ground will only be satisfactory if they can adequately resist the effects of deleterious substances, such as sulphates (see Site preparation and moisture exclusion in section 8.5).

8.4.9   Adequacy of workmanship

It should be remembered that Building Regulations set different standards of workmanship to those imposed by, for example, a building specification. Building Regulations are not concerned with quality or value for money; they are concerned with public health and safety, the conservation of fuel and power, and access and facilities for people in buildings.

Adequacy of workmanship, like that of materials, may be established in a number of ways.

- A British Standard Code of Practice or other equivalent technical specification (e.g. of Member States which are contracting parties to the European Economic Area) may be used. In this context, BS 8000 *Workmanship on Building Sites* may be useful since it gathers together guidance from a number of other BSI Codes and Standards.
- Technical approvals, such as national or European certificates issued by European Technical Approvals issuing bodies, often contain workmanship recommendations. Additionally, it may be possible to use an equivalent technical approval (such as those of a member of EOTA) if this provides an equivalent level of protection and performance. The onus of proof of acceptability rests with the user in this case.
- Workmanship which is covered by a scheme complying with BS EN ISO 9000 *Quality management and quality assurance standards*, will demonstrate an acceptable standard since these schemes relate to processes and products for which there may also be a suitable British or other technical standard. A number of such schemes have been accredited by UKAS. There are also a number of independent schemes for accreditation and registration of installers of materials, products and services and these ensure that work has been carried out to appropriate standards by knowledgeable contractors.

- In some cases past experience of a method of workmanship such as a building in use, may be relied upon to ensure that the method is capable of producing the intended standard of performance.
- Local authorities are empowered under regulation 18 to make such tests of any building work to enable them to establish if the work complies with regulation 7 or any other applicable requirements of Schedule 1. Approved inspectors have similar powers under regulation 11 of the Building (Approved Inspectors etc.) Regulations 2000 (as amended).

8.5   Site preparation and moisture exclusion

8.5.1   Introduction

Part C of Schedule 1 to the 2000 Regulations is concerned with site preparation and resistance to moisture. In addition to moisture exclusion, paragraph C2 contains provisions controlling sites containing dangerous or offensive substances. This replaces section 29 of the Building Act 1984.

The supporting Approved Document C now contains recommendations relating to the control of radon gas in certain areas of England and Wales, and landfill gases on certain sites near waste disposal tips, etc. The guidance on the control of dampness in buildings in Approved Document C now covers damage from condensation and mould growth. This was formerly found in Approved Document F, Ventilation.

Certain provisions regarding the preparation of the site, damp-proofing and weather resistance of floors and walls do not apply to buildings used solely for storage of plant or machinery in which the only persons habitually employed are storemen, etc. Other similar types of buildings where the air is so moisture-laden that any increase would not adversely affect the health of the occupants are also excluded.

8.5.2   Preparation of site – site investigation

In order to prepare a site for construction work it will usually be necessary to carry out a site investigation. This will entail a study of site conditions to determine their probable influence on the design, construction and subsequent performance of a building. The following items may be encountered and are discussed in Approved Document C:

- unsuitable material such as turf and roots;
- mature trees which might affect services, floor slabs, oversite concrete and foundations;
- pre-existing foundations, services and other infra-structure, and buried tanks;
- fill or made up ground;
- contaminants;
- high water table, which might necessitate subsoil drainage to avoid damage to the building.

Additionally, the site investigation will identify the type and nature of the subsoil and its loadbearing capacity which will be of use in the design of foundations in accordance with Part A (see Chapter 7).

The site investigation will normally consist of the following stages:

(1) **Planning stage.** In this stage the objectives, scope and requirements of the investigation are set so as to enable it to be planned and carried out efficiently and so that the required information may be provided.

(2) **Desk study.** A desk study is a review of the historical, geological and environmental information about the site. The information can be obtained much more quickly and cheaply than information from boreholes and trial pits and a great deal of factual information is publicly available in the UK from sources such as geological maps, Ordnance Survey sheets, aerial photographs, computer satellite databases, geological books and records, local authority building control data, mining records, reports of previous site investigations, etc.

(3) **Site reconnaissance or walkover survey.** The site reconnaissance or walkover survey aids the design of the main investigation in that it identifies actual and potential physical hazards. The object of the walkover survey is to check and make additions to the information already collected during the desk study. The site and its surrounding area are visited and covered carefully on foot. When a walkover survey is carried out, it is also possible to gather information from local authorities, local inhabitants and people working in the area, such as builders, electricity and gas workers. On completion of the survey a structured report can be produced from the information gathered at the site and from the local enquiries.

(4) **Main investigation and reporting.** This will usually include an examination of the geotechnical properties of the ground and should be designed to verify and expand information previously collected from the desk study and site walkover survey. The investigation will usually include intrusive and non-intrusive sampling and testing by means of trial pits and boreholes to provide soil parameters for design and construction.

In determining the extent and level of investigation it will be necessary to consider the type of development proposed and the previous use of the land. Typically the site investigation should include:

- susceptibility to groundwater levels and flow;
- underlying geology; and
- ground and hydro-geological properties.

A geotechnical site investigation should:

- identify physical hazards for site development;
- determine an appropriate design; and
- provide soil parameters for design and construction.

Where there is concern that the site might be affected by contaminants, a combined geotechnical and geo-environmental investigation should be considered. Section 2 of Approved Document C *Resistance to contaminants* (which is considered in section 8.5.5) contains guidance on assessing and remediating sites affected by contaminants.

8.5.3   Site investigation – sources of guidance

Comprehensive guidance on site investigations may be found in BS 5930: 1999 *Code of practice for site investigations*.

For low-rise buildings the following documents may also be consulted:

- BRE Digest 322 *Site investigation for low-rise building: procurement*, 1987;
- BRE Digest 318 *Site investigation for low-rise building: desk studies*, 1987;
- BRE Digest 348 *Site investigation for low-rise building: the walk-over survey*, 1989;
- BRE Digest 381 *Site investigation for low-rise building: trial pits*, 1993;
- BRE Digest 383 *Site investigation for low-rise building: soil description*, 1993;
- BRE Digest 411 *Site investigation for low-rise building: direct investigations*, 1995;
- BS 8103: Part 1: 1995 *Structural design for low rise buildings*.

8.5.4   Preparation of site – clearance or treatment of unsuitable material

The ground to be covered by the building is required to be reasonably free from any material that might damage the building or affect its stability. This includes vegetable matter, topsoil and pre-existing foundations. (See Requirement C1(1)).

Decaying vegetable matter could be a danger to health and it could also cause a building to become unstable if it occurred under foundations. Approved Document C, therefore, recommends that the site should be cleared of all turf and vegetable matter at least to a depth to prevent future growth. This might not apply to buildings used solely for storage of plant or machinery in which the only persons habitually employed are storemen, etc. engaged only in taking in, caring for or taking out the goods. Other similar types of buildings where the air is so moisture-laden that any increase would not adversely affect the health of the occupants are also excluded.

Below-ground services (such as foul or surface water drainage) should be designed to resist the effects of tree roots. This can be achieved by making services sufficiently robust or flexible and with joints that cannot be penetrated by roots. Consideration should be given to the removal of roots where they could pose a hazard to below-ground services,

Where a site has been previously built on it will be necessary to consider if it contains any pre-existing foundations, services, buried tanks and any other infra-structure, etc. that could be a danger to persons in and about the building and any land associated with the building. In this context, Approved Document C defines *building and land associated with the building* as 'the building and all the land forming the site subject to building operations which includes land under the building and the land around it which may have an effect on the building or its users'. This definition is further clarified in paragraph 2.11 of Approved Document C, where it refers to the 'area of the site subject to building operations', i.e. 'those parts of the land associated with the building that include the building itself, gardens and other places on the site that are accessible to users of the building and those in and about the building'.

Where shrinkable clays soils are present on a site, the presence of mature trees can exacerbate the tendency of the soil to cause heave and subsidence and this may lead to damage to services, floor slabs and oversite concrete. On such soils, the potential for damage should be assessed and Approved Document C gives guidance, in general terms, on the likely potential for volume change for some commonly occurring clays. Reference should be made to Diagram 1 from Approved Document C (reproduced below) to ascertain the type of clay that might be occurring. When used with Table 1 from Approved Document C (shown on the same diagram) the volume change potential can be approximately assessed.

For more detailed guidance, reference may also be made to BRE Digest 298 *Low-rise building foundations: the influence of trees in clay soils*, 1999.

AD C
Diagram 1   Distribution of shrinkable clays and principal sulphate/sulphide bearing strata in England and Wales

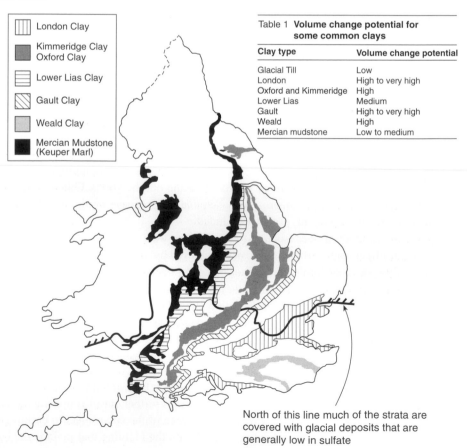

Key:
- London Clay
- Kimmeridge Clay / Oxford Clay
- Lower Lias Clay
- Gault Clay
- Weald Clay
- Mercian Mudstone (Keuper Marl)

Table 1 **Volume change potential for some common clays**

Clay type	Volume change potential
Glacial Till	Low
London	High to very high
Oxford and Kimmeridge	High
Lower Lias	Medium
Gault	High to very high
Weald	High
Mercian mudstone	Low to medium

North of this line much of the strata are covered with glacial deposits that are generally low in sulfate

Sometimes it becomes necessary to remove significant quantities of soil. In these cases reference may also be made to:

- BRE Digest 241 *Low-rise buildings on shrinkable clay soils*: Part 2, 1993.
- BRE Digest 242 *Low-rise buildings on shrinkable clay soils*: Part 3, 1993 (Note: Although this document is referred to in Approved Document C, the author was unable to trace it, even after searching the BRE website);
- the Foundation for the Built Environment (FBE) report *Subsidence damage to domestic buildings: lessons learned and questions remaining*, FBE, 2000.

The National House Building Council (NHBC) Standards Chapter 4.2 *Building near trees*, 2003 can also be consulted to assess the effects of remaining trees on services and building movements close to the building.

Sites which contain fill or made ground can present particular problems related to the compressibility of the ground and its potential for collapse when wetted. Appropriate remedial measures may need to be taken to prevent differential settlement from causing damage to the building. Guidance on these issues may be found in BRE Digest 427 *Low-rise buildings on fill* and BRE Report BR 424 *Building fill: Geotechnical aspects*, 2001.

8.5.5   Resistance to contaminants – introduction

Reasonable precautions must be taken to prevent any contaminants found on or in the ground from causing a danger to health and safety (see requirement C1(2)). This is, of course, the ground covered (or to be covered) by the building and includes any land associated with the building (see definition in section 8.5.4 above).

There is a special definition of *contaminant* for the purposes of requirement C1(2) – any substance which is or could become harmful to persons or buildings including substances that are toxic, corrosive, explosive, flammable or radioactive.

Contaminants that occur on sites can be liquids, solids or gases and can arise out of a previous use of land, especially where this was related to an industrial undertaking. In recent years problems have arisen from the emission of landfill gas from waste disposal sites where it is common for biodegradable waste to be buried. Even sites which have been used for rural purposes, such as agriculture, may be contaminated by pesticides, fertiliser, fuel and oils and decaying matter of biological origin. The author remembers encountering an edge-of-town site contaminated with the carcasses of cattle slaughtered in an outbreak of foot and mouth disease from the 1950s.

Where a site is being redeveloped, knowledge of its previous use, from planning or other local records, may indicate a possible source of contamination. Table 2 to section 2 of AD C (reproduced below) lists a number of site uses that are likely to contain contaminants. It should not be considered to be an exhaustive list. It is derived from the *Industry Profile* guides published by the former Department of the Environment (*Department of the Environment Industry Profiles*, 1996). Each of these profiles considers a different industry which has the potential to cause contamination. The particular contaminant associated with the industry is identified together with details of where it may be found on the site and the routes it might take to migrate to other areas. Further information on the assessment of land affected by contaminants (including a large number of industrial sites where certain contaminants may be found) is presented in Defra/Environment Agency

Contaminated Land Research Report CLR 8 *Potential contaminants for the assessment of land*, 2002 (this is referred to incorrectly in AD C as *Priority* contaminants for the assessment of land).

Additionally, in certain parts of the country, the following naturally occurring contaminants can arise from the underlying geology:

- mining areas:
  - ○ certain heavy metals, such as cadmium and arsenic;
  - ○ gases such as methane and carbon dioxide (mainly from coal mining areas);
- carbon dioxide and methane gases arising from organic rich soils and sediments (such as peat and river silts) (see section 8.5.12).

Further information on these contaminants including their geographical extent, associated hazards, site investigation methods, and protective measures that can be taken can be obtained from the following Environment Agency publications:

- in England – Environment Agency R & D Technical Report P291 *Information on land quality in England: Sources of information (including background contaminants)*;
- in Wales – Environment Agency R & D Technical Report P292 *Information on land quality in Wales: Sources of information (including background contaminants)*.

Other naturally-occurring contaminants include:

- the radioactive gas radon;
- sulfates.

AD C
Table 2   Examples of sites likely to contain contaminants.

Animal and animal products processing works
Asbestos works
Ceramics, cement and asphalt manufacturing works
Chemical works
Dockyards and docklands
Engineering works (including aircraft manufacturing, railway engineering works, shipyards, electrical and electronic equipment manufacturing works)
Gas works, coal carbonisation plants and ancillary by-product works
Industries making or using wood preservatives
Landfill and other waste disposal sites
Metal mines, smelters, foundries, steel works and metal finishing works
Munitions production and testing sites
Oil storage and distribution sites
Paper and printing works
Power stations
Railway land, especially the larger sidings and depots
Road vehicle fuelling, service and repair: garages and filling stations
Scrap yards
Sewage works, sewage farms and sludge disposal sites
Tanneries
Textile works and dye works

Contamination by radon gas and its products of decay has led to concern over the long-term heath of occupants of affected buildings. Measures to protect buildings and occupants against ingress of radon gas are considered in section 8.5.9 below. Sulfate attack affects concrete floor slabs and foundations and can cause failure and disruption. Measures can be taken (such as the use of sulfate-resisting cement) in those areas where naturally occurring sulfates are present. Diagram 1 and Table 1 from Approved Document C (illustrated above) show the principle areas of sulphate-bearing strata in England and Wales. Reference may also be made to BRE Special Digest SD1 *Concrete in aggressive ground*, 2003 where guidance will be found on investigation, concrete specification and design to mitigate the effects of sulfate attack.

8.5.6   Resistance to contaminants – risk assessment

Introduction and general concepts

In order to develop potentially contaminated land safely it is necessary to carry out a risk assessment.

This is normally done by adopting a tiered approach in which an increasing level of detail is required as the tiers are worked through. For a full risk assessment it will be necessary to progress through the following three tiers:

- preliminary risk assessment;
- generic quantitative risk assessment (GQRA); and
- detailed quantitative risk assessment (DQRA).

The need for a risk assessment will usually be identified during the first stages of the site investigation (desk studies and site walkover survey). Once this need has been identified, a preliminary risk assessment must always be undertaken. The extent to which it is necessary to do a more detailed risk assessment will depend on the situation and outcome of the preliminary assessment. This may indicate that it will be necessary to do only one or other (or both) of the more detailed risk assessments.

The general approach to risk assessment described above is based on the concept of the relationship between the *source* of contamination (the contaminants found on or in the ground), the *pathway* taken by the contaminants (e.g. ingestion, inhalation, direct contact, attack on building materials and services) and the *receptor* of those contaminants (i.e. buildings, building materials and services and people). This 'source–pathway–receptor' relationship, or pollutant linkage, is illustrated in Fig. 8.1.

The development of contaminated land has the inevitable consequence of introducing receptors (buildings, building services, building materials and people) onto the development site. In order to mitigate the effects of the pollutants on the receptors the pollutant linkages must be broken. This can be achieved in a number of ways, for example, by:

- using physical, chemical, biological or other processes to treat the contaminant so as to eliminate or reduce its toxicity or harmful properties;
- removing or blocking the contaminant pathway (e.g. by installing barriers to prevent migration or protective layers to isolate the contaminant);

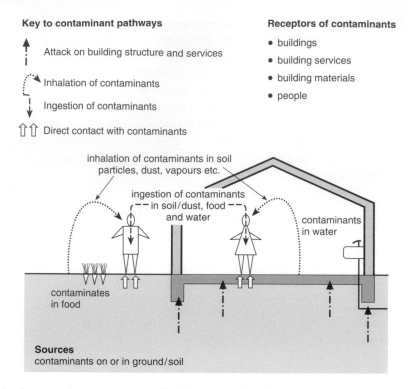

Key to contaminant pathways

Attack on building structure and services

Inhalation of contaminants

Ingestion of contaminants

Direct contact with contaminants

Receptors of contaminants

- buildings
- building services
- building materials
- people

inhalation of contaminants in soil particles, dust, vapours etc.

ingestion of contaminants in soil/dust, food and water

contaminants in water

contaminates in food

Sources
contaminants on or in ground/soil

Fig. 8.1   Source–pathway–receptor – typical site conceptual model.

- removing or protecting the receptor (perhaps by the use of appropriately designed building materials, or by changing the form or layout of the development);
- removing the contaminant by excavating the contaminated material.

Risk assessment stages

For each of the tiers mentioned above (preliminary risk assessment, generic quantitative risk assessment (GQRA) and detailed quantitative risk assessment (DQRA)) AD C recommends that the Defra/Environment Agency Contaminated Land Research Report CLR 11 *Handbook of model procedures for the management of contaminated land*, 2003 should be consulted. At the time that AD C was published this was only a consultation draft. It has since been published (in September 2004) as Contaminated Land Report 11 *Model procedures for the management of land contamination* and it may be of considerable help when developing a site affected by contamination since it describes the stages of risk assessment that should be followed in order to identify risks and make judgements about the consequences of contamination for the affected site. These stages are summarised in Fig. 8.2 and described in more detail below.

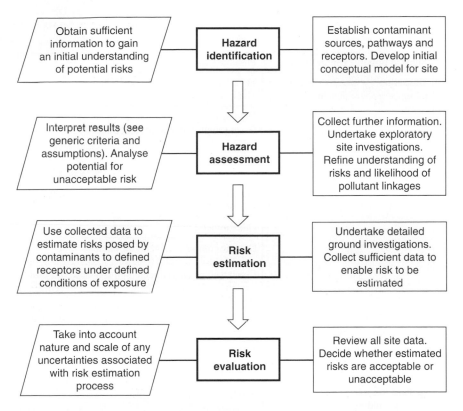

Fig. 8.2 Resistance to contaminants – risk assessment stages.

Hazard identification and assessment

One of the needs of the preliminary site assessment is to provide information on any possible contamination of the site and surrounding area due to its past and present uses (see Table 2 from AD C in section 8.5.5 above for typical uses that might give rise to site contamination). The site walkover survey may reveal signs of possible contaminants. Table 3 from AD C (which is reproduced below) gives some examples of the signs that may indicate particular contaminants; however, this is not a comprehensive list and should be used with caution since it is merely indicative.

Information provided by the desk study and site walkover survey will be of primary importance in the design of the exploratory and detailed ground investigation.

Clearly, the site assessment and risk evaluation should be concentrated on the area of the site subject to building operations. Therefore, the following parts of a site should be remediated to the requirements of the Building Regulations:

- those parts of the land associated with the building itself;
- gardens;
- other places on the site that are accessible to users of the building and those in and about the building.

ADC
Table 3   Examples of possible contaminants.

Signs of possible contaminants	Possible contaminant
Vegetation (absence, poor or unnatural growth)	Metals Metal compounds
	Organic compounds Gases (landfill or natural source)
Surface materials (unusual colours and contours may indicate wastes and residues)	Metals Metal compounds
	Oily and tarry wastes
	Asbestos
	Other mineral fibres
	Organic compounds including phenols
	Combustible material including coal and coke dust
	Refuse and waste
Fumes and odours (may indicate organic chemicals)	Volatile organic and/or sulfurous compounds from landfill or petrol/solvent spillage
	Corrosive liquids
	Faecal animal and vegetable matter (biologically active)
Damage to exposed foundations of existing buildings	Sulfates
Drums and containers (empty or full)	Various

An incremental approach to remediation of a site (perhaps including lower levels of remediation) may be acceptable where part of (or the remainder of) the land associated with the building is accessible to a lesser extent to the user or those in and about the building than the main parts of the buildings and their respective gardens. This could also apply to areas adjacent to such land. This incremental approach may also apply in the case of phased redevelopment of very large sites, where it may be possible to limit remediation to the part of the site that is actually being developed at any particular time. In all cases reliance will be placed on the risk evaluation and remediation strategy documentation in order to demonstrate that restricted remediation is acceptable. It should be noted that this will place the onus on the applicant to show why part of a site may be excluded from particular remediation measures.

The scope of the Building Regulations is limited to matters concerning health, safety, welfare and convenience. Therefore, even if the adjacent land is not subject to Building Regulations, it may still be subject to planning control legislation or to control under Part IIA of the Environmental Protection Act 1990. In fact, a substantial amount of guidance on the assessment of contaminated land has been published to support the implementation of this Act. Most of this guidance is contained in the joint Defra/Environment Agency Contaminated Land Research Reports (CLRs). These include:

- CLR 7 *Assessment of risks to human health from land contamination: an overview of the development of soil guideline values and related research*, 2002;
- CLR 8 *Priority contaminants for the assessment of land*, 2002;
- CLR 9 *Contaminants in soil: collation of toxicological data and intake values for humans*, 2002;
- CLR 10 *The contaminated land exposure assessment model (CLEA): technical basis and algorithms*, 2002.

Some guidance on these reports is given below and an outline of the process is shown in Figure A1 from Appendix A of Approved Document C.

Additionally, if any substance is found which is at variance with any preliminary statements made about the nature of the site, the Planning Authority should be informed before any intrusive investigations are carried out.

Risk estimation and evaluation

During the risk estimation phase, detailed ground investigations will be carried out. These must provide sufficient information for:

- confirmation of a conceptual model for the site;
- the risk assessment itself;
- the design and specification of any remedial works.

The ground investigations are likely to involve collection and analysis, by the use of invasive and/or non-invasive techniques, of:

- soil;
- soil gas;
- surface and groundwater samples.

Since elevated groundwater levels could bring contaminants close to the surface (both beneath the building and in any land associated with the building), an investigation of the groundwater regime, levels and flows is essential for most sites.

Therefore, if land affected by contaminants is to be developed, the health and safety of both the public and workers should be considered. When considering health, risk estimation can be carried out using generic assessment criteria such as contaminant Soil Guideline Values (SGVs) or relevant and appropriate environmental standards. SGVs represent concentrations of contaminant that may pose unacceptable risks to health. The development of SGVs for a range of priority contaminants is described in the Defra/Environment Agency reports mentioned above. There is also a range of corresponding TOX reports which contain the toxicological data used to derive the SGVs. For example, CLR 10 describes the Contaminated Land Exposure Assessment Model (CLEA) for deriving SGVs for three different site uses:

- residential;
- residential with plant uptake; and
- commercial/industrial.

Approved Document C – Annex A: Guidance

RISK ASSESSMENT

START → Define the context & set the objectives → Preliminary risk assessment

Collect more site data and review assessment

Are there potential risks? — Not known / No / Yes

Is further assessment required?

Is assessment using generic criteria appropriate?

Collect more site data and review assessment

Generic quantitative risk assessment

Are there unacceptable risks? — Not known / No / Yes

Is assessment using site specific criteria appropriate?

Collect more site data and review assessment

Detailed quantitative risk assessment

Are there unacceptable risks? — Not known / Yes

NO FURTHER ACTION REQUIRED

OPTIONS APPRAISAL

Define the context & set or refine the objectives → Identification of feasible remediation options

Collect more data, review objectives or monitor condition

Have feasible options been identified? — No / Yes

Detailed evaluation of options

Collect more site data and review assessment

Can the most appropriate option (or combination of options) be selected? — No / Yes

Development of the remediation strategy

Can an appropriate strategy be identified? — No / Yes

Review decisions taken earlier in the process

IMPLEMENTATION OF RISK MANAGEMENT ACTION

Define the context & set or refine the objectives → Preparation of the implementation plan

Adjust plan until agreement is reached

Is the implementation plan agreed with all parties? — No / Yes

Design, implementation and verification of the works

Adjust design and/or arrangements for supervision

Have the works been verified? — No / Yes

Are long term monitoring data required? — Yes / No

Long-term monitoring and maintenance

Are the monitoring data acceptable? — No / Yes

Is further remediation required? — No / Yes

NO FURTHER ACTION REQUIRED

Note: The process may apply to one or more pollutant linkages each of which may follow a different route. For some linkages, it may be possible to stop at an early stage – others will progress all the way through the process. The level of complexity of each stage may also vary and in some cases may be very simple.

This enables the relative importance of each of the pollutant linkages to be considered. As an example of this approach, for a residential site use the assumption is made that residents have private gardens and/or access to community open space close to the home and that some of them may use their gardens to grow vegetables. CLR 10 gives details of the conceptual model underpinning each of the standard land uses.

As an alternative to the generic approach it is also possible to undertake a more site specific quantitative risk assessment using the principles of risk assessment or a risk assessment model. For this, specialist advice should be sought.

Further guidance

For guidance on the investigation of sites potentially affected by contaminants, the following documents may also be consulted:

- Association of Geotechnical and Geoenvironmental Specialists *Guidelines for combined geoenvironmental and geotechnical investigations* (Available from: AGS, Forum Court, 83 Copers Cope Road, Beckenham, Kent, BR3 1NR. Website: www.ags.org.uk/publications/pubcat.cfm);
- BS 5930: 1999 *Code of practice for site investigations*;
- BS 10175: 2001 *Investigation of potentially contaminated land. Code of practice*;
- National Groundwater & Contaminated Land Centre Report NC/99/38/2 *Guide to good practice for the development of conceptual models and the selection and application of mathematical models of contaminant transport processes in the subsurface* (available from website: www.environment-agency.gov.uk/subjects/waterres/groundwater/);
- Environment Agency R & D Technical Report P5–065 *Technical aspects of site investigation, 2000*;
- Environment Agency R & D Technical Report P5–066 *Secondary model procedure for the development of appropriate soil sampling strategies for land contamination*;
- Environment Agency/NHBC R & D Publication 66 *Guidance for the safe development of housing on land affected by contamination*, TSO, 2000.

These documents recommend the adoption of a risk-based approach, so that any hazards that are present can be identified and quantified and an assessment can be made of the nature of the risk they might pose. The design and execution of field investigations are described together with suitable sample distribution strategies, testing and sampling.

8.5.7   Resistance to contaminants – remedial measures

Introduction

If the risk assessment stage results in the identification of unacceptable risks to the particular receptor (e.g. buildings, building services, building materials and people), then appropriate remedial measures will be needed in order to manage these risks. This will mean defining the risk management objectives in terms of the need to break the pollutant linkages (see section 8.5.6 and the text below in this section). Other objectives that will also need to be considered include such items as:

- timescale and cost;
- remedial works;
- planning constraints; and
- sustainability.

The remedial measures that are adopted will depend on the contaminant that has been identified. In general, three generic types of remedial measures can be considered:

- treatment;
- containment; and
- removal.

It should be noted that it may be necessary to obtain a waste management licence from the Environment Agency where the containment or treatment of waste is anticipated. It is also important to consider the effects of building work on sites affected by contaminants. For example, an existing control measure which included a cover system (i.e. containment) could be breached if excavations were carried out for foundations or underground services when an extension was added.

Treatment

Contaminants can be dealt with by a wide range of treatment processes using biological, chemical and physical techniques. These can be carried out either on or off the site and are designed to decrease one or more of the contaminant's features such as mass, concentration, mobility, flux or toxicity. Since the choice of the most appropriate technique is highly site-specific, specialist advice should be sought.

Containment

In general, the term containment means encapsulation of material containing contaminants. However, in the context of building development it is usually taken to mean cover systems, sometimes incorporating vertical barriers in the ground to control lateral migration of contaminants.

In a cover system layers of materials are placed over the site in order to:

- interrupt the pollutant linkage between the contaminants and the receptors;
- sustain vegetation;
- improve geotechnical properties; and
- reduce exposure to an acceptable level.

Some parts of the structure of the building, (foundations, substructure, ground floor, etc.) may assist other containment measures in providing effective protection of health from contaminants. However, the extent to which this is feasible will depend on the circumstances and form of construction.

The following issues need to be addressed when using imported fill and soil for cover systems:

- It should be assessed at source to ensure that it is not contaminated above specified concentrations.
- It should meet required standards for vegetation (see BS 3882: 1994 *Specification for topsoil*).
- Where intermixing of the soil cover with the contaminants in the ground can take place, it will be necessary in the design and dimensioning of the cover system to consider its long-term performance. This may include the need for maintenance and monitoring.
- Gradual intermixing of the soil and contaminants due to natural effects and activities (e.g. by burrowing animals, gardening, etc.) should be taken into account.
- Excavations by householders for garden features, walls, ponds, etc., can penetrate the cover layer leading to possible exposure to contaminants.

Further guidance on the design, construction and performance of cover layers can be found in the Construction Industry Research and Information Association (CIRIA) Special Publication SP124 *Barriers, liners and cover systems for containment and control of land contamination*, 1996.

Removal

Removal means the excavation and safe disposal to a licensed landfill site of the contaminants and contaminated material. This can be achieved by targeting the excavation on contaminant hot spots, or by removing sufficient depth of contaminated material so that a cover system can be accommodated within the planned site levels. Removal may not always be viable since it will depend on the depth and extent of the contaminants on the site and the availability of suitably licensed landfills.

Where removal is incorporated with a subsequent cover system any imported fill should be assessed at source to ensure that there are no materials that will pose unacceptable risks to potential receptors.

Further detailed guidance on treatment, containment and removal is given in the Environment Agency/NHBC R & D Publication 66 referred to in section 8.5.6 and in the following CIRIA publications:

- CIRIA Special Publication SP102 *Decommissioning, decontamination and demolition*, 1995;
- CIRIA Special Publication SP104 *Classification and selection of remedial methods*, 1995;
- CIRIA Special Publication SP105 *Excavation and disposal*, 1995;
- CIRIA Special Publication SP106 *Containment and hydraulic measures*, 1996;
- CIRIA Special Publication SP107 *Ex-situ remedial methods for soils, sludges and sediments*, 1995;
- CIRIA Special Publication SP109 *In-situ methods of remediation*, 1995.

8.5.8   Resistance to contaminants – risks to buildings, building materials and services

Receptors of contaminants include not only people but also buildings, building materials and services on sites. The hazards to these receptors might include:

- **Aggressive substances** – which may affect the long-term durability of construction materials such as concrete, metals and plastics (e.g. organic and inorganic acids, alkalis, organic solvents and inorganic chemicals such as sulfates and chlorides).
- **Combustible fill** – which may lead to subterranean fires, if ignited, and consequent damage to the structural stability of buildings, and the integrity or performance of services (e.g. domestic waste, colliery spoil, coal, plastics, petrol-soaked ground, etc.).
- **Expansive slags** – which may expand some time after deposition (usually when water is introduced onto the site) causing damage to buildings and services (e.g. blast furnace and steel making slag).
- **Contaminant-affected floodwater** – floodwater may be contaminated by substances in the ground, waste matter or sewage. Building elements that are close to or in the ground, such as walls or ground floors may be affected by this contaminated water. The following documents contain guidance on resistant construction:
  - ○ *Preparing for floods: interim guidance for improving the flood resistance of domestic and small business properties*, DTLR, 2002;
  - ○ BRE for Scottish Office *Design guidance on flood damage for dwellings*, TSO, 1996.

Although the main receptors with these hazards are the building, the building materials and the building services, ultimately the health of the occupants may be put at risk. In particular, potable water pipes made of polyethylene may be permeated by hydrocarbons. Reference should be made to Foundation for Water Research report FR0448 *Laying potable water pipelines in contaminated ground: guidance notes*, 1994, where guidance on reducing these risks may be found. Additionally, the Environment Agency document *Assessment and management of risks to buildings, building materials and services from land contamination*, 2001, contains further guidance.

8.5.9   Radon gas contamination – introduction

Radon is a naturally-occurring, colourless and odourless gas which is radioactive. It is formed in small quantities by the radioactive decay of uranium and radium, and thus travels through cracks and fissures in the subsoil until it reaches the atmosphere or enters spaces under or in buildings.

It is recognised that radon gas occurs in all buildings; however, the concentration may vary from below 20 Bq/m^3 (the national average for houses in the UK) to more than 100 times this value. The National Radiological Protection Board (NRPB) has recommended an action level of 200 Bq/m^3 for houses. The lifetime risk of contracting lung or other related cancers at the action level is about 3%. Geographical distribution of houses at or above the action level is very uneven with about two thirds of the total being in Devon and Cornwall.

The DCLG is reviewing the areas where preventative measures should be taken as information becomes available from the NRPB and the British Geological Survey (BGS). This information has been placed in the BRE guidance document *Radon: guidance on protective measures for new dwellings* (BRE Report BR 211: 1999, obtainable from Building Research Establishment, Garston, Watford WD2 7JR), which will be updated as necessary.

A common basis for radiation protection legislation in all Member States of the European Union has been established in a European Council Directive. This Directive has been put into effect in the UK by virtue of the Ionising Radiations Regulations 1999 (SI 1999/3232). These regulations set a national reference level for radon gas and they require employers and self-employed persons responsible for a workplace to measure radon levels on being directed to do so. Reference may also be made to BRE Report BR 293 *Radon in the workplace*, 1995, which provides guidance for existing non-domestic buildings.

The guidance in BR 211 has been developed to show radon protective measures in dwellings. No guidance exists at present for radon protection in workplaces. However, for domestic-sized workplaces with heating and ventilation regimes similar to those in dwellings (such as small office buildings and primary schools) some of the techniques described in BR 211 for installing radon resistant membranes may be suitable. The guidance in BR 211 can also be used as the basis for radon protection of other building types provided that caution is exercised. Good Building Guide 25 *Buildings and radon*, 1996, contains interim guidance on extensions.

Although no guidance is currently available in the 1999 edition of BR 211 for suspended timber ground floors in new dwellings, it is understood that the DCLG is sponsoring research into how this form of construction could provide adequate protection against radon.

BR 211 identifies those areas where either basic or full radon protection is needed by reference to a series of maps derived from:

- statistical analysis of radon measurements of existing houses carried out by the NRPB (Annex A of BR 211); and
- an assessment of geological radon potential prepared by the BGS (Annex B of BR 211).

Use of the maps in accordance with directions given in BR 211 will determine whether basic, full or no protection is needed.

Areas most at risk include parts of:

- Devon
- Cornwall
- Somerset
- Gloucestershire
- Oxfordshire
- Northamptonshire
- Leicestershire and Rutland
- Lincolnshire

- Staffordshire
- Derbyshire
- West Yorkshire
- Northumberland and parts of southern Cumbria
- most of Wales.

As more information becomes available from NRPB it is likely that further areas will be covered by the need for radon precautions. Current information on the areas delineated by DCLG for the purposes of Building Regulations can be obtained from local authority building control officers or from approved inspectors. When it is necessary to make changes to areas delineated as requiring radon protection these will be notified to Building Control Bodies and will be posted on the DCLG website. The results will be published in due course.

8.5.10   Basic protection against radon

Basic protection may be provided by an airtight, and therefore radon-proof, barrier across the whole of the building including the floor and walls. This could consist of:

- polyethylene (polythene) sheet membrane of at least 300 micrometre (1200 gauge) thickness;
- flexible sheet roofing materials;
- prefabricated welded barriers;
- liquid coatings;
- self-adhesive bituminous-coated sheet products;
- asphalt tanking.

It is important to have adequately sealed joints and the membrane must not be damaged during construction. Where possible, penetration of the membrane by service entries should be avoided. With careful design it may be possible for the barrier to serve the dual purpose of damp-proofing and radon protection although the damp-proof course to a cavity wall should be in the form of a cavity tray to prevent radon entering the building through the cavity.

Some typical details are shown in Fig. 8.3.

8.5.11   Full protection against radon

In practical terms, a totally radon-proof barrier may be difficult to achieve.

Therefore, in high-risk areas it is necessary to provide additional secondary protection. This might consist of:

- natural ventilation of an underfloor space by airbricks or ventilators on at least two sides;
- the addition of an electrically operated fan in place of one of the airbricks to provide enhanced sub-floor ventilation;
- a sub-floor depressurisation system comprising a sump located beneath the floor slab, joined by pipework to a fan. It may only be necessary to provide the sump and under-

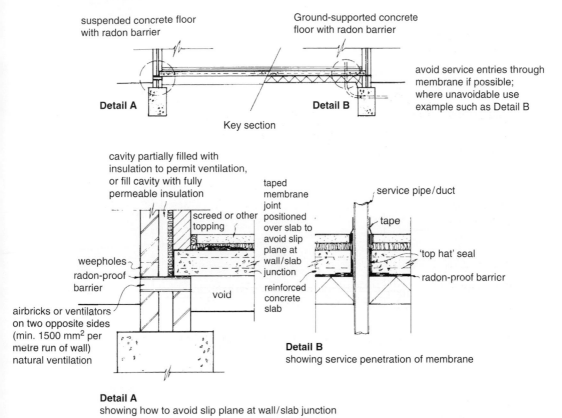

Fig. 8.3　Basic protection against radon.

floor pipework during construction thus giving the owner the option of connecting a fan at a later stage if necessary.

Examples of these methods are shown in Figs 8.3 and 8.4.

It should be noted that the above brief notes on BR 211 are intended to give an idea of the content of that document. Designers of buildings in the delimited areas should consult the full report.

8.5.12　Contamination of landfill gas

Landfill gas is typically made up of 60% methane and 40% carbon dioxide, although small quantities of other gases such as hydrogen, hydrogen sulphide and a wide range of trace organic vapours (called volatile organic compounds or VOCs in Approved Document C) may also be present. The gas is produced by the breakdown of organic material by micro-organisms under oxygen-free (anaerobic) conditions on biodegradable waste materials in landfill sites. Gases similar to landfill gas can also arise naturally from coal strata, river silt, sewage and peat. Additionally, atmospheres that are deficient in methane and oxygen (usually referred to as stythe or black-damp by miners) and which are rich in carbon

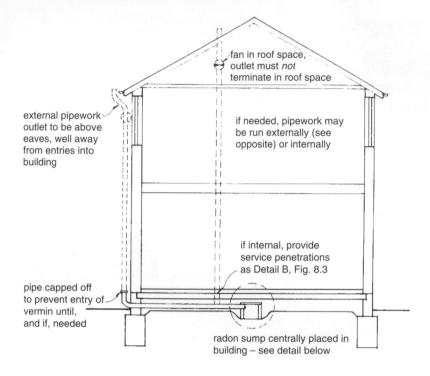

external pipework outlet to be above eaves, well away from entries into building

fan in roof space, outlet must *not* terminate in roof space

if needed, pipework may be run externally (see opposite) or internally

if internal, provide service penetrations as Detail B, Fig. 8.3

pipe capped off to prevent entry of vermin until, and if, needed

radon sump centrally placed in building – see detail below

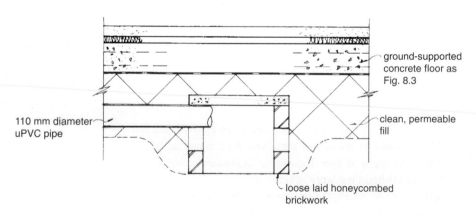

ground-supported concrete floor as Fig. 8.3

110 mm diameter uPVC pipe

clean, permeable fill

loose laid honeycombed brickwork

Fig. 8.4   Additional protection against radon.

dioxide and nitrogen can be produced naturally in coal mining areas. VOCs can also arise as a result of spillages of petrol, oil and solvents.

8.5.13   Properties of landfill gases

The largest component of landfill gas, methane, is a flammable, asphyxiating gas with a flammable range between 5% and 15% by volume in air. If such a concentration occurs within a building and the gas is ignited it will explode. Methane is lighter than air.

The other major component, carbon dioxide, is a non-flammable, toxic gas which has a long-term exposure limit of 0.5% by volume and a short-term exposure limit of 1.5%. It is heavier than air. VOCs are both inflammable and toxic and can also have strong unpleasant odours. A build-up to hazardous levels of any of these gases within buildings will result in harm to health and will compromise safety.

8.5.14   Movement of landfill gases

The proportions of the two main landfill gases and the amount of air mixed with them will largely determine the properties of the landfill gas since they remain mixed and do not separate, although the mixture can remain separate from surrounding air. These landfill gases will migrate from a landfill site as a result of diffusion through the ground and this migration may be increased by rainfall or freezing temperatures as these conditions tend to seal the ground surface. The gases will also follow cracks, cavities, pipelines and tunnels, etc., as these form ideal pathways. Landfill gas emissions can be increased by rapid falls in atmospheric pressure and by a rising water table. Thus, landfill gas may enter buildings and may collect in underfloor voids, drains and soakaways.

8.5.15   Building near landfill sites and on gas contaminated land – risk assessment

The risk assessment stages referred to in section 8.5.6 should be followed for methane and other landfill gases. Further investigations for hazardous soil gases may also be required if the ground to be covered by a building and/or any land associated with the building is:

(a)   on, or within 250 m of a landfill site or within the likely sphere of influence of a landfill; in these cases the policy of the Environment Agency regarding building on or near landfill sites should be followed;

(b)   on a site where biodegradable substances (including made ground and fill) have been deposited on a large scale;

(c)   on a site where the previous use has meant that spillages of petrol, oil and solvents could have taken place (such as vehicle scrap yards);

(d)   in an area where naturally occurring gases (e.g. methane, carbon dioxide, hydrogen sulphide and VOCs) may be present (such as old mining areas and spoil heaps).

In these specific instances Approved Document C recommends use of the following guidance documents covering hazardous soil gases in the contexts referred to in (a) to (d) above:

●   Guidance on the generation and movement of landfill gas as well as techniques for its investigation are given in HMIP Waste Management Paper No 27 *The control of landfill gas*, TSO, 2nd edition, 1991. Complementary guidance is given in a publication by the Chartered Institution of Wastes Management (CIWM) entitled *Monitoring of landfill gas*, 2nd edition, 1998.

- Guidance covering petroleum retail sites may be found in TP 95 *Guidelines for investigation and remediation of petroleum retail sites*, Institute of Petroleum, 1998.
- Guidance on the geographical extent, associated hazards and methods of site investigation of methane, carbon dioxide and oil seeps from natural sources and mining areas can be found in BGS Technical Report WP/95/1 *Methane, carbon dioxide and oil seeps from natural sources and mining areas: characteristics, extent and relevance to planning and development in Great Britain*, 1995. This should be read in conjunction with a report sponsored by the former Department of the Environment entitled *Methane and other gases from disused coal mines: the planning response*, TSO, 1996.
- Additionally, the following three documents published by CIRIA contain relevant information on methane and other gases including means of generation and movement within the ground, detection and monitoring methods, and investigation strategies:
  - CIRIA Report 130 *Methane: its occurrence and hazards in construction*, 1993;
  - CIRIA Report 131 *The measurement of methane and other gases from the ground*, 1993;
  - CIRIA Report 150 *Methane investigation strategies*, 1995.

When a site investigation is carried out for methane and other hazardous gases consideration should be given to the following matters:

- In order to characterise gas emissions fully, measurements should be taken over a sufficiently long period of time (including periods when gas emissions are likely to be higher, e.g. during periods of falling atmospheric pressure).
- It is also important to establish:
  - the concentration of methane and other gases in the ground;
  - the quantity of gas generating materials and their rate of gas generation;
  - gas movement in the ground;
  - gas emissions from the ground surface.

Measurements taken of the surface emission rates and borehole flow rates will give an indication of the gas regime in the ground. For further guidance on this reference should be made to:

- CIRIA Report 151 *Interpreting measurements of gas in the ground*, 1995;
- CIRIA Report 152 *Risk assessment for methane and other gases from the ground*, 1995.

Additionally, the gas regime on the site can be altered by construction activities undertaken as part of building development. For example, surface gas emissions can be increased by the site strip, by piling and by excavations for foundations and below ground services; and dry biodegradable waste can be pushed into moist, gas-active zones by dynamic compaction.

Reference is made in section 8.5.6 above to the use of contaminant Soil Guideline Values (SGVs) when estimating the risk to health and safety of the public on a site. It should be noted that there are no SGVs for methane and other gases.

Therefore, in the context of traditional housing the assessment of gas risks needs to take into account two possible contaminant pathways for human receptors:

- direct entry of gas into the dwelling through the substructure (where it will ultimately build up to hazardous levels); and
- later exposure of the householder in garden areas by:
  o the construction of outbuildings (e.g. garden sheds and greenhouses) and extensions to the dwelling, and
  o the carrying out of excavations for garden features (e.g. ponds).

Guidance on the carrying out of risk assessments for methane and other ground gases can be found in CIRIA Report 152 referred to above, in Contaminated Land Research Report CLR 11 (see section 8.5.6) and in Environment Agency GasSIM – *Landfill gas assessment tool* (this document also includes the assessment of gas emissions from landfill sites).

The following documents describe a range of ground gas regimes (defined in terms of soil gas concentrations of methane and carbon dioxide as well as borehole flow rate measurements), which can also be helpful in assessing gas risks:

- CIRIA Report 149 *Protecting development from methane*, 1995;
- DETR/Arup Environmental PIT Research Report: *Passive venting of soil gases beneath buildings*, 1997.

The above discussion has mainly been in the context of housing development. In the context of non-domestic development the focus might be on the building only, but the general approach is the same.

8.5.16 Building near landfill sites and on gas contaminated land – gas control measures

The investigations into gas risks carried out in accordance with the advice given above may conclude that the risks posed by the gases which are present are unacceptable. If this is the case, then appropriate remedial measures may be required to manage the risks. These can be applied to the building alone or, where the risks on any land associated with the building are deemed unacceptable, this could even mean adopting site-wide gas control measures which could include:

- removal of the material generating the gas; or
- covering together with systems for gas extraction.

CIRIA Report 149 (referred to above) contains further guidance on this. In general, where site-wide gas control measures are thought to be needed then expert advice should be sought.

Gas control measures – dwellings

Practical guidance on construction methods to prevent the ingress of landfill gas in buildings is not given in Approved Document C. Instead, the reader is referred to BRE/Environment Agency Report BR 414 *Protective measures for housing on gas-contaminated land*, 2001, where detailed practical guidance on the construction of protective measures may be found.

Gas control measures for dwellings are normally passive (i.e the gas flow is driven by temperature differences (stack effect) and the effects of wind) and consist of a barrier which is gas resistant, across the entire walls and floor of the dwelling. Below this will be an extraction (or ventilation) layer from which gases can be dispersed and vented into the atmosphere. In order to maximise the driving forces of natural ventilation it will be necessary to consider the design and layout of buildings.

Figure 8.5 gives typical examples of some of the constructional details contained in the BR 414. The DETR/Arup Environmental PIT Research Report *Passive venting of soil gases beneath buildings*, referred to above, can also be used as a guide to the design of a range of commonly used gas control measures.

Gas control measures – non-domestic buildings

In non-domestic buildings gas control measures are based on the same principles as those used for housing. Therefore, the DETR/Arup Environmental report referred to in the previous paragraph can also be used as a design guide. Since the floor areas of non-domestic buildings can be considerably larger than those of dwellings and the adequate dispersal of gas from beneath the floor must be ensured, it is usually necessary to seek expert advice. With larger floor areas, passive systems may not be efficient at ensuring removal of gases; therefore, mechanical systems (which may include monitoring and alarm systems) may be necessary. Such systems also need to be calibrated and continually maintained, so they are more appropriate for non-domestic buildings where there is scope for this. Since special sub-floor ventilation systems are carefully designed to ensure adequate performance, they should not be modified unless a specialist review of the design is undertaken. It should be noted that the use of continuous mechanical ventilation for the removal of landfill gases in dwellings is not recommended since there is a risk of interference by users and maintenance of the system cannot be guaranteed. Consequently, a failure might result in a sharp increase in indoor methane concentration with the possibility of an explosion occurring.

It should be noted that the above brief notes on BR 414 are intended to give an idea of the content of that document. Designers of buildings which are likely to be affected by landfill gas should consult the full report and any of the references mentioned in sections 8.5.12 to 8.5.16.

8.5.17   Subsoil drainage

Subsoil drainage must be provided *if* it is necessary to avoid:

- the passage of moisture from the ground to the inside of the building; or

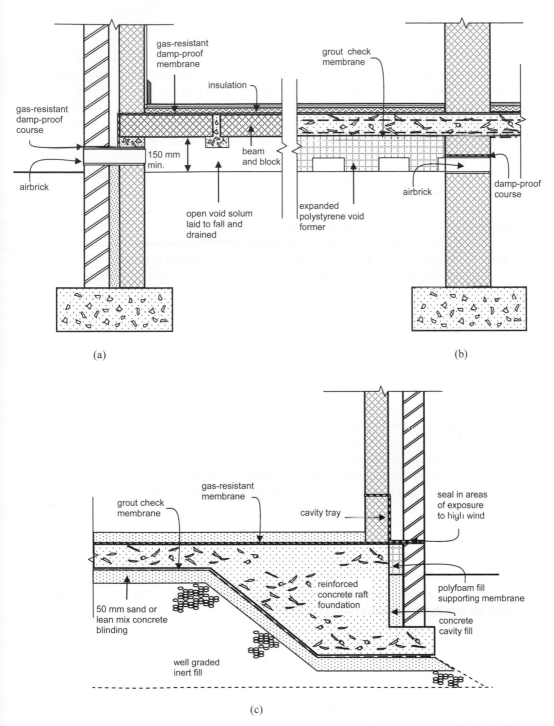

Fig. 8.5   Landfill gas protection details. (a) Beam and block floor with open void. (b) In-situ slab with EPS void former. (c) Raft foundation with membrane on top of slab.

- damage to the fabric of the building. This includes damage to the foundations of the building caused by the transport of water-borne contaminants.

The provisions in AD C assume either that the site of the building will not be subject to general flooding or, if it is, then suitable steps are being taken. Interestingly, although flood resistance is not covered by the Building Regulations 2000 at present, there is a presumption in planning guidance (see Planning Policy Guidance Note PPG 25 *Development and flood risk*, DTLR, 2002) that development should not take place in areas that are at risk of flooding.

However, where local considerations might necessitate building in areas that are prone to flooding the following guidance is offered to mitigate some of its effects:

- The existence of elevated groundwater levels or the flow of subsoil water across a site may be alleviated by the provision of adequate subsoil drainage (see below).
- The creation of blockages in drains and sewers caused by flooding can lead to backflow of sewage into buildings through low sited toilets and gullies, etc. This can be mitigated by the use of anti-flooding devices and non-return valves (see section 13.3.26 of this book or the CIRIA publication C506 *Low cost options for prevention of flooding from sewers*, 1998).
- In areas where the design of the below-ground drainage system is such that foul water drainage also receives rainwater, these systems may surcharge in periods of heavy rainfall. This could lead to increased risks of localised flooding within properties in low-lying areas or in those that contain basements unless preventative measures are taken. Some guidance on protection is given in Approved Document H *Drainage and waste disposal* (see section 13.3.26 of this book).
- The passage of groundwater through a floor can be dealt with using water resistant construction (see section 8.5.19 below).
- The entry of water into under-floor voids can be addressed by making provision for the inspection and clearing out of such locations beneath suspended floors.

The following publications may also be consulted for further guidance on flooding and flood resistant construction:

- *Preparing for floods: interim guidance for improving the flood resistance of domestic and small business properties*, DTLR, 2002;
- BRE for Scottish Office *Design guidance on flood damage for dwellings*, TSO, 1996;
- CIRIA/Environment Agency Flood *Products. Using flood protection products – a guide for home owners*, 2003.

Subsoil water may cause problems where:

- there is a high water table (i.e. within 0.25 m of the lowest floor in the building);
- surface water may enter or adversely affect the building;
- an active subsoil drain is severed during excavations;

- the stability and properties of the ground are adversely affected by groundwater beneath or around the building;
- groundwater flows are altered by general excavation work for foundations and services.

Where problems are anticipated, it will usually be necessary either to drain the site of the building or to design and construct it to resist moisture penetration. Subsoil drain-

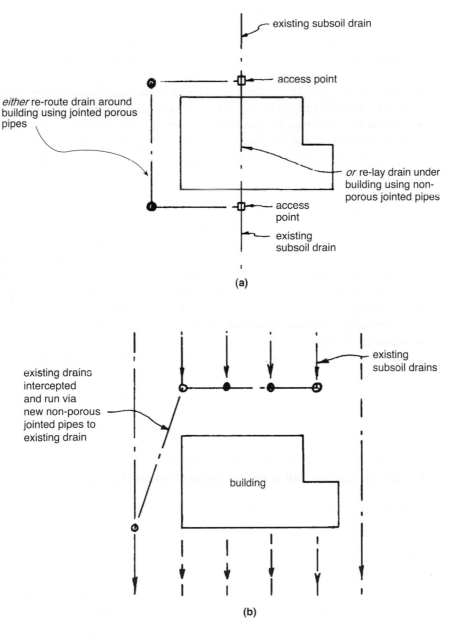

Fig. 8.6   Subsoil drainage. (a) Single subsoil drain. (b) Interception of multiple subsoil drains.

age should also be considered where contaminants are present in the ground, in order to prevent the transportation of water-borne contaminants to the foundations or into the building or its services.

Severed subsoil drains that pass under the building should be intercepted and continued in such a way that moisture is not directed into the building. Figure 8.6 illustrates a number of possible solutions.

8.5.18   Resistance to moisture

Requirement C2 of Schedule 1 to the Building Regulations 2000 deals with resistance to moisture. It requires the floors, walls and roof of a building to adequately protect it and the people who use it from harmful effects caused by:

- moisture from the ground;
- precipitation and wind-driven spray;
- surface and interstitial condensation; and
- water spilt from or associated with sanitary fittings and fixed appliances.

Guidance on the application of Requirement C2 is given in Approved Document C in the following sections:

- Section 4: Floors covering:
    - (a) ground-supported floors, suspended timber ground floors and suspended concrete ground floors when these are exposed to ground moisture;
    - (b) interstitial condensation risk in ground floors and upper floors exposed from below;
    - (c) surface condensation risk and mould growth on any floor type;
- Section 5: Walls;
- Section 6: Roofs.

In Approved Document C the following definitions apply with regard to Requirement C2:

FLOOR – the lower horizontal surface of any space in a building including any surface finish which is laid as part of the permanent construction. This would, presumably, exclude carpets, linoleum, tiles, etc., but would include screeds and granolithic finishes.

GROUNDWATER – Liquid water (i.e. not water vapour, ice or snow, etc.), either flowing through the ground or as a static water table.

INTERSTITIAL CONDENSATION – Water vapour being deposited as liquid water within or between the layers of the envelope of the building.

MOISTURE – Water present as a liquid, gas (e.g. water vapour) or solid (e.g. ice or snow, etc.).

PRECIPITATION – Moisture in any form falling from the atmosphere, such as rain, sleet, snow or hail, etc.

ROOF – Any part of the external envelope of a building that makes an angle of less than 70° to the horizontal.

SPRAY – Wind driven droplets of water blown from the surface of the sea or other bodies of water close to buildings. The salt content of sea spray makes it especially hazardous to many building materials.

SURFACE CONDENSATION – Water vapour being deposited as liquid water on visible surfaces within the building.

VAPOUR CONTROL LAYER – Typically, this is a membrane material which is used in the construction, with the purpose of substantially reducing the transfer of water vapour through any building in which it is incorporated.

WALL – Any opaque part of the external envelope of a building that makes an angle of 70° or more to the horizontal.

8.5.19   Protection of floors next to the ground

Ground floors should be designed and constructed so that:

- The passage of moisture to the upper surface of the floor is resisted (this might not apply to buildings used solely for storage of goods in which the only persons habitually employed were storemen, etc. engaged only in taking in, caring for, or taking out the goods; other similar types of buildings where the air is so moisture-laden that any increase would not adversely affect the health and safety of the occupants might also be excluded).
- They will not be adversely affected by moisture from the ground.
- They will not be adversely affected by groundwater.
- The passage of ground gases is resisted. This relates back to Requirement C1 (2) (see section 8.5.5 above) where floors in certain localities may need to be constructed to resist the passage of hazardous gases such as methane or radon. The remedial measures shown in sections 8.5.10 to 8.5.16 above can function as both a gas resistant barrier and damp-proof membrane if properly detailed.
- The structural and thermal performance of the floor is not adversely affected by interstitial condensation. This also applies to floors exposed from below.
- Surface condensation and mould growth is not promoted under reasonable occupancy conditions. This applies to all floors (not just those next to the ground).

This guidance is illustrated in Fig. 8.7.

Floors next to the ground must:
- not be damaged by moisture or groundwater
- resist passage of ground moisture to upper surface
- resist passage of ground gases
- be designed and constructed so that structural and thermal performance are not affected by interstitial condensation

Floors exposed from below:
- should be designed and constructed so that structural and thermal performance are not affected by interstitial condensation

All floors:
- should not promote surface condensation or mould growth

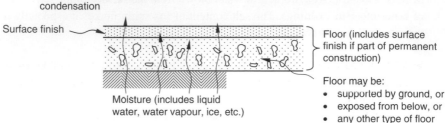

Fig. 8.7 Floors – general guidance.

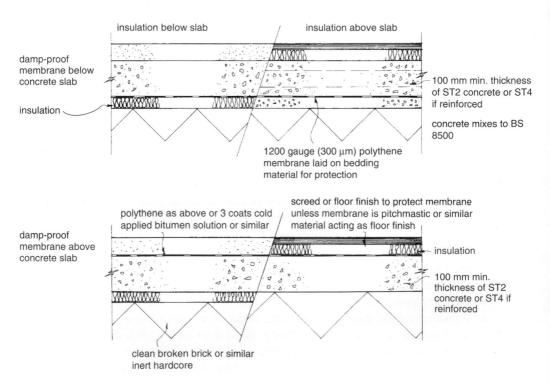

Fig. 8.8 Ground-supported floor.

8.5.20 Floors supported directly by the ground

The requirements mentioned above can be met, for ground-supported floors, by covering the ground with dense concrete incorporating a damp-proof membrane, laid on a hardcore bed. If required, insulation may also be incorporated in the floor construction.

This form of construction is illustrated in Fig. 8.8 and, unless the floor is subjected to water pressure, such as occurs with buildings on very permeable strata like chalk, gravel or limestone (see note on the use of BS 8102 below) the construction of the floor should take into account the following points:

- Well compacted hardcore less than 600 mm thick laid under the floor next to the ground should not contain water-soluble sulfates or deleterious matter in such quantities as might cause damage to the concrete. Broken brick or stone are the best hardcore materials. Clinker is dangerous unless it can be shown that the actual material proposed is free from sulfates, etc., and colliery shales should likewise be avoided. In any event, the builder might well be liable for breach of an implied common law warranty of fitness of materials; see *Hancock* v. *B. W. Brazier (Anerley) Ltd* [1966] 2 All ER 901, where builders were held liable for subsequent damage caused by the use of hardcore containing sulfates.
- A damp-proof membrane (DPM) may be provided above or below the concrete floor slab and should be laid continuous with the damp-proof courses in walls, piers, etc. If laid below the concrete, the DPM should be at least equivalent to 300 μm (1200 gauge) polyethylene (e.g. polythene). It should have sealed joints and should be supported by a layer of material that will not cause damage to the polythene. If a polyethylene sheet membrane is laid above the concrete, there is no need to provide the bedding material. It is also possible to use a three-coat layer of cold-applied bitumen emulsion (or equivalent material with similar moisture and water-vapour resistance) in this position. These materials should be protected by a suitable floor finish or screed. Surface protection does not need to be provided where the membrane consists of pitchmastic or similar material which also serves as a floor finish. There is one particular case where the membrane should be placed below the concrete slab, and that is where the ground contains water-soluble sulfates or other deleterious material that could contaminate the hardcore.
- The minimum thickness for the concrete slab is 100 mm (although the structural design for the slab may require it to be thicker) designed to mix ST2 in BS 8500. Reinforced concrete should be to mix ST4 in BS 8500.
- Insulants located below floor slabs should:
  - ○ have sufficient strength to resist the weight of the slab;
  - ○ be able to carry the anticipated floor loading;
  - ○ be able to support any overloading during construction;
  - ○ if placed below the damp-proof membrane have low water absorption (in order to resist degradation) and, if necessary, be resistant to contaminants in the ground.
- AD C section 4 gives no guidance on the position of the floor relative to outside ground level. Since this type of floor is unsuitable if subjected to water pressure, it is reasonable to assume that the top surface of the slab should not be below outside ground level unless special precautions are taken.

If it is proposed to lay a timber floor finish directly on the concrete slab, it is permissible to bed the timber in a material that would also serve as a damp-proof membrane.

No guidance is given regarding suitable DPM materials. However 12.5 mm of asphalt or pitchmastic will usually be satisfactory for most timber finishes and it may be possible to lay wood blocks in a suitable adhesive DPM. If a timber floor finish is fixed to wooden fillets embedded in the concrete, the fillets should be treated with a suitable preservative unless they are above the DPM (see BS 1282: 1999 *Wood preservatives guidance on choice, use and application*).

Clause 11 of CP 102: 1973 *Protection of buildings against water from the ground* may be used as an alternative to the above. Where groundwater pressure is evident, recommendations may be found in BS 8102: 1990 *Code of practice for protection of structures against water from the ground.*

8.5.21   Suspended timber ground floors

The performance requirements mentioned above may be met for suspended timber ground floors by:

- covering the ground with suitable material to resist moisture and deter plant growth;
- providing a ventilated space between the top surface of the ground covering and the timber;

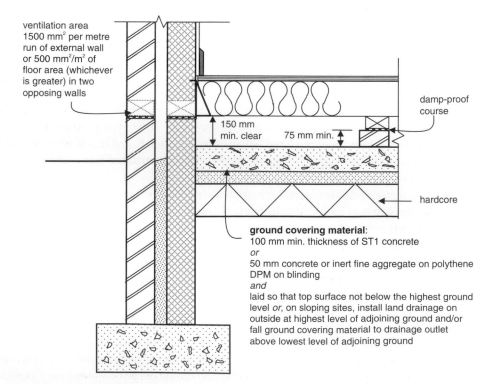

Fig. 8.9   Suspended timber floor.

- isolating timber from moisture-carrying materials by means of damp-proof courses.

A suitable form of construction is shown in Fig. 8.9 and is summarised below:

- The ground surface should be covered with at least 100 mm of concrete to BS 8500 mix ST1, if unreinforced. It should be laid on clean broken brick or similar inert hardcore not containing harmful quantities of water-soluble sulfates or other materials that might damage the concrete. (The Building Research Establishment suggests that over 0.5% of water-soluble sulfates would be a harmful quantity.)

   Alternatively, the ground surface may be covered with at least 50 mm of concrete, as described above, or inert fine aggregate, laid on a polythene DPM as described for ground supported floors above. The joints should be sealed and the membrane should be laid on a protective bed such as sand blinding.
- Since it undesirable for water to collect on top of the ground covering material under a timber floor, the ground-covering material should be laid so that *either* its top surface is not below the highest level of the ground adjoining the building *or*, where the site slopes, it may be necessary to install land drainage on the outside at the highest level of the ground adjoining the building and/or fall the ground-covering material to a drainage outlet above the lowest level of the adjoining ground.
- There should be a space above the top of the concrete of at least 75 mm to any wall-plate and 150 mm to any suspended timber (or insulation where provided). This depth may need to be increased where the building is constructed on shrinkable clays in order to allow for heave.
- There should be ventilation openings in two opposing external walls allowing free ventilation to all parts of the sub-floor. An actual ventilation area equivalent to 1500 mm^2 per metre run of external wall or 500 mm^2/m^2 of floor area (whichever area is greater) should be provided and any ducts needed to convey ventilating air should be at least 100 mm in diameter. Ventilation openings should be fitted with grilles so as to prevent vermin entry but these grilles should not unduly resist the flow of air. It may be difficult, where there is a requirement for level access to the floor, to provide the ventilators in the position shown in Fig. 8.8 since the top surface of the floor may well be nearer to the ground. The problem can usually be solved using offset (periscope) ventilators.
- Damp-proof courses of impervious sheet materials, slates or engineering bricks bedded in cement mortar should be provided between timber members and supporting struc-tures to prevent transmission of moisture from the ground. BS 5628: Part 3: 2001 gives guidance on the choice of suitable materials.
- In areas where water may be spilled (such as bathrooms, utility rooms and kitchens), boards used for flooring should be moisture resistant, irrespective of the storey in which they are located.

   Softwood boarding should be:
   ○ a minimum of 20 mm thick; and either
   ○ from a durable species; or
   ○ treated with a suitable preservative.

Chipboard is particularly susceptible to moisture damage, so where this is used as a flooring material it should be:

    ○ of one of the grades specified in BS 7331: 1990 or BS EN 312 Part 5: 1997 as having improved moisture resistance;

    ○ laid, fixed and jointed in the manner recommended by the manufacturer with the identification marks facing upwards in order to demonstrate compliance.

Again, the recommendations of clause 11 of CP 102: 1973 may be used instead of the above, especially if the floor has a highly vapour-resistant finish.

8.5.22   Suspended concrete ground floors

Moisture should be prevented from reaching the upper surface of the floor and the reinforcement should be protected against moisture if the construction is to be considered satisfactory.

Suitable suspended concrete ground floors may be of pre-cast construction with or without infilling slabs or they may be cast in-situ.

Normally, for in-situ construction the concrete should be at least 100 mm thick (unless required to be thicker by the structural design) and it should contain a minimum of 300 kg of cement per m^3 of concrete. The reinforcement should be protected by at least 40 mm cover.

Pre-cast concrete construction offers another solution and this can be built with or without infilling slabs or blocks. The reinforcement in this case should have at least the thickness of cover required for moderate exposure.

A damp-proof membrane should be provided if the ground below the floor has been excavated so that it is lower than outside ground level and it is not effectively drained. The space between the underside of the floor and the ground should be ventilated. The space should be at least 150 mm in depth (measured from the ground surface to the underside of the floor or insulation, if provided) and the ventilation recommendations should be as for suspended timber floors (see section 8.5.21).

If the building is located in an area where flooding might be a problem, it may be necessary to include a means of inspecting and clearing out the sub-floor voids beneath suspended floors. The DTLR publication *Preparing for floods: interim guidance for improving the flood resistance of domestic and small business properties*, DTLR, 2002, is available from DCLG and offers guidance on preparing for floods.

These recommendations are summarised in Fig. 8.10.

8.5.23   Resistance to damage from interstitial condensation for ground floors and floors exposed from below

Ground floors and floors exposed from below (such as those above an open parking bay under a building or an open passageway) should be designed and constructed so that their structural and thermal performance are not adversely affected by interstitial condensation. No actual guidance is given in Approved Document C, the reader merely being referred to other sources of guidance such as:

• Clause 8.5 and Appendix D of BS 5250: 2002 *Code of practice for the control of condensation in buildings*;

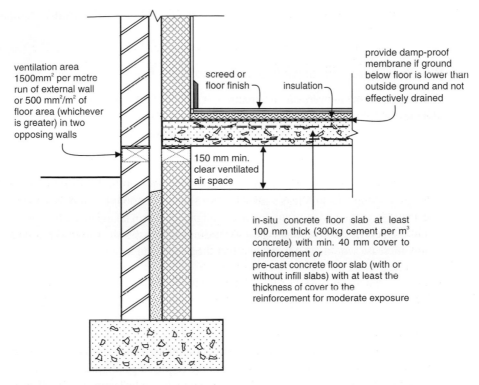

ventilation area 1500mm² per metre run of external wall or 500 mm²/m² of floor area (whichever is greater) in two opposing walls

screed or floor finish

insulation

provide damp-proof membrane if ground below floor is lower than outside ground and not effectively drained

150 mm min. clear ventilated air space

in-situ concrete floor slab at least 100 mm thick (300kg cement per m³ concrete) with min. 40 mm cover to reinforcement *or* pre-cast concrete floor slab (with or without infill slabs) with at least the thickness of cover to the reinforcement for moderate exposure

Fig. 8.10 Suspended concrete ground floors.

- BS EN ISO 13788: 2001 *Hygrothermal performance of building components and building elements. Internal surface temperature to avoid critical surface humidity and interstitial condensation. Calculation methods*; and
- BRE Report BR 262 *Thermal insulation: avoiding risks*, 2002.

8.5.24   Resistance to surface condensation and mould growth in floors

In order to resist surface condensation and mould growth in floors it is necessary to ensure that the surface is maintained above the dewpoint temperature. This will depend on the outside temperature, the temperature of the room in which the floor surface is situated and the relative humidity of the room. It can be affected by ventilation, by thermal bridging of construction elements at junctions and by the degree of thermal insulation provided in the flooring elements.

Therefore, in all floors, care should be taken to design the junctions between the elements so that thermal bridging is avoided. This can be done by following the recommendations in the report *Limiting thermal bridging and air leakage: robust construction details for dwellings and similar buildings*, published by The Stationery Office (TSO, 2001) or by following the guidance of BRE Information Paper IP17/01 *Assessing the effects of thermal bridging at junctions and around openings*.

Additionally, ground floors should be designed and constructed so that the thermal transmittance (U-value) does not exceed 0.7 W/m²K at any point.

8.5.25   Protection of walls against moisture from the ground

The term *wall* means vertical construction, which includes piers, columns and parapets and may include chimneys if they are attached to the building. Windows, doors and other openings are not included, but the joint between the wall and the opening is included.

Walls should be constructed so that:

- The passage of moisture from the ground to the inside of the building is resisted (this might not apply to buildings used solely for storage of goods in which the only persons habitually employed were storemen, etc. engaged only in taking in, caring for, or taking out the goods. Other similar types of buildings where the air is so moisture-laden that any increase would not adversely affect the health and safety of the occupants might also be excluded.)
- They will not be adversely affected by moisture from the ground.
- They will not transmit moisture from the ground to another part of the building that might be damaged.

The requirements mentioned above can be met for internal and external walls by providing a damp-proof course of suitable materials in the required position.

Figure 8.11 illustrates the main provisions, which can be summarised as follows:

- The damp-proof course may be of any material that will prevent moisture movement. This would include bituminous sheet materials, engineering bricks or slates laid in cement mortar, polyethylene or pitch polymer materials.
- The damp-proof course and any damp-proof membrane in the floor should be continuous.
- Unless an external wall is suitably protected by another part of the building, the damp-proof course should be at least 150 mm above outside ground level.
- Where a damp-proof course is inserted in an external cavity wall, the cavity should extend at least 225 mm below the level of the lowest damp-proof course. Alternatively, precipitation (see definition in section 8.5.18) can be prevented from reaching the inner leaf of the wall by the use of a damp-proof tray. This may be particularly useful where a cavity wall is built directly off a raft foundation, ground beam or similar supporting structure, and it is impractical to continue the cavity down 225 mm. Where a cavity tray is inserted, weep holes should provided every 900 mm in the outer leaf in order to allow moisture collecting on the tray to pass out of the wall. In some circumstances, such as above a window or door opening, the cavity tray will not extend the full length of the exposed wall. Here stop ends should be provided to the tray and at least two weep holes should be provided.

For walls subject to groundwater pressure reference should be made to the relevant recommendations of clauses 4 and 5 of BS 8215: 1991 *Code of practice for design and installation of damp-proof courses in masonry construction.* Additionally, BS 8102: 1990

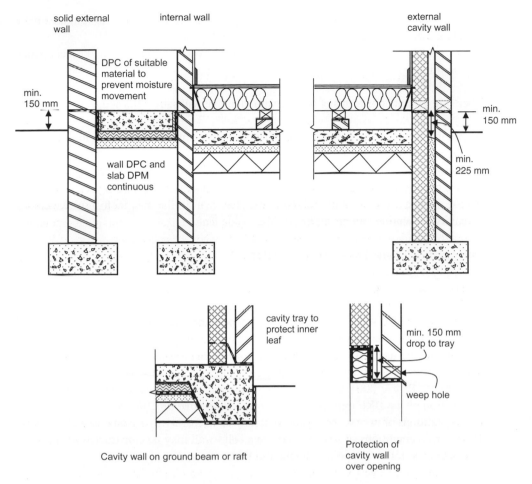

solid external wall

internal wall

external cavity wall

DPC of suitable material to prevent moisture movement

min. 150 mm

wall DPC and slab DPM continuous

min. 150 mm

min. 225 mm

cavity tray to protect inner leaf

min. 150 mm drop to tray

weep hole

Cavity wall on ground beam or raft

Protection of cavity wall over opening

Fig. 8.11   Protection of walls against moisture from the ground.

Code of practice for protection of structures against water from the ground includes recommendations for walls subject to groundwater pressure including basement walls.

8.5.26   Weather resistance of external walls

In addition to resisting ground moisture, external walls should:

- resist the passage of precipitation to the inside of the building (this might not apply to buildings used solely for storage of goods in which the only persons habitually employed were storemen, etc. engaged only in taking in, caring for, or taking out the goods; other similar types of buildings where the air is so moisture-laden that any increase would not adversely affect the health and safety of the occupants might also be excluded);
- not transmit moisture due to precipitation to other components of the building that might be damaged;

- be designed and constructed so as not to allow interstitial condensation to adversely affect their structural and thermal performance;
- not promote surface condensation and mould growth under reasonable occupancy conditions.

There are a number of forms of wall construction that will satisfy the above requirements:

- A solid wall of sufficient thickness holds moisture during bad weather until it can be released in the next dry spell.
- An impervious or weatherproof cladding prevents moisture from penetrating the outside face of the wall.
- The outside leaf of a cavity wall holds moisture in a similar manner to a solid wall, the cavity preventing any penetration to the inside leaf.

These principles are illustrated in Fig. 8.12.

8.5.27   Solid external walls

The thickness of a solid external wall will depend on the type of brick or block used and the severity of exposure to wind-driven rain. This may be assessed for a building in a given area by using BS 8104: 1992 *Code of practice for assessing exposure of walls to wind-driven rain.* Reference may also be made to BS 5628 *Code of practice for use of masonry,* Part 3: 2001 *Materials and components, design and workmanship* and the publication by the BRE entitled *Thermal insulation – avoiding the risks.*

In conditions of *very* severe exposure it may be necessary to use an external cladding. However, in conditions of severe exposure a solid wall may be constructed as shown in Fig. 8.13. The following points should also be considered.

- For brickwork or stonework the wall should be at least 328 mm thick.
- For dense aggregate blockwork the wall should be at least 250 mm thick.
- For lightweight aggregate or aerated autoclaved concrete the wall should be at least 215 mm thick.
- The brickwork or blockwork should be rendered or given an equivalent form of protection.
- Rendering should have a scraped or textured finish and be at least 20 mm thick in two coats. This permits easier evaporation of moisture from the wall.
- The bricks or blocks and mortar should be matched for strength to prevent cracking of joints or bricks and joints should be raked out to a depth of at least 10 mm in order to provide a key for the render.

    Further guidance on mortar mixes is given in BS EN 998: Part 2: 2002 *Specification of mortar for masonry.*
- The render mix should not be too strong or cracking may occur. A mix of 1:1:6 cement: lime: well grade sharp sand is recommended for all walls except those constructed of dense concrete blocks where $1:\frac{1}{2}:4$ should prove satisfactory.

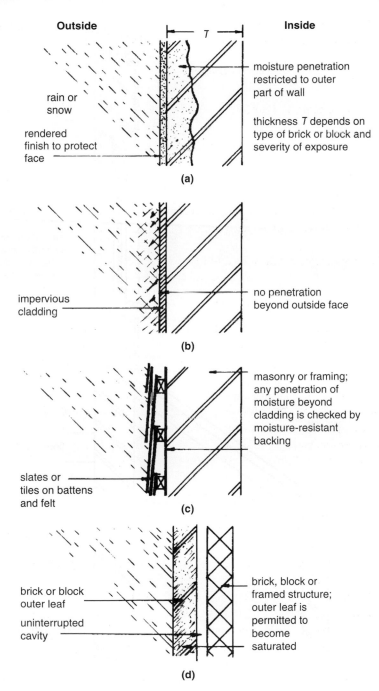

Fig. 8.12   Weather resistance of external walls – principles. (a) Solid external wall. (b) Impervious cladding. (c) Weather-resistant cladding. (d) Cavity wall.

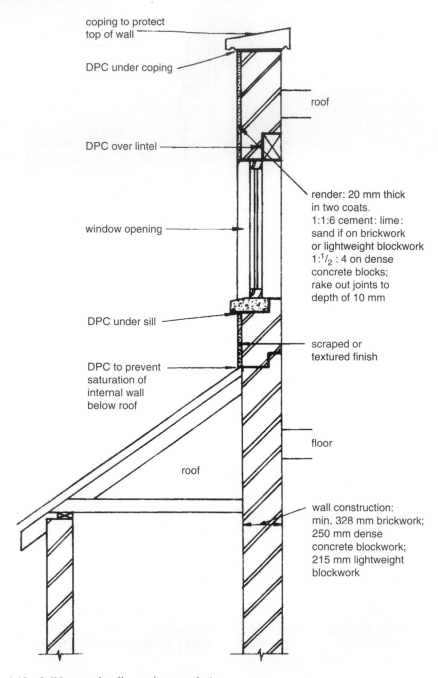

coping to protect top of wall

DPC under coping

roof

DPC over lintel

render: 20 mm thick in two coats. 1:1:6 cement: lime: sand if on brickwork or lightweight blockwork 1:$^1/_2$: 4 on dense concrete blocks; rake out joints to depth of 10 mm

window opening

DPC under sill

scraped or textured finish

DPC to prevent saturation of internal wall below roof

floor

roof

wall construction: min. 328 mm brickwork; 250 mm dense concrete blockwork; 215 mm lightweight blockwork

Fig. 8.13 Solid external walls – moisture exclusion.

Further details of a wide range of render mixes according to the severity of exposure and the type of brick or block may be obtained from BS 5262: 1991 *Code of practice for external renderings.*

- It is, of course, possible to obtain a wide range of premixed and proprietary mortars and renders. These should be used in accordance with the manufacturer's instructions.
- Where the top of a wall is unprotected by the building structure, it should be protected to resist moisture from rain or snow. Unless the protection and joints form a complete barrier to moisture, a damp-proof course should also be provided.
- Damp-proof courses, trays and closers should be provided to direct moisture towards the outside face of the wall in the positions shown in Fig. 8.13.
- Insulation to solid external walls may be provided on the inside or outside of the wall. Externally placed insulation should be protected unless it is able to offer resistance to moisture ingress so that the wall may remain reasonably dry (and the insulation value may not be reduced). Internal insulation should be separated from the wall construction by a cavity to give a break in the path for moisture. (Some examples of external wall insulation are given in Fig. 8.14.)

The performance requirements for solid and cavity external walls can also be met by complying with BS 5628 *Code of practice for use of masonry:* Part 3: 2001 *Materials and components, design and workmanship.*

8.5.28   External cavity walls

In order to meet the performance requirements, an external cavity wall should consist of an internal leaf which is separated from the external leaf by:

- a drained air space; or
- some other method of preventing precipitation from reaching the inner leaf.

An external cavity wall may consist of the following:

- An outside leaf of masonry (brick, block, natural or reconstructed stone).
- Minimum 50 mm uninterrupted cavity. Where a cavity is bridged (by a lintel, etc.) a damp-proof course or tray should be inserted in the wall so that the passage of moisture from the outer to the inner leaf is prevented. This is not necessary where the cavity is bridged by a wall tie, or where the bridging occurs, presumably, at the top of a wall and is then protected by the roof. Cavity walls may also be bridged by cavity barriers and fire stops where other parts of the building regulations require this (such as Part B or Part E). Where an opening is formed in a cavity wall, the jambs should have a suitable vertical damp-proof course or the cavity should be provided with a suitable cavity closure so as to prevent the passage of moisture.
- An inside leaf of masonry or framing with suitable lining.
- In order to ensure structural robustness and weather resistance in the wall, masonry units should be laid on a full bed of mortar and the cross joints should be continuously and substantially filled.
- Where a cavity is only partially filled with insulation, the remaining cavity should be at least 50 mm wide (see Fig. 8.14).

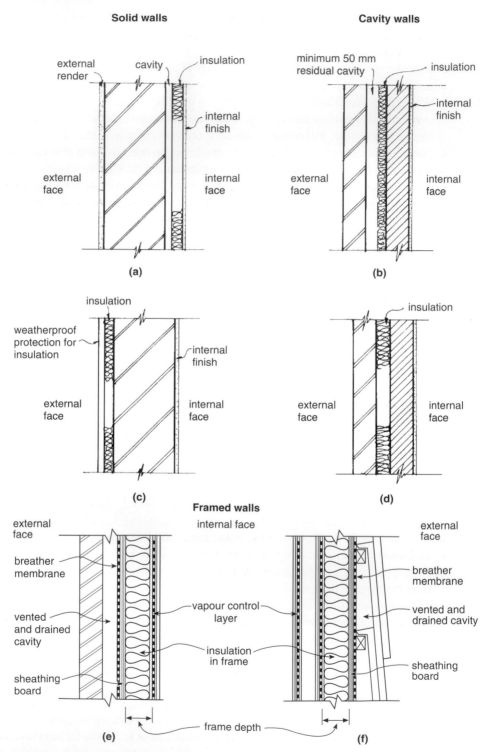

Fig. 8.14 Weather resistance and insulation of external walls. (a) Internal insulation. (b) Partially filled cavity. (c) External insulation. (d) Fully filled cavity. (e) Brick-clad timber-framed wall. (f) Tile-clad timber-framed wall.

Alternatively, the relevant recommendations of BS 5628 *Code of practice for use of masonry:* Part 3: 2001 *Materials and components, design and workmanship* may be followed. Factors affecting rain penetration of cavity walls are indicated in the Code.

8.5.29   Weather resistance and cavity insulation

Since the installation of cavity insulation effectively bridges the cavity of a cavity wall and could give rise to moisture penetration to the inner leaf, it is most important that it be carried out correctly and efficiently. Approved Document C lists a number of British Standards, Codes of Practice and other documents that cover the various materials that may be incorporated into a cavity wall. It also sets out a simple method for determining the minimum width of cavity whether filled or clear for various exposure conditions, various insulating methods and various forms of wall construction. Diagram 12 from Approved Document C is reproduced below. When used with Table 4 (also reproduced) the suitability of a wall for installing insulation may be determined as follows:

(1) From Diagram 12 determine the national exposure (e.g. zone 1, 2, 3 or 4).
(2) Modify this result by either adding one to the exposure zone if local conditions accentuate wind effects (e.g. open hillsides or valleys where the wind is funneled onto the wall), or subtract one from the exposure zone where the walls do not face into the prevailing wind.
(3) Select a form of wall construction from impervious cladding, rendered finish or facing masonry.
(4) Choose an insulation method from built-in full fill, injected fill (UF foam or other material), partial fill, internal insulation or fully filled cavity.
(5) Select the appropriate cavity width by reference to the exposure zone figure.

BRE Report 262 *Thermal insulation: avoiding risks* contains further information regarding the use of Table 4.

It is also possible to determine the suitability of a wall for installing insulation into the cavity by following the calculation or assessment procedure in current British or CEN standards. When partial fill materials are to be used, the residual cavity should not be less than 50 mm nominal.

The following points should also be taken into account:

- Rigid materials (boards or batts) which are built in as the wall is constructed should be the subject of current certification from an appropriate body or a European Technical Approval and the work should be carried out to meet the requirements of that document.
- Urea-formaldehyde foam inserted after the wall has been constructed should comply with BS 5617: 1985 *Specification for urea-formaldehyde (UF) foam systems suitable for thermal insulation of cavity walls with masonry or concrete inner and outer leaves* and should be installed in accordance with BS 5618: 1985 *Code of practice for thermal insulation of cavity walls (with masonry or concrete inner and outer leaves) by filling with urea-formaldehyde (UF) foam systems.* Before work is commenced the wall should be assessed for suitability and the work should be carried out by a person operating

AD C
Diagram 12   UK zones for exposure to driving rain.

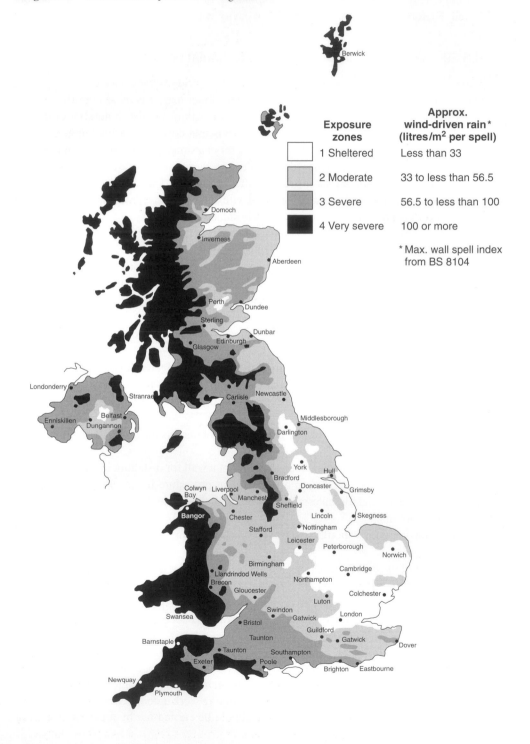

Exposure zones	Approx. wind-driven rain* (litres/m² per spell)
1 Sheltered	Less than 33
2 Moderate	33 to less than 56.5
3 Severe	56.5 to less than 100
4 Very severe	100 or more

* Max. wall spell index from BS 8104

AD C

Table 4   Maximum recommended exposure zones for insulated masonry walls.

Wall construction		Maximum recommended exposure zone for each construction						
		Impervious cladding		Rendered finish		Facing masonry		
Insulation method	Min. width of filled or clear cavity (mm)	Full height of wall	Above facing masonry	Full height of wall	Above facing masonry	Tooled flush joints	Recessed mortar joints	Flush sills and copings
Built-in full fill	50	4	3	3	3	2	1	1
	75	4	3	4	3	3	1	1
	100	4	4	4	3	3	1	2
	125	4	4	4	3	3	1	2
	150	4	4	4	4	4	1	2
Injected fill **not UF foam**	50	4	2	3	2	2	1	1
	75	4	3	4	3	3	1	1
	100	4	3	4	3	3	1	1
	125	4	4	4	3	3	1	2
	150	4	4	4	4	4	1	2
Injected fill **UF foam**	50	4	2	3	2	1	1	1
	75	4	2	3	2	2	1	1
	100	4	2	3	2	2	1	1
Partial fill								
Residual 50 mm cavity	50	4	4	4	4	3	1	1
Residual 75 mm cavity	75	4	4	4	4	4	1	1
Residual 100 mm cavity	100	4	4	4	4	4	2	1
Internal insulation								
Clear cavity 50 mm	50	4	3	4	3	3	1	1
Clear cavity 100 mm	100	4	4	4	4	4	2	2
Fully filled								
Cavity 50 mm	50	4	3	3	3	2	1	1
Cavity 100 mm	100	4	4	4	3	3	1	2

under an Approved Installer Scheme that includes an assessment of capability (see also Chapter 9 *Toxic substances*).

- Other insulating materials inserted after the wall has been constructed should have certification from an appropriate body and be installed in accordance with the appropriate installations code. Before work is commenced the wall should be assessed for suitability and the work should be carried out by a person operating under an Approved Installer Scheme that includes an assessment of capability. Alternatively, the insulating material should be the subject of current certification from an appropriate body or a

European Technical Approval. The work should be carried out to meet the requirements of the relevant document by operatives directly employed by the document holder. Alternatively, they may be employed by an installer approved to operate under the document.

- Where materials are being inserted into a cavity wall of an existing house, the suitability of the wall for filling should be assessed, before installation, in accordance with BS 8208: Part 1: 1985 *Guide to assessment of suitability of external cavity walls for filling with thermal insulation*. Special attention should be given to the condition of the external leaf of the wall (e.g. its state of repair and type of pointing). Some materials that are used to fill existing cavity walls may have a low risk of moisture being carried over to the internal leaf of the wall. In cases where a third party assessment of such a cavity fill material contains a method of assessing the construction of the walls and exposure risk, the procedure set out above may be replaced by that method.

(See Fig. 8.14 for further information regarding the insulation of external walls.)

8.5.30   Impervious claddings for external walls

The principles of external claddings are illustrated in Fig. 8.12(b) and (c) above where it is shown that they should:

- resist the passage of precipitation to the inside of the building;
- not be damaged by precipitation;
- not transmit moisture due to precipitation to any part of the building that might be damaged.

It is possible to design cladding so that it can either hold precipitation (even if wind driven) at the face of the building (impervious cladding), or allow precipitation to penetrate beyond the face but stop it from getting beyond the back of the cladding.

Therefore the cladding should be either:

- jointless (or have sealed joints) and be impervious to moisture (such as sheets of metal, glass, plastic or bituminous materials); or
- have overlapping dry joints and consist of impervious or weather-resisting materials (such as natural stone or slate, cement based products, fired clay or wood).

Dry jointed claddings should be backed by a material (such as sarking felt), which will direct any penetrating moisture to the outside surface of the structure.

Moisture-resisting materials consisting of bituminous or plastic products lapped at the joints are permitted but they should be permeable to water vapour unless there is a ventilated space behind the cladding.

Materials that are jointless or have sealed joints should be designed to accommodate structural and thermal movement.

Dry joints between cladding units should be designed either to resist precipitation or to direct any moisture entering them to the outside face of the structure. The suitability of dry joints will depend on the design of the joint and cladding and the severity of exposure of the building to wind and rain.

All external claddings should be securely fixed and particular care should be taken with detailing and workmanship at the junctions between window and door openings and the cladding since these junctions are particularly susceptible to the ingress of moisture. BS 8000: Part 6: 1990 *Workmanship on building sites. Code of practice for slating and tiling of roofs and claddings* may be of some assistance for claddings containing these materials.

Some materials, such as timber claddings, are subject to rapid deterioration unless properly treated. These materials should only be used as the weather-resisting part of a roof or wall if they can meet the conditions specified in AD Regulation 7 described earlier in this chapter. It should be noted that the weather-resisting part of a roof or wall does not include paint or any surface rendering or coating which does not of itself provide all the weather resistance. Where the cladding is on the façade of a timber-framed building or is itself supported by timber components, the construction should be ventilated so that rapid drying out of any penetrating moisture is ensured. Furthermore, the cladding to framed external walls (see Fig. 8.14(e) and (f)) should be separated from the insulation or sheathing by a vented and drained cavity with a membrane that allows water vapour to pass through it but resists the passage of liquid water on the inside of the cavity.

Insulation may be incorporated into the roof or wall cladding provided that it is protected from moisture (or is unaffected by it). Possible problems may arise due to interstitial condensation and cold bridges in the construction. Further guidance on this may be found in BRE Report BR 292 *Thermal insulation – avoiding risks*.

8.5.31 Impervious claddings for external walls – alternative approach

Relevant recommendations of the following alternative guidance sources can be used instead of the detailed guidance given in Approved Document C:

- British Standard Code of Practice 143 *Code of practice sheet roof and wall coverings* (this includes recommendations for aluminium, zinc, galvanised corrugated steel, copper and semi-rigid asbestos bitumen sheet);
- BS 6915: 2001 *Specification for design and construction of fully supported lead sheet roof and wall coverings*;
- BS 8219 *Profiled fibre cement. Code of practice*;
- BS 8200: 1985 *Code of practice for design of non-loadbearing external vertical enclosures of buildings*;
- BS 8297: 2000 *Code of practice for design and installation of non-loadbearing precast concrete cladding*;
- BS 8298: 1994 *Code of practice for design and installation of natural stone cladding and lining*;
- MCRMA Technical Paper 6 *Profiled metal roofing design guide*, revised edition 1996;
- MCRMA Technical Paper 9 *Composite roof and wall cladding panel design guide*, 1995.

The above documents describe the materials to be used and contain design guidance including fixing recommendations.

8.5.32 Cracking of external walls

This very brief section, which was added into the 2004 edition of Approved Document C, is curious in that it rather states the obvious, that severe rain penetration may occur through cracks in masonry external walls! It goes on to give the reasons for the cracking as being thermal movement in hot weather or subsidence after prolonged droughts and advises that this should be taken into account when designing the building. There are, of course, many causes of cracking in external walls including moisture movement in clay brickwork, cavity wall tie failure, sulfate attack, frost action, swelling and shrinking of clay soils, mining subsidence, faulty drains, etc. In order to assist the reader the following reference sources are given in the Approved Document:

- *BRE Building Elements. Walls, windows and doors*;
- BRE Report BR 292 *Cracking in buildings*;
- BS 5628: Part 3: 2001 *Code of practice for use of masonry. Materials and components, design and workmanship*.

The first two references are both excellent information sources. The author has also found the publication *Common Defects in Buildings* published by The Stationery Office to be invaluable, especially in the diagnosis of defects with particular reference to cracking in external walls. BS 5628 is more relevant to choice of materials.

8.5.33 Detailing of joints between walls and door and window frames

In many cases, the most vulnerable part of the construction regarding moisture penetration due to precipitation is at the junction between the wall and any door, window or other openings into the building. The joint formed at such junctions should:

- resist the passage of precipitation to the inside of the building;
- not transmit moisture due to precipitation to other parts of the building that might be damaged;
- not be damaged by precipitation.

The usual way of satisfying these performance criteria is by the use of suitably placed damp-proof courses to direct moisture to the outside. These should be provided:

- where moisture flowing downward would be interrupted by an obstruction, such as a lintel;
- where a sill under a door, window or other opening (including any associated joint) might not form a complete barrier to the transfer of precipitation;
- where the construction of a reveal at a door, window or other opening (including any associated joint) might not form a complete barrier to the passage of rain or snow.

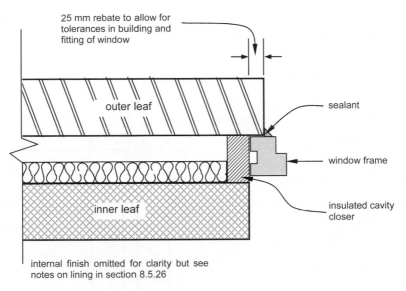

25 mm rebate to allow for tolerances in building and fitting of window

outer leaf

sealant

window frame

inner leaf

insulated cavity closer

internal finish omitted for clarity but see notes on lining in section 8.5.26

Fig. 8.15 Example of checked rebate window reveal detail – areas of severe or very severe exposure to driving rain.

Modern methods of construction often lead to rather wide cavities due to the additional thicknesses of insulation now required by Part L (see Chapter 16) and the need in some cases for a drained 50 mm wide cavity in addition to the insulation where the cavity is only partially filled (see section 8.5.29). In such cases most window and door frames will not fully cover the cavity closer. This will necessitate lining the reveal with plasterboard, dry lining, a thermal backing board or even a support system. The internal reveal may also be wet plastered provided that the plaster is applied to a suitable backing such as expanded metal lathing or similar support system.

In areas of the country where the exposure to driving rain is in zone 4 (see section 8.5.29) additional precautions should be taken to prevent the penetration of precipitation. Approved Document C recommends the use of checked rebates at all window and door frames. In this form of construction (Fig. 8.15) the frame is set back behind the outer leaf of masonry. An alternative to this might be by the use of a proprietary finned cavity closer.

8.5.34 Door thresholds

Approved Document M, *Access to and use of buildings* (2004 edition, see Chapter 17), specifies that in certain circumstances buildings should be provided with accessible thresholds. Clearly, this has implications for the weatherproofing of the building since there is a danger that precipitation and groundwater could enter the building at door openings. In order to prevent this, Approved Document C recommends that the external landing should slope away from the doorway at a fall of between 1 in 40 and 1 in 60. In order to encourage run-off, the door sill should have a maximum slope of 15°. Some additional recommendations are illustrated in Fig. 8.16. Additional advice may be obtained from the following publications:

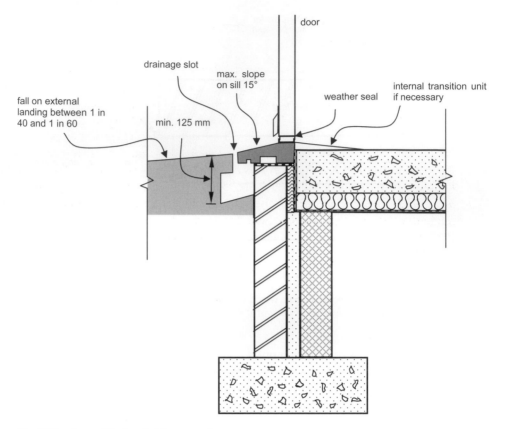

Fig. 8.16 Accessible threshold – exposed areas.

- BRE Good Building Guide 47 *Level external thresholds: reducing moisture penetration and thermal bridging*, 2001;
- *Accessible thresholds in new buildings: guidance for house builders and designers*, TSO, 1999.

8.5.35 Resistance to damage from interstitial condensation for external walls

In general, external walls will satisfy the requirements of the Building Regulations regarding resistance to damage from interstitial condensation if they are designed and constructed in accordance with Clause 8.3 of BS 5250: 2002 *Code of practice for the control of condensation in buildings* and BS EN ISO 13788: 2001 *Hygrothermal performance of building components and building elements. Internal surface temperature to avoid critical surface humidity and interstitial condensation. Calculation methods*.

In swimming pools and other buildings where high levels of moisture are generated, there is a particular risk of interstitial condensation in walls and also roofs (see section 8.5.38) because of the high internal temperatures and humidities that exist. In these cases specialist advice should be sought when these are being designed.

8.5.36   Resistance to surface condensation and mould growth in external walls

In order to resist surface condensation and mould growth in external walls it is necessary to ensure that the surface is maintained above the dewpoint temperature. This will depend on the outside temperature, the temperature of the room in which the wall surface is situated and the relative humidity of the room. It can be affected by ventilation, by thermal bridging of construction elements at junctions and by the degree of thermal insulation provided in the wall elements. Therefore, in external walls, care should be taken to design the junctions between the elements so that thermal bridging is avoided. This can be done by following the recommendations in the report *Limiting thermal bridging and air leakage: robust construction details for dwellings and similar buildings*, published by The Stationery Office (TSO, 2001) or by following the guidance of BRE Information Paper IP17/01 *Assessing the effects of thermal bridging at junctions and around openings*.

Additionally, externals walls should be designed and constructed so that the thermal transmittance (U-value) does not exceed 0.7 W/m^{2}K at any point.

8.5.37   Weather resistance of roofs

The roof of a building should:

- resist the penetration of precipitation to the inside of the building;
- not be damaged by precipitation;
- not transmit precipitation to another part of the building that might be damaged;
- be designed and constructed so as not to allow interstitial condensation to adversely affect its structural and thermal performance;
- not promote surface condensation and mould growth under reasonable occupancy conditions.

The recommendations for impervious external wall claddings mentioned in section 8.5.30 apply equally to roof covering materials.

The performance requirements for external wall and roof claddings can also be met if they comply with:

- British Standard Code of Practice 143 *Code of practice sheet roof and wall coverings* (this includes recommendations for aluminium, zinc, galvanised corrugated steel, copper and semi-rigid asbestos bitumen sheet);
- BS 6915: 2001 *Specification for design and construction of fully supported lead sheet roof and wall coverings*;
- BS 8219 *Profiled fibre cement. Code of practice*;
- BS 8200: 1985 *Code of practice for design of non-loadbearing external vertical enclosures of buildings*;
- MCRMA Technical Paper 6 *Profiled metal roofing design guide*, revised edition 1996;
- MCRMA Technical Paper 9 *Composite roof and wall cladding panel design guide*, 1995.

The above documents describe the materials to be used and contain design guidance including fixing recommendations.

8.5.38   Resistance to damage from interstitial condensation in roofs

This section was originally contained in Part F *Ventilation* of Schedule 1 to the Building Regulations 2000 – requirement F2. With the extension of Part C to cover condensation risk in the 2004 amendment it was logical to transfer these provisions to Part C and they have been omitted from the 2006 edition of Part F.

The guidance given in the current Approved Document C is much less detailed than that originally contained in Approved Document F and is now stated in performance terms and by reference to other sources of guidance where particular design details are illustrated. The current trend of continually referring to additional sources of guidance is making the Approved Documents much less useful to the practitioner as, in many cases, they can only be regarded as a directory and not as a direct source of technical guidance in their own right.

When condensation occurs in roof spaces it can have two main effects:

- the thermal performance of the insulant materials may be reduced by the presence of the water; and
- the structural performance of the roof may be affected due to increased risk of fungal attack.

Approved Document C recommends that interstitial condensation in roofs should be limited such that the thermal and structural performance of the roof will not be adversely affected.

It should be noted that the provisions of AD C apply to roofs of any pitch even though a roof that exceeds 70° in pitch is required to be insulated as if it were a wall. Additionally, small roofs over porches or bay windows, etc., may sometimes be excluded from the requirements of requirement C2(c) if there is no risk to health or safety.

In swimming pools and other buildings where high levels of moisture are generated, there is a particular risk of interstitial condensation in walls and roofs because of the high internal temperatures and humidities that exist. In these cases specialist advice should be sought when these are being designed.

For cold deck roofs (where the insulation is placed at ceiling level and can be permeated by moisture from the building) requirement C2(c) can be met by the provision of adequate ventilation in the roofspace. In such roofs it is obviously essential that moist air is prevented from reaching the roofspace where it might condense within or on the insulation layer and this is particularly important above areas of high humidity such as bathrooms and kitchens. Weak points in the construction where this might occur include:

- gaps and penetrations for pipes and electrical wiring (these should be filled and sealed); and
- loft access hatches (where an effective draught seal should be provided to reduce the inflow of warm moist air).

The requirements can be met for both flat and pitched roofs by following the relevant recommendations of clause 8.4 of BS 5250: 2002 *Code of practice: the control of condensation in* buildings and BS EN ISO 13788: 2001 *Hygrothermal performance of building components and building elements. Internal surface temperature to avoid critical surface humidity and interstitial condensation. Calculation methods.* Further guidance may also be found in the 2002 edition of BRE Report BR 262 *Thermal insulation: avoiding the risks.*

The following guidance is based on that which was originally contained in Approved Document F which was in turn based on the guidance provided in BS 5250. It covers only a few typical examples of commonly found forms of construction where the insulation is placed at ceiling level (cold deck roofs). For full details and for other forms of construction reference should be made to BS 5250, clause 8.4.

Roofs with a pitch of 15° or more

Pitched roofs should be cross-ventilated by permanent vents at eaves level on the two opposite sides of the roof, the vent areas being equivalent in area to a continuous gap along each side of 10 mm width. Ridge ventilation equivalent to a continuous gap of 5 mm should also be provided where the pitch exceeds 35° and/or the spans are in excess of 10 m.

Mono-pitch or lean-to roofs should have ventilation at eaves level as above and also at high level either at the point of junction or through the roof covering at the highest practicable point. The high-level ventilation should be equivalent in area to a continuous gap 5 mm wide (see Fig. 8.17).

In recent years vapour permeable underlays have come onto the market. If these are used without continuous boarding below the tiling battens, it is not necessary to ventilate the roof space below the underlay. However, BS 5250 recommends that, in these circumstances, a ventilated space should be formed above the underlay by using 25 mm battens and counterbattens and provision of ventilation at low level equivalent to a 25 mm continuous gap and at high level equivalent to a 5 mm continuous gap. Simply relying on fortuitous ventilation through the tile/slate joints should not be relied on to ventilate this space adequately.

Roofs with a pitch of less than 15°

In low-pitched roofs the volume of air contained in the void is less and therefore the risk of saturation is greater.

This also applies to roofs with pitch greater than 15° where the ceiling follows the pitch of the roof. The high-level ventilation should be equivalent in area to a continuous gap 5 mm wide (see Fig. 8.18).

Cross-ventilation should again be provided at eaves level but the ventilation gap should be increased to 25 mm width.

8.5.39   Resistance to surface condensation and mould growth in roofs

In order to resist surface condensation and mould growth in roofs and roofspaces, care should be taken to design the junctions between the elements and the details of openings,

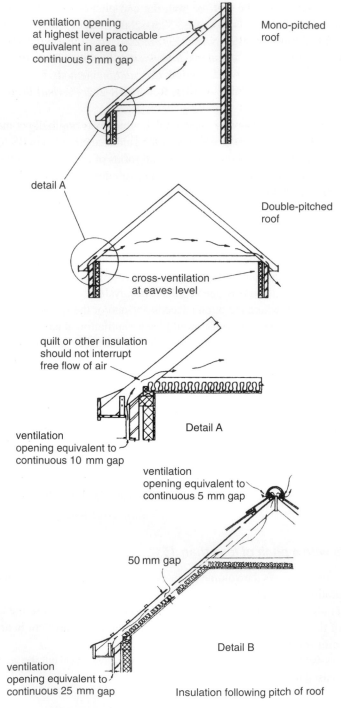

ventilation opening
at highest level practicable
equivalent in area to
continuous 5 mm gap

Mono-pitched
roof

detail A

Double-pitched
roof

cross-ventilation
at eaves level

quilt or other insulation
should not interrupt
free flow of air

Detail A

ventilation
opening equivalent to
continuous 10 mm gap

ventilation
opening equivalent to
continuous 5 mm gap

50 mm gap

Detail B

ventilation
opening equivalent to
continuous 25 mm gap

Insulation following pitch of roof

Fig. 8.17 Roof void ventilation – roofs pitched at 15° or more.

such as windows, so that thermal bridging and air leakage is avoided. This can be done by following the recommendations in the report *Limiting thermal bridging and air leakage: robust construction details for dwellings and similar buildings*, published by The Stationery Office (TSO, 2001) or by following the guidance of BRE Information Paper IP17/01 *Assessing the effects of thermal bridging at junctions and around openings* and/or MCRMA Technical Paper 14 *Guidance for the design of metal roofing and cladding to comply with approved document L2:2001*, 2002.

Additionally, roofs should be designed and constructed so that the thermal transmittance (U-value) does not exceed 0.35 W/m^{2}K at any point.

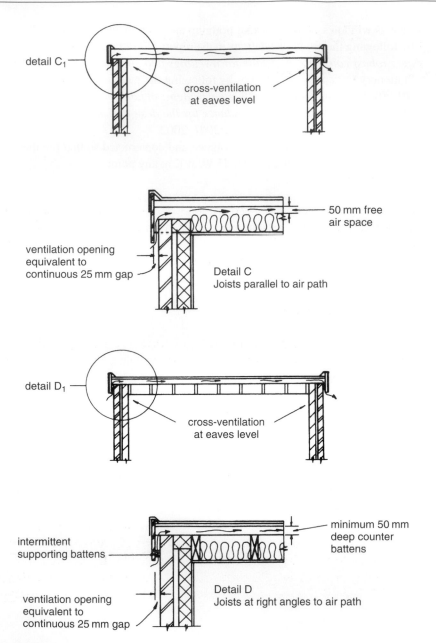

Fig. 8.18   Roof void ventilation – roofs pitched at less than 15°. A free airspace of at least 50 mm should be provided between the roof deck and the insulation. This may need to be formed using counter-battens if the joists run at right angles to the flow of air (see Fig. 8.18). Where it is not possible to provide proper cross-ventilation an alternative form of roof construction should be considered. It is possible to install vapour checks (called vapour control layers in BS 5250) at ceiling level using polythene or foil-backed plasterboard, etc., to reduce the amount of moisture reaching the roof void. This is not acceptable as an alternative to ventilation unless a complete vapour barrier is installed.

9 Toxic substances (Part D)

9.1 Introduction

In recent years there has been evidence to suggest that fumes from urea-formaldehyde foam, when used as a cavity wall filling, can have an adverse effect on the health of people occupying the building. This is still a contentious issue, but has nevertheless become a subject for building control.

9.2 Cavity insulation

Where insulating material is inserted into a cavity in a cavity wall reasonable precautions must be taken to prevent toxic fumes from penetrating occupied parts of the building.

It should be noted that Approved Document D1 does not require total exclusion of formaldehyde fumes from buildings but merely that these should not increase to an irritant concentration.

The inner leaf of the cavity wall should provide a continuous barrier to the passage of fumes and for this purpose it should be of brick or block construction.

Before work is commenced the wall should be assessed for suitability in accordance with BS 8208 *Guide to assessment of suitability of external walls for filling with thermal insulants* Part 1: 1985 *Existing traditional cavity construction.*

The work should be carried out by a person holding a current BSI Certificate of Registration of Assessed Capability for this particular type of work.

The urea-formaldehyde foam should comply with the requirements of BS 5617: 1985 *Specification for urea-formaldehyde (UF) foam systems etc.* and the installation with BS 5618: 1985 *Code of practice for thermal insulation of cavity walls etc.*

10 Sound insulation (Part E)

J.R. Waters

10.1 Introduction

Over recent years, living standards have improved, life styles have altered and as a consequence, sources of the noise which people commonly encounter have changed. Householders are now subjected to higher levels of noise from traffic and neighbourhood sources and this, coupled with higher expectations of indoor environmental quality and reduced tolerance to unwanted noise, has led to increasing dissatisfaction with the standard of sound insulation provided in dwellings. Part E of Schedule 1 to the Building Regulations 2000 (as amended) is aimed at targeting widespread dissatisfaction of occupants living in properties constructed under the preceding legislation that derived from studies undertaken prior to the 1965 Regulations. In addition to annoyance, it should be recognised that domestic noise is also a potential risk to health and whilst difficult to quantify, it is known to be a source of loss of sleep, stress and in extreme cases contributes to assaults on neighbours.

Initial proposals for amending Requirement E included the insulation of the envelope of the building against noise from external sources, but this has not been included in Part E of the 2000 Regulations. Requirements E1 and E2 relate to resistance to the passage of sound between dwellings constructed as part of the same building and within dwellings, and E3 is aimed at limiting reverberation of noise in the common parts of multi-occupancy buildings. The only requirement which does not relate to buildings providing living accommodation is E4, the subject of which is acoustic conditions in schools. This requirement is not addressed in Approved Document E and designers are referred to The Stationery Office publication Building Bulletin 93, *The acoustic design of schools*.

In Approved Document E 1992 it was stated that one way of satisfying the requirements of Part E of Schedule 1 to the 1991 Building Regulations was to build components for which sound insulation was required in accordance with specific 'deemed to satisfy' constructions referred to in the document. Whilst elements constructed to a standard equal to the prescribed constructions may well have provided adequate sound insulation, a fundamental shortcoming of such a procedure is that it makes no allowance for the consequences of bad workmanship other than inspection by building control authorities. Unfortunately, sound insulation is seriously affected by localised construction imperfections and details which on completion may be hidden from view may render a construction unsatisfactory. For this reason, specific performance standards are stipulated in Approved Document E 2003 and there is a requirement for pre-completion testing, by the controlling authority, of a sample of completed units to ensure compliance with the standards. This is a very important aspect of the control procedure, and it places a requirement on the contractor to maintain quality control.

The Building Regulations 2000 (as amended) extend the requirement to provide sound insulation in dwellings beyond the circumstances stipulated in the 1991 Regulations. Prior to the current legislation, the requirement to provide sound insulation was restricted to walls separating houses together with walls and floors separating flats. This has now been extended to include similar levels of insulation between 'rooms for residential purposes' which include hotel rooms, rooms in residential homes and student halls of residence accommodation. In addition, there is now a requirement to restrict the passage of sound between certain walls and between floors within dwelling-houses and other types of accommodation. An additional source of annoyance in multi-occupancy buildings containing dwelling units is noise generated in the common parts of such buildings which tend to be constructed using hard durable noise-reflecting surfaces. As a consequence, the regulations now include a requirement to restrict the level of reverberant sound in such areas.

The current edition of the Approved Document E, *Resistance to the passage of sound*, was published in 2003 and came into effect on 1 July 2003. Amendments to this document, entitled *Amendments 2004 to Approved Document E (Resistance to the passage of sound)* were published in June 2004 and came into effect on 1 July 2004. These two documents have now been consolidated into a single document, published in April 2006. The main features of the 2004 amendments are:

- clarification of some essential definitions;
- consequential changes to other parts of the regulations;
- the correction of minor errors in the 2003 document;
- the inclusion of Robust Details as an alternative to pre-completion testing for demonstrating compliance.

Approved Document E (2003) applies to dwelling-houses, flats and rooms for residential purposes. Amendments 2004 includes the clarification, stated formally in regulation 2, of the definition of a room for residential purposes as follows:

'A "room for residential purposes" means a room, or a suite of rooms, which is not a dwelling-house, or a flat and which is used by one or more persons to live and sleep and includes a room in a hostel, an hotel, a boarding house, a hall of residence or a residential home, whether or not the room is separated from or arranged in a cluster group with other rooms, but does not include a room in a hospital, or other similar establishment, used for patient accommodation and, for the purposes of this definition, a "cluster" is a group of rooms for residential purposes which is:
(a) separated from the rest of the building in which it is situated by a door which is designed to be locked;
and
(b) not designed to be occupied by a single household.'

The consequential changes to other parts of the regulations concern regulations 20A and 12A, which have been modified to allow forms of construction approved by Robust Details Limited to be used as an alternative to pre-completion testing. This is discussed in section 10.9.

For convenience, dwelling-houses, flats and rooms for residential purposes will, for the remainder of this chapter be referred to collectively as 'dwelling places'. Also, each section heading within this chapter is followed by a list of the relevant paragraphs in Approved Document E (2003), modified, as appropriate, by Amendments (2004). However, since material relating to a specific subject may occur in a number of places within the AD, these reference lists may not be totally complete or comprehensive.

10.2   Performance requirements (Approved Document E, Section 0: Performance)

10.2.1   Sound insulation between dwelling places AD E: 0.1–0.8

Requirement E1 of Schedule 1 to the Building Regulations 2000 (as amended) states that the design and construction of houses, flats and rooms for residential purposes should be such that there is reasonable resistance to sound transmission between them and other parts of the same building and other buildings with which they are in direct physical contact. It should be noted that this requirement applies not only to purpose-built units but also to those resulting from the material change of use from some other type of building. Rooms for residential purposes include (see definitions in Chapter 2, section 2.4) rooms in residential homes, student halls of residence and hotels in which people may be expected to sleep.

In circumstances where Requirement E1 applies, regulation 20A of the Building Regulations 2000 (as amended) and regulation 12A of the Building (Approved Inspectors etc) Regulations 2000 (as amended) also apply. These additional regulations impose a requirement to pursue one or other of two courses of action:

(1) To undertake an approved programme of sound insulation testing and provide the appropriate authority with a copy of the test results. In the case of both regulations there are specific time limits within which the results must be presented to the building control body. It is stated in Approved Document E that the normal way of satisfying regulation 20A or 12A will be to undertake this programme of testing (known as pre-completion testing) in accordance with section 1 of the approved document.

(2) Only in the case of a dwelling-house or a building containing flats, to use one or more of the design details approved by Robust Details Limited. This is an alternative to undertaking a programme of testing, and is subject to observing specified procedures.

The requirements for pre-completion testing are considered in section 10.3, and the experimental procedure is discussed in section 10.12 of this chapter. The use of Robust Details is described in section 10.9.

Requirement E1 will be satisfied for separating walls and floors together with stairs which separate dwelling units if they are constructed such that they satisfy the sound insulation values given in Tables 1a and 1b in the Approved Document and also in Table 10.1 below. Separating floors are defined as floors which separate flats or rooms for residential purposes and separating walls are those which separate adjoining dwellinghouses, flats or rooms for residential purposes. It is important to note that sound will not only be transmitted directly through the element in question but also through adjacent, or flanking, elements and that when assessing compliance with Table 10.1, flanking as well as direct transmission must be taken into account. See section 10.14 for explanations of these terms.

Requirement E1 does not relate only to walls and floors which separate dwelling units. If a wall or floor separates a dwelling unit from a space used for communal or non-domestic purposes, e.g. a refuse chute, the level of sound insulation required may be greater than that given in Table 10.1 for separating walls and floors. In such circumstances, the required value of sound insulation will depend on the level of noise generated in the non-domestic space and expert advice should be sought to assess the anticipated level and, if necessary, recommend higher levels of sound insulation. Figure 10.1 summarises the obligations of E1 in terms of elements requiring insulation and the nature of insulation required.

It will be seen from Table 10.1 that the criterion applied to assess airborne sound insulation is a single figure index, $D_{nT,w} + C_{tr}$, which is obtained from a frequency band analysis of the noise. The term 'airborne sound insulation' is explained in section 10.14 and the procedure required for measurement of $D_{nT,w} + C_{tr}$ is described in section 10.12. This index replaces $D_{nT,w}$ which applied in the 1992 Approved Document by the addition of the C_{tr} term. C_{tr} modifies the index by increasing the significance of low frequency noise. The addition of C_{tr} is considered important since the powerful low frequency output of modern musical entertainment systems is such that sound transmission at low frequencies is now a frequent source of considerable annoyance.

The value of 45 dB for airborne sound insulation between purpose-built dwellings that appears in Table 10.1 appears at first sight to represent a reduction in the required standard when compared with the equivalent value of 49 dB which it replaces. However, the old

Table 10.1 Performance standards for separating walls, separating floors and stairs that have a separating function.

	Minimum Value of Airborne Sound Insulation $D_{nT,w} + C_{tr}$ dB	Maximum Value of Impact Sound Insulation $L'_{nT,w}$ dB
Purpose built dwelling-houses and flats		
Walls (inc. walls that separate houses and flats from rooms for residential purposes)	45	–
Floors and stairs	45	62
Dwelling-houses, flats and rooms for residential purposes formed by material change of use		
Walls	43	–
Floors and stairs	43	64
Purpose built rooms for residential purposes		
Walls	43	–
Floors and stairs	45	62

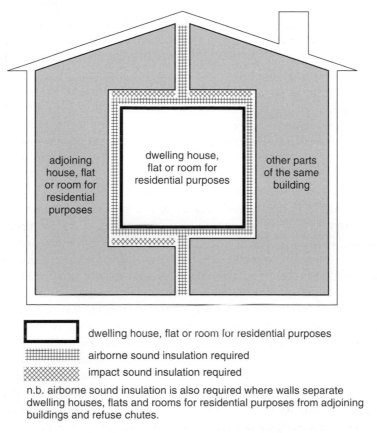

Fig. 10.1   Requirements for protection against sound between dwelling places.

figure of 49 dB can be reduced by 2 dB to 47 dB to allow for the permitted margin of error in the field measurement procedure. Furthermore, the computational effect of including the C_{tr} term is to reduce results calculated from measured values by approximately 5 dB, so that the old figure of 49 dB would, under the current system, be reduced by a further 5 dB to 42 dB. Thus the new requirement of 45 dB may be considered as an increase in the standard of 3 dB, i.e:

$$45 \text{ dB} = (49 \text{ dB} - 2 \text{ dB} - 5 \text{ dB}) + 3 \text{ dB}$$

In the case of a construction with a particularly poor low frequency performance the overall improvement in performance would need to be greater than 3 dB in order to compensate for the effect of C_{tr} at low frequencies.

Required values of the index used to assess the standard of impact sound insulation, $L'_{nT,w}$ are shown in Table 10.1. See section 10.14 for an explanation of the term 'impact sound insulation', and section 10.12 for an explanation of the index $L'_{nT,w}$ and a description of the associated measurement procedure.

There are three circumstances in which it may not be necessary to comply fully with Requirement E1. These are:

- some buildings constructed from prefabricated assemblies;
- school buildings;
- some historic buildings when subject to a material change of use.

Buildings constructed from prefabricated assemblies that are newly made or delivered from stock are subject to the full requirements of Schedule 1 of the building regulations. However, in some cases, e.g. temporary dwelling places, the requirement to provide reasonable sound insulation may vary. Examples referred to in the Approved Document are buildings:

- formed by dismantling, transporting and reassembling sub-assemblies on the same premises;
- constructed from sub-assemblies from other premises or stock manufactured before 1 July 2003.

These buildings would normally be acceptable if they satisfied the requirements of Approved Document E, 1992 edition.

For school buildings, the relevant sections of Building Bulletins 87 and 93 should be consulted, and it would normally be acceptable to implement their recommendations.

In the case of the material change of use of some historic buildings it may not be viable to fully comply with E1 whilst still conserving their special historical features. In such cases the aim should be to improve sound insulation as much as is practically possible without adversely effecting the character of the building or its long-term rate of deterioration. When this approach is adopted, and E1 is not fully complied with, a prominent notice should be displayed in the building indicating the level of insulation obtained by tests carried out using the appropriate procedure. For the purposes of Approved Document E 'historic buildings' include:

- listed buildings;
- buildings in conservation areas;
- buildings of architectural or historic interest and referred to as a material consideration in local authority development plans or in national parks, areas of outstanding national beauty or world heritage sites;
- vernacular buildings of traditional form and construction.

It is suggested in the AD that when assessing the relative merits of reduced sound transmission and conservation, advice should be obtained from the local planning authority's conservation officer.

10.2.2   Sound insulation within dwelling places AD E: 0.9–0.10

Requirement E2 of Schedule 1 to the Building Regulations 2000 (as amended) states that the design and construction of houses, flats and rooms for residential purposes (including extensions) should be such that:

- internal walls between bedrooms and any other rooms;
- internal walls between rooms containing water closets and any other rooms;
- all internal floors,

provide reasonable resistance to the transmission of sound. These requirements are shown diagrammatically in Fig. 10.2.

Requirement E2 will be satisfied if the above elements meet the insulation values quoted in Table 2 of the AD. This simply states that new internal walls and floors in dwelling-houses, flats and rooms for residential purposes, whether purpose-built or formed by material change of use must have a value of airborne sound insulation, $R_w = 40$ dB or above. R_w, the weighted sound reduction index, is a single figure index which results from a frequency band analysis carried out in a laboratory of a sample of the element of construction. It follows therefore that pre-completion testing of internal elements is not necessary but that builders will need to show that elements used satisfy the above criterion. The laboratory measurement of sound insulation is discussed later in section 10.12. There is no requirement to provide impact sound insulation in order to satisfy E2.

As with E1, Requirement E2 extends to dwelling places resulting from the material change of use of a building but unlike E1 the same value applies to both categories of accommodation. Figure 10.2 summarises the obligations of E2 in terms of elements requiring insulation and the nature of insulation required.

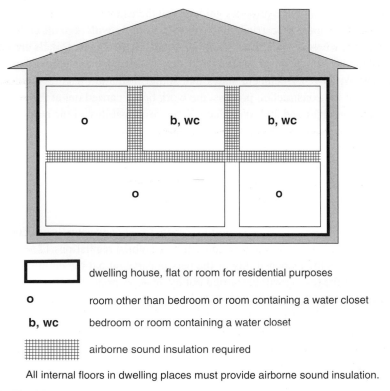

| | dwelling house, flat or room for residential purposes |

o   room other than bedroom or room containing a water closet

b, wc   bedroom or room containing a water closet

airborne sound insulation required

All internal floors in dwelling places must provide airborne sound insulation.

Fig. 10.2   Requirements for protection against sound within dwelling places.

10.2.3 Reverberation in common spaces and acoustic conditions in schools AD E: 0.11–0.12

Requirement E3 of the Building Regulations 2000 (as amended) states that the construction of the common internal spaces in buildings which contain flats or rooms for residential purposes shall be such as to prevent more reverberation than is reasonable. This requirement will be satisfied if sound absorptive measures described in section 7 of the Approved Document are applied. The provision of absorption in common spaces is considered in section 10.10 of this chapter.

Requirement E4 states that the design and construction of all rooms and other places in school buildings shall be such that their acoustic conditions and insulation against disturbance by noise is appropriate to their intended use. It is stated in AD E that the normal way of satisfying these conditions will be to meet the requirements given in section 1 of Building Bulletin 93, *The acoustic design of schools*. See section 10.11 of this chapter.

10.3 Pre-completion testing (Approved Document E, Section 1: Pre-completion testing)

10.3.1 Requirements AD E: 1.1–1.10

Approved Document E provides guidance on the nature and extent of the pre-completion testing that should be carried out in order to comply with regulation 20A, or 12A. This states that where Requirement E1 applies, and where Robust Details are not used, sound insulation testing must be carried out in accordance with an approved procedure.

The Approved Document states that sound insulation testing should be considered as part of the construction process, the work being carried out at the cost of the builder. Regulations 12A and 20A state that it is the responsibility of the builder to ensure that:

- the work is carried out in an approved manner;
- the results are recorded in an approved manner;
- the results are given to the approved authority or inspector within five days of the completion of the work.

However, it is expected that the building control body will actually select which properties to test and the nature of the tests in each case. Whilst regulations 12A and 20A state that the testing is the responsibility of the person carrying out the building work it is recognised that the testing will be carried out by an appropriate testing body. Indeed, for field measurements of sound insulation to be acceptable it will be necessary to demonstrate that the organisation undertaking the work has appropriate accreditation for field measurements, preferably by the United Kingdom Accreditation Service (UKAS), or a European equivalent. Members of the Association of Noise Consultants who are approved under their association's ADE Registration Scheme are also regarded as being suitably qualified.

Testing must be carried out on:

- purpose-built dwelling-houses, flats and rooms for residential purposes;
- dwelling-houses, flats and rooms for residential purposes formed by material change of use;

with the standard to be achieved being that given in Table 10.1. Guidance on appropriate construction details is given in Approved Document E and developers are recommended in the AD that this, or other suitable guidance, is closely followed in order to ensure that the standards set in Table 10.1 are achieved. This is very important since there is no tolerance for measurement uncertainty, which has been built into the values given in the table; and if a test does not reach the prescribed value, by even a very small margin, it will be deemed to be a failure. Where there is doubt as to the required form of construction, specialist advice should be obtained at design stage. The exception to the standards defined in Table 10.1 is when historic buildings are undergoing a material change of use and it is not practicable to achieve the prescribed standard. In this case, appropriate testing is still required with, as mentioned earlier, a notice of the standard achieved posted in a prominent position in the building.

The tests are designed to be carried out on the common wall which separates specific types of rooms and to be undertaken as soon as the dwelling place is, with the exception of decoration, complete. In the case of impact tests, with certain exceptions which are referred to in section 10.12.3, it is important that these are carried out before soft coverings are installed.

10.3.2   Grouping for tests AD E: 1.11–1.18

Not all constructions have to be tested, but those that are must allow inferences of the performance of all constructions in the particular development to be drawn. It is therefore necessary to classify the units into a number of notional groups which are representative of the whole. Houses, flats and rooms for residential purposes must form three different groups.

If there are any significant differences in the method of construction used within each of these groups, then each must be classified as a sub-group. The AD provides guidance, as follows, as to how sub-groups should be determined:

- dwelling-houses (including bungalows) by type of separating wall, and flats and rooms for residential purposes by type of separating wall and separating floor;
- where there are significant differences in flanking elements, e.g. walls, floors and cavities;
- units with specific features which are acoustically unfavourable.

It should be noted that sub-grouping is not necessary if houses, flats and rooms for residential purposes are built with the same separating and flanking constructions and have similar room layouts and dimensions. However, whilst the same principles apply in the case of material change of use as for new buildings, in this case it is probable that significant differences between separating walls, separating floors and flanking constructions will occur more frequently. It follows that more sub-groups will be necessary than in the case of new build work.

10.3.3   Sets of tests AD E: 1.19–1.28

The nature of the tests required between dwelling places depends on the type of accommodation they offer, i.e. there are different requirements for houses and flats. The tests required on a specific type of dwelling place are referred to collectively as a 'set of tests' and the specific requirements for sets of tests between different types of dwelling place are shown in Table 10.2. Note that:

- it is not appropriate to undertake impact tests between houses and bungalows;
- tests are only required in the circumstances identified in the table; and
- the tests referred to in Table 10.2 should each be carried out between a pair of adjacent rooms on opposite sides of the separating wall or floor between them.

Each set of tests should preferably contain individual tests between bedrooms and between living rooms. However, although it is recognised that this may not always be possible when, for example, room types are not mirrored on each side of a separating wall or floor, at least one of the rooms in one pair should be a bedroom and at least one in the other pair should be a living room. Tests should not be carried out between a living space and an adjoining corridor, stairwell or hallway even though the walls between them may need to be constructed to separating wall standard. When the layout of a property is such that there is only one pair of rooms opposite to each other over the complete area of a separating wall or floor, the number of sound insulation tests may be reduced accordingly.

Table 10.2   Requirements for a 'set of tests' between different types of dwelling places.

	Dwelling-houses (including bungalows)	Flats with separating floors but without separating walls	Flats with separating floors and separating walls
Airborne test of walls Where possible between rooms suitable for use as living rooms	✓		✓
Airborne test of walls Where possible between rooms suitable for use as bedrooms	✓		✓
Airborne test of floors Where possible between rooms suitable for use as living rooms		✓	✓
Airborne test of floors Where possible between rooms suitable for use as bedrooms		✓	✓
Impact test of floors Where possible between rooms suitable for use as living rooms		✓	✓
Impact test of floors Where possible between rooms suitable for use as bedrooms		✓	✓
Total number of tests in a set	2	4	6

In the case of rooms for residential purposes, whether in new buildings or constructed by material change of use, the sound insulation of their walls and floors should be tested by application of the measurement principles applicable to dwelling-houses and flats but adapted to suit the individual circumstances. It is sometimes the case that dwelling places are sold before fitting out with internal walls and internal fitments, the example of loft apartments being quoted in the Approved Document. In these situations the dimensions and locations of the principal room are not available but tests should still be undertaken, again applying the principles outlined for dwelling-houses and flats. It is stated in the Approved Document that it is necessary in such cases to make sure that the subsequent work will not be detrimental to sound insulation by application of guidance provided on relevant construction details, see following sections.

10.3.4   Testing programmes AD E: 1.29–1.31

It is expected that the building control authority will determine the nature and extent of the testing programme which is required. However, the Approved Document lays down the overall strategy that should be adopted.

Tests should be carried out between the first pair of properties scheduled for completion/sale in each group or sub-group irrespective of the size of the development, thus providing all parties with an early indication as to the likelihood of any problems. To this end, the frequency of testing should be higher in the early stages of completion than towards its end although in the case of large developments testing should continue for a substantial part of the time that construction is taking place. Notwithstanding the fact that the rate of testing may change during the construction period, the Approved Document states that, assuming that there are no failures, the building control body should require at least one set of tests for every ten dwelling units completed in each group or sub-group. It should be recognised that each set of tests relates to the sound insulation between at least two properties, three in the case of flats with separating walls and floors.

10.3.5   Failure of tests AD E: 1.32–1.39

A set of tests is deemed to have failed if one or more of its individual tests do not satisfy the requirements of Table 10.1. In the event of a failure, remedial action must be taken to rectify the failure between the actual rooms which have failed and also, since the separating element between those rooms may be defective, the developer must demonstrate that other rooms which share the element in question satisfy the performance criteria by:

- further testing between the other rooms; and/or
- application of remedial measures to the other rooms; and/or
- showing that the failure is limited to the rooms initially tested.

If a test between two rooms fails, careful consideration must be given to other properties which incorporate similar elements and forms of flanking construction, and developers will be expected to take the action necessary to satisfy building control authorities that the required standards are being achieved in those properties. Clearly, this will present problems if, by then, some of the properties are occupied. In any event, it is stated in the Approved Document that following the failure of a set of tests, the rate of testing should be increased from what had previously been agreed until the building control body is satisfied that the problems have been overcome.

The AD recommends reference to BRE Information Paper IP 14/02 for guidance relating to the failure of sound insulation tests. It should be recognised that failure may be due not to the construction of the separating wall but rather to excessive flanking transmission, the two causes requiring totally different types of remedial action. The reason for the failure will also indicate which of the other rooms that have not been tested are likely to fail to meet the standard. Such rooms will require remedial treatment. Whenever remedial treatment is applied, the building control body will normally call for further sound insulation testing in order that they can be satisfied with its effectiveness.

10.3.6   Recording of pre-completion test results AD E: 1.41

Regulations 20A and 12A state that the results of sound insulation testing shall be recorded in a manner approved of by the Secretary of State. In order to comply with this requirement, the report on a set of tests must contain at least the following information in the order in which it is listed. The following list is quoted directly from the AD:

'(1) Address of Building.
(2) Type(s) of property. Use the definitions in Regulation 2: dwelling house, flat, room for residential purposes. State if the building is a historic building (see definition in the section on Requirements of this Approved Document).
(3) Date(s) of testing.
(4) Organisation carrying out testing, including:
   (a) name and address;
   (b) third party accreditation number (e.g. UKAS, ANC or European equivalent);
   (c) name(s) of person(s) in charge of test;
   (d) name(s) of client(s).
(5) A statement (preferably in a table) giving the following information:
   (a) the rooms used for each test within the set of tests;
   (b) the measured single-number quantity ($D_{nT,w} + C_{tr}$ for airborne sound insulation and $L'_{nT,w}$ for impact sound insulation) for each test within the set of tests;
   (c) the sound insulation values that should be achieved according to the values set out in Section 0: Performance – Table 1a or 1b, and
   (d) an entry stating 'Pass' or 'Fail' for each test within the set of tests according to the sound insulation values set out in Section 0 Performance – Table 1a or 1b.
(6) Brief details of the test, including:
   (a) equipment;
   (b) a statement that the test procedures in Annex B have been followed. If the procedure could not be followed exactly then the exceptions should be described and reasons given;
   (c) source and receiver room volumes (including a statement on which rooms were used as source rooms);
   (d) results of tests shown in tabular and graphical form for third octave bands according to the relevant part of the BS EN ISO 140 series and BS EN ISO 717 series, including:

(i)   single-number quantities and the spectrum adaptation terms;

(ii) D_{nT} and L'_{nT} data from which the single-number quantities are calculated.'

10.4   Separating walls and their flanking constructions – new buildings
(Approved Document E, Section 2: Separating walls and associated flanking constructions for new buildings)

10.4.1   Construction forms AD E: 2.1–2.17

Section 2 of the Approved Document E gives examples of walls which should achieve the performance standards, for houses and flats, tabulated in section 0: Performance – Table 1a of the AD and in section 10.2 of this chapter. However, as is stated in the document, the actual performance of a wall will depend on the quality of its construction. The information in this section of the AD is provided only for guidance; it is not intended to be exhaustive and alternative constructions may achieve the required standard. Designers are also advised to seek advice from other sources, such as manufacturers, regarding alternative designs, materials or products. The details in no way override the requirement to undergo pre-completion testing. For the provision of walls in 'rooms for residential purposes', see section 10.8 of this chapter.

The separating wall constructions described in the AD are divided into four types.

(1)   Type 1 Solid masonry
      Walls consisting of one leaf of brick or concrete block, plastered on both sides.
(2)   Type 2 Cavity masonry
      Walls consisting of two leaves of brick or concrete block separated by a cavity and usually plastered on both sides.
(3)   Type 3 Masonry between independent panels
      Walls consisting of a layer of brick or block with independent panels on each side.
(4)   Type 4 Framed walls with absorbent material
      Walls consisting of independent frames of, for example, timber with panels, typically of plasterboard, fixed to the outside of each.

Within each wall type, alternatives are presented in rank order such that, as far as possible, the one which should offer the best sound insulation is described first.

It is very important that the junctions between a separating wall and the elements to which it is attached are carefully designed because this will have a significant influence on its overall acoustic effectiveness. Since the level of insulation achieved will be strongly dependent on the extent to which flanking transmission exists, considerable advice is given in the AD on this issue. Guidance is provided on the junctions between separating walls and the following other elements:

*   external cavity walls with masonry inner leaf;
*   external cavity wall with timber framed inner leaf;
*   solid masonry external walls;

- internal walls – masonry;
- internal walls – timber;
- internal floors – timber;
- internal floors – concrete;
- ground floors – timber;
- ground floors – concrete;
- ceiling and roof spaces;
- separating floors Types 1, 2 and 3 (these are considered with separating floors).

However, not all wall/junction combinations are considered. It is stated in the AD that if any element has a separating function, e.g. a ground floor which is also a separating floor for a basement flat, then the separating element requirements take priority.

The following notes should be borne in mind when considering the alternatives which are presented.

(1) The mass per unit area of walls is quoted in kilograms per square metre (kg/m^2) and procedures for calculating it, as described in Annex A of the AD, are outlined in section 10.13. Densities are quoted in kilograms per cubic metre (kg/m^3) and when used for calculating the mass per unit area of brick and blockwork they should take into account moisture content by using data available in the current edition of the CIBSE Part A. Alternatively, manufacturers data relating to mass per unit area may be used where this is available.

(2) The guidance assumes solid blocks which do not have voids within them. The presence of voids may affect sound transmission and relevant information should be sought from manufacturers.

(3) Some of the constructions are described with only wet finishes. If a dry finish is used instead, advice should be sought from the manufacturer.

(4) Drylining laminates of plasterboard with mineral wool may be used whenever plasterboard is recommended. Manufacturer's advice should be obtained for all other drylining laminates. Plasterboard lining should always be fixed in accordance with manufacturer's instructions.

(5) The cavity widths recommended in the guidance are minimum values.

(6) The requirements of Building Regulations Part C (Site preparation and resistance to moisture) and Part L (Conservation of fuel and power) must be considered when applying Part E to junctions between separating walls and ground floors.

(7) The walls described in Tables 10.3, 10.5, 10.6 and 10.7 are only example constructions.

Wall ties in separating and external masonry walls AD E: 2.18–2.24

The sound insulating properties of cavity walls are effected by the coupling effect of wall ties and so it is important that they are used correctly. In AD E, wall ties are defined, in accordance with BS 1243: 1978, *Metal ties for cavity wall construction*, as Type A (butterfly ties) and Type B (double triangle ties) and their layout should be as described in BS 5628–3: 2001, *Code of practice for use of masonry*. However, BS 5628–3 limits the above tie types and spacing to prescribed combinations of cavity widths and minimum masonry leaf thicknesses. It is stated in the AD that tie Type B should be used only in external cavity walls where tie Type A does not satisfy the requirements of Building Regulation Part A

– Structure, and that tie Type B may decrease airborne sound insulation, due to flanking transmission via the external leaf, compared with tie Type A.

An option, which is of value in improving sound insulation, is to select wall ties with an appropriate dynamic stiffness for the cavity width as explained in AD E. The procedure for measuring the dynamic stiffness of wall ties is explained in BRE IP 3/01, *Dynamic stiffness of wall ties used in masonry cavity walls*. It is essential to ensure that, if wall ties are selected on the basis of satisfying dynamic stiffness criteria, they also comply with the requirements of Building Regulation A – Structure.

10.4.2   Design of wall Type 1 – solid masonry AD E: 2.29–2.35

The insulation offered against airborne sound transmission by solid walls is primarily determined by their mass per unit area. Three forms of construction based on solid walls, i.e. wall Type 1, are described in the AD, and also in Table 10.3, and the document states that these should, if built correctly, comply with Requirement E1, see Table 10.1. Examples of each of the three types of wall are provided in the AD and these are also described in Table 10.3 and shown in Fig. 10.3.

Table 10.3   Types and examples of solid masonry separating wall constructions.

	Wall Type		
	Type 1.1	**Type 1.2**	**Type 1.3**
Form			
Description	Dense aggregate concrete block, plastered on both faces	Dense aggregate cast in-situ concrete, plastered on both faces	Brickwork, plastered on both faces
Minimum Mass per unit area	415 kg/m² including plaster	415 kg/m² including plaster	375 kg/m² including plaster
Surface finish	13 mm plaster on both room faces	Plaster on both room faces	13 mm plaster on both room faces
Other factors	Blocks laid flat to full thickness of the wall		Bricks laid 'frog up', coursed with headers
Example constructions that provide the required mass per unit area			
Thickness	215 mm block	190 mm of concrete	215 mm brick
Detail of structure	Dense aggregate concrete block, density 1840 kg/m³	Density of concrete 2200 kg/m³	Density of brick 1610 kg/m³
Coursing	110 mm	n.a.	75 mm
Surface Finish	13 mm lightweight plaster with a minimum mass per unit area of 10 kg/m², on both room faces	13 mm lightweight plaster with a minimum mass per unit area of 10 kg/m², on both room faces	13 mm lightweight plaster with a minimum mass per unit area of 10 kg/m², on both room faces
Further information	215 mm blocks laid flat		

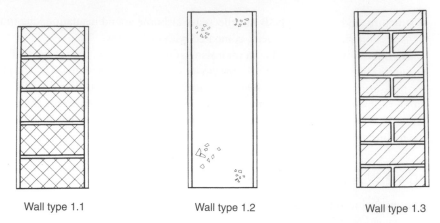

Wall type 1.1   Wall type 1.2   Wall type 1.3

Fig. 10.3   Wall Types 1.1, 1.2 and 1.3.

In order for the performance of solid separating walls to be satisfactory, the junctions between them and their surrounding construction elements must be designed correctly. The requirements detailed in the AD are explained below. See also section 10.4.7.

Junctions with external cavity walls which have masonry inner leaves AD E: 2.36–2.39

- There are no restrictions on the outer leaf of the wall.
- Unless the cavity is fully filled with mineral wool, expanded polystyrene beads or some other suitable insulating material (the AD states that manufacturer's advice should be sought regarding alternative suitable materials), the cavity should be stopped with a flexible closer as shown in Fig. 10.4.
- The separating wall should be joined to the inner leaf of the external wall either by bonding with at least half of the bond provided by the separating wall, or by using tied construction with the inner leaf abutting the separating wall see Fig. 10.5. See, also, Building Regulation Part A – Structure.
- The mass per unit area of the inner leaf of the external wall should be at least 120 kg/m² excluding its finish in order to reduce the effects of flanking transmission. This minimum mass requirement is waived if there is no separating floor and there are openings in the external wall which are at least 1 m high, within 0.7 m of the separating wall and on both sides of the separating wall at every storey. Note, however, that if openings in the external wall are placed close to and on either side of a separating wall, some sound will be transmitted out of the opening on one side of the separating wall and back in through the opening on the other side. This may affect the result of a pre-completion test.

Junctions with external cavity walls which have timber framed inner leaves AD E: 2.40–2.41

- There are no restrictions on the outer leaf of the wall.
- The cavity should be stopped with a flexible closer as shown in Fig. 10.4.

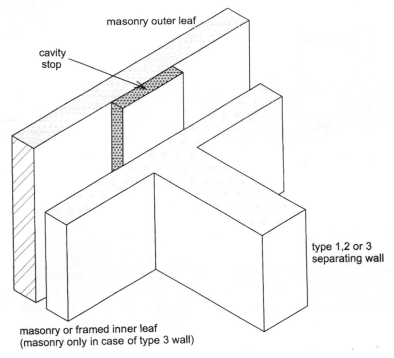

cavity
stop

masonry outer leaf

type 1,2 or 3
separating wall

masonry or framed inner leaf
(masonry only in case of type 3 wall)

Fig. 10.4 Provision of a flexible cavity stop at junction of a Type 1, 2 or 3 separating wall and an external cavity wall.

- The inner leaf frame of the external wall should abut the separating wall and be tied to it vertically with ties at not greater than 300 mm centres.
- The finish of the inner leaf of the external wall should be one layer of plasterboard with a mass per unit area of at least 10 kg/m^2 and having all joints sealed with tape or caulked with appropriate sealant. Two layers of plasterboard, each with a mass per unit area of at least 10 kg/m^2, are required if there is a separating floor.

Junctions with solid external masonry walls AD E: 2.42

No guidance is provided in the AD. Designers are advised to seek specialist advice.

Junctions with internal framed walls AD E: 2.43

No restrictions are imposed.

Junctions with internal masonry walls AD E: 2.44

Internal masonry walls should abut the separating wall and have a mass per unit area of at least 120 kg/m^2 excluding finish, see Fig. 10.5.

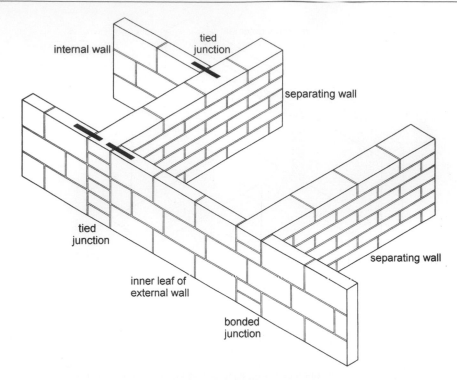

Fig. 10.5   Junctions between separating wall and other masonry walls.

Junctions with internal timber floors AD E: 2.45

Floor joists should not be built into this type of wall. Instead, they should be supported on joist hangers.

Junctions with internal concrete floors AD E: 2.46–2.48

- The requirements for solid concrete slabs are shown in Fig. 10.6.
- Hollow core concrete plank floors and concrete beam with infill block floors should not be continuous through Type 1 separating walls.
- If concrete beam with infill block floors are built into separating walls, then the blocks in the floor must fill the space between the beams where they penetrate the wall.

Junctions with timber ground floors AD E: 2.49

Ground floor joists should not be built into this type of wall. Instead, they should be supported on joist hangers.

Junctions with concrete ground floors AD E: 2.51–2.53

- For suspended concrete ground floors and solid slabs laid directly onto the ground, see Fig. 10.6.

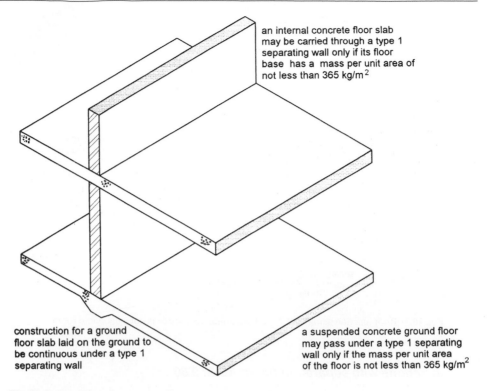

an internal concrete floor slab may be carried through a type 1 separating wall only if its floor base has a mass per unit area of not less than 365 kg/m^2

construction for a ground floor slab laid on the ground to be continuous under a type 1 separating wall

a suspended concrete ground floor may pass under a type 1 separating wall only if the mass per unit area of the floor is not less than 365 kg/m^2

Fig. 10.6 Conditions for concrete floors to pass through Type 1 separating walls.

- Hollow core concrete plank floors and concrete beam with infill block floors should not be continuous under Type 1 separating walls.

Junctions with ceilings and roofs AD E: 2.55–2.59

- The separating wall should continue to the underside of the roof with a flexible closer, which is suitable for use as a fire stop, at its junction with the roof, see Fig. 10.7.
- If the roof or loft space is not a habitable room, then the mass per unit area of the separating wall may be reduced to 150 kg/m^2 above the ceiling which separates the roof or loft space from the uppermost habitable rooms. For this reduction to apply, the mass per unit area of the ceiling must be at least 10 kg/m^2 with sealed joints around the perimeter of the ceiling, see Fig. 10.7.
- If lightweight aggregate blocks with a density of less than 1200 kg/m^3 are used above ceiling level, one side of the wall should be sealed with cement paint or plaster skim.
- Cavities in external walls should be closed at their eaves since this will have the effect of reducing any sound transmission through the cavity from the roof space. A flexible material should be used for the closure with no rigid joint between the two leaves. It is important to avoid a rigid connection between the two leaves and if a rigid material is used it should only be rigidly attached to one leaf, see BRE BR 262, *Thermal insulation: avoiding risks*, section 2.3.

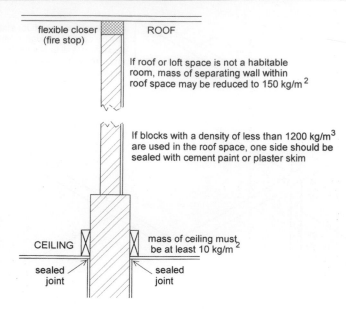

Fig. 10.7   Roof junction using wall Type 1, assuming roof space is not a habitable room.

Junctions with separating floors AD E: 2.60

See section 3 of the AD (section 10.5 of this chapter) for details of junctions between separating floors and Type 1 separating walls.

10.4.3   Design of wall Type 2 – cavity masonry AD E: 2.61–2.72

The insulation offered against airborne sound transmission by cavity masonry walls depends on the mass per unit area of the masonry, the cavity width and also the nature of the connection there is between the two leaves by, for example, wall ties and foundations.

Four types of cavity masonry separating wall are described in Approved Document E, which according to the document should, if built correctly, comply with Requirement E1, see Table 10.4. However, only two of these are appropriate for general application, two being restricted to situations where there is a step or stagger in the separating wall. Examples of each of the four types of wall are provided in the AD and these are described in Table 10.5. For each of these constructions it is important to note that:

- cavity widths are minimum values;
- Type A wall ties should be used to connect the leaves of the walls;
- blocks without voids are used (manufacturers advice should be sought regarding the use of blocks with voids).

In order for the performance of cavity masonry separating walls to be satisfactory, the junctions between them and their surrounding construction elements must be designed correctly. The requirements detailed in the AD are explained below. See also section 10.4.7.

Table 10.4 Types of cavity masonry separating wall constructions.

	Wall Type			
	Type 2.1	**Type 2.2**	**Type 2.3**	**Type 2.4**
Description of wall	Two leaves of dense aggregate concrete block with 50 mm cavity, plaster on both room faces	Two leaves of lightweight aggregate block with 75 mm cavity, plaster on both room faces	Two leaves of lightweight aggregate block with 75 mm cavity and step/stagger. Plasterboard on both room faces	Two leaves of aircrete block with 75 mm cavity and step/stagger. Plasterboard or plaster on both room faces
Limitations			**Should only be used where there is a step and/or stagger of at least 300 mm**	**Should only be used in constructions without separating floors and where there is a step and/or stagger of at least 300 mm**
Minimum mass per unit area	415 kg/m² including plaster	300 kg/m² including plaster	290 kg/m² including plasterboard	150 kg/m² including finish
Minimum cavity width	50 mm	75 mm	75 mm	75 mm
Surface finish	13 mm plaster on both room faces	13 mm plaster on both room faces	Plasterboard on both room faces, minimum mass per unit area of each sheet 10 kg/m²	Plasterboard on both room faces, minimum mass per unit area of each sheet 10 kg/m² or, 13 mm plaster on both room faces.
Other factors			The lightweight blocks should have a density in the range 1350 to 1600 kg/m³ The composition of the lightweight aggregate blocks contributes to performance of this construction with a plasterboard finish. Denser blocks may not give equivalent performance Increasing size of step or stagger in the separating wall tends to increase airborne sound insulation	Increasing size of step or stagger in the separating wall tends to increase airborne insulation

Table 10.5 Examples of cavity masonry separating wall constructions which provide the required mass per unit area.

	Wall Type			
	Type 2.1	**Type 2.2**	**Type 2.3**	**Type 2.4**
Description of wall	Two leaves of dense aggregate concrete block with 50 mm cavity, plaster on both room faces	Two leaves of lightweight aggregate block with 75 mm cavity, plaster on both room faces	Two leaves of lightweight aggregate block with 75 mm cavity and step/stagger. Plasterboard on both room faces	Two leaves of aircrete block with 75 mm cavity and step/stagger. Plasterboard or plaster on both room faces
Limitations			**Should only be used where there is a step and/or stagger of at least 300 mm**	**Should only be used in constructions without separating floors and where there is a step and/or stagger of at least 300 mm**
Thickness of block leaves	100 mm	100 mm	100 mm	100 mm
Description of blocks	Dense aggregate concrete block, density 1990 kg/m³	Lightweight aggregate block, density 1375 kg/m³	Lightweight aggregate block, density 1375 kg/m³	Aircrete block, density 650 kg/m³
Coursing	225 mm	225 mm	225 mm	225 mm
Surface finish	13 mm lightweight plaster with a minimum mass per unit area of 10 kg/m², on both room faces	13 mm lightweight plaster with a minimum mass per unit area of 10 kg/m², on both room faces	Plasterboard on both room faces. Minimum mass per unit area of each sheet 10 kg/m²	Plasterboard on both room faces. Minimum mass per unit area of each sheet 10 kg/m²

Junctions with external cavity walls which have masonry inner leaves AD E: 2.73–2.76

- There are no restrictions on the outer leaf of the wall.
- Unless the cavity is fully filled with mineral wool, expanded polystyrene beads or some other suitable insulating material (the AD states that manufacturer's advice should be sought regarding alternative suitable materials), the cavity should be stopped with a flexible closer as shown in Figs 10.4 and 10.8. Note the nature of the cavity closer for staggered wall construction in Fig. 10.8.
- The separating wall should be joined to the inner leaf of the external wall either by bonding with at least half of the bond provided by the separating wall, or by using tied construction with the inner leaf of the external wall abutting the separating wall. See Building Regulation Part A – Structure.
- The mass per unit area of the inner leaf of the external wall should be at least 120 kg/m² excluding its finish for Type 2.2 separating walls in order to reduce the effects of flanking transmission. This requirement does not apply to Type 2.1, 2.3 and 2.4 walls unless there is also a separating floor in which case it applies to all Type 2 separating walls.

Junctions with external cavity walls which have timber framed inner leaves AD E: 2.77–2.78

- There are no restrictions on the outer leaf of the wall.
- The cavity should be stopped with a flexible closer as shown in Fig. 10.4.
- The inner leaf frame of the external wall should abut the separating wall and be tied to it vertically with ties at not greater than 300 mm centres.
- The finish of the inner leaf of the external wall should be one layer of plasterboard with a mass per unit area of at least 10 kg/m² with all joints sealed with tape or caulked with appropriate sealant. Two layers of plasterboard, each with a mass per unit area of at least 10 kg/m², are required if there is a separating floor.

Junctions with solid external masonry walls AD E: 2.79

No guidance is provided in the AD. Designers are advised to seek specialist advice.

Junctions with internal framed walls AD E: 2.80

No restrictions are imposed.

Junctions with internal masonry walls AD E: 2.81–2.83

Internal masonry walls which abut a Type 2 separating wall, see Fig. 10.8, should have a mass per unit area of at least 120 kg/m² excluding finish. Also, where there is a separating floor, internal masonry walls should have a mass of at least 120 kg/m² excluding finish. In the case of Type 2.3 and 2.4 separating walls (which are only applicable if there is a step

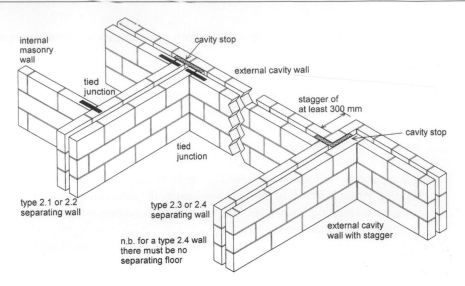

Fig. 10.8 Junctions of cavity separating walls.

or stagger), and when there is no separating floor, there is no minimum mass requirement for internal masonry walls.

Junctions with internal timber floors AD E: 2.84

Floor joists should not be built into this type of separating wall. They should be supported on joist hangers.

Junctions with internal concrete floors AD E: 2.85

Internal concrete floors should generally be built into Type 2 separating walls and continue to the face of the cavity, see Fig. 10.9. Such floors should not bridge the cavity.

Junctions with timber ground floors AD E: 2.86

Ground floor joists should not be built into this type of separating wall. They should be supported on joist hangers.

Junctions with concrete ground floors AD E: 2.88–2.89

- Concrete floor slabs laid directly onto the ground should not be continuous beneath Type 2 separating walls but should abut them as shown in Fig. 10.9.
- Suspended concrete ground floors should not be continuous beneath Type 2 separating walls but should be built in and continue to the face of the cavity, see Fig. 10.9. Floors should not bridge the cavity.

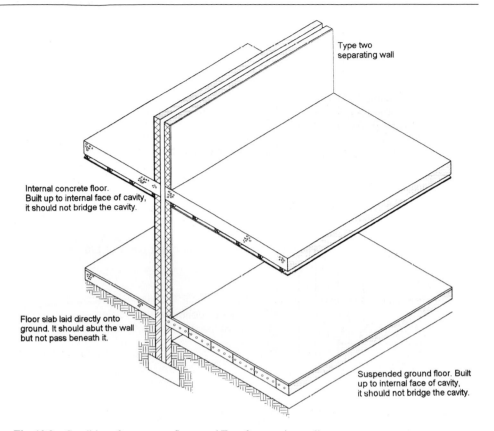

Fig. 10.9   Conditions for concrete floors and Type 2 separating walls.

Junctions with ceilings and roofs AD E: 2.91–2.95

- The separating wall should continue to the underside of the roof with a flexible closer, which is suitable for use as a fire stop, at its junction with the roof, see Fig. 10.10.
- If the roof or loft space is not a habitable room, then the mass per unit area of the separating wall may be reduced to 150 kg/m² above the ceiling which separates the roof or loft space from the uppermost habitable rooms. For this reduction to apply, the mass per unit area of the ceiling must be at least 10 kg/m², all joints within the ceiling and around its perimeter must be sealed and the wall above the ceiling must remain a cavity wall, as shown in Fig. 10.10.
- If lightweight aggregate blocks with a density of less than 1200 kg/m³ are used above ceiling level, one side of the wall should be sealed with cement paint or plaster skim.
- Cavities in external walls should be closed at their eaves since this will have the effect of reducing any sound transmission through the cavity from the roof space. A flexible material should be used for the closure with no rigid joint between the two leaves. It is important to avoid a rigid connection between the two leaves and if a rigid material is used it should only be rigidly attached to one leaf, see BRE BR 262, *Thermal insulation: avoiding risks*, section 2.3.

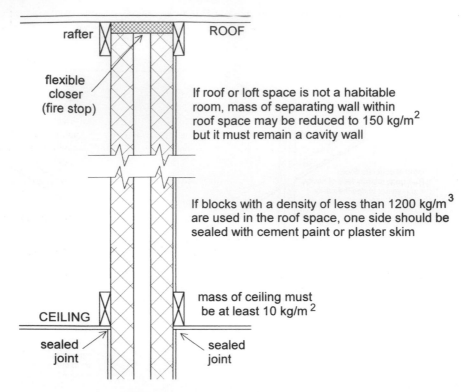

rafter

flexible
closer
(fire stop)

ROOF

If roof or loft space is not a habitable
room, mass of separating wall within
roof space may be reduced to 150 kg/m^2
but it must remain a cavity wall

If blocks with a density of less than 1200 kg/m^3
are used in the roof space, one side should be
sealed with cement paint or plaster skim

mass of ceiling must
be at least 10 kg/m^2

CEILING

sealed
joint

sealed
joint

Fig. 10.10   Roof junction using wall Type 2, assuming roof space is not a habitable room.

Junctions with separating floors AD E: 2.96

See section 3 of the AD (section 10.5 of this chapter) for details of junctions between separating floors and Type 2 separating walls.

10.4.4   Design of wall Type 3 – masonry walls between independent panels AD E: 2.97–2.107

The independent panels used in this type of construction are typically of plasterboard fixed to timber studs and the insulation such walls offer to airborne sound transmission depends on the mass per unit area of the masonry and the independent panels. It is also influenced by the degree of isolation between the masonry core and the panels. Separating walls of this type are capable of providing very good airborne and impact sound insulation.

Three types of masonry/independent panel walls are described in Approved Document E, which according to the document should, if built correctly, comply with Requirement E1, see Table 10.6. Examples of each of the three types of wall are provided in the AD and these are also described in Table 10.6. The essential components of this form of construction are a masonry core of either solid or cavity construction with independent panels on each side of it. The panels, and any framing which is associated with them, must not be in contact with the masonry core.

Table 10.6　Types and examples of masonry with independent panels separating wall constructions.

	Wall Type		
	Type 3.1	**Type 3.2**	**Type 3.3**
Form Description	Solid masonry core of dense aggregate concrete block, independent panels on both room faces	Solid masonry core of lightweight concrete block, independent panels on both room faces	Cavity masonry core of brickwork or blockwork, 50 mm cavity, independent panels on both room faces
Minimum mass per unit area of core	300 kg/m^2	150 kg/m^2	No minimum specified
Cavity width	n.a.	n.a.	50 mm minimum
Panels	Independent panels on both room faces	Independent panels on both room faces	Independent panels on both room faces
Other factors	Structural requirements determine minimum core width. Ref: Part A Structure	Structural requirements determine minimum core width. Ref: Part A Structure	Structural requirements determine minimum core width. Ref: Part A Structure
Example construction Thickness of core	140 mm	140 mm	Two leaves, each one at least 100 mm thick
Cavity width	n.a.	n.a.	50 mm minimum
Detail of core	Dense aggregate concrete block, density 2200 kg/m^3	Lightweight concrete block, density 1400 kg/m^3	Concrete block, density not specified
Coursing	110 mm	225 mm	Not specified
Form of panels	Independent panels, each consisting of two sheets of plasterboard with staggered joints Mass per unit area of each panel 20 kg/m^2	Independent panels, each consisting of two sheets of plasterboard joined by a cellular core Mass per unit area of each panel 20 kg/m^2	Independent panels, each consisting of two sheets of plasterboard joined by a cellular core Mass per unit area of each panel 20 kg/m^2

For each of these constructions:

- cavity widths in masonry cores are minimum values;
- Type A wall ties should be used to connect the leaves of cavity masonry cores; and
- a stringent specification for the construction of the independent panels should be adhered to as follows:
  - (a) the independent panels must each have a mass per unit area of at least 20 kg/m^2 excluding the mass of any supporting framework;

(b) panels to consist of two or more layers of plasterboard with staggered joints, or composite panels formed by two sheets of plasterboard separated by a cellular core; and,

(c) a gap of at least 35 mm between panels and masonry core if the panels are not supported on a frame, or a gap of at least 10 mm between the frame and the masonry core if the panels are supported on a frame.

In order for the performance of masonry with independent panel separating walls to be satisfactory, the junctions between them and their surrounding construction elements must be designed correctly. The requirements detailed in the AD are explained below. See also Table 10.8 and section 10.4.7 regarding correct and incorrect construction procedures.

Junctions with external cavity walls which have masonry inner leaves AD E: 2.108–2.112

- There are no restrictions on the outer leaf of the wall.
- Unless the cavity is fully filled with mineral wool, expanded polystyrene beads or some other suitable insulating material (the AD states that manufacturers' advice should be sought regarding alternative suitable materials), the cavity should be stopped with a flexible closer as shown in Figs 10.4 and 10.11.
- There should be a bonded or tied connection between the core of the separating wall and the inner leaf of the external wall. See Fig. 10.11 for detail of tied connection.
- The inner leaf of the external wall should be lined in the same way as the separating wall, see Fig. 10.11.

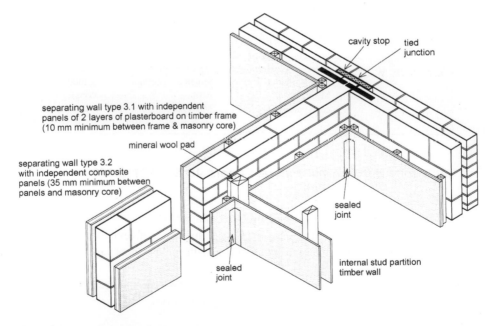

Fig. 10.11   Junctions using Type 3 separating walls.

- The mass per unit area of the inner leaf of the external wall should be at least 120 kg/m² excluding its finish, but if there is no separating floor and the inner leaf of the external wall is lined with independent panels in the same way as the separating wall there is no minimum mass requirement for the inner leaf.
- Where the mass per unit area of the inner leaf of the external wall is at least 120 kg/m² excluding its finish, there is no separating floor and wall Type 3.1 or 3.3 is used, the inner leaf of the external wall may be finished with plaster or plasterboard with a mass per unit area of not less than 10 kg/m².

Junctions with external cavity walls which have timber framed inner leaves AD E: 2.113

No guidance is provided in the AD. Designers are advised to seek specialist advice.

Junctions with solid external masonry walls AD E: 2.114

No guidance is provided in the AD. Designers are advised to seek specialist advice.

Junctions with internal framed walls AD E: 2.115–2.117

Load bearing internal framed walls should be fixed to the masonry core using a continuous pad of mineral wool as shown in Fig. 10.11, whereas non-load bearing internal walls should be butted to the independent panels. The joints between all internal wall panels should be either sealed with tape or caulked with appropriate sealant.

Junctions with internal masonry walls AD E: 2.118

Internal walls of masonry construction should not abut Type 3 separating walls.

Junctions with internal timber floors AD E: 2.119–2.120

Floor joists should not be built into this type of wall. They should be supported on joist hangers and the spaces at the wall surface between the joists should be sealed to the full depth of the joists with timber blocking.

Junctions with internal concrete floors AD E: 2.121–2.122

For separating wall Types 3.1 and 3.2 (solid), internal concrete floor slabs may only be carried through the solid masonry core if the floor base has a mass per unit area of 365 kg/m² or more. For separating wall Type 3.3 (cavity), internal concrete floors should normally be built into the wall and continue to the face of the cavity. Such floors should not bridge the cavity.

Junctions with timber ground floors AD E: 2.123–2.124

Floor joists should not be built into this type of wall. They should be supported on joist hangers and the spaces at the wall surface between the joists should be sealed to the full depth of the joists with timber blocking.

Junctions with concrete ground floors AD E: 2.127–2.131

- A concrete floor slab laid on the ground may be continuous under the solid core of a Type 3.1 or 3.2 (solid core) separating wall.
- A suspended concrete floor may pass under the masonry core of a Type 3.1 or 3.2 (solid core) separating wall only if the mass per unit area of the floor is 365 kg/m^2 or more.
- Hollow core concrete plank floors and concrete beam with infill block floors should not be continuous under the masonry core of a Type 3.1 or 3.2 (solid core) separating wall.
- Concrete floor slabs laid directly onto the ground should not be continuous beneath Type 3.3 (cavity core) separating walls but should abut them in a manner similar to that shown in Fig. 10.9 for cavity walls.
- Suspended concrete ground floors should not be continuous beneath Type 3.3 (cavity core) separating walls but should be built in and continue to the face of the cavity in a manner similar to that shown in Fig. 10.9 for cavity walls. The floor should not bridge the cavity.

Junctions with ceilings and roofs AD E: 2.133–2.139

- The masonry core should continue to the underside of the roof with a flexible closer, which is also suitable for use as a fire stop, at its junction with the roof. Also, the independent panels should be either sealed with tape or caulked with appropriate sealant at their junction with the ceiling, see Fig. 10.12.
- Cavities in external walls should be closed at their eaves since this will have the effect of reducing any sound transmission through the cavity from the roof space. A flexible material should be used for the closure with no rigid joint between the two leaves. It is important to avoid a rigid connection between the two leaves and if a rigid material is used it should only be rigidly attached to one leaf, see BRE BR 262, *Thermal insulation: avoiding risks*, section 2.3.
- If the roof or loft space is not a habitable room and the mass per unit area of the ceiling is at least 10 kg/m^2 and it has sealed joints, then the independent panels may be omitted in the roof space. Also, in the case of wall Types 3.1 and 3.2 the mass per unit area of the separating wall may be reduced to 150 kg/m^2 above the ceiling, see Fig 10.12. However, in the case of wall Type 3.3, the cavity masonry core must continue to the underside of the roof.
- If lightweight aggregate blocks with a density of less than 1200 kg/m^3 are used above ceiling level, one side of the wall should be sealed with cement paint or plaster skim.

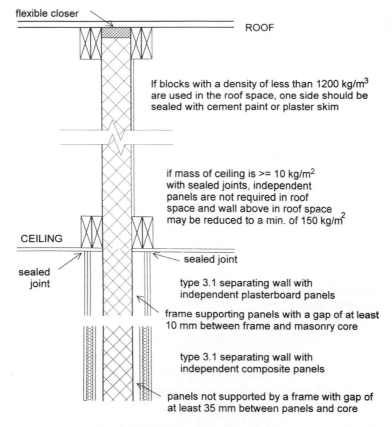

flexible closer

ROOF

If blocks with a density of less than 1200 kg/m³ are used in the roof space, one side should be sealed with cement paint or plaster skim

if mass of ceiling is >= 10 kg/m² with sealed joints, independent panels are not required in roof space and wall above in roof space may be reduced to a min. of 150 kg/m²

CEILING

sealed joint

sealed joint

type 3.1 separating wall with independent plasterboard panels

frame supporting panels with a gap of at least 10 mm between frame and masonry core

type 3.1 separating wall with independent composite panels

panels not supported by a frame with gap of at least 35 mm between panels and core

Fig. 10.12 Roof junction using wall Type 3.1 or 3.2, assuming roof space is not a habitable room.

Junctions with separating floors AD E: 2.140

See section 3 of the AD (section 10.5 of this chapter) for details of junctions between separating floors and Type 3 separating walls.

10.4.5 Design of wall Type 4 – framed walls with absorbent material AD E: 2.141–2.147

Walls of this type consist of two independent frames of timber or steel with panels fixed to the outside of each and sound absorbent material such as mineral wool placed in the void between the two panels. The insulation they offer to airborne sound transmission depends on the mass per unit area of the panels, the extent to which the frames are isolated from each other and on the absorption properties of the void between them.

One type of framed wall consisting of timber frames and plasterboard lining is described in Approved Document E, see Table 10.7, and its construction form is shown in Fig. 10.13. According to the document this wall should, if built correctly, comply with Requirement E1. Designers are advised in the AD to seek advice from manufacturers for steel framed alternatives.

Table 10.7   One type of framed wall with absorbent material.

Form	Wall Type 4.1
Structure	A two leaf frame
Lining	Each lining to consist of two or more layers of plasterboard with staggered joints. Minimum mass per unit area of each lining: 10 kg/m²
Width	Distance between inside lining faces: 200 mm minimum
Form of absorbent material	Unfaced mineral wool batts or quilt with a minimum density of 10 kg/m³. These may be wire reinforced
Thickness of absorbent material	25 mm min. if suspended in cavity between frames 50 mm min. if fixed to one frame 25 mm min. per batt or quilt if one is fixed to each frame
Other information	Plywood sheathing may be used in the cavity as necessary for structural reasons

Notes:

(1)   See Fig. 10.13 for details associated with this type of wall.

(2)   A masonry core may be incorporated, if this is necessary for structural reasons, but it may only be connected to one frame.

In order for the performance of frame and absorbent infill separating walls to be satisfactory, the junctions between them and their surrounding construction elements must be designed correctly. The requirements detailed in the AD are explained below. See also Table 10.8 and section 10.4.7 regarding correct and incorrect construction procedures.

Junctions with external cavity walls which have masonry inner leaves AD E: 2.148

No guidance is provided in the AD. Designers are advised to seek specialist advice.

Junctions with external cavity walls which have timber framed inner leaves AD E: 2.149–2.150

- No restrictions are imposed on the form of the outer leaf.
- The cavities in the external wall and the separating wall should be stopped with flexible closers as shown in Fig. 10.13.
- The finish of the inner leaf of the external wall, assuming there is no separating floor, should consist of one layer of plasterboard with a mass per unit area of at least 10 kg/m², all joints being sealed with tape or caulked with appropriate sealant. If there is a separating floor, two layers of plasterboard, each with a mass per unit area of at least 10 kg/m², should be provided.

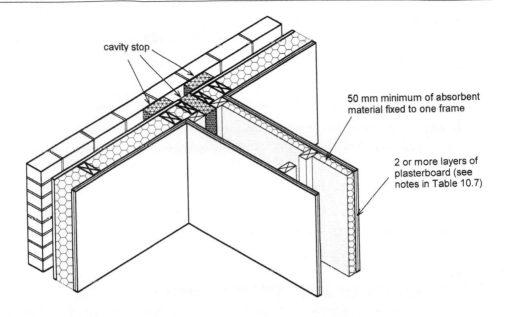

cavity stop

50 mm minimum of absorbent
material fixed to one frame

2 or more layers of
plasterboard (see
notes in Table 10.7)

Alternative locations for absorbent material (see Table 10.7)

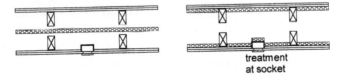

treatment
at socket

Fig. 10.13   Details of timber framed separating wall with absorbent infill.

Junctions with solid external masonry walls AD E: 2.151

No guidance is provided in the AD. Designers are advised to seek specialist advice.

Junctions with internal framed walls AD E: 2.152

No restrictions are imposed on internal framed walls meeting a Type 4 separating wall.

Junctions with internal masonry walls AD E: 2.153

No restrictions are imposed on internal masonry walls meeting a Type 4 separating wall.

Junctions with internal timber floors AD E: 2.154

In order to prevent flanking transmission, air paths into the cavity from floor voids must be blocked by solid timber blockings, continuous ring beams or joists.

Junctions with internal concrete floors AD E: 2.155

No guidance is provided in the AD. Designers are advised to seek specialist advice.

Junctions with timber ground floors AD E: 2.156

In order to prevent flanking transmission, air paths into the cavity from floor voids must be blocked by solid timber blockings, continuous ring beams or joists.

Junctions with concrete ground floors AD E: 2.158

A concrete ground floor slab laid directly on the ground may be of continuous construction under a Type 4 separating wall whereas a suspended concrete ground floor may only pass beneath a Type 4 wall if the floor has a mass per unit area of not less than 365 kg/m^2.

Junctions with ceilings and roofs AD E: 2.160–2.163

- The separating wall should preferably continue to the underside of the roof, a flexible closer should be provided at its junction with the roof and the separating wall linings should be sealed with tape or caulked with appropriate sealant at their junction with the ceiling.
- If the roof or loft space is not a habitable room and the ceiling has sealed joints and a mass per unit area of at least 10 kg/m^2, then within the roof space, either:
  (a) the lining on each of the two frames may be reduced to two layers of plasterboard, each sheet with a mass per unit area of at least 10 kg/m^2; or
  (b) the wall may be reduced to one frame with two layers of plasterboard, each sheet with a mass per unit area of at least 10 kg/m^2, on both sides of the frame. In this case, the cavity must be closed at ceiling level in such a way that the two frames are not rigidly connected together.
- Cavities in external walls should be closed at their eaves. A flexible material should be used for the closure with no rigid joint between the two leaves. It is important to avoid a rigid connection between the two leaves and if a rigid material is used it should only be rigidly attached to one leaf, see BRE BR 262, *Thermal insulation: avoiding risks*, section 2.3.

Junctions with separating floors AD E: 2.164

See section 3 of the AD (section 10.5 of this chapter) for details of junctions between separating floors and Type 4 separating walls.

10.4.6   Walls separating habitable rooms from other parts of a building AD E: 2.25–2.28

Buildings containing flats and rooms for residential purposes often contain non-habitable spaces juxtaposed to those which are occupied. The AD stipulates the sound insulation provided by the walls of refuse chutes in terms of mass, including finishes, per unit area.

If the refuse chute is separated from a habitable room or kitchen its mass per unit area should be 1320 kg/m^2 or more, whereas if it is a non-habitable room the mass per unit area may be reduced to 220 kg/m^2.

Another common source of noise in multi-occupancy buildings is that transmitted from corridors. The AD states that in order to control noise from this source, walls between corridors and flats should be constructed to the standard used for separating walls, see sections 10.4.2 to 10.4.5. A weak point in the sound insulation provided from corridor noise is that transmitted through doors. For this reason, the document states that doors to corridors should have good sealing around their perimeters (including, if practical, their thresholds), and a mass per unit area of at least 25 kg/m^2, or a weighted sound reduction index of at least 29 dB. The term sound reduction index is described in section 10.13 and the relevant measurement standards are listed in section 10.11.4 (AD E B3.9). In 'noisy' parts of a building, noise should be contained by a lobby, two doors in series (i.e. such that each door is passed through in turn) or a high performance door set and if this is impossible, flats in the vicinity should be provided with this type of protection at their entrances. It is very important when considering the sound insulation of doors to bear in mind also the requirements of Building Regulations Part B – Fire safety and Part M – Access and facilities for disabled people.

10.4.7   Construction procedures

The AD provides information regarding correct and incorrect construction procedures which will influence the sound insulation offered by separating walls. This is summarised in Table 10.8.

10.5   Separating floors and their flanking constructions – new buildings
(Approved Document E, Section 3: Separating floors and associated flanking constructions for new buildings)

10.5.1   Construction forms AD E: 3.1–3.7 & 3.9

Section 3 of the Approved Document E gives examples of floors which should achieve the performance standards for houses and flats tabulated in section 0: Performance – Table 1a of the AD and in section 10.2 of this chapter. However, as is stated in the document, the actual performance of a floor will depend on the quality of its construction. The information in this section of the AD is provided only for guidance; it is not intended to be exhaustive and alternative constructions may achieve the required standard. Designers are also advised to seek advice from other sources such as manufacturers regarding alternative designs, materials or products. The details in no way override the requirement to undergo pre-completion testing. For the provision of floors in 'rooms for residential purposes', see section 10.8 of this chapter.

The separating floor constructions described in the AD are divided into three types:

Table 10.8   Construction procedures relating to the sound insulation of separating walls.

Separating Wall Type	Correct Procedure	Procedures which must be avoided
1, 2, 3, 4	Control flanking transmission between separating walls and connected walls and floors as described in the text	
	Ensure external cavity walls are stopped with flexible closers at junctions with separating wall. See text	
1, 2, 3	Fill and seal all masonry joints with mortar	
	Ensure flue blocks will not adversely affect sound insulation. Ensure that a suitable finish is used over flue blocks (see BS 1289–1:1986 and obtain manufacturer's advice)	
1, 2, 4	Stagger the positions of sockets on opposite sides of separating walls. In the case of Type 4 walls, use a similar thickness of cladding material behind socket boxes	In the case of Type 1 and 2 walls, do not use deep sockets and chases in the wall and do not locate them back to back. In the case of Type 4 walls do not locate sockets back to back or chase plasterboard and a minimum of 150 mm edge to edge stagger is recommended between sockets
1, 2		Do not create a junction between Type 1 and 2 walls in which the cavity is bridged by the solid wall
		Do not attempt to convert a Type 2 separating wall into a Type 1 wall by inserting mortar or concrete into the cavity
1	Lay bricks 'frog up'	
	Use bricks and blocks which extend to the full thickness of the wall	
2	Keep cavity leaves separate below ground floor level	Do not change to a solid wall in the roof space. A rigid connection between the leaves will adversely affect performance
		Do not build a cavity wall off a continuous concrete floor slab
3	Fix panels or supporting frames to ceiling and floor only	Do not fix, tie or otherwise connect the free standing panels to the masonry core wall
	Tape and seal all joints	
4	Ensure that fire stops in the cavity between frames are either flexible or fixed to one frame only	If the two leaves have to be connected for structural purposes, do not use ties of greater cross section than 40 mm by 3 mm. Fix them to the studwork at or just below ceiling level. Do not set the ties at closer than 1.2 m centres
	Ensure that each layer of plasterboard is independently fixed to the frame	

- **Type 1 Concrete base with ceiling and soft floor covering.**
- **Type 2 Concrete base with ceiling and floating floor.**
  Floors of this type require one of three types of floating floor. These are defined by the terms (a), (b) and (c).
- **Type 3 Timber frame base with ceiling and platform floor.**

The three floor types are shown in Fig. 10.14 and within each floor type, alternatives are presented in a ranking order such that the one which should offer the best sound insulation is described first.

  The following notes should be borne in mind when considering the alternatives which are presented.

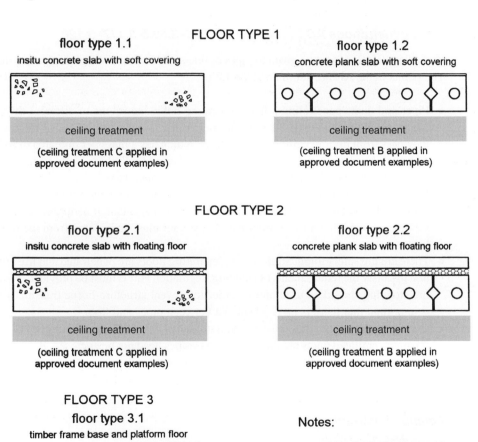

FLOOR TYPE 1

floor type 1.1
insitu concrete slab with soft covering

(ceiling treatment C applied in approved document examples)

floor type 1.2
concrete plank slab with soft covering

(ceiling treatment B applied in approved document examples)

FLOOR TYPE 2

floor type 2.1
insitu concrete slab with floating floor

(ceiling treatment C applied in approved document examples)

floor type 2.2
concrete plank slab with floating floor

(ceiling treatment B applied in approved document examples)

FLOOR TYPE 3

floor type 3.1
timber frame base and platform floor

(ceiling treatment A applied in approved document example)

Notes:

1. Alternative floating floors are shown in Figure 10.18

2. Ceiling treatments are shown in Figure 10.15

Fig. 10.14   Alternative types of floor construction.

- The mass per unit area of floors is quoted in kilograms per square metre (kg/m²) and this may be obtained from manufacturers' data. Alternatively it may be obtained using a procedure described in Annex A of the AD, and outlined in section 10.12 of this chapter. Densities of materials are quoted in kilograms per cubic metre (kg/m³).
- Acoustically, the mass of a bonded screed acts with the slab on to which it is laid and therefore, where appropriate, may be taken into account when calculating the mass per square metre of a floor, but the mass of a floating screed does not, and therefore must not be taken into account.
- Solid in-situ concrete and hollow plank floors are considered in the guidance notes but beam and block floors are not. It is suggested that designers take advice from manufacturers regarding the sound insulation properties of this latter type of floor.

Floor penetrations AD E: 3.41–3.43 & 3.79–3.82 & 3.117–3.120

Particular attention must be given to pipes or ducts which pass through separating floors. The points in the AD are summarised as follows.

- Pipes and ducts penetrating a floor separating habitable rooms in different dwelling places should be enclosed for their full height, i.e. for the full height of the room if they pass from floor to ceiling, with material which has a mass per unit area of at least 15 kg/m². Not less than 25 mm of unfaced wool or mineral fibre should be used either to wrap the pipe or duct, or to line the inside of the enclosure.
- If the floor incorporates a floating layer, a gap of approximately 5 mm should be left between this and the enclosure with the gap filled with sealant or neoprene. The enclosure may extend down to the floor base but if so it must be isolated from the floating layer.
- There are fire safety implications where separating floors are penetrated and fire protection to satisfy the requirements of Building Regulation Part B must be provided. Flexible fire stopping should be used and, in order to prevent structure-borne transmission, it is important that there is no rigid contact between pipes and floor.
- Ducts containing gas pipes must be ventilated at each floor. Gas pipes may be housed in separate ventilated ducts or, alternatively, gas pipes need not be enclosed. It is essential that relevant codes and standards are complied with to ensure gas safety. Reference: The Gas Safety (Installation and Use) Regulations 1998 (SI 1998 2451).

Ceiling construction AD E: 3.17–3.22

The type of ceiling provided has a significant bearing on the acoustic performance of a floor and so, since alternative ceiling treatments may be combined with each type of floor, the three ceiling treatments described in the AD are defined by the letter A, B or C and each floor type is qualified by the letter A, B or C depending on its ceiling treatment. The AD states that the ceiling treatments, as described in Table 10.9 and shown in Fig. 10.15, are ranked such that A should provide the highest level of sound insulation and, further, that if a ceiling treatment of a higher ranking than one applied in a guidance example is used, then providing that there is no significant flanking transmission, the result should be improved sound insulation from the complete floor.

Table 10.9   Ceiling treatments for separating wall constructions.

Ceiling treatment	Form	Specification
A	Independent ceiling with absorbent material	At least two layers of plasterboard with staggered joints and total mass per unit area of at least 20 kg/m²
		Mineral wool laid in cavity above ceiling to provide sound absorption. Minimum thickness 100 mm, minimum density 10 kg/m³
		See note below regarding fixings
B	Plasterboard on proprietary resilient bars with absorbent material to fill ceiling void	Single layer of plasterboard with mass per unit area of at least 10 kg/m² Plasterboard fixed using proprietary resilient bar. For concrete floor, fix bar to timber battens. Bar should be fixed in accordance with manufacturers instructions
		Fill ceiling void with mineral wool with density of at least 10 kg/m³
C	Plasterboard on timber or plasterboard on resilient channels with absorbent material to fill ceiling void	Single layer of plasterboard with mass per unit area of at least 10 kg/m² Plasterboard should be fixed using proprietary resilient channels or timber battens Fill ceiling void with mineral wool with density of at least 10 kg/m³ if resilient channels are used

Notes:

(a)   If using ceiling treatment A with floor types 1 or 2, the ceiling should be attached to independent joists supported only by surrounding walls with a clearance of a minimum of 100 mm between top of ceiling plasterboard and underside of base floor. If using floor type 3, the ceiling should be attached to independent joists supported by surrounding walls with extra support from resilient hangers attached directly to the floor. In this case there should be a clearance of a minimum of 100 mm between the top of the ceiling joists and underside of base floor.

(b)   Light fittings recessed into ceilings may reduce sound insulation.

(c)   See BRE BR 262, *Thermal Insulation: Avoiding Risks*, section 2.4, regarding heat emission from electrical cables which may be covered by absorbent material.

Junctions and flanking transmission AD E: 3.10

It is very important that the junctions between a separating floor and the elements to which it is attached are carefully designed because this will have a significant influence on its overall effectiveness. Since the level of sound insulation achieved will be strongly dependent on the extent to which flanking transmission exists, considerable advice is given in the AD on this issue. Guidance is provided on the junctions between separating floors and the following other elements:

- external cavity walls with masonry inner leaf;
- external cavity wall with timber framed inner leaf;
- solid masonry external walls;
- internal walls – masonry;
- internal walls – timber;
- floor penetrations (see above);
- separating walls Type 1, 2, 3 and 4 (these being relevant to the design of flats).

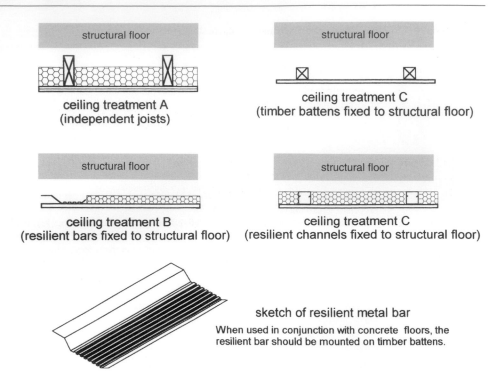

Fig. 10.15   Alternative ceiling treatments.

However, not all floor/junction combinations are considered. It is stated in the AD that if any element has a separating function, e.g. a ground floor which is also a separating floor for a basement flat, then the separating element requirements should take precedence.

10.5.2   Design of floor Type 1 – concrete base with ceiling and soft floor covering AD E: 3.23–3.30

It is the large mass of the concrete slab together with that of the ceiling which provides airborne sound insulation from this type of floor. For impact sound insulation it relies on the soft covering preventing the contact of hard surfaces.

To be acceptable, a soft covering should be a resilient material, or have a resilient base, with a thickness, when uncompressed, of 4.5 mm or more. An alternative approach is to measure the 'weighted reduction of impact sound pressure level', ΔL_w, of the floor covering in accordance with the procedure described in Appendix B3 of the AD, see section 10.12.4 of this chapter (Floor coverings and floating floors). To be acceptable the measured value of ΔL_w should be 17 dB or more.

Two variations of floor Type 1 are described in Approved Document E, which according to the document should, if built correctly, comply with Requirement E1, see Table 10.10. The first floor has ceiling treatment C and hence is referenced 1.1C, whereas the second has ceiling treatment B and hence is referenced 1.2B. In the case of floor type 1.1C, a pre-cast floor slab may be acceptable if it is fully sealed and bonded around the perimeter.

Table 10.10   Examples of concrete base with ceiling and soft covering floor constructions.

	Floor Type	
	1.1C	**1.2B**
Description	Solid concrete slab, cast in-situ with or without permanent shuttering. Soft floor covering and ceiling treatment C	Hollow or solid concrete planks. Soft floor covering and ceiling treatment B
Minimum mass per unit area of concrete base	365 kg/m² including any bonded screed. Permanent shuttering of solid concrete or metal may also be included	365 kg/m² including any bonded screed
Floor covering	Soft floor covering. This is essential	Soft floor covering. This is essential
Ceiling treatment	C or better is essential	B or better is essential
Other requirements		Use regulating floor screed
		All joints between and around planks to be fully grouted to ensure complete air tightness

In order for the performance of Type 1 floors to be satisfactory, the junctions between them and their surrounding elements of constructions must be designed correctly. The requirements detailed in the AD are explained below. See also Table 10.14 and section 10.5.5 regarding correct and incorrect construction procedures.

Junctions with external cavity walls which have masonry inner leaves AD E: 3.31–3.35

- There are no restrictions on the outer leaf of the wall.
- Unless the cavity is fully filled with mineral wool, expanded polystyrene beads or some other suitable insulating material (the AD states that manufacturers' advice should be sought regarding alternative suitable materials), the cavity should be stopped with a flexible closer as shown in Fig. 10.16. The flexible closer must be protected from the effects of moisture and it is essential that adequate drainage is provided (note the provision of a flexible cavity tray, or an equivalent, as shown in Fig. 10.16).
- The mass per unit area of the inner leaf of the external wall should be at least 120 kg/m² excluding its finish in order to reduce the effects of flanking transmission.
- The floor base, but not its screed, should continue to the face of the cavity without bridging the cavity and if floor Type 1.2B is used, with the planks laid parallel to the wall, the first joint between planks should be 300 mm or more from the cavity face, as shown in Fig. 10.16.
- The use of wall ties in external masonry walls is considered in section 2 of the AD. See section 10.4.1 (wall ties) of this chapter.

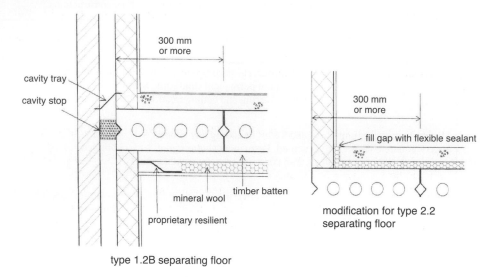

type 1.2B separating floor

Fig. 10.16   Details of junctions between Type 1 and 2 separating floors and external cavity walls.

Junctions with external cavity walls which have timber framed inner leaves AD E: 3.36

- There are no restrictions on the outer leaf of the wall.
- It is assumed that the cavity will not be filled and so the cavity should be stopped with a flexible closer.
- The finish of the inner leaf should consist of two layers of plasterboard each with a mass per unit area of at least 10 kg/m² with all joints sealed with tape or caulked with appropriate sealant.

Junctions with solid external masonry walls AD E: 3.37

No guidance is provided in the AD. Designers are advised to seek specialist advice.

Junctions with internal framed walls AD E: 3.38

No restrictions are imposed.

Junctions with internal masonry walls AD E: 3.39–3.40

The floor base should be continuous through or above such walls and any internal load-bearing wall or any other internal wall rigidly connected to a separating floor should have a mass per unit area of at least 120 kg/m² excluding finish.

Junctions with solid masonry (Type 1) separating walls AD E: 3.44–3.45

Floors bases of Type 1.1C should pass through Type 1 separating walls. Floor bases of Type 1.2B should not be continuous through Type 1 separating walls. In neither case should any screed penetrate the separating wall. See Fig. 10.17 for examples and requirements of both types of construction.

Junctions with cavity masonry (Type 2) separating walls AD E: 3.46–3.48

- The mass per unit area of any leaf, excluding finish, that supports or adjoins the floor should be 120 kg/m^2 or more.
- The floor base, but not its screed, should continue to the face of the cavity without bridging the cavity and if floor Type 1.2B is used, with the planks laid parallel to the wall, the first joint between planks should be 300 mm or more from the cavity face as shown in Fig. 10.17.

Junctions with masonry between independent panels (Type 3) separating walls AD E: 3.49–3.54

Assuming the separating wall has a solid core, i.e. wall Types 3.1 and 3.2:

- Floor bases of Type 1.1C should pass through the wall. Floor bases of Type 1.2B should not be continuous through this type of separating wall. In neither case should any screed penetrate the separating wall. Construction should be similar to that shown in Fig. 10.17 for solid masonry separating walls.
- If floor Type 1.2B is used in conjunction with wall Type 3.2, with the planks laid parallel to the wall, the first joint between planks should be 300 mm or more from the centreline of the masonry core.

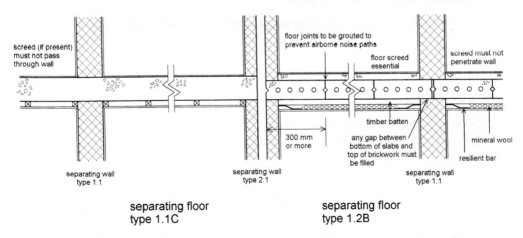

Fig. 10.17 Details of junctions between Type 1 separating floors and Type 1 and 2 separating walls.

Assuming the separating wall has a cavity core, i.e. wall Type 3.3:

- The mass per unit area of any leaf that is supporting or adjoining the floor, excluding finish, should be 120 kg/m² or more.
- The floor base, but not its screed should continue to the face of the cavity without bridging the cavity and if floor Type 1.2B is used, with the planks laid parallel to the wall, the first joint between planks should be 300 mm or more from the cavity face of the adjacent leaf of the masonry core. Construction should be similar to that shown in Fig. 10.17 for cavity masonry separating walls.

Junctions with framed with absorbent material (Type 4) separating walls AD E: 3.55

No guidance is provided in the AD. Designers are advised to seek specialist advice.

10.5.3   Design of floor Type 2 – concrete base with ceiling and floating floor AD E: 3.56–3.60

The complete floor construction consists of a concrete floor base on top of which is a resilient layer. On top of the resilient layer there is a solid 'floating' layer and below the concrete floor base is a ceiling which, as has previously been defined, may be classified as Type A, B or C. The floating floor consists of the floating layer **and** the resilient layer.

The sound insulation offered by floors of this type depends on the mass of the concrete base, the floating layer and the ceiling. It is also improved by the isolation of the floating layer and the ceiling. Impact sound is reduced at source by the floating layer. However, even if resistance to airborne sound only is required, the full construction should still be used.

Floating floors AD E: 3.62–3.66

Examples of three types of floating floors are presented in the AD. These are described in Table 10.11 and shown in Fig. 10.18.

Floor, floating floor, ceiling combinations AD E: 3.67–3.68

Two variations of floor Type 2 are described in Approved Document E, which according to the document should, if built correctly, comply with Requirement E1, see Table 10.12. The first floor has ceiling treatment C and hence is referenced 2.1C, whereas the second has ceiling treatment B and hence is referenced 2.2B. In the case of floor Type 2.1C, a pre-cast floor slab may be acceptable if it is fully sealed and bonded around the perimeter.

When considering the alternatives in Table 10.12, it should be borne in mind that in the AD, alternatives are ranked such that, as far as possible, the one placed first should give the best performance. Also, the ceiling treatment specified is that which is required to meet the requirements of the AD. Since, as already stated, the ceilings are ranked in descending order of performance from A to C, substitution of Type A or B ceilings in Floor 2.1 or a Type A ceiling in Floor 2.2 should give improved sound insulation.

Table 10.11   Alternative types of floating floors as described in the Approved Document.

| | Floating Floor Type | | |
	(a)	**(b)**	**(c)**
Description	Timber raft floating layer supported on a resilient layer	Sand and cement screed floating layer supported on a resilient layer	A floating floor designed to satisfy a performance criterion
Floating layer	Raft of board material with bonded edges, such as t. & g. timber with mass per unit area of not less than 12 kg/m², fixed to 45 mm × 45 mm battens Raft laid on resilient layer without any fixings. Do not lay battens along joints in resilient layer	65 mm of sand and cement or suitable proprietary screed product Screed to have mass per unit area of 80 kg/m² or more Resilient layer must be protected while screed is laid, e.g. by using a 20–50 mm wire mesh	The floating floor should consist of a rigid board above a resilient and/or damping layer which provides a weighted reduction in impact sound pressure level, ΔL_w, of 29 dB or more ΔL_w is the improvement in impact sound insulation obtained in a laboratory by installing a floating floor over a test floor Laboratory measurement should be in accordance with the procedure described in section 10.11.4
Resilient layer	Mineral wool with density of 36 kg/m³ and thickness of at least 25 mm. The layer of mineral wool may have a paper faced under side	Mineral wool with density of 36 kg/m³ and thickness of at least 25 mm. The layer of mineral wool to be paper faced on upper side to prevent wet screed entering the resilient layer, or, a layer meeting the specific dynamic stiffness criterion of 15 MN/m³ and min. thickness of 5 mm under specified load. See BS 29052–1: 1992	

Note: Designers are advised to take advice from manufacturers on proprietary screed products and the performance and installation of proprietary floating floors.

In order for the performance of Type 2 floors to be satisfactory, the junctions between them and their surrounding elements of constructions must be designed correctly. The requirements detailed in the AD are explained below. See also Table 10.14 and section 10.5.5 regarding correct and incorrect construction procedures.

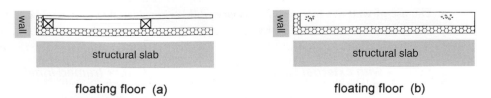

Fig. 10.18   Alternative floating floor constructions.

Table 10.12   Examples of concrete base with ceiling and floating floor constructions.

	Floor Type	
	2.1C	**2.2B**
Description	Floating floor	Floating floor
	Solid concrete slab, cast in-situ with or without permanent shuttering Ceiling treatment C	Hollow or solid concrete planks Ceiling treatment B
Minimum mass per unit area of concrete base	300 kg/m² including any bonded screed. Permanent shuttering of solid concrete or metal may also be included	300 kg/m² including any bonded screed
Floor covering	Floating floor (a), (b) or (c) is essential	Floating floor (a), (b) or (c) is essential
Ceiling treatment	C or better is essential	B or better is essential
Other requirements	Regulating floor screed is optional	Use regulating floor screed
		All joints between and around planks to be fully grouted to ensure complete air tightness

Junctions with external cavity walls which have masonry inner leaves AD E: 3.69–3.73

- There are no restrictions on the outer leaf of the wall.
- Unless the cavity is fully filled with mineral wool, expanded polystyrene beads or some other suitable insulating material (the AD states that manufacturers' advice should be sought regarding alternative suitable materials), the cavity should be stopped with a flexible closer as shown in Fig. 10.16. The flexible closer must be protected from the effects of moisture and it is essential that adequate drainage is provided (note the provision of a flexible cavity tray, or an equivalent, as shown in Fig. 10.16).
- The mass per unit area of the inner leaf of the external wall should be at least 120 kg/m² excluding its finish in order to reduce the effects of flanking transmission.
- The floor base, but not its screed, should continue to the face of the cavity without bridging the cavity and if floor Type 2.2B is used, with the planks laid parallel to the wall, the first joint between planks should be 300 mm or more from the cavity face, as shown in Fig. 10.16.
- Treatment of the intersection of the floating layer and the wall should be as shown in Figs 10.16 and 10.19.
- The use of wall ties in external masonry walls is considered in section 2 of the AD. See section 10.4.1 of this chapter.

Junctions with external cavity walls which have timber framed inner leaves AD E: 3.74

- There are no restrictions on the outer leaf of the wall.

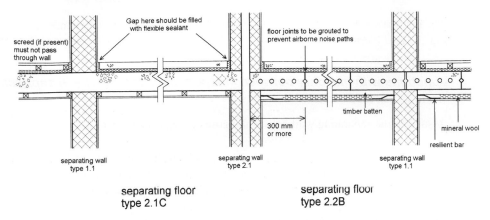

Fig. 10.19   Details of functions between Type 2 separating floors and Type 1 and 2 separating walls.

- It is assumed that the cavity will not be filled and so the cavity should be stopped with a flexible closer.
- The finish of the inner leaf should consist of two layers of plasterboard each with a mass per unit area of at least 10 kg/m^2 with all joints sealed with tape or caulked with appropriate sealant.

Junctions with solid external masonry walls AD E: 3.75

No guidance is provided in the AD. Designers are advised to seek specialist advice.

Junctions with internal framed walls AD E: 3.76

No restrictions are imposed.

Junctions with internal masonry walls AD E: 3.77–3.78

The floor base should be continuous through or above such walls and any internal load bearing wall or any other internal wall rigidly connected to a separating floor should have a mass per unit area of at least 120 kg/m^2 excluding finish.

Junctions with solid masonry (Type 1) separating walls AD E: 3.83–3.84

Floors bases of Type 2.1C should pass through Type 1 separating walls. Floor bases of Type 2.2B should not be continuous through Type 1 separating walls. In neither case should any screed penetrate the separating wall. See Fig. 10.19 for examples and requirements of both types of construction.

Junctions with cavity masonry (Type 2) separating walls AD E: 3.85–3.86

The floor base, but not its screed, should continue to the face of the cavity without bridging the cavity and if floor Type 2.2B is used, with the planks laid parallel to the wall, the first joint between planks should be 300 mm or more from the cavity face as shown in Fig. 10.19.

Junctions with masonry between independent panels (Type 3) separating walls AD E: 3.87–3.92

Assuming the separating wall has a solid core, i.e. wall Types 3.1 and 3.2:

- Floor bases of Type 2.1C should pass through the wall. Floor bases of Type 2.2B should not be continuous through this type of separating wall. In neither case should any screed penetrate the separating wall. Construction should be as shown in Fig. 10.20.
- If floor Type 2.2B is used in conjunction with wall Type 3.2, with the planks laid parallel to the wall, the first joint between planks should be 300 mm or more from the centreline of the masonry core.

Assuming the separating wall has a cavity core, i.e. wall Type 3.3:

- The mass per unit area of any leaf that is supporting or adjoining the floor, excluding finish, should be 120 kg/m^2 or more.
- The floor base, but not its screed should continue to the face of the cavity without bridging the cavity and if floor type 2.2B is used, with the planks laid parallel to the wall, the first joint between planks should be 300 mm or more from the cavity face of the adjacent leaf of the masonry core.

Junctions framed with absorbent material (Type 4) separating walls AD E: 3.93

No guidance is provided in the AD. Designers are advised to seek specialist advice.

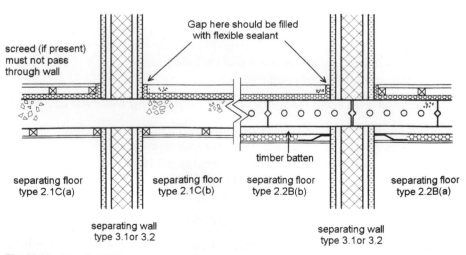

Fig. 10.20   Details of junctions between Type 2 separating floors and Type 3 separating walls.

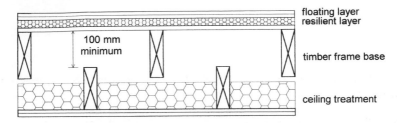

floating layer
resilient layer

100 mm
minimum

timber frame base

ceiling treatment

Fig. 10.21   Construction form of floor Type 3.1A.

10.5.4   Design of floor Type 3 – timber frame base and platform floor with ceiling treatment AD E: 3.94–3.102

Floors of this type consist of a structural timber base comprising of boarding supported on timber joists upon which is a floating layer resting on a resilient layer. The platform floor consists of the floating layer and the resilient layer. Beneath the structural floor is a Type A ceiling treatment. The construction form is shown in Fig. 10.21. In order to obtain good insulation against both airborne and impact sound transmission it is essential that there is good acoustic isolation between the platform floor and the structural base, and between the structural base and the ceiling. The platform floor is important because it reduces the effects of impact noise at source. However, even if resistance to airborne sound only is required, the full construction should still be used.

There are fewer variations of this floor than there are of floor Types 1 and 2 and only one construction form is described in the AD. According to the document this should, if built correctly, comply with Requirement E1. Since this example, which is described in Table 10.13 and shown in Fig. 10.21, utilises ceiling treatment A it is referred to as separating floor Type 3.1A.

In order for the performance of Type 3 floors to be satisfactory, the junctions between them and their surrounding elements of constructions must be designed correctly. The requirements detailed in the AD are explained below. See also Table 10.14 and section 10.5.5 regarding correct and incorrect construction procedures.

Junctions with external cavity walls which have masonry inner leaves AD E: 3.103–3.108

- There are no restrictions on the outer leaf of the wall.
- Unless the cavity is fully filled with mineral wool, expanded polystyrene beads or some other suitable insulating material (the AD states that manufacturers' advice should be sought regarding alternative suitable materials), the cavity should be stopped with a flexible closer.
- The AD states that it is necessary to line the internal face of the masonry inner leaf of a cavity wall with independent panels as is described for Type 3 separating walls. However, if the mass per unit area of the inner leaf is greater than 375 kg/m², the independent panels need not be provided.
- The ceiling, which consists of independent joists supporting plasterboard, must continue to the masonry of the inner leaf. However, where the ceiling passes above the independent panels it should be sealed with tape or caulked with appropriate sealant.

Table 10.13   Example of timber frame base with ceiling and platform floor.

Floor Type 3.1A	
Description	Timber joists supporting a deck and platform floor (consisting of a floating layer and a resilient layer) together with ceiling treatment A beneath the structural floor
Structural floor	Timber joists selected to satisfy structural requirements together with a deck which has a minimum mass per unit area of 20 kg/m^2
Floating layer	A minimum of two layers of board each with a minimum thickness of 8 mm to provide a total mass per unit area of at least 25 kg/m^2. The layers should be fixed together ensuring staggered joints and laid loose on the resilient layer
	Two example floating layer constructions are provided in Approved Document E. These are as follows:
	1. 18 mm of timber or wood based board with tongued and grooved edges and glued joints. These should be spot bonded to a substrate of 19 mm thick plasterboard with staggered joints to give a total mass per unit area of at least 25 kg/m^2
	2. Two layers of cement bonded particle board glued and screwed together with staggered joints. The resulting platform must have a total thickness of 24 mm and a mass per unit area of at least 25 kg/m^2
Resilient layer	Mineral wool, which may be paper faced on its underside, with a thickness of at least 25 mm and density in the range 60 to 100 kg/m^3
	If the material chosen is towards the lower end of the above range, impact sound insulation is improved but it may result in what is described in the document as a 'soft' floor, i.e. one which may be considered to respond excessively to fluctuating loads. It is suggested in the document that this may be overcome by providing additional support via a timber batten which is fixed to the walls and has a foam strip along its top
Ceiling treatment	Ceiling treatment A must be used in conjunction with this floor

- The method of connecting the floor base, i.e. the floor joists, to the external wall is not prescribed but it is emphasised that there must be no air paths connecting the floor to the wall cavity. This would pose a source of flanking transmission which it would be difficult to remedy at a later date.
- The use of wall ties in external masonry walls is considered in section 2 of the AD. See section 10.4.1 of this chapter.

Junctions with external cavity walls which have timber framed inner leaves AD E: 3.109–3.112

- There are no restrictions on the outer leaf of the wall.
- It is assumed that the cavity will not be filled and so the cavity should be stopped with a flexible closer.
- The finish of the inner leaf should consist of two layers of plasterboard each with a mass per unit area of at least 10 kg/m^2 with all joints sealed with tape or caulked with appropriate sealant.

Table 10.14   Construction procedures relating to the sound insulation of separating floors.

Element	Correct procedure	Procedures which must be avoided
Floor Types 1, 2 and 3	Seal the perimeter of independent ceilings with tape or sealant Give extra attention to workmanship and detailing at perimeter and where there are floor penetrations, to reduce flanking transmission and prevent air paths Ensure notes on junction construction are complied with to reduce flanking effects	Do not create a rigid or direct connection between an independent ceiling and its floor base
Floor Types 1 and 2	Fill all joints between floor components to avoid air paths between floors Build concrete separating floors (or floor bases) into all of their masonry perimeter walls Ensure that all gaps between heads of masonry walls and undersides of concrete floors are filled with mortar	Do not allow a floor base to bridge across a cavity in a masonry cavity wall
Floor Type 1	Fix or glue the soft covering to the floor	A soft floor covering must be used. Do not use ceramic floor tiles, wood blocks or any other non-resilient floor finish rigidly connected to the floor base
Floor Type 2	Leave small gap of size recommended by manufacturers between floating layer and wall at room edge and fill with flexible sealant Leave gap of about 5 mm between skirting and floating layer and fill with flexible sealant Lay rolls or sheets of resilient materials with lapped joints or tightly butted and taped joints Prevent screed entering resilient layers by laying fibrous materials with paper facing uppermost	Do not bridge between the floating layer and the base or surrounding walls, e.g. with services or fixings which penetrate the resilient layer between floating layer and floor base or walls Do not allow the floating layer to bridge to the floor base or walls, e.g. through a void in the resilient layer
Floor Type 3	With respect to the platform floor: (a) Ensure that the resilient layer has the correct density, and that it is adequate to carry the applied load (b) Use a resilient material, e.g. expanded or extruded polystyrene strip around the perimeter to ensure that a gap is maintained between wall and floating layer during construction. The strip should be about 4 mm higher than top surface of the floating layer. The gap may be filled with a flexible sealant (c) Lay sheets of resilient materials with tightly butted and taped joints	Do not bridge between the floating layer and the timber frame base or surrounding walls, e.g. with services or fixings which penetrate the resilient layer between the floating layer and the timber base or walls

LEEDS COLLEGE OF BUILDING
LIBRARY

- The method of connecting the floor base, i.e. the floor joists, to the external wall is not prescribed but where floor joists are perpendicular to the wall, the spaces between the floor joists should be sealed to the full depth of the floor with timber blocking.
- The junction between the ceiling and wall lining should be sealed with tape or caulked with appropriate sealant.

Junctions with solid external masonry walls AD E: 3.113

No guidance is provided in the AD. Designers are advised to seek specialist advice.

Junctions with internal framed walls AD E: 3.114–3.115

- Where floor joists are perpendicular to the internal framed wall, the spaces between the floor joists should be sealed to the full depth of the floor with timber blocking.
- The junction between the ceiling and the wall should be sealed with tape or caulked with appropriate sealant.

Junctions with internal masonry walls AD E: 3.116

No guidance is provided in the AD. Designers are advised to seek specialist advice.

Junctions with solid masonry (Type 1) separating walls AD E: 3.121–3.122

- When floor joists are to be supported on this type of wall, they should not be built into the wall but must be supported on joist hangers.
- The junction between the ceiling and the wall should be sealed with tape or caulked with appropriate sealant.

Junctions with cavity masonry (Type 2) separating walls AD E: 3.123–3.126

- When floor joists are to be supported on this type of wall, they should not be built into the wall but must be supported on joist hangers.
- The AD states that it is necessary to line the adjacent leaf of the cavity wall (i.e. the leaf which is nearest to the room with the Type 3 floor) with independent panels as is described for Type 3 separating walls. However, if the mass per unit area of the adjacent leaf is greater than 375 kg/m^2, the independent panels need not be provided.
- The ceiling, which consists of independent joists supporting plasterboard, should continue to the masonry of the wall. However, where the ceiling passes above the independent panels it should be sealed with tape or caulked with appropriate sealant.

Junctions with masonry between independent panels (Type 3) separating walls AD E: 3.127–3.128

- When floor joists are to be supported on this type of wall, they should not be built into the wall but must be supported on joist hangers.
- The ceiling, which is supported on independent joists, should pass through the independent panels to the masonry core. The junction where the ceiling passes over the independent panels should be sealed with tape or caulked with appropriate sealant.

Junctions with timber frames with absorbent material (Type 4) separating walls AD E: 3.129–3.130

- Where the floor joists are perpendicular to the separating wall, the spaces between the floor joists should be sealed to the full depth of the floor with timber blocking.
- The junction between the ceiling and the wall lining should be sealed with tape or caulked with appropriate sealant.

10.5.5   Construction procedures

The AD provides information regarding correct and incorrect construction procedures which will influence the sound insulation offered by separating floors. This is summarised in Table 10.14.

10.6   Dwelling-houses and flats formed by material change of use (Approved Document E, Section 4: Dwelling-houses and flats formed by material change of use)

10.6.1   General requirements AD E: 4.1–4.21

When a building is converted to dwelling places, consideration must be given to satisfying the requirements of Part E. The complexities of sound transmission in a building that is subject to conversion are considerable, and a superficial or incomplete approach to possible solutions is fraught with danger. High-quality acoustic advice is almost always essential at an early stage of the design.

Section 4 of Approved Document E describes forms of treatment to walls, floors, stairs and their associated junctions which it may be appropriate to apply when converting existing buildings by material change of use to provide dwelling houses and flats. Rooms for residential purposes are considered in section 6 of the AD, see section 10.8 of this chapter. The point is made in the AD that existing components may already satisfy the requirements of Table 1a of the document, which are outlined in section 10.2, and it is suggested that this would be so if the construction of a component, its junction detailing and flanking construction was sufficiently similar to that of one of the example walls or floors presented for new buildings. It is also stated that in some circumstances the example of separating walls and floors (including their flanking constructions), described in sections 2 and 3 of the AD for new buildings, can be used as guidance for the work that may be necessary to bring existing components up to the required standard. The importance of the control of flanking transmission, see section 10.6.6, in instances of change of use is also emphasised in the AD.

The construction of an existing component may be such that a designer has little idea as to whether it will comply with the requirements of Table 1a of the AD. To help in this case, treatments to existing components are provided in the AD as follows:

- walls, one form of treatment;
- floors, two forms of treatments;
- stairs, one form of treatment.

The nature of each of these is explained below. It is important to note that the AD states that the example treatments 'can be used to increase sound insulation', but not that if constructed correctly, treated components will achieve the required standard. This is because the level of insulation reached will depend heavily on the existing construction. It is also stated in the AD that the information provided is only guidance and that other designs, materials or products may be used to achieve the required performance standard. Designers are recommended to seek advice from manufacturers and/or other expert sources.

It is suggested in the AD that, with respect to requirements for mass, for example, an existing component may be sufficiently similar to comply with the required performance requirements if its mass was within 15% of that of an equivalent component recommended for the construction of new buildings. This may be the case, but due to uncertainty about the workmanship and density of material used throughout the component, additional treatment may still be advisable.

In view of the complex nature of the construction forms resulting from conversion work, it may be necessary to seek specialist acoustic advice. The AD cites, as an example of a construction which will require special attention, the consequences of constructing a wall or floor across an existing continuous floor or wall such that the original floor or wall becomes a flanking element. The nature of the building may make the simple isolation of such flanking components technically difficult and in such circumstances and when significant additional loads are imposed on the building as a result of floor and wall treatment, structural advice should be sought.

The AD recommends that the following work should be undertaken to existing floors before the recommended treatments to improve sound insulation are applied:

- Since gaps in existing floorboards are a potential source of airborne sound transmission they should be covered with hardboard or filled with sealant.
- Replacement of floorboards. If required, these should be at least 12 mm thick and mineral wool with a density of not less than 10 kg/m^2 should be laid between joists to a depth of not less than 100 mm.
- The mass per unit area of concrete floors should be increased to at least 300 kg/m^2, air gaps within them should be sealed and a regulating screed provided if required.

- Existing lath and plaster ceilings should be retained as long as they comply with Building Regulation B – Fire Safety, whereas other ceilings should be modified such that they consist of at least two layers of plasterboard with staggered joints and a total mass per unit area of 20 kg/m^2.

Sound transmission from corridors and, particularly, through doors in corridors is often a serious source of annoyance in dwellings formed by material change of use. It is suggested in the AD that the separating walls described in this section, i.e. section 4 of the AD, should be used to control sound insulation and flanking transmission between houses or flats formed by material change of use and corridors. The AD states that corridor doors and doors to noisy parts of a building should be constructed in the same way as the equivalent doors in a new building, as described in section 10.4.6.

10.6.2   Wall Treatment 1: independent panel(s) with absorbent material AD E: 4.22–4.24

This treatment consists of adding a panel, or panels, to an existing wall which has poor sound insulating properties and the result may be similar to the 'masonry wall with independent panels' as described for new constructions but with insulation in the void between the two. The existing wall will offer some sound insulation and this will be enhanced by the provision of an independent panel, the isolation of that panel and the provision of absorbent material between the two.

The guidance suggests that if the original wall is of masonry construction, is 100 mm or more thick and is plastered on both sides, one independent panel will suffice, but in the case of different types of existing wall, a panel should be built on each side of the wall.

The specification of Wall Treatment 1 is as described in Table 10.15 and shown in Fig. 10.22. See also Table 10.19 and section 10.6.7 regarding correct and incorrect construction procedures.

Table 10.15   Construction form of Wall Treatment 1.

	Description
Construction form	Plasterboard supported on a timber frame or freestanding panels consisting of plasterboard sandwiching a cellular core
Panels	The mass per unit area of each panel should be at least 20 kg/m^2 excluding the mass of any framework
	If supported on a frame, panels should consist of at least two layers of plasterboard with staggered joints. Alternatively, freestanding panels, e.g. two sheets of plasterboard separated by a cellular core, may be used
Spacing	There must be a gap of at least 35 mm between the inside face of the panels and the surface of the existing wall. The panels may be free standing, but if they are supported on a frame, there must also be a gap of at least 10 mm between the frame and the wall
Mineral wool	Mineral wool with a minimum density of 10 kg/m^2 and minimum thickness of 35 mm to be located in the cavity formed between the panel and the existing wall

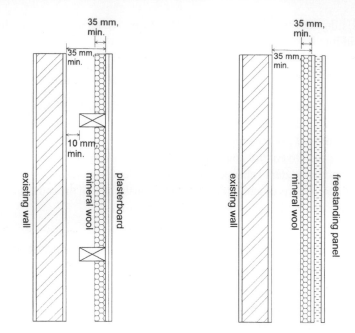

Fig. 10.22   Construction form of Wall Treatment 1.

10.6.3   Floor Treatment 1: independent ceiling with absorbent material AD E: 4.26–4.29

This treatment consists of adding an additional independent ceiling comprising plasterboard, joists and absorbent infill to an existing conventional floor comprising joists, boarding and plasterboard. The mass of the new ceiling, its absorbent infill together with its acoustic isolation will add significantly to the airborne and impact sound insulation offered by the original ceiling. Further, for good insulation, the construction should be made as airtight as possible.

The specification of Floor Treatment 1 is as described in Table 10.16. See also Table 10.19 and section 10.6.7 regarding correct and incorrect construction procedures.

The point is made in the AD that adoption of this procedure significantly reduces the floor to ceiling height in the converted building and that this should be borne in mind at design stage, i.e. the reduction will be a minimum of 125 mm plus the thickness of the plasterboard and even more if the joists need to be deeper than 100 mm. This can present a problem if window heads are close to the height of the original ceiling and a method of solution is proposed. This, together with the wall junction treatment that should be used when this procedure is employed are shown in Fig. 10.23. See also Table 10.19 and section 10.6.7 regarding correct and incorrect construction procedures.

Table 10.16   Construction form of Floor Treatment 1.

	Description
Construction form	Independent joist and plasterboard ceiling with absorbent mineral wool infill constructed below a conventional timber joist floor
Independent ceiling	Two or more layers of plasterboard having staggered joints and a mass per unit area of at least 20 kg/m²
Mineral wool	Mineral wool with a minimum density of 10 kg/m³ and minimum thickness of 100 mm to be located between the joists in the void between the new and old ceilings
Ceiling support	Independent joists fixed to surrounding walls
	There should be either:
	(a) no further support between top of the joists and underside of the existing floor; or
	(b) additional support by resilient hangers fixed directly to the underside of the existing floor
Existing ceiling	Upgrade as necessary to provide two or more layers of plasterboard having staggered joints and a mass per unit area of at least 20 kg/m²

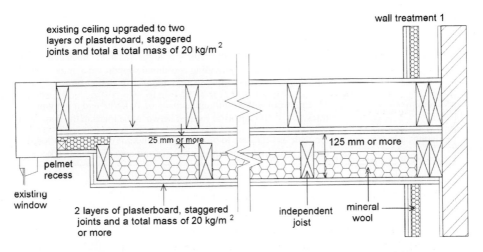

Fig. 10.23   Construction details of Floor Treatment 1 showing junction with Wall Treatment 1.

10.6.4   Floor Treatment 2: platform floor with absorbent material AD E: 4.31–4.33

This treatment consists of adding a platform floor, i.e. a floating layer supported on a resilient layer, to an existing conventional floor comprising timber joists, boarding and plasterboard ceiling. The increase in the total mass of the floor, the action of the resilient layer and the absorbent material should all serve to increase the airborne and impact sound resistance of the floor.

The specification of Floor Treatment 2 should be as described in Table 10.17 and the construction form together with a junction treatment recommended in the AD are shown in Fig. 10.24. See also Table 10.19 and section 10.6.7 regarding correct and incorrect construction procedures.

Table 10.17 Construction form of Floor Treatment 2.

	Description
Construction form	A floating layer laid on a resilient layer which has first been laid onto the boarding of a conventional timber joist floor
Floating layer	A minimum of two layers of board each with a minimum thickness of 8 mm to provide a total mass per unit area of at least 25 kg/m². The layers should be fixed together, for example by spot bonding or glueing and screwing, ensuring staggered joints and then laid loose on the resilient layer
	Two examples of how a platform floor/floating layer of this type could be constructed are provided in the Approved Document, see new floor Type 3.1A
Resilient layer	Mineral wool, which may be paper faced on its underside, with a thickness of at least 25 mm and density in the range 60 to 100 kg/m³
	If the material chosen is towards the lower end of the above range, sound insulation is improved but it may result in what is described in the document as a 'soft' floor, i.e. one which may be considered to respond excessively to fluctuating loads. It is suggested in the document that this may be overcome by providing additional support via a timber batten which is fixed to the walls and has a foam strip along its top
Mineral wool	Mineral wool with a minimum density of 10 kg/m³ and minimum thickness of 100 mm to be located between the joists in the floor cavity
Existing ceiling	Upgrade as necessary to provide two or more layers of plasterboard having staggered joints and a mass per unit area of at least 20 kg/m²

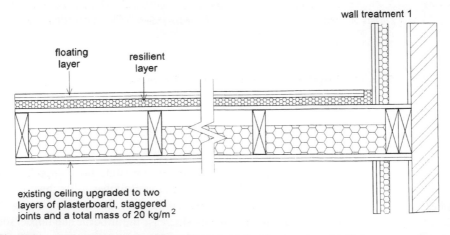

Fig. 10.24 Construction details of Floor Treatment 2 showing junction with Wall Treatment 1.

Table 10.18   Construction form of stair treatment.

	Description
Construction form	The addition of a soft layer over the existing treads and an independent ceiling lined with absorbent material beneath the stairs. A cupboard enclosure may be provided beneath the stairs
Soft covering	At least 6 mm thick, the soft covering must be securely fixed in order that it does not become a safety hazard
If there is an under stairs cupboard	The stair within the cupboard should be lined with plasterboard which has a mass per unit area of at least 10 kg/m². This should be mounted on battens, the space between the battens being filled with mineral wool with a density of at least 10 kg/m³. See Fig. 10.25
	The cupboard should be built with a small heavy well fitted door and walls consisting of two layers of plasterboard, or its equivalent, each layer of which has a mass per unit area of 10 kg/m²
If there is NOT an under stairs cupboard	In this case an independent ceiling as recommended in Floor Treatment 1 should be constructed beneath the stair

10.6.5   Stair treatment: stair covering and independent ceiling with absorbent material AD E: 4.35–4.38

This treatment consists of adding a soft layer over the treads, which reduces impact noise at source, and constructing an independent ceiling lined with absorbent material beneath the stairs. It is the mass of the stair and independent ceiling, the isolation of the independent ceiling and the presence of the absorbent material which provides sound insulation. This is improved by constructing a cupboard enclosure beneath the stairs. It may well be necessary to consider the influence of stairs when constructing dwellings by material change of use since when stairs provide a separating function they must make the same contribution to sound insulation as separating floors.

The specification of the stair treatment is as described in Table 10.18 and its construction form, together with associated treatment of the space beneath the stair as recommended in the AD, are shown in Fig. 10.25. If a staircase performs a separating function, reference should be made to Building Regulation Part B – Fire Safety.

10.6.6   Specific junction requirements in the event of change of use AD E: 4.39–4.50

- Advice is provided in the AD for abutments of floor treatments:
  - (a) In the case of floating floors, e.g. Floor Treatment 2, the resilient layer should be turned vertically upwards at room edges to ensure that there is no contact between the floating layer and the wall and a gap of about 5 mm, filled with flexible sealant, should be left between the floating layer and the skirting.
  - (b) The junctions between new ceilings and walls should be sealed with tape or caulked with appropriate sealant, and detailed as indicated in Figs 10.23 and 10.24.

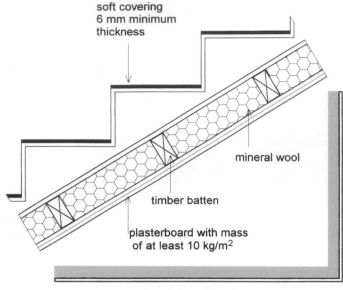

Fig. 10.25   Stair treatment assuming an under stair cupboard.

- Junctions of separating walls and floors with other elements, e.g. external or other walls or floors, are a potential source of significant flanking transmission. As stated in the AD, it may be necessary in such circumstances to seek expert advice regarding the diagnosis and control of flanking transmission. If there is significant flanking transmission via adjoining walls, this can be reduced by lining all adjoining masonry walls with:
  (a) an independent layer of plasterboard; or
  (b) a laminate of plasterboard and mineral wool.
  The AD suggests seeking manufacturers' advice for other dry-lining laminates but does not provide guidance as to the thickness of plasterboard which is appropriate since this will depend on the structure of the adjoining wall. Indeed, the document states that if the adjoining masonry wall has a mass per unit area of more than 375 kg/m², the lining may not significantly improve the insulation and so its use may not be appropriate.
- The requirements in the AD for penetrating services, i.e. piped services and ducts, passing through separating floors between flats in conversion buildings are essentially the same as for penetrating services in new buildings which are explained in section 10.5.1.

10.6.7   Construction procedures

The AD provides information regarding correct and incorrect construction procedures which will influence the sound insulation provided when undertaking material change of use. This is summarised in Table 10.19.

Table 10.19 Construction procedures relating to material change of use.

Element	Correct procedure	Procedures which must be avoided
Wall Treatment 1	Ensure independent panel (and frame if present) have no contact with existing wall Seal perimeter of independent panel with tape or sealant	Do not tightly compress absorbent material as this may bridge the cavity
Floor Treatments 1 and 2	Apply appropriate remedial work to the existing construction	
Floor Treatment 1	Seal perimeter of any new or independent ceiling with tape or sealant	Do not create a rigid or direct connection between an independent ceiling and its floor base Do not tightly compress the absorbent material as doing so may bridge the cavity
Floor Treatment 2	Use correct density of resilient layer and ensure it can carry anticipated load Allow for movement of materials, e.g. expansion of chipboard after laying, to maintain isolation Carry resilient layer up to room edges to isolate floating layer from wall surfaces Leave a gap of approximately 5 mm between skirting and floating layer and fill with flexible sealant Lay resilient layers in sheets with joints tightly butted and taped	Do not bridge between the floating layer and the timber base or surrounding walls, e.g. with services or fixings which penetrate the resilient layer between the floating layer and timber base or walls
Junctions with floor penetrations	Seal joints between casings and ceilings with tape or sealant Leave a gap of approximately 5 mm between casing and floating layer and fill with flexible sealant.	

10.7 Internal walls and floors for new buildings (Approved Document E, Section 5: Internal walls and floors for new buildings)

10.7.1 General requirements AD E: 5.1, 5.2 & 5.9–5.12

Requirement E2 of Schedule 1 to the Building Regulations 2000 (as amended) relates to protection against noise *within* dwelling places. Unlike Requirement E1 which relates to protection against noise *between* dwelling places, the performance of components within dwelling places is not assessed by pre-completion testing but simply by demonstrating that they meet the laboratory sound insulation values which are tabulated in Table 2 of section 0: Performance of the AD, see section 10.2.2 of this chapter. This requirement relates only to airborne sound insulation and advice to occupiers wishing to improve impact sound insulation should be to provide carpets or other soft coverings to floor surfaces. The examples in the AD are for guidance, they are not exhaustive and other designs, materials or products may be used to provide the desired performance.

The following points should be considered when designing internal walls and floors.

- If a door assembly in an internal wall offers a lower level of insulation than the wall in which it is located this will reduce the overall insulation offered. The avoidance of air paths by good perimeter sealing is an essential way of reducing sound transmission, and in addition the AD recommends the use of door sets.
- Stairs provide an important route of sound transmission between floors within dwellings, and if not enclosed the overall airborne sound insulation will be such that the potential of the associated floor will not be realised. However, as stated in the AD, the floor must still be constructed to comply with Requirement E2.
- The requirements for internal noise control should be borne in mind by designers who ought to ensure that noise sensitive rooms such as bedrooms are not constructed immediately adjacent to noise source rooms. The AD refers designers to BS 8233:1999, *Sound Insulation and Noise Reduction for Buildings – Code of Practice*.
- The sealing of walls, floors and gaps around doors has the potential to reduce air supply and in this context the requirements of Building Regulation Part F – Ventilation and Building Regulation Part J – Combustion Appliances and fuel storage systems must be taken into account.

10.7.2   Internal walls AD E: 5.4, 5.5 & 5.8

Four examples of different types of construction which should meet the requirements for internal walls are described in the AD. These are presented in the document, as follows, as wall Types A to D in a ranking order such that, as far as possible, the ones which give the best sound insulation appear first.

- Type A   Timber or metal frames with plasterboard linings on each side.
- Type B   Timber or metal frames with plasterboard linings on each side and absorbent infill material.
- Type C   Concrete block wall with plaster or plasterboard finish on both sides.
- Type D   Aircrete block wall with plaster or plasterboard finish on both sides.

Type A and B internal walls AD E: 5.17–5.18

These types of walls are similar, not only in construction form but in the ways in which they insulate against airborne sound transmission. The mass of the panels and the provision of absorbent infill between them determines their resistance to airborne sound transmission which is also influenced by cavity width and material with which the frame is constructed. The construction of the examples described in the AD is as outlined in Table 10.20.

Table 10.20   Internal walls Type A and B.

Element	Wall Type A	Wall Type B
Frame	Timber or metal	Timber or metal
Linings	Linings to be provided on each side of the frame	Linings to be provided on each side of the frame
	Each lining to consist of two or more layers of plasterboard, each sheet of which has a mass per unit area of 10 kg/m^2 or more	Each lining to consist of a single layer of plasterboard which has a mass per unit area of 10 kg/m^2 or more
Distance between linings	A minimum of 75 mm if fixed to a timber frame	A minimum of 75 mm if fixed to a timber frame
	A minimum of 45 mm if fixed to a metal frame	A minimum of 45 mm if fixed to a metal frame
Absorbent layer	Not required	Unfaced wool batts or quilt with: • thickness of 25 mm or more; and • density of 10 kg/m^3 or more suspended in the cavity The absorbent layer may be wire reinforced
Other requirements	All joints to be well sealed	All joints to be well sealed

Type C and D internal walls AD E: 5.19–5.20

These types are similar, not only in construction form but in the ways in which they insulate against airborne sound transmission. The mass of the panels determines their resistance to airborne sound transmission. The construction of the examples described in the AD is as outlined in Table 10.21.

10.7.3   Internal floors AD E: 5.6, 5.7 & 5.8

Three examples of different types of construction which should meet the requirements for internal floors are described in the AD. These are presented in the document, as follows, as floor Types A to C in a ranking order such that, as far as possible, the ones which give the best sound insulation appear first.

- Type A   Concrete planks.
- Type B   Concrete beams with infilling blocks.
- Type C   Timber or metal joists with board and plasterboard surfaces.

Type A and B internal floors AD E: 5.21–5.22

These types of floors are similar, not only in construction form but in the ways in which they insulate against airborne sound transmission. The mass of the planks (Type A) or beams and infilling blocks (Type B) and screed determines their resistance to airborne

Table 10.21   Internal walls Type C and D.

Element	Wall Type C	Wall Type D
Core structure	Concrete block wall	Aircrete block wall
Required mass per unit area	A mass per unit area, excluding finish, of 120 kg/m^2 or more	A mass per unit area, including finish, of • 90 kg/m^2 or more for plaster finish • 75 kg/m^2 or more for plasterboard finish
Surface finish	Plaster or plasterboard on both sides	Plaster or plasterboard on both sides
Other requirements	All joints to be well sealed	All joints to be well sealed
Restrictions	No specific restrictions	This type of wall should: • not be used as a load-bearing wall rigidly connected to a separating floor • not be rigidly connected to the separating floors described in the AD (see guidance relating to separating floors) • only be used with the separating walls described in the AD where there is no minimum mass requirement on the internal masonry walls (e.g. with separating wall Types 2.3 and 2.4 when there is no separating floor – see guidance relating to separating walls).

sound transmission. The provision of a soft covering, such as carpet, will improve impact sound insulation by reducing impact noise at its source. The construction of the examples described in the document is as outlined in Table 10.22 and shown in Fig. 10.26.

Table 10.22   Internal floors Type A and B.

Element	Floor Type A	Floor Type B
Structure	Concrete planks	Concrete beams with infilling blocks, bonded screed and ceiling
Required mass per unit area	Mass per unit area of 180 kg/m^2 or more	Mass per unit area of concrete beams and blocks to be 220 kg/m^2 or more
Screed	The provision of a regulating screed is optional	A bonded screed is required. If a sand and cement screed is used, it should have minimum thickness of 40 mm. If a proprietary product is used, manufacturers advice should be sought regarding its thickness
Ceiling finish	The provision of a ceiling finish is optional	Ceiling finish C or better, as defined in section 3, of the AD, is required. See section 10.5.1 of this chapter for details
Other requirements	Although not stated in the AD, floor joints should be fully grouted to ensure air tightness	

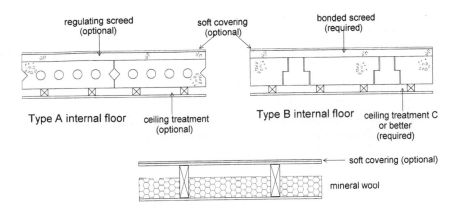

Fig. 10.26   Internal floors Type A, B and C.

Type C internal floor AD E: 5.23

Resistance to airborne sound transmission is determined by the joist and board construction, the ceiling and the absorbent material used. The provision of a soft covering, such as carpet, will improve impact sound insulation by reducing impact noise at its source. The construction form of the example described in the document is as outlined in Table 10.23 and shown in Fig. 10.26.

Table 10.23   Internal floor Type C.

Element	Floor Type C
Structure	Timber or metal joists supporting timber or wood based boarding, a plasterboard ceiling and absorbent material between the joists
Required mass per unit area	The timber or wood based boarding of the floor surface should have a mass per unit area of not less than 15 kg/m^2
Ceiling finish	A ceiling consisting of a single layer of plasterboard. Mass per unit area of not less than 10 kg/m^2. Normal method of fixing
Absorbent layer	Mineral wool laid in cavity between joists. Thickness not less than 100 mm and density not less than 10 kg/m^3
Other requirements	See BRE BR 262 *Thermal Insulation: Avoiding risks*, section 2.4 regarding heat emission from electrical cables which may be covered by any material, including a sound absorbing layer, which may act as thermal insulation

10.7.4   Internal wall and floor junctions AD E: 5.13–16

Guidance on the form of junctions between separating walls and internal floors, and separating floors and internal walls is provided in sections 2 and 3 respectively of the AD and this is considered in sections 10.4 and 10.5 of this chapter. When separating elements are present, the junctions between them and internal walls and floors should always be constructed in accordance with this guidance.

Whenever internal walls or floors are constructed, care should be taken to ensure that there are no air paths, i.e. direct or indirect routes of air passage, between rooms by filling all gaps around the wall or floor.

10.8   Rooms for residential purposes (Approved Document E, Section 6: Rooms for residential purposes)

10.8.1   General requirements AD E: 6.1–6.3 & 6.16–6.18

'Rooms for residential purposes' are defined in section 10.1 of this chapter. Typical examples are rooms used by one or more persons for living and sleeping in a hotel, hostel, boarding house or residential home, but not in a hospital.

Section 6 of the AD gives examples of walls and floors which it states should achieve the performance standards laid out in section 0: Performance – Table 1b of the AD and section 10.2 of this chapter although, as is stated in the document, the actual performance of a wall or floor will depend on the quality of its construction. The information in this section of the AD is provided only for guidance. It is not exhaustive, and designers are recommended also to seek advice from other sources, such as manufacturers, regarding alternative designs, materials or products. Note that although Robust Details may be used, they do not automatically demonstrate compliance for spaces of this type, and so pre-completion testing is always required.

It is not only sound insulation but room layout and building services which determine internal noise levels and these factors may be particularly significant in the confined spaces of 'rooms for residential purposes'. The requirements for internal noise control should be borne in mind by designers who ought to endeavour to ensure that noise sensitive rooms such as bedrooms are not constructed immediately adjacent to noise source rooms. The AD refers designers to BS 8233: 1999 *Sound insulation and noise reduction for buildings – code of practice* and the BRE/CIRIA Report: *Sound control for homes*.

10.8.2   New buildings containing rooms for residential purposes

Separating walls AD E: 6.4

Examples of alternative types of separating walls suitable for new buildings are presented in section 2 of the AD and described in section 10.4 of this chapter. It is stated in the AD that where buildings contain rooms for residential purposes, of the examples given, the most appropriate choices are:

- wall Type 1 (solid masonry), where the in-situ plastered finishes associated with each of the variants 1.1, 1.2 and 1.3 may be substituted for by a single sheet of plasterboard on each face provided that the sheets each have a mass per unit area of 10 kg/m^2 or more;
- wall Type 3 (masonry between independent panels) but only Types 3.1 and 3.2 which have solid masonry cores.

The AD states that wall Type 2 (cavity masonry) and wall Type 4 (framed wall with absorbent material) may be used but particular care must be taken to maintain isolation between the leaves and specialist advice may be required.

Separating floors AD E: 6.8

Examples of alternative types of separating floors suitable for new buildings are presented in section 3 of the AD and described in section 10.5 of this chapter. It is stated in the AD that where the buildings contain rooms for residential purposes, of the examples given, the most appropriate choice is one of the following sub-groups of floor Type 1 (concrete base with soft covering):

- floor Type 1.1C Solid concrete slab, cast in-situ with or without permanent shuttering. Soft floor covering and ceiling treatment C;
- floor Type 1.2B Hollow or solid concrete planks. Soft floor covering and ceiling treatment B.

The AD states that floor Type 2 (concrete base with ceiling and floating floor) and floor Type 3 (timber frame base with ceiling and platform floor) may be used but their floating floors and ceilings must not be continuous across the walls which separate rooms for residential purposes. Designers are advised that this type of construction may require specialist advice.

10.8.3   Corridor walls and doors AD E: 6.5–6.7

The walls between rooms for residential purposes and corridors should be built to the specification, as described in section 6 of the AD, of the separating walls and their junction details. However, a weak point in the sound insulation provided from corridor noise is that transmitted through doors. For this reason, the document states that doors to corridors should have good sealing around their perimeters (including, if practical, their thresholds), and a mass per unit area of at least 25 kg/m^2, or alternatively, a doorset with a weighted sound reduction index of at least 29 dB. The term sound reduction index is described in section 10.13 and the relevant measurement standards are listed in section 10.11.4.

Particular attention should be paid to potentially noisy places such as bars, which ideally should be separated from the rest of the building by a lobby, two doors in series or a high performance doorset. If this provision can not be made to the noisy room, it should be applied to rooms for residential purposes in its vicinity.

The requirements of Building Regulation Part B – Fire safety and Building Regulation Part M – Access and facilities for disabled people must be taken into account when selecting doors for buildings of this type.

10.8.4   Material change of use – rooms for residential purposes AD E: 6.9–6.10

Section 6 of the AD considers the treatments it may be appropriate to apply to walls, floors, stairs and their associated junctions when converting an existing building by material change of use to provide rooms for residential purposes. The point is made in the AD that existing components may already satisfy the requirements of 'Table 1b of Section 0: Performance' of the AD (see Table 10.1 of this chapter) without the need for remedial work. It is suggested that this could be so if the construction (including its flanking construction) were similar to one of the examples listed in section 10.8.2 for 'rooms for residential purposes' in a new building. For walls, this includes wall Types 1.1, 1.2, 1.3, 3.1 and 3.2 which all have a solid masonry core, and for floors it includes floor Types 1.1C and 1.2B, which are both concrete floors. If the existing wall or floor has a mass per unit area that is within 15% of the mass per unit area of the equivalent construction for a new building, then the existing structure may achieve the required performance standard. However, for an existing structure it will often be difficult to obtain the necessary details of the design, the properties of the materials and the quality of workmanship and, even if the mass per unit area is within the 15% criterion, it may still be advisable to provide additional treatment.

If it has been decided that an existing component requires additional treatment, use may be made of the floor, wall and stair treatments recommended for houses and flats resulting from material change of use. These treatments are described in section 4 of the AD and in section 10.6 of this chapter. The design of separating components and their associated junctions to provide adequate sound insulation in such circumstances may be of a complex nature and developers are advised in the AD that specialist advice may be required when undertaking this type of work.

In order to demonstrate compliance with the relevant requirement, the finished structure will be subject to pre-completion testing. It may, therefore, be both helpful and economical for the builder or designer, before commencing building work, to obtain the results of a preliminary sound transmission test on the existing structure to ascertain the extent of the need for additional treatment.

10.8.5   Junctions AD E: 6.11–6.15

As with other forms of dwelling place, in order for separating elements to fulfil their full potential it is essential that flanking transmission is restricted by careful junction design and selection of flanking elements. The AD states that in the case of new buildings, the guidance relating to junction design and flanking transmission provided in sections 2 and 3 of the document for separating walls and floors respectively of houses and flats should be adhered to when building 'rooms for residential purposes'. This guidance is described in sections 10.4 and 10.5 of this chapter.

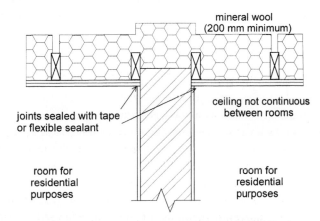

Fig. 10.27   Ceiling void/roof space detail – rooms for residential purposes only.

Similarly, the guidance relating to junction and flanking details for houses and flats formed by material change of use in section 4 of the AD should be applied when creating 'rooms for residential purposes' by material change of use. This guidance is explained in section 10.6 of this chapter.

A relaxation to the guidance for houses and flats is that where a solid masonry Type 1 separating wall is used, the wall need not be continuous through a ceiling void or roof space to the underside of a structural floor or roof, subject to the following conditions:

- there is a ceiling consisting of two or more layers of plasterboard with a total mass per unit area of not less than 20 kg/m²;
- there is a layer of mineral wool in the roof void which is not less than 200 mm thick and has a density of not less than 10 kg/m³; and
- the ceiling is not perforated.

Also, the structure of the building should be such that neither the ceiling joists or plaster-board sheets are continuous across the wall separating 'rooms for residential purposes' with sealed joints as shown in Fig. 10.27. As stated in the AD, this construction detail may only be used if Building Regulation Part B – Fire Safety and Building Regulation Part L – Conservation of Fuel and Power are satisfied.

10.9   Robust Details – an alternative to pre-completion testing

10.9.1   Introduction

Requirement E1 of Schedule 1 to the Building Regulations 2000 stipulates that a pre-defined level of acoustic performance must be provided for the walls, floors and stairs that separate adjoining dwelling places. The AD lays down the minimum acceptable performance standard in terms of specific values of airborne and impact sound insulation, as shown in Table 10.1. The forms of construction described in the AD and in sections 10.4 to 10.8 of this book are suggestions which, if followed, may be expected to provide the required performance. However, these constructions are not 'deemed to satisfy' provisions. The formal 'deemed to satisfy' provision remains the performance standard stated in Table 10.1. As this can only be guaranteed by actual measurement on site of the finished construction, the regulations specify a programme of pre-completion testing, the details of which are described in sections 10.3 and 10.11. However, if the construction of a separating element and its associated details were good enough to offer a potential performance well above the standards laid down in Table 10.1, then it may be argued that the need for pre-completion testing is an unnecessary expense and may be waived. This is the theory behind Robust Details, and Amendments 2004 to Approved Document Part E allow for this possibility in certain circumstances.

10.9.2   Robust Details

The purpose of Robust Details is to provide construction designs with a sound insulation performance that is consistently greater than that required to satisfy requirement E1. To ensure that an installed separating structure will provide a performance, in terms of sound insulation, that is greater than the required standard is difficult since there are inherent variations in the composition of the materials forming the components of the structure, as well as variations in the quality of workmanship. Thus, several examples of a particular design will, when measured in a finished building, give rise to a spread of values, some of which may fail. It follows that in order to be assured that a separating component complies with the prescribed performance criteria in all but a very small number of cases, the average (i.e. design) performance will have to be significantly higher than the minimum requirement. The purpose of a Robust Detail, therefore, is to provide a design which, on average, has a measured performance that:

- has an average performance sufficiently above the minimum requirement to ensure that the lowest values in a normal distribution of its measured results will rarely, if ever, fall below the minimum requirement;
- be capable of consistently exceeding the performance requirement of AD E;
- be practical to construct on site and be reasonably tolerant to standards of workmanship.

In order to be accepted and approved as a Robust Detail, a particular form of construction must pass a thorough assessment and approval process. This includes an initial set of at least eight on-site sound transmission tests to demonstrate the feasibility of the construction. If this first 'Stage A' assessment is successful then the proposal will be registered as a 'candidate robust detail'. A subsequent 'Stage B' application should be supported by a further 22 on-site tests, which if successful will qualify the construction to be accepted for publication as a 'robust detail'. Approved constructions are published in the Robust Details Part E Resistance to the Passage of Sound Handbook (see www.robustdetails.com), and include the following generic types:

- Separating walls: Masonry   Timber   Steel
- Separating floors: Concrete   Timber   Steel-concrete composite

It is important to ensure that a published Robust Detail continues to deliver the required sound insulation performance in practice. Thus, an important aspect of the Robust Details scheme is the monitoring of the performance of constructed Robust Details in completed buildings. This is undertaken by acoustic consultants working for Robust Details Ltd on a percentage of plots registered. It includes visual inspections of work in progress and acoustic testing of completed properties.

10.9.3   Conditions relating to the utilisation of a Robust Detail

In order to avoid pre-completion testing by using a Robust Detail, four conditions must be satisfied as follows:

- The building work must consist of the erection of new attached dwelling-houses, or a new building containing flats.
- The person carrying out the building work must notify the building control body that design details approved by Robust Details Ltd are to be used in specified locations in the building. If building control is being carried out by the local authority, notification must be given no later than the date on which notice of commencement of construction is given under regulation 15(1) of the Building Regulations 2000. In the case of Approved Inspectors, notification must normally be given prior to commencement of the work.
- The notification must specify the unique identification numbers, issued by Robust Details Ltd, which attach to the specified use of the design details.
- The building work must be carried out in accordance with the design details specified in the notification.

Note that in the case of a 'room for residential purposes', although it is permissible to use a Robust Detail in the construction, the waiver on site measurement does not apply, and pre-completion testing is still required.

Failure to adhere, precisely and completely, to all these conditions may invalidate the use of Robust Details. Failure could include one or more of:

- late notification;
- failure to specify the part or parts of the building where the Robust Details will be used;
- failure to provide a means to identify the Robust Detail, either by its unique reference number as included on the purchase statement or by providing a copy of the detail;
- failure to carry out the work in accordance with the design details.

In the case of the last of these (but not the first three), it may be possible for the builder to take remedial action, and then to request the building control body to accept the improved construction.

If the building control body is not satisfied that all the conditions have been met, or that there has been a failure in meeting one or more parts of the required conditions, then the building control body may determine that the exemption from pre-completion testing no longer applies and that all relevant parts of the building in question should be subjected to pre-completion testing in the usual way.

10.9.4   Procedure for using Robust Detail

The Robust Details scheme is operated by Robust Details Ltd. The procedure to be followed by a builder is as follows:

- Purchase a copy of the *Robust Details Handbook*, and decide which detail or details are to be used.
- Obtain and complete a plot registration form, listing all plots and the codes of the proposed Robust Details (see Table 10.24 for an example).
- Submit the completed form to Robust Details Ltd together with the appropriate fee.

On receipt of the registration form, Robust Details Ltd will allocate a unique plot registration number to each plot. They will issue a purchase statement, a copy of the check list for each Robust Detail and a compliance certificate for each plot. It is then necessary to:

- Pass the purchase statement to the building control body – this must be done before starting building work, otherwise pre-completion testing will be required (if some building work has commenced, none of which affects the construction of the Robust Detail, a building control body may still accept the purchase statement and permit registration, but only at its discretion).
- Sign the compliance certificate for each home as soon as the Robust Detail work has been completed, and make this available to the building control body for review.

10.9.5   Potential failure of Robust Details after completion

The usual method of establishing compliance with requirement E1 is by means of measurement, on site, of the finished structure. The assumption which underpins Robust Details is that their acoustic performance will exceed the standard required by the Building Regulations, as stated in Table 10.1, by a significant margin, and that therefore pre-completion testing is unnecessary. However, the use of a Robust Detail is not in itself a guarantee that requirement E1 has been satisfied. What, then, happens in the case of a Robust Detail which, after completion and acceptance by the building control body, is subjected to a sound transmission test and found to fail? In these circumstances, the results of the pre-completion test take precedence over the use of a Robust Detail in determining

Table 10.24   Example of the information given in a Plot Registration Form.

Block Number (if applicable)	Plot Number(s)	Quantity of Plots	Plot Type (H or F)	Wall RD Type	Floor RD Type
1	1–10	10	H	E-WM-1	n/a
3	16–19	4	F	E-WM-1	E-FC-1
5	25–28	4	F	E-WM-1	PCT

whether or not requirement E1 had been satisfied, and it would be open to the owner or occupier of the building to request remedial action from the builder. It would not be a defence for the builder to show that he had correctly carried out a design detail approved by Robust Details Ltd.

10.10 Reverberation in common parts of buildings (Approved Document E, Section 7: Reverberation in the common internal parts of buildings containing flats or rooms for residential purposes)

10.10.1 General requirements AD E: 7.1–7.9

The common parts of buildings containing flats and/or 'rooms for residential purposes' tend to be constructed with hard durable surface finishes, which are easily maintained. Unfortunately, such surfaces lack the soft open texture which efficiently absorbs sound and so the level of reflected, or reverberated, sound tends to be high in such places. Requirement E3 of Part E of the Building Regulations 2000 (as amended) states that the design and construction of the common parts of such buildings shall be such as to prevent more reverberation around them than is reasonable.

Fortunately, it is relatively easy to increase sound absorption and hence reduce reverberant noise levels by surface treatment. Procedures for determining the amount of additional absorption which is required in corridors, hallways, stairwells and entrance halls providing access to flats and 'rooms for residential purposes' are described in section 7 of Approved Document E. The document differentiates between different types of common space as shown in Table 10.25.

The types of space referred to in Table 10.25 are frequently interconnected and in such circumstance the guidance for each space should be followed individually.

The AD describes two methods, A and B, for calculating how much sound absorption is required and the procedure for the use of each method is explained in sections 10.10.3 and 10.10.4 with an example of their application in section 10.10.5. However, whichever procedure is adopted, it is essential that any material chosen to line the internal surfaces complies with the requirements of Building Regulation Part B – Fire Safety and it is recommended in the AD that if designers require guidance regarding the provision of additional sound absorbent material, this should be sought at an early stage of the design process.

Table 10.25 Limiting dimensions of common areas.

Type of space	Ratio of longest to shortest floor dimensions
corridor, hallway	greater than three
entrance hall	three or less
stairwell	not defined in terms of ratio of dimensions

10.10.2 Sound absorption properties of materials

When sound energy strikes a surface, some passes through it, some is absorbed by it and some is reflected. The absorbent coefficient, α, of a material is defined as:

$$\alpha = \frac{\text{sound energy not reflected}}{\text{sound energy incident}}$$

It follows that if no sound is reflected $\alpha = 1$ and if none is absorbed or transmitted then $\alpha = 0$. In theory, all building materials have absorption coefficients between zero and one, values close to zero representing reflective surfaces with low absorption power and vice versa. Sound absorption is highly dependent on frequency. Some of the types of material used to line rooms have good absorption power in the mid to high frequencies of the audible spectrum whereas for others it is highest at the lower end of the spectrum. For this reason it is usual to quote one third octave or one octave band values of absorption coefficients measured across the audible spectrum. Octave band values of absorption coefficient for a range of materials are shown in Table 10.28 below.

10.10.3 Calculation of additional absorption – Method A
AD E: 7.10–7.12

This method makes use of a procedure for classifying sound absorbers which is described in EN ISO 11654: 1997. To apply this procedure, absorption coefficients are measured at one-third octave bands and the arithmetic mean of the three one-third octave bands within each octave band is taken to obtain the 'practical sound absorption coefficient', α_{pi}, a one-octave band average value where i is the ith octave band. The one-octave band values are then compared with a reference curve which is 'shifted' in increments of 0.05 dB until the sum of the adverse deviations does not exceed 0.01 dB. The value of the reference curve at 500 Hz, α_w, is then taken as a single figure weighted average absorption coefficient of the material in question. The material is then given a classification letter according to its value of α_w as shown in Table 10.26 which is an extract from EN ISO 11654: 1997.

Method A involves covering a prescribed area with an absorbent material of a defined classification as follows.

Entrance halls, corridors, hallways

Cover an area, at least as great as the floor area with an absorber which is Class C or better. The document states that it will normally be expedient to cover the ceiling with the additional absorptive material.

Stairwells and stair enclosures

Calculate the combined area of:

- stair treads;

Table 10.26   Sound absorption classes defined in EN ISO 11654: 1997.

Sound absorption class	α_w
A	0.90; 0.95; 1.00
B	0.80; 0.85
C	0.60; 0.65; 0.70; 0.75
D	0.30; 0.35; 0.40; 0.45; 0.50; 0.55
E	0.25; 0.20; 0.15
Not classified	0.10; 0.05; 0.00

- upper surface of intermediate landings;
- upper surface of landings (excluding ground floor); and
- ceiling area on the top floor,

and cover an area which is equal to or larger than the calculated area with a Class D absorber, or cover an area which is at least half of the calculated area with an absorber of Class C or better. In either case the sound absorbing material should be distributed equally between floor levels. It is stated in the AD that it will normally be appropriate to cover the:

- underside of intermediate landings;
- underside of the other landings; and
- ceiling area on the top floor.

The materials used to increase sound absorption tends to be light and porous, lacking the robustness which is required to withstand day-to-day wear and tear and vandalism in common areas where they may come into contact with human beings. This is why it is recommended in the AD that they are used to cover ceilings and the undersides of landings despite the fact that they would be equally effective on other surfaces. The AD states that the use of proprietary acoustic ceilings is appropriate. These have the benefits of being available with a variety of surface finishes and are lightweight and very efficient absorbers of sound.

10.10.4 Calculation of additional absorption – Method B
AD E: 7.13–7.21

This method takes into account the actual absorption power of the surfaces of the enclosure prior to the provision of additional absorbent material and allows the amount of additional material which is required to be calculated. It is stated in the AD that it is only intended for corridors, hallways and entrance halls (not stairwells) but that it offers a more flexible approach which may require less additional absorption than Method A. It is also a more efficient approach in that the additional absorption can be directed at the frequencies at which it is most needed.

For a particular surface of area S m^2 and absorption coefficient α, its absorption area A is defined as the product of S and α and it has units of m^2. This may be considered as the hypothetical area of a material which is a perfect absorber, i.e. one with $\alpha = 1$, which would provide the same sound absorption as the actual material with area S.

If an enclosure has n internal surfaces, its total absorption area, A_T is defined as the sum of the absorption areas of its individual components, such that:

$$A_T = \alpha_1 S_1 + \alpha_2 S_2 + \alpha_3 S_3 + \ldots + \alpha_n S_n$$

It follows that A_T may be considered as the hypothetical area of a material which is a perfect absorber which would provide the same total sound absorption as the sum of the actual materials which are present.

Requirement E3 will be satisfied when the total absorption area stipulated in Table 10.27 for different types of common area has been provided at each octave band from 250 Hz to 4000 Hz.

Absorption coefficient data which should be accurate to two decimal places may be obtained from a variety of sources. For generic materials, data is provided in Table 7.1 of the AD which is reproduced as Table 10.28. It is stated in the AD that this may be supplemented by other published data. Whilst published data will be necessary and appropriate for generic surface materials not provided specifically to increase the absorption area of the enclosure, materials which are selected for their absorption properties will usually be trade products and their absorption coefficients should be available from their manufacturers or suppliers. These should be established by measuring one-third octave band values of absorption coefficients in accordance with BS EN 20354:1993 and then converting them to one-octave band values of 'practical sound absorption coefficient', α_{pi}, using the procedure described EN ISO 11654: 1997, see section 10.10.3.

When calculating the area and absorption coefficient of the material necessary to satisfy the requirements of Table 10.27, calculation steps should be rounded to two decimal places.

10.10.5 Example of absorption calculation

There follows an example of the application of Methods A and B to calculate the extra absorption required for an entrance hall, see Fig. 10.28.

Table 10.27   Total absorption areas for common areas using Method B.

Type of common area	Minimum total absorption area required	Location of additional absorptive material
Entrance halls	0.20 m^2 per cubic metre of volume	To be distributed over the available surfaces
Corridors, hallways	0.25 m^2 per cubic metre of volume	To be distributed over one or more of the surfaces

Table 10.28 Copy of Table 7.1 from Approved Document E providing absorption coefficients for commonly used materials.

Material	Sound absorption coefficient, α in octave frequency bands (Hz)				
	250	500	1000	2000	4000
Fair-faced concrete or plastered masonry	0.01	0.01	0.02	0.02	0.03
Fair-faced brick	0.02	0.03	0.04	0.05	0.07
Painted concrete block	0.05	0.06	0.07	0.09	0.08
Windows, glass facade	0.08	0.05	0.04	0.03	0.02
Doors (timber)	0.10	0.08	0.08	0.08	0.08
Glazed tile/marble	0.01	0.01	0.01	0.02	0.02
Hard floor coverings (e.g. lino, parquet) on concrete floor	0.03	0.04	0.05	0.05	0.06
Soft floor coverings (e.g. carpet) on concrete floor	0.03	0.06	0.15	0.30	0.40
Suspended plaster or plasterboard ceiling (with large airspace behind)	0.15	0.10	0.05	0.05	0.05

Application of Method A

This requires covering an area equal or greater than the floor area with a Class C, or better, absorber.

An appropriate solution would therefore be to cover $4.0 \times 5.0 = 20$ m², e.g. the ceiling with an appropriate material, probably acoustic ceiling tiles, of Class C or better.

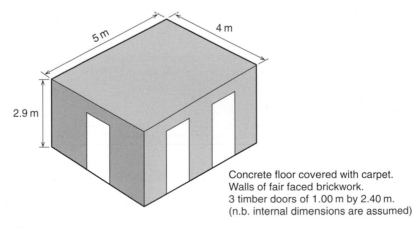

Concrete floor covered with carpet.
Walls of fair faced brickwork.
3 timber doors of 1.00 m by 2.40 m.
(n.b. internal dimensions are assumed)

Fig. 10.28 Entrance hall relating to example question.

Application of Method B

Assuming that the designer has decided to cover the entire ceiling with absorptive material the necessary calculations are as described below.

Stage 1

Calculate area of each internal surface and establish the absorption coefficient from Table 10.28 for each surface which is not being selected for its sound absorbing properties.

Surface	area (m²)	Absorption coefficient for each octave band				
		250 (Hz)	500 (Hz)	1000 (Hz)	2000 (Hz)	4000 (Hz)
floor (concrete, carpet covered)	20	0.03	0.06	0.15	0.30	0.40
doors (timber)	7.2	0.10	0.08	0.08	0.08	0.08
walls (exc. doors) (fair-faced brick)	45	0.02	0.03	0.04	0.05	0.07
ceiling (yet to be determined)	20					

Note that the carpeted floor provides good absorption at the higher frequency octave bands. This reduces the absorption required at these frequencies from the ceiling covering.

Stage 2

Calculate the absorption area of floor, doors and walls (the product of area and absorption coefficient) and then establish the overall absorption area of these surfaces at each octave band by summing.

Surface	Absorption area (m²)				
	250 (Hz)	500 (Hz)	1000 (Hz)	2000 (Hz)	4000 (Hz)
floor	0.60	1.20	3.00	6.00	8.00
doors	0.72	0.58	0.58	0.58	0.58
walls (exc. doors)	0.90	1.35	1.80	2.25	3.15
sum of absorption areas	2.22	3.13	5.38	8.83	11.73

Stage 3

Calculate the total absorption area, A_T, required at each octave band, i.e. $0.2 \times$ volume $= 0.2 \times 5 \times 4 \times 2.9 = 11.6$ m².

Stage 4

Calculate the additional absorption area required from ceiling by subtracting the values found at Stage 2 from that found at Stage 3:

	Absorption area (m²)				
	250 (Hz)	**500 (Hz)**	**1000 (Hz)**	**2000 (Hz)**	**4000 (Hz)**
Additional area required from ceiling	9.38	8.47	6.22	2.77	–0.13

The negative value at 4000 Hz indicates that sufficient absorption is provided at this octave band by the floor, doors and walls.

Stage 5

Calculate the absorption coefficient required for the ceiling by dividing the required absorption area by the area of the ceiling.

	Absorption area (m²)				
	250 (Hz)	**500 (Hz)**	**1000 (Hz)**	**2000 (Hz)**	**4000 (Hz)**
Required absorption coefficient of ceiling material	0.47	0.42	0.31	0.14	any value

Select a patent ceiling product which has absorption coefficients greater than those in the above table.

10.10.6 Report on compliance AD E: 7.22

Demonstration of compliance with Requirement E3 may take the form of an annotated drawing or a report. The information which should be included, quoted directly from the AD, is as follows.

'(1) A description of the enclosed space (entrance hall, corridor, stairwell etc.).
(2) The approach used to satisfy Requirement E3, Method A or B.
   • With Method A, state the absorber class and the area to be covered.
   • With Method B, state the total absorption area of additional absorptive material used to satisfy the Requirement.
(3) Plans indicating the assignment of the absorptive material in the enclosed space.'

10.11 School acoustics (Approved Document E, Section 8: Acoustic conditions in schools)

Requirement E4 states that the design and construction of all rooms and other places in school buildings shall be such that their acoustic conditions and insulation against disturbance by noise is appropriate to their intended use. It is stated in AD E that the normal way of satisfying these conditions will be to meet the requirements given in section 1 of

Building Bulletin 93, *The acoustic design of schools* with respect to sound insulation, reverberation time and internal ambient noise levels. This document is produced by DfES and published by The Stationery Office, ISBN No. 0 11 271105 7.

10.12 Calculation of sound transmission indices (Approved Document E, Annex E: Procedures for sound insulation testing)

10.12.1 Overview of procedures for establishing sound insulation values from field measurements

Field measurement of airborne sound insulation for compliance with Requirement E1 requires calculation of the 'weighted standardised level difference' $D_{nT,w}$ and the 'spectrum adaptation term' C_{tr}. Initially, a noise source is located in one of two rooms between which there is a separating element and microphones are located in this and the room on the other side of the separating element as shown in Fig. 10.29. On activation of the noise source, noise passes from the source room to the receiving room by the direct and indirect routes indicated on the figure and the level difference between them, D, is measured in accordance with BS EN ISO 140–4:1998 and as described in section 10.12.3. The level measured at the microphone in the receiving room will be determined not only by the energy passing directly to the microphone but also by noise energy reflected within the room which in turn will influence the calculated level difference D.

To take account of the reflected energy in the receiving room, a noise source is activated in the receiving room and then switched off. The sound pressure then falls, and the time taken for the sound to decay by 60 dB is established. This is known as the reverberation time and the standards relevant to reverberation time measurement are referred to in section 10.12.3. The level difference is then normalised by calculating the value D_{nT}:

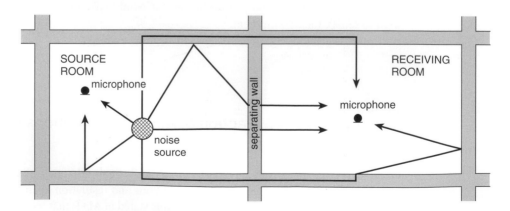

Fig. 10.29 Measurement of airborne sound insulation.

$$D_{nT} = D + 10 \, \text{Log} \left(\frac{T}{T_0}\right) \, \text{dB with } T_0 = 0.5 \text{ s}$$

where T is the reverberation time of the receiving room, and T_0 is a reference value. D_{nT} is known as the 'standardised level difference' and is the level difference which would have been measured if the receiving room had a reverberation time of 0.5 s. Hence all level differences are normalised to a receiving room reverberation time of 0.5 s. D_{nT} is measured using one-third-octave bands and these values are combined to give a single value, the weighted standardised level difference $D_{nT,w}$, using a procedure described in BS EN ISO 717–1:1997.

The spectrum adaptation term, C_{tr}, is also a single number value measured in decibels and calculated from the spectrum D_{nT} using BS EN ISO 717–1:1997. If a separating element provides poor insulation at low frequencies, C_{tr} will be large and negative and so when added to $D_{nT,w}$ for the purposes of Requirement E1, its effect will be to reduce the overall magnitude of the insulation value.

Field measurement of impact sound insulation for compliance with Requirement E1 requires calculation of the 'weighted standardised impact sound pressure level', $L'_{nT,w}$. Initially a tapping machine is located on top of the separating floor and a microphone(s) is located in the room beneath it, see Fig. 10.30. A tapping machine is a device with five small hammers which consecutively strike the floor producing impacts of predetermined momentum. On activation of the tapping machine, the impact sound pressure level, L_i is measured in the room beneath the floor (the receiving room) in accordance with BS EN ISO 140–7:1998. As with airborne sound transmission, the level measured at the microphone in the receiving room will be influenced by noise energy reflected within the room, and so reverberation time is also measured when assessing impact sound insulation. In this case, the standardisation is also to the equivalent of a room with a reverberation time of 0.5 s. to give the 'standardised impact sound pressure level' L'_{nT}:

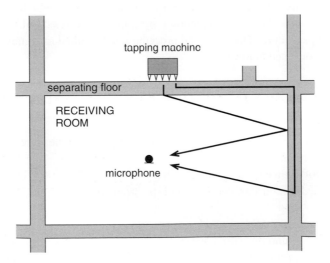

Fig. 10.30   Measurement of impact sound transmission.

$$L'_{nT} = L_i + 10 \, \text{Log} \left(\frac{T}{T_0} \right) \, \text{dB with } T_0 = 0.5 \, \text{s}.$$

L'_{nT} is measured using one-third-octave bands and these values are combined to give a single value, the weighted standardised impact sound pressure level, $L'_{nT,w}$ using a procedure described in BS EN ISO 717–2:1997.

10.12.2 General requirements AD E: B1.4, B2.1 & B3.2

It is the responsibility of the person undertaking the building work to arrange for the necessary sound insulation testing to be carried out by a testing organisation with appropriate third party accreditation. The AD states that the organisation undertaking the testing should preferably have United Kingdom Accreditation Service (UKAS) accreditation, or a European equivalent. Members of the Association of Noise Consultants who are approved under their association's ADE Registration Scheme are also regarded as being suitably qualified. It is also a requirement that a valid traceable calibration certificate exists for the instrumentation used for the measurement and that it has been tested within the preceding two years.

Measurements are based on the BS EN ISO 140 series of standards and rating is based on the BS EN ISO 717 series of standards and as stated in the AD it is important when calculating test results that rounding does not occur until required by the standard in question.

10.12.3 Field measurement of the sound insulation of separating walls and floors AD E: 0.1

Requirement E1 is satisfied if separating elements comply with Tables 1a and 1b of the AD following pre-completion testing, see section 10.2 of this chapter.

Regulations 20A and 12A apply to building work to which E1 applies and these regulations state that pre-completion testing must be carried out in accordance with an approved procedure. The approved procedure is described in Annex B2 of AD E and its content is outlined below.

Airborne insulation of separating walls or floors AD E: B2.2–B2.8

Procedure:
Measure in accordance with BS EN ISO140–4:1998 (all measurements and calculations carried out in one-third-octave bands).
Rate performance in accordance with BS EN ISO 717–1:1997 (in terms of the weighted standardised level difference $D_{nT,w}$ and spectrum adaptation term C_{tr}).

(1) Measurements using a single sound source
The average sound pressure level (SPL) is measured in one-third-octave bands in source and receiving rooms for each source position using either fixed microphone positions, and averaging values on an energy basis, or a moving microphone.

The differences between average source room SPL measurements in adjacent one-third-octave bands should not exceed 6 dB. If this criterion is not satisfied, the instrumentation should be adjusted and measurement repeated until it is. When it is satisfied, the average SPL in the receiving room and hence the level difference should be established.

The sound source must not be moved or its output level adjusted until measurements in the source and receiving rooms are complete.

The procedure should be repeated with the sound source moved to another position in the source room. At least two source positions should be used and the level differences obtained at each position should be averaged to obtain the level difference D defined in BS EN ISO 140–4:1998.

(2) Measurements using simultaneously operating multiple sound sources
The average SPL is measured in one-third-octave bands in source and receiving rooms with the multiple sources operating simultaneously. Measurement is undertaken using either fixed microphone positions, and averaging values on an energy basis, or a moving microphone.

The differences between average source room SPL measurements in adjacent one-third-octave bands should not exceed 6 dB. If this criterion is not satisfied, the instrumentation should be adjusted and measurement repeated until it is. When it is satisfied, the average SPL in the receiving room and hence the level difference, D defined in BS EN ISO 140–4:1998 may be established.

Impact sound transmission through separating floors AD E: B2.9

Procedure:
Measure in accordance with BS EN ISO140–7:1998 (all measurements and calculations carried out in one-third-octave bands).
Rate performance in accordance with BS EN ISO 717–2:1997 (in terms of the weighted standardised impact sound pressure level, $L'_{nT,w}$).

Measurement of reverberation time AD E: B2.10

In order to calculate the weighted standardised level difference $D_{nT,w}$, and the weighted standardised impact sound pressure level, $L'_{nT,w}$ measurements have to be corrected to take into account the degree of reverberation in the receiving room. This requires measurement of its reverberation time which is defined as the time taken for the SPL to decrease by 60 dB after a sound source is switched off. The measurement standards referenced above refer to ISO 354 for the method of measuring reverberation time. Guidance in BS EN ISO 140–7:1998 with respect to positioning of equipment and number of measurements should be adhered to.

Room types and sizes AD E: B2.11

The types of rooms to be used for testing are considered in section 10.3. They should have volumes of 25 m³ or more and if not, the volumes of the rooms used should be reported.

Room conditions for tests AD E: B2.12–B2.17

The following conditions should apply.

(1) Rooms (or available spaces in the case of properties sold before fitting out) used for testing should be completed but unfurnished.
(2) Floors without soft coverings should be used for impact tests. Exceptions are separating floors Type 1 and structural concrete floor bases which have integral soft coverings. If soft coverings have been laid on other types of floor, these should be removed and if that is not possible at least half of the floor should be exposed, with the tapping machine (this is required for impact transmission measurement) used on the exposed part of the floor.
(3) When a pair of rooms used for airborne sound insulation measurement are of different volumes, the sound source should be activated in the larger room.
(4) All doors and windows should be closed and the doors of fitments such as kitchen units and cupboards on all walls should be empty and have open doors.

Accuracy of measurement AD E: B2.18–B2.19

SPL and reverberation time measurements should be to an accuracy of 0.1 dB and 0.01 s. respectively.

Measurement using moving microphones AD E: B2.20–B2.21

If a moving microphone is employed, measurements should be taken with it centred at two or more positions. Due to the transient nature of the reverberation time measurement process, fixed microphone positions as opposed to moving microphones should be used when measuring reverberation times.

10.12.4 Laboratory measurement procedures AD E: B1.2 & B3.1–B3.2

Laboratory testing may be required to establish the performance of building elements in order to comply with Requirement E2, components such as wall ties and floating floors to which Requirement E1 applies as well as to assess the performance of proposed novel constructions. Appropriate laboratory testing procedures are described in Annex B3 of the AD and are outlined below.

No rounding in calculations associated with sound insulation test results should take place until required by the relevant BS EN ISO 140 and BS EN ISO 717 series standards.

Floor coverings and floating floors: AD E: B3.3–B3.6

Procedure:
Test in accordance with BS EN ISO 140–8:1998 (using a test floor with a thickness of 140 mm).
Rate performance in accordance with BS EN ISO 717–2:1997.

The AD states that text has been omitted from BS EN ISO 140–8:1998 and that for the purposes of the document, section 6.2.1 of BS EN ISO 140–8:1998 should be disregarded and section 5.3.3 of BS EN ISO 140–7:1998 referred to instead.

BS EN ISO-8:1998 refers to ISO 354 for the method of measuring reverberation time. It is stated in the AD that guidance in BS EN ISO 140–8:1998 with respect to positioning of equipment and number of required decay measurements should be adhered to.

When assessing Category II specimens, defined in BS EN ISO 140–8:1998 as large specimens including rigid homogeneous surface materials (e.g. floating floors), the value ΔL_w should be measured with and without the floor carrying a defined load. ΔL_w is a measure of the improvement in impact sound insulation provided by installing a floor covering or floating floor over a laboratory test floor.

Dynamic stiffness AD E: B3.7–B3.8

Procedure for resilient layer:
Measure in accordance with BS EN 29052–1:1992 (use sinusoidal signals method with no pre-compression of specimens).
Procedure for wall ties:
Measure in accordance with BRE Information Paper IP 3/01.

Airborne sound insulation of wall and floor elements AD E: B3.9

Procedure:
Measure in accordance with BS EN ISO 140–3:1995.
Rate performance in accordance with BS EN ISO 717–1:1997
(determine weighted sound reduction index R_w).
Sound reduction index and weighted sound reduction index are described in section 10.14.

Flanking laboratory measurement AD E: B3.10–B3.14

Although tests in flanking laboratories are very useful because they include the effects of transmission through direct and flanking routes, they may not be used to indicate compliance with Requirement E1 since this relates to field performance.

Flanking laboratory measurements are a useful way of assessing the probable performance in practice of novel or alternative constructions and it is stated in the AD that if a test construction provides airborne insulation, $D_{nT,w} + C_{tr}$, value of 49 dB or more when measured in a flanking laboratory, this may be taken as an indication that a construction which is identical in all significant details may achieve a $D_{nT,w} + C_{tr}$ value of 45 dB or more if built in the field. Note first paragraph of this section.

It is also stated in the AD that if a test construction provides impact sound transmission, $L'_{nT,w}$, of 58 dB or less measured in a flanking laboratory, this may be taken as an indication that a construction which is identical in all significant details may achieve an $L'_{nT,w}$ value of 62 dB or less if built in the field. Note first paragraph of this section.

A standard for laboratory measurement of flanking transmission is being developed and construction details of a flanking laboratory are available from The Acoustics Centre, BRE, Garston, Watford WD25 9XX.

10.12.5 Test report information

Field test report information AD E: B4.1

The information to be provided in a test report relating to tests done in compliance with regulations 20A or 12A is itemised in section 1 of AD E, see section 10.3.6 of this chapter. It is suggested in the AD that, although not mandatory, it may be useful to also provide the following information which is quoted directly from the document:

'(1) sketches showing the layout and dimensions of rooms tested;
(2) description of separating walls, external walls, separating floors, and internal walls and floors including details of materials used for their construction and finishes;
(3) mass per unit area in kg/m2 of separating walls, external walls, separating floors, and internal walls and floors;
(4) dimensions of any step and/or stagger between rooms tested;
(5) dimensions and position of any windows or doors in external walls.'

Laboratory test report information AD E: B4.2

It is stated that the following information, which is quoted directly from the AD, should be provided in test reports:

'(1) Organisation conducting test, including:
    (a) name and address;
    (b) third party accreditation number (e.g. UKAS or European equivalent);
    (c) Name(s) of person(s) in charge of test.
(2) Name(s) of client(s).
(3) Date of test.
(4) Brief details of test, including:
    (a) equipment;
    (b) test procedures.
(5) Full details of the construction under test and the mounting conditions.
(6) Results of test shown in tabular and graphical form for third-octave bands according to the relevant part of the BS EN ISO 140 series and BS EN ISO 717 series, including:
    (a) single-number quantity and the spectrum adaptation terms;
    (b) data from which the single-number quantity is calculated.'

10.13 The calculation of mass (Approved Document E, Annex A: Method for calculating mass per unit area)

The mass per unit area, expressed in kilograms per square metre may be obtained from manufacturers' data. Alternatively, it can be calculated for walls and floors by the following methods as described in Annex A of AD E.

10.13.1 Walls AD E: A2–A4

In the case of a brick or block wall, mortar beds and perpends may constitute a considerable proportion of the total volume and, since the density of mortar may be significantly different to that of the blocks, account must be taken of this when calculating the mass per unit area. The procedure used is to define the 'co-ordinating area' as shown in Fig. 10.31, and to calculate the mass per unit area, M_A from:

$$M_A = \frac{\text{mass of co-ordinating area}}{\text{co-ordinating area}} = \frac{M_B + \rho_m(T\,d(L+H-d)+V}{LH}\,\text{kg/m}^2$$

where, in the above equation:

M_B = brick or block mass (kg)
ρ_m = density of mortar (kg/m³)
T = brick or block thickness (m) (excluding any finish)
d = mortar joint thickness (m)
L = co-ordinating length (m)
H = co-ordinating height (m)
V = volume of any mortar filled void, e.g. a frog

In using the above equation the following points should be noted.

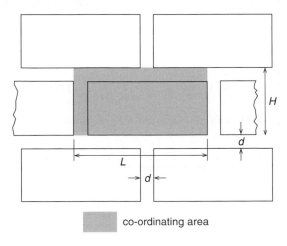

co-ordinating area

Fig. 10.31 Co-ordinating area used for calculation of mass per unit area.

- The brick or block unit must penetrate the full thickness (excluding finishes) of the wall.
- Since density is a function of moisture content, the mass of brick/block and density of mortar should be taken at the appropriate moisture content. The AD makes reference to Table 3.2 of CIBSE Guide A (1999) for this information.
- The above formula calculates the mass per unit area of a single leaf without surface finish. If a finish, e.g. plaster, is to be applied and/or if it is a cavity wall with another leaf, the mass per unit area of each component must be added together to obtain the total mass per unit area of the wall.
- Manufacturers' data should be used to obtain the mass per unit area of surface finishes.

Example calculation

Assume that a concrete block has the following dimensions:

thickness   = 0.120 m
length   = 0.440 m
height   = 0.290 m
mass of block   = 14 kg

and it is to be used on edge with mortar which has a density of 1800 kg/m^3 and a thickness of 0.010 m (10 mm).

In this case: M_B = 14 kg
                ρ_m = 1800 kg/m^3
                T = 0.120 m
                d = 0.010 m
                L = 0.450 m
                H = 0.300 m
                V = 0 (it is assumed that there is no frog)

and substituting into the above equation gives a mass of 115.5 kg/m^2.

The AD makes reference to 'simplified equations'. These, however, are simply representations of the equation shown above with substitution for all variables for the particular case with the exception of the mass of the block. In this format the above equation would be written as:

$$\text{Mass per unit area} = 7.4\,M_B + 11.8 \text{ kg/m}^2$$

Various values of M_B may now be substituted into the equation to establish the mass per unit area.

10.13.2 Floors AD E: A5

The mass per unit area of a flat solid concrete slab floor, M_F, may be obtained by multiplying its density, ρ_c, by its thickness, T:

$$M_F = \rho_c T$$

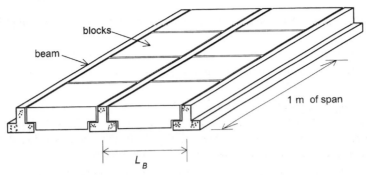

Fig. 10.32   Beam and block floor.

In the case of a conventional beam and block floor, as shown in Fig. 10.32, the mass per unit area, M_F, can be obtained from:

$$M_F = \frac{(M_{beam,\,1m} + M_{block,\,1m})}{L_B}$$

where, in the above equation:

$M_{beam,1m}$	=	the mass of 1 m of the length of the beam
$M_{block,1m}$	=	the mass of the blocks spanning between two consecutive beams for 1 m of the length of the beam
L_B	=	the dimension defined in Fig. 10.32

No other examples are given in the AD. Designers are advised in the AD to seek advice from manufacturers on the mass per unit area of other floor types.

10.14 Explanation of important terms

Absorption of sound

When sound strikes a surface, a fraction is reflected back from the surface and the remainder is absorbed at the surface. Sound absorption, which may be thought of as the conversion of sound energy to heat, may be caused by a number of mechanisms. The absorption coefficient is the fraction of the incident energy which is not reflected from the surface and has a value between zero and one, the extreme values representing perfectly reflecting and perfectly absorbing surfaces respectively.

Airborne sound insulation

Airborne sound is sound, from a source such a speech or loudspeakers, which travels through air. Elements reducing its transmission are therefore providing airborne sound insulation. Airborne sound may pass from one room to another by:

- direct transmission: sound which passes directly through the element separating two rooms;
- flanking transmission: sound which passes indirectly between rooms, e.g. via elements which abut the separating element.

Dwelling places

Within this chapter, dwelling-houses, flats and rooms for residential purposes are, for convenience, referred to collectively as 'dwelling places'.

Frequency

Sound results from cyclic variations in air pressure, the frequency of which is measured in the units of hertz and given the symbol Hz. The frequency in hertz refers to the number of pressure variation cycles per second. The human ear can detect sound within a range of frequencies from about 20 Hz to 20,000 Hz. An octave band contains all frequencies from a lower limiting frequency to an upper limiting frequency of twice the lower limit. One-third-octave bands result from dividing an octave band into three contiguous bands in which the upper limiting frequency in each sub band is its lower limiting frequency multiplied by $2^{1/3}$.

Impact sound insulation

Impact sound is sound created by the impact of objects, e.g. footsteps, directly with part of the building structure. If a component is provided which reduces impact sound transmission it is providing impact sound insulation. When sound is transmitted via the structural components of a building it is known as structure borne sound.

Impact sound pressure level

The standardised impact sound pressure level, L'_{nT}, is obtained from frequency band measurements of the impact sound generated by a standard test. The weighted standardised impact sound pressure level, $L'_{nT,w}$, is a single number expression of impact sound level measured in decibels and obtained from L'_{nT}. These indices are described in section 10.11.

Level difference

The standardised level difference, D_{nT}, is obtained from frequency band measurements of the sound pressure level difference between two rooms, and the weighted standardised level difference, $D_{nT,w}$, is a single number expression of level difference obtained from D_{nT}. The spectrum adaptation term, C_{tr}, is a single number modification to $D_{nT,w}$. These indices are described in section 10.11.

Separating floors and walls

These are the elements between dwelling places. Separating floors separate flats and 'rooms for residential purposes'. Separating walls separate dwelling-houses, flats and 'rooms for residential purposes'.

Sound reduction index

The sound reduction index, R, of a building element is a laboratory measured value relating to its airborne sound insulating properties. Sound reduction index is measured in accordance with BS EN ISO 140–3:1995 in one-third-octave bands across the audible frequency range and is expressed in decibels. The weighted sound reduction index, R_w, is a single number quantity which expresses the airborne sound insulation of a building element. It is derived from the sound reduction index in accordance with BS EN ISO 717–1:1997.

Units of measurement

The decibel, given the symbol dB, is a convenient unit for representing magnitude of sound. Sound pressure level, which is used to measure the magnitude of sound in building acoustics, is measured in decibels.

11 Ventilation (Part F)

J.R. Waters

11.1 Introduction

Ventilation is the replacement of polluted air within a building with fresh air from outside the building. It is also possible to recirculate polluted air provided it has been treated in some way, but 100% recirculation is not normally acceptable in buildings. The process of ventilation consumes energy, both because it may be necessary to heat or cool the incoming fresh air and because it may also involve the operation of energy consuming equipment, e.g. electric fans. Because there are energy implications in providing ventilation, Part L, Conservation of Fuel and Power, imposes some limitations. The relevant ones are:

- a requirement to make buildings air tight in order to limit accidental infiltration of fresh air through the building fabric;
- limitations on fan power in mechanically ventilated systems.

Both of these limitations tend to reduce the fresh air supply to a building. Thus the provisions of Part F become particularly important in providing a counterbalance to Part L, and ensuring that there is sufficient ventilation for the purposes of health and safety.

11.2 Definition and interpretation of terms

The principal terms are described in the glossary of Part F (p. 32). They are presented here, defined as they are used in Part F, together with other significant and relevant terms, in overall alphabetic order.

AIR PERMEABILITY – Air permeability is the air leakage rate per unit envelope area at the test reference pressure differential of 50 Pascal.

The envelope area of the building (or its measured part) is the total area of all floors, walls and ceilings bordering the space being tested, and includes walls and floors below external ground level. Overall internal dimensions are used to calculate this area, with no subtractions for the area of the junctions where internal elements meet the external envelope.

AIR TIGHTNESS – Air tightness is the ability of the building envelope to resist air infiltration. An air tight building has a low air permeability, and therefore a low rate of infiltration.

AUTOMATIC CONTROL – This is a non-manual control that operates a ventilation device in response to a signal from a sensor which detects some property of the air or the air flow.

BACKGROUND VENTILATOR – A small ventilation opening designed to provide controllable whole building ventilation.

BASEMENT FOR DWELLINGS – A usable and habitable part of a dwelling that is wholly or partly below ground level. A cellar or other space used for purposes other than habitation (e.g. storage, heating plant, etc.) is not a basement.

BATHROOM – Any room containing a bath or shower and which may also include sanitary accommodation.

CELLAR – See the definition of basement for dwellings.

CLOSABLE – A ventilation opening which may be opened and closed either manually or automatically.

COMMON SPACES – Spaces where large numbers of people are expected to gather, such as shopping malls and cinema/theatre foyers. However, for the purposes of Part F, spaces used solely or principally for circulation (e.g. corridors, lift lobbies, etc.) are *not* regarded as common spaces.

CONTINUOUS OPERATION – A mechanical ventilation device that runs all the time. The air flow rate through such a device may vary.

EQUIVALENT AREA – The area of a sharp-edged orifice through which air would pass at the same volume flow rate as an actual opening when both are subjected to the same applied pressure difference.

EXTRACT VENTILATION – The intermittent or continuous extraction of air, either naturally or mechanically, to the outside from a space (or spaces) where large quantities of a pollutant (e.g. water vapour) are generated in order to maintain acceptable pollutant levels in the space itself and also to prevent spread of the pollutant to the rest of the building.

FREE AREA – The geometric open area of a ventilator.

GROSS INTERNAL VOLUME – The total internal volume of the heated space, including the volume of all contents, including furniture, internal walls and floors, etc.

HABITABLE ROOM – A room which is used for dwelling purposes but which is not solely a kitchen, utility room, bathroom, cellar or sanitary accommodation.

HISTORIC BUILDING – These normally include:

- listed buildings;
- buildings situated in conservation areas;

- buildings of architectural and historical interest and which are referred to as a material consideration in a local authority development plan;
- buildings of architectural and historical interest within national parks, areas of outstanding natural beauty, and world heritage sites.

INFILTRATION – The uncontrolled exchange of air between the inside and outside of a building through unintentional openings in the external envelope.

INTERMITTENT OPERATION – A mechanical ventilation device that does not run all the time. The intermittent operation may be controlled manually or automatically.

MANUAL CONTROL – A ventilation device that is adjusted (e.g. on or off, opened or closed, up or down, etc.) manually by occupants.

MECHANICAL VENTILATION – A ventilation system in which the driving force is provided by fans or any other mechanical devices which require the supply of energy.

MIXED-MODE VENTILATION – A ventilation system which is designed to use both naturally driven and mechanically driven air flows.

NATURAL VENTILATION – A ventilation system in which the air flow is generated by natural forces such as the buoyancy effect or wind pressures.

OCCUPIABLE ROOM – A room in a building other than a dwelling that is occupied by people. However this does *not* include any of the following:

- bathroom;
- sanitary accommodation;
- utility room;
- rooms or spaces used solely or principally for circulation;
- rooms or spaces used solely or principally for building services plant;
- rooms or spaces used solely or principally for storage.

PASSIVE STACK VENTILATION (PSV) – A ventilation device consisting of ducts connecting terminals in the ceiling of a room (or at high level) to terminals on or above the roof that extract air by a combination of stack effect and wind pressure effects at roof level.

PERMANENT – A ventilation opening which is permanently fixed in the open position.

PURGE VENTILATION – Purge ventilation (also known as rapid ventilation) is ventilation of rooms or spaces at a high rate to rapidly dilute high concentrations of pollutants. Such pollutants are usually released by occasional activities (e.g. painting and decorating) or by accidental spillage. Purge ventilation is usually manually controlled and may be provided by natural or mechanical means.

PURPOSE-PROVIDED VENTILATION – That part of the ventilation of a building provided by ventilation devices designed into the building, e.g. background ventilators, PSV, extract fans, mechanical ventilation, air conditioning, etc.

SANITARY ACCOMMODATION – A space containing one or more water closets or urinals. Sanitary accommodation containing more than one cubicle counts as a single space provided there is free circulation of air throughout the space.

STACK EFFECT – The pressure differential between the inside and outside of a building due to differences in air pressure caused by inside/outside temperature difference.

UTILITY ROOM – A room containing a sink or other feature that may reasonably be expected to produce water vapour in significant quantities.

VENTILATION – The supply and removal of air (naturally or mechanically) to or from a space or spaces in a building. It includes both purpose-provided ventilation and infiltration.

VENTILATION OPENING – Any means of purpose-provided ventilation, permanent or closable, which opens directly to external air, including the openable parts of a window, a louvre, a background ventilator, a door which opens directly to external air, etc.

WET ROOM – A room used for domestic activities (e.g. cooking, clothes washing, bathing) which give rise to significant production of airborne moisture. For the purposes of Part F, sanitary accommodation is regarded as a wet room.

WHOLE BUILDING VENTILATION – Nominally continuous ventilation of rooms or spaces at a relatively low rate to dilute, disperse and remove pollutants, and to supply fresh outdoor air. It does not include removal of pollutants by extract ventilation, purge ventilation or infiltration.

11.3 General principles

The Part F requirement and its associated guidance are based on some general principles regarding ventilation. These principles can be grouped under two main headings, one concerning the purposes of ventilation and the other to do with the means of ventilation.

11.3.1 The purposes of ventilation

The general purpose of ventilation is to remove contaminated indoor air and to replace it with fresh outdoor air. It is assumed throughout Part F that outdoor air is relatively uncontaminated, of reasonable quality and with low levels of pollutants. If this is not the case, then it is the responsibility of the designers of a project to make appropriate provision.

The main purposes of ventilation may be listed as one or more of the following:

- The removal of CO_2 and the provision of fresh air for occupants to breathe properly.

- The dilution and removal of airborne pollutants, including odours. This includes not only the pollutants generated physiologically by normal human activity, but those released from the materials and products used in construction, decoration and furnishing. Activities within a building, especially non-dwellings, may also add pollutants to the indoor air (e.g. photocopiers, printers, etc.).
- Removal of excess moisture in order to control the humidity of the indoor air.
- Removal of excess heat in order to control temperature (see Part L).
- Provision of air for fuel-burning appliances (see Part J).

The principal requirements and provisions of Part F do not specify maximum permissible levels for pollutants, it being assumed that if the recommendations for fresh air supply rates and the sizing of ventilation openings are followed, then in normal circumstances indoor pollutant levels will be kept sufficiently low. However, as an alternative, Part L does permit the evaluation of ventilation rates based on pollutant concentrations (see section 11.8, Performance-based ventilation). This type of approach may become necessary if circumstances are not normal, or when it is known that the occupants may be exposed to unacceptable levels of specific pollutants.

11.3.2   Ventilation strategies, types and systems

Part F identifies three broad strategies for ventilation:

- extract ventilation;
- whole building ventilation;
- purge ventilation.

These ventilation strategies may be provided by three broad types of ventilation:

- natural ventilation;
- mechanical ventilation;
- mixed-mode (or hybrid) ventilation.

There are many systems and devices that can be used to give effect to a particular choice of ventilation strategy and type. Whichever system is chosen, it is now considered important to ensure that the ventilation is controllable so that it achieves reasonable indoor air quality without wasting energy. The efficiency of a ventilation system in maintaining air quality can be assessed using the concept of ventilation effectiveness, but there is no requirement in Part F 2006 to achieve a specific value of this parameter.

11.4   Part F – The requirement and its applicability

The requirement is:

MEANS OF VENTILATION – There shall be adequate means of ventilation provided for people in the building.

This requirement applies to all buildings, but with some exceptions. Part F does not apply to a building or a space within a building:

- into which people do not go; or
- which is used solely for storage; or
- which is a garage used solely in connection with a single dwelling.

In general terms the requirement may be achieved by a ventilation system (or systems) which:

- extracts, before they are generally widespread, water vapour and/or hazardous pollutants from areas where they are produced in significant quantities;
- rapidly dilutes or preferably extracts water vapour and/or pollutants produced in habitable rooms, occupiable rooms and sanitary accommodation;
- makes available over long periods a minimum supply of outdoor air for occupants and to disperse where necessary residual water vapour and/or pollutants;
- is designed, installed and commissioned to perform in a way that is not detrimental to the health of the people in the building;
- is installed to facilitate maintenance where necessary.

It should be noted that the guidance given in the AD does not specifically take account of the products of tobacco smoke. If tobacco smoke is likely to be an issue, this must be dealt with separately.

11.5   The ventilation of dwellings

There are four main ways of complying with the requirement:

- by following the rules set out in section 11.5.1 and by providing the minimum ventilation rates in Tables 11.1 and 11.2; or
- for dwellings without basements, by following the rules set out in section 11.5.2 and providing the system requirements set out in Tables 11.3; or
- for dwellings with basements, by following the requirements set out in section 11.5.7; or
- by using other ventilation systems provided it can be demonstrated that they meet the pollutant criteria set out in Table 11.8 for performance-based ventilation.

Each of these is now considered in turn.

11.5.1   Compliance using ventilation rates

It is necessary to provide all of the following:

- Extract ventilation to outside in each kitchen, utility room, bathroom and room with sanitary accommodation, with minimum air flow rates as specified in Table 11.1.

Table 11.1 Minimum extract ventilation rates.

Room	Intermittent extract	Continuous extract	
	Minimum rate (ls⁻¹)	Minimum high rate (ls⁻¹)	Minimum low rate (ls⁻¹)
Kitchen, sited adjacent to hob	30	13	The total extract rate (room plus rest of dwelling) must be at least the whole building ventilation rate as in Table 11.2
Kitchen, not sited adjacent to hob	60	13	
Utility room	30	8	
Bathroom	15	8	
Sanitary accommodation	6	6	

Table 11.2 Minimum whole building ventilation rates.

	Number of bedrooms in dwelling				
	1	2	3	4	5
Whole building ventilation rate (ls⁻¹)	13	17	21	25	29

Notes:
(a)   In addition, the minimum ventilation should be not less than 0.3 ls⁻¹ per m² of the total internal floor area (i.e. ground floor plus upper floors).
(b)   These minimum rates are based on two occupants in the main bedroom and a single occupant in all other bedrooms. If a greater occupancy level is expected, add 4 ls⁻¹ per additional occupant.

- Minimum whole building ventilation rates as shown in Table 11.2.
- In each habitable room not provided with extract ventilation, purge ventilation capable of extracting a minimum of four air changes per hour per room directly to outside.

The airflow rates in Tables 11.1 and 11.2 are for the performance of the complete installation. On-site measurement of the performance is not required, except where necessary to balance the system, but equipment should be tested according to the appropriate standards (see section 11.5.6).

11.5.2   Compliance using system specification for dwellings without basements

This method has five steps:

(1)   Select one out of the four ventilation systems described below, and comply with the minimum system specifications in the tables relevant to each system.
(2)   For all four systems, make at least minimum provision for purge ventilation, as described in section 11.5.3.
(3)   Choose at least minimum areas for background ventilators from Table 11.3, and select acceptable ventilator locations as described in section 11.5.4.
(4)   Follow the guidance on ventilation controls in section 11.5.5.
(5)   Follow the guidance on performance test methods, see section 11.5.6.

Table 11.3   Minimum total equivalent ventilator area, mm².

Total floor area (m²)	Number of bedrooms				
	1	**2**	**3**	**4**	**5**
Up to 50	25 000	35 000	45 000		
51–60	25 000	30 000	40 000		
61–70	30 000	30 000	30 000	All	All
71–80	35 000	35 000	35 000	45 000	55 000
81–90	40 000	40 000	40 000		
91–100	45 000	45 000	45 000		
Greater than 100	Add 5000 mm² for every extra 10 m² floor area				

Notes:

(a)   The equivalent area of a background ventilator should be determined at 1 Pa pressure difference using an approved test method.

(b)   The table assumes two occupants in the main bedroom and one in all other bedrooms. For each extra occupant use the value for one extra bedroom. For more than five bedrooms add 10 000 mm² per bedroom.

(c)   The areas given in this table are the total for the whole dwelling.

SYSTEM 1 – background ventilators and intermittent extract fans

This system is illustrated in Fig. 11.1. Minimum equivalent ventilator areas are given in Table 11.3, with the additional following information on intermittent extract, background ventilators and single-sided ventilation.

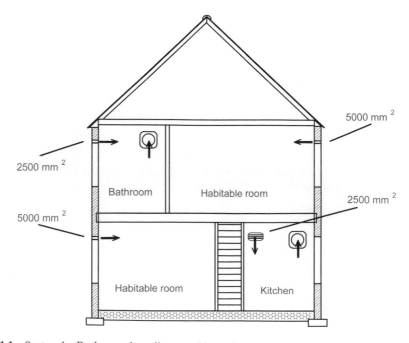

Fig. 11.1   System 1 – Background ventilators and intermittent extract fans.

INTERMITTENT EXTRACT

- Intermittent extract rates are given in Table 11.1. For sanitary accommodation only, purge ventilation provisions (windows) as given in section 11.5.3 can be used provided security is not an issue.
- Instead of a conventional intermittent fan, a continuously running single-room heat-recovery ventilator could be used in a wet room. It should use the minimum high rate in Table 11.1 and 50% of this rate as the minimum low rate. No background ventilator is required in the same room as a single-room heat-recovery ventilator. Also, the total equivalent background ventilator area given below can be reduced by 2500 mm² for each room containing a single-room heat-recovery ventilator. Continuously running fans should be quiet so as not to discourage their use by occupants.

BACKGROUND VENTILATORS

- For dwellings with more than one exposed façade:
  - For multi-storey dwellings, and single-storey dwellings *more than* four storeys above ground level, use the ventilator areas in Table 11.3 unchanged.
  - For single-storey dwellings *up to* four storeys above ground level, add 5000 mm² to all the ventilator areas in Table 11.3.

DWELLINGS WITH ONLY A SINGLE EXPOSED FAÇADE

Cross ventilation is not possible in this type of dwelling with this type of ventilation system. An alternative is required, as shown in Fig. 11.2. For all single and multi-storey

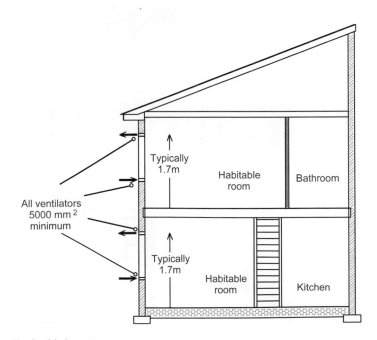

Fig. 11.2   Single-sided ventilation.

dwellings, background ventilators should be located at both high and low positions in the exposed façade as follows:

- high-level ventilator at least 1.7 m above floor level;
- low-level ventilator at least 1.0 m below the high level ventilator;
- total equivalent area at the high level as given in Table 11.3;
- in addition, the same total equivalent area at the low level.

Arrange the habitable rooms to be on the exposed façade, and ensure that these rooms have a maximum depth of 6 m.

SYSTEM 2 – passive stack ventilation (PSV)

The system is shown in Fig. 11.3. The minimum internal dimensions of the stack ducts should be taken from either column 1 or column 2 of Table 11.4. The dwelling will also require background ventilators to provide inlet air to replace the air extracted via the stacks. The required equivalent area of the background ventilators is found in three stages as follows:

(1) Determine from Table 11.3 the equivalent ventilator area for the dwelling.
(2) Make an allowance for the total air flow through all PSV units. As an approximation under normal conditions it may be assumed that each PSV unit provides an equivalent ventilator area of 2500 mm^2.

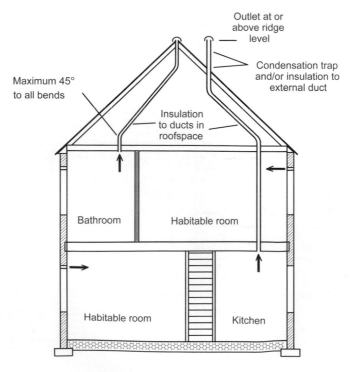

Fig. 11.3   System 2 – Passive stack ventilation.

Table 11.4   Minimum dimensions for passive stack ventilators.

Room	Internal duct diameter (mm)	Internal cross sectional area (mm^2)
Kitchen	125	12 000
Utility room	100	8 000
Bathroom	100	8 000
Sanitary accommodation	80	5 000

Notes:

(a)   An open-flued appliance may provide sufficient extract ventilation for the room in which it is located when in operation, and can be arranged to provide sufficient ventilation when not firing. For instance, the provisions would be adequate if: (i) the solid fuel appliance is a primary source of heating, cooking or hot water production, or (ii) the open-flued appliance has a flue of free area at least equivalent to a 125 mm diameter duct and the appliance's combustion air inlet and dilution inlet are permanently open, so that either there is a path with no control dampers which could block the flow, or the ventilation path can be left open when the appliance is not in use. Note that PSV, in common with extract fans, depressurises the internal space, and so also creates the danger of spillage into the dwelling from open flues.

(b)   For sanitary accommodation only, as an alternative, the purge ventilation provisions given in section 11.5.3 can be used when security is not an issue.

(c)   In addition, use this procedure to calculate sizes of background ventilators if both PSV (or open-flued appliances as described in note (a) above) and intermittent extract fans are used in different rooms in the same dwelling. Pay due regard to the danger of spillage from any open-flued appliances.

(3)   The equivalent area is the greater of:
   (a)   the difference stage 1 – stage 2;
   (b)   the total maximum cross-sectional area of all the PSV ducts.

See section 11.9 for details of the design and installation of PSV. Note also that for dwellings with a single exposed façade, all habitable rooms should be on that exposed façade.

SYSTEM 3 – continuous mechanical extract

This is illustrated in Fig. 11.4. The procedure is in three stages:

(1)   Determine the whole building ventilation rate from Table 11.2. Note that no allowance is made for infiltration because the extract system lowers the pressure in the dwelling and limits the exit of air through the building fabric.

(2)   Calculate the whole-dwelling air extract rate at maximum operation by summing the individual room rates for 'minimum high rate' from Table 11.1. For sanitary accommodation only, as an alternative, the purge ventilation provisions given in section 11.5.3 can be used where security is not an issue. In this case, the 'minimum high rate' for sanitary accommodation should be omitted from this stage of the calculation.

(3)   The required extract rates are as follows:
   • The maximum rate (e.g. boost) should be at least the greater of stage 1 and stage 2. The maximum individual room extract rates should be at least the 'minimum high rate' given in Table 11.1.
   • The minimum rate should be at least the whole building ventilation rate in stage 1.

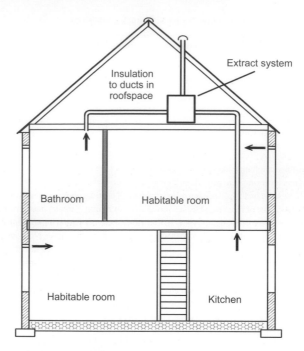

Fig. 11.4 System 3 – Continuous mechanical extract.

This system could comprise either a central extract system or individual room fans or a combination of both. In all cases, fans should operate quietly at their minimum (i.e. normal) rate so as not to discourage their use. To ensure that the system performs to specification, extract terminals should be located to minimise wind effects. Possible solutions include ducting to a sheltered façade or installing constant volume flow rate units. Further guidance may be found at www.est.org.uk.

If a single-room heat-recovery ventilator (SRHRV) is used to ventilate a habitable room, with the rest of the dwelling provided with continuous mechanical extract, the airflow rates are determined as follows:

- Determine the whole building ventilation rate from Table 11.2.
- Calculate the room supply rate required for the SRHRV from:

$$\text{SRHRV supply rate} = \frac{\text{Whole building ventilation rate} \times \text{room volume}}{\text{Total volume of all habitable rooms}}$$

Then follow stages 1 to 3 for sizing the mechanical extract for the rest of the dwelling, except that in stage 1 the supply rate of the SRHRV should be subtracted from the value given in Table 11.2.

Background ventilators to provide make-up air will be needed depending on the air permeability of the dwelling. As the air permeability is unlikely to be known at the design stage, it is recommended that controllable background ventilators with a minimum equivalent area of 2500 mm² are fitted in each room (except wet rooms) from which air is extracted. If this approach causes difficulties (e.g. on a noisy site), specialist advice may be required.

SYSTEM 4 – continuous mechanical supply and extract with heat recovery (MVHR)

This is illustrated in Fig. 11.5. The procedure is in three stages:

(1) Determine the whole building ventilation rate from Table 11.2. Allow for infiltration by subtracting from this value $X\%$ of the gross internal volume of the dwelling heated space, where $X = 4\%$ for multi-storey dwellings and $X = 6\%$ for single-storey dwellings.

(2) Calculate the whole dwelling air extract rate at maximum operation by summing the individual room rates for 'minimum high rate' from Table 11.1. For sanitary accommodation only, as an alternative, the purge ventilation provisions given in section 11.5.3 can be used where security is not an issue. In this case, the 'minimum high rate' for sanitary accommodation should be omitted from this stage of the calculation.

(3) The required air flow rates are as follows:
- The maximum rate (e.g. boost) should be at least the greater of stage 1 and stage 2. The maximum individual room extract rates should be at least the 'minimum high rate' given in Table 11.1.
- The minimum air supply rate should be at least the whole building ventilation rate in stage 1.

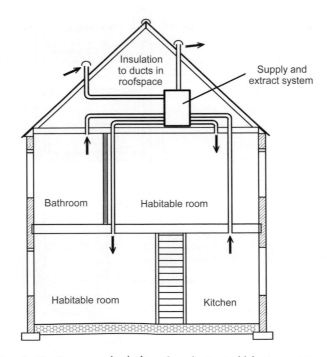

Fig. 11.5   System 4 – Continuous mechanical supply and extract with heat recovery.

11.5.3   Purge ventilation provisions for all four ventilation systems

Purge ventilation must be provided. For each habitable room with an external wall, the necessary dimensions of windows and doors are as follows (more detailed guidance is given in BS 5925:1991):

WINDOWS:

- For a hinged or pivot window that opens 30° or more, or for sliding sash windows, the openable area should be at least one twentieth of the floor area of the room.
- For a hinged or pivot window that opens between 15° and 30°, the openable area should be at least one tenth of the floor area of the room.
- If there is more than one openable window in a room, the required area (i.e. one twentieth or one tenth of the floor area) is calculated according to the opening angle of the largest window. The openable area is then the sum of the areas of the openable parts of all windows in the room, whatever their opening angle.
- Refer to AD B for the sizing of escape windows, and use the larger of the AD B or AD F provisions.
- The method of measuring the openable area of windows is illustrated in Fig. 11.6.

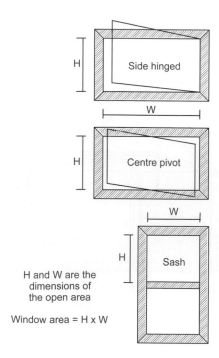

Fig. 11.6   Purge ventilation – calculation of opening area.

EXTERNAL DOORS, INCLUDING PATIO DOORS:

- For an external door, the height × width of the openable part should be at least one twentieth of the floor area of the room.
- If the room contains more than one external door, the areas of all openable areas may be added together.
- If the room contains a combination of at least one external door and at least one openable window, the areas of all the opening parts may be added to achieve at least one twentieth of the floor area of the room.

11.5.4   Location of all ventilation devices in rooms

Mechanical (intermittent and continuous) extract or supply

- Cooker hoods should be 650 mm to 750 mm above the hob surface (or follow manufacturer's instructions).
- Mechanical extract terminals and extract fans should be placed as high as practicable and preferably less than 400 mm below the ceiling.
- Mechanical supply terminals should be located and directed so as to avoid draughts.
- Where ducts, etc. are provided in a dwelling with a protected stairway, precautions may be necessary to avoid the possibility of the system allowing smoke or fire to spread into the stairway (see Approved Document B).
- The fans or terminals should be located in the following rooms:
  - **System 1:** Extract should be from each wet room.
  - **System 3:** Extract should be from each wet room.
  - **System 4:** Extract should be from each wet room. In addition, air should normally be supplied to each habitable room. The total supply airflow should usually be distributed in proportion to the habitable room volumes. Recirculation by the system of moist air from wet rooms to habitable rooms should be avoided.

Passive stack ventilation

- PSV extract terminals should be located in the ceiling or on a wall less than 400 mm below the ceiling. There should be no background ventilators within the same room as a PSV extract terminal. For open-flued appliances, room air supply is necessary as given in Approved Document J.
- Where PSV is provided in a dwelling with a protected stairway, precautions may be necessary to avoid the possibility of the system allowing smoke or fire to spread into the stairway (see Approved Document B).

Background ventilators

Background ventilators should be located according to the chosen ventilation system.

- **System 1:** Located in all rooms, with dimensions such that they have a minimum equivalent area of:

- ○ 5000 mm² in all habitable rooms with an external wall;
- ○ 2500 mm² in all wet rooms with an external wall, and with a total equivalent area not less than that given in Table 11.3.

For habitable rooms without an external wall, follow the appropriate guidance in section 11.5.8 or 11.5.9 below.

For wet rooms without an external wall, follow the guidance for mechanical intermittent extract in section 11.5.5 below.

Where background ventilators and individual fans are fitted in the same room, they should be a minimum of 0.5 m apart.

- **System 2:** Located in all rooms, except within the same room as a passive stack extract ventilator, with dimensions such that they have a minimum equivalent area of:
  - ○ 5000 mm² in all habitable rooms with an external wall, and with a total equivalent area not less than that given in Table 11.4.
- **System 3:** Located in each habitable room. See the description of system 3 in section 11.5.2.
- **System 4:** No background ventilators.

In addition, background ventilators should be:

- **All systems:** Located so as to avoid draughts, typically 1.7 m above floor level. For system 1, if the dwelling has a single exposed façade, the low ventilators should be below this level (see section 11.5.2).
- **Systems 1 and 2:** If the dwelling has more than one exposed façade, airflow should be maximised by encouraging cross ventilation. To do this, similar equivalent areas of background ventilator should be located on opposite (or if that is not possible, adjacent) façades.

For systems 1 and 2 the background ventilators have been sized for the winter period. Additional ventilation may be needed during warm weather because stack driving pressures are reduced. The provisions for purge ventilation (e.g. windows) may be used. However, for dwellings designed to a high air-tightness standard, additional background ventilation may be necessary, or systems 3 or 4 may be more appropriate. Specialist advice may be required.

Purge ventilation

The location of ventilation devices is not critical.

Air transfer between rooms

To ensure good transfer of air throughout the dwelling, there should be an undercut of minimum area 7600 mm² in all internal doors above floor finish. This corresponds to an undercut of 10 mm on a 760 mm wide door.

11.5.5   Controls for ventilation devices

Mechanical intermittent extract

Intermittent extract can be controlled:

- manually; or
- automatically; or
- by a combination of manual and automatic control.

In kitchens, any automatic control must have manual over-ride to allow the occupant to turn the extract on.

In any room without an openable window, the extract fan must have at least a 15 minute over-run.

In a room with no natural light, the extract fan could be controlled from the main room light switch.

Sensors for operating automatic controls could be based on:

- humidity, but not in sanitary accommodation where odour is the main pollutant;
- occupancy/usage;
- moisture or pollutant release.

Mechanical continuous supply or extract, and passive stack ventilation

These systems should be set up to operate without occupant intervention. However, they may have manual controls to allow the occupant to select a 'boost' rate. The system must always provide at least the minimum whole building ventilation rate as given in Table 11.2.

In kitchens, any automatic control must have manual over-ride to allow the occupant to turn the extract on.

For sensors, see mechanical intermittent extract above.

Background ventilator controls

These can be controlled manually or automatically. Trickle ventilators should be installed with the following in mind:

- They should be mounted preferably at least 1.7 m above floor level.
- If located above a window, they should be mounted either in the frame above the glass or directly through the wall'.
- They should have a flap to shut off the ventilation in adverse weather conditions.
- If provided with pressure-difference regulated automatic control, they should have a flap with manual over-ride to allow the occupant to fully close the ventilator.
- If provided with an automatic control which is not pressure-difference regulated, they should have a flap with manual over-ride to allow the occupant to fully open as well as fully close the ventilator.

Openable windows with a night latch position are not recommended as a means of providing background ventilation.

Purge ventilation

In dwellings, purge ventilation should be manually operated.

Accessibility of controls

Where manual controls are provided, they should be within reasonable reach of occupants. If necessary, pull cords, operating rods, etc. may be used to achieve this.

Compliance with Requirement N3 'Safe opening and closing of windows etc.', which for the purposes of Part F also applies to dwellings, is required. The relevant guidance in Approved Document N should be followed.

11.5.6   Performance test methods

The principal standards for performance testing are:

- BS EN 13141–1:2004 *Ventilation for buildings. Performance testing of components/ products for residential ventilation. Externally and internally mounted air transfer devices.*
- BS EN 13141–3:2004 *Ventilation for buildings. Performance testing of components/ products for residential ventilation. Range hoods for residential use.*
- BS EN 13141–4:2004 *Ventilation for buildings. Performance testing of components/ products for residential ventilation. Fans used in residential ventilation systems.*
- BS EN 13141–6:2004 *Ventilation for buildings. Performance testing of components/ products for residential ventilation. Exhaust ventilation system packages used in a single dwelling.*
- BS EN 13141–7:2004 *Ventilation for buildings. Performance testing of components/ products for residential ventilation. Performance testing of mechanical supply and exhaust ventilation units [including heat recovery] for mechanical ventilation systems intended for single family dwellings.*
- prEN 13141–8:2004 *Ventilation for buildings. Performance testing of components/products for residential ventilation. Performance testing of unducted mechanical supply and exhaust ventilation units [including heat recovery] for mechanical ventilation systems intended for a single room.*

Other standards are in preparation. Full details are given in Table 1.6 and section 4 of the Approved Document.

11.5.7   Compliance for dwellings with basements

Guidance depends on whether or not the basement is connected to the dwelling by a large permanent opening. In this context, a large permanent opening would typically be an open stairway.

A basement connected by a large permanent opening to the rest of the dwelling above ground

Follow the guidance in sections 11.5.2 to 11.5.6 for a multi-storey dwelling without a basement. If the basement has only a single façade while the rest of the dwelling has more than one exposed façade, systems 3 and 4 are preferred, and systems 1 and 2 should only be used with expert advice.

A basement that is not connected by a large permanent opening to the rest of the dwelling above ground

Consider the above-ground dwelling and the below-ground basement separately:

- For the part of the dwelling above ground, ventilate as for a dwelling without a basement, sections 11.5.2 to 11.5.6. If this part has no bedrooms, assume it has one bedroom when determining rates from Tables 11.2 and 11.3.
- For the basement, ventilate as for a single-storey dwelling, sections 11.5.2 to 11.5.6. If this basement has no bedrooms, assume it has one bedroom when determining rates from Tables 11.2 and 11.3.

For a dwelling that comprises a basement only, use the guidance for a single-storey dwelling above ground.

11.5.8 Ventilation of a habitable room through another room

A habitable room not containing openable windows may be ventilated through another habitable room provided there is adequate provision for purge ventilation and background ventilation for both rooms. This requires all three of the following:

- a permanent opening (or openings) between the two rooms of minimum area one tenth or one twentieth of the total floor area of both rooms, calculated according to the rules for purge ventilation in section 11.5.3;
- one or more ventilation openings in the outer room of minimum area one tenth or one twentieth of the total floor area of both rooms, also calculated according to the rules for purge ventilation in section 11.5.3;
- a background ventilator between the outer room and the outside of at least 8000 mm^2 equivalent area.

See Fig. 11.7.

11.5.9 Ventilation of a habitable room through a conservatory

A habitable room not containing openable windows may be ventilated through a conservatory provided there is adequate provision for purge ventilation and background ventilation for both rooms. This requires all four of the following:

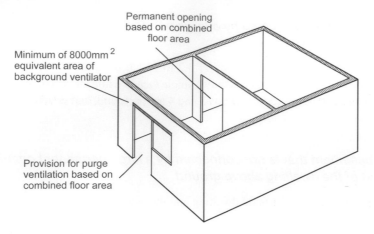

Fig. 11.7   Ventilation through a habitable room.

- an opening (or openings), which must be closable, between the room and the conservatory of minimum area one tenth or one twentieth of the total floor area of both spaces, calculated according to the rules for purge ventilation in section 11.5.3;
- a background ventilator between the room and the conservatory of at least $8000 \, mm^2$ equivalent area;
- one or more ventilation openings in the conservatory of minimum area one tenth or one twentieth of the total floor area of both rooms, also calculated according to the rules for purge ventilation in section 11.5.3;
- a background ventilator between the conservatory and the outside of at least $8000 \, mm^2$ equivalent area.

See Fig. 11.8.

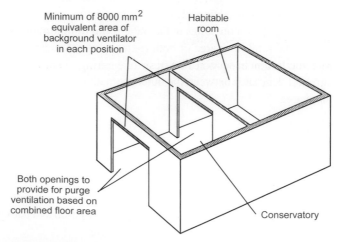

Fig. 11.8   Ventilation through a conservatory.

11.5.10 Example calculations

Appendix C of AD F gives detailed example calculations.

11.6   The ventilation of buildings other than dwellings

The guidance in the Approved Document is grouped into:

- offices;
- car parks;
- all other building types.

However, compared to dwellings, there is relatively little specific guidance in the Approved Document itself. Most of the detailed guidance is to be found in other publications to which AD F refers.

11.6.1 General requirements for all building types

The ventilation provision in all buildings should take account of:

- cooling;
- protection of fresh air supplies;
- the danger of legionella contamination;
- the health implications of recirculated air;
- access for maintenance.

Cooling

The ventilation provisions of AD F do not include an allowance for cooling to remove excess heat. This is covered in AD L2A.

Protection of fresh air supplies

Provision should be made to protect the fresh air entering a ventilation system from harmful contaminants. See section 11.11.

Legionella

Measures for avoiding the danger of legionella contamination are given in:

- *Legionnaires' disease – the control of legionella bacteria in water systems*, HSE (paragraphs 79 to 144);
- *Minimising the risk of Legionnaires' disease*, CIBSE TM 13.

Recirculated air

Information and guidance is given in:

- *Approved code of practice and guidance L24*, Workplace (Health, safety and welfare) Regulations 1992, HSE (paragraph 32).

Access for maintenance

In order to allow for essential maintenance, it is necessary to provide:

- access for the purpose of replacing filters, fans and coils;
- access points for cleaning ductwork;
- adequate space in a central plant room for the maintenance of plant.

The minimum space requirements in a central plant room are:

- for passageways, 2 m wide by 0.6 m high;
- for crawl spaces, 1.4 m high by 1.1 m wide, with equipment requiring attention ideally not less than 0.69 m above floor level.

Dimensions greater than these minima may be necessary for opening access doors, etc.

Commissioning

Commissioning is dealt with in Approved Document L2A. There is no additional requirement in AD F.

11.6.2   Extract ventilation rates

The Approved Document gives minimum extraction rates applicable primarily to offices but which may also be applied to other buildings where there are similar activities or similar types and levels of contaminant release. Minimum rates are specified for individual rooms and for the whole building, as shown in Table 11.5.

11.6.3   Offices

The Approved Document identifies four methods of complying with the requirement:

- by providing specified extract rates, whole-building ventilation rates and purge ventilation, or
- by following guidance on the design of the ventilation systems, or
- by using approaches as described in other authoritative and acceptable documents, or
- by demonstrating that the proposed ventilation systems can satisfy the requirement by meeting specified moisture and air quality criteria.

Table 11.5   Minimum ventilation rates.

Room	Minimum room extract rate
Rooms containing printers and photocopiers in substantial use (more than 30 minutes per hour)	20 ls^{-1} per machine during use If the operator(s) is in the room continuously, use the greater of the extract rate and the whole building ventilation rate
Office sanitary accommodation and washrooms	Intermittent air extract rate of: 15 ls^{-1} per shower or bath 6 ls^{-1} per WC or urinal
Food and beverage preparation areas, but not commercial kitchens	Intermittent air extract rate of: 15 ls^{-1} with microwave ovens and beverages only 30 ls^{-1} adjacent to a hob with cooker(s) 60 ls^{-1} elsewhere with cooker(s) All extract should operate while food and/or beverage preparation is in progress
Specialist buildings and spaces, including commercial kitchens and sports centres	See section 11.6.4
Whole building	**Minimum outdoor air supply rate**
Total outdoor air supply rate – no smoking or significant pollutant sources	10 ls^{-1} per person

Although these are presented as four independent approaches, they are to some degree linked.

Specification of extract rates and whole-building ventilation rate

The following, which are based on the control of body odours and when no other significant pollutants are present, should be provided:

- **Extract rates:** The minimum extract rates are given in Table 11.5.
- **Whole-building ventilation rate:** The minimum supply rate of outdoor air is given in Table 11.5. This assumes there is no smoking.
- **Purge ventilation:** Purge ventilation is required in each office, and should be sufficient to reduce pollutants to an acceptable level before the space is occupied. Purged air must be exhausted to the outside and not recirculated.

If there are significant levels of other pollutants, these rates will have to be increased in accordance with an approved calculation method such as that given in CIBSE Guide A.

Compliance by following guidance on ventilation system design

For all systems, the minimum extract rates and the minimum whole-building ventilation rate as given in Table 11.5 should be provided. The three types of ventilation system for achieving this are:

- natural ventilation;
- mixed-mode ventilation;
- mechanical ventilation.

NATURAL VENTILATION – Approved guidance is given in:

- *Natural ventilation in non-domestic buildings* CIBSE Application Manual AM 10:2005.

This provides guidance on:

- ventilation provisions;
- location of ventilators in rooms.

It also provides guidance on controls for ventilators in rooms, including:

- extract;
- whole-building ventilation;
- purge ventilation.

In addition, the following should also be noted:

- Extract ventilators should be located as high as practicable and preferably less than 400 mm below ceiling level. This helps to remove pollutants from the breathing zone and improves the extraction of buoyant pollutants and water vapour.
- Passive stack ventilation (PSV) may be used instead of a mechanical extract fan to office sanitary accommodation, washrooms and food preparation areas.
- PSV extract terminals should be located in the ceiling.
- When an open-flued appliance is provided in a building with mechanical extract, spillage of flue gases must be avoided whether or not the extract fan is running. See Approved Document J for guidance.
- Controls for extract fans may be manual or automatic. For a room with no openable window, the extract fan should have a 15 minute over-run.
- Controls for PSV may be manual and/or operated automatically by a sensor or controller.
- Readily accessible controls must be provided, particularly over-ride controls for the occupants.

MECHANICAL VENTILATION

A mechanical system must satisfy the following:

- The minimum ventilation rates should be as given in Table 11.5.
- Extract ventilators should be located as high as practicable and preferably less than 400 mm below ceiling level. This helps to remove pollutants from the breathing zone and improves the extraction of buoyant pollutants and water vapour.

- Controls for extract fans may be manual or automatic. For a room with no openable window, the extract should have a 15 minute over-run.
- Readily accessible controls must be provided, particularly over-ride controls for the occupants.

ALTERNATIVE APPROACHES

Acceptable alternative approaches include following the recommendations in:

- *Mixed mode ventilation* CIBSE Application Manual AM 13: 2000;
- CIBSE Guide A and CIBSE Guide B2.

11.6.4   Other building types

With the exception of common spaces and sanitary accommodation, guidance for all other building types is to be found not in AD F but in other publications. These are identified in Table 11.6, and the publications themselves are as follows:

(1) CIBSE Guide B2: 2001
(2) *The Welfare of Farm Animals (England) Regulations*, SI 2000 No. 1870
(3) *The Welfare of Farm Animals (England) (Amendment) Regulations*, SI 2002 No. 1646 and SI 2003 No. 299
(4) *Buildings and structures for agriculture*, BS 5502
(5) *General ventilation in the workplace – guidance for employers*, HSE guidance note HSG 202
(6) *Specification for safety aspects in the design, construction and installation of refrigeration appliances and systems*, BS 4434: 1989
(7) *Ventilation of kitchens in catering establishments*, HSE catering information sheet No. 10, 2000
(8) *The main health and safety law applicable to catering*, HSE information sheet No. 11, 2000
(9) *Designing energy efficient multi-residential buildings*, Energy efficiency best practice in housing, good practice guide GPG 192
(10) *Court standards and design guide*, Department of Constitutional Affairs, 2004
(11) Factories Act
(12) Health and safety at work etc. Act
(13) National Health Service database
(14) Health technical memorandum (HTM) 03
(15) Health building notes (HBN), various
(16) *Manual of recommended practice, Industrial ventilation*, 24th edition, ACGIH
(17) *An introduction to local exhaust ventilation*, HS(G) 37
(18) *Maintenance, examination and testing of local exhaustion ventilation*, HS(G) 54
(19) *COSHH essentials*, HS(G) 193
(20) BS 5454: 2000
(21) National offender management service (NOMS), Technical services, Home Office
(22) *Ventilation of school buildings*, Building Bulletin 101, DfES, and Education (school premises) Regulations

Table 11.6 Sources of guidance for other building types.

Building type	Source of guidance																					
	1	2	3	4	5	6	7	8	9	10	11	12	13	14	15	16	17	18	19	20	21	22
Animal husbandry	✓	✓	✓	✓																		
Assembly halls	✓																					
Atria	✓																					
Broadcasting studios	✓																					
Building services plant rooms					✓																	
Call centres	✓																					
Catering, including commercial kitchens	✓						✓	✓														
Cleanrooms	✓																					
Communal residential buildings	✓								✓													
Computer rooms	✓																					
Courtrooms										✓												
Darkrooms – photographic	✓																					
Dealing rooms	✓																					
Factories and warehouses	✓										✓	✓										
High-rise non-domestic buildings	✓																					

Source of guidance

Building type	1	2	3	4	5	6	7	8	9	10	11	12	13	14	15	16	17	18	19	20	21	22
Horticulture	✓																					
Hospitals and healthcare buildings	✓												✓	✓	✓							
Hotels	✓																					
Industrial ventilation																✓	✓	✓	✓			
Laboratories	✓																			✓		
Museums, libraries and art galleries																						
Plant rooms	✓																				✓	
Prison cells																						
Schools and educational buildings																						✓
Shops and retail premises	✓																					
Sports centres and swimming pools	✓																					
Standards rooms	✓																					
Transportation buildings and facilities	✓																					

11.6.5   Car parks

In car parks, the objective is to limit the concentration of carbon monoxide (CO) arising from vehicle exhausts. There are limits on both the long-term average and the short-term peak concentrations as follows:

- **maximum permissible long-term concentration of CO:** 30 parts per million averaged over an 8-hour day;
- **maximum permissible short-term concentration of CO:** 90 parts per million averaged over any 15-minute period.

The short-term concentrations are most likely to occur in the vicinity of ramps and exits.
   Ventilation systems must be designed and must operate in such a way as to keep concentrations within the permissible limits. Compliance may be demonstrated by showing by means of appropriate calculations that the limits will not be exceeded. If this is impractical or inappropriate, compliance may be demonstrated by following the guidance on natural or mechanical ventilation.

Naturally ventilated car parks

Well distributed permanent natural ventilation must be provided. At each and every car park level openings must:

- have an aggregate equivalent area of at least one twentieth of floor area at that level;
- be distributed so that at least 25% of the opening area is on each of two opposing walls.

Mechanically ventilated car parks

Either

- provide permanent natural ventilation openings of equivalent floor area at least one fortieth of the floor area, and also a mechanical ventilation system capable of extracting not less than 3 ach (air changes per hour); or
- for basement car parks, provide a mechanical ventilation system capable of extracting not less than 6 ach (air changes per hour).

In addition to whichever of these alternative is chosen, at ramps and exits provide local ventilation capable of extracting not less than 10 ach (air changes per hour).
   Note that Approved Document B deals with the ventilation of car parks for the purpose of fire risk management.
   Further guidance is in:

- *Code of practice for ground floor, multi-storey and underground car parks*, Association of petroleum and explosives administration;
- CIBSE Guide B2.

11.7   Work on existing buildings

If a building is subject to material change of use, then Part F applies to the building or that part of the building which has been subjected to a change of use. For all other cases of work on an existing building:

- Part F applies to the building work which has been carried out, and
- the rest of the building should not be made worse in relation to the requirements of Part F than it was before the work was carried out.

11.7.1   Replacement windows and background ventilation

It should be remembered that windows are a controlled fitting. This means that:

- In addition to complying with Part F, the replacement work should comply with Parts L and N.
- Once completed, the building work should not result in a worse level of compliance with other applicable Parts of Schedule 1 than existed before the work began, particularly Parts B and J.

If the original windows were fitted with trickle ventilators, the replacement windows should also include them. They should be sized as set out in Table 11.7. Even if the original windows did not have trickle ventilators, it would still be good practice to fit them (or an equivalent device) in all replacement windows in order to improve air quality and to assist in the control of condensation. Such ventilators should be provided with accessible controls.

When windows are replaced as part of a material change of use, the relevant guidance for new buildings should be followed, i.e. section 11.5 for dwellings, or section 11.6 for buildings other than dwellings.

Table 11.7   Minimum background ventilation provision.

Building type	Room type	Minimum equivalent area of opening
Dwellings	Habitable rooms	5000 mm²
	Kitchen, utility room and bathroom (with or without WC)	2500 mm²
Buildings other than dwellings	Occupiable rooms with floor area up to 10 m²	2500 mm²
	Occupiable rooms with floor area greater than 10 m²	250 mm² per square metre of floor area
	Kitchens (domestic type)	2500 mm²
	Bathrooms and shower rooms	2500 mm²
	Sanitary accommodation and/or washing facilities	2500 mm² per WC

In all cases, the ventilation opening should be controllable and be at least the same size as that originally provided. Where the size of any existing ventilation opening is not known, or if there was no ventilation opening, the minimum area as shown in Table 11.7 should be provided. As a general rule, a night latch setting on a window is not considered acceptable as an alternative to a trickle ventilator. However, exceptions to this rule are possible if the window cannot reasonably accommodate a trickle ventilator, either because of its design or because it is small.

11.7.2   Addition of a habitable room that is not a conservatory to an existing dwelling

The general ventilation rate for the additional room and if necessary adjoining rooms can be achieved by either:

- providing background ventilators; or
- providing a single-room heat-recovery ventilator in the additional room.

When background ventilators are provided, there are three possibilities:

- If the additional room is connected to an existing habitable room which now has no windows opening to the outside, the guidance in section 11.5.8 must be followed.
- If the additional room is connected to an existing habitable room which still has windows opening to the outside but has a total background ventilator equivalent area less than 5000 mm², the guidance in section 11.5.8 must be followed.
- If the additional room is connected to an existing habitable room which still has windows opening to the outside and has a total background ventilator equivalent area of at least 5000 mm², then there should be:
  ○ background ventilators of at least 8000 mm² equivalent area between the rooms, and also
  ○ background ventilators of at least 8000 mm² equivalent area between the additional room and the outside.

When a single-room heat-recovery ventilator is provided in the additional room, the fresh air supply rate to that room is found from:

$$\text{Supply rate} = \frac{\text{Volume of additional room}}{\text{Total volume of all habitable rooms}} \times \text{whole building ventilation rate}$$

where the whole building ventilation rate is obtained from Table 11.2.

The ventilation provision in the additional room must also comply with the requirements for purge ventilation, location of ventilation devices and controls as described in sections 11.5.4 to 11.5.6 inclusive.

11.7.3   Addition of a wet room to an existing dwelling

Whole building and extract ventilation

There are four alternatives:

(1) Provide intermittent extract as described for intermittent extract in section 11.5.2 for system 1, using data from Table 11.3.
(2) Provide a single-room heat-recovery ventilator as described for intermittent extract in section 11.5.2 for system 1.
(3) Provide passive stack ventilation as described for passive stack ventilation in section 11.5.2 for system 2, using data from Table 11.4.
(4) Provide a continuous extract fan as described in section 11.5.2 for system 3.

The ventilation provision in the wet room must also comply with the requirements for purge ventilation, location of ventilation devices and controls as described in sections 11.5.4 to 11.5.6 inclusive.

11.7.4   Addition of a conservatory

Conservatories up to and including 30 m² floor area are exempt from the regulations. Above 30 m² floor area, the general ventilation of the conservatory and, if relevant or necessary, adjoining rooms can be achieved with background ventilators. Follow the guidance in section 11.5.9 whatever the ventilation provisions in the existing room adjoining the conservatory.

The ventilation provision in the conservatory must also comply with the requirements for purge ventilation, location of ventilation devices and controls as described in sections 11.5.4 to 11.5.6 inclusive.

11.7.5   Historic buildings

No specific requirements are given in AD F. However, it is stated that the aim should be to improve ventilation to the extent that it is necessary without:

- prejudicing the character of the building;
- making alterations that risk increasing the long-term deterioration of the building.

Further details and references are given in Chapter 16, sections 16.7.1.2 and 16.9.1.2.

11.8   Performance based ventilation

Appendix A of AD F details the assumptions used to establish the guidance given in the main body of the Approved Document and its tables. Data is given for dwellings and for buildings other than dwellings. If, in a particular building project, there is the possibility that these assumptions are not applicable, particularly if pollutant concentrations may be exceeded or if there are significant sources of other pollutants, then special ventilation provision will have to be made.

11.8.1 Dwellings

The primary assumptions are in terms of three criteria:

- that there should be no visible mould on external walls due to excessive levels of humidity in the indoor air; and
- that bio-effluents (i.e. body odours) will be controlled by a fresh air supply rate not less than 3.5 ls^{-1} per person, and
- the maximum permissible concentrations of certain common indoor pollutants are not exceeded.

For convenience, the maximum permissible pollutant concentrations for dwellings are listed in Table 11.8.

11.8.2 Buildings other than dwellings

The primary assumptions are in terms of three criteria:

- In the absence of tobacco smoke or other excessive pollutants, a fresh air supply rate of 10 ls^{-1} will be sufficient to avoid health effects and to control bio-effluents (body odour).
- That there should be no visible mould on external walls due to excessive levels of humidity in the indoor air.
- The maximum permissible concentrations of certain common indoor pollutants are not exceeded.

For convenience, the maximum permissible pollutant concentrations for buildings other than dwellings are listed in Table 11.9.

Table 11.8 Maximum concentrations of pollutants – dwellings.

Pollutant	Maximum concentration	Averaging time
Nitrogen dioxide	288 µg/m³ (150 ppb)	1 hour
Nitrogen dioxide	40 µg/m³ (20 ppb)	Long-term average
Carbon monoxide	100 mg/m³ (90 ppm)	15 minutes
Carbon monoxide	60 mg/m³ (50 ppm)	30 minutes
Carbon monoxide	30 mg/m³ (25 ppm)	1 hour
Carbon monoxide	10 mg/m³ (10 ppm)	8 hours
Total organic volatile compounds (TOVC)	300 µg/m³	8 hours

Table 11.9 Maximum concentrations of pollutants – buildings other than dwellings.

Pollutant	Maximum concentration	Averaging time
Nitrogen dioxide	288 µg/m³ (150 ppb)	1 hour
Nitrogen dioxide	40 µg/m³ (21 ppb)	Long-term average
Carbon monoxide	100 mg/m³ (90 ppm)	15 minutes
Carbon monoxide	60 mg/m³ (50 ppm)	30 minutes
Carbon monoxide	30 mg/m³ (25 ppm)	1 hour
Carbon monoxide	10 mg/m³ (10 ppm)	8 hours
Carbon monoxide for occupational exposure	35mg/m³ (30 ppm)	8 hours
Total organic volatile compounds (TOVC)	300 µg/m³	8 hours
Ozone	100 µg/m³	At any time

11.9 Passive stack ventilation system design

The main recommendations are as follows.

Figure 11.3 shows the design layouts which should be suitable for the majority of dwellings of up to four storeys. It should be noted that:

- The preferred position for the outlet terminal is the ridge of the roof.
- A tile ventilator is acceptable if it is within 0.5 m of the ridge.
- A duct that penetrates the roof must extend to at least ridge level.
- Ducts from kitchens, bathrooms utility rooms or WC's must be separate, with no branches or common outlets.
- Each duct should have a maximum of two bends, each with a maximum change of direction of 45° or less, and the bends should be swept and not sharp.
- If the dwelling is near a much taller building (i.e. more than 50% taller), PSV should *not* be installed if the distance between them is less than five times the difference in height.

Recommendations for the components of the PSV system are:

- Ceiling extract grilles and all outlet terminals should have a free area not less than the duct cross-sectional area.
- Rigid and flexible ducts are acceptable – flexible ducts should be fully extended but not taut.
- Duct cross-sectional areas should be maintained throughout their run, especially at junctions with terminals, etc.
- Any part of a duct in an unheated space must be insulated with at least 25 mm of insulation with a thermal conductivity of 0.04 $Wm^{-1}K^{-1}$, or material giving an equivalent degree of insulation.
- If a duct extends above roof level, it should be either insulated or fitted with a condensation trap just below roof level.

- The outlet terminal should:
  - ○ not allow ingress of birds or large insects;
  - ○ not allow rain to enter the duct and into the dwelling;
  - ○ ensure that any condensate due to moisture in the outward airflow runs off onto the roof and not into the dwelling.
- Roof terminal design is particularly important, and careful attention should be paid to their selection.

Other points to be noted are:

- Fire precautions – see Approved Document B.
- Noise – in noisy areas, sound attenuation in the duct may be necessary or at least desirable.

11.10 Installation of fans in dwellings

Appendix E of AD F gives some general advice on the suitability of different fan types and their installation. However, in addition, manufacturer's technical data and installation instructions should be carefully examined.

11.11 Ingress of external pollution

Appendix F of AD F gives guidance on strategies for minimising the intake of external air pollutants into a building. The basic principle is to locate the air intake points as far as is possible from the pollutant source or sources. In urban areas where there are often multiple sources, including traffic and industry, the solution depends on the exact location of the building, not only in relation to the sources but in relation to other buildings. Exhaust outlets must also be carefully located to avoid pollution near pedestrians and to avoid short-circuiting to ventilation intakes. Two sources of information are recommended:

- *Minimising pollution at air intakes*, Technical Memorandum TM 21, CIBSE;
- Liddament, M. W., Chapter 13 'Ventilation strategies' in *Indoor air quality handbook*, McGraw-Hill, 2000.

12 Hygiene (Part G)

12.1 Introduction

When first introduced, Part G of Schedule 1 to the Building Regulations 1985 consisted of four requirements grouped under the title 'Hygiene'.

The first of these requirements (G1 – Food storage) required that dwellings be provided with adequate food storage accommodation. Since most people have refrigerators or deep freezers today this regulation has become outdated and food storage is no longer controlled by the building regulations.

Consequently, the former regulation G4 (Sanitary conveniences and washing facilities) was renumbered G1 in the 1991 Regulations to fill the void left by the now defunct food storage requirements. This convention has been continued in the 2000 Regulations

The remaining regulations in Part G relate to Bathrooms (G2) and Hot water storage (G3).

12.2 Sanitary conveniences and washing facilities

Adequate sanitary conveniences (i.e. closets and urinals) situated in purpose-built accommodation or bathrooms, must be provided in buildings. This requirement replaces section 26 of the Building Act 1984.

Additionally, adequate washbasins with suitable hot and cold water supplies must be provided in rooms containing water-closets or in adjacent rooms or spaces.

These sanitary conveniences and washbasins must be separated from places where food is prepared and must be designed and installed so that they can be cleaned effectively.

It may be noted that section 66 of the 1984 Act enables the local authority to serve a notice on an occupier requiring him to replace any closet provided for his building which is not a water-closet. The notice can only be served where the building has a sufficient water supply and a sewer available. Where a notice requiring closet conversion is served, the local authority must bear half the cost of carrying out the work.

A satisfactory level of performance will be achieved if:

- sufficient numbers of the appropriate type of sanitary convenience are provided depending on the sex and age of the users of the building
- washbasins with hot and cold water supply are provided either in or adjacent to rooms containing water-closets.

Both sanitary conveniences and washbasins should be sited, designed and installed so as not to be a health risk.

12.2.1   Provision of sanitary conveniences and washbasins

The following definitions apply in AD G1.

SANITARY CONVENIENCE – Closets and urinals.

SANITARY ACCOMMODATION – A room containing closets or urinals. Other sanitary fittings may also be present. Sanitary accommodation containing more than one cubicle may be treated as a single room provided there is free air circulation throughout the room.

WATER-CLOSET – Defined by section 126 of the Building Act 1984 as a closet which has a separate fixed receptacle connected to a drainage system and separate provision for flushing from a supply of clean water, either by the operation of mechanism or by automatic action.

AD G1 also permits the use of a chemical or other means of treatment where drains and water supply are not available. It is not clear whether earth-closets would be permitted, but on normal principles of interpretation it is unlikely that they would be.

Houses, flats and maisonettes should have at least one closet and one washbasin. This also applies to houses in multiple occupation (houses where the occupants are not part of a single household), if the facilities are available for the use of all the occupants.

In other types of buildings the scale of provision and the siting of appliances may be the subject of other legislation as follows:

● Workplace (Health, Safety & Welfare) Regulations 1992. Approved Code of Practice & Guidance.

(This document is not referred to in the current edition of ADGI because it was published after that document. However, it repeals those provisions of the Offices, Shops and Railway Premises Act 1963 and the Factories Act 1961 which are referred to in ADGI.)

● The Food Hygiene (General) Regulations 1970
● Part M of Schedule 1 to the 2000 Regulations (Access and facilities for disabled people).

The requirement to provide satisfactory sanitary conveniences can also be met, subject to other legislation, by referring to the relevant clauses of BS 6465 *Sanitary installations*, Part 1: 1984 which contains details of the scale of provision, selection and installation of sanitary appliances.

A room or space containing closets or urinals should be separated by a door from any area in which food is prepared or washing up done. AD G1 makes it clear, therefore, that a separate lobby is not required.

Additionally, washbasins should be placed:

- in the room containing the closet; or
- in the room or space immediately leading to the room containing the closet provided it is not used for food preparation; or
- in the case of dwellings, in the room or space adjacent to the room containing the closet.

In this last case it is unclear whether or not the space may be used for the preparation of food, but in all probability it is not.

Closets, urinals and washbasins should have smooth, readily-cleaned, non-absorbent surfaces.

Any flushing apparatus should be capable of cleansing the receptacle effectively. The receptacle should only be connected to a flush pipe or branch discharge pipe.

Any washbasins required by the provisions of regulation G4 should have a supply of hot water from a central source or unit water heater and a piped cold water supply.

12.2.2   Discharge from sanitary conveniences and washbasins

Water-closets should discharge via a trap and branch pipe to a soil stack pipe or foul drain.

In recent years a system of waste disposal has been developed in which the discharge from a waste appliance is fed into a macerator. The liquified contents are then pumped via a small bore pipe to the normal foul drainage system. A closet is permitted to be connected to such a system provided:

- a closet discharging directly to a gravity system is also available; and
- the macerator system is the subject of a current European Technical Approval issued by a member body of the European Organisation for Technical Approvals e.g. the British Board of Agrément. The conditions of use must be in accordance with the terms of the ETA.

Urinals which are fitted with flushing apparatus should have an outlet fitted with an effective grating and trap and should discharge via a branch pipe to a soil stack pipe or foul drain (see Approved Document H1 and Chapter 13 for details of drainage).

Washbasins should discharge via a trap and branch discharge pipe to a soil stack. If on the ground floor, it is permissible to discharge the basin to a gulley or direct to a drain.

12.3   Bathrooms

Dwellings are required to be provided with a bathroom containing a fixed bath or shower. Hot and cold water must also be supplied to the bath or shower. This requirement replaces section 27 of the Building Act 1984.

The foregoing requirements apply to dwellings (i.e. houses, flats and maisonettes) and houses in multiple occupation (houses where the occupants are not part of a single household). In the latter case the facility should be available to all the occupants.

The hot and cold water supplies should be piped to the bath or shower and hot water may come from a central source such as a hot water cylinder or from a unit water heater.

The discharge from the bath or shower should be via a trap and waste pipe to a gulley, soil stack pipe or foul drain direct (see Approved Document H1 and Chapter 13 for details of drainage).

A bath or shower may be connected via a macerator system provided it complies with a current European Technical Approval.

12.4   Hot water storage

A hot water storage system incorporating a hot water storage vessel which is not vented to the atmosphere must be installed by a competent person and adequate precautions must be taken to:

- prevent the water temperature exceeding 100°C; and
- ensure that any hot water discharged from safety devices is conveyed safely to a disposal point where it is visible but will not be a danger to users of the building.

The above requirements do not apply to space heating systems, systems which heat or store water for industrial processes and systems which store 15 litres or less of water.

Approved Document G3 describes the provisions for an unvented hot water storage system. In such a system, the stored hot water is heated in a closed vessel. Without adequate safety devices an uncontrolled heat input would cause the water temperature to rise above the boiling point of water at atmospheric pressure (100°C). At the same time the pressure would increase until the vessel burst. This would result in an almost instantaneous conversion of water to steam with the large increase in volume producing a steam explosion.

Water for domestic use is required at temperatures below 100°C, therefore, an explosion cannot occur if the water is released at these temperatures, however great the pressure, hence the precautions required by regulation G3 to prevent the water temperature exceeding 100°C.

The term 'domestic hot water' is defined in AD G3 as water which has been heated for washing, cooking and cleaning purposes. The term is used irrespective of the type of building in which an unvented hot water storage system is installed.

Figure 12.1 illustrates the three independent levels of protection which should be provided for each source of energy supply to the stored water. These are:

- Thermostatic control (see Part L, Chapter 16).
- Non self-resetting thermal cut-outs to BS 3955: 1986 (electrical controls) or BS 4201: 1979 (for gas burning appliances).
- One or more temperature operated relief valves to BS 6283 *Safety devices for use in hot water systems* Part 2: 1991 or Part 3: 1991.

These safety devices are required for both directly and indirectly heated unvented hot water storage systems. The safety devices are designed to work in sequence as the temperature rises. All three means of protection would have to fail for the water temperature to exceed 100°C.

AD G3 provides separate recommendations for smaller (usually domestic) systems (not exceeding 500 litres capacity with a heat input below 45 kW) in section 3. Systems which exceed 500 litres capacity or have a heat input in excess of 45 kW are dealt with in section 4.

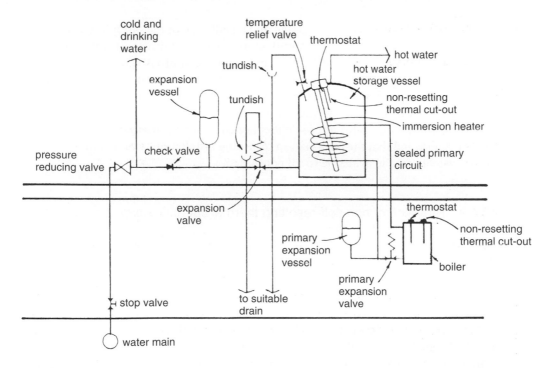

Fig. 12.1   Directly and indirectly heated unvented hot water storage system.

12.4.1   Section 3 hot water storage systems

Generally, a system covered by section 3 of AD G3 should be in the form of a unit or package which is:

- approved by a member body of the European Organisation for Technical Approvals (EOTA) operating a technical approvals scheme which ensures that the relevant requirements of regulation G3 will be met (e.g. the British Board of Agrément); or
- approved by a certification body having accreditation from the National Accreditation Council for Certification Bodies (NACCB). This would include testing to the requirements of an appropriate standard to ensure compliance with regulation G3 (e.g. BS 7206: 1990 *Specification for unvented hot water storage units and packages*); or
- independently assessed to clearly demonstrate an equivalent level of verification and performance to those above.

This means that the system should be factory made and supplied either as a *unit* (fitted with all the safety protection devices mentioned above and incorporating any other operating devices to stop primary flow, prevent backflow, control working pressure, relieve excess pressure and accommodate expansion fitted to the unit by the manufacturer) or as a *package* in which the safety devices are fitted by the manufacturer but the operating devices are supplied in kit form to be fitted by the installer.

This approach ensures that the design and installation of the safety and operating devices are carried out by the manufacturer who is conversant with his own equipment and can control the training and supervision of his staff.

It should be noted that where a system is subject to the above approvals it is unlikely to need site inspection by the building control authorities.

This may not be the case in other situations.

The recommendations for approval mentioned above ensure that the system is fit for its purpose and that the information regarding installation, maintenance and use of the system is made available to all concerned.

12.4.2   Provision of non self-resetting thermal cut-outs

Storage systems may be heated directly or indirectly. In an unvented, indirectly heated system (see Fig. 12.1) the non self-resetting thermal cut-out should be wired up to a motorised valve or other approved device or should shut off the flow to the primary heater. These devices should be subject to the same approvals as the units or packages.

Sometimes a unit system may incorporate a boiler. In this case the thermal cut-out may be located on the boiler.

In many cases an indirect system will also contain an alternative direct method of water heating (such as an immersion heater). This alternative heating source will also need to be fitted with a non self-resetting thermal cut-out. The non self-resetting thermal cut-out should be connected to the direct heat source or the indirect primary flow control device in accordance with BS 7671 Requirements for Electrical Installations. IEE Wiring Regulations.

12.4.3   Provision of temperature relief valves

Whether the unit or package is directly or indirectly heated the temperature relief valve should be situated directly on the storage vessel in order to prevent the stored water exceeding 100°C.

BS 6283 requires that each valve be marked with a discharge rating (in kW). This rating should never be less than the maximum power input to the vessel which the valve protects. More than one valve may be needed.

Valves should also comply with the following.

- They should not be disconnected except for replacement.
- They should not be relocated in any other position.
- The valve connecting boss should not be used to connect any other devices or fittings.
- They should discharge through a short length of metal pipe (D1) which is of at least the same bore as the valve's nominal outlet size.

    The discharge should either be direct or by way of a manifold which is large enough to take the total discharge of all the pipes connected to it. It should then continue via an air break to a tundish which is located vertically as near as possible to the valve.

It may be possible to provide an equivalent degree of safety using other safety devices but these would need to be assessed in a similar manner to units or packages (see above).

12.4.4   Installation

The installation of the system should be carried out by a competent person.

This means a person who holds a current Registered Operative Identity Card for the installation of unvented domestic hot water storage systems issued by:

- the Construction Industry Training Board (CITB); or
- the Institute of Plumbing; or
- the Association of Installers of Unvented Hot Water Systems (Scotland and Northern Ireland); or
- designated Registered Operatives who are employed by companies included on the list of Approved Installers published by the BBA up to 31 December 1991; or
- an equivalent body.

12.4.5   Discharge pipes

Discharge pipe D1 (see above and Fig. 12.2) is usually supplied by the storage system manufacturer. (If not it should be fitted by the installer of the system.)

In either case the tundish should be:

- vertical;
- located in the same space as the unvented hot water storage system; and
- fitted within 500 mm of the safety device.

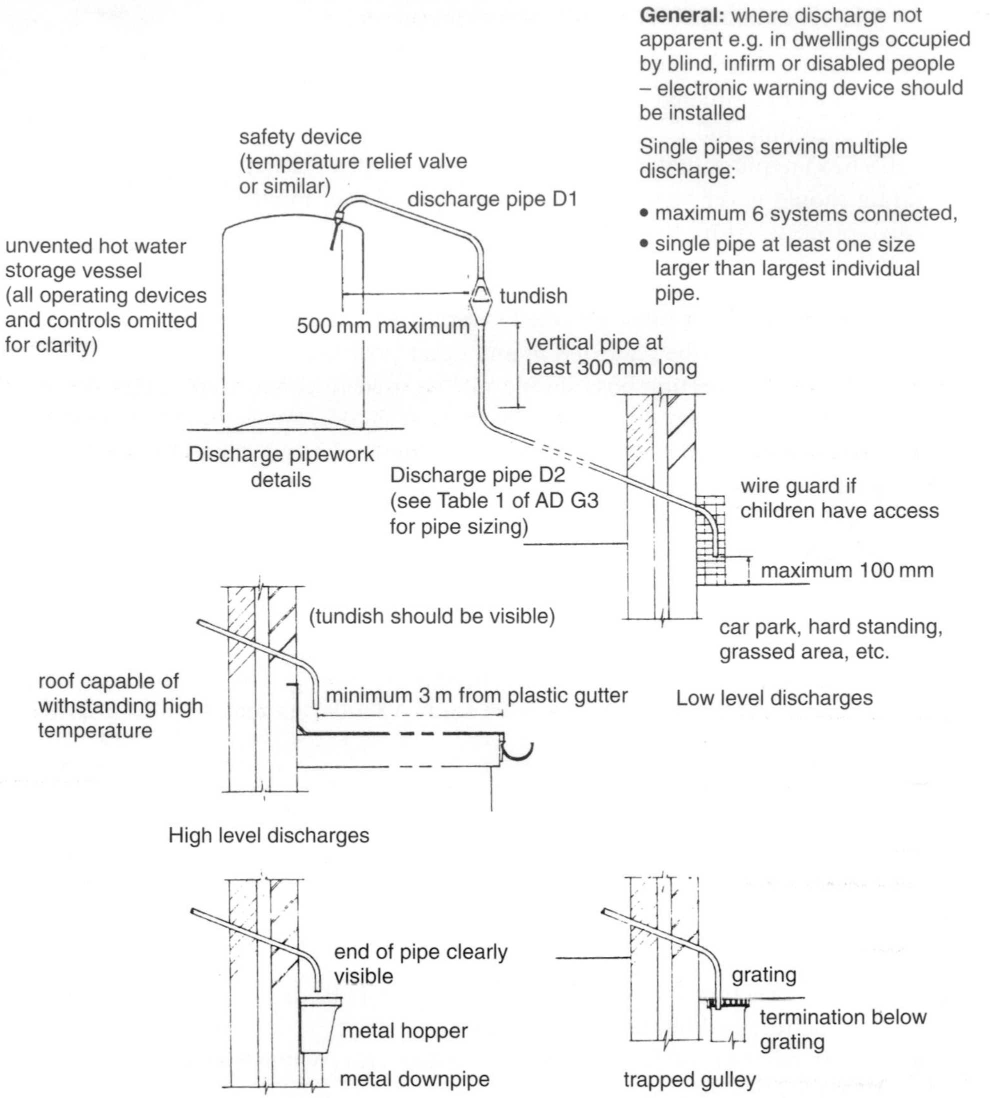

General: where discharge not apparent e.g. in dwellings occupied by blind, infirm or disabled people – electronic warning device should be installed

Single pipes serving multiple discharge:

- maximum 6 systems connected,
- single pipe at least one size larger than largest individual pipe.

safety device (temperature relief valve or similar)

discharge pipe D1

unvented hot water storage vessel (all operating devices and controls omitted for clarity)

tundish

500 mm maximum

vertical pipe at least 300 mm long

Discharge pipework details

Discharge pipe D2 (see Table 1 of AD G3 for pipe sizing)

wire guard if children have access

maximum 100 mm

(tundish should be visible)

car park, hard standing, grassed area, etc.

roof capable of withstanding high temperature

minimum 3 m from plastic gutter

Low level discharges

High level discharges

end of pipe clearly visible

metal hopper

metal downpipe

grating

termination below grating

trapped gulley

Fig. 12.2   Discharge pipes.

Discharge pipe D2 (see Fig. 12.2) from the tundish should comply with the following:

- it should terminate in a safe place where it cannot present a risk of contact to users of the building;
- it should be laid to a continuous fall;
- the discharge should be visible at either tundish or final outlet, but preferably at both of these locations (Fig. 12.2 shows possible discharge arrangements);

- it should be at least one pipe size larger than the nominal outlet size of the safety device unless its total equivalent hydraulic resistance exceeds that of a straight pipe 9 m long (bends will increase flow resistance, therefore Table 1 from AD G3 is reproduced below and shows how to calculate the minimum size of discharge pipe D2); and
- it should have a vertical section of pipework at least 300 mm long below the tundish before any changes in direction.

12.4.6   Alternative approach

Discharge pipes may also be sized in accordance with BS 6700: 1987 *Specification for design, installation, testing and maintenance of services supplying water for domestic use within buildings and their curtilages*, Appendix E, section E2 and Table 21.

AD G3, Section 3

Table 1   Sizing of copper discharge pipe 'D2' for common temperature relief valve outlet sizes

Valve outlet size	Minimum size of discharge pipe D1*	Minimum size of discharge pipe D2* from tundish	Maximum resistance allowed, expressed as a length of straight pipe (i.e. no elbows or bends)	Resistance created by each elbow or bend
$G\frac{1}{2}$	15 mm	22 mm	up to 9 m	0.8 m
		28 mm	up to 18 m	1.0 m
		35 mm	up to 27 m	1.4 m
$G\frac{3}{4}$	22 mm	28 mm	up to 9 m	1.0 m
		35 mm	up to 18 m	1.4 m
		42 mm	up to 27 m	1.7 m
G1	28 mm	35 mm	up to 9 m	1.4 m
		42 mm	up to 18 m	1.7 m
		54 mm	up to 27 m	2.3 m

* see 3.5, 3.9, 3.9(a) and Fig. 12.2

Worked example:
The example below is for a $G\frac{1}{2}$ temperature relief valve with a discharge pipe (D2) having 4 No. elbows and length of 7 m from the tundish to the point of discharge.

From Table 1:
Maximum resistance allowed for a straight length of 22 mm copper discharge pipe (D2) from a $G\frac{1}{2}$ temperature relief valve is: 9.0 m
Subtract the resistance for 4 No. 22 mm elbows at 0.8 m each = 3.2 m

Therefore the maximum permitted length equates to: 5.8 m

5.8 m is less than the actual length of 7 m therefore calculate the next largest size.

Maximum resistance allowed for a straight length of 28 mm pipe (D2) from a $G\frac{1}{2}$ temperature relief valve equates to: 18 m

Subtract the resistance for 4 No. 28 mm elbows at 1.0 m each = 4 m

Therefore the maximum permitted length equates to: 14 m

As the actual length is 7 m, a 28 mm (D2) copper pipe will be satisfactory.

12.4.7   Section 4 hot water storage systems

Systems within the scope of section 4 exceed 500 litres in capacity or have a power input of more than 45 kW. Generally they will be individual designs for specific projects and therefore, not systems appropriate for EOTA or NACCB certification. Nevertheless, these systems should still conform to the same general safety recommendations as in section 3 including design by an appropriately qualified engineer and installation by a competent person.

Systems with a storage vessel of more than 500 litres capacity but with a power input of not more than 45kW should have safety devices conforming to BS 6700: 1987 (section two, clause 7) or other equivalent practice specifications which recommend a similar operating sequence for the safety devices to prevent the stored water temperature exceeding 100°C.

An unvented hot water storage vessel with a power input which exceeds 45 kW should also have an appropriate number of temperature relief valves which:

- either comply with BS 6283: Parts 2 or 3 or equivalent giving a combined discharge rating at least equivalent to the power input; or
- are equally suitable and marked with the set temperature in °C and a discharge rating marked in kW, measured in accordance with Appendix F of BS 6283: Part 2: 1991 or Appendix G of BS 6283: Part 3: 1991 and certified by a member of EOTA such as BBA or another recognised testing body (e.g. the Associated Offices Technical Committee, AOTC).

The temperature relief valves should be factory fitted to the storage vessel and the sensing element located as described in paragraph 3.5 of AD G3.

The non self-resetting thermal cut-outs should be installed in the system as described in Section 3 and the discharge pipes should also comply with that section.

13 Drainage and waste disposal (Part H)

13.1 Introduction

This chapter describes Part H of Schedule 1 to the Building Regulations 2000 (as amended) and the associated Approved Document H. Together, these documents cover:

- Foul water drainage (H1);
- Wastewater treatment systems and cesspools (H2);
- Rainwater drainage (H3);
- Building over existing sewers (H4);
- Separate systems of drainage (H5); and
- Solid waste storage (H6).

Part H was substantially revised in 2002 and the current edition of Approved Document H draws together not only guidance on the drainage items listed above but also a certain amount of information on legislation related to drainage and waste disposal under the following headings:

- Repairs, alterations and discontinued use of drains and sewers (Appendix H1-B)
- Adoption of sewers and connection to public sewers (Appendix H1-C)
- Maintenance of wastewater treatment systems and cesspools (Appendix H2-A)
- Relevant waste collection legislation (Appendix H6-A).

13.2 Repairs, alterations and discontinued use of drains and sewers

Reconstruction and alteration to existing drains and sewers is deemed to constitute a material alteration of a controlled service under the Building Regulations and should be carried out to the same standards as new drains and sewers. Therefore, where new drainage is connected to existing pipework, the following points should be considered.

- Existing pipework should not be damaged, (e.g. use proper cutting equipment when breaking into existing drain runs).
- The resulting joint should be watertight, (e.g. by making use of purpose made repair couplings).

- Care should be taken to avoid differential settlement between the existing and new pipework, (e.g. by providing proper bedding of the pipework).

Even though the Building Regulations do not cover requirements for ongoing maintenance or repair of drains or sewers, sewerage undertakers and local authorities have a variety of powers under other legislation to make sure that drains, sewers, cesspools, septic tanks and settlement tanks do not deteriorate to the extent that they become a risk to public health and safety. This includes powers to ensure that:

- adequate maintenance is carried out
- repairs and alterations are properly carried out
- disused drains and sewers are sealed.

Requirements for inspection, maintenance, repairs and alterations

Section 48 of the Public Health Act 1936 (*power of relevant authority to examine and test drains etc. believed to be defective*) enables a local authority to examine and test any sanitary convenience, drain, private sewer or cesspool where it feels that it has reasonable grounds for believing that the drain is in such a condition:

- as to be prejudicial to health or a nuisance; or
- is so defective as to admit subsoil water (where the drain or private sewer connects indirectly with a public sewer).

Similar powers exist to enable sewerage undertakers to examine and test drains and private sewers under section 114 of the Water Industry Act 1991 (*power to investigate defective drain or sewer*).

Section 59 of the Building Act 1984 (*drainage of building*) allows a local authority to require a building owner to carry out remedial works on soil pipes, drains, cesspools or private sewers where these are deemed to be:

- insufficient for adequately draining the building
- prejudicial to health or a nuisance
- so defective as to admit subsoil water.

Section 59 also applies to disused cesspools, septic tanks or settlement tanks where these are considered to be prejudicial to health or a nuisance. The local authority can require the owner or occupier to fill or remove the tank or otherwise render it innocuous.

Under section 60 of the of the Building Act 1984 a pipe for conveying rainwater from a roof may not be used for conveying soil or drainage from a sanitary convenience, or as a ventilating shaft to a foul drain. The practical effect of this provision is that all rainwater pipes must be trapped before entering a foul drain.

Section 61 of the Building Act 1984 (*Repair etc. of drain*), requires any person intending to repair, reconstruct or alter a drain to give 24 hours notice to the local authority of their intention to carry out the works. This does not apply in an emergency; however such work must not be subsequently covered over without giving 24 hours notice. Free access must also be given to the local authority to inspect the works.

Section 17 of the Public Health Act 1961 (*power to repair drains etc. and to remedy stopped-up drains*) provides a swift procedure whereby local authorities may repair or clear blockages on drains or private sewers which have not been properly maintained. The repairs etc. must not cost more than £250 and can only be carried out after a notice has been served on the owner from whom costs can be recovered.

Section 50 of the Public Health Act 1936 (*overflowing and leaking cesspools*) allows the local authority to take action against any person who has caused by their action, default or sufferance, a septic tank, settlement tank or cesspool to leak or overflow. The person can be required to carry out repairs or to periodically empty the tank. This does not apply to the overflow of treated effluent or flow from a septic tank into a drainage field, provided the overflow is not prejudicial to health or a nuisance. It should be noted that under this section action can be taken against a builder who had caused the problem, as well as against the owner.

Sealing and/or removal of disused drains and sewers

Disused drains and sewers can be prejudicial to health in that they harbour rats, allow them to move between sewers and the surface, and may collapse causing possible subsidence. Therefore local authorities have a number of powers to control the sealing and removal of such drains and sewers as follows.

- Where a person carries out work which results in any part of a drain becoming permanently disused, under section 62 of the Building Act 1984 (*disconnection of drain*) a local authority may require the drain to be sealed at such points as it directs.
- Section 82 of the Building Act 1984 (*notices about demolition*), allows the local authority to require any person demolishing a building to remove or seal any sewer or drain to which the building was connected (see Chapter 1, section 1.6).
- A local authority can also use its powers under section 59 of the Building Act 1984 (see above) to require an owner of a building to remove or otherwise render innocuous any disused drain or sewer which is a health risk.

Disused drains or sewers should be disconnected from the sewer system as near as possible to the point of connection. Care should be taken not damage any pipe which is still in use and to ensure that the sewer system remains watertight. Disconnection is usually carried out by removing the pipe from a junction and placing a stopper in the branch of the junction fitting. If the connection is to a public sewer the sewerage undertaker should be consulted.

Shallow drains or sewers (i.e. less than 1.5 m deep) in open ground should, where possible, be removed. To ensure that rats cannot gain access, other pipes should be grout filled and sealed at both ends and at any point of connection. Larger pipes

(225 mm diameter or greater) should be grout filled to prevent subsidence or damage to buildings or services in the event of collapse.

Pollution of watercourses and ground water

Under Section 85 (*offences of polluting controlled waters*) of the Water Resources Act 1991 the Environment Agency have powers to prosecute anyone causing or knowingly permitting pollution of any stream, river, lake etc. or any groundwater. They also have powers under section 161A (*notices requiring persons to carry out anti-pollution works and operations*) of the Water Resources Act 1991 (as amended by the Environment Act 1995) to take action against any person causing or knowingly permitting a situation in which pollution of a stream, river, lake etc. or groundwater, is likely. Such a person can be required to carry out works to prevent the pollution.

Control over solid waste storage

With regard to solid waste storage, all dwellings are now required to have satisfactory means of storing solid waste and the provision of sections 23(1) and (2) of the Building Act 1984 which required satisfactory means of access for removal of refuse have been replaced by paragraph H4 of Schedule 1 to the Building Regulations 2000 (as amended). This paragraph of the regulations must be read in light of other legislative provisions in respect of refuse disposal. In particular, sections 45 to 47 of the Environmental Protection Act 1990 should be referred to (see Chapter 5) since those sections deal with the removal of refuse and allied matters. Thus, under section 45 of the 1990 Act a duty is placed on the local authority to collect all household waste in their area, while sections 46 and 47 make provision for the removal of trade and other refuse. Section 23(3) of the Building Act 1984 requires the local authority's consent to close or obstruct the means of access by which refuse is removed from a house.

13.3   Sanitary pipework and drainage

Paragraph H1 of Schedule 1 to the Building Regulations 2000 (as amended) requires that an adequate system of drainage must be provided to carry foul water from appliances in a building to one of the following, listed in order of priority:

- a public sewer; or
- a private sewer communicating with a public sewer; or
- a septic tank which has an appropriate form of secondary treatment or another wastewater treatment system; or
- a cesspool.

Movement to a lower level in the order of priority may only be on the grounds of reasonable practicability. For example, if no public or private sewer was available within a reasonable distance then a septic tank might be a suitable alternative.

FOUL WATER is defined as waste water which comprises or includes:

- waste from a sanitary convenience, bidet or appliance used for washing receptacles for foul waste, or
- water which has been used for food preparation, cooking or washing.

Where it is proposed to divert water that has been used for personal washing or for the washing of clothes, linen or other articles to a collection system for reuse, then the provisions of requirement H1 will not apply.

Further guidance on the meaning of SANITARY CONVENIENCE is given in the guidance to Approved Document G4 where it is defined as a closet or urinal.

FOUL WATER OUTFALL may be a foul or combined sewer, cesspool, septic tank or holding tank. This term is not specifically defined in AD H1; however the term is inferred from the description of Performance on page 6.

The requirements of Paragraph H1 may be met by any foul water drainage system which:

- conveys the flow of foul water to a suitable foul water outfall;
- reduces to a minimum the risk of leakage or blockage;
- prevents the entry of foul air from the drainage system to the building, under working conditions;
- is ventilated;
- is accessible for clearing blockages, and
- does not increase the vulnerability of the building to flooding.

AD H1 sets out detailed provisions in two sections. Section 1 deals with sanitary pipework (i.e. above ground foul drainage) and is applicable to domestic buildings and small non-domestic buildings. Section 2 deals with foul drainage (i.e. below ground foul drainage). There is also an appendix (H1-A) which contains additional guidance for large buildings. Complex systems in larger buildings should follow the guidance in BS EN 12056 *Gravity drainage systems inside buildings*.

13.3.1   Above-ground foul drainage

A number of terms are used throughout AD H1. These are defined below and illustrated in Fig. 13.1. It should be noted that these definitions do not appear in the AD.

DISCHARGE STACK – A ventilated vertical pipe which carries soil and waste water directly to a drain.

VENTILATING STACK – A ventilated vertical pipe which ventilates a drainage system either by connection to a drain or to a discharge stack or branch ventilating pipe.

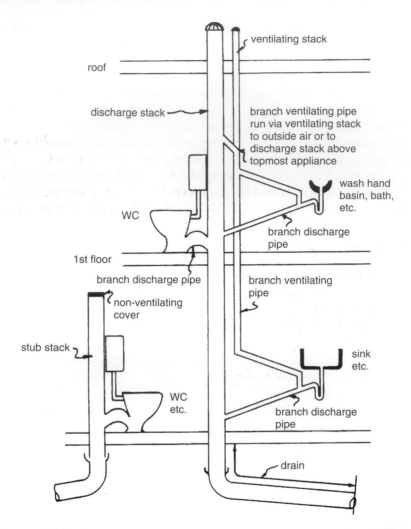

Fig. 13.1 Definitions.

BRANCH DISCHARGE PIPE (sometimes referred to as a BRANCH PIPE) – The section of pipework which connects an appliance to another branch pipe or a discharge stack if above the ground floor, or to a gully, drain or discharge stack if on the ground floor.

BRANCH VENTILATING PIPE – The section of pipework which allows a branch discharge pipe to be separately ventilated.

STUB STACK – An unventilated discharge stack.

A drainage system, whether above or below ground, should have sufficient capacity to carry the anticipated flow at any point. The capacity of the system, therefore, will depend on the size and gradient of the pipes whereas the flow will depend on the type, number and grouping of appliances. Table 13.1 below is based on information from BS EN 12056 and Table A2 of AD HI, and gives the expected flow rates for a range of appliances.

Since sanitary appliances are seldom used simultaneously, the normal size of discharge stack or drain will be able to take the flow from quite a large number of appliances. Table A1 of AD H1 is reproduced below and is derived from BS EN 12056. It shows the approximate flow rates from dwellings and is based on an appliance grouping per household of 1 WC, 1 bath, 1 or 2 washbasins and 1 sink.

The guidance given in section 1 of AD H1 is applicable for WCs with major flush volumes of 5 litres or more. WCs with flush volumes of less than 5 litres may give rise to an increased risk of blockages, however BS EN 12056 contains guidance on the design of sanitary pipework suitable for WCs with flush volumes as low as 4 litres.

13.3.2   Pipe sizes

Since individual manufacturer's pipe sizes will vary, the sizes quoted in AD H1 are nominal and give a numerical designation in convenient round numbers. Similarly, equivalent pipe sizes for individual pipe standards are given in the standards listed in AD H Tables 4, 7 and 14 reproduced below.

Table 13.1   Appliance flow rates.

Appliance	Flow rate (litres/sec)
WC (9 litre washdown)	2.3
Washbasin	0.6
Sink	0.9
Bath	1.1
Shower	0.1
Washing machine	0.7
Urinal (per person unit)	0.15
Spray tap basin	0.06
Dishwashing machine	0.25

AD H1, section 1

Table 1   Flow rates from dwellings

Number of dwellings	Flow rate (litres/sec)
1	2.5
5	3.5
10	4.1
15	4.6
20	5.1
25	5.4
30	5.8

13.3.3 Trap water seals

Trap water seals are provided in drainage systems to prevent foul air from the system entering the building. All discharge points into the system should be fitted with traps and these should retain a minimum seal of 25 mm or equivalent under test and working conditions.

Traditionally the 'one pipe' and 'two pipe' systems of plumbing have required the provision of branch ventilating pipes and ventilating stacks unless special forms of trap are used. The 'single-stack' system of plumbing obviates the need for these ventilating pipes and is illustrated in Fig. 13.2. Table 13.2 below, which is based on Table 1 and Table A3 of AD H1, gives minimum dimensions of pipes and traps where it is proposed to use appliances other than those shown in Fig. 13.2.

It is permissible to reduce the depth of trap seal to 38mm where washing machines, dishwashers, baths or showers discharge directly to a gully. Additionally, traps used on appliances with flat bottom (trailing waste) discharge which discharge to a gully with a grating may also have a water seal of not less than 38 mm.

It should be stressed that the minimum pipe sizes given above relate to branch pipes serving a single appliance. Where a number of appliances are served by a single branch pipe which is unventilated, the diameter of the pipe should be at least the size given in Table 2 to section 1 of AD HI, which is reproduced below.

If it is not possible to comply with the figures given in Table 13.1, Fig. 13.2 or Table 2, then the branch discharge pipe should be ventilated in order to prevent loss of trap seals. This is facilitated by means of a *branch ventilating pipe* which is connected to the discharge pipe within 750 mm of the appliance trap. The branch ventilating pipe may be run direct to outside air, where it should finish at least 900 mm above any opening into the building which is nearer than 3m, or, it may be connected to the ventilating stack or stack vent above the 'spillover' level of the highest appliance served. In this case it should have a continuous incline from the branch discharge pipe to the point of connection with the stack (see Fig. 13.3).

Where a branch ventilating pipe serves only one appliance it should have a minimum diameter of 25 mm. This should be increased to 32 mm diameter if the branch ventilating pipe is longer than 15 m or contains more than five bends.

Table 13.2   Minimum dimensions of branch pipes and traps.

Appliance	Minimum diameter of pipe and trap (mm)	Depth of trap seal (mm)
Bidet	32	75
Shower Food waste disposal unit Urinal bowl Sanitary towel macerator Washing machine Dishwashing machine	40	75
Industrial food waste disposal unit	50	75
Urinal stall (1 to 6 person position)	65	50

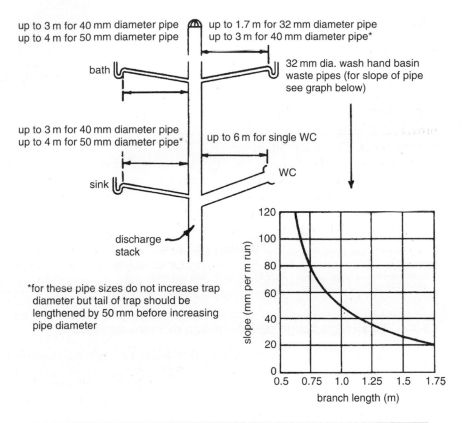

up to 3 m for 40 mm diameter pipe — up to 1.7 m for 32 mm diameter pipe
up to 4 m for 50 mm diameter pipe — up to 3 m for 40 mm diameter pipe*

bath

32 mm dia. wash hand basin
waste pipes (for slope of pipe
see graph below)

up to 3 m for 40 mm diameter pipe
up to 4 m for 50 mm diameter pipe*

up to 6 m for single WC

sink

WC

discharge
stack

*for these pipe sizes do not increase trap
diameter but tail of trap should be
lengthened by 50 mm before increasing
pipe diameter

slope (mm per m run)

branch length (m)

Appliance	Minimum diameter of pipe and trap (mm)	Depth of trap seal	Slope (mm/m)
Sink	40	75	18–90
Bath	40	50	18–90
WC – outlet < 80 mm	75	50	18
WC – outlet > 80 mm	100	50	18
washbasin	32	75*	See graph above

*Depth of seal may be reduced to 50 mm only with flush grated wastes
without plugs on spray tap basins

Fig. 13.2 Single stack system – design limits.

As appliance traps present an obstacle to the normal flow in a pipe they may be subject to periodic blockages. It is important, therefore, that they be fitted immediately after an appliance and either be removable or be fitted with a cleaning eye. Where a trap forms an integral part of an appliance (such as in a WC pan), the appliance should be removable.

AD H1 Section 1

Table 2   Common Branch discharge pipes (unventilated).

Appliance	Max no. to be connected	Max length of branch pipe (m)	Min size of pipe (mm)	Gradient limits (mm fall per metre)
WC outlet > 80 mm	8	15	100	18^2 to 90
WC outlet < 80 mm	1	15	75^3	18 to 90
Urinal – bowl		3^1	50	
Urinal – trough		3^1	65	18 to 90
Urinal – slab[4]		3^1		
Washbasin or bidet	3	1.7	30	18 to 22

Notes:
[1] Should be as short as possible to prevent deposition
[2] May be reduced to 9 mm on long drain runs where space is restricted, but only if more than one WC is connected
[3] Not recommended where disposal of sanitary towels may take place via the WC, as there is an increased risk of blockages
[4] Slab urinals longer than seven, persons should have more than one outlet.

13.3.4   Branch discharge pipes – design recommendations

In addition to size and gradient there are other design recommendations for branch discharge pipes that should be adhered to for efficient operation, and in order to prevent loss of trap seals.

Branch pipes should only discharge into another branch pipe, a discharge stack or a gully. Gullies are usually at ground floor level but may be situated in a basement and are only permitted to take wastewater. It is not permissible to discharge a branch pipe into an open hopper. Branch pipes to ground floor appliances may also discharge into a stub stack or directly to a drain.

In high buildings especially, back-pressure may build up at the foot of a discharge stack and may cause loss of trap seal in ground floor appliances. Therefore, the following recommendations should be followed.

- For multi-storey buildings up to five storeys high there should be a minimum distance of 750 mm between the point of junction of the lowest branch discharge pipe connection and the invert of the tail of the bend at the foot of the discharge stack. This is reduced to 450 mm for discharge stacks in single dwellings up to three storeys high (see Fig. 13.4).
- For appliances above ground floor level the branch pipe should only be run to a discharge stack, to another branch pipe or to a stub stack (but see also section 13.3.8 below for more information on stub stacks).
- Ground floor appliances may be run to a separate drain, gully or stub stack. (A gully connection should be restricted to pipes carrying waste water only.) They may also be run to a discharge stack in the following circumstances:
  - (a) in buildings up to five storeys high – without restriction;
  - (b) in buildings with six to twenty storeys – to their own separate discharge stack;

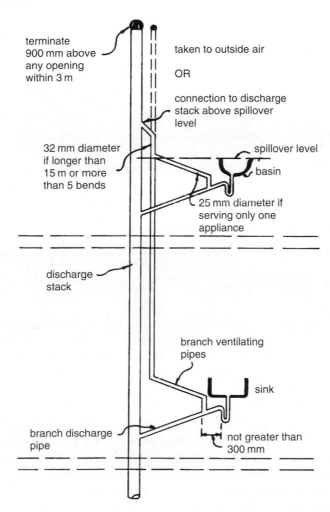

terminate
900 mm above
any opening
within 3 m

taken to outside air

OR

connection to discharge
stack above spillover
level

32 mm diameter
if longer than
15 m or more
than 5 bends

spillover level

basin

25 mm diameter if
serving only one
appliance

discharge
stack

branch ventilating
pipes

sink

branch discharge
pipe

not greater than
300 mm

Fig. 13.3　Branch ventilating pipes.

(c)　in buildings over 20 storeys – ground and first floor appliances to their own separate discharge stack; (see Fig. 13.5).

Back-pressure and blockages may occur where branches are connected so as to be almost opposite one another. This is most likely to occur where bath and WC branch connections are at or about the same level. Figure 13.6 illustrates ways in which possible cross flows may be avoided.

Additionally, a long vertical drop from a ground floor water closet to a drain may cause self-syphonage of the WC trap. To prevent this the drop should not exceed 1.3 m from floor level to invert of drain (see Fig. 13.7).

Similarly, there is a chance of syphonage where a branch discharge pipe connects with a gully. This can be avoided by terminating the branch pipe above the water level but below the gully grating or sealing plate (see Fig. 13.7).

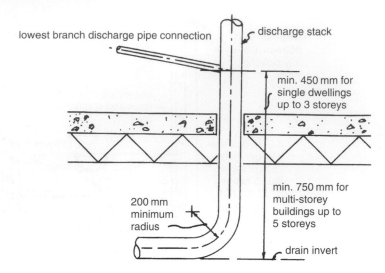

Fig. 13.4   Connection of lowest branch to discharge stack.

Self-syphonage can also be prevented by ensuring that bends in branch discharge pipes are kept to a minimum. Where bends are unavoidable they should be made with as large a radius as possible. Junctions on branches should be swept in the direction of flow with a minimum radius of 25 mm or should make an angle of 45° with the discharge stack. Where a branch diameter is 75 mm or more the sweep radius should be increased to 50 mm (see Fig. 13.6). Branch pipes up to 40 mm diameter joining other branch pipes which are 100 mm diameter or greater should, where possible, connect to the upper part of the pipe wall of the larger branch.

Branch discharge pipes should be fully accessible for clearing blockages. Additionally rodding points should be provided so that access may be gained to any part of a branch discharge pipe which cannot be reached by removing a trap or an appliance with an integral trap.

13.3.5   Drainage of condensate from boilers

It is permissible to connect condensate drainage from boilers to sanitary pipework. The connecting pipework should have a minimum diameter of 22 mm and should pass through a 75 mm condensate trap. This can be by means of an additional trap provided externally to the boiler to achieve the 75 mm seal. If this is the case, an air gap should be provided between the boiler and the trap. The following recommendations should also be observed.

- For preference, the connection should be made to an internal stack with a 75 mm condensate trap.
- Any connection made to a branch discharge pipe should be downstream of any sink waste connection.
- All sanitary pipework receiving condensate should be made of materials which can resist a pH value of 6.5 and lower.

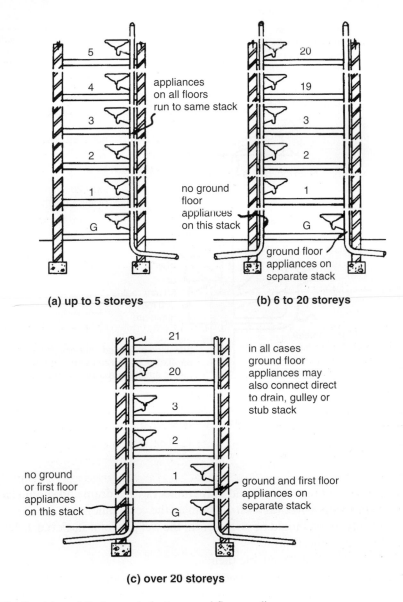

Fig. 13.5 Provision of discharge stacks to ground floor appliances.

- The installation should follow the guidance in BS 6798 *Specification for installation of gas-fired hot water boilers of rated input not exceeding 60 kW.*

13.3.6 Discharge stacks – design recommendations

The satisfactory performance of a discharge stack will be ensured if it complies with the following rules.

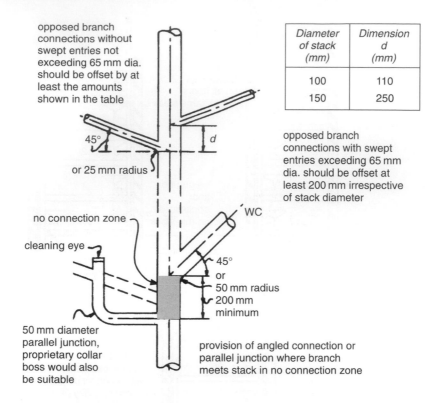

Diameter of stack (mm)	Dimension d (mm)
100	110
150	250

opposed branch connections without swept entries not exceeding 65 mm dia. should be offset by at least the amounts shown in the table

45°

or 25 mm radius

opposed branch connections with swept entries exceeding 65 mm dia. should be offset at least 200 mm irrespective of stack diameter

no connection zone

cleaning eye

WC

45° or 50 mm radius 200 mm minimum

50 mm diameter parallel junction, proprietary collar boss would also be suitable

provision of angled connection or parallel junction where branch meets stack in no connection zone

Fig. 13.6   Avoidance of cross flows in discharge stacks.

- The foot of the stack should only connect with a drain and should have as large a radius as possible (at least 200 mm at the centreline).
- Ideally, there should be no offsets in the wet part of a stack (i.e. below the highest branch connection).
- If offsets are unavoidable then:
    - (a) buildings over three storeys should have a separate ventilation stack connected above and below the offset; and
    - (b) buildings up to three storeys should have no branch connection within 750 mm of the offset.
- The stack should be placed inside a building, unless the building has not more than three storeys. This rule is intended to prevent frost damage to discharge stacks and branch pipes.
- The stack should comply with the minimum diameters given in Table 3 to section 1 of AD Hl (see below). Additionally, the following minimum internal diameters for discharge stacks also apply:
    - (a) serving urinals – 50 mm,
    - (b) serving closets with outlets less than 80 mm – 75 mm, and
    - (c) serving closets with outlets greater than 80 mm – 100 mm.

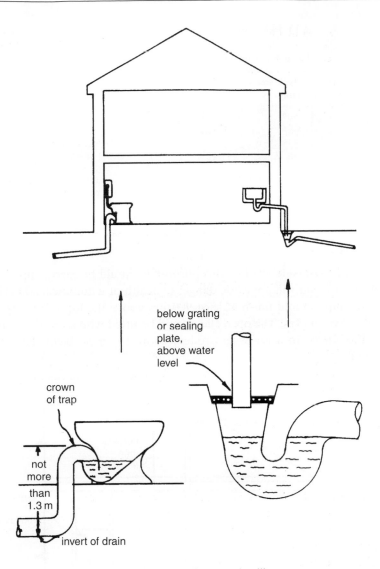

crown
of trap

below grating
or sealing
plate,
above water
level

not
more
than
1.3 m

invert of drain

Fig. 13.7 Ground floor connections for water closets and gullies

- The diameter of a discharge stack should not reduce in the direction of flow and the internal diameter of the stack should not be less than that of the largest trap or branch discharge pipe.
- Adequate access points for clearing blockages should be provided and all pipes should be reasonably accessible for repairs. Rodding points in stacks should be above the spillover level of appliances.

13.3.7 Discharge stacks – ventilation recommendations

In order to prevent the loss of trap seals it is essential that the air pressure in a discharge stack remains reasonably constant. Therefore, the stack should be

AD H1, section 1

Table 3   Minimum diameters for discharge stacks.

Stack size (mm)	Max capacity (litres/sec)
50*	1.2
65*	2.1
75†	3.4
90	5.3
100	7.2

Note
* No wcs.
† Not more than 1 syphonic wc with 75 mm outlet.

ventilated to outside air. For this purpose it should be carried up to such a height that its open end will not cause danger to health or a nuisance. AD HI recommends that the pipe should finish at least 900 mm above the top of any opening into the building within 3 m. The open end should be fitted with a durable ventilating cover (see Fig. 13.8). In areas where rodent control is a problem the cover should be metallic.

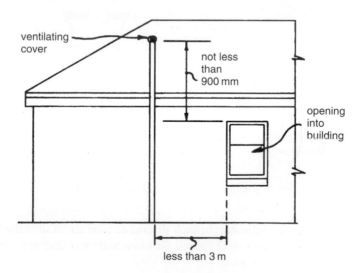

Fig. 13.8   Termination of discharge stacks.

The dry part of a discharge stack above the topmost branch, which serves only for ventilation, may be reduced in size in one and two storey houses to 75 mm diameter.

It is permissible to terminate a discharge stack inside a building if it is fitted with an air admittance valve. This valve allows air to enter the pipe but does not allow foul air to escape. It should comply with prEN 12380 *Ventilating pipework, air admittance valves* and should not adversely affect the operation of the underground drainage system which normally relies on ventilation from the open stacks to the sanitary pipework.

Air admittance valves should also be:

- located in areas which have adequate ventilation
- accessible for maintenance
- removable to give access for clearing blockages.

Air admittance valves should not be used:

- in dust laden atmospheres
- outside buildings
- where there is no open ventilation on a drainage system or through connected drains – other means to relieve positive pressures should be considered.

Some underground drains are subject to surcharging. Where this is the case the discharge stack should be ventilated by a pipe of not less than 50 mm diameter connected at the base of the stack above the expected flood level. This would also apply where a discharge pipe is connected to a drain near an intercepting trap.

13.3.8   Stub stacks

There is one exception to the general rule that discharge stacks should be ventilated. This involves the use of an unvented stack (or *stub stack*). A stub stack should connect to a ventilated discharge stack or a ventilated drain which is not subject to surcharging and should comply with the dimensions given in Fig. 13.9. It is permissible for more than one ground floor appliance to connect to a stub stack.

13.3.9   Dry ventilating stacks

Where an installation requires a large number of branch ventilating pipes and the distance to a discharge stack is also large it may be necessary to use a dry ventilating stack.

It is normal to connect the lower end of a ventilating stack to a ventilated discharge stack below the lowest branch discharge pipe and above the bend at the foot of the stack or to the crown of the lowest branch discharge pipe connection provided that it is at least 75 mm diameter.

Ventilating stacks should be at least 32 mm in diameter if serving a building containing dwellings not more than ten storeys high. For all other buildings reference should be made to BS EN 12056 *Gravity drainage systems inside buildings*.

13.3.10   Greywater recovery systems

Greywater is defined in the Water Regulations Advisory Scheme leaflet No. 09-02-04 *Reclaimed water systems. Information about installing, modifying or maintaining*

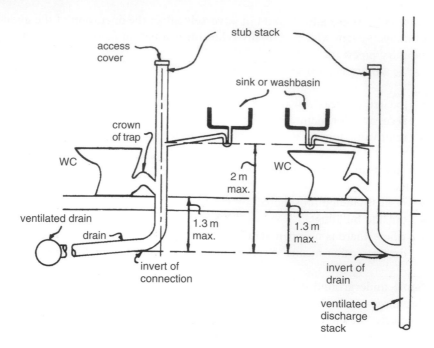

Fig. 13.9 Stub stacks.

reclaimed water systems as – 'water originating from the mains potable water supply that has been used for bathing or washing, washing dishes or laundering clothes'.

Such water can be used for irrigation and for other purposes such as toilet flushing or car washing; however care must be taken to prevent contamination of potable water supplies or accidental misuse due to the greywater being mistaken for potable water.

Approved Document H1 gives very little guidance on the use of greywater other than a passing reference to the leaflet mentioned above. It is more concerned with the identification of the pipes conveying the greywater and the suitability of storage systems. Accordingly, it recommends that all sanitary pipework carrying greywater for reuse should be clearly marked with the word 'GREYWATER' in accordance with the Water Regulations Advisory Scheme leaflet No. 09-02-05 *Marking and identification of pipework for reclaimed greywater systems.*

Guidance on the provision of external tanks for the storage of greywater is given in section 13.5 below.

13.3.11 Materials for above-ground drainage systems

Table 4 to section 1 of AD H1, which is reproduced below, gives details of the materials that may be used for pipes, fittings and joints in above ground drainage systems. The following matters should also be addressed when considering which materials to use in a system of sanitary pipework:

AD H1, section 1

Table 4 Materials for sanitary pipework.

Material	British Standard
Pipes	
Cast Iron	BS 416, BS EN 877
Copper	BC EN 1254, BS EN 1057
Galvanised steel	BS 3868
PVC-U	BS EN 1329
Polypropylene (PP)	BS EN 1451
ABS	BS EN 1455
Polyethylene (PE)	BS EN 1519
Styrene Copolymer blends (PVC + SAN)	BS EN 1565
PVC-C	BS EN 1566
Traps	BS EN 274, BS 3943

Note: Some of these materials may not be suitable for carrying trade effluent or condensate from boilers

- pipes of different metals should be separated where necessary by non-metallic material to prevent electrolytic corrosion;
- pipes should be adequately supported without restricting thermal movement;
- care should be taken to ensure continuity of any electrical earth bonding;
- care should be taken where pipes pass through fire separating elements (see Part B of Schedule 1 to the Building Regulations 2000 and Approved Document B);
- light should not be visible through the pipe wall when sanitary pipework is connected to WCs as this is believed to encourage damage by rodents.

13.3.12 Workmanship

Workmanship should be in accordance with BS 8000 *Workmanship on Building Sites* Part 13: *Code of practice for above ground drainage*.

13.3.13 Test for airtightness

In order to ensure that a completed installation is airtight it should be subjected to a pressure test of air or smoke of at least 38 mm water gauge for a maximum of three minutes. A satisfactory installation will maintain a 25 mm water seal in every trap. PVC-U pipes should not be smoke tested.

13.3.14 Alternative method of design

The requirements of the 2000 Regulations for above-ground drainage can also be met by following the relevant recommendations of BS EN 12056 *Gravity drainage systems inside buildings*. These are:

- in Part 1 *General and performance requirements* – clauses 3 to 6;
- in Part 2 *Sanitary pipework, layout and calculation*, clauses 3 to 6 and national annexes NA to NG (System III is traditionally in use in the UK);
- in Part 5 *Installation and testing, instructions for operation, maintenance and use*, clauses 4 to 6, 8, 9 & 11.

For vacuum drainage systems, designers should follow the guidance in BS EN 12109 *Vacuum drainage systems inside buildings*.

13.3.15   Below-ground foul drainage

Section 2 of AD H1 gives guidance on the construction of underground drains and sewers from buildings to the point of connection to a suitable outfall. This may be an existing sewer, a wastewater treatment system or a cesspool and includes any drains or sewers outside the curtilage of the building.

Section 2 also gives guidance in Appendix H1-B on the repair, alteration and discontinued use of drains and sewers and in Appendix H1-C on the adoption of sewers and connection to public sewers.

In most modern systems of underground drainage foul water and rainwater are carried separately. However, some public sewers are on the combined system taking foul and rainwater in the same pipe. The provisions of AD HI will apply equally to combined systems although pipe gradients and sizes may have to be adjusted to take the increased flows. In some circumstances separate drainage should still be provided on a development even though the outfall of the drainage system is to a combined sewer (see Requirement H5 in section 13.9.2 below). Combined systems should never discharge to a cesspool or septic tank.

13.3.16   Foul water outfalls and connections with sewers

Ideally, foul drainage from a development should connect to a public foul or combined sewer. Section 106 of the Water Industry Act 1991 gives the owner or occupier of a building the right to connect to a public sewer subject to the following conditions:

- where separate foul and surface water sewers are provided the connections must match this appropriately and proof of connectivity will be needed by the Building Control Body;
- the manner of the connection must not prejudice the public sewer system; and
- 21 days notice of intention to connect must be given to the sewerage undertaker.

Section 107 of the Water Industry Act 1991 allows the sewerage undertaker to make the connection and recover reasonable costs from the developer. Alternatively, the sewerage undertaker may allow the developer to carry out the work under its supervision.

Drain connections (drains to drains, drains to public or private sewers, and private sewers to public sewers) should be made obliquely, or in the direction of flow. Connections should be made using prefabricated components and where holes are cut to make the connection, these should be drilled to avoid damaging the pipe. Sometimes, in making a connection, it is preferable to remove a section of pipe and insert a junction. Repair couplings should be used for this to ensure a watertight joint. The coupling should be carefully packed to avoid differential settlement with adjacent pipes.

Where a sewer serves more than one property it should be kept as far away as is practicable from the position where a future extension might be built.

The degree to which it is possible to connect to a public sewer may, to a certain extent, depend on the size of the development. For example, for a small development, it may be reasonable to connect to a public sewer up to 30 m from the development provided that the developer has the right to construct drainage over any intervening private land. This might necessitate the provision of a pumping installation where the levels do not permit drainage by gravity (see below section 13.3.28). The economies of larger developments may make it feasible to connect to a public sewer which is some distance away.

It is also possible, for developments which comprise more than one curtilage, for the developer to requisition a sewer from the sewerage undertaker. This may be done under section 98 of the Water Industry act 1991. In constructing the sewer, the sewerage undertaker may use its rights of access to private land, however the person requisitioning the sewer may be required to contribute towards its cost over a 12 year period.

It may be possible to connect to an existing private sewer that connects with a public sewer where it is not reasonably practicable to connect directly to a public sewer. In such a case permission will need to be granted by the owner(s) of the private sewer and it should be in a satisfactory condition and have sufficient capacity to take the increased flows.

A wastewater treatment system or cesspool should only be provided where it is not reasonably practicable to connect to a sewer as described above.

13.3.17 Design and performance factors

The performance of a below ground foul drainage system depends on the drainage layout, provision for ventilation, the pipe cover and bedding, the pipe sizes and gradients, the materials used and the provisions for clearing blockages.

DRAINAGE LAYOUT – The drainage layout should be kept as simple as possible with pipes laid in straight lines and to even gradients. The number of access points provided should be limited to those essential for clearing blockages. If possible, changes of gradient and direction should be combined with access points, inspection chambers or manholes.

A slight curve in a length of otherwise straight pipework is permissible provided the line can still be adequately rodded. Bends should only be used in or close to inspection chambers and manholes, or at the foot of discharge or ventilating stacks. The radius of any bend should be as large as practicable.

In commercial hot food premises, drains serving kitchens should be fitted with a grease separator in compliance with prEN 1825: *Installations for separation of grease*, Part 1: *Principles of design, performance and testing, marking and quality control* and Part 2: *Selection of nominal size, installation and maintenance*, unless other effective means of grease removal are provided.

VENTILATION – It is important to ventilate an underground foul drainage system with a flow of air. Ventilated discharge pipes may be used for this purpose and should be positioned at or near the head of each main run. An open ventilating pipe (without an air admittance valve) should be fitted on any drain run fitted with an intercepting trap (especially on sealed systems) and on drains subject to surcharge. Ventilating pipes should not finish near openings in buildings (see section 13.3.3 above).

PIPE COVER AND BEDDING – The degree of pipe cover to be provided will usually depend on:

- the invert level of the connections to the drainage system
- the slope and level of the ground
- the necessary pipe gradients
- the necessity for protection to pipes.

In order to protect pipes from damage it is essential that they are bedded and backfilled correctly. The choice of materials for this purpose will depend mainly on the depth, size and strength of the pipes used. If the limits of cover cannot be attained it may be possible to choose another pipe strength and bedding class (see also **BS EN 1295-1:1998** *Structural design of buried pipelines under various conditions of loading*) or provide special protection (see section 13.3.20 below).

Pipes used for underground drainage may be classed as rigid or flexible. Flexible pipes will be subject to deformation under load and will therefore need more support than rigid pipes so that the deformation may be limited.

13.3.18   Rigid pipes

Tables 8 and 9 of AD HI are set out below and contain details of the limits of cover that need to be provided for rigid clay and concrete pipes in any width of trench. For details of the bedding classes referred to in the Tables, see Figs. 13.10 and 13.11.

The backfilling materials should comply with the following.

(1)   Granular material for rigid pipes should conform to BS EN 1610 Annex B Table B.15. The granular material should be single sized or graded from 5 mm up to:
- 10 mm for 100 mm pipes
- 14 mm for 150 mm pipes
- 20 mm for pipes from 150 mm to 600 mm diameter
- 40 mm for pipes more than 600 mm diameter.

The compaction fraction maximum should be 0.3 for class N or B and 0.15 for class F.

(2) Selected fill should be free from stones larger than 40 mm, lumps of clay over 100 mm, timber, frozen material or vegetable matter.

(3) It is possible that groundwater may flow in trenches with granular bedding. Provisions may be required to prevent this.

(4) Socketed pipes used with class D bedding should have holes formed in the trench bottom under the sockets to give a clearance of at least 50 mm. The holes should be as short as possible.

(5) Sockets for pipes used with class F or N bedding should be at least 50 mm above the floor of the trench.

AD H1, section 2

Table 8   Limits of cover for class 120 Clayware pipes in any width of trench.

Nominal size	Laid in fields	Laid in light roads	Laid in main roads
100 mm	0.6 m–8 + m	1.2 m–8 + m	1.2 m–8 m
225 mm	0.6 m–5 m	1.2 m–5 m	1.2 m–4.5 m
400 mm	0.6 m–4.5 m	1.2 m–4.5 m	1.2 m–4 m
600 mm	0.6 m–4.5 m	1.2 m–4.5 m	1.2 m–4 m

Notes:
1. All pipes assumed to be Class 120 to BS EN 295, other strengths and sizes of pipe are available, consult manufacturers;
2. Bedding assumed to be Class B with bedding factor of 1.9, guidance is available on use of higher bedding factors with clayware pipes
3. Alternative designs using different pipe strengths and/or bedding types may offer more appropriate or economic options using the procedures set out in BS EN 1295;
4. Minimum depth in roads set to 1.2 m irrespective of pipe strength.

AD H1, section 2

Table 9   Limits of cover for class M Concrete pipes in any width of trench.

Nominal size	Laid in fields	Laid in light roads	Laid in main roads
300 mm	0.6 mm–3 m	1.2 m–3 m	1.2 m–2.5 m
450 mm	0.6 m–3.5 m	1.2 m–3.5 m	1.2 m–2.5 m
600 mm	0.6 m–3.5 m	1.2 m–3.5 m	1.2 m–3 m

Notes:
1. All pipes assumed to be Class M to BS 5911, other strengths and sizes of pipe are available, consult manufacturers;
2. Bedding assumed to be Class B with bedding factor of 1.9;
3. Alternative designs using different pipe strengths and/or bedding types may offer more appropriate or economic options using the procedures set out in BS EN 1295;
4. Minimum depth in roads set to 1.2 m irrespective of pipe strength.

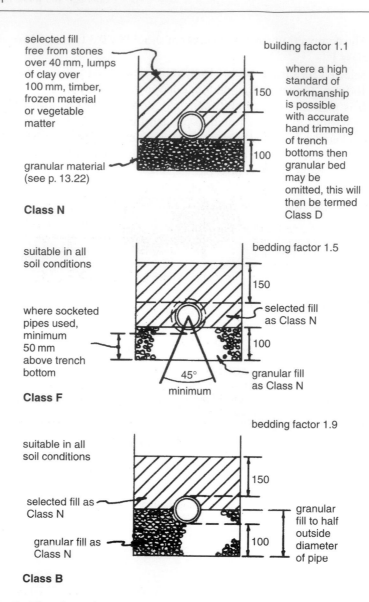

Fig. 13.10 Bedding classes for rigid pipes.

13.3.19 Flexible pipes

Flexible pipes should be provided with a minimum depth of cover of 900 mm under any road. This may be reduced to 600 mm in fields and gardens. The maximum permissible depth of cover is 7 m. Figure 13.11 shows typical bedding and backfilling details for flexible pipes. Table 10 of AD H1 gives limits of cover for thermoplastics pipes in any width of trench. Where flexible pipes have less than the minimum cover depths given in the tables they should be protected where necessary, as shown in Fig. 13.12.

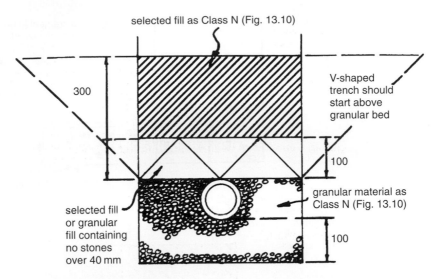

Fig. 13.11 Bedding for flexible pipes.

13.3.20 Special protection to pipes

Where pipes have less than the minimum cover recommended in Tables 8, 9 or 10 they should be bridged by reinforced concrete cover slabs resting on a flexible filler with at least 75 mm of granular fill between the top of the pipe and the underside of the flexible filler (see Fig. 13.12). Where it is necessary to backfill a trench with concrete to protect adjacent foundations (see section 13.3.25 below), the pipes should be surrounded in concrete to a thickness of at least 100 mm. Expansion joints should also be provided at each socket or sleeve joint face (see Fig. 13.12).

PIPE SIZES AND GRADIENTS – Drains should be laid to falls and should be large enough to carry the expected flow. The rate of flow will depend on the type,

AD H1, section 2

Table 10 Limits of cover for thermoplastics (nominal ring stiffness SN4) pipes in any width of trench

Nominal size	Laid in fields	Laid in light roads	Laid in main roads
100 mm–300 mm	0.6 m–7 m	0.9 m–7 m	0.9 m–7 m

Notes:
1. For drains and sewers less than 1.5 m deep where there is a risk of excavation adjacent to the drain, a special calculation is necessary, see BS EN 1295, paragraph NA. 6.2.3
2. All pipes assumed to be in accordance with the relevant standard listed in Table 7 of AD H1 section 2 with nominal ring stiffness SN4, other strengths and sizes of pipe are available, consult manufacturers;
3. Bedding assumed to be Class S2 with 80% compaction and average soil conditions;
4. Alternative designs using different pipe strengths and/or bedding types may offer more appropriate or economic options using the procedures set out in BS EN 1295;
5. Minimum depth in roads is set to 1.5 m irrespective of pipe strength, to cover loss of side support from parallel excavations.

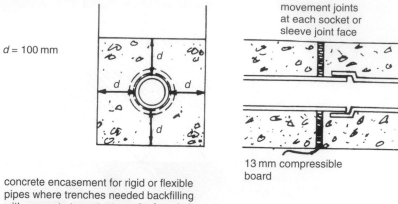

$d = 100$ mm

concrete encasement for rigid or flexible
pipes where trenches needed backfilling
with concrete to protect nearby foundations
(see also Fig. 13.15)

movement joints
at each socket or
sleeve joint face

13 mm compressible
board

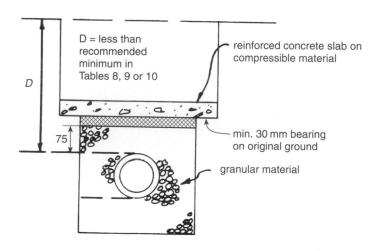

D = less than
recommended
minimum in
Tables 8, 9 or 10

reinforced concrete slab on
compressible material

min. 30 mm bearing
on original ground

granular material

protection to shallow pipes

Fig. 13.12 Special protection to pipes.

number and grouping of appliances that are connected to the drain (see Table A1,
and Table 13.1, section 13.3.1). The capacity will depend on the diameter and
gradient of the pipes.

Table 6 to section 2 of AD HI gives recommended minimum gradients for different sized foul drains and shows the maximum capacities they are capable of
carrying. The Table is set out below.

As a further design guide Diagram 9 from AD HI is reproduced below. This gives
discharge capacities for foul drains running at 0.75 proportional depth.

A drain serving more than one property (i.e. a sewer) should normally have a
minimum diameter of 100 mm if serving no more than ten dwellings. Sewers serving
more than ten dwellings should normally have a diameter of at least 150 mm.

AD H1, section 2

Table 6 Recommended minimum gradients for foul drains.

Peak flow (litres/sec)	Pipe size (mm)	Minimum gradient (l: . . .)	Maximum capacity (litres/sec)
< 1	75	1:40	4.1
	100	1:40	9.2
> 1	75	1:80	2.8
	100	1:80*	6.3
	150	1:150†	15.0

Notes:
* Minimum of 1 wc.
† Minimum of 5 wcs.

If a drain is carrying only foul water it may have a minimum diameter of 75 mm. This is increased to a minimum of 100 mm if the drain is carrying effluent from a WC or trade effluent. Where foul and rainwater drainage systems are combined, the capacity of the system should be large enough to take the combined peak flow (see Rainwater drainage, section 13.6 below).

MATERIALS – Table 7 to section 2 of AD HI, which is reproduced below, gives details of the materials that may be used for pipes, fittings and joints in below-ground foul drainage systems. Joints should remain watertight under working and test conditions and nothing in the joints, pipes or fittings should form an obstruction inside the pipeline. To avoid damage by differential settlement pipes should have flexible joints appropriate to the material of the pipes.

To prevent electrolytic corrosion, pipes of different metals should be separated where necessary by non-metallic material.

PROVISIONS FOR CLEARING BLOCKAGES – Every part of a drainage system should be accessible for clearing blockages. The type of access point chosen and its siting and spacing will depend on the layout of the drainage system and the depth and size of the drain runs. A drainage system designed in accordance with the provisions of AD H1 should be capable of being rodded by normal means (i.e. not by mechanical methods).

13.3.21 Access points

Four types of access points are described in AD H1.

- Rodding eyes (or points). These are extensions of the drainage system to ground level where the open end of the pipe is capped with a sealing plate.
- Access fittings. Small chambers situated at the invert level of a pipe and without any real area of open channel.
- Inspection chambers. Chambers having working space at ground level.
- Manholes. Chambers large enough to admit persons to work at drain level.

AD HI, section 2

Diagram 9 Discharge capacities of foul drains running 0.75 proportional depth.

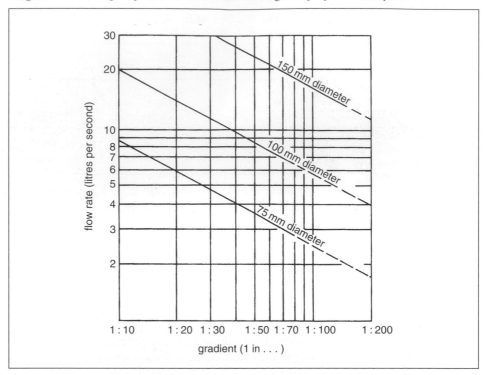

AD H1, section 2

Table 7 Materials for below ground gravity drainage.

Material	British Standard
Rigid pipes	
Vitrified clay	BS 65, BS EN 295
concrete	BS 5911
grey iron	BS 437
ductile iron	BS EN 598
Flexible pipes	
UPVC	BS EN 1401 [+]
PP	BS EN 1852 [+]
Structure Walled Plastic pipes	BS EN 13476
[+] Application area code UD should normally be specified Note: Some of these materials may not be suitable for conveying trade effluent	

Some typical access point details are illustrated in Fig. 13.13.

Whatever form of access point is used it should be of sufficient size to enable the drain run to be adequately rodded. Tables 11 and 12 to section 2 of AD HI set out the maximum depths and minimum internal dimensions for

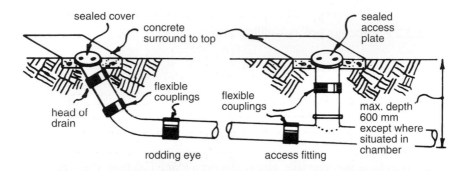

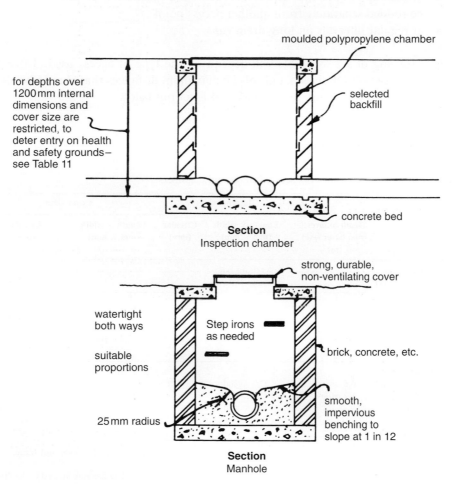

Fig. 13.13 Access points.

each type of access point. Where a large number of branches enter an inspection chamber or manhole the sizes given in Tables 11 and 12 may need to be increased. It is usual to allow 300 mm for each branch connection (thus a 1200 mm long manhole could cater for up to four branch connections on each side). The Tables are set out below.

13.3.22   Access points – siting and spacing

Access points should be provided:

- at or near the head of any drain run
- at any change of direction or gradient
- at a junction, unless each drain run can be rodded separately from another access point
- at a change of pipe size, unless this occurs at a junction where each drain run can be rodded separately from another access point
- at regular intervals on long drain runs.

The spacing of access points will depend on the type of access used. Table 13 to section 2 of AD HI gives details of the maximum distances that should be allowed for drains up to 300 mm in diameter and is set out below.

AD H1, section 2

Table 11   Minimum dimensions for access fittings and chambers.

Type	Depth to invert from cover level (m)	Internal sizes		Cover sizes	
		Length × width (mm × mm)	Circular (mm)	Length × width (mm × mm)	Circular (mm)
Rodding eye		As drain but min 100			Same size as pipework[1]
Access fitting					
small 150 diam	0.6 or less,			150 × 100[1]	Same size as
150 × 100	except where	150 × 100	150	225 × 100[1]	access fitting
large 225 × 100	situated in a chamber	225 × 100	225		
Inspection chamber					
shallow	0.6 or less	225 × 100	190[2]	–	190[1]
	1.2 or less	450 × 450	450	Min 430 × 430	430
deep	> 1.2	450 × 450	450	max 300 × 300[3]	Access restricted to max 350[3]

Notes:
[1] The clear opening may be reduced by 20 mm in order to provide proper support for the cover and frame.
[2] Drains up to 150 mm.
[3] A larger clear opening cover may be used in conjunction with a restricted access. The size is restricted for health and safety reasons to deter entry.

AD H1, section 2

Table 12 Minimum dimensions for manholes.

Type	Size of largest pipe (DN)	Min internal dimensions[1]		Min clear opening size[1]	
		Rectangular length and width	Circular diameter	Rectangular length and width	Circular diameter
Manhole < 1.5 m deep to soffit	150	750 × 675[7]	1000[7]	750 × 675[2]	na[3]
	225	1200 × 675	1200	1200 × 675[2]	
	300	1200 × 750	1200		
	> 300	1800 × (DN + 450)	The larger of 1800 or (DN + 450)		
> 1.5 m deep to soffit	225	1200 × 1000	1200	600 × 600	600
	300	1200 × 1075	1200		
	375–450	1350 × 1225	1200		
	> 450	1800 × (DN + 775)	The larger of 1800 or (DN + 775)		
Manhole shaft[4] > 3.0 m deep to soffit of pipe	Steps[5]	1050 × 800	1050	600 × 600	600
	Ladder[5]	1200 × 800	1200		
	Winch[6]	900 × 800	900	600 × 600	600

Notes:
[1] Larger sizes may be required for manholes on bends or where there are junctions.
[2] May be reduced to 600 by 600 where required by highway loading considerations, subject to a safe system of work being specified.
[3] Not applicable due to working space needed.
[4] Minimum height of chamber in shafted manhole 2 m from benching to underside of reducing slab.
[5] Min clear space between ladder or steps and the opposite face of the shaft should be approximately 900 mm.
[6] Winch only – no steps or ladders, permanent or removable.
[7] The minimum size of any manhole serving a sewer (i.e. any drain serving more than one property) should be 1200 mm × 675 mm rectangular or 1200 mm diameter.

Where an access point is provided to a sewer (i.e. serving more than one property) it should be positioned so that it is both accessible and apparent for use in emergencies. Typically, it could be positioned in a highway, public open space, unfenced front garden or shared and unfenced driveway.

13.3.23   Access points – construction

Generally, access points should:

- be constructed of suitable and durable materials
- exclude subsoil or rainwater
- be watertight under working and test conditions.

Table 14 to section 2 of AD HI is shown below and lists materials which are suitable for the construction of access points.

AD H1, section 2

Table 13   Maximum spacing of access points in metres.

| From | To | Access Fitting | | Junction | Inspection chamber | Manhole |
		Small	Large			
Start of external drain[1]		12	12	—	22	45
Rodding eye		22	22	22	45	45
Access fitting small 150 diam 150 × 100 large 225 × 100		— —	— —	12 45	22 22	22 45
Inspection chamber		22	45	22	45	45
Manhole		—	—	—	45	90[2]

Note
[1] Stack or ground floor appliance
[2] May be up to 200 for man-entry size drains and sewers

AD H1, section 2

Table 14   Materials for access points.

Material	British Standard
1 Inspection chambers and manholes Clay bricks and blocks Vitrified clay Concrete precast in situ Plastics	 **BS 3921** BS EN 295, BS 65 BS 5911 BS 8110 BS 7158
2 Rodding eyes and access fittings (excluding frames and covers)	as pipes see Table 7 ETA Certificates

Inspection chambers and manholes should fulfil the following.

- Have smooth impervious surface benching up to at least the top of the outgoing pipe to all channels and branches. The purpose of benching is to direct the flow into the main channel and to provide a safe foothold. For this reason the benching should fall towards the channel at a slope of 1 in 12 and should be rounded at the channel with a minimum radius of 25 mm (see Fig. 13.13 above).
- Be constructed so that branches up to and including 150 mm diameter discharge into the main channel at or above the horizontal diameter where half-round open

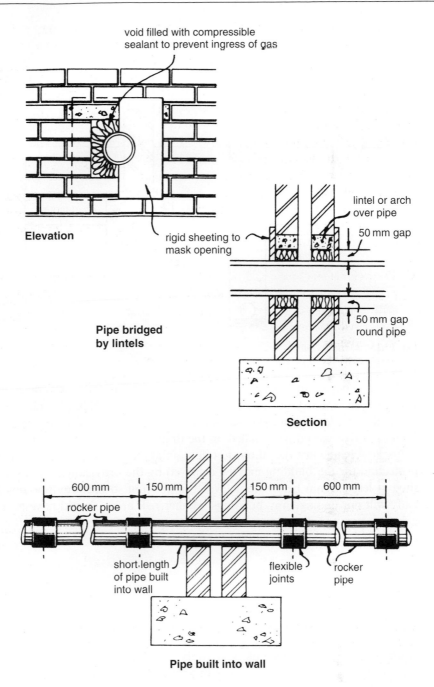

void filled with compressible
sealant to prevent ingress of gas

Elevation

rigid sheeting to
mask opening

**Pipe bridged
by lintels**

lintel or arch
over pipe

50 mm gap

50 mm gap
round pipe

Section

600 mm

150 mm

150 mm

600 mm

rocker pipe

short length
of pipe built
into wall

flexible
joints

rocker
pipe

Pipe built into wall

Fig. 13.14 Drains passing through foundations.

some low lying sites properties may be at increased risk of flooding if the ground
level of the site (or the level of a basement) is below the level at which the drainage
connects to the public sewer. The sewerage undertaker should consulted in such
cases to determine the extent and frequency of the likely surcharge.

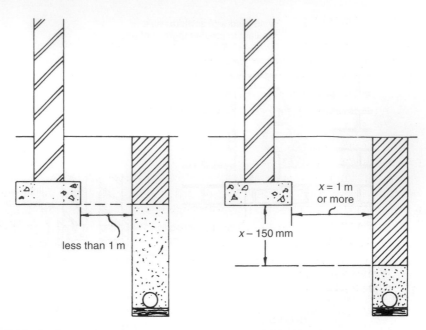

less than 1 m

$x = 1$ m or more

$x - 150$ mm

Fig. 13.15   Drain trenches.

Where a basement contains sanitary appliances and the sewerage undertaker considers that the risk of flooding due to surcharging is high, the drainage from the basement should be pumped (see section 13.3.28 below). For low risks, an anti-flooding valve should be installed on the drainage from the basement.

For low lying sites (i.e. those not containing basements) where the risk is low, protection for the building may be achieved by the provision of an external gully sited at least 75 mm below floor level in a position so that any flooding from the gully will not damage any buildings. Higher risk areas should have anti-flooding valves or pumped drainage systems (see section 13.3.28 below).

Anti-flooding valves should:

- be of the double valve type
- be suitable for foul water
- have a manual closure device
- comply with prEN 13564 *Anti flooding devices for buildings*.

Normally, a single valve should serve only one building and information about the valve should be provided on a notice inside the building. The notice should indicate the location of any manual override and include necessary maintenance information.

Some parts of the drainage system may be unaffected by surcharging. These parts should by-pass any protective measures and should discharge by gravity.

13.3.27   Special protection – rodent control

Generally, rodent infestation (especially by rats) is on the increase. Since rats use drains and sewers as effective communication routes, on previously developed sites the local authority should be consulted to ascertain if any special rodent control precautions are thought necessary.

Special precautions could include the following.

- By providing inspection chambers with screwed access covers on the pipework instead of open channels. These should only be used in inspection chambers where maintenance can be carried out from the surface without personnel entry.
- Intercepting traps may also be provided as in the past, although they do increase the incidence of blockages unless adequately maintained. Trap stoppers should be of the locking type and should be easy to remove and replace after clearing any blockage. They should always be replaced after maintenance operations and should only be used in inspection chambers where maintenance can be carried out from the surface without personnel entry.
- A number of different kinds of rodent barriers might be considered. These include enlarged sections of discharge stacks which prevent rats climbing, flexible downward facing fins in discharge stacks and one way valves in underground drainage.
- The provision of metal cages on ventilator stack terminals and fixed plastic covers or metal gratings on gullies to discourage rats from leaving the drainage system.

13.3.28   Pumping installations

Reference has been made above to the use of pumping installations where gravity drainage is impracticable or where protection is required against flooding due to surcharge in downstream sewers.

AD H1 gives details of packaged pumping systems for use both inside and outside buildings as follows.

Inside buildings

Floor mounted units available for use in basements should comply with BS EN 12050 *Wastewater lifting plants for buildings and sites – principles of construction and testing*. The pumping installation itself should be designed in accordance with BS EN 12056:2000 *Gravity drainage systems inside buildings*: Part 4: *Effluent lifting plants, layout and calculation*.

Outside buildings

Package pumping installations for use outside buildings are also available. The pumping installation should be designed in accordance with BS EN 752 *Drain and sewer systems outside buildings*: Part 6: 1998 *Pumping installations*.

Foul water drainage pumping installations should comply with the following.

- To allow for disruption in service, the effluent receiving chamber should be sized to contain 24-hour inflow.
- For domestic use the minimum daily discharge of foul drainage should be taken as 150 litres per person per day.
- For non-domestic uses the capacity of the receiving chamber should be based on the calculated daily demand of the water intake for the building (and should be assessed on a pro-rata basis where only a proportion of the foul sewage is pumped).
- For all pumped systems the controls should be arranged to optimise pump operation.

13.3.29   Workmanship

In general, workmanship should be in accordance with BS 8000 *Workmanship on building sites* Part 14: *Code of practice for below ground drainage*.

In particular, drains and sewers which are left open during construction should be covered when work is not in progress to prevent entry by rats.

A number of measures are necessary to protect drains during construction work. For example, drains can be damaged by construction traffic and heavy machinery. Barriers should be provided where necessary to keep traffic away from the line of the sewer and heavy materials should not be stored over drains or sewers. Additionally, piling works can cause damage to drains and sewers unless certain precautions are taken. This would include carrying out a survey to establish the exact location of any drain runs and connections before piling commences. Piling should not be carried out where the distance from the outside of the sewer to the outside of the pile is less than two times the diameter of the pile.

13.3.30   Alternative method of design

Additional information on the design and construction of building drainage which meets the requirements of the 2000 Regulations may be found in the relevant parts of:

- BS EN 12056: *Gravity drainage systems inside buildings*. This standard also describes the discharge unit method of calculating flows;
- BS EN 752 *Drain and sewer systems outside buildings*: Part 3:1997 *Planning*, Part 4:1997 *Hydraulic design and environmental aspects* and Part 6:1998 *Pumping installations*;
- BS EN 1610:1998 *Construction and testing of drains and sewers*;
- BS EN 1295: Part 1:1998 *Structural design of buried pipelines under various conditions of loading*;
- BS EN 1091:1997 *Vacuum sewerage systems outside buildings*; and
- BS EN 1671:1997 *Pressure sewerage systems outside buildings*.

BS EN 752 together with BS EN 1610 and BS EN 1295 contain additional information about design and construction.

13.4   Wastewater treatment systems and cesspools

Any septic tank and its form of secondary treatment, other wastewater treatment system or cesspool must be sited and constructed so that:

- it is not prejudicial to health;
- it will not contaminate any watercourse, underground water or water supply;
- it is accessible for emptying and maintenance;
- it will continue to function in the event of a power failure to a standard sufficient for the protection of health, where this is relevant (i.e. where a power supply is needed for normal operation of the system).

Furthermore, any septic tank, holding tank which is part of a wastewater treatment system or cesspool must be:

- adequately ventilated
- of adequate capacity
- constructed to be impermeable to liquids.

Since all wastewater treatment systems and cesspools rely on adequate maintenance in order to continue to operate in a safe and healthy manner the Regulations require that maintenance instructions be provided in the form of a durable notice which must be affixed in a suitable place in the building. Examples of a typical notices are given in the text below.

It should be noted that the use of non-mains foul drainage should only be considered where connection to mains drainage is not practicable and any discharge from a wastewater treatment system is likely to require consent from the Environment Agency. Contact with the Environment Agency should be made as early as possible in the design process (usually when the planning process is being initiated and before a Building Regulation application is made for non-mains drainage). This will determine whether a consent to discharge is required and what parameters apply, which in turn can have an impact on the type of system that may be installed. Further guidance may be obtained from Pollution Prevention Guideline No 4 *Disposal of sewage where no mains drainage is available*: Environment Agency 1999.

13.4.1   Wastewater treatment systems

A wastewater treatment system typically includes a septic or settlement tank which provides primary treatment to the effluent from a building. This is likely to be the most economic form of treating wastewater for one to three dwellings. The discharge from the tank can, however, still be harmful, therefore there is a need for a system of drainage which completes the treatment process after the effluent has passed through the tank, thus providing a means of secondary treatment. The term 'wastewater treatment system' can also include small sewage treatment works (see section 13.4.3 below).

In the past, the Regulations have tended to concentrate on the design and construction of the means of primary treatment but have failed to provide guidance on the ultimate means of disposal of the effluent. This is an area where considerable research and development has taken place in recent years and guidance is now given on the design and construction of drainage fields and mounds, and on constructed wetlands and reedbeds. The performance of a wastewater treatment system will depend on the capacity, siting, design and construction of both the septic or settlement tank and the drainage field or other means of secondary treatment.

13.4.2   Primary treatment systems – septic tanks and settlement tanks

Capacity

The primary treatment system should have sufficient capacity and should provide suitable conditions for the settlement, storage and partial decomposition of solid matter in the wastewater from the building. It should also be sited and constructed so as to prevent overloading of the receiving water.

For up to four users, a minimum capacity of $2.7\,m^3$ (2700 litres) below the level of the inlet is set for septic tanks and settlement tanks in order to reduce danger of overflowing and malfunctioning. This size should be increased by $0.18\,m^3$ (180 litres) for each additional user.

Siting and construction

Septic tanks should be designed and constructed to prevent leakage of contents and the ingress of subsoil water. They should also be provided with adequate ventilation, which should be kept away from buildings. Therefore, they should be kept at least 7 m from any habitable parts of buildings, preferably on a downslope.

Septic tanks must be periodically desludged and cleaned. This is usually carried out mechanically using a tanker. Because of the length of piping involved it is necessary that the cesspool or tank be sited within 30 m of a vehicular access; however, where the invert level of the tank is more than 3 m below the vehicle access level the 30 m distance will need to be reduced accordingly. Emptying and cleaning should not involve the contents being taken through a dwelling or place of work, and there should be a clear route for the hose so that the emptying and cleaning can be carried out without creating a hazard for the building's occupants. Access covers for emptying and cleaning should be sufficiently durable to resist the corrosive nature of the tank contents and should be designed to prevent unauthorised access (by being lockable or otherwise engineered to prevent personnel entry).

Tanks should also be constructed of materials which are impervious to the contents and to ground water. This would include engineering brickwork in 1:3 cement mortar at least 220 mm thick and concrete at least 150 mm thick (C/25/P mix to BS 5328), roofed with heavy concrete slabs. Prefabricated cesspools and tanks are available made of glass reinforced plastic, polyethylene or steel. These should follow the guidance in BS EN 12566 *Small wastewater treatment plants less than 50 PE*: Part 1: 2000 *Prefabricated septic tanks*. Care should be exercised over the stability of these tanks.

The inlet and outlet of the tank should be provided with access for sampling and inspection of the contents, and be designed to avoid excessive disturbance of the surface scum or settled contents by incorporating at least two chambers operating in sequence. The velocity of flow into the tank can be limited by laying the last 12 m of the incoming drain at a gradient of 1 in 50 or flatter for all pipes up to 150 mm in diameter. Alternatively, a dip pipe inlet may be provided (see Fig. 13.16) where the tank width does not exceed 1200 mm.

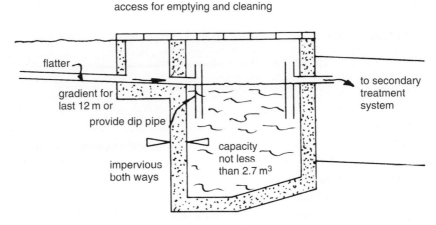

Fig. 13.16 Septic tank.

Septic tank maintenance

It is essential that building owners are kept informed about the maintenance requirements of septic tanks. Septic tanks need to be inspected monthly to make sure that they are working properly. This involves an inspection of the inlet chamber and the outlet from the tank to ensure that the effluent is flowing freely. Additionally, the effluent at the outlet should be clear.

If these conditions are not met the tank should be emptied by a licensed contractor who will make a charge. It may be more economical to take out an annual maintenance contract with a suitable contractor. Septic tanks should be emptied annually and it is usual to leave a small amount of sludge to act as an anaerobic seed.

Failure to adequately maintain the tank may result in solids being carried into the drainage field or mound. The sediments deposited may block the pores in the soil necessitating early replacement of the field or mound and in exceptional circumstances this can even render the site unsuitable for future use as a drainage field or mound.

A durable notice should be fixed within the building describing the necessary maintenance. Fig. 13.17 below is an example of a typical notice.

```
┌─────────────────────────────────────────────────────────────────────────┐
│                    Wastewater Treatment System                          │
│                                                                         │
│  Details of Necessary Maintenance                                       │
│                                                                         │
│  Address of Property   _____  │
│                                                                         │
│  Location of treatment system  _____  │
│                                                                         │
│  The foul drainage system from this property discharges to a septic tank and a │
│  [insert type of secondary treatment]                                   │
│  _____                                              │
│                                                                         │
│  The tank requires monthly inspections of the outlet chamber or distribution box to observe that the │
│  effluent is free-flowing and clear.                                    │
│                                                                         │
│  The septic tank requires emptying at least every 12 months by a licensed contractor. │
│  The [insert type of secondary treatment] should be [insert details of maintenance of secondary │
│  treatment].                                                            │
│                                                                         │
│  The owner is legally responsible to ensure that the system does not cause pollution, a health hazard │
│  or a nuisance.                                                         │
└─────────────────────────────────────────────────────────────────────────┘
```

Fig. 13.17   Septic tank – typical maintenance notice.

13.4.3   Primary treatment systems – packaged treatment works

Packaged treatment works, which are engineered to treat a given hydraulic and organic load to a higher standard than septic tanks, use prefabricated components which can be installed with a minimum amount of site work. They are normally more economic than septic tanks for larger developments and can also discharge direct to a suitable watercourse. They should be considered where there are space limitations or where other options are not possible. AD H2 does not really deal with such installations in any detail since specialist knowledge is needed in their detailed design and installation. However, it does recommend that the discharge from the treatment plant should be sited at least 10 m from watercourses or any other buildings. Furthermore, since many of these systems are powered by electricity it is important that the system should be able to adequately function for up to six hours without power or have an uninterruptible power supply.

Guidance on packaged treatment works may be obtained from BS 6297: 1983 *Code of practice for design and installation of small sewage treatment works and cesspools*. Additionally, packaged treatment works should be type-tested in accordance with BS 7781 or otherwise tested by a notified body.

The guidance regarding maintenance requirements mentioned above in connection with septic tanks also applies generally to packaged treatment works; however there will be variations in maintenance needs depending on the type of plant installed. The manufacturer's instructions regarding maintenance and inspection should always be adhered to. A durable notice should be fixed within the building describing the necessary maintenance. Fig. 13.18 below is an example of a typical notice.

Wastewater Treatment System

Details of Necessary Maintenance

Address of Property   _____

Location of treatment system   _____

The foul drainage system from this property discharges to a packaged treatment works.

Maintenance is required [insert frequency] and should be carried out by the owner in accordance with the manufacturer's instructions.

The owner is legally responsible to ensure that the system does not cause pollution, a health hazard or a nuisance.

Fig. 13.18   Septic tank – typical maintenance notice.

13.4.3   Secondary treatment systems – drainage fields and drainage mounds

Drainage fields and drainage mounds are used to provide secondary treatment to the discharge from a septic tank or packaged treatment plant. Drainage fields normally consist of below ground irrigation pipes which allow the partially treated effluent from the septic tank to percolate into the surrounding soil. Further biological treatment takes place naturally in the aerated soil layers. They may be used in subsoils with good percolation characteristics on sites which are not prone to flooding or waterlogging at any time of the year. Drainage mounds consist of drainage fields placed above the ground surface thus providing an aerated soil layer to treat the discharge. On sites where there is a high water table or impervious ground where occasional waterlogging is possible, drainage mounds could be used.

It should be noted that drainage fields and mounds are not permitted by the Environment Agency in prescribed Zone 1 groundwater source protection zones.

Siting

Care has to be taken with siting in order to protect underground water sources and watercourses and to ensure that the system will operate effectively. Therefore, a drainage field or mound serving a wastewater treatment system or septic tank should be sited:

- not less than 10 m from any permeable drain or watercourse;
- not less than 50 m from the abstraction point of any groundwater supply;
- away from any Zone 1 groundwater protection zone;
- not less than 15 m from any building;
- far enough away from other drainage fields, mounds or soakaways so that the overall soakage capacity of the ground is not exceeded;
- on the downslope side of any groundwater sources.

The disposal area should be isolated and should not contain any access roads, driveways or paved areas. Additionally, no water supply pipes or other underground services should be located within the disposal area other than those required by the disposal system itself.

Ground conditions and percolation

Some indication of the likely percolation characteristics of a site may be gained by taking a sample and observing the nature of the subsoil. Table 13.3 below gives an indication of the likely percolation characteristics of different subsoil colours and types. Percolation characteristics should be ascertained under both summer and winter conditions. This usually takes the form of a preliminary assessment followed by a percolation test.

Table 13.3   Likely percolation characteristics of different soil types.

Likely percolation characteristics	Soil colour	Likely soil type
Well drained and well aerated	Brown, yellow or reddish	Sand, gravel, chalk, sandy loam, clay loam
Poorly drained or saturated	Grey, blue	Sandy clay, silty clay, clay
Indicative of periodic saturation	Grey or brown mottling	Sandy clay, silty clay, clay

The preliminary assessment should involve:

- consultation with the Environment Agency and the local authority to determine the possible suitability of the site;
- an assessment of the on-site natural vegetation (most plants generally grow best on well drained land, the presence of certain plants may indicate wet or boggy conditions etc.);
- a determination of the position of the standing ground water table.

The standing ground water table is determined by excavating a trial hole. This should be at least $1\,m^2$ in area and should be at least 2 m deep or 1.5 m below the invert of the proposed drainage field pipework. The ground water table should be at least 1m below the invert level of the proposed effluent distribution pipes in both summer and winter.

The preliminary assessment should be followed by a percolation test of the proposed disposal area. Fig. 13.19 below illustrates the three stages of the percolation test. Where deep drains are needed the 300 mm hole should be excavated at the base of a wider excavation to allow room for working. Alternatively a modified test procedure can be adopted using a 300 mm earth auger bored vertically into the ground. All debris should be removed from the hole before the test is carried out.

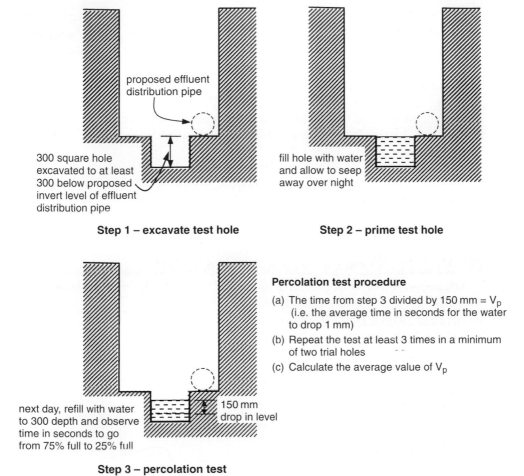

proposed effluent distribution pipe

300 square hole excavated to at least 300 below proposed invert level of effluent distribution pipe

Step 1 – excavate test hole

fill hole with water and allow to seep away over night

Step 2 – prime test hole

Percolation test procedure

(a) The time from step 3 divided by 150 mm = V_p (i.e. the average time in seconds for the water to drop 1 mm)

(b) Repeat the test at least 3 times in a minimum of two trial holes

(c) Calculate the average value of V_p

next day, refill with water to 300 depth and observe time in seconds to go from 75% full to 25% full

150 mm drop in level

Step 3 – percolation test

Fig. 13.19   Percolation test method.

The test should only be carried out when weather conditions are suitable (i.e. not during heavy rain, severe frost or drought).

Disposal via a drainage field should only be considered when:

- the percolation tests give values of V_p between 12 and 100 sec/mm; and
- the preliminary site assessment report is favourable; and
- the trial hole tests give acceptable results.

The values of V_p given above ensure that untreated effluent cannot percolate too rapidly into the ground water. Where V_p is outside the quoted range, treatment via a drainage field is unlikely to be successful. In these circumstances it may still be possible to use a septic tank provided that an alternative method of secondary treatment can be used to treat the effluent from the septic tank. It might then be possible to take the final discharge to a soakaway.

Design and construction

The main features of a typical drainage field are illustrated in Fig. 13.20, whilst Fig. 13.21 shows the main features of a drainage mound. Both are designed to ensure aerobic contact between the liquid effluent and the subsoil.

Pipes for drainage fields should be:

- perforated and laid in trenches of uniform gradient not exceeding 1 in 200;
- laid on a 300 mm layer of clean shingle or broken stone graded between 20 mm and 50 mm;
- laid at a minimum depth of 500 mm below the surface of the ground;
- laid in a continuous loop fed from an inspection chamber sited between the septic tank and the drainage field.

Pipe trenches for drainage fields should be:

- filled to a level of 50 mm above the pipe with the shingle or broken stone and covered with a layer of geotextile to prevent entry of silt;
- topped up with soil to ground level;
- between 300 mm and 900 mm wide;
- at least 2 m from other trenches in the same drainage field.

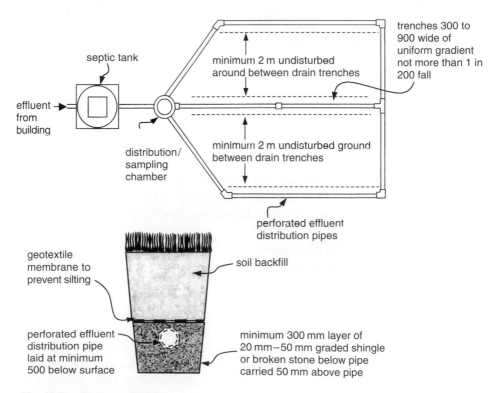

Fig. 13.20   Drainage field.

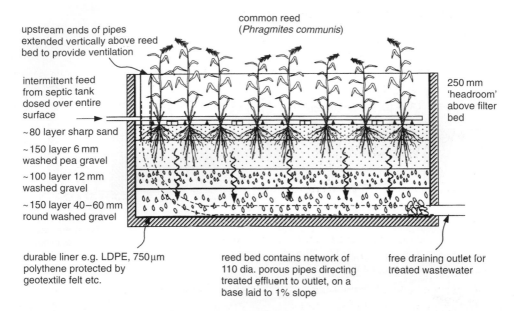

upstream ends of pipes
extended vertically above reed
bed to provide ventilation

common reed
(*Phragmites communis*)

intermittent feed
from septic tank
dosed over entire
surface

250 mm
'headroom'
above filter
bed

~80 layer sharp sand

~150 layer 6 mm
washed pea gravel

~100 layer 12 mm
washed gravel

~150 layer 40–60 mm
round washed gravel

durable liner e.g. LDPE, 750 µm
polythene protected by
geotextile felt etc.

reed bed contains network of
110 dia. porous pipes directing
treated effluent to outlet, on a
base laid to 1% slope

free draining outlet for
treated wastewater

Fig. 13.22   Typical vertical flow reed bed treatment system.

A typical vertical flow reed bed treatment system is shown in Fig. 13.22.

Horizontal flow systems

In a horizontal flow system, wastewater is continuously fed in from the upstream end and passes over the full width of a gravel bed to an outlet at the downstream end. Horizontal flow systems have the disadvantage of being oxygen-limited and therefore incapable of fully treating concentrated effluents, especially those containing high levels of ammonia. They also require a relatively level site but have lower maintenance needs than vertical flow systems since only one bed is needed. A typical horizontal flow reed bed system is illustrated in Fig. 13.23.

The guidance provided in AD H2 on vertical and horizontal flow reed bed systems is extremely limited and has been enhanced in the above notes and illustrations by reference to BRE Good Building Guide 42 (GBG 42) to which reference should be made. There are many other forms of constructed wetland treatment systems available and being developed both nationally and internationally. GBG 42 contains a number of references to such systems which will normally require specialist advice.

Maintenance of constructed wetlands

It is essential that building owners are kept informed about the maintenance requirements where constructed wetlands are provided. The main maintenance

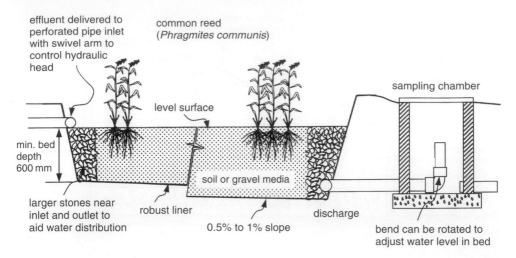

Fig. 13.23 Typical horizontal flow reed bed.

requirements are weeding, annual reed cutting and general grounds-care around the system. There will also be the need for periodic resting where multiple beds are provided in horizontal flow systems. Horizontal flow reed beds require less routine maintenance than vertical flow beds and where weeds are a problem the bed can be flooded to kill off the weeds. Full details of guidance on maintenance of reedbeds can be found in GBG 42 Part 2. A durable notice should be fixed within the building describing the necessary maintenance. Fig. 13.24 below is an example of a typical notice.

Wastewater Treatment System

Details of Necessary Maintenance

Address of Property _____

Location of treatment system _____

The foul drainage system from this property discharges to a [*insert type of primary treatment*] and a constructed wetland.

The [*insert type of primary treatment*] requires [*insert details of maintenance of the primary treatment*].

The constructed wetland system requires [*insert details of maintenance of the constructed wetland*].

Fig. 13.24 Constructed wetland – typical maintenance notice.

13.4.5 Cesspools

Where no other drainage disposal option is available it may be acceptable to provide a cesspool. Quite simply, a cesspool is a watertight underground tank provided for the storage of raw sewage. No treatment is involved.

Siting

Cesspools should be sited:

- on sloping ground away from and lower than nearby buildings;
- below, and at least 7 m from, the habitable parts of buildings;
- within 30 m of a vehicle access;
- at such levels that emptying and cleaning can be carried out without creating a hazard for the building's occupants and without the contents being taken through a dwelling or place of work.

Access for emptying and cleaning may be through a covered space which may be lockable.

Design and construction

The minimum capacity of the cesspool measured below the level of the inlet should be $18 \, m^3$ (18,000 litres) based on two users. This capacity should be increased by $6.8 \, m^3$ (6800 litres) for each additional user.

Additionally, cesspools should:

- have no openings except for the inlet from the drain, access for emptying and cleaning and ventilation;
- prevent leakage of the contents and ingress of subsoil water;
- be ventilated;
- be provided with access for emptying, cleaning, and inspection at the inlet.

Access covers for emptying and cleaning should be sufficiently durable to resist the corrosive nature of the tank contents and should be designed to prevent unauthorised access (by being lockable or otherwise engineered to prevent personnel entry).

Cesspools should also be constructed of materials which are impervious to the contents and to ground water. This would include engineering brickwork in 1:3 cement mortar at least 220 mm thick and concrete at least 150 mm thick (C/25/P mix to BS 5328), roofed with heavy concrete slabs. Prefabricated cesspools are available made of glass reinforced plastics, polyethylene or steel. These should follow the guidance in BS EN 12566 *Small wastewater treatment plants less than 50 PE:* Part 1: 2000 *Prefabricated septic tanks*. Care should be exercised over the stability of these tanks. Fig. 13.25 illustrates a typical cesspool designed for two users.

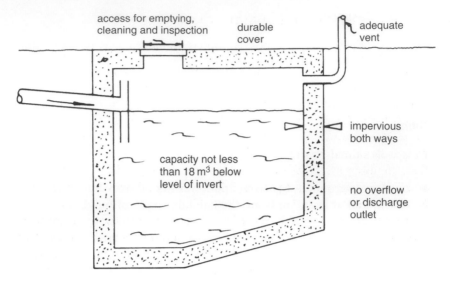

access for emptying,
cleaning and inspection

durable
cover

adequate
vent

impervious
both ways

capacity not less
than 18 m³ below
level of invert

no overflow
or discharge
outlet

Fig. 13.25 Cesspools.

Maintenance of cesspools

It is essential that building owners are kept informed about the maintenance requirements where cesspools are provided. Cesspools should be inspected every two weeks for overflow and emptied as necessary. A typical emptying frequency is once per month and this can be estimated by assuming a filling rate of 150 litres per person per day. Cesspools which do not fill within the expected period may be leaking and should be checked out. A durable notice should be fixed within the building describing the necessary maintenance. Fig. 13.26 below is an example of a typical notice.

Cesspool foul drainage system

Details of Necessary Maintenance

Address of Property   _____

Location of treatment system   _____

The foul drainage system from this property is served by a cesspool

The system should be emptied approximately every *[insert design emptying frequency]* by a licensed contractor and inspected fortnightly for overflow.

The owner is legally responsible to ensure that the system does not cause pollution, a health hazard or a nuisance.

Fig. 13.26 Cesspool – typical maintenance notice.

13.5   Greywater and rainwater storage tanks

AD H2 concludes with limited guidance information on tanks for storage of greywater or rainwater for reuse within the building. The guidance does not apply to water butts used for storing rainwater for use in gardens.

Reclaimed water systems aid water conservation by reducing the amount of mains supply water used in houses and commercial buildings. Clearly there is a potential for reclaimed water systems to contaminate potable mains water supplies by inadvertent cross connection or backflow. Therefore, it is essential that water installations comply with the Water Supply (Water Fittings) Regulations 1999 in England and Wales, the Water Byelaws 1999 in Scotland or the Water Regulations in Northern Ireland.

Greywater should only be used for irrigation purposes and even then, certain precautions should be taken to prevent possible health problems occurring. Section 9 of the Water Regulations Advisory Scheme leaflet 09-02-04 *Reclaimed Water Systems. Information about installing, modifying or maintaining reclaimed water systems* gives general tips on reclaimed water uses and treatment. It also includes a great deal more information than can be found in AD H2, including a definition of greywater. This means water originating from the mains potable water supply that has been used for bathing or washing, washing dishes or laundering clothes.

Therefore, greywater and rainwater tanks should be:

- ventilated and prevent ingress of subsoil water or leakage of the contents;
- fitted with an anti-backflow device on any overflow connected to a drain or sewer to prevent contamination should a surcharge occur in the drain or sewer;
- provided with access for emptying and cleaning.

Access covers for emptying and cleaning should be sufficiently durable to resist the corrosive nature of the tank contents and should be designed to prevent unauthorised access (by being lockable or otherwise engineered to prevent personnel entry).

13.5.1   Alternative approach

Requirement H2 of Schedule 1 to the Building Regulations 2000 may also be met by complying with the relevant recommendations of BS 6297: 1983 *Code of practice for the design and installation of small sewage treatment works and cesspools*. These are – sections one, two, four and the appendices and clauses 6 – 11 of section three.

13.6   Rainwater drainage

Paragraph H3 of Schedule 1 to the Building Regulations 2000 requires that:

- any system carrying rainwater from the roof of a building is adequate; and
- paved areas around the building are so constructed as to be adequately drained.

It should be noted that only the following paved areas are covered by the Regulations:

- those which provide access for disabled people in accordance with requirement M2 (see Chapter 17 below);
- those which provide access to or from a place used for the storage of solid waste (see requirement H6(2) below); and
- those which give access to the building where this is intended to be used in common by the occupiers of one or more other buildings.

Rainwater from the roof of the building and any relevant paved areas must be taken to one of the following, listed in order of priority:

- an adequate soakaway or some other adequate infiltration system; or
- a watercourse; or
- a sewer.

Movement to a lower level in the order of priority may only be on the grounds of reasonable practicability since it is the purpose of the Regulation to encourage drainage connections to other than surface water sewers where this is technically feasible. This is an attempt to lessen the effect of flash flooding occurring in times of exceptionally heavy rainfall. This requirement does not apply to the gathering of rainwater for re-use.

The requirements of Paragraph H3 will be met if the following are fulfilled.

- Rainwater from paved areas and roofs is carried away from the relevant surface by a drainage system or some other appropriate means.
- Any rainwater drainage system:
  (a) conveys the flow of rainwater to a suitable outfall (soakaway, watercourse, surface water or combined sewer);
  (b) reduces to a minimum the risk of leakage or blockage;
  (c) is accessible for clearing blockages.
- Rainwater soaking into the ground (to a soakaway etc.) is sufficiently distributed so as not to damage the foundations of the building or any adjacent structure.

The emphasis in H3 on infiltration systems to dispose of rainwater means that it is essential to ensure adequate distribution of rainwater where it will not harm the structure of the building. Rainwater or surface water should never be discharged to a cesspool or septic tank.

13.6.1 Gutters and rainwater pipes

A rainwater drainage system should be capable of carrying the anticipated flow at any point in the system. The flow will depend on the area of roofs to be drained and on the intensity of the rainfall. For eaves gutters, a design rainfall intensity of 0.021 litres/second/m^2 (i.e. 75 mm of rainfall in any one hour) should be assumed in design

AD H3, Section 1

Diagram 1   Rainfall intensities for design of gutters and rainfall pipes (litres per second per square metre)

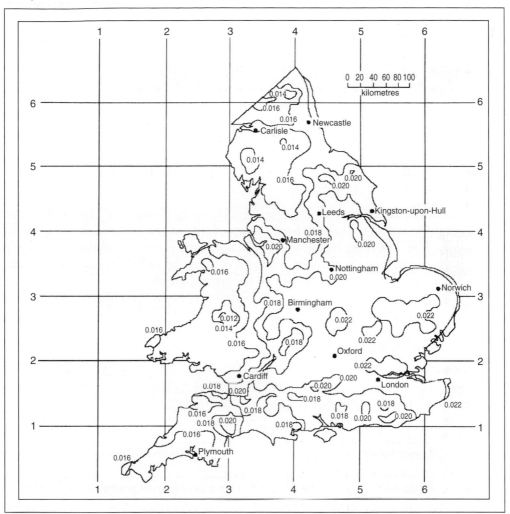

calculations. For valley gutters, parapet gutters, siphonic systems and other rain gathering systems the rainfall intensity should be obtained from Diagram 1 of AD H3 which is reproduced above.

For some roof designs incorporating valley gutters, parapet gutters or drainage from flat roofs it is possible that intense rainfall could cause overtopping of the construction resulting in water entering the building causing damage, wetting of insulation etc. The design of such systems should be carried out in accordance with BS EN 12056 *Gravity drainage systems inside buildings*.

The ultimate capacity of gutters and rainwater pipes depends on their length, shape, size and gradient and on the number, disposition and design of outlets. AD

AD H3, section 1

Table 2   Gutter sizes and outlet sizes.

Max effective roof area (m²)	Gutter size (mm dia)	Outlet size (mm dia)	Flow capacity (litres/sec)
6.0	—	—	—
18.0	75	50	0.38
37.0	100	63	0.78
53.0	115	63	1.11
65.0	125	75	1.37
103.0	150	89	2.16

Note
Refers to nominal half round eaves gutters laid level with outlet at one end sharp edged.
Round edged outlets allow smaller downpipe sizes.

H3 contains design data for half-round gutters up to 150 mm in diameter. They are assumed to be laid level and to have a sharp-edged outlet at one end only. Table 2 to section 1 of AD H3 is reproduced above and gives gutter and outlet sizes for the drainage of different roof areas for lengths of gutter up to 50 times the water depth. The gutter capacity should be reduced for greater lengths.

The maximum roof areas given in Table 2 are the largest effective areas which should be drained into the gutters given in the table. The effective area of a roof will depend on whether the surface is flat or pitched. Table 1 to section 1 of AD H3 shows how the effective area may be calculated for different roof pitches.

AD H3, section 1

Table 1   Calculation of area drained.

Type of surface	Effective design area (m²)
1 flat roof	plan area of relevant portion
2 pitched roof at 30° pitched roof at 45° pitched roof at 60°	plan area of portion × 1.29 plan area of portion × 1.50 plan area of portion × 1.87
3 pitched roof over 70° or any wall	elevational area × 0.5

Gutters should also be fitted so that any overflow caused by abnormal rainfall will be discharged clear of the building. Additional outlets may be necessary on flat roofs, valley gutters and parapet gutters to avoid over-topping.

Where it is not possible to comply with the conditions assumed in Table 2, further guidance is given in AD H3.

- Where an end outlet is not practicable the gutter should be sized to take the larger of the roof areas draining into it.
- If two end outlets are provided they may be 100 times the depth of flow apart.
- It may be possible to reduce pipe and gutter sizes if:
  (a)   the gutter is laid to fall towards the nearest outlet; or
  (b)   a different shaped gutter is used with a larger capacity than the half round gutter; or
  (c)   a rounded outlet is used.

In these cases reference should be made to the following parts of **BS EN 12056**:

- Part 3: *Roof drainage layout and calculation*, clauses 3 to 7
- Annex A and National Annexes
- Part 5: *Installation, testing instructions for operation and maintenance and use*, clauses 3, 4, 6 & 11.

Rainwater pipes should comply with the following rules.

- Discharge should be to a drain, gully, other gutter or surface which is drained.
- Any discharge into a combined system of drainage should be through a trap (e.g. into a trapped gully).
- Rainwater pipes should not be smaller than the size of the gutter outlet. Where more than one gutter serves a rainwater pipe the pipe should have an area at least as large as the combined areas of the gutter outlets and be large enough to take the flow from the whole contributing area.
- Discharge from a rainwater pipe onto a lower roof or paved area should be via a pipe shoe to divert water away from the building.
- Where a single downpipe serves a roof with an effective area greater than $25\,m^2$ and discharges onto a lower roof, a distributor pipe should be fitted to the shoe to ensure that the width of flow at the receiving gutter is great enough to prevent overtopping of the gutter.

13.6.2   Siphonic roof drainage systems

Using siphonic action to accelerate the flow of water from the gutters to the below ground drainage system enables small-diameter pipes to achieve high rates of discharge. This permits a reduction in the number of downpipes that need to be provided when compared with traditional systems of roof drainage, and this is particularly useful in large single-storey commercial buildings with restricted numbers of columns.

For the siphonic action to start, the pipework must be airtight. This requires special rainwater outlets with baffle plates, correctly dimensioned pipes and fully sealed joints. Additionally, care should be taken that other trades do not connect their internal drainage pipes into a siphonic system, and thus break the vacuum.

The following considerations should also be taken into account.

- Possible surcharging in the downstream drainage system as this can cause reductions in flow rates in downpipes.
- The time taken to prime the siphonic action may be excessive where long gutters are specified. In this case, overflow arrangements to prevent gutters from over-topping, should be provided.

More information on the design of siphonic systems of roof drainage may be obtained from Report SR 463 *Performance of syphonic drainage systems for roof gutters* published by Hydraulics Research Ltd. Reference should also be made to BS EN 12056: Part 3.

13.6.3   Eaves drop systems

In modern buildings it is normal for rainwater from roofs to be collected by a system of guttering and downpipes to be transmitted to a system of below-ground drainage. In fact, the requirement of H3 for adequacy or rainwater systems means that an eaves drop system (where rainwater is allowed to fall from the roof freely to the ground) can be a perfectly acceptable solution provided that the following design considerations are taken into account.

- The fabric of the building should be protected against ingress of water caused by splashing against the external walls.
- The entry of water into doorways and windows should be prevented.
- Persons should be protected from falling water in doorways etc.
- Splashback caused by water hitting the ground should be prevented from affecting people and the fabric of the building (e.g. by providing a gravel layer or angled concrete apron to deflect water away).
- Foundations should be protected from concentrated discharges which occur at valleys, valley gutters or from excessive flows caused by large roofs (where the area of roof per unit length of eaves is high).

13.6.4   Rainwater recovery systems

In order to conserve water supplies, it is possible to collect rainwater for re-use within the building provided that the following considerations are taken into account.

- Storage tanks should follow the guidance given in section 13.5 above.
- Pipework, valves and washouts used for recovered water should be clearly identified on marker plates in accordance with the recommendations the Water Regulations Advisory Scheme leaflet 09-02-04 *Reclaimed Water Systems. Information about installing, modifying or maintaining reclaimed water systems* where further guidance on the use of rainwater recovery systems will be also be found.

13.6.5   Materials

Materials used should be adequately strong and durable. Additionally:

- Gutters should have watertight joints under working conditions.
- Downpipes placed inside a building should be capable of withstanding the test for airtightness described in section 13.3.13 above.
- Gutters and rainwater pipes should be adequately supported with no restraint on thermal movement.
- Pipes and gutters of different metals should be separated by non-metallic material to prevent electrolytic corrosion.
- Siphonic roof drainage pipework should be designed to resist negative pressures.

13.6.6   Alternative method of design

The requirements of the 2000 Regulations for rainwater drainage can also be met by following the relevant recommendations of BS EN 12056 *Gravity drainage systems inside buildings*. These are:

- In Part 3 *Rainwater drainage, layout and calculation*, clauses 3 to 7.
- Annex A and National Annexes.
- In Part 5 *Installation and testing, instructions for operation, maintenance and use*, clauses 3, 4, 6, & 11.

13.6.7   Drainage of paved areas

Section 2 of AD H3 contains information on the design of rainwater drainage systems for paved areas around buildings and small car parks up to 4000 m². For the design of systems serving larger catchment areas the guidance in BS EN 752 *Drain and sewer systems outside buildings* Part 4: 1998 *Hydraulic design and environmental aspects* should be followed. Rainfall intensities of 0.014 litres/sec/m² (i.e. 50 mm of rainfall in any one hour) are assumed for normal situations. More accurate local figures can be obtained from Diagram 2 of Section 2 to AD H3 which is reproduced below. In very high risk areas where ponding could lead to flooding of buildings, drainage of paved areas should be designed in accordance with BS EN 752: Part 4.
  AD H3 describes three methods for draining paved areas:

- Allow pavings to drain freely onto adjacent pervious surfaces
- Use pervious paving
- Use impervious paving discharging to gullies or channels connected to a drainage system.

13.6.8   Design of freedraining surfaces

Surface water should not be allowed to soak into ground where the conditions are not suitable; however, paved areas do not always have to be served by underground

AD H3, Section 2

Diagram 2 Rainfall intensities for design of drainage from paved areas and underground rainwater drainage (litres per second per square metre)

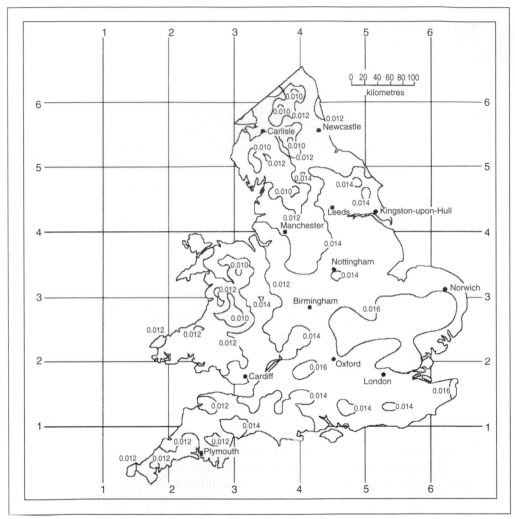

drainage systems. It is acceptable for paths, driveways and other narrow areas of paving to be freedraining to pervious areas (e.g. grassland) if the following conditions are met.

- Water should be directed away from buildings where foundations could be damaged. This can be achieved by suitable surface gradients (e.g. where ground levels would cause water to collect along the wall of a building a reverse gradient could be created at least 500 mm wide to divert water away).
- Impervious surfaces should have a cross fall of at least 1 in 60 to permit rapid draining. The fall across a path should not exceed 1 in 40.

- Where paving drains onto adjacent ground it should be finished flush with, or above the level of the surrounding ground to permit the water to run off.
- The soakage capacity of the ground should not be overloaded and where the adjacent ground is not sufficiently permeable to take the flow it may be necessary to provide filter drains (see section 13.7.4 below).

13.6.9   Pervious paving

As an alternative, and where it is not possible to drain large paved areas to adjacent pervious surfaces, it may be possible to construct pervious paving to deal with surface drainage. Pervious paving is made up of a porous or permeable surface material placed onto a granular layer which acts as a storage reservoir, retaining peak water flows until soakage into the underlying subsoil takes place. The storage layer should be designed on a similar basis to the design of the storage volume in a soakaway (see section 13.7.4 below).

On steeply sloping surfaces it will be necessary to check that the water level can rise sufficiently in the storage reservoir to enable its full capacity to be used. It will also be necessary to check that water is not inadvertently accumulating around the building foundations.

Pervious paving, on flat or sloping sites, may even be used where infiltration drainage is not possible (see section 13.7.4). In this case an impermeable barrier is placed below the storage layer to act as a detention tank or pond prior to discharge of the stored water to a drainage system (see section 13.6.10 below).

Pervious paving should not be used:

- where excessive amounts of sediment are present since these can enter the pavement and block the pores;
- in oil contaminated areas or where run-off may be contaminated with pollutants.

More information on the design of pervious paving can be found on pages 64 to 66 of CIRIA report C522: *Sustainable urban drainage systems – design manual for England and Wales.*

13.6.10   Paving connected to drainage system

Where it is not possible for the paving to be freedraining or for pervious paving to be used, impervious paving should be used in conjunction with gullies or channels connected to a drainage system. Gullies should comply with the following guidance.

- Be provided as necessary at low points to ensure that ponding does not occur.
- Be provided at intermediate positions so that individual gullies are not over-loaded and channels do not have excessive depths of flow.
- Have their gratings set about 5 mm below the surrounding paving to allow for settlement of the paving.

Since it is possible that drainage from pavings may encourage silt and grit to enter the drainage system this should be intercepted by providing suitably sized gully pots or catchpits.

13.6.11   Alternative method of design

The requirements of the 2000 Regulations for drainage of pavings can also be met by following the relevant recommendations of BS EN 752 *Drain and sewer systems outside buildings*. These are:

- in Part 4, *Hydraulic design and environmental considerations*, clause 11.
- National Annexes ND and NE.

13.7   Rainwater drainage below ground

13.7.1   Connections and outlets

Section 3 of AD H3 deals specifically with drainage systems carrying only rainwater. Where practicable, surface water drainage should discharge to a soakaway or other infiltration system. Discharge to a watercourse is the next best option but the consent of the Environment Agency may be required and they may put a limit on the rate of discharge, although this can be attenuated by the use of detention ponds or basins. Where these forms of outlet are not practicable, discharge should be made to a suitable sewer.

Combined systems (those carrying both foul and rainwater) are permitted by some drainage authorities where allowance is made for the additional capacity. Where a combined system does not have sufficient capacity, rainwater will need to be taken via a separate system to its own outfall. Even where a sewer is operated as a combined system and has sufficient capacity it may still be necessary to provide separate systems of drainage to the building in accordance with the provisions of Requirement H5 (see below section 13.9.2). Surface water drainage connected to a combined system should have traps on all inlets.

Pumped systems of surface water drainage may be needed where there is a tendency to surcharging or gravity connections are impracticable (see also section 13.3.28 above).

The design information contained below is suitable for the drainage of small impervious catchment areas up to 2 hectares with an assumed design rainfall intensity of 0.014 litres/sec/m^2 for normal situations. Rainfall intensity may also be obtained from Diagram 2 of AD H3 illustrated in section 13.6.7 above. Where it is intended to drain larger areas than 2 hectares or where low levels of surface flooding could cause flooding of buildings, reference should be made to BS EN 742: Part 4.

With the exception of pipe gradients and sizes, the recommendations given above for below ground foul drainage (materials, bedding and backfilling, clearance of blockages, workmanship and testing and inspection) apply equally to rainwater drainage below ground.

13.7.2 Pipe sizes and gradients

Drains should be laid to falls and should be large enough to carry the expected flow. The rate of flow will depend on the area of the surfaces (including paved or other hard surfaces) being drained. The capacity will depend on the diameter and gradient of the pipes. The minimum permitted diameter of any rainwater drain is 75 mm. Surface water sewers (i.e. drains serving more than one building) should have a minimum diameter of 100 mm. Diagram 3 to Section 3 of AD H3 is reproduced below and gives discharge capacities for rainwater drains running full where it will be seen that the capacity increases proportionately with the pipe diameter and gradient.

AD H3, Section 3

Diagram 3   Discharge capacities of rainwater drains running full.

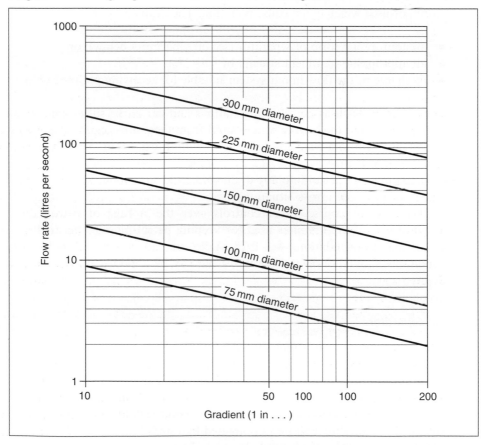

In general, the minimum permitted gradient of a pipe is related to its diameter as shown in Table 13.4 below.

Table 13.4 Pipe sizes and minimum gradients.

Pipe diameter mm	Minimum gradient
75	1 in 100
100	1 in 100
150	1 in 150
225	1 in 225
Over 225	See BS EN 752: Part 4

13.7.3 Contaminated runoff

It is an offence under section 85 (*offences of polluting controlled waters*) of the Water Resources Act 1991 to discharge any polluting or noxious material into coastal or underground water, or a watercourse. Since most surface water sewers discharge to watercourses, separate drainage systems should be provided where materials are stored or used which could cause pollution. The drainage system should include:

- an appropriate form of separator (see oil separators below); or
- an appropriate treatment system; or
- discharge of the flow into a system suitable for receiving polluted effluent.

Certain areas, such as petrol filling stations and car parks suffer from leakage or spillage of oil and the surface water runoff from these areas can find its way via the drainage system to a watercourse where pollution can occur. Since it is an offence under section 111 (*restrictions on the use of public sewers*) of the Water Industry Act 1991 to discharge petrol into any drain connected to a public sewer, oil separators should be provided in risk situations.

There are, of course, other controls over the storage of petrol, due to its combustible nature. Premises used for keeping petrol must be licensed under the Petroleum (Consolidation) Act 1928. A licence can be granted with or without conditions. Guidance on the storage of oil may be obtained from the Health and Safety Executive.

Oil separators

The type of oil separator that should be provided will depend on the risk of contamination represented by the site.

For comparatively low risk areas, such as paved areas around buildings and car parks, a bypass separator should be provided with a nominal size (NSB) of 0.0018 times the contributing area and a silt storage volume in litres equal to $100 \times$ NSB. Bypass separators treat all flows generated by rainfall rates of up to 5 mm/hr, thus accounting for 99% of all rainfall events. Flows above this rate are allowed to bypass the separator.

For fuel storage and other high risk areas, full retention separators should be provided with nominal size (NS) equal to 0.018 times the contributing area and a silt storage volume in litres equal to $100 \times$ NS. Full retention separators treat the full flow that can be delivered by the drainage system which is normally equivalent to the flow generated by a rainfall intensity of 50 mm/hr.

Separators should:

- be Class 1 when discharging to infiltration devices or surface water sewers (i.e. designed to achieve a concentration of less than 5mg/litre of oil under standard test conditions);
- be leaktight and adequately ventilated;
- have inlet arrangements that avoid directing the inflow to the surface of the water already in the separator;
- comply with the requirements of the Environment Agency and prEN 858 *Installations for separation of light liquids (e.g. petrol or oil)*;
- comply with the requirements of the licensing authority where the Petroleum Act applies;
- be regularly maintained to ensure continued effectiveness. It is normal for routine inspections to be carried out every six months including the completion of a log which details the inspection date, depth of oil and any cleaning undertaken; and
- be provided with sufficient access points to allow for inspection and cleaning of all internal chambers.

More information on the provision of oil separators may be found in the Environment Agency publication – Pollution Prevention Guidelines 3 (PPG 3) *Use and design of oil separators in surface water drainage systems.*

13.7.4   Infiltration drainage systems

Infiltration drainage systems are designed to return rainwater from roofs and pavings to the ground in the vicinity of the building, without involving connection to sewers or watercourses. They include such devices as soakaways, swales, infiltration basins, filter drains and detention ponds (but see the comments on these below). AD H3 gives a very brief summary of the various infiltration devices which are available; however the information provided is too brief to be of any real use for the designer. The notes which follow have been enhanced using the various documents referred to in the text. These reference sources are essential for anyone seriously interested in infiltration drainage systems.

It is not always possible to provide infiltration drainage to a building. For example, infiltration devices should not be provided in the following situations.

- Within 5 m of a road or building.
- In areas of unstable land. (Annex 1 of Planning Policy Guidance Note 14 warns against the use of infiltration systems in areas subject to landslip).
- In ground with a high water table (i.e. where the water table reaches the base of the device at any time of the year.
- Where ground water source or resource might be polluted by the presence of contamination in the runoff.

- At such a distance from drainage fields, drainage mounds or other soakaways so that the overall soakage capacity of the ground would be exceeded and the effectiveness of any drainage field would be impaired.

Soakaways

Soakaways should be designed to store the immediate surface water runoff and allow for its efficient infiltration into the surrounding soil. Stored water must be discharged sufficiently quickly to provide the necessary capacity to receive runoff from a subsequent rainfall event. The time taken for discharge depends upon the soakaway shape and size, and the infiltration characteristics of the surrounding soil.

Soakaways serving catchment areas of less than $100\,m^2$ are usually built as square or circular pits filled with rubble or lined with dry-jointed masonry or pre-cast perforated concrete ring units surrounded by suitable granular backfill. For drained areas above $100\,m^2$, soakaways can be lined pits or of trench type and usually a depth of 3 to 4 m is adequate if ground conditions allow. Trench soakaways are cheaper to dig with readily available excavating equipment.

Although the design of soakaways should be carried out by considering storms of different durations over a ten year period in order to determine the maximum storage volume, for small soakaways serving $25\,m^2$ or less, a design rainfall of 10 mm in five minutes can be taken to represent the worst case. For soakaways serving larger areas reference should be made to **BRE Digest 365** *Soakaway design* or BS EN 752: Part 4. Where the percolation characteristics of the ground are marginal it may still be possible to use soakaways in conjunction with overflow drains.

The percolation test described in AD H2 (see Fig. 13.19 above) may be carried out to determine the capacity of the soil to receive infiltration. The value of V_p from the percolation test may be used in the equation below to determine the soil infiltration rate:

$$f = 10^{-3}/2V_p$$

where f = the soil infiltration rate

Therefore, assuming a value of V_p of 20,

$$f - 1/1000 \times 2 \times 20 = 0.000025\,m/sec$$

The storage volume of the soakaway should be able, during storm conditions, to accommodate the difference between the inflow volume and the outflow volume.

The inflow volume is simply calculated by considering the design rainfall depth during a storm multiplied by the drainage area. Therefore, if a rainfall depth of 10mm is considered over an area of $25\,m^2$ in a five minute period:

$$\text{the inflow volume} = 0.01 \times 25 = 0.25\,m^3$$

The outflow volume (O) is calculated from the equation:

$$O = a_{s50} \times f \times D$$

Where:
a_{s50} is the area of the side of the storage volume when filled to 50% of its effective depth, and D is the duration of the storm in minutes.

Using the figures from the example given above and assuming a soakaway 2m deep and 2m x 1m in area with the inlet 1m below ground:

$$O = 4 \times 0.000025 \times 5 \times 60 = 0.03\,m^3$$

Therefore the difference between the inflow volume and the outflow volume equals the storage volume $= 0.25 - 0.03 = 0.22\,m^3$

The actual volume of soakaway below inlet $= 1 \times 1 \times 2 = 2\,m^3$ which is more than adequate.

Swales

Swales are simply grass-lined channels with shallow side-slopes used to carry rainwater from a site. They can also control the flow and quality of surface runoff and allow a certain amount of the flow to infiltrate into the ground. To increase the infiltration and detention capacity of swales they can be provided with low check dams across their width. To prevent overtopping during wet spells, it is possible to provide an overflow at one end discharging into another form of infiltration device or watercourse. They can be used to treat runoff from small residential developments, parking areas and roads.

Infiltration basins

These are dry grass-lined basins for storage of surface runoff that are free from water under dry weather flow conditions. They can be designed to manage water quantity and quality and are used to encourage surface water infiltration into the ground.

Filter drains

Otherwise known as french drains, filter drains consist of geotextile-lined trenches filled with gravel, sometimes containing perforated pipes to assist drainage. They are designed so that most of the flow enters the filter drain directly from the runoff or is discharged into it through other drains from where it infiltrates into the ground.

Detention ponds

The term 'detention pond' appears to be a mistake in the AD since the reference material given for this section in AD H3 (*Sustainable urban drainage systems – a design manual for England and Wales* published by CIRIA) makes reference to 'detention basins' and 'retention ponds' but not 'detention ponds'.

According to this reference source '*detention basins are vegetated depressions. They are formed below the surrounding ground, and are dry except during and immediately following storm events. Detention basins only provide flood storage to attenuate flows. Extending the detention times improves water quality by permitting the settlement of course silts*'.

On the other hand, '*retention ponds are permanently wet ponds with rooted wetland and aquatic vegetation – mainly around the edge. The retention time of several days provides better settlement conditions than offered by extended detention ponds and provides a degree of biological treatment*'. The description given in AD H3 could apply to either or both of the above.

13.7.5   Alternative method of design

Requirement H3 can also be met by following the relevant recommendations of BS EN 752 *Drain and sewer systems outside buildings*.

These are:

- in Part 4, *Hydraulic design and environmental considerations*, clauses 3 to 12;
- National Annexes NA, NB and ND to NI.

Additionally, detailed information about design and construction can be found in:

- BS EN 1295, *Structural design of buried pipelines under various conditions of loading:* Part 1: 1998 *General requirements*; and
- BS EN 1610: 1998 *Construction and testing of drains and sewers*.

13.8   Building over existing sewers

13.8.1   Introduction

Control of building works over or near existing sewers has long been subject to control in England and Wales. Until the coming into effect of the first amendment to the 2000 Regulations on 1 April 2002 this control was exercised by local authorities through the medium of section 18 of the Building Act 1984. The first amendment introduced a new Building Regulation requirement H4, which replaced section 18 and can be administered by both local authorities and approved inspectors, although the substance of section 18 has been little altered by the change. What has altered is the substantial amount of guidance provided by Approved Document H4.

13.8.2   Interpretation

The following terms apply in AD H4:

DISPOSAL MAIN – Any pipe, tunnel or conduit used for the conveyance of effluent to or from a sewage disposal works, which is not a public sewer.

MAP OF SEWERS – Any records kept by a sewerage undertaker under section 199 of the Water Industry Act 1991.

13.8.3   Building over sewers

H4 requires that where it is intended to:

- erect a building; or
- extend a building; or
- carry out works of underpinning to a building

near to or over a drain, sewer or disposal main, then the work must be carried out so that it is not detrimental to the building or extension or to the continued maintenance of the drain, sewer or disposal main.

H4 is limited to work carried out:

- near to or over a drain, sewer or disposal main which is shown on any map of sewers; or
- which will result in interference with the use of, or obstruction of any person's access to, any drain, sewer or disposal main shown on any map of sewers.

In order to meet the requirements of H4 it is necessary to ensure the following.

(1)   That the work of building, extending or underpinning:
- is expedited so as not to overload or otherwise damage the drain, sewer or disposal main both during construction and after it is completed; and
- will not prevent reasonable access to any manhole or inspection chamber situated on the drain, sewer or disposal main.
(2)   That where the drain, sewer or disposal main needs to be replaced:
- a satisfactory diversionary route can be provided; or
- the building or extension will not unduly obstruct the replacement work if the current alignment is maintained.
(3)   That if the drain, sewer or disposal main fails, the risk of damage to the building will not be excessive. To assess the risk of damage to the building it is necessary to consider:
- the nature of the ground;
- the location, construction and condition of the drain, sewer or disposal main;
- the nature, volume and pressure of the flow in the drain, sewer or disposal main; and
- the design and construction of the building's foundations.

13.8.4   Application

The provisions of H4 apply where it is intended to erect, extend or underpin a building that is situated over, or within 3 m of the centreline of, an existing drain, sewer or disposal main shown on the sewer records of the sewerage undertaker, even if the sewer is not a public sewer.

The public have access to copies of sewer record maps during normal office hours, these being held by both sewerage undertakers and local authorities.

13.8.5   Consultation

When it is proposed to carry out any work to which H4 applies, the developer should always consult the owner of the drain or sewer (for public sewers this would be the sewerage undertaker) unless, of course, the developer is also the owner. In the case of public sewers the sewerage undertaker should be able to provide useful information regarding the age, location, condition and depth to invert of the sewer. They may also be able to arrange an inspection and if a public sewer needs to be repaired or replaced, they will carry out this work. The sewerage undertaker should also be consulted where it is proposed to build or extend over a sewer that is later intended for adoption.

In order to ensure compliance with H4 it will be necessary to apply to the relevant building control body (local authority or approved inspector) so that the works can be properly controlled. This will involve the carrying out of further consultations as follows.

- If using the local authority you must deposit full plans. This enables the local authority to carry out its duties under Regulation 14A of the Building Regulations 2000 to consult the sewerage undertaker as soon as practicable after the plans have been deposited. The local authority is not permitted to pass the plans or issue a completion certificate until the consultation has taken place (the sewerage undertaker has up to 15 days to reply) and it must have regard to the views expressed by the sewerage undertaker.
- If using an approved inspector, he must consult the sewerage undertaker where an initial notice or amendment notice is to be given (or has been given). The consultation must take place at the following stages:
  - (a) before or as soon as is practicable after giving an initial notice or an amendment notice;
  - (b) before giving a plans certificate (whether or not this is combined with an initial notice); and
  - (c)   before giving a final certificate.

  Additionally, he must allow the sewerage undertaker up to 15 working days to comment, and have regard to the views it expresses, before giving a plans certificate or final certificate to the local authority.

13.8.6   Building near drains or sewers in risk situations

Unless special measures are taken, buildings should not be constructed or extended over or within 3 m of any of the following:

- drains or sewers in poor condition (pipes which are cracked, fractured, misaligned or more than 5% deformed);

- drains or sewers constructed from brick or other masonry;
- rising mains (except those used only to drain the building);

since failure of the drain or sewer would expose the building to a high level of risk.

Additionally, certain soil types (fine sands, fine silty sands, saturated silts and peat) are easily eroded by groundwater leaking into drains or sewers. Therefore, failure of a drain or sewer could result in erosion of soil from around the foundations thereby exposing the building to undue risk. Where such soils are present, buildings should not be constructed or extended over or within 3 m of any drain or sewer to which H4 applies unless special measures are taken in the design and construction of the foundations to mitigate the effect of drain or sewer failure. Special measures are not needed if the invert of the drain or sewer is:

- above the level of the foundations; and
- above the level of the groundwater; and
- no deeper than 1 m.

13.8.7 Access for maintenance

Fig 13.27 below gives details of the precautions that should be taken to ensure that sewers remain accessible when buildings are constructed over or within 3m of them. The following main points should be observed.

- Do not construct a building or extension over a manhole, inspection chamber or access fitting on a sewer (i.e. a drain serving more than one property).
- Locate access points to sewers where they are accessible and apparent for use in emergency. Where this provision is already met by the existing sewer, do not construct a building or extension which would remove this provision unless a satisfactory alternative on the line of the sewer can be agreed with the sewer owner.
- Ensure that a satisfactory diversionary route is available at least 3 m from the building to allow the drain or sewer to be reconstructed without affecting the building. Where existing drains or sewers more than 1.5 m deep have access for mechanical excavators, ensure that the diversionary route also has such access.
- Unless the sewer owner agrees, the length of drain or sewer under a building should not exceed 6 m in length.
- Do not build over or near an existing sewer more than 3 m deep or more than 225 mm diameter without the sewer owner's permission.

13.8.8 Protection of drains and sewers

Approved Document H4 contains details of protection which should be provided both during construction of the building over the drain or sewer, and subsequently to prevent damage by settlement of the building. Details of other protection

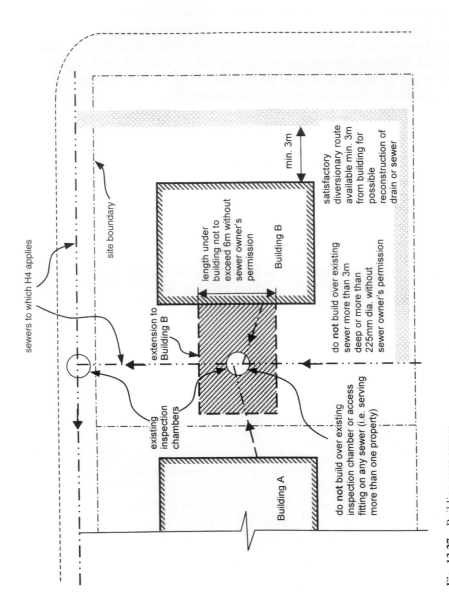

Fig. 13.27 Building over sewers.

measures for below ground drainage also apply to drains or sewers covered by the following notes. They may be found in sections 13.3.20 (pipes); 13.3.25 (drains); 13.3.26 (surcharging); and 13.3.27 (rodent control).

Protection during construction

During construction activities drains and sewers should be protected from damage by:

- providing barriers to keep construction traffic and heavy machinery away from the line of the sewer;
- not storing heavy materials over drains or sewers.

Piling works present a special risk and care should be taken to avoid damage to drains and sewers in the vicinity of such activities. The following precautions should be taken:

- a survey should be carried out to establish the position of the drain or sewer;
- where piling will take place within 1 m of a drain or sewer trial holes should be excavated to establish its exact position and the location of any connections;
- piling should not be carried out where the distance from the outside of the pile to the outside of the drain or sewer is less than twice the pile diameter.

Protection from settlement

Drains or sewers passing under buildings should comply with the following guidance.

- Provide at least 100 mm of granular or other flexible filling round the pipe.
- Where excessive subsidence is possible provide additional flexible joints or adopt other solutions (e.g. suspended drainage).
- Provide special protection where the crown of the pipe is within 300 mm of the underside of the slab (see section 13.3.20 above).
- For drains or sewers less than 2 m deep, increase the depth of the foundations in the vicinity of the drain or sewer so that it may pass through the wall.
- For drains or sewers greater than 2 m deep, design the foundations as a lintel spanning the drain or sewer. The 'lintel' should extend at least 1.5 m on either side of the pipe and should be designed so that no loads are transmitted to the drain or sewer.
- Where the drain or sewer passes through a wall or foundation follow the guidance given in section 13.3.25 above and shown in Fig. 13.14.
- Trenches for drains and sewers should only be excavated below the level of the building foundations if the precautions described in section 13.3.25 and illustrated in Fig.13.15 are taken.

13.9 Separate systems of drainage

13.9.1 Introduction

Control over the provision of separate systems of drainage was first introduced into the Building Regulations with the coming into force of the first amendment to the 2000 Regulations on 1 April 2002. The provisions are aimed at helping to minimise the volume of rainwater which enters the public foul sewer system since this can lead to overloading of the capacity of sewers and treatment works, and can cause flooding. At the date of this edition, this requirement sits uncomfortably beside six local Acts of Parliament which cover broadly the same area of control and are listed in Appendix A. A consultative document published by the Office of the Deputy Prime Minister in July 2002 aims to repeal these six Acts and will probably come into force some time in 2003.

13.9.2 Requirement H5

Any system for discharging water to a sewer which is provided to take rainwater from roofs or from paved areas around buildings covered by the requirements of H3 (see section 13.6), must be separate from that provided for the conveyance of foul water from the building.

For H5 to apply:

- the drainage system must be provided in connection with the erection or extension of a building; and
- it must be reasonably practicable for the system to discharge directly or indirectly to a sewer for the separate conveyance of surface water.

Additionally, the sewer must be:

- shown on a map of sewers (see definition in section 13.8.2 above); or
- under construction either by the sewerage undertaker or by some other person (although in this case the sewerage undertaker must have agreed in advance to adopt the drain or sewer in accordance with section 104 of the Water Industry Act 1991).

13.9.3 Meeting the requirement

The requirements of H5 can be met in either of two ways:

- by connecting to separate public sewers which are already in existence; or
- by providing separate drainage systems on the site of the building which will later be connected to separate public sewers which are under construction at the time of the building works.

13.9.4   Provision where separate sewer systems already exist

Where the sewerage undertaker has provided separate sewer systems, the owner or occupier of a building has a right to connect to the public sewers (see section 106 of the Water Industry Act 1991) provided that the following restrictions are observed.

- The surface water drainage from the building must be connected to the appropriate public surface water sewer.
- The foul water drainage from the building must be connected to the appropriate public foul water sewer.
- The way in which the connection is made must not be prejudicial to the public sewer system.
- 21 days notice must be given to the sewerage undertaker of the intention to make the connection.

It is normal for the sewerage undertaker to carry out the work of making the connection and recover its reasonable costs from the developer (see section 107 of the Water Industry Act 1991). Alternatively the developer may be permitted to carry out the work under the supervision of the sewerage undertaker.

13.9.5   Provision where separate sewer systems are proposed

Separate sewer systems should still be provided to drain the building even if only a combined system exists at the time of the building works, provided that separate public sewers are under construction by the sewerage undertaker, or by some other person for later adoption by the sewerage undertaker. Depending on the respective programmes for the building works and the public sewer construction, it may be necessary initially to connect the separated site drainage to the existing combined sewer. Later reconnection to the separate sewer systems can be made when these are completed thus minimising disruption to the building occupiers.

13.9.6   Dealing with contaminated surface water

It should be noted that the necessity to connect to a separate surface water sewer would only apply if the surface water was uncontaminated. Drainage from areas where materials are stored could contaminate runoff and lead to pollution if discharged to a surface water sewer. The alternative of discharging such contaminated water to a foul sewer needs to be discussed with the sewerage undertaker (see section 106 of the Water Industry Act 1991 and notes above) whose consent is required. It will also be necessary to consult the sewerage undertaker when connecting such contaminated runoff via a new foul sewer to an existing combined sewer, if it is intended that this will eventually be reconnected to a foul sewer that is proposed or under construction.

13.10   Solid waste storage

13.10.1   Introduction

The efficacy of the refuse storage system is dependent on its capacity and ease of collection by the waste collection authority. Under section 46 (*Receptacles for household waste*) and section 47 (*Receptacles for commercial or industrial waste*) of the Environmental Protection Act 1990, the waste collection authority has powers to specify the type and number of receptacles which should be provided and the position where the waste should be placed for collection. Therefore it is important that consultations take place with the waste collection authority to establish its specific requirements regarding the storage and collection of waste.

The opportunity has been taken in AD H6 to give general recommendations regarding the separate storage of waste for recycling. This is interesting since the Building Regulations do not cover the recycling of household or other waste. However, there are moves afoot to amend sections 46 and 47 of the Environmental Protection Act 1990 to allow for separate storage, and of course, there are a number of national initiatives on recycling and waste reduction which the AD is attempting to support. From a legal standpoint it is unlikely that the recommendations can, in fact, be enforced.

13.10.2   Requirement H6

Buildings are required to have:

- adequate means of storing solid waste;
- adequate means of access for the users of the building to the place of storage; and
- adequate means of access from the place of storage to:
  - (a)   a collection point where one has been specified by the waste collection authority under section 46 (for household waste) or section 47 (for commercial waste) of the Environmental Protection Act 1990; or
  - (b)   to a street (in the case of no collection point being specified).

The requirements of paragraph H6 may be met by providing solid waste storage facilities which are:

- large enough, bearing in mind the requirements of the waste collection authority for the number and size of receptacles (this relates to the quantity of refuse generated and the frequency of removal, see sections 46 and 47 of the Environmental Protection Act 1990);
- designed and sited so as not to present a health risk; and
- sited so as to be accessible for filling by people in the building and for removing to the access point specified by the waste collection authority.

13.10.3   Domestic buildings – storage capacity

Assuming weekly collection, dwellinghouses, flats and maisonettes up to four storeys high should have, or have access to, a location large enough to accommodate at least two movable, individual or communal containers, which meet the requirements of the waste collection authority.

The location should cater for separated waste (i.e. one container taking waste for recycling and another taking all other waste). The combined capacity of the two containers should not be less than $0.25\,m^3$ per dwelling (or such other capacity as is agreed with the waste collection authority). If the waste collection authority does not provide weekly collections then larger capacity containers or more individual containers will need to be provided.

The size of the location will depend on whether this is based on the provision of communal or individual storage. Where individual storage is provided for each dwelling this should be an area with dimensions of at least $1.2\,m \times 1.2\,m$. The waste collection authority should be consulted regarding space requirements for communal storage areas.

Dwellings in buildings above the fourth storey may either:

- share a container fed by a chute for non-recyclable waste, plus be provided with separate storage for waste which is to be recycled; or
- be provided with storage compounds or rooms for both types of waste if suitable management arrangements can be assured for conveying the waste to the place of storage.

For large blocks, recyclable waste can also be dealt with by providing 'Residents Only' recycling centres (places where residents can bring their own recyclable waste for storage in large containers, such as bottle banks).

13.10.4   Siting of waste containers and storage areas

Waste containers should comply with the following rules with regard to siting.

- For new buildings it should be possible to take a container to a collection point without taking it through a building. (It is permissible to pass through a porch, garage, carport or other covered space.) Buildings should not be extended or converted in such a way as to remove such an access facility where it already exists.
- Waste containers and chutes should not be sited more than 25 m from the waste collection point specified by the waste collection authority.
- Householders should not be required to carry refuse more than 30 m to a storage area for a waste container or chute (excluding any vertical distance).
- For waste containers with capacities up to 250 litres:
  - (a) steps should be avoided wherever possible on the route between the container store and the waste collection point and where unavoidable they should be restricted to three in number;

(b)   ideally, slopes should not be greater than 1 in 12 but where this is unavoidable they should be of restricted length and not in a series.

- For waste containers with capacities greater than 250 litres the storage area
should be located so that steps are avoided altogether.
- The waste collection authority should be consulted to ensure that the collection
point can be accessed by its normal size of waste collection vehicle.
- External locations for waste containers should be sited in shade or under shelter
away from windows or ventilators.
- Waste storage areas should not obstruct pedestrian or vehicle access routes to
buildings.

13.10.5   Design of waste containers and storage areas

Enclosures, compounds or storage rooms

Enclosures, compounds and storage rooms should comply with the following.

- Be designed to allow room for filling and emptying.
- Have a clear space of 150 mm provided between and around containers.
- Be permanently ventilated at top and bottom.
- Have paved impervious floors.
- Be secure to prevent access by vermin (unless, in the case of compounds, the
refuse is stored in secure containers with close fitting lids).
- If enclosing communal containers, have:
  (a)   clear headroom of 2 m;
  (b)   provision for washing down and draining the floor into a drainage system
designed to receive polluted effluent;
  (c)   gullies incorporating traps which maintain their seals even after prolonged
periods of disuse.
- If enclosing individual containers, be sufficiently high to allow the lid to be
opened for filling.
- Where storage rooms are provided these should contain separate rooms for
recyclable and non-recyclable waste.

A waste storage facility which is located in a publicly accessible area or in an open
area around a building (such as a front garden) should be provided with an
enclosure or shelter.

Refuse chutes

For high rise domestic developments where refuse chutes are provided, AD H6
recommends that they should be at least 450 mm in diameter and constructed with:

- smooth, non-absorbent inner surfaces
- close fitting access doors at each storey containing a dwelling
- ventilation at top and bottom.

Alternatively, refuse chutes may be designed in accordance with the relevant clauses in BS 5906: 1980 *Code of practice for storage and on-site treatment of solid waste from buildings*. Figure 13.28 is based on the recommendations of BS 5906: 1980 and illustrates a typical refuse chute installation.

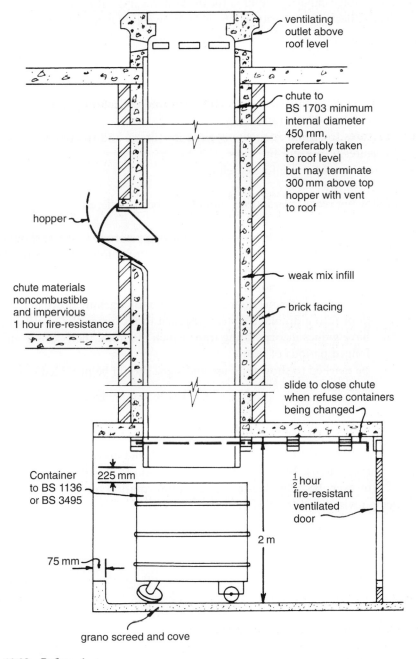

ventilating
outlet above
roof level

chute to
BS 1703 minimum
internal diameter
450 mm,
preferably taken
to roof level
but may terminate
300 mm above top
hopper with vent
to roof

hopper

weak mix infill

brick facing

chute materials
noncombustible
and impervious
1 hour fire-resistance

slide to close chute
when refuse containers
being changed

Container
to BS 1136
or BS 3495

225 mm

½ hour
fire-resistant
ventilated
door

2 m

75 mm

grano screed and cove

Fig. 13.28   Refuse chutes.

13.10.6   Non-domestic buildings

In the development of non-domestic buildings special problems may arise. It is therefore essential to consult the refuse collection authority for their requirements with regard to the following.

- The storage capacity required for the volume and nature of the waste produced. (The collection authority will be able to give guidance as to the size and type of container they will accept and the frequency of collection).
- Storage method. This may include details of any proposed on-site treatment and should be related to the future layout of the development and the building density.
- Location of storage and treatment areas and collection points, including access for vehicles and operatives.
- Measures to ensure adequate hygiene in storage and treatment areas.
- Measures to prevent fire risks.
- Segregation of waste for recycling.

The following recommendations should also be considered.

- Rooms and compounds provided for the open storage of waste should be secure to prevent access by vermin (unless in the case of compounds, the refuse is stored in secure containers with close fitting lids).
- Waste storage areas should:
  (a) have an impervious floor and provision for washing down and draining the floor into a drainage system designed to receive polluted effluent;
  (b) have gullies incorporating traps which maintain their seals even after pro-longed periods of disuse;
  (c) be marked to show their use and signs should be provided to indicate their location.

13.10.7   Alternative approach

As an alternative to the recommendations listed above for waste disposal, it is permissible to use BS 5906: 1980 especially clauses 3 to 10, 12 to 15 and Appendix A. However, BS 5906 does not contain information on recycling. It is currently being revised and the new edition is expected to contain such information.

14 Combustion appliances and fuel storage systems (Part J)

14.1 Introduction

Part J of Schedule 1 to the Building Regulations 2000 (as amended) is concerned with the safe installation and use of combustion appliances in buildings. In 2002 Part J was extended to include the protection of liquid fuel storage systems (oil and liquefied petroleum gas) from fire in neighbouring buildings and protection against pollution resulting from spills from oil storage tanks.

It is limited to fixed appliances burning solid fuel, oil or gas and to incinerators. (It is assumed that this means incinerators that burn solid fuel, oil or gas since no further guidance is given.) This excludes all electric heating appliances and small portable heaters such as paraffin stoves.

In order that it may function safely a combustion appliance needs an adequate supply of combustion air and it must be capable of discharging the products of combustion to outside air. This must be achieved without allowing noxious fumes to enter the building and without causing damage by heat or fire to the fabric of the building.

14.1.1 The Clean Air Acts 1956–1993

As a direct result of the great London smog of 1952 and after the Report of the Committee on Air Pollution (Cmnd. 9322, November 1954) the Clean Air Act 1956 was passed to give effect to some of the Committee's recommendations. The 1956 Act, amended in 1968, was consolidated in the Clean Air Act 1993. Its main provisions may be summarised briefly, and should be borne in mind in considering the effect of Part J of Schedule 1 to the 2000 Regulations.

The Act makes it an offence to allow the emission of *dark smoke* from a chimney, but certain special defences are allowed, e.g. unavoidable failure of a furnace. DARK SMOKE is defined as smoke which appears to be as dark as, or darker than, shade 2 on the Ringelmann Chart. Regulations made under the Act amplify its provisions in relation to industrial and other buildings, and it should be noted that the prohibition on dark smoke applies to all buildings, railway engines and ships. However, its chief effect is on industrial and commercial premises.

House chimneys rarely emit dark smoke, but local authorities may, by order confirmed by the Secretary of State for the Environment, declare *smoke control areas*. In a smoke control area the emission of smoke from chimneys constitutes an offence, although it is a defence to prove that the emission of smoke was not caused by the use of any fuel other than an authorised fuel. Regulations prescribe the following authorised fuels: anthracite; briquetted fuels carbonised in the process of manufacture; coke; electricity; low temperature carbonisation fuels; low volatile steam coals; and fluidised char binderless briquettes.

The 1956 Act, as amended, provides for the payment of grants by local authorities, and Exchequer contributions, towards the cost of any necessary adaptation or conversion of fireplaces to smokeless forms of heating in private dwellings in smoke control areas.

14.1.2   Combustion appliances and fuel storage systems

Although Part J applies to any combustion installation and liquid fuel storage system within the limits on application, the guidance in Approved Document J deals mainly with domestic installations, such as those which comprise space and water heating systems and cookers and their flues, and their attendant oil and LPG fuel storage systems.

Therefore, the guidance is concerned only with combustion installations having the following power ratings:

- solid fuel installations of up to 50 kW rated output;
- gas installations of up to 70 kW net (77.7 kW gross) rated input; and
- oil installations of up to 45 kW rated heat output;

and with fuel storage installations with the following capacities:

- heating oil storage installations with capacities up to 3500 litres; and
- liquefied petroleum gas (LPG) storage installations with capacities up to 1.1 tonne.

It should be noted, however, that no upper size limit is specified on the application of paragraph J5 of Part J of Schedule 1 to the Building Regulations 2000 (which deals with the protection of liquid fuel storage systems).

There are no specific references to incinerators or to the installation of appliances with a higher rating than those given above in AD J even though these are covered by the requirements of Part J and it is evident that specialist guidance will usually be needed (since these will almost invariably be installed under the supervision of a heating engineer). Some larger installations can be shown to comply by adopting the relevant recommendations of:

- CIBSE Design Guide Volume B, and
- practice standards produced by the British Standards Institute and the Institution of Gas Engineers.

14.2   Interpretation

Compared with the previous edition of AD J in which only one definition appeared, in an attempt to provide clarity, the current edition contains no fewer than 39!

APPLIANCE COMPARTMENT means an enclosure specifically constructed or adapted to accommodate one or more gas or oil-fired appliances.

BALANCED COMPARTMENT refers to a method of installing an open-flued appliance into a compartment so that it is sealed from the remainder of the building and whose ventilation is so arranged in conjunction with the appliance flue as to achieve a balanced flue effect (see below for definitions of BALANCED FLUED APPLIANCE and OPEN FLUED APPLIANCE).

BALANCED FLUED APPLIANCE means a combustion appliance that draws its combustion air from a point immediately adjacent to the point where it discharges its combustion products. The inlet and outlet are so arranged as to substantially balance any wind effects. Balanced flues can run vertically, but they usually discharge horizontally through the external wall on which the appliance is situated.

BOUNDARY. This definition should not be confused with the definition of relevant boundary given in AD B (see Chapter 7) and applies solely for the purposes of AD J. What is referred to here is the boundary of the land or buildings belonging to and under the control of the building owner. Some sections of AD J relate to the distance that outlets from chimneys and flues can be from the boundary. In these cases the measurements may be taken up to the centreline of adjacent routes or waterways. Other sections refer to the distance that oil and LPG storage tanks must be from the boundary. In these cases the measurements may only be taken to the physical boundaries of the site.

BUILDING CONTROL BODY may be either the local authority or an approved inspector. These are fully described in Chapters 3 and 4.

CAPACITY of an oil tank means its nominal volume as stated by the manufacturer. This is usually about 97% of the volume of liquid required to totally fill it.

CHIMNEY includes a wall or walls enclosing one or more flues (see Fig. 14.1). (The chimney for a gas appliance may be referred to as the flue in the gas industry.)

COMBUSTION APPLIANCE means an apparatus which burns fuel to generate energy for space heating, water heating, cooking etc. (e.g. boilers, warm air heaters, water heaters, fires, stoves and cookers), but not including fuel delivery or heat distribution systems.

DECORATIVE FUEL EFFECT (DFE) FIRES are described in BS 5871: Part 3. These are gas-fired imitations, which can be substituted for the solid fuel appliances in open fires. Where suitable, they can also be used in flueboxes designed for gas appliances only. Common designs include beds of artificial coals shaped to fit into a fireplace recess or baskets of artificial logs for use in larger fireplaces or under canopies (see Fig. 14.2(a)).

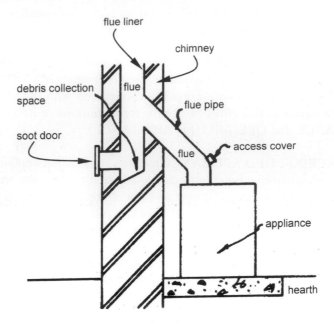

Fig. 14.1   Interpretation – chimneys and flues.

DESIGNATION SYSTEM for the performance characteristics of a chimney or its components means designations which are referred to throughout AD J and details of the designation system are contained in BS EN 1443:1999. This allows the performance characteristics to be expressed by means of a code such as *Chimney EN 1443 T400 P1 S W 1 R22 C50*. In this example the various parts of the code stand for:

Chimney EN 1443 – the number of corresponding European standard
T400 – the temperature class (i.e. a product with a normal working temperature of 400°C, see Table 1 of EN 1443)
P1 – the pressure class N or P or H (negative pressure, positive pressure or high positive pressure)
S – sootfire resistance class S or O (S – chimney has sootfire resistance, O – no sootfire resistance)
W – resistance to condensate class (W – chimneys operating under wet conditions, D – dry conditions)
1 – corrosion resistance class (1 – gas fuel, 2 – oils with sulphur content less than 0.2% and natural wood, 3 – oils with sulphur content over 0.2 % and solid mineral fuels and peat)
R22 – the thermal resistance
C50 – distance to combustible material (i.e. 50 mm)

Sometimes the designation is shortened, e.g. clay ceramic flue liners with the designation EN 1457 T600 N2 S D 3 can be described as Class A1N2.

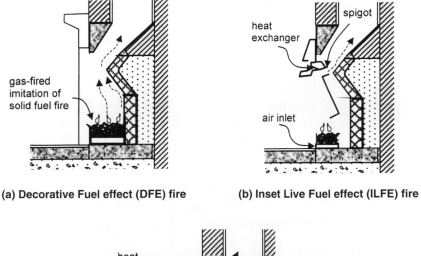

(a) Decorative Fuel effect (DFE) fire   **(b) Inset Live Fuel effect (ILFE) fire**

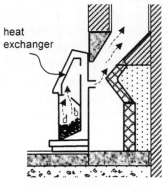

(c) Radiant convector gas fire

Fig. 14.2   Interpretation – gas fires.

DRAUGHT BREAK is an opening into any part of the flue serving an open flued appliance, formed by a factory-made component. This can allow dilution air to be drawn into a flue or can be used to lessen the effects of down-draught on combustion in the appliance.

DRAUGHT DIVERTER is a form of draught break which allows the appliance to operate without interference from down-draughts that may occur in adverse wind conditions and excessive draught.

DRAUGHT STABILISER is a factory made counter-balanced flap device usually mounted in the fluepipe or chimney, but sometimes located on the appliance, which admits air to the flue from the same space as the combustion air. It is designed to prevent excessive variations in the draught.

FACTORY-MADE METAL CHIMNEYS are prefabricated chimneys that are usually manufactured as sets of components for assembly on site. Commonly available types range from single-walled metal chimneys suitable for some gas appliances to chimneys with insulation sandwiched between an inner liner and an outer metal wall designed for oil or solid fuel use. They are also known as system chimneys.

FANNED DRAUGHT INSTALLATION. Sometimes known as forced draught appliances, these incorporate a fan to enable the proper discharge of the flue gases. The fan may be separately installed in the flue or may be an integral part of the combustion appliance. Fans can be installed in most oil-fired and many gas-fired boilers and may either extract flue gases from the combustion chamber or may cause the flue gases to be displaced from the combustion chamber if the fan is supplying it with air for combustion. Depending on the location of the fan, flues in fanned draught installations can run horizontally or vertically and can be at higher or lower pressures than their surroundings.

FIREPLACE RECESS means a structural opening formed in a wall or chimney breast, from which a chimney leads and including a hearth at its base. For closed appliances such as stoves, cookers or boilers a simple structural opening may be suitable. For accommodating open fires it will usually be necessary to form a gather to reduce the recess to the size of the flue. Fireplace recesses are often lined with firebacks to accommodate inset open fires and lining components and decorative treatments may be fitted around openings to reduce the opening area. It is this finished fireplace opening area which determines the size of flue required for an open fire in such a recess (see Fig. 14.14 in section 14.7.1).

FIRE WALL is a means of shielding a fuel tank from the heat of a fire. For LPG tanks, a fire wall will lengthen the path that has to be travelled by gas accidentally leaking from the tank or fittings. This will allow it more time to disperse safely, before reaching a hazard such as other potential ignition sources, an opening in a building or a boundary.

FLUE means a passage conveying the products of combustion to the external air (see Fig. 14.1).

FLUEBLOCK CHIMNEY means a chimney constructed from a set of factory-made components. Flueblock chimneys may be made from precast concrete, clay or other masonry units, designed for assembly on site to provide a complete chimney with the performance appropriate for the intended appliance. Two types of common systems are available:

- for use solely with gas burning appliances, and
- for solid fuel burning appliances (sometimes called chimney block systems).

FLUE BOX means a factory-made unit (usually of metal), which is designed to accommodate a gas burning appliance in conjunction with a factory-made chimney. (See also PREFABRICATED APPLIANCE CHAMBER below.)

FLUELESS APPLIANCE means an appliance, which is designed to be used without being connected to a flue. Its products of combustion mix with the surrounding room air and are eventually ventilated from the room to the outside. Examples include gas cookers, gas instantaneous water heaters and some types of gas space heaters.

FLUE LINER is the wall of the chimney that is in contact with the products of combustion. This could be a concrete flue liner, the inner liner of a factory-made chimney system or a flexible liner fitted into an existing chimney (see Fig. 14.1).

FLUE OUTLET is that part of the combustion installation where the products of combustion are discharged from the flue to the outside air, such as the top of a chimney pot or flue terminal.

FLUEPIPE (see Fig. 14.1) means a pipe that connects a combustion appliance to a flue in a chimney (sometimes called a CONNECTING FLUEPIPE). It may be either single-walled (bare or insulated) or double-walled. (The term FLUEPIPE is also used to describe the tubular components from which some factory made chimneys for gas and oil appliances are made or from which plastic flue systems are made.)

HEARTH is a base on which a combustion appliance is placed (see Fig. 14.1). It provides safe isolation between the appliance, and people, combustible parts of the building fabric and soft furnishings. The exposed surface of the hearth usually extends beyond the appliance and provides a region which can be kept clear of anything at risk of fire. The hearth may be constructed of:

- thin insulating board;
- a substantial thickness of material such as concrete; or
- some intermediate form of construction depending on the weight and downward heat emission characteristics of the appliance(s) upon it.

For solid fuel open fires the substantial thickness of material necessary may be provided by a constructional hearth (often as part of the building structure, floor slab etc.), on which may be placed a decorative superimposed hearth to provide the clear surface.

HEAT INPUT RATE. For a gas appliance, this is the maximum rate of energy flow that could be provided by the prevailing rate of fuel flow into the appliance if the fuel were to be burned with full oxidation. It is calculated as the rate of fuel flow to the appliance multiplied by either the fuel's gross or net calorific value. The gross calorific value takes account of the latent heat due to the condensation of water in combustion products and allows this to be included in the heat obtained from the fuel (such as in a gas condensing boiler). It is thus a larger figure than the net heat input rate. Either heat input rating can be used for any given appliance; however, it is now usual to express the rating of a gas appliance as a net heat input rate (kW (net)).

INDEPENDENTLY CERTIFIED means that the product conforms to a product certification scheme that has been approved by an independent certification body. Such schemes certify compliance with the requirements of a recognised document that is appropriate to the purpose for which the material is to be used. Materials which are not so certified may still conform to a relevant standard. Many certification bodies, which approve such schemes, are accredited by UKAS.

INSET LIVE FUEL EFFECT (ILFE) FIRES are described in BS 5871: Part 2. These gas fires stand fully or partially within a fireplace recess or suitable fluebox and give the impression of an open fire. The appliance covers the full height of the fireplace opening so that air only enters through purpose designed openings and the flue gases only discharge through the spigot (see Fig. 14.2(b)).

INSTALLATION INSTRUCTIONS means the manufacturer's instructions, to enable installers to correctly install and test appliances and flues and to commission them into service.

NATURAL DRAUGHT flue – this is the traditional concept for a flue whereby the draught which takes the flue gases up the flue to outside air relies on the difference between the temperature of the gases within the flue and the temperature of the ambient air. Draught increases with the height of the flue and a satisfactory natural draught requires an essentially vertical run of flue. This concept can be contrasted with balanced flue appliances which are designed to discharge directly through the wall adjacent to the appliance (see above).

NON-COMBUSTIBLE means capable of being classed as non-combustible:

- when subjected to the non-combustibility test of BS 476, Part 4: 1970 (1984) *Non-combustibility test for materials*; and
- any material which when tested to BS 476, Part 11: 1982 (1988) *Method for Assessing the Heat Emission from Building Materials* does not flame nor cause any rise in temperature on either the centre (specimen) or furnace thermocouples.

(See also Chapter 7, Fire, section 7.18.5.)

NOTIFIED BODY for the purposes of the Gas Appliances (Safety) Regulations (1995) means a body that:

- is approved by the Secretary of State for Trade and Industry as being competent to carry out the required Attestation procedures for gas appliances and whose name and identification number has been notified by him/her to the Commission of the European Community and to other member states in accordance with the Gas Appliances (Safety) Regulations 1995;
- has been similarly approved for the purposes of the Gas Appliances Directive by another member state and whose name and identification number has been notified to the Commission and to other member states pursuant to the Gas Appliances Directive.

OPEN FLUED APPLIANCE is, for example, the traditional open fire or stove that draws its combustion air from the room or space in which it is installed and which requires a flue to discharge its products of combustion to the outside air (see Fig. 14.3).

PREFABRICATED APPLIANCE CHAMBER means a set of factory-made pre-cast concrete components designed to provide a FIREPLACE RECESS (see above). It is normal for the chamber to be positioned against a wall and it may be designed to support a chimney. The chamber and chimney can be enclosed to create a false chimney breast. (See also FLUE BOX above.)

RADIANT CONVECTOR GAS FIRES, CONVECTOR HEATERS AND FIRE/ BACK BOILERS, are described in BS 5871: Part 1. These stand in front of a closure plate which is fitted to the fireplace opening of a fireplace recess or suitable fluebox. The appliance covers the full height of the fireplace opening so that air only enters through purpose designed openings and the flue gases only discharge through the flue spigot (see Fig. 14.2(c)).

RATED HEAT INPUT (or rated input) for a gas appliance means the maximum heat input rate at which it can be operated. This will be declared on the appliance data plate and for gas appliances it is now usual to express this rating as a net value (kW (net)) although the gross value (kW (gross)) was used until recently. (For details of net and gross values see HEAT INPUT RATE above.)

RATED HEAT OUTPUT for an oil appliance is the maximum declared energy output rate (kW) as declared on the appliance data plate. For a solid fuel appliance, this is the maximum manufacturers' declared energy output rate (kW) for the appliance. This may be different for different fuels.

ROOM-SEALED APPLIANCE means an appliance whose combustion system is sealed from the room in which the appliance is situated. The appliance obtains combustion air either from a ventilated uninhabited space within the building or directly from the open air outside the building. The products of combustion will be vented directly to open air outside the building (see Fig. 14.3).

THROAT means a narrowing part of the flue between a fireplace recess and its chimney. Throats can be formed from prefabricated components or can be built in brickwork by a process of corbelling.

14.3   Rules for measurement

When measuring the size of a duct or flue (to establish the area, diameter etc. for the purposes of the approved document guidance) the dimensions should be taken at right angles to the direction of gas flow. Minimum requirements for flue sizes are given in the text below and where offset components are used they should not reduce the flue area to less than the quoted figures.

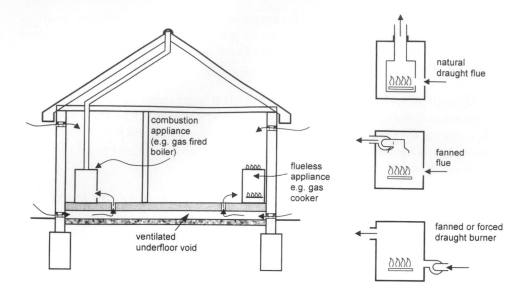

Open flued appliances

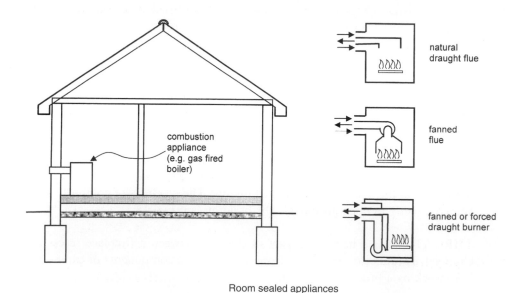

Room sealed appliances

Fig. 14.3   Types of combustion installation.

14.4   Checking the condition of combustion installations before use

Combustion installations must be designed and constructed in accordance with the requirements of Part J. Additionally, before being used they should be inspected and tested to prove that they are in fact in compliance. This applies not only to new build but also to repairs, refurbishment and re-use of existing flues (see section 14.8.3 below for details of this). Responsibility for such proof of compliance rests with the person carrying out the work. This could be a specialist firm working directly for the client, a developer, a main contractor or even a sub-contractor working for the main contractor.

Proving compliance can involve a number of steps, which should be documented in the form of a report drawn up by a 'specialist' firm for the main contractor, client or developer.

To be acceptable to the Building Control Body, the specialist firm would probably need to be a registered member of one of the following organisations although there is no clear guidance in the AD to this effect:

- The Council for Registered Gas Installers (CORGI) contactable at www.corgi-gas.com;
- The Heating Equipment Testing and Approval Scheme (HETAS) contactable at www.hetas.co.uk;
- The Oil Firing Technical Association for the Petroleum Industry (OFTEC) contactable at www.oftec.org;
- The National Association of Chimney Sweeps (NACS) contactable at www.chimneyworks.co.uk;
- The National Association of Chimney Engineers (NACE) contactable at www.nace.org.uk.

The Building Control Body can ask for this report as a way of proving compliance. An example of a completed checklist for such a report is given in Appendix A to AD J, which is reproduced below. The report will need to show that materials and components have been used which are suitable for the intended application and that flues have passed appropriate tests.

In addition to checking a combustion installation for compliance, where a material change of use takes place in a building (e.g. conversion to flats, see Chapter 2) the fire resistance of the existing chimney walls should be checked and improved as necessary. This can be done by applying additional layers of non-combustible material to the existing chimney walls.

14.5   Air supply

Paragraph J1 of Part J requires that combustion appliances be installed so that an adequate supply of air is provided for combustion of the fuel, to prevent overheating and for efficient operation of the flue.

14.5.1   Air supply – general provisions

Where combustion appliances are installed in a building, Paragraph J1 can be met if provisions are made to enable the admission of sufficient air for:

AD J, Appendix A

Hearths, Fireplaces, flues and chimneys

This checklist can help you to ensure hearths, fireplaces, flues and chimneys are satisfactory. If you have been directly engaged, copies should also be offered to the client and to the Building Control Body to show what you have done to comply with the requirements of Part J. If you are a sub-contractor, a copy should be offered to the main contractor.

1. **Building address, where work has been carried out** . . . .
. . . .
. . . .

2. **Identification of hearth, fireplace, chimney or flue**	*Example:* *Fireplace in lounge*	*Example:* *Gas fire in rear addition bedroom*	*Example:* *Small boiler room*
3. **Firing capability: solid fuel/gas/oil/all**	*All*	*Gas only*	*Oil only*
4. **Intended type of appliance.** **State type or make. If open fire give finished fireplace opening dimensions**	*Open fire* *480 W × 560 H (mm)*	*Radiant/convector fire 6 kW input*	*Oil fired boiler 18 kW output (pressure jet)*
5. **Ventilation provision for the appliance:** **State type and area of permanently open air vents**	*2 through wall ventilators each 10,000 mm^2 (100 cm^2)*	*Not fitted*	*Vents to outside: Top 9,900 mm^2 Bottom 19,800 mm^2*
6. **Chimney or flue construction**			
(a) **State the type or make and whether new or existing**	*New. Brick with clay liners*	*Existing masonry*	*S.S. prefab to BS 4543-2*
(b) **Internal flue size (and equivalent height, where calculated – natural draught gas appliances only)**	*200 mm Ø*	*125 mm Ø (H_e = 3.3 m)*	*127 mm Ø*
(c) **If clay or concrete flue liners used confirm they are correctly jointed with socket end uppermost and state jointing materials used**	*Sockets uppermost Jointed by fire cement*	*Not applicable*	*Not applicable*
(d) **If an existing chimney has been refurbished with a new liner, type or make of liner fitted**	*Not applicable to BS 715*	*Flexible metal liner*	*Not applicable*
(e) **Details of flue outlet terminal and diagram reference** **Outlet Detail:**	*Smith Ltd Louvred pot 200 mm Ø*	*125 mm Ø GC1 terminal*	*Maker's recommended terminal*
Complies with:	*As Diagram 2.2 AD J*	*As BS 5440-1: 2000 Figure C.1*	*As Diagram 4.2 AD J*
(f) **Number and angle of bends**	*2 × 45°*	*2 × 45°*	*1 × 90° Tee*
(g) **Provision for cleaning and recommended frequency**	*Sweep annually via fireplace opening*	*Annual service by CORGI engineer*	*Sweep annually via base of Tee and via appliance*
7. **Hearth. Form of construction. New or existing?**	*New. Tiles on concrete floor. 125 mm thick. As Diagram 2.9 AD J*	*Existing hearth for solid fuel fire, with fender*	*New. Solid floor Min 125 mm concrete above DPM. As Diagram 4.3 AD J*
8. **Inspection and testing after completion** **Tests carried out by:** **Tests (Appx E in AD J 2002 ed) and results**	*Inspected and tested by J Smith, Smith Building Co*	*Tested by J Smith, CORGI Reg no. 12345*	*Tested by J Smith, The Oil Heating Co*
Flue visual	*Not possible, bends*	*Not possible, bends*	*Checked to Section 10, BS 7566: Part 3:*
Inspection sweeping	*OK*	*Not applicable*	*1992 – OK*
coring ball	*OK*	*Not applicable*	
smoke	*OK*	*Not applicable*	*OK*
Appliance (where included) spillage	*Not included*	*OK*	*OK*

I/We the undersigned confirm that the above details are correct. In my opinion, these works comply with the relevant requirements in Part J of Schedule 1 to the Building Regulations.

Print name and title . . . . Profession . . . .

Capacity . . . (e.g. 'Proprietor of Smith's Flues', Authorising Engineer for Brown plc) . . . . Tel no . . . .

Address . . . . Postcode . . . .

Signed . . . . Date . . . .

Registered membership of . . . (e.g. CORGI, OFTEC, HETAS, NACE, NACS) . . . .

- proper combustion of the fuel;
- proper operation of any flues, (or for flueless appliances, safe dispersal of the products of combustion to the outside air); and
- cooling control systems and/or to make sure that appliance casings do not become too hot to touch, where this is deemed necessary.

AD J gives a range of air vent sizes, which vary with the type of fuel being burned. These are described below and should be read with the following general notes.

- The figures given are for single combustion appliances only, and will need to be increased if more than one appliance is installed in a room (e.g. where a kitchen contains an open-flued boiler and a flueless appliance such as a cooker).
- Where an open-flued appliance is installed in a room it will receive a small amount of combustion air from infiltration through the building fabric. Depending on the type of appliance installed this will need to be supplemented by permanently open air vents.
- Where an open-flued appliance is installed in an appliance compartment:
  - (a)   all the air necessary for combustion and proper operation of the flue must be supplied through adequately sized permanent vents which may be situated in an outside or internal wall;
  - (b)   where cooling air is needed, the compartment should be large enough to allow air to circulate via high and low level vents;
  - (c)   the appliance and ventilation system manufacturer's instructions should be followed where appliances are to be installed within balanced compartments since special provisions will be necessary.
- Ventilation direct to outside air should be provided for other rooms or spaces within the building where:
  - (a)   a room-sealed appliance takes its combustion air from another space (e.g. a roof void);
  - (b)   a flue has a permanent opening to another space (e.g. where it feeds a secondary flue in the roof void).

In the case of ventilation via roof voids, the ventilation provisions contained in Approved Document F (see Chapter 11) would normally be adequate, where the combustion installation serves a dwelling.

In other cases, the room or space from which the combustion air is obtained should have air vent openings direct to outside air of at least the same size as the internal openings serving the appliance, although air vents for flueless appliances should always open direct to outside air (i.e. not through an adjoining space). Figure 14.4 below gives examples of locations for permanent air vent openings.

Where permanently open air vents are called for (see guidance related to specific fuels below) they should be:

- non-adjustable;
- appropriately sized to admit the correct amount of air taking into account their free area (or equivalent free area, see Fig. 14.5) and any obstructions such as

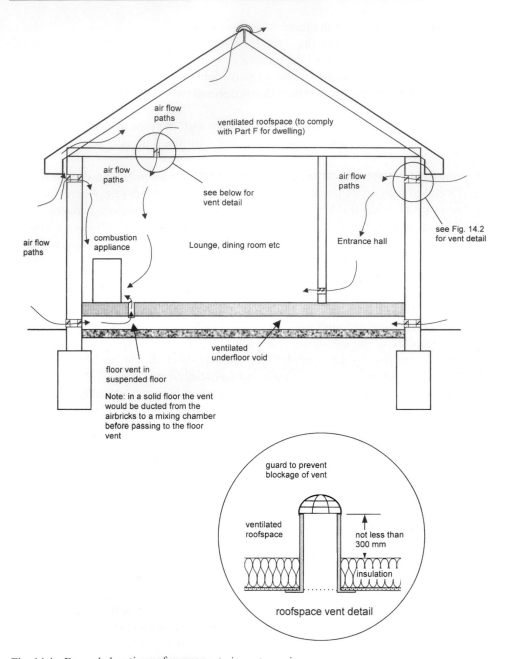

Fig. 14.4   Example locations of permanent air vent openings.

grilles and anti-vermin mesh (which should have aperture dimensions no smaller than 5 mm);

- positioned so that they are unlikely to become blocked;
- located so that occupants are not provoked to seal them against cold draughts and noise (e.g. place vents close to appliances; draw air from hallways or other intermediate spaces; place air vents next to ceilings to ensure good mixing of

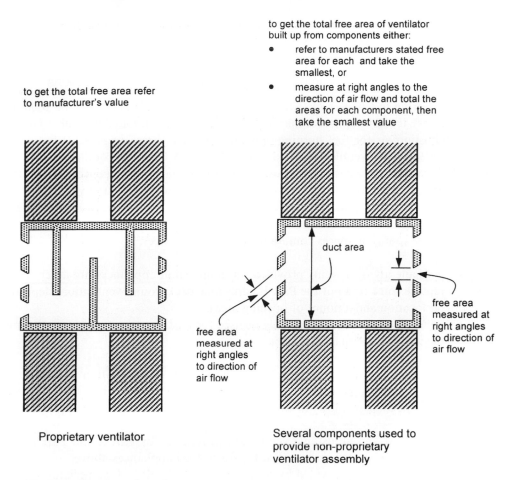

to get the total free area of ventilator built up from components either:

- refer to manufacturers stated free area for each and take the smallest, or

- measure at right angles to the direction of air flow and total the areas for each component, then take the smallest value

to get the total free area refer to manufacturer's value

duct area

free area measured at right angles to direction of air flow

free area measured at right angles to direction of air flow

Proprietary ventilator

Several components used to provide non-proprietary ventilator assembly

Fig. 14.5   Free area of ventilators.

incoming cold air; install noise attenuated ventilators to cut down on unwanted external noise; place ventilators outside fireplace recesses and beyond the hearths of open fires where dust or ash might be disturbed by draughts).

Permanently open air vents should not be installed in fire-resisting walls. Although external walls are excluded from this provision, this exclusion will not apply to parts of external walls shielding LPG tanks. Additionally, vents should not be sited in fireplace recesses unless expert advice has been sought.

Ventilation via permanently open vents can be provided in a number of ways. Where proprietary components or assemblies are used the manufacturer will usually be able to give a value for the free area (or equivalent free area). Where this is not available (e.g. some air bricks, grilles or louvres) it will be necessary to calculate the free area as shown in Fig. 14.5 by aggregating the individual apertures and taking the smallest free area in the assembly (or the overall duct area if this is less).

Where buildings have air tight membranes in their floors (such as radon or landfill gas barriers, see Chapter 8) ventilation ducts or vents should be installed so as not to compromise the effectiveness of the membrane.

14.5.2   Air supply – compliance with other parts of the Regulations

By referring to Chapter 11 (which deals with Part F, Ventilation of the 2000 Regulations), it will be apparent that buildings need to be ventilated not only to provide combustion air for fuel burning appliances (where these are present) but also to provide general ventilation for health reasons. A possible conflict might occur between the two forms of ventilation since the guidance in Approved Document F allows the background ventilation to be adjustable, whereas Part J ventilation must be permanently open. Previous editions of Part J have been unsuccessful in addressing this conflict, however the 2002 edition gives the following guidance.

Where rooms or spaces contain open-flued appliances:

- permanently open vents provided for combustion appliances are acceptable to replace some or all of the Part F adjustable background ventilation (depending on location and amount of opening area);
- adjustable Part F vents can only be used for Part J combustion air ventilation if the are fixed permanently open.

Where rooms or spaces contain flueless appliances (see definition in section 14.2 above) it may be necessary to provide permanent, adjustable and rapid ventilation (e.g. an openable window) to comply with Parts F and J. For such appliances:

- permanent and adjustable ventilation provisions for Part J and Part F compliance can be used as described for open-flued appliances above;
- rapid ventilation provided by opening elements for Part F compliance can also be accepted for Part J compliance provided that the minimum opening areas are achieved.

Where mechanical extract ventilation is provided for Part F compliance, dangerous conditions can be created where open-flued appliances are also present due to the spillage of flue gases (even where the fans and appliances are in different rooms). Approved Document F (see Chapter 11 section 11.3.5 above) contains recommendations designed to avoid flue gas spillage, for different types of combustion appliances. They should be read in conjunction with the following.

- Specialist advice may be needed for commercial and industrial installations, regarding the possible need for the interlocking of gas heaters and any mechanical ventilation systems.
- Suitable spillage tests for gas appliances may be provided by appliance manufacturers in their installation instructions. Alternatively, the procedure given in BS 5440: *Installation and maintenance of flues and ventilation for gas appliances of rated input not exceeding 70 Kw net (1st, 2nd and 3rd family gases)*, Part 1: 2000 *Specification for installation and maintenance of flues* can also be used. The object

of the test is to check for spillage when appliances are subjected to the greatest possible depressurisation. To cater for this the following should be taken into account.

(a)   All external doors, windows and other adjustable ventilators to outside should be closed.

(b)   Several tests may be necessary to demonstrate the safe operation of the appliance with reasonable certainty since various combinations of fans in operation and open internal doors will be possible, and the specific combination causing the greatest depressurisation at the appliance will depend on the circumstances in each case. One test should of course be carried out with the door leading into the room of installation closed and all fans in that room switched on.

(c)   The effect of ceiling fans should be taken into account during the tests.

(d)   It is important to consider all fans which might be in use and not only the obvious ones, such as those on view in kitchens. Others include fans installed in domestic appliances such as tumble dryers, fans fitted to other open-flued combustion appliances, and fans installed to draw radon gas from the ground below a building (see also *BRE Good Building Guide GBG 25* and Chapter 8 section 8.5.5 above).

14.5.3   Air supply – specific provisions relating to appliances burning solid fuel

In addition to the general recommendations given above, appliances burning solid fuel with rated outputs up to 50 kW should have permanently open air vents at least as great as the sizes shown in Table 2.1 to Approved Document J, which is reproduced below. It should be noted that where an appliance is installed that is capable of burning a range of different solid fuels the ventilation requirements should suit the fuel that produces the greatest heat output.

Manufacturers' installation instructions should be followed where these vary from the Table recommendations (e.g. they may specify even larger areas of permanently open air vents or, in the case of a cooker, omit to specify a rated output).

14.5.4   Air supply – specific provisions relating to appliances burning gas

Appliances burning gas with rated inputs up to 70 kW (net) should comply with the general recommendations given above. Additionally, the amount of permanently open air vents which should be provided will vary with the type of appliance (and whether it is room-sealed, open-flued or flueless). The various combinations are discussed below.

Flued decorative fuel effect (DFE) fires

These are defined in section 14.2 above. Any room or space containing a DFE fire should have ventilation provided in accordance with Table 14.1 below.

AD J Section 2

Table 2.1   Air supply to solid fuel appliances.

Type of appliance	Type and amount of ventilation (1)	
Open appliance, such as an open fire with no throat, e.g. a fire under a canopy as in Diagram 2.7	Permanently open air vent(s) with a total free area of at least 50% of the cross sectional area of the flue	
Open appliance, such as an open fire with a throat as in Diagram 2.6 and 2.13	Permanently open air vent(s) with a total free area of at least 50% of the throat opening area (2)	
Other appliance, such as a stove, cooker or boiler, with a flue draught stabiliser	Permanently open air vent(s) as below: (3)	
		Total free area
	First 5 kW of appliance rated output	300 mm^2/kW
	Balance of rated output	850 mm^2/kW
Other appliance, such as a stove, cooker or boiler, with no flue draught stabiliser	A permanent air entry opening or openings with a total free area of at least 550 mm^2 per kW of appliance rated output above 5 kW	

Notes:
1. Divide the area given in mm^2 by 100 to find the corresponding area in cm^2
2. For simple open fires (see Fig. 14.15) the requirement can be met with room ventilation areas as follows:
   Nominal fire size (fireplace opening size)   500 mm   450 mm   400 mm   350 mm
   Total free area of permanently open air vents   20,500 mm^2 18,500 mm^2 16,500 mm^2 14,500 mm^2
3. Example: an appliance with a flue draught stabiliser and a rated output of 7 kW would require a free area of:
   [5 × 300] + [2 × 850] = 3200 mm^2

Table 14.1   Supply of combustion air to flued decorative fuel effect fires.

Type of appliance	Type of ventilation
1 DFE fire in a fireplace recess with a throat	Air vent free area of at least 10,000 mm^2 (100 cm^2)
2 DFE fire in a fireplace with no throat (e.g. under a canopy)	Air vent free area sized as for a solid fuel fire (see Table 2.1 to AD J section 2 above)
3 DFE fire with rating not exceeding 7 kW (net)	Permanently open air vents not necessary for appliances certified by a Notified Body (see section 14.2 above) as having a flue gas clearance rate (without spilling) not exceeding 70 m^3/hour

Flued appliances other than DFE fires

These include inset live fuel effect fires (ILFE), radiant convector gas fires, convector heaters and fire/back boilers (all as defined in section 14.2 above). All these combustion appliances come in both room-sealed and open-flued variants. Table 14.2 gives the free areas of permanently open vents for these appliances.

Table 14.2 Supply of combustion air to gas appliance installations (other than DFE fires or flueless appliances).

Location/type of appliance	Amount/type of ventilation
1 Appliance in a room or space:	Ventilation direct to outside air
(a) Open-flued	Permanently open vents of at least 500 mm^2 per kW (net) of rated input over 7 kW (net)
(b) Room-sealed	No vents needed
2 Appliance in appliance compartment:	Ventilation via adjoining room or space
(a) Open-flued	From adjoining room or space to outside air: permanently open vents of at least 500 mm^2 per kW (net) of rated input over 7 kW (net) Between adjoining room or space and appliance compartment, permanently open vents: at high level – 1000 mm^2 per kW input (net) at low level –2000 mm^2 per kW input (net)
(b) Room-sealed	Between adjoining space and appliance compartment, permanently open vents: both high and low levels – 1000 mm^2 per kW input (net)
3 Appliance in appliance compartment:	Ventilation direct to outside air:
(a) Open-flued	Permanently open vents: at high level – 500 mm^2 per kW input (net) at low level – 1000 mm^2 per kW input (net)
(b) Room-sealed	Permanently open vents: both high and low levels – 500 mm^2 per kW input (net)

Example calculation
An open-flued boiler with a rated input of 20 kW (net) is installed in a boiler room (appliance compartment, row 3 above) ventilated directly to the outside. The design of the boiler is such that it requires cooling air in these circumstances.
The cooling air will need to be exhausted via a high level vent.
From above, the area of ventilation needed = 20 kW × 500 mm^2/kW = 10,000 mm^2.
A low level vent will need to be provided to allow cooling air to enter, as well as admitting the air needed for combustion and the safe operation of the flue.
From above the area of ventilation needed = 20 kW × 1000 mm^2/kW = 20,000 mm^2.
These ventilation areas can be converted to cm^2 by dividing the results given above in mm^2 by 100.
The calculated areas are the free areas of the vents (or equivalent free areas for proprietary ventilators) described in section 14.5.1 above.

Air supply to flueless gas appliances

Flueless appliances are designed to be used without being connected to a flue. The products of combustion mix with the surrounding room air and are eventually ventilated from the room to the outside. Examples include gas cookers, gas instantaneous water heaters and some types of gas space heaters.

It will usually be necessary to comply with Part F (see Chapter 11) regarding background, rapid and extract ventilation as well as the permanent ventilation provisions in Part J.

Table 14.3 below gives details of the amounts of permanent ventilation that

Table 14.3 Ventilation of flueless gas appliances.

Type of flueless appliance	Maximum rated heat input of appliance	Volume of room, space or internal space[1] (m^3)	Free area of permanently open air vents (mm^2)
1 Cooker, oven, hob, grill or combination of these	Not applicable	Up to 5 5 to 10 over 10	10,000 5000[3] permanently open vent not needed
2 Instantaneous water heater	11 kW (net)	5 to 10 10 to 20 over 20	10,000 5000 permanently open vent not needed
3 Space heater[2]:			
(a) not in an internal space[4]	0.045 kW (net) per m^3 volume of room or space	All room sizes	10,000 **plus** 5500 per kW (net) greater than 2.7 kW
(b) in an internal space[5]	0.09 kW (net) per m^3 volume of room or space	All room sizes	10,000 **plus** 2750 per kW (net) greater than 5.4 kW

Notes:
[1] In this Table 'internal space' means e.g. a hallway or landing (a space which communicates with several other rooms spaces).
[2] For LPG fired space heaters which conform to BS EN 449: 1997 *Specification for dedicated liquified petroleum gas appliances. Domestic flueless space heaters (including diffusive catalytic combustion heaters)*, follow the guidance in BS 5440: Part 2: 2000.
[3] If the room or space has a door direct to outside air, no permanently open air vent is needed.
[4] A space heater is to be installed in a lounge measuring $5 \times 4 \times 2.5 = 50$ m^3. Its maximum rated input should not exceed $50 \times 0.045 = 2.25$ kW (net).
[5] A space heater installed in a hallway to provide background heating has a rated input of 6 kW (net). It will need to be provided with $10,000 + 2750 \times (6 - 5.4) = 11,650$ mm^2 of permanently open ventilation.

should be provided in different circumstances. These are in addition to any openable elements or extract ventilation needed to comply with Part F. The Table should be read in conjunction with the general provisions listed above. Where a gas point is installed in a room which is intended to be used with a flueless appliance, the room should have the ventilation provisions required for the intended appliance. The ventilation provision should be calculated on the basis that the largest rated appliance consistent with the Table 14.3 recommendations will be installed in the room.

A flueless instantaneous water heater should never be installed in a room or space which has a volume of less than 5m^3.

14.5.5 Air supply – specific provisions relating to appliances burning oil

The guidance in Approved Document J is relevant to combustion appliances burning oils meeting the BS 2869: 1998 (or equivalent) specifications for Class C2 (Kerosene), and Class D (Gas oil). Appliances burning oil with rated outputs up to 45 kW should comply with the general recommendations given above. In rooms such as bathrooms and bedrooms, where there is an increased risk of carbon monoxide poisoning, open-flued oil-fired combustion appliances should not be installed. Instead, room-sealed appliances could be installed.

Table 14.4 below gives details of the amounts of permanent ventilation that should be provided in different circumstances. Manufacturers' installation instructions should be followed where these require greater areas or permanent ventilation than shown in the table recommendations.

Table 14.4 Supply of combustion air to oil-fired appliance installations.

Location/type of appliance	Amount/type of ventilation
1 Appliance in a room or space:	Ventilation direct to outside air
(c) Open-flued	Permanently open vents of at least $550\,mm^2$ per kW of rated output over $5\,kW^1$
(d) Room-sealed	No vents needed
2 Appliance in appliance compartment:	Ventilation via adjoining room or space
(c) Open-flued	From adjoining room or space to outside air: permanently open vents of at least $550\,mm^2$ per kW of rated output over 5 kW Between adjoining room or space and appliance compartment, permanently open vents: at high level – $1100\,mm^2$ per kW output at low level – $1650\,mm^2$ per kW output
(d) Room-sealed	Between adjoining space and appliance compartment, permanently open vents: both high and low levels – $1100\,mm^2$ per kW output
3 Appliance in appliance compartment:	Ventilation direct to outside air:
(c) Open-flued	Permanently open vents: at high level – $550\,mm^2$ per kW output at low level – $1100\,mm^2$ per kW output
(d) Room-sealed	Permanently open vents: both high and low levels – $550\,mm^2$ per kW output

Notes:
[1] Increase the area of permanent ventilation by a further $550\ mm^2$ per kW output if appliance fitted with draught break.

Example calculation
An open-flued boiler with a rated output of 15 kW is installed in a cupboard (appliance compartment, row 2 above) ventilated via an adjacent room. Since the boiler output exceeds 5 kW, permanent ventilation openings will be needed in the adjacent room in addition to the vents between the cupboard and the room designed to provide combustion and cooling air.
Area of permanent vents to outside air needed in adjacent room $= (15\,kW - 5\,kW) \times 550\,mm^2/kW = 5500\,mm^2$.
The cooling air will need to be exhausted via a high level vent.
From above, the area of ventilation needed $= 15\,kW \times 1100\,mm^2/kW = 16{,}500\,mm^2$.
A low level vent will need to be provided to allow cooling air to enter, as well as admitting the air needed for combustion and the safe operation of the flue.
From above the area of ventilation needed $= 15\,kW \times 1650\,mm^2/kW = 24{,}750\,mm^2$
These ventilation areas can be converted to cm2 by dividing the results given above in mm^2 by 100.
The calculated areas are the free areas of the vents (or equivalent free areas for proprietary ventilators) described in section 14.5.1 above.

14.6 Discharge of products of combustion

Paragraph J2 of Part J requires that combustion appliances have adequate provision for the discharge of the products of combustion to the outside air.

In general, this means that the combustion installation must enable normal operation of the appliances without a hazard to health being created by the products of combustion.

Apart from flueless appliances, this is achieved by providing each combustion appliance with a suitable flue which discharges to outside air.

14.6.1 Provision of flues and chimneys

The guidance in AD J is based on the provision of a separate flue for each appliance. Although every solid fuel appliance should be connected to its own flue, for oil and gas-fired installations it is possible to connect more than one appliance to a single flue by following the alternative guidance in:

- BS 5410: *Code of practice for oil firing,* Part 1: 1977 *Installations up to 44 kW output capacity for space heating and hot water supply purposes*, AMD 3637, for oil-fired installations; and
- BS 5440: *Installation and maintenance of flues and ventilation for gas appliances of rated input not exceeding 70 Kw net (1st, 2nd and 3rd family gases)*, Part 1: 2000 *Specification for installation and maintenance of flues*, for gas-fired installations.

A chimney is defined above as a structure consisting of a wall or walls enclosing one or more flues and the old practice of building chimneys with unlined flues became obsolete many years ago.

Interestingly, AD J now contains provisions for the repair and testing of flues in existing chimneys when these are brought back into use or re-use. In many cases this will involve relining the old flues.

14.6.2 Chimneys, flues and flue pipes – general provisions

AD J gives a range of flue sizes, and a number of specific recommendations, which vary with the type of fuel being burnt. These are described below and should be read with the following general notes.

Liners to masonry chimneys

Liners to masonry chimneys should be installed in accordance with manufacturer's instructions and the following notes:

- the flue should be formed with appropriate components, keeping joints to a minimum and avoiding cutting;
- matching factory-made components should be used for bends and offsets;
- liners should be built into the chimney with sockets or rebate ends uppermost to

keep moisture and other condensates in the flue (this also prevents condensate from running out of the joints where it might adversely affect any caulking material);
- joints between liners should be sealed with fire cement, refractory mortar or installed in accordance with manufacturer's instructions;
- the space between the liners and the masonry should be filled with weak mortar or insulating concrete, with mixes such as:
  - (a) 1 : 20 ordinary Portland cement : suitable lightweight expanded clay aggregate (minimally wetted);
  - (b) 1 : 6 ordinary Portland cement : vermiculite;
  - (c) 1 : 10 ordinary Portland cement : perlite.

The following liners for masonry chimneys are suitable for all fuels.

- Liners as described in BS EN 1443: 1999 with performance at least equal to the designation T450 N2 S D 3 (see section 14.2 above), such as:
  - (a) clay flue liners with rebated or socketed joints which meet the requirements for Class A1 N2 or Class A1 N1 to BS EN 1457: 1999
  - (b) concrete flue liners independently certified as meeting the requirements for the classification Type A1, Type A2, Type B1 or Type B2 as described in prEN 1857(e18) January 2001; or
  - (c) other products that are independently certified as meeting the BS EN 1443 criteria.
- Imperforate clay pipes with sockets for jointing to comply with BS 65: 1991 (1995) Specification for vitrified clay pipes, fittings, joints and ducts, also flexible mechanical joints for use solely with surface water pipes and fittings, AMD 8622.

Construction of flueblock chimneys

For all fuels, flueblock chimneys should:

- be constructed using factory-made components suitable for the intended application;
- be installed in accordance with manufacturer's instructions;
- have joints sealed in accordance with the flueblock manufacturer's instructions;
- have bends and offsets formed only with matching factory-made components;
- be constructed using:
  - (a) liners as described in BS EN 1443: 1999 with performance at least equal to the designation T450 N2 S D 3 (see section 14.2 above), such as:
    - (i) clay flueblocks meeting the requirements for the class FB1 N2 to BS EN 1806: 2000 *Chimneys. Clay/ceramic flue blocks for single wall chimneys. Requirements and test methods*; or
    - (ii) other products that are independently certified as meeting the BS EN 1443 criteria;
  - (b) blocks lined as for masonry chimneys above (and independently certified as suitable for the purpose).

Condensates in flues

The products of combustion of all fuels contain considerable quantities of water vapour which can condense on the inside of the flue and cause damage if not satisfactorily controlled. Some modern appliances (called condensing appliances) are designed to extract the latent heat of vaporisation from the condensate to improve efficiency.

In the case of chimneys that do not serve condensing appliances, satisfactory control of condensation can be achieved by insulating flues so that flue gases do not condense under normal conditions of operation.

For chimneys serving condensing appliances, satisfactory control of condensation can be achieved by:

- using lining components that are impervious to condensates and suitably resistant to corrosion;
- making appropriate drainage provisions and avoiding ledges, crevices etc;
- providing for the disposal of condensate from condensing appliances.

Plastic fluepipe systems

Under certain circumstances it is possible to use plastic fluepipe systems. For example, with condensing boiler installations, the fluepipes should be:

- supplied by or specified by the appliance manufacturer, and
- approved by a Notified Body or independently certified as being suitable for purpose.

Design of flues serving natural draught open-flued appliances

Flue systems serving natural draught open-flued appliances rely on the stack effect (the natural buoyancy of hot air) to ensure that the products of combustion are successfully discharged to outside air. Therefore, the system should offer as little resistance as possible to the passage of the flue gases and it should be possible to inspect and sweep the whole flue after the appliance (if any) has been installed. This can be done by minimising changes in direction and avoiding long horizontal sections of flue. The following design factors illustrated in Fig. 14.6 below should be considered:

- build flues so that they are as straight and vertical as possible;
- restrict horizontal sections of flue to connections to appliances with rear outlets and do not allow these to be longer than 150 mm;
- if bends have to be included in the flue make sure that they are angled at not greater than 45° to the vertical;
- have no more than four 45° bends between the appliance outlet and the flue outlet, and not more than two of these between:
  (a) adjacent cleaning access points, or
  (b) a cleaning access point and the flue outlet.

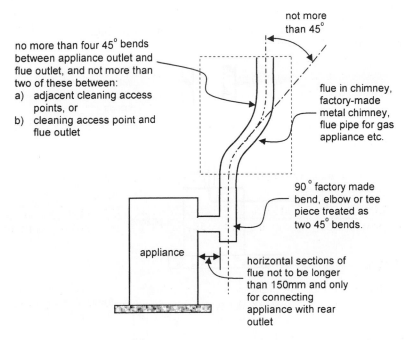

build flues so that they are as
straight and vertical as possible

not more
than 45°

no more than four 45° bends
between appliance outlet and
flue outlet, and not more than
two of these between:
a) adjacent cleaning access
    points, or
b) cleaning access point and
    flue outlet

flue in chimney,
factory-made
metal chimney,
flue pipe for gas
appliance etc.

90° factory made
bend, elbow or tee
piece treated as
two 45° bends.

appliance

horizontal sections of
flue not to be longer
than 150mm and only
for connecting
appliance with rear
outlet

Fig. 14.6   Configuration of natural draught flues serving open-flued appliances.

It should be noted that a 90° factory-made bend, elbow or tee piece should be treated as two 45° bends.

Openings into flues for inspection and cleaning

No flue should communicate with more than one room or internal space in a building except for the purposes of:

- inspection or cleaning, or
- the fitting of a draught diverter, draught stabiliser, draught break or explosion door.

Openings for inspection and cleaning should:

- be formed using purpose factory-made components which are compatible with the flue system;
- have an access cover with the same level of gas-tightness as the flue system and an equal thermal insulation level;
- allow easy passage of the sweeping brush.

Covers should be non-combustible unless fitted to a combustible fluepipe (e.g. a plastic fluepipe). A chimney which cannot be cleaned through an appliance should be provided with:

(a) suitably sized openings for cleaning provided at a sufficient number of locations in the chimney (see Fig. 14.7); and

(b) a debris collecting space with access for emptying.

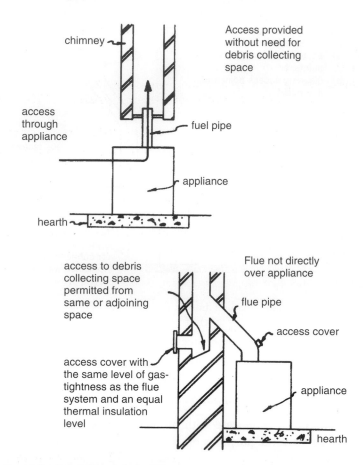

Fig. 14.7   Access for inspection and cleaning.

For appliances burning gas, the debris collection space should be provided at the base of a flue unless it is a factory-made metal chimney with a flue box, is lined or is constructed of flue blocks. To achieve this provide a space which:

• has a volume of at least 12 litres and a depth of at least 250 mm below the point where flue gases discharge into the chimney;

• is readily accessible for clearing debris, (e.g. by removing the appliance).

For radiant convector gas fires, convector heaters, fire/back boilers and inset live fuel effect fires, provide a clearance of at least 50 mm between the end of the appliance flue outlet and any surface.

Connecting fluepipes

A connecting fluepipe is used to connect a combustion appliance to a flue in a chimney. Suitable components for constructing connecting fluepipes include:

- cast iron, in accordance with BS 41: 1973 (1988) *Specification for cast iron spigot and socket flue or smoke pipes and fittings*;
- mild steel complying with BS 1449: *Steel plate, sheet and strip. Carbon and carbon manganese plate, sheet and strip*. Part 1: 1991 *General specifications*, with a flue wall thickness of at least 3 mm;
- stainless steel pipes at least 1 mm thick as described in BS EN 10088: *Stainless Steels:* Part 1: 1995. *List of stainless steels*, grades 1.4401, 1.4404, 1.4432 or 1.4436;
- vitreous enamelled steel complying with BS 6999: 1989 *Specification for vitreous enamelled low carbon steel flue pipes, other components and accessories for solid fuel burning appliances with a maximum rated output of 45 kW*;
- other fluepipes which have been independently certified to have the necessary performance designation for suitable use with the intended appliance.

Where spigot and socket fluepipes are used the sockets should be placed uppermost to contain moisture and other condensates in the flue. In order to achieve gas-tight joints it is usual either to use proprietary jointing accessories or, where appropriate, to pack the joints with noncombustible rope and fire cement.

Inspection and testing of flues

As explained in section 14.4 above, combustion installations should be inspected and tested before use to prove that they are in compliance with the Building Regulations. For flues this entails carrying out checks to show that they are:

- free from obstructions;
- satisfactorily gas-tight; and
- constructed of materials and components of the correct size to suit the intended application.

If building work also includes installation of the combustion appliance, tests can be carried out which involve firing up the appliance. These should include:

- tests for gas-tightness of the flue and at joints between the flue and the combustion appliance outlet;
- spillage tests as part of the commissioning process to check compliance with:
  - (a) Part J2 and Part L (see Chapter 16); and
  - (b) The Gas Safety (Installation and Use) Regulations 1998 (see Chapter 5).

It will normally be necessary for a suitably qualified person to prepare a report showing that the above considerations have been taken into account. An example checklist for such a report is given in section 14.4 above). For more information on methods of checking compliance with Regulation J2 see the section on *repair and reuse of existing flues* in section 14.8.3 below.

Dry lining around fireplace openings

It is usual to finish off around a fireplace opening with decorative treatment, such as a fireplace surround, masonry cladding or dry lining. Care must be taken to avoid the creation of any gaps that could allow flue gases to escape from the fireplace opening into the void behind the decorative treatment, by applying a suitable sealant at the junction. The sealant should be able to remain in place despite any relative movement between the decorative treatment and the fireplace recess.

14.6.3   Access to combustion appliances for maintenance purposes

Combustion appliances should be provided with a permanent means of access for maintenance purposes. The means of access should suit the location of the appliance, e.g. where an appliance is installed in a roof space it would probably be necessary to provide an access walkway.

14.6.4   Factory-made metal chimneys

These are defined in section 14.2 above and consist of prefabricated chimneys that are usually manufactured as sets of components for assembly on site.

Specification of factory-made metal chimneys

Where it is proposed to use component systems they should be independently certified as complying with:

- BS 4543 *Factory-made insulated chimneys,*
  - (a)   Part 1: 1990 (1996) *Methods of test,* (this part was withdrawn in April 2000 and partially replaced by BS EN 1859:2000);
  - (b)   Part 2: 1990 (1996) *Specification for chimneys with stainless steel flue linings for use with solid fuel fired appliances;* and
  - (c)   Part 3: 1990 (1996) *Specification for chimneys with stainless steel flue lining for use with oil-fired appliances.*

Such component systems should be installed in accordance with the relevant recommendations of:

- BS 7566: *Installation of factory-made chimneys to BS 4543 for domestic appliances,*
  - (a)   Part 1: 1992 (1998) *Method of specifying installation design information;*
  - (b)   Part 2: 1992 (1998) *Specification for installation design;*
  - (c)   Part 3: 1992 (1998) *Specification for site installation;*
  - (d)   Part 4: 1992 (1998) *Recommendations for installation design and installation.*

Gas and oil-fired appliances with flue gas temperatures not exceeding 250°C using twin wall component systems (oil-fired) or single wall component systems (gas-fired) should comply with BS 715: 1993. They should be installed in accordance with BS 5440 *Installation and maintenance of flues and ventilation for gas appliances of rated input not exceeding 70 Kw net (1st, 2nd and 3rd family gases)*, Part 1: 2000 *Specification for installation and maintenance of flues.*

Apart from the above, any other chimney system can be used provided that it is independently certified as being suitable for its purpose and it is installed in accordance with BS 7566 or BS 5440 as appropriate.

Using factory-made metal chimneys – general precautions

Where a factory-made metal chimney passes through a fire compartment wall or floor it must comply with the requirements of Part B (see Chapter 7) regarding the integrity of the compartmentation. Additionally, it may meet the requirements if:

- it is surrounded in non-combustible material having at least half the fire resistance required for the compartment wall or floor; or
- it has an appropriate level of fire resistance.

No combustible material should be placed nearer to the outer surface of the chimney than the distance (X) derived from the test procedures specified in BS 4543 *Factory-made insulated chimneys*, Part 1: 1990 (1996) *Methods of test*. The distance X will be the flue manufacturer's declared minimum distance as derived from the test. Additionally, a chimney passing through a cupboard, storage space or roof space may be separated from that space by a guard of suitable imperforate material provided that no combustible material is enclosed within the guard, and the distance between the inside of the guard and the outside of the chimney is not less than the distance (X) specified above.

Although it is not permissible to take a connecting fluepipe serving a solid fuel appliance (or an oil-fired appliance with flue gas temperatures which exceed 250°C) through a roof space, partition, internal wall or floor, it is permissible to do this for a factory-made metal chimney if the following precautions are taken:

- when passing through a wall, provide sleeves so that thermal movement is allowed to take place without damaging the flue or the building;
- do not conceal joints between chimney sections within ceiling joist spaces or walls, since this might prevent the flue from being checked for gas-tightness.

The chimney should be designed and constructed so that the appliance can be changed at a later date without the need for the chimney to be dismantled.

Like any other flue, factory-made metal chimneys should be guarded if they present a burn hazard which is not immediately apparent to people or they are so sited as to be at risk of damage.

14.6.5   Flues – specific provisions relating to appliances burning solid fuel

In addition to the general recommendations given above, appliances burning solid fuel with rated outputs up to 50 kW should comply with the following specific recommendations.

Size of flues for appliances burning solid fuel

Fluepipes and flues should:

- be at least the size shown in Table 2.2 to Approved Document J, which is reproduced below;
- have the same diameter or equivalent cross sectional area as that of the appliance flue outlet;
- not be smaller than the size recommended by the appliance manufacturer; and
- where a multifuel appliance is installed, be sized to accommodate burning the fuel that requires the largest flue.

Table 2.2 should used in conjunction with the following notes.

- A fireplace with an opening larger than 500 mm × 550 mm should be provided with a flue with a cross sectional area equal to 15% of the total face area of the fireplace opening.
- Specialist advice should be sought when proposing to construct flues having an area of:
  (a)   more than 15% of the total face area of the fireplace openings; or
  (b)   more than 120,000 mm^2 (0.12 m^2).
- Where a fireplace is exposed on two or more sides the area of the opening should be calculated from the formula:
  Fireplace opening area (mm^2) = L × H, where
  L = total horizontal length of fireplace opening in mm, and
  H = height of fireplace opening in mm
  For examples, see Fig. 14.8 below.

Outlets of flues for appliances burning solid fuel

Flue outlets should be located above the roof of the building where, whatever the wind conditions, the products of combustion can discharge freely and will not present a fire hazard.

   Where wind exposure, surrounding tall buildings, high trees or high ground could have adverse effects on flue draught, these chimney heights and/or separations may need to be increased.

   The outlet of any flue should be at least:

- 1 m above the highest point of contact between the flue and the roof, for roofs pitched at less than 10° (i.e. flat roofs);

AD J Section 2

Table 2.2 Size of flues in chimneys.

Installation (1)	Minimum flue size
Fireplace with an opening of up to 500 mm × 550 mm	200 mm diameter or rectangular/square flues having the same cross sectional area and a minimum dimension not less than 175 mm
Fireplace with an opening in excess of 500 mm × 550 mm or a fireplace exposed on two or more sides	See Paragraph 2.7. If rectangular/square flues are used the minimum dimension should not be less than 200 mm
Closed appliance of up to 20 kW rated output which: (a) burns smokeless or low volatiles fuel (2); or (b) is an appliance which meets the requirements of the Clean Air Act when burning an appropriate bituminous coal (3)	125 mm diameter or rectangular/square flues having the same cross sectional area and a minimum dimension not less than 100 mm for straight flues or 125 mm for flues with bends or offsets
Other closed appliance of up to 30 kW rated output burning any fuel	150 mm diameter or rectangular/square flues having the same cross sectional area and a minimum dimension not less than 125 mm
Closed appliance of above 30 kW and up to 50 kW rated output burning any fuel	175 mm diameter or rectangular/square flues having the same cross sectional area and a minimum dimension not less than 150 mm

Notes:
1. Closed appliances include cookers, stoves, room heaters and boilers.
2. Fuels such as bituminous coal, untreated wood or compressed paper are not smokeless or low volatiles fuels.
3. These appliances are known as 'exempted fireplaces'.

- 2.3 m measured horizontally from the roof surface for roofs pitched at 10° or more;
- 1 m above the top of any openable part of a dormer window, rooflight, or similar opening which is in a roof or external wall and is not more than 2.3 m horizontally from the top of the flue; and
- 600 mm above the top of any part of an adjoining building which is not more than 2.3 m horizontally from the top of the flue.

Additionally, if the flue passes through the roof within 2.3 m of the ridge and both slopes are at 10° or more to the horizontal, the top of the flue should be not less than 600 mm above the ridge (see Fig. 14.9).

Where flues discharge on or close to roofs with readily ignitable surfaces (e.g. roofs covered in thatch or shingles), the clearances shown in Fig. 14.9 might be insufficient to avoid a fire hazard. In such cases the outlet of any flue should be:

- at least 2.3 m measured horizontally from, and at least 1.8 m above, the roof surface;

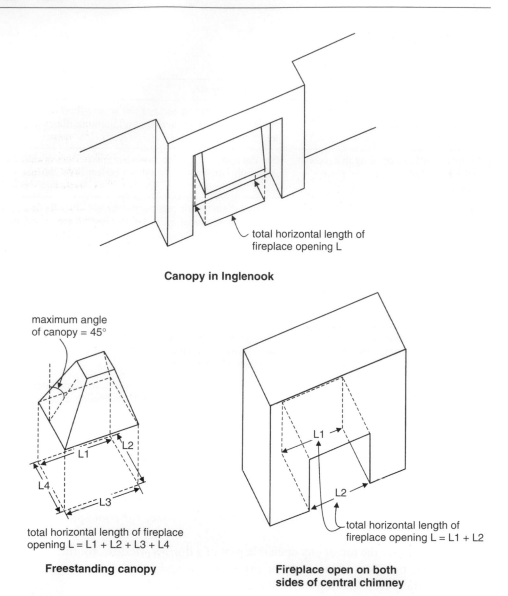

Canopy in Inglenook

total horizontal length of
fireplace opening L

maximum angle
of canopy = 45°

L1   L2

L4

L3

total horizontal length of fireplace
opening L = L1 + L2 + L3 + L4

Freestanding canopy

L1

L2

total horizontal length of
fireplace opening L = L1 + L2

**Fireplace open on both
sides of central chimney**

Fig. 14.8 Calculating fireplace opening.

- at least 600 mm above the ridge and at least 1800 mm vertically above the roof
  surface.

Fig. 14.10 gives details of these clearances.

Heights of flues for appliances burning solid fuel

The height of flue needed to ensure that sufficient draught is provided to clear the
products of combustion, will depend on:

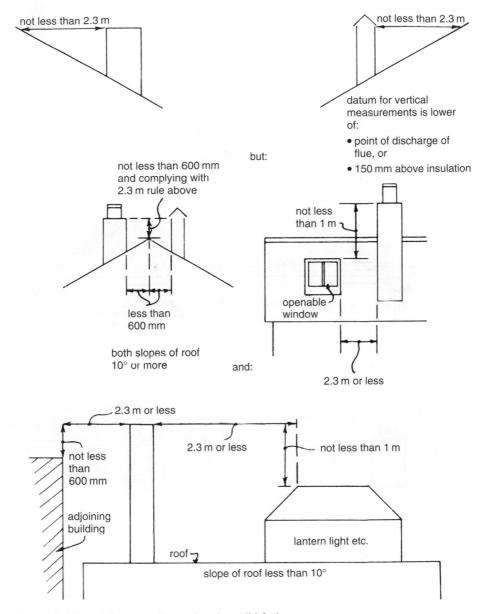

Fig. 14.9   Flue outlets – appliances burning solid fuel.

- the type of the appliance;
- the height of the building;
- the type of flue and the number of bends in it; and
- a careful assessment of local wind patterns.

It is possible that a flue height of 4.5 m could be satisfactory if the guidance on the position of flue outlets in sections 14.6.2 and 14.6.5 is followed.

Alternatively, the calculation procedure shown in BS 5854:1980 (1996): *Specifi-*

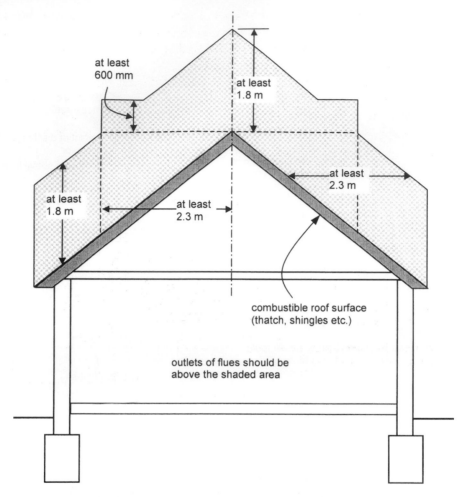

at least
600 mm

at least
1.8 m

at least
2.3 m

at least
1.8 m

at least
2.3 m

combustible roof surface
(thatch, shingles etc.)

outlets of flues should be
above the shaded area

Fig. 14.10   Flue outlets from solid fuel appliances above easily ignitable roof surfaces.

cation for installation in domestic premises of gas-fired ducted-air heaters of rated input not exceeding 60 kW, AMD 8130, can be used to decide whether or not a chimney design will provide sufficient draught.

When serving an open fire, the flue height is measured vertically from the highest point at which air can enter the fireplace (e.g. the top of the fireplace opening or, for a fire under a canopy, the bottom of the canopy) to the level at which the flue discharges into the outside air.

14.6.6   Flues – specific provisions relating to appliances burning gas

In addition to the general recommendations given above, appliances burning gas with rated inputs up to 70 kW (net) should comply with the following specific recommendations.

Additionally, for new appliances of known type, satisfactory provision for chimneys and flues will be achieved by:

- using factory-made components which have been independently certified (when tested to an appropriate BS EN European chimney standard) as complying with the performance requirements corresponding to the designations given in Table 3.2 from AD J (reproduced below); and
- installing these components in accordance with the appliance and component manufacturer's installation instructions and the guidance on location and shielding of fluepipes, connecting fluepipe components, masonry and flueblock chimneys and factory-made metal chimneys given in the text below.

AD J Section 3

Table 3.2 Minimum performance designations for chimney and fluepipe components for use with new gas appliances.

Appliance type		Minimum designation (See Notes)
Boiler: open-flued	natural draught	T250 N2 O D 1
	fanned draught	T250 P2 O D 1
	condensing	T250 P2 O W 1
Boiler: room-sealed	natural draught	T250 N2 O D 1
	fanned draught	T250 P2 O D 1
Gas fire Radiant/convector, ILFE or DFE		T300 N2 O D 1
Air heater	natural draught	T250 N2 O D 1
	fanned draught	T200 P2 O D 1
	SE – duct	T450 N2 O D 1

Notes:
1. The designation of chimney products is described in section 14.2 above. The BS EN for the product will specify its full designation and marking requirements.
2. These are default designations. Where appliance manufacturer's installation instructions specify a higher designation, this should be complied with.

Connecting DFE, ILFE and other gas appliances to flues

The building provisions needed to safely accommodate each of the main categories of gas appliance will differ for each type. Provided that the safety of the installation can be assured, it is permissible to install gas fires into fireplaces with flues designed for solid fuel appliances. Certain types of gas fire may also be installed in fireplaces with flues designed specifically for gas appliances. Reference to the Gas Appliances (Safety) Regulations 1995 will show that it is a requirement for particular combinations of appliance, flue box (if required) and flue to be selected from those stated in the manufacturer's instructions as having been shown to be safe by a Notified Body (see section 14.2 above).

Size of natural draught flues for open-flued appliances burning gas

Flues for gas-fired appliances should:

- be at least the size shown in Table 3.1 to Approved Document J (see below), where the builder is responsible for providing (or refurbishing) the flue but is not responsible for supplying the appliance;
- for a connecting fluepipe, have the same diameter or equivalent cross sectional area as that of the appliance flue outlet (the chimney flue should also have the same cross sectional area as the appliance flue outlet);
- be sized in accordance with the appliance manufacturer's installation instructions for appliances that are CE marked as complying with the Gas Appliances (Safety) Regulations 1995.

AD J Section 3

Table 3.1   Size of flues for gas fired appliances.

Intended installation	Minimum flue size	
Radiant/Convector gas fire	New flue: Circular Rectangular	 125 mm diameter 16,500 mm² cross sectional area with a minimum dimension of 90 mm
	Existing flue: Circular Rectangular	 125 mm diameter 12,000 mm² cross sectional area with a minimum dimension of 63 mm
ILFE fire or DFE fire within a fireplace opening up to 500 mm × 550 mm	Circular or Rectangular	Minimum flue dimension of 175 mm
DFE fire installed in a fireplace with an opening in excess of 500 mm × 550 mm	Calculate in accordance with Fig. 14.8 and the notes referring to similar sized openings for appliances burning solid fuel in section 14.6.5	

Outlets of flues for appliances burning gas

Flue outlets should be located externally so as to permit dispersal of the products of combustion and, for balanced flues, the intake of air. Suitable positions for outlets for both balanced and open-flued appliances are shown in Diagram 3.4. Minimum separation distances of outlets from various elements of buildings are given in the *Table to Diagram 3.4*. Both Diagram 3.4 and its accompanying Table are reproduced below. It should be noted that Diagram 3.4 and its Table are substantially the same as Figure C.1 and Table C.1 from BS 5440: *Installation and maintenance of flues and ventilation for gas appliances of rated input not exceeding 7 kW net (1st, 2nd and 3rd family gases): Part 1: 2000 Specification for installation and maintenance of flues.* The version of Diagram 3.4 shown in this chapter has been corrected for typographical errors found in the original in AD J.

Where a flue passes through a roof and the outlet is near to a roof window it should not be sited nearer to the window than:

- for a flat roof – 600 mm
- for a pitched roof – 600 mm if sited alongside or above, and 2000 mm if below the window.

AD J Section 3

Diagram 3.4   Location of outlets from flues serving gas appliances.

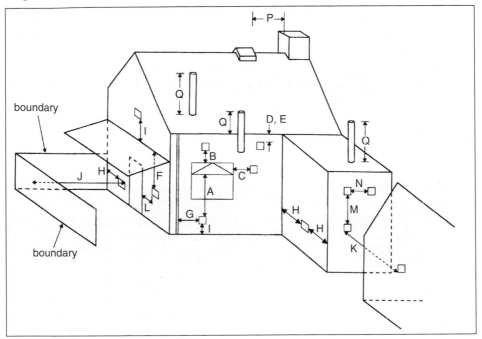

Where they are at significant risk of blockage flue outlets should be suitably protected as follows.

- For flues serving natural draught open-flued appliances consideration should be given to the following:
  - (a)   Protect with a suitable outlet terminal if the flue is less than 170 mm diameter. Suitable terminals are specified in BS 715: 1993 *Specification for metal flue pipes, fittings, terminals and accessories for gas-fired appliances with a rated input not exceeding 60 kW*, AMD 8413 and BS 1289: 1986 *Flue blocks and masonry terminals for gas appliances*, Part 1: 1986 *Specification for precast concrete flue blocks and terminals*.
  - (b)   For flues of over 170 mm diameter local conditions should be taken into account when assessing the risk of blockage. This could include the risk of blockage from nesting squirrels or jackdaws in areas where these creatures are prevalent. Protective cages designed for use with solid fuel appliances with mesh sizes between 6 mm and 25 mm could be used provided that the total free area of the outlet openings in the cage is at least twice the cross sectional area of the flue.
- Any flue outlet should also be protected with a guard to prevent it being damaged and to protect people who might otherwise come in contact with it.
- Flue outlets in vulnerable positions (e.g. in reach of a balcony, veranda, window

AD J Section 3

Table to Diagram 3.4 Location of outlets from flues serving gas appliances.

		Minimum separation distances for terminals in mm			
Location		**Balanced flue**		**Open flue**	
		Natural draught	Fanned draught	Natural draught	Fanned draught
A	Below an opening (1)	Appliance rated heat input (net)	300	(3)	300
		0–7 kW 300 > 7–14 kW 600 > 14–32 kW 1500 > 32 kW 2000			
B	Above an opening (1)	0–32 kW 300 > 32 kW 600	300	(3)	300
C	Horizontally to an opening (1)	0–7 kW 300 > 7–14 kW 400 > 14 kW 600	300	(3)	300
D	Below gutters, soil pipes or drain pipes	300	75	(3)	75
E	Below eaves	300	200	(3)	200
F	Below balcony or car port roof	600	200	(3)	200
G	From a vertical drain pipe or soil pipe	300	150 (4)	(3)	150
H	From an internal or external corner or to a boundary alongside the terminal (2)	600	300	(3)	200
I	Above ground, roof or balcony level	300	300	(3)	300
J	From a surface or a boundary facing the terminal (2)	600	600	(3)	600
K	From a terminal facing the terminal	600	1200	(3)	1200
L	From an opening in the car port into the building	1200	1200	(3)	1200
M	Vertically from a terminal on the same wall	1200	1500	(3)	1500
N	Horizontally from a terminal on the same wall	300	300	(3)	300
P	From a structure on the roof	Not applicable	Not applicable	1500 mm if a ridge terminal. For any other terminal, as given in BS 5440-1:2000	N/A
Q	Above the highest point of intersection with the roof	Not applicable	Site in accordance with manufacturer's instructions	Site in accordance with BS 5440-1:2000	150

Notes:
1. An opening here means an openable element, such as an openable window, or a fixed opening such as an air vent. However, in addition, the outlet should not be nearer than 150 mm (fanned draught) or 300 mm (natural draught) to an opening into the building fabric formed for the purpose of accommodating a built in element, such as a window frame.
2. Boundary as defined in section 4.2. Smaller separations to the boundary may be acceptable for appliances that have been shown to operate safely with such separations from surfaces adjacent to or opposite the flue outlet.
3. Should not be used.
4. This dimension may be reduced to 75 mm for appliances of up to 5 kW input (net).

or from the ground) should be designed so that the entry of any matter which might restrict the flue is prevented.

Heights of natural draught flues for open-flued appliances burning gas

The height of flue needed to ensure that sufficient draught is provided to clear the products of combustion, will depend on:

- the type of the appliance;
- the height of the building;
- the type of flue and the number of bends in it; and
- a careful assessment of local wind patterns.

The requirement for sufficient flue height can be met:

- for appliances that are CE marked as complying with the Gas Appliances (Safety) Regulations 1995, by following the appliance manufacturer's installation instructions;
- for older appliances that are not CE marked (but have manufacturer's installation instructions) by following:
  - (a) the guidance in BS 5871: *Specification for installation of gas fires, convector heaters, fire/back boilers and decorative fuel effect gas appliances:* Part 3: 2001 *Decorative fuel effect gas appliances of heat input not exceeding 20 kW (2nd and 3rd family gases)*, AMD 7033, for decorative fuel effect fires; or
  - (b) the calculation procedure BS 5440: Part 1: 2000, for appliances other than decorative fuel effect fires.

Components for connecting fluepipes for appliances burning gas

Suitable components for constructing connecting fluepipes include:

- any components given in section 14.6.2 'Connecting fluepipes' above;
- sheet metal fluepipes complying with BS 715: 1993 *Specification for metal flue pipes, fittings, terminals and accessories for gas-fired appliances with a rated input not exceeding 60 kW*, AMD 8413; or
- fibre cement pipes as described in BS 7435 *Fibre cement flue pipes, fittings and terminals,* Part 1: 1991 (1998) *Specification for light quality fibre cement flue pipes, fittings and terminals:* Part 2 1991 *Specifications for heavy quality cement flue pipes, fittings and terminals*; or
- any other material or component that has been independently certified as suitable for this purpose.

Construction of flueblock chimneys for installations burning gas

In addition to the general recommendations for flueblock chimneys (see section 14.6.2 'Construction of flueblock chimneys' above), for gas fired installations, flueblock chimneys:

- should be supported and restrained in accordance with manufacturer's instructions where they are not intended to be bonded into surrounding masonry;
- may be constructed from factory-made flueblock systems (consisting of, for example, straight blocks, lintel blocks, offset blocks, transfer blocks, recess units and jointing materials) complying with:
  - (a) BS 1289: *Flue blocks and masonry terminals for gas appliances* Part 1: 1986 *Specification for Precast Concrete Flue Blocks and Terminals*, or
  - (b) BS EN 1806: 2000 (with a performance class of at least FB4 N2).

14.6.7   Flues – specific provisions relating to appliances burning oil

In addition to the general recommendations given above, appliances burning oil with rated outputs up to 45 kW should comply with the following specific recommendations.

Size of flues for appliances burning oil

Unlike previous editions of AD J, no specific minimum flue sizes are given in the 2002 edition. This could reflect the different types of oil-fired appliances that are available (and their different modes of operation, e.g. pressure jet or vaporising) since they are likely to have different discharge velocities. Therefore, in general, flues should be sized to suit the appliance being installed, so that adequate discharge velocities are achieved to prevent flow reversal problems and excessive flow resistances are avoided.

Oil-fired appliances can be connected to discharge:

- via a connecting fluepipe to a flue in a chimney or flueblock chimney; or
- to a balanced flue (or a flue designed to discharge through or adjacent to a wall).

In the first case the connecting fluepipe should be the same size as the appliance flue outlet and any flue in a chimney should have the same cross sectional area as the appliance flue outlet. If the make and model of the appliance is known when the chimney (or flueblock chimney) is being designed and built then the flue can be made the same size as the appliance flue outlet. Otherwise, the flue can be made large enough to allow the later insertion of a suitable flexible flue liner matching the appliance to be installed.

In all cases, the flue size should always be provided to suit the appliance manufacturers' installation instructions.

Outlets of flues for appliances burning oil

Flue outlets should be located externally so as to permit:

- dispersal of the products of combustion;
- correct operation of a natural draught flue; and
- for balanced flues, the intake of air.

Suitable positions for outlets for both balanced and open-flued appliances are shown in Diagram 4.2. Minimum separation distances of outlets from various elements of buildings are given in the Table to Diagram 4.2. Both Diagram 4.2 and its accompanying Table are reproduced below.

AD J Section 4

Diagram 4.2   Location of outlets from flues serving oil-fired appliances.

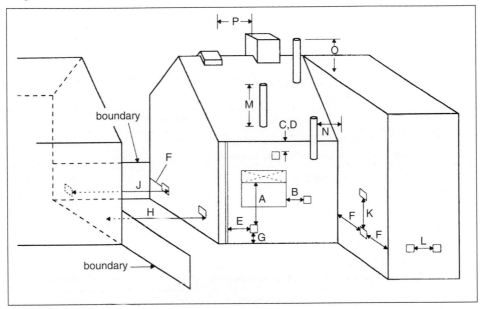

The minimum separation distances given in Table to Diagram 4.2 may need to be increased where:

- local factors such as wind patterns, might disrupt operation of the flue; or
- the height of a natural draught flue may be insufficient to disperse the products of combustion from an open-flued appliance.

Flue outlets should be protected with guards to prevent them being damaged and to protect people who might otherwise come in contact with them. Flue outlets in vulnerable positions (e.g. in reach of a balcony, veranda, window or from the ground) should be designed so that the entry of any matter which might restrict the flue is prevented.

Appliances burning oil – the effect of flue gas temperatures on the design of chimneys and fluepipes

In order to provide a satisfactory chimney or fluepipe for an oil-fired appliance it is necessary to establish whether the flue gas temperature is above or below 250°C as measured by a suitable method (e.g. as in *OFTEC Standards A 100 or A 101*).

AD J Section 4

Table to Diagram 4.2   Location of outlets from flues serving oil-fired appliances.

Minimum separation distances for terminals in mm		
Location of outlet (1)	**Appliance with pressure jet burner**	**Appliance with vaporising burner**
A Below an opening (2, 3)	600	should not be used
B Horizontally to an opening (2, 3)	600	should not be used
C Below a plastic/painted gutter, drainage pipe or eaves if combustible material protected (4)	75	should not be used
D Below a balcony or a plastic/painted gutter, drainage pipe or eaves without protection to combustible material	600	should not be used
E From vertical sanitary pipework	300	should not be used
F From an external or internal corner or from a surface or boundary alongside the terminal	300	should not be used
G Above ground or balcony level	300	should not be used
H From a surface or boundary facing the terminal	600	should not be used
J From a terminal facing the terminal	1200	should not be used
K Vertically from a terminal on the same wall	1500	should not be used
L Horizontally from a terminal on the same wall	750	should not be used
M Above the highest point of an intersection with the roof	600 (6)	1000 (5)
N From a vertical structure to the side of the terminal	750 (6)	2300
O Above a vertical structure which is less than 750 mm (pressure jet burner) or 2300 mm (vaporising burner) horizontally from the side of the terminal	600 (6)	1000 (5)
P From a ridge terminal to a vertical structure on the roof	1500	should not be used

Notes:
1. Terminals should only be positioned on walls where appliances have been approved for such configurations when tested in accordance with BS EN 303-1:1999 or OFTEC standards OFS A100 or OFS A101.
2. An opening means an openable element, such as an openable window, or a permanent opening such as a permanently open air vent.
3. Notwithstanding the dimensions above, a terminal should be at least 300 mm from combustible material, e.g. a window frame.
4. A way of providing protection of combustible material would be to fit a heat shield at least 750 mm wide.
5. Where a terminal is used with a vaporising burner, the terminal should be at least 2300 mm horizontally from the roof.
6. Outlets for vertical balanced flues in locations M, N and O should be in accordance with manufacturer's instructions.

As a guide to establishing the likely flue gas temperatures for a particular appliance the following notes may prove useful.

- Since flue gas temperatures depend on the appliance type and its age, older and re-used appliances will probably have flue gas temperatures in excess of 250°C.
- Modern appliances bearing the CE Mark (showing compliance with the *Boiler (Efficiency) Regulations 1993*) will usually have flue gas temperatures which do not exceed 250°C.
- Manufacturers of appliances should be able to supply information on flue gas temperatures for individual appliances, in their installation instructions.
- The Oil Firing Technical Association for the Petroleum Industry (OFTEC) can be contacted to obtain information on individual appliances. They may be contacted at Century House, 100 High Street, Banstead, Surrey, SM7 2NN.

Where information is unobtainable, the flue gas temperature for a particular appliance should be assumed to be greater than 250°C.

Design of flues for oil-fired appliances with flue gas temperatures exceeding 250°C

Where the flue gas temperature exceeds 250°C, a satisfactory design for the flue could be achieved by discharging the appliance:

- via a connecting fluepipe, masonry or flueblock chimney designed for use with solid fuel;
- to a suitable factory-made metal chimney designed for flue gas temperatures exceeding 250°C (see section 14.7.3 below);
- via other products which have been independently certified as being suitable for this purpose.

Design of flues for oil-fired appliances with flue gas temperatures not exceeding 250°C

Where the flue gas temperature does not exceed 250°C, a satisfactory design for the flue could be achieved:

- for any appliances (new or existing), by:
  - (a) following the guidance for installations where the flue gas temperature exceeds 250°C listed above; and
  - (b) the provisions for location and shielding of fluepipes and connecting flue-pipe components, outlined in section 14.7.3 below; or
- for new appliances of known type, by:
  - (a) using factory-made components which have been independently certified (when tested to an appropriate BS EN European chimney standard) as complying with the performance requirements corresponding to the designations given in Table 4.1 from AD J (reproduced below); and

AD J Section 4

Table 4.1 Minimum performance designations for chimneys and fluepipe components for use with new oil-fired appliances with flue gas temperature less than 250°C.

Appliance type	Minimum designation (See Notes)	
Boiler – pressure jet (including combination)	T160 P2 O D 1	Class C2 oil
Boiler – condensing Cooker – pressure jet	T160 P2 O D 2	Class D oil
Cooker – vaporising burner	T160 N2 O D 1	Class C2 oil
Room heater – vaporising burner	T160 N2 O D 2	Class D oil

Notes:
3. The designation of chimney products is described in section 14.2 above. The BS EN for the product will specify its full designation and marking requirements.
4. These are default designations. Where appliance manufacturer's installation instructions specify a higher designation, this should be complied with.

(b)   installing these components in accordance with the appliance and component manufacturer's installation instructions and the guidance on location and shielding of fluepipes, connecting fluepipe components, flueblock chimneys and factory-made metal chimneys given in section 14.7.3 below.

Components for connecting fluepipes for appliances burning oil

Suitable components for constructing connecting fluepipes include:

- any components given in section 14.6.2 'Connecting fluepipes' above;
- any components given in section 14.6.5 'Connecting fluepipes' above; or
- any other material or component that has been independently certified as suitable for this purpose.

Construction of flueblock chimneys for installations burning oil

In addition to the general recommendations for flueblock chimneys (see section 14.6.2 'Construction of flueblock chimneys' above), for oil fired installations, flueblock chimneys:

- should be supported and restrained in accordance with manufacturer's instructions where they are not intended to be bonded into surrounding masonry;
- may be constructed from factory-made flueblock systems (consisting of, for example, straight blocks, lintel blocks, offset blocks, transfer blocks, recess units and jointing materials) complying with:
  - (a)   BS 1289: *Flue blocks and masonry terminals for gas appliances* Part 1: 1986 *Specification for precast concrete flue blocks and terminals*, or
  - (b)   BS EN 1806: 2000 (with a performance at least equal to the Table 4.1 of AD J Section 4 above, designations for oil-fired appliances).

14.7   Protection of building against fire and heat

Paragraph J3 of Part J requires that the construction of fireplaces and chimneys and the installation of combustion appliances and fluepipes must be carried out so as to reduce to a reasonable level the risk of people suffering burns or the building catching fire in consequence of their use.

In general, this means that the combustion installation must enable normal operation of the appliances without danger being caused through damage to the fabric of the building by heat or fire. Additionally, the combustion installation should undergo inspection and testing to ensure that it is suitable for its intended purpose.

Therefore, there are implications for the design and construction of hearths, fireplaces, chimneys and flues to ensure that they are:

- of sufficient size
- constructed of suitable materials
- suitably isolated from any adjacent combustible materials.

14.7.1   Protection of the building against fire and heat – appliances burning solid fuel

Hearths for appliances burning solid fuel – general provisions

In most cases constructional hearths should be provided where an appliance burning solid fuel is to be installed. However, where it can be independently certified that the appliance cannot cause the temperature of the hearth to exceed 100°C (and the appliance is not designed to stand in an appliance recess), it is possible to provide a hearth made of non-combustible board/sheet material or tiles at least 12 mm thick.

Constructional hearths should be constructed of solid non-combustible material at least 125 mm thick (including the thickness of any non-combustible floor under the hearth).

Constructional hearths built in connection with a fireplace recess should:

- extend within the recess to the back and jambs of the recess;
- project at least 500 mm in front of the jamb; and
- extend outside the recess to at least 150 mm beyond each side of the opening.

If not built in connection with a fireplace recess, the plan dimensions of the hearth should be such as to accommodate a square of at least 840 mm. See Fig. 14.11 for details.

Hearths for appliances burning solid fuel – proximity of combustible materials

Hearths are provided to prevent combustion appliances setting fire to the building fabric and furnishings and to limit the risk of people being accidentally burnt.

Minimum plan dimensions

with fireplace recess

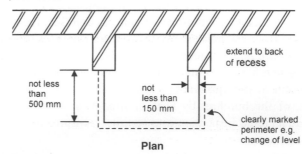

extend to back
of recess

not less
than
500 mm

not
less than
150 mm

clearly marked
perimeter e.g.
change of level

Plan

without fireplace recess

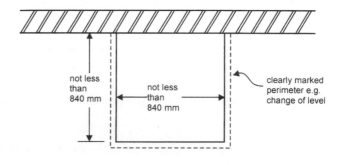

not less
than
840 mm

not less
than
840 mm

clearly marked
perimeter e.g.
change of level

Plan

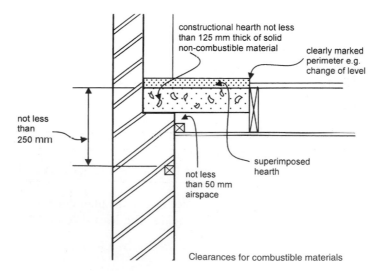

constructional hearth not less
than 125 mm thick of solid
non-combustible material

clearly marked
perimeter e.g.
change of level

not less
than
250 mm

superimposed
hearth

not less
than 50 mm
airspace

Clearances for combustible materials

Fig. 14.11   Hearths for appliances burning solid fuel.

Therefore, they should be separated from adjacent combustible materials and should be satisfactorily delineated from surrounding floor finishes (carpets etc.) as follows.

- Combustible material should not be placed under a constructional hearth for a solid fuel appliance within a vertical distance of 250 mm from the upper surface of the hearth, unless there is an airspace of at least 50 mm between the combustible material and the underside of the hearth (see Fig. 14.11).
- Where a superimposed hearth has been placed onto a constructional hearth, combustible material placed on or beside the constructional hearth should not extend under the superimposed hearth by more than 25 mm or closer to the appliance than 150 mm (see Fig. 14.12).
- Ensure that the hearth (superimposed or constructional) is suitably delineated to discourage combustible floor finishes from being laid too close to the appliance, by marking the edges or providing a change of level.
- Position the appliance on the hearth such that combustible material cannot be laid closer to the base of the appliance than:
  - (a) *at the front*, 300 mm if the appliance is an open fire or stove which can, when opened, be operated as an open fire, or 225 mm in any other case;
  - (b) *at the back and sides*, 150 mm or in accordance with the recommendations below which relate to distance from hearth to walls.

If any part of the back or sides of the appliance lies within 150 mm horizontally of the wall, then the wall should be of solid non-combustible construction at least 75 mm thick from floor level to a level of 300 mm above the top of the appliance and 1200 mm above the hearth.

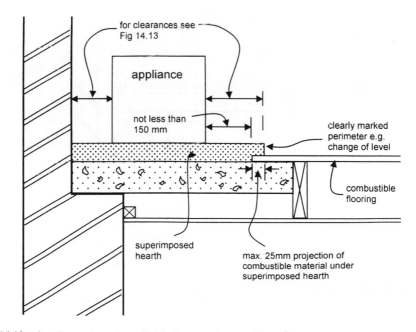

Fig. 14.12   Appliances burning solid fuel – superimposed hearths.

If, however, any part of the back or sides of the appliance lies within 50 mm of the wall, then the wall should be of solid non-combustible construction at least 200 mm thick from floor level to a level of 300 mm above the top of the appliance and 1200 mm above the hearth (see Fig. 14.13). Where the hearth itself is at least 150 mm from an adjacent wall there is no requirement for protection of the wall. It should be noted that these thicknesses of solid non-combustible material can be substituted by thinner material if the same overall level of protection can be achieved.

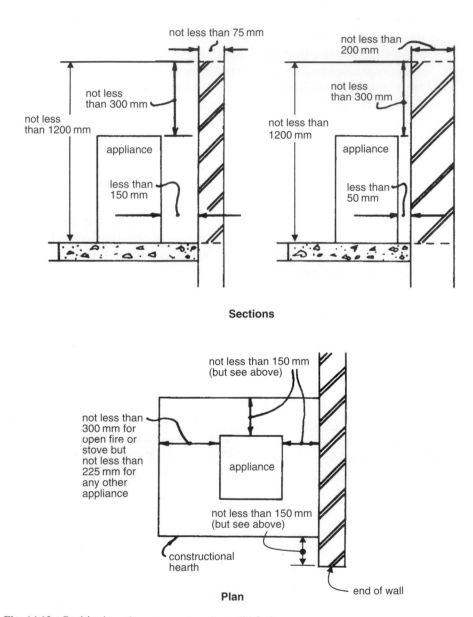

Fig. 14.13   Positioning of appliances burning solid fuel.

Fireplace recesses for appliances burning solid fuel – general provisions

Fireplace recesses need to be constructed so that they protect the building from the risk of fire. Traditionally, fireplace recesses have been constructed in masonry or concrete and guidance on these forms is still available in AD J. It is also possible to construct prefabricated factory-made appliance chambers and the 2002 edition of AD J contains guidance on these for the first time.

Fireplace recesses constructed in masonry or concrete should have a jamb on each side at least 200 mm thick, a solid back wall at least 200 mm thick, or a cavity wall back with each leaf at least 100 mm thick. These thicknesses are to run the full height of the recess. However, if a fire-place recess is in an external wall the back may be a solid wall of not less than 100 mm thickness. Similarly, if part of a wall acts as the back of two recesses on opposite sides of the wall, it may be a solid wall not less than 100 mm thick. It is assumed that this latter exemption does not apply to a wall separating buildings or dwellings within a building since the requirements for chimney walls (see below) specify a minimum thickness of 200mm in these circumstances (see Fig. 14.14).

Fireplace recesses constructed from prefabricated factory-made appliance chambers use components made from insulating concrete with a density between 1200 kg/m^2 and 1700 kg/m^2. AD J lays down the minimum thicknesses for the various components as follows:

- base – 50 mm
- side sections (forming walls on each side if the chamber) – 75 mm
- back section (forming the rear of the chamber) – 100 mm
- top slab, lintel or gather (forming the top of the chamber) – 100 mm.

These components should be supplied as sets which must be assembled and jointed in accordance with the manufacturer's instructions.

Fireplace recesses for appliances burning solid fuel – linings

In most cases a fireplace recess will need to be lined so that it is able to provide an acceptable setting for the installation of an appliance (such as an inset open fire). There are a great many lining components on the market and one example of a traditional fireplace lining and throat forming lintel is shown in Fig. 14.15. In the example, the tapered gather is formed by a throat forming lintel. Gathers are needed to ensure the proper working of the flue and can be formed in other ways, such as by:

- using a combined prefabricated lintel and gather unit built into the fireplace recess; or
- traditional corbelling of the masonry; or
- using a prefabricated appliance chamber incorporating the gather; or
- using a suitable canopy (see Fig. 14.8 above).

Recesses built in masonry

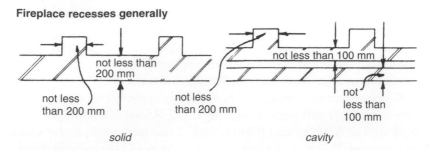

Fireplace recesses generally

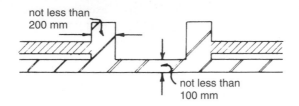

Recesses in external wall

Back-to-back recesses

Fig. 14.14   Fireplace recesses – appliances burning solid fuel.

Fireplace recesses for appliances burning solid fuel – proximity of combustible material

Combustible materials should not be placed where they could be ignited by heat dissipating through the walls of the fireplace recess. This means that combustible material should be at least:

- 200 mm from the *inside* surface of the fireplace recess; or
- 40 mm from the *outside* surface (although this does not apply to floorboards, skirting boards, dado and picture rails, mantelshelves and architraves).

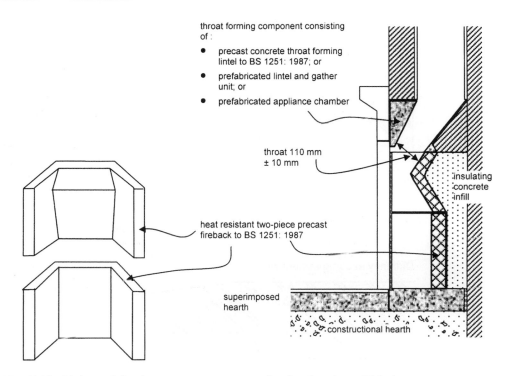

throat forming component consisting
of :

- precast concrete throat forming
  lintel to BS 1251: 1987; or
- prefabricated lintel and gather
  unit; or
- prefabricated appliance chamber

throat 110 mm
± 10 mm

insulating
concrete
infill

heat resistant two-piece precast
fireback to BS 1251: 1987

superimposed
hearth

constructional hearth

Fig. 14.15   Lining and fireplace components to open fireplace burning solid fuel.

Masonry and flueblock chimneys for appliances burning solid fuel

A masonry chimney provides structural support to suitably caulked flue liners and flueblocks, and fire protection to the building, by being constructed of bricks, medium weight concrete blocks or stone set in suitable mortar. The thickness of the masonry will vary with the type of fuel. Therefore, for appliances burning solid fuel, if a chimney is built of masonry and is lined as described above, any flue in that chimney (including a flue composed of flueblocks) should be:

- surrounded and separated from any other flue in that chimney by at least 100 mm thickness of solid masonry material, excluding the thickness of any flue lining material;
- separated by at least 200 mm of solid masonry material from another compartment of the same building, another building or another dwelling;
- separated by at least 100 mm of solid masonry material from the outside air (see Fig. 14.16).

Masonry and flueblock chimneys for appliances burning solid fuel – proximity of combustible material

Combustible materials should not be placed where they could be ignited by heat dissipating through the walls of flues. This means that combustible material should be at least:

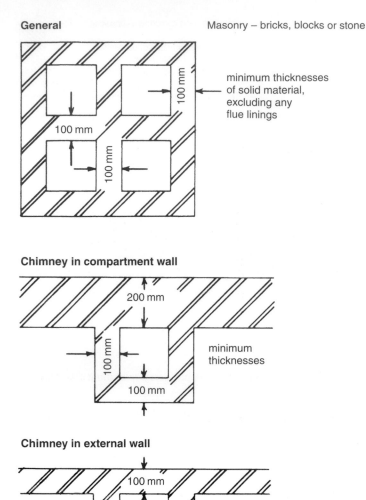

Fig. 14.16   Masonry and flueblock chimneys – wall thicknesses.

- 200 mm from the *inside* surface of a flue; or
- 40 mm from the *outside* surface of the chimney (although this does not apply to floorboards, skirting boards, dado and picture rails, mantelshelves and architraves).

No metal fastening in contact with combustible material should be placed within 50 mm of the inside of the flue (see Fig. 14.17).

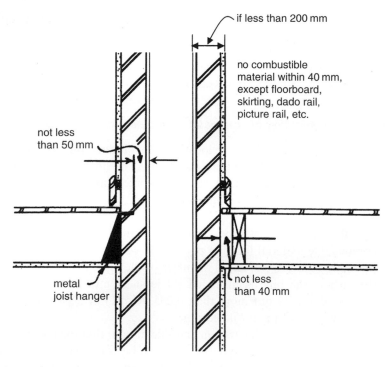

Fig. 14.17 Proximity of combustible materials – appliances burning solid fuel.

Connecting fluepipes serving appliances burning solid fuel

Connecting fluepipes serving appliances burning solid fuel should not pass through any roof space, partition, internal wall or floor (although they are allowed to pass through the wall of a chimney and a floor supporting a chimney!) and should only be used to connect an appliance to a chimney (see Fig. 14.18). The obvious intention is that connecting fluepipes should be as short as possible and should only be used to connect an appliance to a proper chimney. They should also be guarded if:

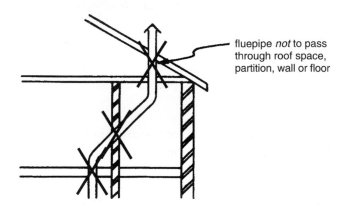

Fig. 14.18 Connecting the fluepipes for appliances burning solid fuel.

- the burn hazard they present is not immediately apparent to people; or
- they are so sited as to be at risk of damage.

A horizontal connection is permitted to connect a back outlet appliance to a chimney but this should not exceed 150 mm in length (see Fig. 14.6).

Placing and shielding of connecting fluepipes for appliances burning solid fuel

A connecting fluepipe should be located so that it does not risk igniting combustible material. Horizontal and sloping runs should be limited in length and where the connecting fluepipe is adjacent to combustible material, the following examples show how the risk could be minimised.

- By allowing sufficient separation, e.g. setting a minimum distance from any combustible material forming part of the wall or partition of at least:
  - (a)   three times the external diameter of the pipe; or
  - (b)   where the pipe is insulated (at least 12 mm of insulation with thermal conductivity not more than 0.065 W/mK) at least 0.75 times the outside diameter of the insulated pipe.
- By a combination of separation and shielding so that the pipe is at least 1.5 times its external diameter from the combustible material, and a shield of non-combustible material is placed so that there is an airspace of at least 12 mm between the shield and the combustible material. The non-combustible shield may either:
  - (a)   extend past the fluepipe by at least 1.5 times its external diameter; or
  - (b)   be positioned to ensure that the minimum distance from the fluepipe to any combustible material is at least three times the pipe diameter (see Fig. 14.19).
- By using a factory-made metal chimney and following the guidance in section 14.6.4 above.

14.7.2   Protection of the building against fire and heat – appliances burning gas

Hearths for appliances burning gas

The decision as to the type of hearth to provide (or indeed whether one is needed at all) will depend on the type of appliance installed. For example, a hearth is not required:

- if the appliance is installed so that no part of any flame or incandescent material is less than 225 mm above the floor; or
- the manufacturer's instructions state that a hearth is not required.

For a back boiler behind a gas fire, the hearth should be constructed of solid non-combustible material:

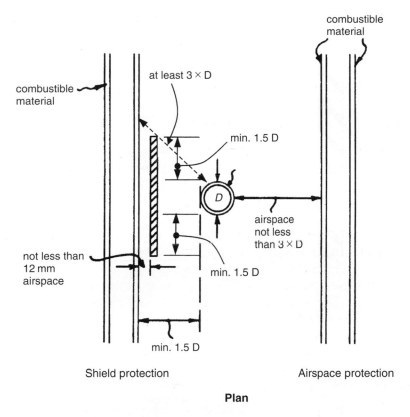

at least 3 × D

combustible material

min. 1.5 D

D

airspace not less than 3 × D

not less than 12 mm airspace

min. 1.5 D

min. 1.5 D

combustible material

Shield protection

Airspace protection

Plan

Fig. 14.19 Protection of combustible materials next to uninsulated fluepipes for appliances burning solid fuel.

- at least 125 mm thick (i.e. as for solid fuel appliances); or
- at least 25 mm thick on 25 mm non-combustible supports.

It should extend at least 150 mm beyond the back and sides of the back boiler and extend forward in front as required for the type of fire fitted (see Fig. 14.20).

The hearth for a decorative fuel effect fire (DFE) or inset live fuel effect fire (ILFE) should consist of a top layer of non-combustible, non-friable material at least 12 mm thick. The extent of any projections will depend on the type of appliance and whether it is freestanding or situated in a fireplace recess (see Fig. 14.21). The edges of the hearth should be designed to be apparent to building occupiers so as to discourage the laying of combustible floor finishes too close to the appliance. This could be achieved by providing a change in levels.

Shielding of appliances burning gas

Gas-fired appliances should be positioned so that the possibility of accidental contact is minimised, and separated from combustible materials as shown in Fig. 14.22 by a non-combustible surface such as:

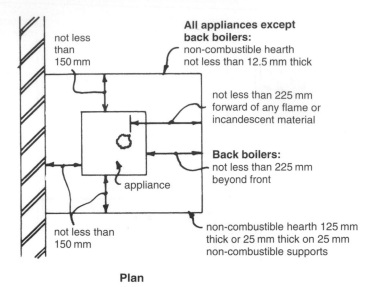

not less
than
150 mm

**All appliances except
back boilers:**
non-combustible hearth
not less than 12.5 mm thick

not less than 225 mm
forward of any flame or
incandescent material

Back boilers:
not less than 225 mm
beyond front

appliance

not less than
150 mm

non-combustible hearth 125 mm
thick or 25 mm thick on 25 mm
non-combustible supports

Plan

Unless:

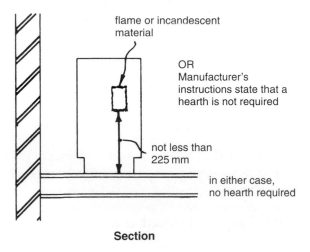

flame or incandescent
material

OR
Manufacturer's
instructions state that a
hearth is not required

not less than
225 mm

in either case,
no hearth required

Section

Fig. 14.20   Hearths for appliances burning gas.

- a shield of non-combustible material at least 25 mm thick, or
- at least 75 mm airspace.

Alternatively, for appliances that are CE marked as compliant with the Gas
Appliances (Safety) Regulations 1995 the manufacturer's instructions regarding
shielding should be followed.

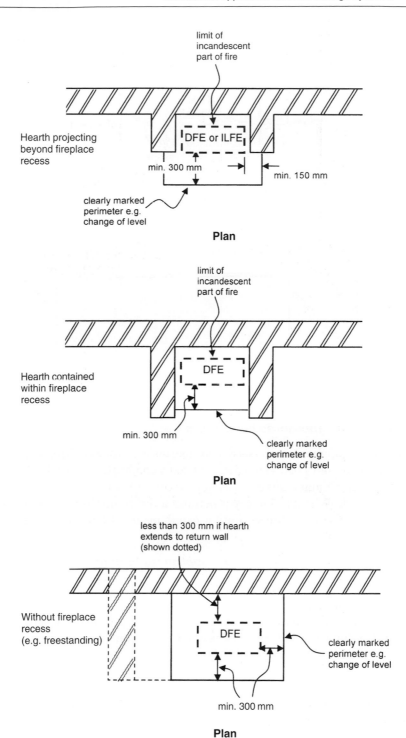

Fig. 14.21 Hearths for decorative fuel effect (DFE) and inset live fuel effect (ILFE) fires burning gas.

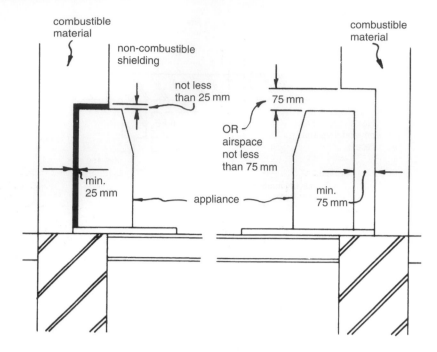

Fig. 14.22   Shielding of appliances burning gas.

Chimneys for appliances burning gas

Where gas appliances are served by masonry chimneys there should be at least 25 mm of masonry between the flues and any combustible material. Similarly, where a flueblock chimney serves a gas appliance the flueblock walls should be at least 25 mm thick. Where a chimney penetrates a fire compartment wall or floor it must also comply with the fire separation requirements of Part B (see Chapter 7).

Placing and shielding of flues for appliances burning gas

Connecting flues and factory-made chimneys complying with BS 715: 1993 *Specification for metal flue pipes, fittings, terminals and accessories for gas-fired appliances with a rated input not exceeding 60 kW*, serving appliances burning gas should be placed as shown in Fig. 14.23 to ensure the following.

- Every part of the flue is at least 25 mm from any combustible material. The distance is measured from the outer surface of the flue wall, or the outer surface of the inner wall for multi-walled products.
- Where the flue passes through a roof, floor or wall formed of combustible materials (other than a compartment roof, floor or wall), it is enclosed in a sleeve of non-combustible material and there is at least 25 mm airspace between the flue and the sleeve. The airspace could be wholly or partially filled with non-combustible insulating material.

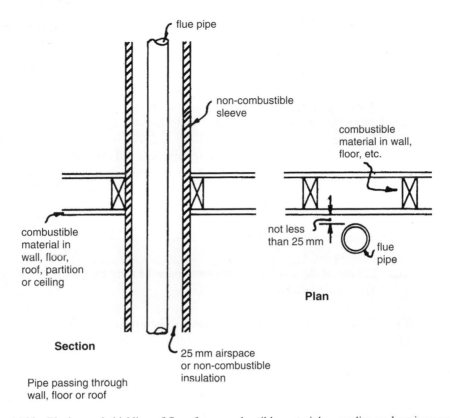

Fig. 14.23 Placing and shielding of flues from combustible materials – appliances burning gas.

Factory-made chimneys complying with BS 4543: 1996 *Factory-made insulated chimneys* should also be separated from combustible materials. This is dealt with in section 14.6.4 above. Factory-made chimneys and connecting fluepipes should also be guarded if:

- the burn hazard they present is not immediately apparent to people; or
- they are so sited as to be at risk of damage.

14.7.3   Protection of the building against fire and heat – appliances burning oil

Hearths for appliances burning oil

The basic function of a hearth is to stop the building from catching fire. Where oil is the fuel it is customary to place a non-combustible tray on top of the hearth to collect any spilled fuel although this is not a health and safety provision.

The decision as to the type of hearth to provide will depend on the temperature reached by the floor below the hearth as a result of the operation of the appliance as follows.

- For hearth temperatures which are unlikely to exceed 100°C (as shown by a suitable test procedure such as that contained in *OFTEC Oil-Fired Appliance Standards OFS A 100 and OFS A 101*) it is usual to provide a rigid, imperforate, non-absorbent, non-combustible sheet, (e.g. a steel tray) which can be part of the appliance. No other special measures are necessary.
- Where hearth temperatures could exceed 100°C, the guidance on hearths for appliances burning solid fuel should be followed (see section 14.7.1 above).

The hearth surface surrounding the appliance should be kept clear of any combustible material by maintaining the following minimum clearances:

- at the back and sides – 150 mm (clear space or distance to a suitably heat resistant wall);
- at the front –
  (a)   150 mm; or
  (b)   225 mm if the appliance provides space heating by means of visible flames or radiating elements.

The edges of the hearth should be designed to be apparent to building occupiers so as to discourage the laying of combustible floor finishes too close to the appliance. This could be achieved by providing a change in levels.

Shielding of appliances burning oil

Combustible materials adjacent to oil-fired appliances may only need special protection if they are subjected to temperatures which exceed 100°C. In these cases they should be separated from combustible materials by a non-combustible surface such as:

- a shield of non-combustible material at least 25 mm thick; or
- at least 75 mm airspace.

Alternatively, shielding would not usually be needed for appliances that are independently certified as having surface temperatures of not more than 100°C during normal operation.

Chimneys and flues for appliances burning oil

Where oil-fired appliances are served by masonry chimneys there should be at least 25 mm of masonry between the flues and any combustible material. Similarly, where a flueblock chimney serves an oil-fired appliance the flueblock walls should be at least 25 mm thick.

Where a chimney penetrates a fire compartment wall or floor it must also comply with the fire separation requirements of Part B (see Chapter 7).

Flues which are likely to serve appliances burning Class D oil (gas oil) should be made of materials which are resistant to acids of sulphur.

Placing and shielding of flues for appliances burning oil

Where flue gas temperatures are unlikely to exceed 250°C, the building fabric may be protected from heat dissipation from connecting flues and factory-made chimneys complying with BS 715: 1993 serving appliances burning oil by making sure of the following.

- Every part of the flue is at least 25 mm from any combustible material. The distance is measured from the outer surface of the flue wall, or the outer surface of the inner wall for multi-walled products.
- Where the flue passes through a roof, floor or wall formed of combustible materials (other than a compartment roof, floor or wall), it is enclosed in a sleeve of non-combustible material and there is at least 25 mm airspace between the flue and the sleeve. The airspace could be wholly or partially filled with non-combustible insulating material.

Factory-made chimneys complying with BS 4543: 1996 *Factory-made insulated chimneys* should also be separated from combustible materials. This is dealt with in section 14.6.4 above. Factory-made chimneys and connecting fluepipes should also be guarded if:

- the burn hazard they present is not immediately apparent to people (e.g. when they cross intermediate floors which are not visible from the appliance); or
- they are so sited as to be at risk of damage.

Where a flue assembly for a room-sealed appliance passes through a combustible wall it should be:

- surrounded by insulating material with a thickness of at least 50 mm; and
- provided with a minimum clearance of 50 mm, from any combustible wall cladding to the edge of the flue outlet.

14.8   Repair and re-use of existing flues

14.8.1   Introduction

Until the coming into force of the current edition of Part J on 1 April 2002 it was possible to open up disused flues in buildings undergoing refurbishment for the purposes of connecting new appliances without having to comply with the Building Regulations. In many cases this led to the installation of appliances burning inappropriate fuels for the old flue and, of course, the use of flues which were in a poor state of repair. Inevitably this resulted in danger to the health and safety of building occupants from fire and noxious fumes.

14.8.2   Repairs to flues

Under Regulations 2 and 3 of the Building Regulations 2000, building work to which Part J imposes requirements, constitutes work to a controlled service or fitting. This means that the following types of work must be notified to a Building Control Body (local authority or approved inspector) and will need to comply with Part J.

- Renovation, refurbishment or repair work involving the provision of a new or replacement flue liner (e.g. the insertion of new linings such as rigid or flexible prefabricated components; or a cast in situ liner that significantly alters the flue's internal dimensions);
- Proposals to bring a flue in an existing chimney back into use or to re-use a flue with a different type or rating of appliance.

The Building Control Body will need to be assured that any altered flues comply with the Regulations. This will involve the inspection and testing of the altered flue by a competent person and is described more fully below.

14.8.3   Re-lining and repairing existing flues

General provisions

In many older buildings with flues designed for use with open fires, flues are likely to be larger than normal and unlined (or they may originally have been parged with lime mortar which has subsequently deteriorated). Since oversize flues can be unsafe, lining may be necessary in order to reduce the flue area to suit the intended appliance.

It is usually possible to refurbish old or defective flues by relining them using materials and components designed to suit the appliance and/or fuel which it is intended to burn. Before being relined, flues should be swept to remove deposits.

Of course, it may be the case that a chimney has been relined in the past using a metal lining system to suit a particular appliance. Where a decision is taken to replace the appliance, the metal liner should also be replaced unless it is in good condition or it can be shown that it has recently been installed. Details of the lining materials and components that are suitable for the various appliances and fuels are considered below.

Chimneys can be relined using independently certified flexible metal flue liners which have been specifically designed to suit the particular fuel being burnt. Such flue liners should only be used for relining purposes and should not be used as the primary liner of a new chimney.

Relining flues serving appliances burning solid fuel

Liners for existing flues serving appliances burning solid fuel should have an independently certified performance at least equal to that corresponding to the BS EN 1443 designation of T450 N2 S D 3 i.e.:

T450 – a product with a normal working temperature of 450°C, see Table 1 of EN 1443

N2 – the pressure class (N = negative pressure)

S – chimney has sootfire resistance

D – resistance to condensate class (D – dry conditions)

3 – corrosion resistance class (3 – Oils with sulphur content over 0.2% and solid mineral fuels and peat).

Examples of liners which comply with this include the following, provided that they are independently certified as being suitable for appliances burning solid fuel:

- factory-made flue lining systems (e.g. double skin flexible stainless steel);
- cast in situ flue relining systems (materials and installation procedures must be independently certified as suitable); and
- any other systems meeting the BS EN 1443 designation and independent certification criteria.

Relining flues serving appliances burning gas

Liners for existing flues serving appliances burning gas can be:

- any of the liners described in section 14.6.2 above as being suitable for all fuels;
- any of the liners described in the section on relining flues serving appliances burning solid fuel immediately above;
- flexible stainless steel liners independently certified as complying with BS 715: 1993 and installed in accordance with BS 5440: Part 1: 2000;
- any other lining systems independently certified as being fit for purpose.

When installing flexible metal flue liners the following points should be noted:

- install the liner in one complete length (without joints in the chimney);
- leave the space between the liner and the chimney empty (other than for sealing at top and bottom) unless the manufacturer's instructions say otherwise;
- double skin flexible flue liners should be installed in accordance with manufacturer's instructions.

Relining flues serving appliances burning oil

Liners for existing flues serving appliances burning oil, with flue gas temperatures which are expected to exceed 250°C can be:

- any of the liners described in section 14.6.2 above as being suitable for all fuels;
- any of the liners described in the section on relining flues serving appliances burning solid fuel immediately above;
- flexible stainless steel liners independently certified as complying with BS 715: 1993 and installed in accordance with BS 5440: Part 1: 2000;
- any other lining systems independently certified as being fit for purpose.

Liners for existing flues serving appliances burning oil, with flue gas temperatures which are unlikely to exceed 250°C can be:

- any of the liners described above for flue gas temperatures exceeding 250°C;
- any other lining systems independently certified as being fit for purpose;
- for new appliances of known type, flue lining systems which have been independently certified as complying with the performance requirements corresponding to the designations given in Table 4.1 from AD J (see section 14.6.7 above).

When installing flexible metal flue liners the following points should be noted:

- install the liner in one complete length (without joints in the chimney);
- leave the space between the liner and the chimney empty (other than for sealing at top and bottom) unless the manufacturer's instructions say otherwise;
- double skin flexible flue liners should be installed in accordance with manufacturer's instructions.

14.8.4   Checking and testing refurbished and repaired flues

Reference has been made in section 14.4 above regarding the need to inspect and test combustion installations before they are first used. This requirement also applies to flues which have been refurbished or repaired in line with the guidance given in this section. The tests, inspections and report should be done by a specialist firm and copies should be supplied to the main contractor, client or developer, and the Building Control Body.

Guidance on methods of checking and testing for compliance with Regulation J2 (*Discharge of products of combustion*) is given in Appendix E of AD J. The guidance applies to new, and existing re-used or relined natural draught flues intended for open flued appliances.

The described procedures cover flues in chimneys, connecting fluepipes, and flue gas passages in appliances, to ensure that these are acceptably gastight and free of obstruction. Furthermore, when an appliance is commissioned to check for compliance with Part L (see Chapter 16) and as required by the Gas Safety (Installation and Use) Regulations 1998, it will also be necessary to carry out appliance performance tests (including flue spillage tests to check for compliance with J2).

It is not necessary to wait for all the building work to be complete before carrying out tests on flues etc. In fact, the most appropriate time to carry out a test may be before the application of plaster finishes or drylining to the structure of a chimney, because at such times possible smoke leakage will be unobscured by surface finishes.

14.9   Test methods

14.9.1   Tests on existing flues

Flues in existing chimneys may suffer from the following defects:

- obstruction caused by bird nests, soot, tar and debris resulting from deterioration of the structure (e.g. pieces of brickwork and chimney pot, and decaying flue lining materials);
- leakage of flue gases as a result of holes or cracks appearing in the structure and linings, particularly at joints;
- decay of the exposed part of the chimney above the roof.

The following methods can be used to determine the state of repair of a flue before it is brought back into re-use.

- Sweeping – to clean the flue and show that it is substantially free from obstructions. It will also enable better visual inspection and testing of the flue. Some deposits (such as tar from burning wood) may be especially hard to dislodge and should be removed. Examine the debris that comes down the chimney when sweeping, to see if it contains excessive quantities of flue lining or brick. These are signs that further repairs may be necessary.
- Visual inspection of the accessible parts – to identify the following.
  - (a) Deterioration in the structure, connections or linings which could affect the gastightness and safe performance of the flue and its associated combustion appliance. Examine the exterior of the chimney (including the part in the roofspace) and the interior of the flue. Look for the presence of smoke or tar stains on the exterior of a chimney/breast. These are signs of leaks and could indicate possible damage.
  - (b) Modifications made whilst the flue was no longer in service. Look out for the presence of ventilator terminals. These could be incompatible when the flue is used with the intended appliance.
  - (c) Correct specification and size of the lining for the proposed new application.
  - (d) Freedom from restriction in the flue. A visual inspection may be sufficient where the full length of the flue can be seen. In other cases it may be better to carry out a coring ball test.
- Smoke testing to check the operation and gastightness of the flue.

14.9.2   Tests on new masonry and flueblock chimneys

Flues in new masonry and flueblock chimneys should be checked to demonstrate that they have been correctly constructed, are free of restrictions and are acceptably gas tight. The following defects may occur.

- Incorrect installation of flue liners. Checks should be made during construction that liners are installed with sockets facing upwards and joints are sealed so that moisture and condensate will be contained in the chimney.
- Obstructions (particularly at bends), caused by:
  - (a) debris left during construction;
  - (b) excess mortar falling into the flue;
  - (c) jointing material extruded from between liners and flueblocks.

Before bringing a new flue into use its condition should be checked as follows:

- carry out a visual inspection of the accessible parts to check that the lining, liners or flueblocks are correctly specified and of the correct size for the proposed application;
- carry out a coring ball test, or sweep the flue to remove flexible debris if a visual examination cannot confirm that the flue is free from restrictions;
- carry out a smoke test to confirm the correct operation and gastightness of the flue.

14.9.3   Tests on new factory-made metal chimneys

Newly completed factory-made metal chimneys should follow the checklist for visual inspection given in section 10 of BS7566 *Installation of factory-made chimneys to BS 4543 for Domestic appliances:* Part 3:1992 (1998) *Specification for site installation.* This covers the following:

- chimney route – to be in accordance with installation design;
- openings for testing, cleaning and maintenance – must be accessible;
- components, joints, connections, locking bands, etc. – to be secure;
- combustible material – to be at least the manufacturer's specified distance X mm from the chimney (see section 14.6.4 above);
- fire-stops, fire-stop spacers and ceiling supports – to be in position;
- weatherproofing where the chimney penetrates the roof – use only cover flashing and sealing material specified by either the installation designer or the chimney manufacturer;
- enclosure – to be clear of all rubbish and extraneous matter;
- aerials, etc. – not to be attached to the chimney structure.

Additional checks or particular variants may be included in manufacturer's installation instructions. A smoke test should be carried out after the inspection.

14.9.4   Tests on relined flues

Apply the same tests for freedom from restrictions and gastightness as for newly built flues. A flue for a gas appliance which has been relined with a flexible metal liner (see section 14.8.3 above) may be assumed to be unobstructed and acceptably gastight. (The use of a coring ball or inappropriate sweeps brushes can seriously damage a flexible metal flue liner.)

14.9.5   Testing appliances

Where the building work involves the complete installation of combustion appliance and flue system, the entire system should be tested for gastightness in addition to testing the flue separately as above. Appropriate spillage test procedures for different fuels are given in:

- BS 5440: Part 1: 2000 – gas appliances;
- BS 5410 Part 1: 1997 – oil fired appliances;
- BS 6461: *Installation of chimneys and flues for domestic appliances burning solid fuel (including wood and peat): Part 1: 1984 (1998) Code of practice for masonry chimneys and flue pipes* – solid fuel appliances.

14.9.6   Test procedures for flues

Coring ball test

In this test a heavy ball, (about 25 mm less in diameter than the flue) is lowered on a rope from the flue outlet to the bottom of the flue. On encountering any obstruction the flue should be unblocked and the test repeated. It should be carried out before smoke testing.

The test can be used for:

- proving the minimum diameter of circular flues;
- checking for obstructions in square flues (but it will not detect obstructions in the corners unless a purpose made coring ball or plate for rectangular flues is used).

The test is not applicable to fluepipes and, as explained above, should not be used with flexible metal flue liners.

Smoke testing

Smoke tests are carried out in order to check that flue gases can rise freely through the flue and to identify any faults that would cause the flue gases to escape into the building (e.g. incorrectly sealed joints or damage to the flue).

Two types of smoke test are described in Appendix E of AD J. Smoke Test 1 is the most stringent and should be used to test flues serving appliances burning solid fuel or oil, since it tests the gastightness of the whole flue. It can be used for flues serving appliances burning gas if there is any doubt about the state of the flue. Smoke Test 2 is used exclusively for appliances burning gas and does not involve sealing the flue.

These tests are in addition to any spillage test carried out when the appliance is commissioned. Where an approved flue or relining system is installed it is possible that other tests could be a requirement of the installation procedure.

Smoke Test 1 is carried out as follows:

- close all doors and windows in the room served by the flue;
- warm the flue to establish a draught (e.g. with a blow lamp or electric heater);
- place and ignite a suitable number of flue testing smoke pellets at the base of the flue, (e.g. in the fireplace recess or in the appliance, where fitted);
- when smoke starts to form, seal off the base of the flue or fireplace opening, or close the appliance if fitted, so that smoke can only enter the flue (e.g. close off the recess opening with a board or plate, sealed at the edges or, where an appliance is fitted, close its doors, ashpit covers and vents);

- establish that there is a free flow of smoke from the flue outlet or terminal;
- seal the top of the flue;
- check the entire length of the flue to establish that there is no significant leakage (see also notes below on checking flues for leakage);
- allow the test to continue for at least 5 minutes;
- remove the closures at the top and bottom.

Smoke Test 2 is carried out as follows:

- close all doors and windows in the room served by the flue;
- warm the flue to establish a draught (e.g. with a blow lamp or electric heater);
- place and ignite a suitable flue testing smoke pellet at the base of the flue, (e.g. in the fireplace recess or in the intended position of the appliance);
- partially close off the opening between the recess and the room with a board to leave an air entry gap of about 25 mm at the bottom;
- establish that smoke is issuing freely from the flue outlet or terminal and not to spilling back into the room;
- check the entire length of the flue to establish that there is no significant leakage inside or outside the building.

Smoke Tests 1 and 2 conform to the recommendations for testing given in BS 6461: Part 1: 1984 (1998) and BS 5440: Part 1: 2000.

Further guidance on smoke testing of flues

The following notes should read in conjunction with the procedures for smoke testing referred to above.

(1) Flues should be warmed for a minimum of 10 minutes in order to establish a draught. Large or cold flues may take longer than this.
(2) Where an appliance is fitted it should not be under fire at the time of carrying out the test.
(3) When being tested, smoke should only emerge from the flue under test. Smoke issuing from any other flue would indicate leakage between the flues.
(4) If leaks do occur during the test they may be quite remote from the area of the fault. This may make it difficult to pinpoint the exact location. For example, where a chimney is located on a gable wall smoke can track down the verge and emerge from under the bargeboard overhang, or it may issue from window reveals where it has penetrated into the cavity of the wall
(5) Smoke pellets create a significantly higher pressure than that required in the product standards for natural draught chimneys and for flues having a gas tightness designation of N1 (negative pressure chimney tested at a pressure of 40 Pa). BS EN 1443 permits flues to this designation to have a leakage rate of up to 2 litres/second per m^2 of flue wall area. Therefore, when assessing the extent of any smoke leakage seen during the test the following points should be considered:

- wisps of smoke seen on the outside of the chimney or near joints between chimney sections are not necessarily indicative of a fault;
- evidence of forceful plumes, or large volumes of smoke could indicate a major fault (e.g. an incorrectly made connection or joint, or a damaged section of chimney).

Therefore, evidence of smoke leakage during a test may not necessarily indicate failure and it can be a matter of expert judgement as to whether this is, in fact, the case. Where leakage is established it will be necessary to conduct an investigation and carry out any remedial action before repeating the smoke test.

14.10   Provision of information

It has been shown in the preceding parts of this chapter that safe installation of a combustion appliance depends on making sure that the appliance is compatible with the hearth, fireplace, flue or chimney to which it is connected. It is vital therefore, that anyone carrying out work to such an installation is aware of its performance capabilities so that an appliance is not connected to an inappropriate flue etc.

Accordingly, paragraph J4 of Part J requires that where a hearth, fireplace, flue or chimney is provided or extended, a durable notice containing information on the performance capabilities of the installation must be affixed in a suitable place in the building.

14.10.1   Notice plates for hearths, fireplaces, chimneys and flues

A robust and indelibly marked notice plate containing information essential to the correct application and use of the combustion installation should be permanently posted up in an unobtrusive but obvious position in the building where any of the following are being provided or extended:

- a hearth and fireplace (including a flue box); and/or
- a flue or chimney (including the provision of a flue as part of the refurbishment work).

A suitable location in the building for the notice plate might be next to the:

- electricity consumer unit; or
- chimney or hearth described on the plate; or
- water supply stop-cock.

The notice plate should convey the following information:

- location of the hearth, fireplace (or flue box) or the location of the start of the flue;
- category of flue and the generic types of appliances that the flue can safely accommodate;

- type and size of the flue (or liner for relined flues) and the name of the manufacturer;
- date of installation.

An example of a typical notice plate is given in Diagram 1.9 from Section 1 of AD J, which is reproduced below. Where a chimney product has had its performance characteristics assessed in accordance with a European Standard (EN) and it is supplied or marked with a designation as described in section 14.2 above, this designation may be included on the notice plate at the option the installer (see Diagram 1.9).

AD J Section 1

Diagram 1.9   Example notice plate for hearths and flues.

Essential information	**IMPORTANT SAFETY INFORMATION** **This label must not be removed or covered**
	Property address .... *20 Main Street* *New Town* The hearth and chimney installed in the .... *lounge* are suitable for .... *decorative fuel effect gas fire* Chimney liner .... *double skin stainless steel flexible, 200 mm* *diameter* Suitable for condensing appliance ...... *no* Installed on .... *date*
Optional additional information	Other information (optional) .... *Designation of stainless steel liner stated by* *manufacturer to be T450 N2 S D 3* *e.g. installer's name, product trade names,* *installation and maintenance advice,* *European chimney product designation,* *warnings on performance limitations of* *imitation elements e.g. false hearths*

14.11   Alternative means of compliance

As has been shown in Chapter 2, it is possible to achieve the levels of performance described in the Approved Documents (and which are needed to satisfy the requirements of the Building Regulations) by using other sources of guidance, such as British and European Standards. AD J lists a number of alternative guidance sources according to the fuel being burnt. These are outlined below.

Alternative sources of guidance for appliances burning solid fuel

For appliances burning solid fuel the requirements of the Regulations will be met if the relevant recommendations of the following publications are adopted:

- BS 6461: *Installation of chimneys and flues for domestic appliances burning solid fuel (including wood and peat). Code of practice for masonry chimneys and flue pipes.* Part 1: 1984 (1998); and

- BS 7566: *Installation of factory-made chimneys to BS 4543 for domestic appliances* Parts 1 to 4: 1992 (1998); and
- BS 8303: *Installation of domestic heating and cooking appliances burning solid mineral fuels.* Parts 1 to 3: 1994.

Alternative sources of guidance for appliances burning gas

For appliances burning gas the requirements of the Regulations will be met if the relevant recommendations of the following publications are adopted:

- BS 5440: *Installation and maintenance of flues and ventilation for gas appliances of rated input not exceeding 70 kW net (1st, 2nd and 3rd family gases),* Part 1: 2000 *Specification for installation and maintenance of flues;* Part 2: 2000 *Specification for installation and maintenance of ventilation for gas appliances.*
- BS 5546: 2000 *Specification for installation of hot water supplies for domestic purposes, using gas-fired appliances of rated input not exceeding 70 kW.*
- BS 5864: 1989 *Specification for installation in domestic premises of gas-fired ducted-air heaters of rated input not exceeding 60 kW.*
- BS 5871: *Specification for installation of gas fires, convector heaters, fire/back boilers and decorative fuel effect gas appliances,* Part 1: 2001 *Gas fires, convector heaters and fire/back boilers and heating stoves (1st, 2nd and 3rd family gases);* Part 2: 2001 *Inset live fuel effect gas fires of heat input not exceeding 15 kW and fire/back boilers (2nd and 3rd family gases);* Part 3: 2001 *Decorative fuel effect gas appliances of heat input not exceeding 20 kW (2nd and 3rd family gases).*
- BS 6172: 1990 *Specification for installation of domestic gas cooking appliances (1st, 2nd and 3rd family gases).*
- BS 6173: 2001 *Specification for installation of gas-fired catering appliances for use in all types of catering establishments (2nd and 3rd family gases).*
- BS 6798: 2000 *Specification for installation of gas-fired boilers of rated input not exceeding 70 kW net.*

Alternative sources of guidance for appliances burning oil

For appliances burning oil the requirements of the Regulations will be met if the relevant recommendations of the following publication are adopted:

- BS 5410: *Code of practice for oil firing,* Part 1: 1997 *Installations up to 45 kW output capacity for space heating and hot water supply purposes.*

14.12   Protection of liquid fuel storage systems

Under the requirements of Paragraph J5 of Part J, liquid fuel storage systems and the pipes connecting them to combustion appliances must be constructed and separated from buildings and the boundary of the premises so as to reduce to a reasonable level the risk of the fuel igniting in the event of fire in adjacent buildings or premises.

This requirement is limited in application to the following:

- fixed oil storage tanks with capacities in excess of 90 litres and their connecting pipes; and
- fixed liquefied petroleum gas (LPG) storage installations with capacities in excess of 150 litres and associated connecting pipes

provided that these are located outside the building, serving fixed combustion appliances (including incinerators) inside the building.

14.12.1   Limitations on the Approved Document guidance

It will be noticed that there are no upper limits set on the size of the liquid fuel storage systems described above and that the requirements apply to all building types and all types of tanks (even those that are buried). The guidance in AD J, however is restricted to the following types of installation:

- oil storage systems with above-ground or semi-buried tanks of up to 3500 litres capacity, used exclusively for heating oil burning:
  (a)   Class C2 oil (kerosene); or
  (b)   Class D oil (gas oil);
- LPG storage systems of up to 1.1 tonne capacity comprising one tank standing in the open air, or LPG storage systems consisting of sets of cylinders, under the conditions described below.

For oil storage tanks with capacities exceeding 3500 litres, AD J recommends that advice should be sought from the Fire Authority regarding suitable fire precautions. Further guidance may also be obtained from BS 5410: Part 1: 1997 for oil storage systems, and from Code of Practice 1: *Bulk LPG storage at fixed installations* Part 1: 1998 *Design, installation and operation of vessels located above ground*, published by the LP Gas Association for LPG storage systems.

14.12.2   Protection of oil storage tanks against fire

Oil and LPG fuel storage installations (including the pipework connecting them to combustion appliances in the buildings they serve) should be located and constructed so that they are reasonably protected from fires which may occur in buildings or beyond boundaries. To this end Table 5.1 from Section 5 of AD J (which is reproduced below) contains guidance on measures that can be taken to protect the tank from a fire occurring within the building. Additionally the guidance offered would help to reduce the risk of fuel storage fires igniting buildings although this is thought (in BS 5410) to be unlikely to happen.

In addition to the measures described in Table 5.1, the following issues should also be addressed.

AD J Section 5

Table 5.1 Fire protection for oil storage tanks.

Location of tank	Protection usually satisfactory
Within a building	Locate tanks in a place of special fire hazard which should be directly ventilated to outside. Without prejudice to the need for compliance with all the requirements in Schedule 1, the need to comply with Part B should particularly be taken into account
Less than 1800 mm from any part of a building	(a) Make building walls imperforate (1) within 1800 mm of tanks with at least 30 minutes fire resistance (2) to internal fire and construct eaves within 1800 mm of tanks and extending 300 mm beyond each side of tanks with at least 30 minutes fire resistance to external fire and with non-combustible cladding; or (b) Provide a fire wall (3) between the tank and any part of the building within 1800 mm of the tank and construct eaves as in (a) above. The fire wall should extend at least 300 mm higher and wider than the affected parts of the tank
Less than 760 mm from a boundary	Provide a fire wall between the tank and the boundary or a boundary wall having at least 30 minutes fire resistance to fire on either side. The fire wall or the boundary wall should extend at least 300 mm higher and wider than the top and sides of the tank
At least 1800 mm from the building and at least 760 mm from a boundary	No further provisions necessary

Notes:
1. Excluding small openings such as air bricks etc.
2. Fire resistance in terms of insulation, integrity and stability.
3. Fire walls are imperforate non-combustible walls or screens, such as masonry walls or steel screens.

- To prevent the storage installation becoming overgrown by weeds, ensure that the ground under the tank is hard surfaced with 100 mm concrete or paving slabs at least 42 mm thick extending at least 300 mm beyond the edge of the tank (or the face of the external skin of the tank if it is of the integrally bunded type).
- Ensure that the recommended fire walls are structurally stable and do not pose a danger to people coming near them. Since the Building Regulations do not apply to freestanding walls of this description, other guidance should be sought. The former DTLR has published a guide to the construction of garden walls entitled *Your Garden Walls. Better Safe than Sorry*. This is obtainable from DTLR Free Literature, PO Box 236, Wetherby, West Yorkshire, LS23 7NB. Guidance will be found in this publication relating wall thickness to height.

A way to protect the oil supply pipework is to install it so that it is resistant to the effects of fire and to fit a proprietary fire valve system. More information on this can be gained by reading sections 8.2 and 8.3 of BS 5410: Part 1: 1997.

14.12.3   Protection of LPG storage installations against fire

Because of the potential hazards presented by liquefied petroleum gas (LPG) installations, they are controlled by legislation enforced by the Health and Safety Executive or its agents. There is a considerable amount of legislation dealing with the safe use of LPG in different branches of industry and compliance is usually demonstrated by constructing an LPG storage installation in accordance with an appropriate industry Code of Practice, prepared in consultation with the HSE. Therefore, the amount of building work that is needed in order for an installation to comply will depend on:

- the capacity of the installation;
- whether or not tanks are installed above or below ground; and
- the nature of the premises served.

However, by following the guidance in Approved Document J as outlined in this section, and the relevant guidance in Approved Document B (see Chapter 7), no further building work will normally be necessary to comply with other legislation if the tank stands in the open air and the installation has a capacity not exceeding 1.1 tonne.

LPG tank location

Since liquefied petroleum gas is heavier than air the LPG tank should be installed outdoors and not within an open pit or bund which would trap the vapour and not allow it to disperse. In fact, safe dispersal is essential should accidental venting or leakage of the tank occur and this necessitates adequate separation of the tank from buildings, boundaries and any fixed sources of ignition to reduce the risk of fire spread. Safe separation distances from buildings, boundaries and fire walls are shown in Fig. 14.24 and Fig. 14.25 below. It also follows that drains, gullies and cellar hatches should be protected from gas entry if they are within the separation distance shown in the Figures.

Fire walls

Fire walls can be freestanding walls which separate the tank from the building, the boundary or a fixed source of ignition. They can also be part of the building itself or a boundary wall belonging to the property. All these options are illustrated in Fig 14.24 and Fig. 14.25 together with their respective separation distances.

Acceptable fire walls should:

- be imperforate and constructed of masonry, concrete or similar materials;
- have fire resistance in terms of stability, integrity and insulation (see Chapter 7 for details of these concepts) of at least:
    - (a)   30 minutes if a freestanding or boundary wall
    - (b)   60 minutes if part of a building

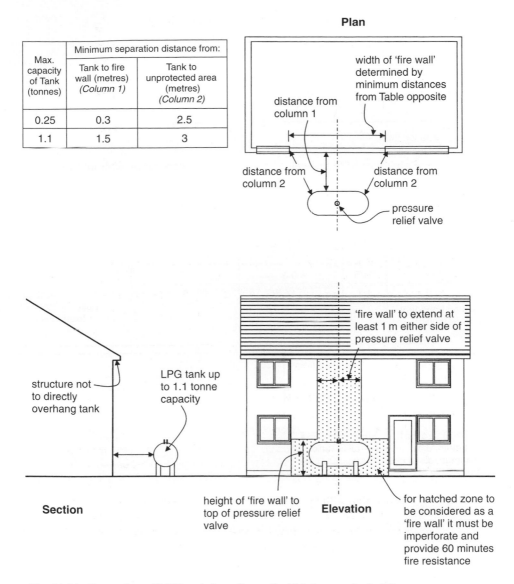

Max. capacity of Tank (tonnes)	Minimum separation distance from:	
	Tank to fire wall (metres) *(Column 1)*	Tank to unprotected area (metres) *(Column 2)*
0.25	0.3	2.5
1.1	1.5	3

Plan

width of 'fire wall' determined by minimum distances from Table opposite

distance from column 1

distance from column 2

distance from column 2

pressure relief valve

'fire wall' to extend at least 1 m either side of pressure relief valve

structure not to directly overhang tank

LPG tank up to 1.1 tonne capacity

Section

height of 'fire wall' to top of pressure relief valve

Elevation

for hatched zone to be considered as a 'fire wall' it must be imperforate and provide 60 minutes fire resistance

Fig. 14.24 Separation of LPG tank from fire wall which is part of a building.

- only be built on one side of a tank (to assist ventilation and dispersal of the gas in the event of a leak);
- be at least as high as the pressure relief valve on the tank;
- extend horizontally so that the separation shown in Figs. 14.24 and 14.25 is maintained.

Sets of cylinders – location and support

LPG storage installations that consist of sets of cylinders should be adequately supported and sufficiently far enough away from openings into the building to not

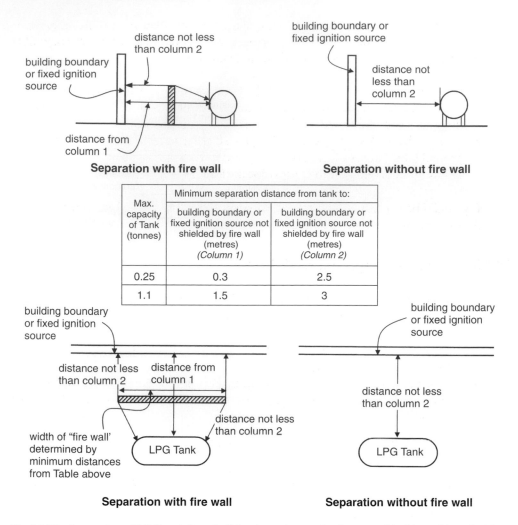

	Minimum separation distance from tank to:	
Max. capacity of Tank (tonnes)	building boundary or fixed ignition source not shielded by fire wall (metres) (Column 1)	building boundary or fixed ignition source not shielded by fire wall (metres) (Column 2)
0.25	0.3	2.5
1.1	1.5	3

Fig. 14.25 Separation of LPG tank from building boundary or fixed source of ignition with and without fire wall.

present a risk to the health and safety of the occupants. In terms of support they should:

- stand upright;
- be readily accessible;
- be reasonably protected from physical damage; and
- not obstruct exit routes from the building.

It is usual to secure the cylinders by means of chains or straps against a wall outside the building at ground level, in a well-ventilated position. They should be placed on a firm, level base (e.g. concrete at least 50 mm thick or paving slabs bedded in mortar).

In terms of separation distances, these should be sufficient to prevent the entry of gas into the building should venting or a leak occur. Examples of acceptable separation distances include the following from AD J:

- from openings through the wall of the building (doors, windows, airbricks etc.) and from heat sources such as flue terminals or tumble-dryer vents – at least 1 m horizontally and 300 mm vertically measured from the nearest cylinder valve;
- from drains without traps, unsealed gullies, cellar hatches and similar entries at ground level – at least 2 m horizontally measured from the nearest cylinder valve, unless an intervening wall is provided at least 250 mm high.

Fig. 14.26 below shows how sets of cylinders should be located in relation to any doors, windows or other openings into the building.

14.13   Protection against pollution

Under the requirements of Paragraph J6 of Part J, oil storage tanks and the pipes connecting them to combustion appliances, must:

- be constructed and protected so as to reduce to a reasonable level the risk of oil escaping and causing pollution; and
- have a durable notice affixed in a prominent position that contains information on how to respond to an escape of oil so that the risk of pollution is reduced to a reasonable level.

This requirement is limited in application to the following:

- fixed oil storage tanks with capacities of 3500 litres or less and their connecting pipes which are:
  (a)   located outside the building; and
  (b)   serve fixed combustion appliances (including incinerators) inside a building which is used wholly or mainly as a private dwelling;
- above-ground storage systems (i.e. it does not apply to buried systems).

In addition to paragraph J6 of Part J of the Building regulations, there are controls exercised over storage tanks in England under the provisions of the Control of Pollution (Oil Storage) (England) Regulations 2001 (SI 2001/2954). These regulations apply to a wide range of oil storage installations but they do not apply to any premises which are wholly or mainly used as private dwellings where the storage capacity of the oil installation is less than 3500 litres. These regulations are policed by the Environment Agency in the event of any oil pollution.

14.13.1   Provisions to prevent oil pollution

As has been stated above, Paragraph J6 applies to oil storage tanks with capacities of 3500 litres or less serving combustion appliances in buildings used

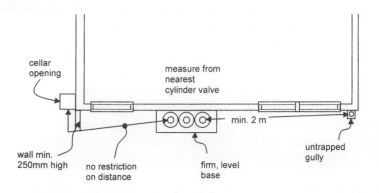

Elevation

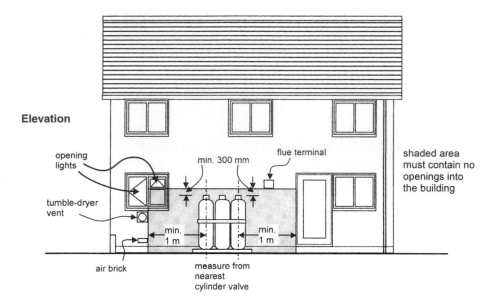

Fig. 14.26 Location and separation of LPG cylinders.

wholly or mainly as private dwellings. In these circumstances, secondary containment (the provision of bunds etc. see below) should be provided where there is considered to be a significant risk of oil pollution, i.e. if the oil storage installation:

- has a total capacity exceeding 2500 litres; or
- is closer than 10 m from inland freshwaters (see definition below) or coastal waters; or
- is located where an open drain or loose fitting manhole cover could be affected by spillage; or
- is closer than 50 m from sources of drinkable water (e.g. wells, bore-holes or springs); or

- is situated where oil spillage from the installation could reach inland freshwaters or coastal waters by running across hard ground; or
- is located where tank vent pipe outlets are not visible from the intended filling point (possibly resulting in accidental overfilling of the tank).

INLAND FRESHWATERS mean streams, rivers, reservoirs and lakes, and the ditches and ground drainage (including perforated drainage pipes) that run into them.

Secondary containment, when considered necessary, can be provided by means of:

- integrally bunded prefabricated tanks; or
- bunds constructed from masonry or concrete.

General advice on the construction of masonry or concrete bunds can be obtained from the Environment Agency in their publication – Pollution Prevention Guidelines PPG2 – *Above Ground Oil Storage Tanks*. They can also supply specific advice for bunds constructed of masonry or concrete in their publications – *Masonry Bunds for Oil Storage Tanks and Concrete Bunds for Oil Storage Tanks*. All these guidance documents are available free of charge from DMS Mailing House, Fax No. 0151 604 1222.

When constructing bunds the following points should also be taken into account.

- They should have a capacity of at least 110% of the largest tank they contain, whether they are part of a prefabricated tank system or constructed on site.
- Where the walls of a chamber or building enclosing a tank act as a bund for the purposes of this section, any door through such walls should be above bund level.
- Where the bund has a structural role as part of a building, specialist advice should be sought.

Labelling of oil storage installations

A label should be placed on an oil storage installation in a prominent position, to advise on what to do in the event of an oil spill. It should contain the Environment Agency's Emergency Hotline number 0800 80 70 60.

Useful addresses

Reference is made throughout this chapter to a number of organisations concerned with various aspects of the design, construction and use of combustion installations and who were involved in the development of the 2002 edition of Approved Document J, either directly or as consultees. Contact details are given in the list below.

ACE (Amalgamated Chimney Engineers): White Acre, Metheringham Fen, Lincoln LN4 3AL

Tel 01526 32 30 09 Fax01526 32 31 81

BFCMA (British Flue and Chimney Manufacturers Association): Henley Road, Medmenham, Marlow, Bucks SL7 2ER

Tel 01491 57 86 74 Fax 01491 57 50 24

info@feta.co.uk www.feta.co.uk

BRE (Building Research Establishment Ltd.): Bucknalls Lane, Garston, Watford, Hertfordshire WD25 9XX

Tel 01923 66 4000 Fax 01923 66 4010

enquiries@bre.co.uk www.bre.co.uk

BSI (British Standards Institution): 389 Chiswick High Road, London W4 4AL

Tel 020 8996 9000 Fax 020 8996 7400

www.bsi-global.com

CIBSE (Chartered Institution of Building Services Engineers): 222 Balham High Road, London SW12 9BS

Tel 020 8675 5211 Fax 020 8675 5449

www.cibse.org

CORGI (The Council for Registered Gas Installers): 1, Elmwood, Chineham Business Park, Crockford Lane, Basingstoke, Hampshire RG24 8WG

Tel 01256 37 22 00 Fax 01256 70 81 44

www.corgi-gas.com

Environment Agency: Rio House, Waterside Drive, Aztec West, Almondsbury, Bristol BS32 4UD

Tel 0845 9333111 Fax 01454 624 409

www.environment-agency.gov.uk

(Publication enquiries to: Tel 01454 624 411 Fax 01454 624 014)

Environment Agency Emergency Hotline 0800 80 70 60

HETAS (Heating Equipment Testing and Approval Scheme): PO Box 37, Bishops Cleeve, Gloucestershire, GL52 4TB

Tel 01242 673257 Fax 01242 673463

www.hetas.co.uk

HSE (Health and Safety Executive): Rose Court, 2 Southwark Bridge, London SE1 9HS

Tel 020 7717 6000 Fax 020 7717 6717

www.hse.gov.uk

Gas safety advice line 0800 300 363

IGasE (Institution of Gas Engineers): 21 Portland Place, London W1B 1PY

Tel 020 7636 6603 Fax 020 7636 6602

www.igaseng.com

LP Gas Association: Pavilion 16, Headlands Business Park, Salisbury Road, Ringwood, Hampshire BH24 3PB

Tel 01425 461612 Fax 01425 471131

www.lpga.co.uk

NACE (National Association of Chimney Engineers): PO Box 5666, Belper, Derbyshire, DE56 0YX

Tel 01773 599095 Fax 01773 599195

www.nace.org.uk

NACS (National Association of Chimney Sweeps): Unit 15, Emerald Way, Stone Business Park, Stone, Staffordshire, ST15 0SR

Tel 01785 811732 Fax 01785 811712

nacs@chimneyworks.co.uk

www.chimneyworks.co.uk

NFA (National Fireplace Association): 6th Floor, McLaren Building, 35 Dale End, Birmingham B4 7LN

Tel 0121 200 13 10 Fax 0121 200 13 06

www.nationalfireplaccassociation.org.uk

OFTEC (Oil Firing Technical Association for the Petroleum Industry): Century House, 100 High Street, Banstead, Surrey, SM7 2NN.

Tel 01737 37 33 11 Fax 01737 37 35 53

enquiries@oftec.org www.oftec.org

SFA (Solid Fuel Association): 7 Swanwick Court, Alfreton, Derbyshire, DE55 7AS

Tel 0800 600 000 Fax 01773 834 351

sfa@solidfuel.co.uk www.solidfuel.co.uk

15 Protection from falling, collision and impact (Part K)

15.1 Introduction

The control of stairways, ramps and guards has always formed an important part of Building Regulations since stairways represent, in many cases, the only way out of a building in the event of fire and there is a need to make buildings accessible to disabled people. Additionally, provisions are necessary to protect people from the risk of falling when they use exposed areas such as landings, balconies and accessible roofs.

The 1998 edition of Approved Document K introduced new provisions governing:

- the safe use of vehicle loading bays,
- measures to reduce the risk of collisions with open windows, skylights and ventilators; and
- measures designed to reduce the risk of injury when using various types of sliding or powered doors and gates,

mainly in order to ensure compliance with the Workplace (Health, Safety and Welfare) Regulations 1992.

Therefore, compliance with the revised regulations in Part K prevents action being taken against the occupier under the Workplace Regulations when the building is eventually in use. This involves considering aspects of design which will affect the way a building is used and applies only to workplaces. Although dwellings are excluded from the changes, in mixed use developments the requirements for the non-domestic part of the use would apply to any shared parts of the building (such as common access staircases and corridors). With flats the situation may be less clear for although certain sections of Part K do not apply to dwellings, it is still necessary for people such as cleaners, wardens and caretakers to work in the common parts. Therefore, the requirements of the Workplace Regulations may still apply even though the Building Regulations do not. Additionally, it is now necessary to provide safe access to areas used exclusively for maintenance purposes.

15.2 Stairways, ramps and ladders

Stairways, ramps and ladders which form part of the building must be designed, constructed and installed so that people may move safely between levels, in or about the building. Regulation K1 applies to all areas of a building which need to be accessed, including those used only for maintenance.

It should be noted that compliance with Regulation K1 prevents action being taken against the occupier of a building under regulation 17 of the Workplace (Health, Safety and Welfare) Regulations 1992 when the building is eventually in use. (Regulation 17 relates specifically to permanent stairs, ladders and ramps on routes used by people in places of work and includes access to areas used for maintenance.)

15.3   Application

The provisions contained in AD K1 regarding stairs, ramps and ladders only apply if there is a change in level of:

- more than 600 mm in dwellings; or
- two or more risers in other buildings (the difference in levels in this case will depend on the recommended height for the risers), or 380 mm where there is no stair.

Since stairs, etc. must provide safe access for people, an acceptable level of safety in a building will be dictated by the circumstances. Therefore, the standard in a dwelling will be lower than that recommended for a public building because there are likely to be less people in the dwelling and they will be familiar with the stairs. Similarly, the standard of access to maintenance areas will be lower than that recommended for normal use to reflect the greater care expected of those gaining access.

Outside stairways and ramps, (e.g. entrance steps), are covered by the regulations if they form part of the building (obviously, the proximity of the steps to the building and the way they are associated with it will dictate whether or not they are part of the building, in most cases). Therefore, steps in paths leading to the building would not be covered by Part K. These access routes may, of course, need to comply with other parts of the regulations if:

- they form part of a means of escape in case of fire (see Approved Document B: Fire safety);
- they are intended for use by disabled people (see Approved Document M: Access and facilities for disabled people).

In general, normal access routes in assembly buildings (e.g. sports stadia, theatres, cinemas etc.), should follow the guidance in AD K. Where special consideration needs to be given to guarding spectator areas or there are steps in gangways serving these areas then it may be preferable to follow more specialised guidance contained in the following:

- BS 5588 *Fire Precautions in the Design, Construction and Use of Buildings*, Part 6:1991 *Code of practice for places of assembly* – for new assembly buildings.

- *Guide to fire precautions in existing places of entertainment and like premises*, Home Office 1990 – for existing assembly buildings.
- *Guide to Safety at Sports Grounds*, The Stationery Office 1997 – for stands at sports grounds.

15.4 Interpretation

A number of definitions are given which apply generally throughout AD K:

CONTAINMENT – A barrier to prevent people from falling from a floor to the storey below.

FLIGHT – The section of a ramp or stair running between landings with a continuous slope or series of steps.

RAMP – A slope which is steeper than 1 in 20 intended to enable pedestrians or wheelchair users to get from one floor level to another.

A number of definitions are given which are specific to stairways and ladders:

STAIR – Steps and landings designed to enable pedestrians to get to different levels.

PRIVATE STAIRS – Stairs in or serving only one dwelling.

INSTITUTIONAL AND ASSEMBLY STAIRS – Stairs serving places where a substantial number of people gather.

OTHER STAIRS – Stairs serving all other buildings apart from those referred to above.

GOING – The distance measured in plan across the tread less any overlap with the next tread above (see Fig. 15.1).

RISE – The vertical distance between the top surfaces of two consecutive treads (see Fig. 15.1).

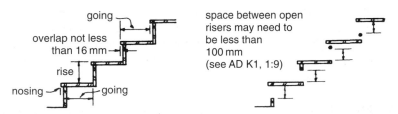

Fig. 15.1 Rise and going.

PITCH AND PITCH LINE – These terms are illustrated in AD K1 but are not defined; however

PITCH LINE – May be defined as a notional line connecting the nosings of all treads in a flight, including the nosing of the landing or ramp at the top of the flight. The line is taken so as to form the greatest possible angle with the horizontal, subject to the special recommendations for tapered treads (see below).

PITCH – May be defined as the angle between the pitch line and the horizontal (see Fig. 15.2).

ALTERNATING TREAD STAIRS – Stairs with paddle shaped treads where the wider portion alternates from one side to the other on each consecutive tread.

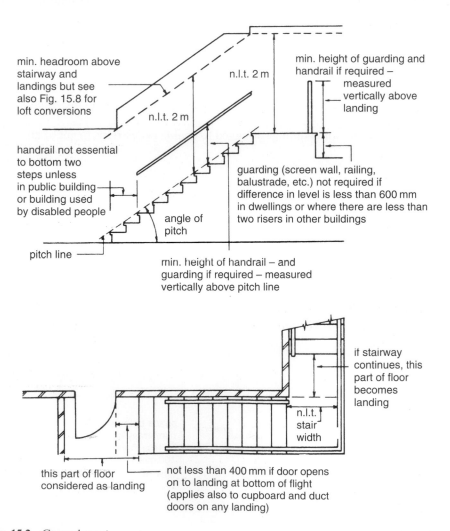

Fig. 15.2 General requirements.

HELICAL STAIRS – Stairs which form a helix round a central void.

SPIRAL STAIRS – Stairs which form a helix round a central column.

LADDER – A series of rungs or narrow treads used as a means of access from one level to another which is normally used by a person facing the ladder.

TAPERED TREAD – A tread with a nosing which is not parallel to the nosing of the tread or landing above it.

15.5   General recommendations for stairways and ramps

15.5.1   Landings

As a general rule a landing should be provided at the top and bottom of every flight or ramp. Where a stairway or ramp is continuous, part of the floor of the building may count as a landing. The going and width of the landing should not be less than the width of the flight or ramp.

Landings should be level and free from permanent obstructions. (This would allow, for example, the placing of a temporary barrier such as a child's safety gate between a landing and a flight.) A door is permitted to swing across a landing at the bottom of a flight or ramp but only if it leaves an area 400 mm wide across the full width of the flight or ramp. This rule also applies to cupboard and duct doors, but in this case they are permitted to open over any landing (including a landing at the top of a flight). Approved Document B contains details of restrictions on the use of cupboards situated in means of escape staircases.

A landing of firm ground or paving at the top or bottom of an external flight or ramp may slope at a gradient of not more than 1 in 20 (see Fig. 15.2).

15.5.2   Handrails

Stairs and ramps should have a handrail on at least one side. Where the width of the stair or ramp is 1m or more, then a handrail should be fixed on both sides.

Handrails are not required:

- beside the bottom two steps in a stairway unless it is in a public building or is intended for use by disabled people (see Approved Document M and Chapter 17).
- where the rise of a ramp is 600 mm or less.

Handrails should provide firm support and grip and be fixed at a height of between 900 mm and 1 m vertically above the pitch line or floor. They can also form guarding where the heights can be matched. Further details regarding handrails for disabled people can be found in Chapter 17 (see also Fig. 15.2).

15.5.3   Headroom

Clear headroom of 2 m should be provided over the whole width of any stairway, ramp or landing. There are reduced dimensions for headroom over stairs in loft conversions (see section 15.6.5, stairs to loft conversions).

Headroom is measured vertically from the pitch line, or where there is no pitch line, from the top surface of any ramp, floor or landing (see Fig. 15.2).

15.5.4   Width

AD K contains no recommendations for minimum stair or ramp widths; however there may be width recommendations in other Approved Documents. Reference should be made to AD B: Fire safety (Chapter 7) and AD M: Access and facilities for disabled people (Chapter 17), where minimum width guidance can be found.

15.6   Stairway recommendations

15.6.1   Rules applying to all stairways

- In any stairway there should not be more than thirty-six rises in consecutive flights, unless there is a change in the direction of travel of at least 30° (see Fig. 15.3).
- For any step the sum of twice its rise plus its going ($2R + G$) should not be more than 700 mm nor less than 550 mm. This rule is subject to variation at tapered steps, for which there are special rules.
- The rise of any step should generally be constant throughout its length and all steps in a flight should have the same rise and going.
- Open risers are permitted in a stairway but for safety the treads should overlap each other by at least 16 mm.
- Each tread in a stairway should be level.

Tapered treads should comply with the following rules:

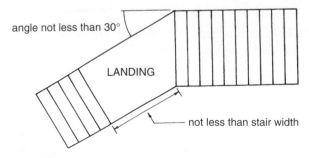

change of direction required if more than 36 risers in consecutive flights

Fig. 15.3   Length of flights.

- The minimum going at any part of a tread within the width of a stairway should not be less than 50 mm.
- The going should be measured:
  (a) if the stairway is less than 1m wide, at the centre point of the length or deemed length of a tread; and
  (b) if the stairway is 1 m or more wide, at points 270 mm from each end of the length or deemed length of a tread. (When referring to a set of consecutive tapered treads of different lengths, the term 'deemed length' means the length of the shortest tread. This term is not used in AD K1.) (See Fig. 15.4.)
- All consecutive tapered treads in a flight should have the same taper.
- Where stairs contain straight and tapered treads the goings of the tapered treads should not be less than those of the straight flight.
- In order to prevent small children from becoming trapped between the treads of open riser staircases there should be no opening in a riser of such size as to allow the passage of a sphere of 100 mm diameter. This rule applies to all stairways which are likely to be used by children under the age of five years. (See Fig. 15.1.)

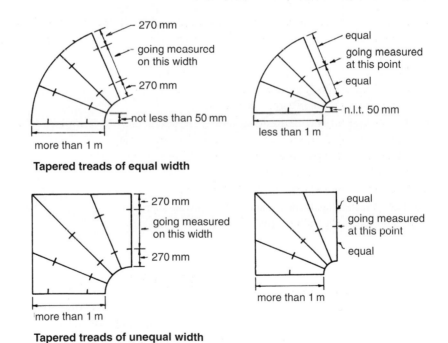

Fig. 15.4   Tapered treads.

15.6.2   Rules applying to private stairways

There are a number of recommendations in AD K which control the steepness, rise and going of private stairs as follows.

- The height of any rise should not be more than 220 mm.
- The going of any step should generally not be less than 220 mm (but see the rules relating to tapered treads above).
- The pitch should not be more than 42°. This means that it is not possible to combine a maximum rise with a minimum going.

Figure 15.5 illustrates the practical limits for rise and going. This figure also incorporates the $(2R + G)$ relationship mentioned above.

The rules governing rise and going for private stairs will also satisfy the recommendations contained in AD M for access for disabled people.

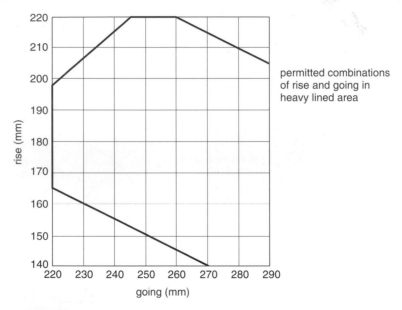

permitted combinations of rise and going in heavy lined area

Fig. 15.5   Practical limits for rise and going – private stairways.

15.6.3   Rules applying to institutional and assembly stairs

Institutional buildings are usually occupied by young children, old people or people with physical or mental disabilities. It is necessary, therefore, that staircases should be of slacker pitch than in other types of buildings and further recommendations regarding rise and going may be found in AD M. Additionally, they may need to comply with width recommendations in AD B or AD M. Assembly buildings contain large numbers of people and similar considerations will apply.

- The height of any rise should not be more than 180 mm.
- Subject to the rules governing tapered treads, the going of any step should generally not be less than 280 mm. This may be reduced to 250 mm if the floor area of the building served by the stairway is less than 100 m².

Figure 15.6 illustrates the practical limits for rise and going. This figure also incorporates the $(2R + G)$ relationship mentioned above.

- In order to maintain sightlines for spectators in assembly buildings, gangways are permitted to be pitched at up to 35°.
- There should not be more than sixteen risers in a single flight where a stairway serves an area used for assembly purposes. (See also Rules applying to other stairs, below.)
- The width of a stairway in a public building which is wider than 1800 mm should be sub-divided with handrails or other suitable means so that the sub-divisions do not exceed 1800 mm in width.

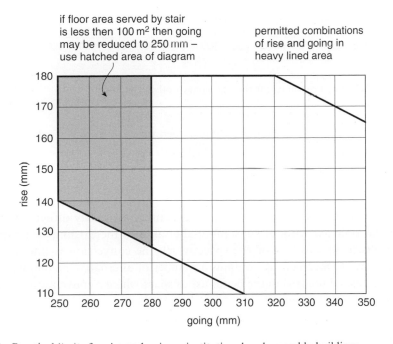

Fig 15.6   Practical limits for rise and going – institutional and assembly buildings.

15.6.4   Rules applying to other stairs

These recommendations apply to all stairs other than those in private dwellings, institutional or assembly buildings. Many of the buildings covered by these recommendations will also need to be accessible to disabled people and should, therefore, conform to the guidance given in AD M.

- The height of any rise should not be more than 190 mm.
- The going of any step should generally not be less than 250 mm (but see the rules relating to tapered treads above).

Figure 15.7 illustrates the practical limits for rise and going. This figure also incorporates the $(2R + G)$ relationship mentioned above.

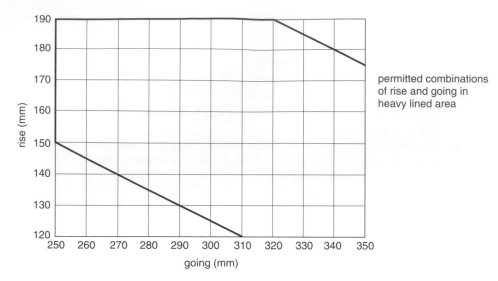

Fig. 15.7   Practical limits for rise and going – other buildings.

- There should not be more than sixteen risers in a single flight where a stairway serves an area used for shop purposes.

15.6.5   Stairs to loft conversions

When carrying out loft conversions it is often extremely difficult to fit a conventional staircase without substantial alteration to the existing structure. This often results in serious loss of the space which the conversion is intended to provide. Approved Document K contains a number of alternative recommendations which are intended to assist in the better use of space where a conventional staircase would prove difficult to install.

Headroom

Where there is insufficient height to achieve the recommended 2 m headroom over a stairway, 1.9 m at the centre of the stair is acceptable reducing to 1.8 m at the side (see Fig. 15.8).

Ladders

Fixed ladders may be installed for access in a loft conversion where there is insufficient room to install a conventional staircase without alteration to the existing space if they conform to the following:

- there should be fixed handrails on both sides;
- the ladder should only serve one habitable room;
- retractable ladders are not acceptable for means of escape in case of fire.

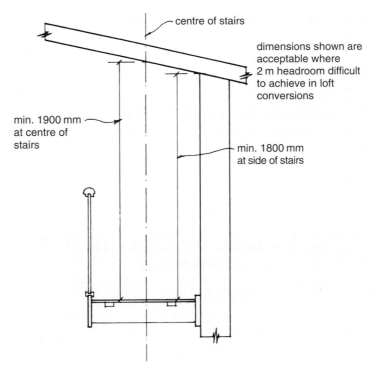

Fig. 15.8   Reduced headroom over stairs – loft conversions.

The definition of habitable room is not included in AD K and it varies from one Approved Document to another. The only safe conclusion which can be drawn is that bathrooms or WCs are not habitable rooms. Therefore, it would seem that a ladder could be used to access one habitable room and a bathroom or WC, but see Alternating tread stairs below.

It would also seem to be the case that retractable ladders can be used where the roof of a bungalow is converted to provide a two-storey house. These houses do not need to have means of escape to a stairway since escape from a suitable first floor window is acceptable in AD B. (See AD B, Section 1, 1.1 and Chapter 7 above).

15.6.6   Spiral and helical stairs

Generally stairs designed in accordance with BS 5395 *Stairs, Ladders and Walkways*, Part 2: 1984 *Code of practice for the design of helical and spiral stairs* will satisfy the recommendations of AD K. It is permissible to provide stairs with lesser goings for conversion work if space is limited, but the stair should only serve one habitable room (and, perhaps a bathroom or WC). The degree of variation from the norm is, however, not specified.

15.6.7   Alternating tread stairs

When first introduced a few years ago, these stairs caused quite a controversy since the treads are not of uniform width and the staircase is steeper than a conventional flight. They rely on a certain degree of familiarity on the part of the user since it is necessary to start the ascent or descent on the correct foot or disaster will ensue!

Alternating tread stairs should only be installed for loft conversion work where insufficient space is available to accommodate a conventional staircase and they should:

- be in one or more straight flights;
- provide access to only one habitable room plus bathroom or WC provided it is not the only WC in the dwelling;
- be fitted with handrails on both sides;
- contain treads which have slip-resistant surfaces;
- have uniform steps with parallel nosings;
- have a minimum going of 220 mm and a maximum rise of 220 mm when measured over the wider part of the tread;
- conform to the recommendations regarding maximum gap sizes for open riser stairs.

A typical design for an alternating tread stair is shown in Fig. 15.9.

15.7   Ramp recommendations

In addition to the general recommendations outlined above, section 2 of AD K contains the following particular guidance for ramps:

- the slope should not be greater than 1 in 12;
- there should be no permanent obstructions placed across any ramp.

15.8   Guarding of stairways, ramps and landings

Guarding should generally be provided at the sides of every flight, ramp or landing. However, guarding need not be provided in dwellings where there is a drop of 600 mm or less, or where there are fewer than two steps in other buildings.

The rules which apply to prevent small children from being trapped in open riser staircases also apply to guarding; that is, there should be no opening of such size as to allow the passage of a sphere of 100 mm diameter. This relates to the guarding on any staircase except that which is in a building which is unlikely to be used by children under the age of five years. It may be difficult in certain circumstances to decide where this exemption should apply. For example, some public houses have a policy of not admitting young children whereas others actually encourage them (accompanied by their parents, of course!). Guarding should also be designed so

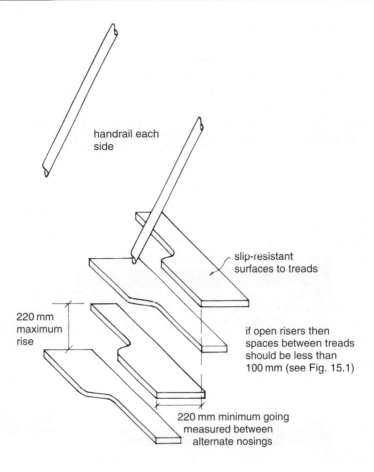

handrail each
side

slip-resistant
surfaces to treads

220 mm
maximum
rise

if open risers then
spaces between treads
should be less than
100 mm (see Fig. 15.1)

220 mm minimum going
measured between
alternate nosings

Fig. 15.9 Alternating tread stairs.

that it cannot easily be climbed by small children. (This might preclude the use of horizontal 'ranch' style balustrading, for example.)

Certain minimum heights for the guarding to flights, ramps and landings are given in AD K2/3, Section 3, Diagram 11. These are illustrated in Fig. 15.10 below and it should be noted that the guarding should be strong enough to resist at least the horizontal forces given in BS 6399: Part 1:1996.

15.9 Access to maintenance areas

It is important to establish how frequently access will be required to areas needing maintenance. If this is likely to be more than once per month then permanent stairs or ladders (such as those suggested in AD K1 for private stairs in dwellings) may need to be provided. Alternatively, the requirement may be satisified by following the guidance in BS 5395: Part 3: 1985 *Code of Practice for the Design of Industrial*

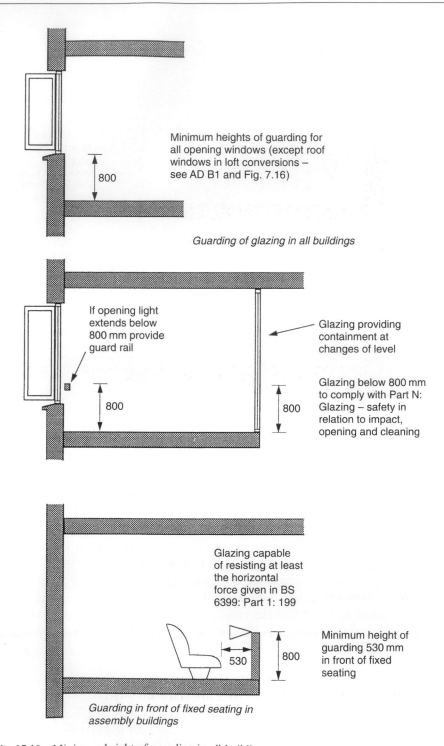

Minimum heights of guarding for all opening windows (except roof windows in loft conversions – see AD B1 and Fig. 7.16)

800

Guarding of glazing in all buildings

If opening light extends below 800 mm provide guard rail

Glazing providing containment at changes of level

800

800

Glazing below 800 mm to comply with Part N: Glazing – safety in relation to impact, opening and cleaning

Glazing capable of resisting at least the horizontal force given in BS 6399: Part 1: 199

530

800

Minimum height of guarding 530 mm in front of fixed seating

Guarding in front of fixed seating in assembly buildings

Fig. 15.10 Minimum height of guarding in all buildings.

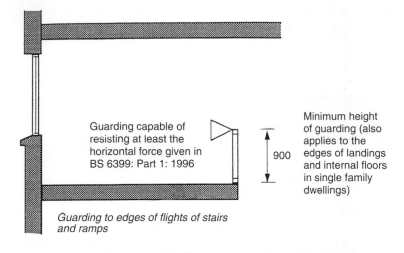

Guarding capable of resisting at least the horizontal force given in BS 6399: Part 1: 1996

900

Minimum height of guarding (also applies to the edges of landings and internal floors in single family dwellings)

Guarding to edges of flights of stairs and ramps

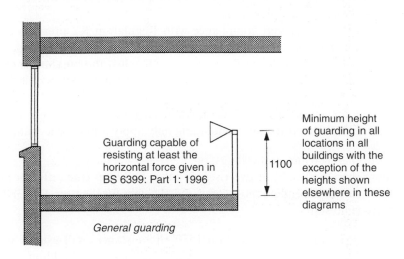

Guarding capable of resisting at least the horizontal force given in BS 6399: Part 1: 1996

1100

Minimum height of guarding in all locations in all buildings with the exception of the heights shown elsewhere in these diagrams

General guarding

Fig. 15.10   continued.

Type Stairs, Permanent Ladders and Walkways. Less frequent access to maintenance areas may be achieved using portable ladders.

15.10   Alternative approach to stairway design

It is permissible to use other sources of guidance when designing stairs.

- BS 5395 *Stairs, ladders and walkways*, Part 1: 1977 *Code of practice for the design of straight stairs* contains recommendations which will meet the steepness requirements of AD K.

LEEDS COLLEGE OF BUILDING
LIBRARY

- Wood stairs designed in accordance with BS 585, Part 1: 1989 will offer reasonable safety to users.
- Stairs, ladders or walkways in industrial buildings should follow the recommendations of BS 5395, Part 3: 1985 *Code of practice for the design of industrial stairs, permanent ladders and walkways* or BS 4211: 1987 *Specification for ladders for permanent access to chimneys, other high structures, silos and bins.*

15.11   Protection from falling

The following areas of the building must be provided with barriers to protect people in or about the building from falling.

- Stairways and ramps which form part of the building.
- Floors, balconies, and any roof to which people normally have access.
- Light wells, basements or similar sunken areas which are connected to the building.

Regulation K2 applies to all areas of a building which need to be accessed, including those used only for maintenance.

It should be noted that compliance with Regulation K2 prevents action being taken against the occupier of a building under Regulation 13 of the Workplace (Health, Safety and Welfare) Regulations 1992 when the building is eventually in use. (Regulation 13 relates to requirements designed to protect people from the risk of falling a distance likely to cause personal injury.)

The recommendations contained in AD K2 regarding the provision of pedestrian barriers to provide protection from falling only apply if there is a change in level of:

- more than 600 mm in dwellings; or
- more than the height of two risers in other buildings (the difference in levels in this case will depend on the recommended height for the risers), or 380 mm where there is no stair.

As in the case of stairs (see section 15.3) circumstances will usually dictate what constitutes an acceptable level of safety for the guarding in a building. Therefore, the standard in a dwelling may be lower than that recommended for a public building because there are likely to be less people in the dwelling and they will be familiar with its layout, etc. Similarly, the standard of access to maintenance areas will be lower than that recommended for normal use to reflect the greater care expected of those gaining access.

The provisions for guarding contained in AD K1 are extended by Approved Document K2/3 to cover the edges of any part of:

- a balcony, gallery, roof (including rooflights or other openings), floor (including the edge below any window openings) or other place to which people have access;
- any light well, basement or sunken area adjoining a building;
- a vehicle park (but not, of course, on any vehicle access ramps).

Guarding is not needed where it would obstruct normal use (for example, at the edges of loading bays).

15.11.1   Guarding recommendations

Where guarding is provided to meet the requirements of AD K2, then it should:

- have a minimum height as shown in Fig. 15.10;
- be capable of resisting the horizontal force given in BS 6399: Part 1: 1996;
- have any glazed part designed in accordance with the recommendations of AD N; Glazing – safety in relation to impact, opening and cleaning (see Chapter 18);
- have no opening of such size as to permit the passage of a 100 mm diameter sphere if it is in a building which is likely to be used by children under 5 years of age.

Guarding can consist of a wall, balustrade, parapet or similar barrier but should be designed so that it cannot be easily climbed by small children (i.e. horizontal rails should be avoided) if they are likely to be present in the building.

15.11.2   Guarding recommendations for maintenance areas

It is important to establish how frequently access will be required to areas needing maintenance. If this is likely to be more than once per month then guarding provisions such as those suggested for dwellings (see Fig. 15.10) will be satisfactory. For less frequent access it may only be necessary to provide temporary guarding and in many circumstances the posting of warning notices may be sufficient to satisfy the requirement. Building Regulations, of course, do not cover the design and installation of temporary guarding or warning signs although they are covered by the Construction (Design and Management) Regulations 1994. (For information on signs see the Health and Safety (Signs and Signals) Regulations 1996.)

15.12   Vehicle barriers and loading bays

Vehicle ramps and any levels in a building to which vehicles have access are required to have barriers in order to protect people in or about the building. Additionally, vehicle loading bays must either be constructed in such a way as to protect people in them from collision with vehicles or they must contain features which achieve the same end.

15.12.1   Vehicle barriers

If the perimeter of any roof, ramp or floor to which vehicles have access, forms part of a building, it should have barriers to protect it, provided that it is level with or above any adjacent floor, ground or vehicular route.

Vehicle barriers can be formed by walls, parapets, balustrading or similar obstructions and should be at least the heights shown in Fig. 15.11. Barriers should be capable of resisting the horizontal forces as set out in BS 6399 *Loading for buildings*, Part 1: 1984 *Code of practice for dead and imposed loads.*

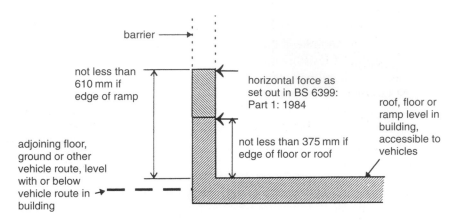

Fig. 15.11   Vehicle barriers – any building.

15.12.2   Loading bays

Loading bays for less than three vehicles should be provided with:

- one exit point (e.g. steps) from the lower level, ideally near to the centre of the rear wall; or
- a refuge into which people can go if in danger of being struck or crushed by a vehicle.

Larger loading bays (i.e. for three or more vehicles) should be provided with:

- one exit point at each side, or
- a refuge.

See Fig. 15.12 for details.

15.13   Protection from collision with open windows, skylights or ventilators

Regulation K4 requires that provision be made to prevent people who are moving in or about a building from colliding with open windows, skylights or ventilators. This requirement does not apply to dwellings but it does apply to all parts of other buildings including, in a limited form, to areas of the building used exclusively for maintenance purposes.

Again, compliance with this regulation prevents action being taken against the

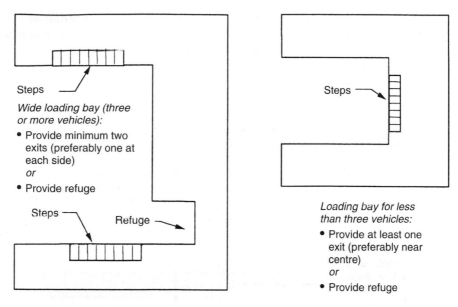

Fig. 15.12   Vehicle loading bays.

occupier of a building under Regulation 15(2) of the Workplace (Health, Safety and Welfare) Regulations 1992 (Regulation 15(2) also relates to projecting windows, skylights and ventilators) when the building is eventually in use.

Since it is desirable to leave windows, etc. open for ventilation purposes, this should be possible without causing danger to people who might collide with them. Generally, this can be achieved by:

- Providing windows, skylights and ventilators so that projecting parts are kept away from people moving in and about the building; or
- Providing features which guide people away from these projections.

In certain special cases (e.g. in spaces used only for maintenance purposes) it is reasonable to expect the exercise of greater care by those gaining access, therefore less demanding provisions than those for normal access could satisfy the regulation.

15.13.1   Avoiding projecting parts

AD K4 makes it clear that the requirements of Regulation K4 do not apply to the opening parts of windows, skylights or ventilators:

- which are at least 2 m above ground or floor level; or
- which do not project more than about 100 mm internally or externally into spaces in or about the building where people are likely to be present.

Since the siting of opening lights above 2 m may not always be possible, an acceptable solution might be to fit projecting lights with restraint straps designed to restrict the projection to about 100 mm. The restraint should only be removed for maintenance and cleaning purposes. An ordinary casement stay would probably not be considered suitable for this purpose. Any such solution should, of course, be checked for acceptability with the building control authority (local authority or approved inspector).

Alternative solutions given in AD K4 also include:

(a)   Marking the projection with a feature such as a barrier or rail about 1100 mm high to prevent people walking into it; or

(b)   Guiding people away from the projection by providing ground or floor surfaces with strong tactile differences or suitable landscaping features.

These solutions are shown in Fig. 15.13 below.

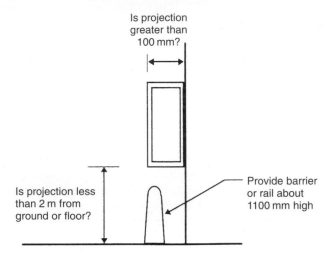

Fig. 15.13(a)   Marking projections – barriers.

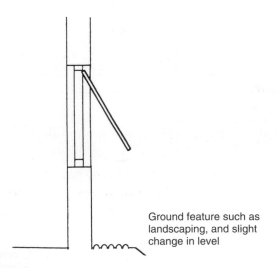

Fig. 15.13(b)   Marking projections – ground features.

15.13.2 Maintenance areas

Where such areas are used infrequently, projecting parts of windows, skylights and ventilators should be made easier to see by being clearly marked, in order to satisfy Regulation K4.

It should be noted that there are other provisions in the Building Regulations which relate to safety on common circulation routes in buildings. In particular, reference should be made to Approved Document B – Fire safety (see section 7.9.3 for guidance on the clear widths of escape routes), and Approved Document M – Access and facilities for disabled people (see section 17.5.5 below for guidance on the avoidance of hazards on circulation routes).

15.14   Safe use of doors

In order that they may be safely used, provisions must be made to prevent doors or gates:

- which slide or open upward, from falling onto any person; and
- which are powered, from trapping anyone.

Furthermore, powered doors and gates must be openable in the event of a power failure and all swing doors and gates are required to be designed to allow a clear view of the space on either side.

Regulation K5 does not apply to dwellings, or to any door or gate which is part of a lift. Compliance with this regulation prevents action being taken against the occupier of a building under Regulation 18 (requirements for doors and gates) of the Workplace (Health, Safety and Welfare) Regulations 1992 when the building is eventually in use.

15.14.1   Door and gate safety features

Typically, the following safety features will satisfy the requirements of Regulation K5 by preventing the opening and closing of doors and gates from presenting a safety hazard:

- Vision panels should be provided in doors and gates on traffic routes and in those which can be pushed open from either side (i.e. double swing doors) unless they can be seen over (e.g. lower than about 900 mm for a person in a wheelchair). For example, this means providing such doors and gates with a glazed panel giving a zone of visibility between 900 mm and 1500 mm from floor level (see also Fig. 17.3 in section 17.6 below).
- Sliding doors and gates should be prevented from leaving the end of the suspension track by a stop or other suitable means. Additionally, if the suspension system fails or the rollers leave the track they should be prevented from falling by means of a retaining rail.

- Upward opening doors and gates should be designed to prevent them from falling and causing injury.
- If doors and gates are power operated the following safety features are examples of those which could be incorporated:
  - (a) devices to prevent injury to people from being struck or trapped (e.g. doors with pressure sensitive edges)
  - (b) readily accessible and identifiable stop switches
  - (c) manual or automatic opening provisions in the event of power failure if this is necessary for health and safety reasons.

Reference should also be made to other provisions in the Building Regulations which relate to the design and use of doors in buildings. In particular, see Approved Document B – Fire safety (see section 7.22.3 above for guidance on the provision of doors on escape routes) and Approved Document M – Access and facilities for disabled people (see sections 17.6, 17.6.1 and 17.6.2 below for guidance on the design of internal and external doors).

16 Conservation of fuel and power (Part L)

J.R. Waters

16.1 Introduction

National and international concerns with respect to fuel resources, atmospheric pollution, the greenhouse effect and global warming have led to a generally accepted need to conserve energy. This problem is being treated with increasing urgency at all levels of society, and a particular consequence is that relevant legislation is subject to regular and frequent change and expansion. This is particularly true of the requirements for the conservation of fuel and power, which are dealt with by Part L of Schedule 1 of the Building Regulations. The documents described in this chapter were those current at the time of writing, but because of the rate of change in this area, the reader must be aware that new or amended documents may have been published since.

The 2006 edition of Part L has been drafted to take into account two main objectives. The first is to contribute to the Government's aim to reduce carbon dioxide emissions into the atmosphere. In the UK, approximately half of the these emissions arise from buildings, and so more stringent measures for the conservation of fuel and power in buildings are likely to make a major contribution to a reduction in the overall UK carbon dioxide emissions. The second objective is for Part L to be compatible with the European Energy Performance of Buildings Directive (EPBD). This sets, in several ways, minimum performance standards for buildings.

The main changes from the previous (2002) version of Part L are:

- the removal of the traditional elemental method for demonstrating compliance;
- the introduction of the calculation of carbon dioxide emissions as the criterion of compliance;
- an extension of the scope of Part L with respect to work on existing buildings;
- the introduction of new parts to regulations 17 and 20, as described in The Building and Approved Inspectors (Amendment) Regulations 2006 (SI 2006 No. 652), to provide the statutory basis for the new approach;
- a split into four separate Approved Documents, the so-called 'first tier' documents;
- the reliance on a range of so-called 'second tier' documents, which provide essential support to the Approved Documents and are referenced from them;
- frequent reference to a large number of other supporting documents.

The Approved Documents contain a substantial amount of material that is common to all four of them. This includes:

- the definitions of key terms and the interpretation of new ones;
- statements of relevant new clauses in the Building Regulations;
- the formal Part L requirement of Schedule 1;
- general exemptions;
- tables of the self-certification schemes and exemptions from various requirements, as given in Schedules 2A and 2B;
- certain aspects of general guidance which are common to all buildings.

For convenience, this common material is gathered together and presented first.

16.2   Definition and interpretation of terms

Definitions are given in section 5 of AD L1A, AD L2A and AD L2B and in section 4 of AD L1B. They are collected together here in overall alphabetic order.

AIR PERMEABILITY – The air leakage rate per unit envelope area at the test reference pressure differential of 50 Pascal.

The envelope area of the building (or its measured part) is the total area of all floors, walls and ceilings bordering the space being tested, and includes walls and floors below external ground level. Overall internal dimensions are used to calculate this area, with no subtractions for the area of the junctions where internal elements meet the external envelope.

BCB – A building control body, local authority or an approved inspector.

BER – The Building Emission Rate –the predicted CO_2 emission rate from a building other than a dwelling.

CHANGE TO A BUILDING'S ENERGY STATUS – Any change that results in a building becoming a building to which the energy efficiency requirements of the regulations apply, where previously it was not.

CONSEQUENTIAL IMPROVEMENTS – Those energy efficiency improvements required by regulation 17D. Consequential improvements are additional to the principal works, and are improvements to an existing building in order to raise its energy efficiency standards to those required by the current Part L.

CONSERVATORY – An extension that:

- has not less than three quarters of its roof area and not less than one half of its external wall area made from translucent material; and
- is thermally separated from the building to which it is attached by walls, windows and doors with U-value and draught-stripping provisions as least as good as provided elsewhere in the building.

CPSU – Combined primary storage unit.

DAYLIT SPACE – Any space:

- within 6 m of a window wall, provided that the glazing area is at least 20% of the internal area of the window wall; or
- below rooflights and similar provided that the glazing area is at least 10% of the floor area.

The normal light transmittance of the glazing should be at least 70%, or, if the light transmittance is reduced below 70%, the glazing qualifying area should be increased proportionately.

DER – the Dwelling Emission Rate – the predicted CO_2 emission rate from a dwelling.

DESIGN AIR PERMEABILITY – The value of air permeability selected by the designer of a dwelling for use in the preliminary calculation of the DER (for dwellings), or the BER (for buildings other than dwellings).

DISPLAY LIGHTING – Lighting intended to highlight displays of exhibits or merchandise, or lighting used in spaces for public leisure and entertainment such as dance halls, auditoria, conference halls, restaurants and cinemas.

DISPLAY WINDOW – An area of glazing, including glazed doors, intended for the display of products or services on offer within the building, positioned:

- at the external perimeter of the building;
- at an access level and immediately adjacent to a pedestrian thoroughfare.

There should be no permanent workspace within one glazing height of the perimeter. Glazing more than 3 m above such an access level should not be considered part of a display window except:

- where the products on display require a greater height of glazing;
- in existing buildings, when replacing display windows that already exist to a greater height;
- in cases of building work involving changes to the façade and glazing requiring planning consent, where planners should have discretion to require a greater height of glazing. For example, the greater height may be required to fit in with surrounding buildings or to match with the character of the existing façade.

It is expected that display windows will be found in the usage classes A1, A2, A3 and D2 as shown in Table 16.1.

DWELLING – A self-contained unit designed to accommodate a single household.

DWELLING TYPE – A group of dwellings on a site having the same generic form (detached, semi-detached, terrace, ground mid or top floor flat) and where the same construction methods have been used for the main elements (walls, floors, roofs, etc.).

Table 16.1   Building classes.

Class	Usage
A1	Shops, including retail-warehouses, undertakers, showrooms, post offices, hairdressers and shops for the sale of cold food for consumption off the premises
A2	Financial and professional services, including banks, building societies, estate and employment agencies and betting offices
A3	Food and drink, including restaurants, pubs, wine bars and shops for sale of hot food for consumption off the premises.
D2	Assembly and leisure buildings, including cinemas, concert halls, bingo halls, casinos, etc.

Note that the use of a consistent set of accredited details can indicate the use of the same construction method, and that small variations in floor area do not constitute a different dwelling type.

EMERGENCY ESCAPE LIGHTING – That part of emergency lighting that provides illumination for the safety of people leaving an area or attempting to terminate a dangerous process before leaving an area.

ENERGY EFFICIENCY REQUIREMENTS – The requirements of regulations 4A, 17C and 17D, and of Part L of Schedule 1.

EPBD – Energy in Buildings Performance Directive.

FIT-OUT WORK – That work needed to complete the partitioning and building services within the external fabric of the building (the shell) to meet the specific needs of incoming occupiers. Fit-out work can be carried out in whole or in parts:

- in the same project and time frame as the construction of the building shell; or
- at some time after the shell has been completed.

FIXED BUILDING SERVICES – Any part of, or any controls associated with:

- fixed internal or external lighting systems, but does not include emergency escape lighting or specialist process lighting; or
- fixed systems for heating, hot water service, air conditioning or mechanical ventilation.

HIGH-USAGE ENTRANCE DOOR – A door to an entrance, primarily for the use of people, that is expected to experience large traffic volumes, and where robustness and/or powered operation is the primary performance requirement. To qualify as a high-usage entrance door, the door should be equipped with automatic closers, and except where operational requirements preclude, be protected by a lobby.

LARGE EXTENSION – An extension is considered to be large when its total useful floor area is both:

- greater than 100 m^2; and
- greater than 25% of the total useful floor area of the existing building.

NCM – The National Calculation Method. Information is available at www.ncm.bre.co.uk.

PRINCIPAL WORKS – The works necessary to achieve the client's purposes in extending a building and/or increasing the installed capacity of any fixed building services. The value of the principal works is the basis for determining a reasonable provision of consequential improvements.

RENOVATION – In relation to a thermal element, renovation means the provision of a new layer in the thermal element or the replacement of an existing layer, but excludes decorative finishes, and 'renovate' shall be construed accordingly.

ROOM FOR RESIDENTIAL PURPOSES – A room, or a suite of rooms, which is not a dwelling-house or a flat, and which is used by one or more persons to live and sleep and includes a room in a hostel, a hotel, a boarding house, a hall of residence or a residential home, whether or not the room is separated from or arranged in a cluster group with other rooms, but does not include a room in a hospital, or other similar establishment, used for patient accommodation and, for the purposes of this definition, a 'cluster' is a group of rooms for residential purposes which is:

- separated from the rest of the building in which it is situated by a door which is designed to be locked; and
- not designed to be occupied by a single household.

SAP 2005 – The Government's Standard Assessment Procedure for Energy Rating of Dwellings, 2005 edition.

SBEM AND iSBEM – SBEM is the Simplified Building Energy Model, which is a computer based. iSBEM is a user interface to the SBEM.

SEDBUK – The Seasonal Efficiency of a Domestic Boiler in the UK. See sedbuk.com for details.

SIMPLE PAYBACK – The amount of time it will take to recover an initial investment in an energy-saving measure through energy savings. It is found by dividing the marginal initial cost of implementing the energy efficiency measure by the value (excluding VAT) of the annual energy savings achieved by that measure. Note that:

- The marginal additional cost is the additional cost (materials and labour) of incorporating the energy saving measure (e.g. additional insulation) and not the whole cost of the work.
- The cost of implementing the measure should be based on prices current at the date the proposals are made known to the BCB and be confirmed in a report signed by a suitably qualified person.

- The annual energy savings should be estimated using SAP 2005 (for dwellings) or SBEM (for buildings other than dwellings).
- For the purposes of AD L1B, the following energy prices (current in 2005) should be used when evaluating the energy savings:
  - mains gas 1.63 p/kWh;
  - electricity 3.65 p/kWh (a weighted combination of peak and off peak tariffs);
  - heating oil 2.17 p/kWh;
  - LPG   3.71 p/kWh.
- For the purposes of AD L2B, the following energy prices should be used when evaluating the energy savings:
  - mains gas 1.45 p/kWh;
  - electricity 5.00 p/kWh;
  - heating oil 1.90 p/kWh;
  - LPG   3.39 p/kWh.

If desired, higher energy prices may be used, such as those prevailing when application for Building Regulations approval is made.

SPECIALIST PROCESS LIGHTING – Lighting intended to illuminate specialist tasks within a space, rather than the space itself. It could include theatre spotlights, projection equipment, lighting in TV and photographic studios, medical lighting in operating theatres and doctors' and dentists' surgeries, illuminated signs, coloured or stroboscopic lighting and art objects with integral lighting such as sculptures, decorative fountains and chandeliers.

SUBSTANTIALLY GLAZED SPACE – An extension that has:

- a large area of glazing, but does not satisfy the definition for a conservatory; and
- is thermally separated from the building by walls, windows and doors with U-value and draught-stripping provisions as least as good as provided elsewhere in the building.

In many cases, a substantially glazed space can be treated as a conservatory.

TER – Target Emission Rate – the target (i.e. maximum permissible) CO_2 emission rate from a building.

THERMAL ELEMENT – In paragraph 2A of regulation 2, 'thermal element' means a wall, floor or roof (but does not include windows, doors, roof windows or roof-lights) that separates a thermally conditioned part of the building ('the conditioned space') from:

- the external environment (including the floor); or
- in the case of walls or floors, another part of the building which is:
  - unconditioned;
  - an extension falling within class VII in Schedule 2; or
  - where this paragraph applies, conditioned to a different temperature, and includes all parts of the element between the surface bounding the conditioned space and the external environment or other part of the building as the case may be.

In paragraph 2B of regulation 2:

- Paragraph 2A(b)(iii) only applies to a building that is not a dwelling, where the other part of the building is used for a purpose that is not similar or identical to the purpose for which the conditioned space is used.

TOTAL USEFUL FLOOR AREA – The total area of all enclosed spaces measured to the internal face of the external walls, that is to say it is the gross floor area as measured in accordance with the guidance issued to surveyors by the RICS. In this convention:

- the area of sloping surfaces such as staircases, galleries, raked auditoria and tiered terraces should be taken as their area on plan; and
- areas that are not enclosed such as open floors, covered ways and balconies are excluded.

16.3   Additions to the regulations with particular relevance to Part L

Changes or additions to regulations 4, 17 and 20 have been made which are directly relevant to Part L.

16.3.1   New regulations 4A and 4B, requirements relating to thermal elements and change of energy status

Where work is to be carried out on an existing building, new regulations 4A and 4B apply, as follows:

- **Regulation 4A:**
  (1) Where a person intends to renovate a thermal element, such work shall be carried out as is necessary to ensure that the whole thermal element complies with the requirements of paragraph L1(a)(i) of Schedule 1.
  (2) Where a thermal element is replaced, the new thermal element shall comply with the requirements of paragraph L1(a)(i) of Schedule 1.
- **Regulation 4B:**
  (1) Where there is a change to a building's energy status, such work, if any, shall be carried out as is necessary to ensure that the building complies with the applicable requirements of Part L of Schedule 1.
  (2) In this regulation 'building' means the building as a whole or parts of it that have been designed or altered to be used separately.

16.3.2   Additions to regulation 17 – energy performance of buildings

The introduction of carbon emissions as the criterion for compliance has necessitated the addition of new parts to regulation 17. These apply to all buildings, and are restated in full here.

REGULATION 17A – METHODOLOGY OF CALCULATION OF THE ENERGY PERFORMANCE OF BUILDINGS

The Secretary of State shall approve a methodology of calculation of the energy performance of buildings.

One set of methodologies has been approved for dwellings, and a different set for buildings which are not dwellings. It is possible that techniques for calculating the energy performance of buildings will continue to evolve, and that other methodologies will be approved in the future.

REGULATION 17B – MINIMUM ENERGY PERFORMANCE REQUIREMENTS FOR BUILDINGS

The Secretary of State shall approve minimum energy performance requirements for new buildings, in the form of target CO_2 emission rates, which shall be based upon the methodology approved pursuant to regulation 17A.

REGULATION 17C – NEW BUILDINGS

Where a building is erected, it shall not exceed the target CO_2 emission rate for the building that has been approved pursuant to regulation 17B.

Taken together, regulations 17B and 17C mean that the limiting factor on the design and operation of a building is the amount of CO_2 released into the atmosphere, and not the total amount of energy consumed. This in turn means that the source of energy is crucial. Buildings which are designed to rely on energy sources with low emission rates per unit of consumed energy will be better able to meet the target.

REGULATION 17D – CONSEQUENTIAL IMPROVEMENTS TO ENERGY PERFORMANCE

(1) Paragraph (2) applies to an existing building with a total useful floor area over 1000 m² where the proposed building work consists of or includes:
     (a) an extension;
     (b) the initial provision of any fixed building services; or
     (c) an increase to the installed capacity of any fixed building services.
(2) Subject to paragraph (3), where this regulation applies, such work, if any, shall be carried out as is necessary to ensure that the building complies with the requirements of Part L of Schedule 1.
(3) Nothing in paragraph (2) requires work to be carried out if it is not technically, functionally and economically feasible.

Regulation 17D means that if any building above a useful floor area of 1000 m² is extended, or its fixed services installed or altered, then the whole building must be brought up to a standard compatible with the current Part L requirements for that building.

REGULATION 17E – INTERPRETATION

In this part 'building' means the building as a whole or parts of it that have been designed or altered to be used separately.

16.3.3 Additions to regulation 20 – pressure testing, commissioning and CO_2 emission rate calculations

Regulations 20B, 20C and 20D impose the statutory requirements for pressure testing, commissioning and CO_2 emission rate calculations. In summary, these regulations are as follows.

REGULATION 20B – PRESSURE TESTING
This applies to the erection of a building for which Part L imposes a requirement to limit heat gains and losses through thermal elements and other parts of the building fabric. Pressure testing must be carried out and recorded in the approved manner, and either:

(a) the results and data must be submitted to the local authority within seven days of the final test; or
(b) the local authority may accept evidence that the test met the requirement in the form of a certificate by a person registered for the purpose by the British Institute of Non-destructive Testing [1].

REGULATION 20C – COMMISSIONING
This applies whenever Part L imposes a requirement to provide and commission energy efficient fixed building services with effective controls. That is to say, it applies whenever Schedule 1 imposes a requirement, but not when the work consists only of work described in Schedule 2B (Table 16.3). Commissioning must be carried out according to the approved procedure, and notice given to the local authority either:

(a) not later than the date on which the notice required by regulation 15(4) is required to be given; or
(b) where that regulation does not apply, not more than 30 days after completion of the work.

REGULATION 20D – CO_2 EMISSION RATE CALCULATIONS
This regulation expands on the requirement stated in regulation 17C, and is particularly important as it gives specifications for the calculation and reporting of CO_2 emission rates. It is stated that it is necessary to obtain:

(a) the target CO_2 emission rate for the building; and
(b) the calculated CO_2 emission rate for the building as constructed.

The person carrying out the building work must, within the same time limits imposed by regulation 20B, either provide the target and calculated CO_2 emission rates as evidence of compliance, or provide an appropriate certificate by a person who is registered by a specified organisation (currently either FAERO Ltd [2] or BRE Certification Ltd [3]).

16.4 Part L – Exemptions from Part L

Buildings and works which have overall exemption from the Building Regulations are covered by regulation 9 and Schedule 2. Other buildings which are normally exempt from the requirements of Part L are those buildings which are:

- listed, located in a conservation area, or included in the schedule of ancient monuments, provided compliance with energy efficiency requirements would unacceptably alter their character or appearance;
- used primarily or solely as places of worship;
- stand-alone buildings other than dwellings with a total useful floor area of less than 50 m^2;
- temporary buildings with a planned time use of two years or less, industrial sites, workshops and non-residential agricultural buildings with low energy demand.

Examples of buildings which are industrial sites and workshops with low energy demand include buildings, or parts of buildings designed to be used separately, whose purpose is to accommodate industrial activities in spaces where the air is not conditioned. Such activities might include hot processes (e.g. foundries, forges, etc.), chemical processes, food and drinks packaging, heavy engineering, storage and warehouses. In each case the air is not fully heated or cooled, although there may be some local conditioning appliances to serve people at work stations, etc. An example of non-residential agricultural building or parts of a building with low energy demand could be one that is heated for a few days each year (e.g. for plants to germinate) but is otherwise unheated.

Further statements of exemptions, etc. which are particularly relevant to Part L are given in Schedules 2A and 2B. Tables 16.2 and 16.3 give these Schedules in full.

Table 16.2   Schedule 2A – Self-certification schemes and exemptions from requirement to give building notice or deposit full plans.

Type of work	Person carrying out work
1. Installation of a heat-producing gas appliance.	A person, or an employee of a person, who is a member of a class of persons approved in accordance with regulation 3 of the Gas Safety (Installation and Use) Regulations 1998 [4].
2. Installation of a heating or hot-water service system connected to a heat-producing gas appliance, or associated controls.	A person registered by CORGI Services Limited [5] in respect of that type of work.
3. Installation of: (a) an oil-fired combustion appliance which has a rated heat output of 100 kW or less and which is installed in a building with no more than three storeys (excluding any basement) or in a dwelling; (b) oil storage tanks and the pipes connecting them to combustion appliances; or (c) heating and hot-water service systems connected to an oil-fired combustion appliance.	An individual registered by Oil Firing Technical Association Limited [6], NAPIT Certification Limited [7] or Building Engineering Services Competence Accreditation Limited [8] in respect of that type of work.
4. Installation of: (a) a solid-fuel burning combustion appliance which has a rated heat output of 50 kW or less which is installed in a building with no more than three storeys (excluding any basement); or (b) heating and hot-water service systems connected to a solid fuel burning combustion appliance.	A person registered by HETAS Limited [9], NAPIT Certification Limited [7] or Building Engineering Services Competence Accreditation Limited [8] in respect of that type of work.

(Contd).

Table 16.2 (*Contd*).

Type of work	Person carrying out work
5. Installation of a heating or hot-water service system, or associated controls, in a dwelling.	A person registered by Building Engineering Services Competence Accreditation Limited [8] in respect of that type of work.
6. Installation of a heating, hot-water service, mechanical ventilation or air conditioning system, or associated controls, in a building other than a dwelling.	A person registered by Building Engineering Services Competence Accreditation Limited [8] in respect of that type of work.
7. Installation of an air conditioning or ventilation system in an existing dwelling, which does not involve work on systems shared with other dwellings.	A person registered by CORGI Services Limited [5] or NAPIT Certification Limited [7] in respect of that type of work.
8. Installation of a commercial kitchen ventilation system which does not involve work on systems shared with parts of the building occupied separately.	A person registered by CORGI Services Limited [5] in respect of that type of work.
9. Installation of a lighting system or electric heating system, or associated electrical controls.	A person registered by The Electrical Contractors Association Limited [10] in respect of that type of work.
10. Installation of fixed low or extra-low voltage electrical installations.	A person registered by BRE Certification Limited [3], British Standards Institution [11], ELECSA Limited [12], NICEIC Group Limited [13] or NAPIT Certification Limited in respect of that type of work.
11. Installation of fixed low or extra-low voltage electrical installations as a necessary adjunct to or arising out of other work being carried out by the registered person.	A person registered by CORGI Services Limited [5], ELECSA Limited [12], NAPIT Certification Limited [7], NICEIC Group Limited [13] or Oil Firing Technical Association Limited [6] in respect of that type of electrical work.
12. Installation, as a replacement, of a window, rooflight, roof window or door (being a door which together with its frame has more than 50% of its internal face area glazed) in an existing building.	A person registered under the Fenestration Self-Assessment Scheme by Fensa Limited [14], or by CERTASS Limited [15] or the British Standards Institution [11] in respect of that type of work.
13. Installation of a sanitary convenience, washing facility or bathroom in a dwelling, which does not involve work on shared or underground drainage.	A person registered by CORGI Services Limited [5] or NAPIT Certification Limited [7] in respect of that type of work.
14(1) Subject to paragraph (2), any building work, other than the provision of a masonry chimney, which is necessary to ensure that any appliance, service or fitting which is installed and which is described in the preceding entries in column 1 above, complies with the applicable requirements contained in Schedule 1.	The person who installs the appliance, service or fitting to which the building work relates and who is described in the corresponding entry in column 2 above.
14(2) Paragraph (1) does not apply to: (a) building work which is necessary to ensure that a heat-producing gas appliance complies with the applicable requirements contained in Schedule 1 unless the appliance: (i) has a rated heat output of 100 kW or less; and (ii) is installed in a building with no more than three storeys (excluding any basement), or in a dwelling; or (b) the provision of a masonry chimney.	

Table 16.3   Schedule 2B – Descriptions of work where no building notice or deposit of full plans is required.

Type of work	Description

1 Work consisting of	(a) replacing any fixed electrical equipment which does not include the provision of:
	    (i) any new fixed cabling;
	    (ii) a new consumer unit;
	(b) replacing a damaged cable for a single circuit only;
	(c) refixing or replacing enclosures of existing installation components, where the circuit protective measures are unaffected;
	(d) providing mechanical protection to an existing fixed installation, where the circuit protective measures and current-carrying capacity of conductors are unaffected by the increased thermal insulation;
	(e) installing or upgrading main or supplementary equipotential bonding;
	(f) in heating and cooling systems:
	    (i) replacing control devices that utilise existing fixed control wiring or pneumatic pipes;
	    (ii) replacing a distribution system output device;
	    (iii) providing a valve or pump;
	    (iv) providing a damper or fan;
	(g) in hot-water service systems, providing a valve or pump;
	(h) replacing an external door (where the door together with its frame has not more than 50% of its internal face area glazed);
	(i) in existing buildings other than dwellings, providing fixed internal lighting where no more than 100 m² of the floor area of the building is to be served by the lighting.

2 Work which	(a) is not in a kitchen, or a special location;
	(b) does not involve work on a special installation, and
	(c) consists of:
	    (i) adding light fittings and switches to an existing circuit; or
	    (ii) adding socket outlets and fused spurs to an existing ring or radial circuit.

3 Work on	(a) telephone wiring or extra-low voltage wiring for the purposes of communications, information technology, signalling, control and similar purposes, where the wiring is not in a special location; (b) equipment associated with the wiring referred to in sub-paragraph (a); (c) pre-fabricated equipment sets and associated flexible leads with integral plug and socket connections.

Note: In Table 16.3 (Schedule 2B):

'kitchen' means a room or part of a room which contains a sink and food preparation facilities;

'special installation' means an electric floor or ceiling heating system, an outdoor lighting or electric power installation, an electricity generator, or an extra-low voltage lighting system which is not a pre-assembled lighting set bearing the CE marking referred to in regulation 9 of the Electrical Equipment (Safety) Regulations 1994 [16];

'special location' means a location within the limits of the relevant zones specified for a bath, a shower, a swimming or paddling pool or a hot air sauna in the Wiring Regulations, 16th edition, published by the Institution of Electrical Engineers and the British Standards Institution as BS 7671: 2001 and incorporating amendments 1 and 2.

16.5   Part L – The requirement and general applicability

The formal requirement for Part L, as stated in Part L of Schedule 1, is given in Table 16.4. In general, the energy efficiency requirements of the Building Regulations (i.e. Part L) apply to:

- any new building;
- the extension of any existing building;

Table 16.4 Part L – Conservation of fuel and power – requirements.

Reasonable provision shall be made for the conservation of fuel and power in buildings by:
(a) limiting gains and losses:
    (i) through thermal elements and other parts of the building fabric; and
    (ii) from pipes, ducts and vessels used for space heating, space cooling and hot-water services;
(b) providing and commissioning energy efficient fixed building services with effective controls; and
(c) providing to the owner sufficient information about the building, the fixed building services and their maintenance requirements so that the building can be operated in such a manner as to use no more fuel and power than is reasonable in the circumstances.

Table 16.5 List of Approved or 'First Tier' Documents.

Reference	Title	Date of Publication
Approved Document L1A	Conservation of fuel and power in new dwellings (2006 edition)	March 2006 Third impression
Approved Document L1B	Conservation of fuel and power in existing dwellings (2006 edition)	March 2006 Third impression
Approved Document L2A	Conservation of fuel and power in new buildings other than dwellings (2006 edition)	March 2006 Third impression
Approved Document L2B	Conservation of fuel and power in existing buildings other than dwellings (2006 edition)	March 2006 Second impression with revisions

- the carrying out of work to or in connection with any such building or extension if it:
  ○ is a roofed construction having walls,
  ○ uses energy to condition the indoor climate, and
  ○ does not fall within the list of exemptions.

Methods by which the requirement set out in Table 16.4 may be satisfied are given in the Approved Documents, of which there are four, as shown in Table 16.5. Each of the Approved Documents gives details of more specific cases of applicability and exemptions, and each re-states:

- the new regulations 17A, 17B, 17C, 17D, 17E, 20B, 20C and 20D;
- the Part L requirement as stated in Table 16.4;
- Schedules 2A and 2B.

Common to all four Approved Documents is a short section on technical risk.

16.5.1 Technical risk

This states that building work must satisfy all the technical requirements set out in regulations 4A, 4B, 17C, 17D and Schedule 1 of the Building Regulations. When considering the incorporation of energy efficiency measures, it is further stated that the following other parts are also of particular relevance:

- Part B: Fire safety;
- Part C: Site preparation and resistance to moisture;
- Part E: Resistance to the passage of sound;
- Part F: Ventilation;
- Part J: Combustion appliances and fuel storage systems;
- Part P: Electrical safety.

The inclusion of an energy efficiency measure should not give rise to excessive technical risk. General guidance on avoiding risks in the application of thermal insulation may be found in BR 262 [17].

16.6   Part L1A – Conservation of fuel and power in new dwellings

16.6.1   Section 0: General guidance

This section includes:

- types of work covered by AD L1A;
- conservatories and substantially glazed spaces;
- technical risk – see section 16.5.1;
- demonstrating compliance.

16.6.1.1   *Types of work covered by AD L1A*

The guidance covers:

(1) New dwellings created through new construction works.
(2) A new building containing multiple dwellings, but *not* the common areas in such a building.
(3) A new building which contains both living accommodation and space used for commercial purposes, subject to the following conditions:
- the commercial part could revert to domestic use on change of ownership;
- there is direct access between the living accommodation and the commercial space;
- both are contained within the same thermal envelope;
- the living accommodation occupies a substantial portion of the total area of the building. This rules out the case of a small manager's or caretaker's flat in a large non-domestic building. It is implied that the building must primarily be conceived of as being a dwelling in which part of the space is used commercially.
(4) A new dwelling or dwellings constructed as part of a larger building that contains other types of accommodation. The non-dwelling parts of such a building (common areas, commercial or retail space) must be considered using AD L2A.

Document AD L1A does not apply to:

(1) Rooms for residential purposes (see section 16.2 for definition). This includes nursing homes, student hostels and similar types of accommodation. These are classified as 'buildings other than dwellings', to which AD L2A applies.
(2) The common areas in buildings containing multiple dwellings. Although the individual dwellings fall within AD L1A, the common areas do not. If they are heated, the guidance in AD L2A should be used; if they are unheated it would be reasonable to provide fabric elements that meet the standards given in section 16.8.

16.6.1.2   Conservatories and substantially glazed spaces

A conservatory is attached to a dwelling, but outside the external fabric of the building, and therefore thermally separated from the building. If a conservatory is built as part of a new dwelling, the performance of the dwelling must be assessed as if the conservatory were not there.

If there is no effective thermal separation between a conservatory and a dwelling, then, by definition, the space is not a conservatory, and must be included as an integral part of the dwelling when assessing compliance.

It should also be noted that:

- A conservatory built at ground level with a floor area of 30 m^2 or less is currently exempt from the Building Regulations.
- A new conservatory to either a new or an existing dwelling is subject to the guidance given in AD L1B.
- A substantially glazed space may be considered in the same way as a conservatory.

16.6.1.3   Demonstrating compliance

In order to demonstrate compliance it is necessary to meet all of the following five criteria.

- **Criterion 1:** The predicted rate of CO_2 emissions from the dwelling as built is less than or equal to the target emission rate from the dwelling.

$$DER \leq TER$$

- **Criterion 2:** The performance of the building fabric and the fixed building services should be no worse than the design limits specified in section 16.6.3.
- **Criterion 3:** The dwelling has appropriate passive control measures to limit the effect of solar gains on indoor temperatures in summer.
- **Criterion 4:** The performance of the dwelling as built is consistent with the predicted DER.
- **Criterion 5:** The necessary conditions for energy efficient operation of the dwelling are put in place.

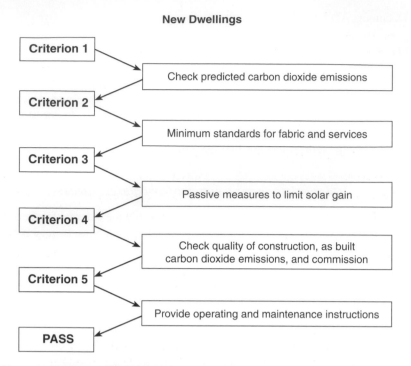

Fig. 16.1   New dwellings – the five criteria.

These criteria, which are illustrated in Fig. 16.1, are dealt with in turn in detail in subsequent sections of the AD L1A. However, it should be noted that they are not wholly independent of each other. In particular, air permeability and pressure testing come under criterion 4, but the result affects the calculations necessary for criterion 1.

16.6.2   Section 1: Design standards – criterion 1

Section 1 of AD L1A discusses criteria 1, 2 and 3. We begin with criterion 1.

16.6.2.1   *The TER (Target Emission Rate) and the DER (Dwelling Emission Rate)*

The essential feature of criterion 1 is the calculation of CO_2 emission rates. The calculation must be carried out using an approved calculation tool, and currently there are two, as shown in Table 16.6. The TER and DER calculations are carried out as follows:

(1) Calculate the TER from a notional dwelling of the same size and shape as the actual dwelling.
(2) Calculate a preliminary value of the DER from the actual dwelling, based on the plans and specifications of the dwelling as it is to be constructed. Although not essential, this should be done to enable both the builder and the BCB to discover whether or not the dwelling is likely to be compliant.

Table 16.6 Approved calculation tools for dwelling.

Official tools		
Individual dwelling	Up to but no greater than 450 m²	Government Standard Assessment Procedure SAP 2005
Individual dwelling	Greater than 450 m²	Simplified Building Energy Model SBEM
Commercial implementations		
Elmhurst Energy Systems	www.elmhurstenergy.co.uk	EES SAP Calculator
National Energy Services	www.nher.co.uk	NHER Plan Assessor
Northgate Land and Property Solutions	www.northgate-is.com	MAXIM 5
ECMK	www.ecmk.co.uk	
Survey Design Associates	www.superheat.com	SuperHeat
BuildDesk	www.builddesk.co.uk	BuildDesk
JPA Technical Literature	www.techlit.co.uk	JPA Designer

(3) Calculate a final value of the DER from the actual dwelling as constructed, allowing for any changes in performance specifications made during construction, and using measured values of air permeability, ductwork leakage and fan performance.

(4) Provide the BCB with either

(a) a notice specifying the TER and the final as constructed DER; or

(b) a certificate stating that the dwelling will satisfy criterion 1 if it was constructed according to an accompanying list of specifications. This certificate must be issued by a person who is registered for CO_2 emission rate calculations by FAERO Ltd [2] or by BRE Certification Ltd [3].

In order to calculate the TER, the construction of the notional dwelling must be assumed to be according to the reference values given in Appendix R of SAP 2005. No variation of these values is allowed except for, in certain circumstances, the air permeability (see section 1.6.5.6 below). Table 16.7 gives these reference values. The notional dwelling is evaluated using the appropriate calculation tool, and from the results of the calculation the following two parameters must be extracted:

- C_H – The CO_2 emissions arising from heating and hot water (including pumps and fans);
- C_L – The CO_2 emissions arising from internal fixed lighting.

The TER is then calculated using the formula:

$$TER = (C_H \times \text{fuel factor} + C_L) \times (1 - \text{improvement factor})$$

Table 16.7   Reference values for TER calculation.

Element or system	Value
Size and shape	Same as actual dwelling
Opening areas – windows and doors	25% of total floor area, or, if the total exposed façade area is less than 25% of the total floor area, the total exposed façade area. This includes one opaque door of area 1.85 m²; any other doors are fully glazed.
Walls	$U = 0.35 \, \text{Wm}^{-2}\text{K}^{-1}$
Floors	$U = 0.25 \, \text{Wm}^{-2}\text{K}^{-1}$
Roofs	$U = 0.16 \, \text{Wm}^{-2}\text{K}^{-1}$
Opaque door	$U = 2.00 \, \text{Wm}^{-2}\text{K}^{-1}$
Windows and glazed door	$U = 2.00 \, \text{Wm}^{-2}\text{K}^{-1}$, double glazed, low-E hard coated. Frame factor 0.7 Solar energy transmittance 0.72 Light transmittance 0.80
Living area fraction	Same as actual dwelling
Shading and orientation	All glazing orientated east/west; average overshading
Number of sheltered sides	2
Allowance for thermal bridging	$0.11 \times$ total exposed surface area, WK^{-1}
Ventilation system	Natural ventilation with intermittent extract fans
Air permeability	10 m³/m².h at 50 Pa
Chimneys	None
Open flues	None
Extract fans	Three for dwellings with floor area greater than 80 m² Two for smaller dwellings
Primary heating fuel for space and water	Mains gas
Heating system	Boiler and radiators; water pump in heated space
Boiler	SEDBUK 78%, room sealed, fanned flue
Heating system controls	Programmer + room thermostat + TRVs, boiler interlock
Hot-water system	Stored hot water, heated by boiler, separate time control for space and water heating
Hot-water cylinder	150 litre cylinder insulated with 35 mm of factory applied foam
Primary water-heating losses	Primary pipework not insulated, cylinder temperature controlled by thermostat
Secondary space heating	10% electric
Low-energy light fittings	30% of fixed outlets

Reproduced with permission from SAP 2005

In this formula, the improvement factor is a constant, currently taken to be 0.2 (i.e. 20%). However, the fuel factor must be selected from Table 16.8, and there are a number of rules governing the selection. These are:

Table 16.8 The fuel factor, C_H for the most common fuels.

Heating fuel	Fuel factor
Mains gas	1.00
LPG	1.10
Oil	1.17
Grid electricity (for direct acting, storage and electric heat pump systems)	1.47
Solid mineral fuel	1.28
Renewable energy, including bio-fuels such as wood pellets	1.00
Solid multi-fuel	1.00

- When all the heating appliances in the dwelling burn the same fuel, then the fuel factor for that fuel must be chosen.
- If an individual appliance burns only one fuel, then the fuel factor for that specific fuel is the appropriate one.
- If a heating appliance is classed as multi-fuel and the dwelling is *not* in a Smoke Control Area, use the multi-fuel factor.
- If a heating appliance is classed as multi-fuel and the dwelling *is* in a Smoke Control Area, then if the appliance *is* approved for use in a Smoke Control Area, use the multi-fuel factor.
- If a heating appliance is classed as multi-fuel and the dwelling *is* in a Smoke Control Area, then if the appliance is *not* approved for use in a Smoke Control Area, use the solid mineral fuel factor.
- If a dwelling has more than one heating system served by different fuels, then choose the fuel factor for either:
  - mains gas if any of the heating appliances are fired by mains gas, or
  - the fuel used in the main heating system.
- Where a dwelling is served by a community heating scheme, choose the fuel factor for the principal fuel used by that scheme.

For buildings containing multiple dwellings, a floor-area-weighted average TER can be calculated as follows:

$$TER_{av} = \frac{TER_1 \times A_1 + TER_2 \times A_2 + TER_3 \times A_3 + \ldots}{A_1 + A_2 + A_3 + \ldots}$$

where TER_1 and A_1 are respectively the TER and floor area of dwelling 1, and so on.

The calculation of the DER must be carried out with the same calculation tool that was used for the TER, but using values for the actual building. However, there are two exceptions, secondary heating and lighting.

Regarding secondary heating, when the DER is calculated, it is necessary to assume that a proportion of the space heat demand will be met by a secondary heating appliance. The proportion that must be used depends on the particular combination of primary and secondary systems, and the principal types are shown in Table 16.9. For other types,

Table 16.9 Secondary heating.

Main heating system	Fraction from secondary
Gas, oil or solid fuel	0.10
Heat pump	0.10
Electric storage heaters, non-integrated and without fan assisted	0.15
Electric storage heaters, non-integrated and with fan assisted	0.10
Electric integrated storage or direct acting	0.10
Electric room heaters	0.20
Other electric systems (except CPSU)	0.10
Community heating	0

Adapted with permission from SAP 2005.

consult SAP 2005. The details of the secondary heating appliance to be used in the calculation are:

- where a secondary heating appliance is fitted, the efficiency of the actual appliance with its appropriate fuel;
- where a chimney or flue is provided but no appliance is installed, the presence of one of the following appliances must be assumed:
  - if a gas point is located adjacent to the hearth, a decorative fuel effect fire open to the chimney or flue with an efficiency of 20%, or
  - if there is no gas point, then an open fire with an efficiency of 37% burning multi-fuel, unless the dwelling is in a Smoke Control Area in which case the fuel must be assumed to be smokeless solid mineral fuel.

If neither of the above apply, it must be assumed that the secondary heating appliance is an electric room heater.

Regarding lighting, for the purposes of the DER calculation, in all cases it must be assumed that 30% of the lighting provision is low-energy lighting, even if more is provided. This means that the result of the final calculation of DER cannot be improved by increasing the proportion of low energy lighting.

16.6.2.2 The TER and DER for buildings containing multiple dwellings

In the case of a building containing multiple dwellings, such as a terrace or a block of flats, criterion 1 can be tested by comparing the TER and DER of each dwelling separately in turn, or the average TER for all dwellings in the building can be compared with the average DER. In each case, the floor-area-weighted-average must be used. The formula for the average TER is given above and a similar formula must used for the average DER:

$$DER_{av} = \frac{DER_1 \times A_1 + DER_2 \times A_2 + DER_3 \times A_3 + \ldots}{A_1 + A_2 + A_3 + \ldots}$$

16.6.2.3   *Strategies for achieving criterion 1*

There can be no absolute certainty that a particular dwelling will satisfy criterion 1, i.e. that the DER is less than the TER, until the final calculation of DER using measured values of air permeability, ductwork leakage and fan performance has been completed. However, there are a number of design features and strategies which can improve the chances of success. Some of these could include:

- adopting roof, wall or floor constructions with U-values significantly lower than the reference values;
- installing windows and doors with lower U-values than the reference values, and/or restricting their total area to below 25% of the total floor area;
- achieving, by attention to construction detailing, an air permeability significantly less than 10 $m^3/h.m^2$;
- adopting low- or zero-energy supply systems, as discussed in 'Low or Zero Carbon Energy Sources – Strategic Guide' [18].

The effectiveness of each of these or any other strategies depends on the overall design of the building, making it difficult to formulate rules for general guidance.

16.6.3   Section 1: Design standards – criterion 2

Criterion 1 does not lay down specific minimum standards for the thermal performance of a dwelling and its installed services (e.g. by specifying maximum permitted U-values). This is intended to allow a substantial measure of design flexibility which, in practice, allows some trade-off between some aspects of the design. However, in addition to criterion 1, and in order to prevent this flexibility being used unwisely, criterion 2 imposes limits on design flexibility. These limits apply to:

- insulation standards (U-values) of the external envelope;
- air permeability of the external envelope;
- fixed heating and hot-water systems (including insulation of pipework, etc.);
- mechanical ventilation and cooling;
- fixed internal and external lighting.

The details are as follows.

16.6.3.1   *Design limits for the external envelope – U-values*

Table 16.10 gives limiting (i.e. maximum) U-values which should not be exceeded. In this table, U_{av} is the area weighted average U-value of the particular element:

$$U_{av} = \frac{U_1A_1 + U_2A_2 + U_3A_3 + \ldots}{A_1 + A_2 + A_3 + \ldots}$$

Table 16.10   Limiting U-values – dwellings

Element	Maximum U_{av} (Wm^{-2}K^{-1})	Maximum U_{part} (Wm^{-2}K^{-1})
Wall	0.35	0.70
Floor	0.25	0.70
Roof	0.25	0.35
Windows, roof windows, rooflights and doors	2.20	3.30

where U_1, A_1, etc. refer to all the individual elements of that type, and U_{part} refers to that part of a particular element which has an unusually high U-value. It should be noted that:

- although these limiting maxima are described in the AD as reasonable, in practice it would be advisable to treat them as absolute;
- if one part of an element, for example part of the external wall, has a high U-value (say, close to the maximum of 0.70 Wm^{2}K^{-1}), then the U-value of the other parts of the external wall would have to have a sufficiently low U-value to bring the average down to the permitted maximum of 0.35 W m^{2}K^{-1};
- the figures given in Table 16.10 for the U-values of roof windows and rooflights assume that they are in the vertical position; if they are not mounted vertically, an adjustment to the U-value must be applied (see section 16.10) before making a comparison with the limiting maximum value in the table.

The U-values themselves must be measured or calculated according to an approved method. Section 16.10 gives details.

16.6.3.2   Design limits for the external envelope – air permeability

The limiting (i.e. maximum) value for the air permeability is 10m^3/h.m^2, which is the same as the reference value that must be used in the calculation of TER. Although this is not a mandatory maximum, in practice it would be advisable to treat it as such, particularly as it may be necessary to achieve a lower measured value in order to ensure that the final DER is less than the TER. Values of air permeability down to 5m^3/h.m^2 or less should normally be achievable if the guidance given in 'Limiting thermal bridging and air leakage' [19] is followed.

16.6.3.3   Design limits for fixed building services

These include:

- heating and hot-water systems;
- insulation of pipes and ducts;
- mechanical ventilation;
- mechanical cooling;
- fixed internal lighting;
- fixed external lighting.

16.6.3.4   Heating and hot-water systems and insulation of pipes and ducts

Heating and hot-water systems must meet the minimum standards set out in the *Domestic Heating Compliance Guide* [20]. This sets minimum standards for the efficiency of heating and hot-water appliances, their controls, and the required insulation standards of pipes and ducts. It covers most energy sources, including gas, LPG, oil, electricity, solid fuel and solar energy. It also covers most of the systems that are likely to be met, including conventional wet systems, air systems, community heating, under-floor heating, heat pump systems, solar systems, etc. It may also be useful to consult the HVAC Guide to Part L [21]. In most conventional domestic central-heating systems fired by oil or natural gas, the minimum boiler seasonal efficiency or SEBUK (see www.sedbuk.com) is 86%. (For oil systems, this is relaxed to 85% until the end of March 2007.)

16.6.3.5   Mechanical ventilation and cooling systems

Mechanical ventilation systems, whether ventilating the whole dwelling or a part of it, must:

- be not worse than the systems described in *Energy efficient ventilation in dwellings – a guide for specifiers,* GPG 268 [22]; and
- have specific fan powers and heat recovery efficiency not worse than those given in Table 16.11.

Note also that mechanical ventilation systems must satisfy the requirements of Part F.

In the case of mechanical cooling, if a fixed household air conditioner is installed, it should have an energy efficiency classification not worse than class C in Schedule 3 of the labelling scheme adopted under the Energy Information (Household Air Conditioners) (No. 2) Regulations 2005 [23].

16.6.3.6   Design limits for fixed internal and external lighting

Fixed internal lighting must include a reasonable proportion of low-energy light fittings. In this context, a low-energy light fitting includes all the components of the fitting (including control gear, shade and/or diffuser, etc.) plus the lamp itself. The lamp must have a luminous efficacy greater than 40 lumens per circuit-watt, where circuit-watt refers to the power consumed by the lamp and all its associated components, including control gear

Table 16.11   Performance limits for mechanical ventilation systems.

System type	Minimum performance
Specific fan power for continuous supply only	0.8 litres/s per watt
Specific fan power for continuous extract only	0.8 litres/s per watt
Specific fan power for balanced systems	2.0 litres/s per watt
Heat recovery efficiency	66%

and power factor correction device. In general, only discharge lamps, such as fluorescent or compact fluorescent types can meet this standard, whereas all filament lamps such as GLS tungsten and tungsten halogen lamps do not. A reasonable proportion would be to provide:

- at least one low-energy light fitting per 25 m^2 of floor area (excluding garages); or
- at least one low-energy light fitting per four fixed light fittings.

Areas of a dwelling that are frequented only occasionally, such as cupboards and storage areas, do not count. Note also that mains-frequency fluorescent lights can give rise to stroboscopic effects with rotating or oscillating machinery, and are therefore not recommended for areas where this might occur, such as garages. Further information on low-energy lighting is given in GIL 20 [24].

Fixed external lighting is defined as lighting fixed to the external surface of a dwelling and supplied from the occupier's electrical system. It does not include the lighting of common areas and access ways in buildings containing multiple dwellings. Reasonable provision would be one of the following:

- either to ensure that the lamp capacity does not exceed 150 W per light fitting and that the external lighting switches off automatically when there is enough daylight or, if there is insufficient daylight, when it is not required;
- or to ensure that the external light fittings have sockets that can only be used with lamps whose efficacy is greater than 40 lumens per circuit-watt.

16.6.4   Section 1: Design standards – criterion 3

The current editions of AD L1A and SAP 2005 do not directly take account of energy used in dwellings for mechanical cooling. Nevertheless, criterion 3 requires that provision should be made to limit internal temperature rise due to solar gains in summer. This is intended to eliminate, or at least minimise, the need for mechanical cooling. Reasonable provision for limiting solar gains can be demonstrated by following the procedure in Appendix P of SAP 2005 and hence demonstrating that the dwelling will not have a high risk of internal temperatures.

Suggested strategies for limiting solar gains include:

- choosing window size and orientation to minimise solar gains;
- protecting windows with solar control measures such as shading, solar control glazing, etc.;
- providing the capability for adequate ventilation at night as well as during the day;
- choosing a structure (e.g. predominantly brick, concrete or similar masonry products) with high thermal capacity.

However, these strategies can conflict with other requirements, for example:

- a desire to take advantage of solar gains in winter;

- the need to provide adequate daylight, and hence to avoid excessive use of electric lighting;
- if the dwelling has a high thermal capacity structure, the need to increase the capacity of the heating system to provide adequate performance during pre-heating from cold.

General design guidance is given by *Reduced overheating – a designers guide* [25].

16.6.5   Section 2: Quality of construction and commissioning – criterion 4

This section of the Approved Document deals with criterion 4 and sets minimum standards for two aspects of the completed dwelling, that is the quality of construction of the building fabric (i.e. the external envelope) as built, and the commissioning of the heating and hot-water systems as installed. The first of these, the building fabric, is to ensure that:

- thermal insulation is reasonably continuous over the whole building envelope;
- the air permeability of the external envelope is within reasonable limits;
- that it has been adequately demonstrated that both of these have been achieved.

Recommended construction details for achieving both continuity of insulation and a sufficiently low air permeability are given in reference [19]. Dwellings which use these details are treated differently from those that do not.

16.6.5.1   *Continuity of Insulation*

At the design stage, if details given in reference [19] or any other approved details have been used, then the design will be considered reasonable. Otherwise it is necessary to demonstrate that the design details will be at least as good as the approved details. This requires an assessment of thermal bridging using the techniques given in reference [26], and described in brief in section 16.10.

Whatever design details have been used, it is also necessary to demonstrate that site inspection of construction procedures is sufficient to ensure that:

- the design details have been correctly executed;
- reasonable continuity of insulation has been achieved; and
- reasonable avoidance of thermal bridging has been achieved.

This can be done by submitting a suitable report, and reference [27] provides checklists suitable for dwellings using approved details.

16.6.5.2   *Required number of pressure tests to determine air permeability*

Pressure testing is a mandatory requirement, but need only be carried out on a sample of the dwellings in a particular development. In all cases it is the right of the BCB to select the sample to be tested. The rules for selection depend on whether or not approved construction details have been used.

Table 16.12   Frequency of pressure tests – dwellings not using accredited details.

Number of occurrences of a particular dwelling type on a development or in a block of flats	Number of tests to be carried out
Four or less	One test of the dwelling type
More than four but less than 40	Two tests of the dwelling type
More than 40	**Either**: at least 5% of the dwelling type **Or:** at least 2% of the dwelling type if the first five tested achieved the design air permeability

For dwellings that have been designed and constructed using approved construction details, an air pressure test must be carried out:

- on each development of houses, on one unit of each dwelling type, selected from the first completed batch of each dwelling type;
- for every block of flats (even if there are other identical blocks on the same site) on one unit of each dwelling type;
- if there is more than one dwelling type in a particular development, on one unit of each type.

For dwellings that have *not* been designed and constructed using approved construction details, the required number of air pressure tests on a particular development or a block of flats is given in Table 16.12.

16.6.5.3   *Compliance conditions for an air pressure test*

Compliance is demonstrated if both:

- the measured air permeability is less than or equal to the limit value of $10 \, \text{m}^3/\text{h.m}^2$; and
- when the measured air permeability is used to calculate the DER, the DER is less than or equal to the TER, i.e.:

$$\text{DER} \leq \text{TER}$$

This means that, even if the measured air permeability is less than the limit value of $10 \, \text{m}^3/\text{h.m}^2$, it will nevertheless count as a fail if, when used in the DER calculation, the resulting DER is greater than the TER. In some cases, the air permeability may need to be significantly less than $10 \, \text{m}^3/\text{h.m}^2$ in order to produce a low enough result for DER. Note that the design value chosen for the air permeability for the preliminary calculation of DER ceases to be relevant once a measured value from the completed dwelling becomes available.

16.6.5.4 Failure of an air pressure test and its consequences

When a dwelling fails an air pressure test, remedial work and a repeat test become necessary, but the remedial strategy will depend on the nature of the failure. An air pressure test may fail for one of the following three reasons:

(1) The measured air permeability exceeds the limit of 10 m³/h.m² but the resulting DER is satisfactory.
(2) The measured air permeability is satisfactory but the resulting DER exceeds the TER.
(3) The measured air permeability exceeds the limit of 10 m³/h.m² and also the resulting DER exceeds the TER.

For case (1), it is essential to improve the air tightness of the external envelope in order to get within the limiting permeability. This should be possible, because the target figure of 10 m³/h.m² is not excessive and should be easily attainable with proper attention to design details and their execution on site. A permeability above the target is likely to indicate a major omission in sealing, draught proofing or membrane fixing, though it may not be easy to find the error. However, if remedial work and a repeat test show that the air permeability has been sufficiently reduced to satisfy the limit value, then the recalculated DER must also have been reduced and must therefore still be satisfactory.

Case (2) is most likely to arise if the design air permeability was chosen to be less than 10 m³/h.m² in order to ensure the predicted DER was below the TER. In this case remedial action to reduce the air permeability is possible, but may be more difficult. Other features of the dwelling should also be considered for treatment. For example, improvements to the U-values of some components by adding insulation will reduce the DER and may bring it within the TER without making changes to reduce the air permeability.

Case (3) is similar to case (1), and improvements to the air tightness of the external envelope are essential. If, after re-test, it is found that the air permeability satisfies the limit value, but the DER is still too high, case (3) then becomes an example of case (2) and must be treated accordingly.

If, after failing an initial air permeability test, a dwelling is found to be satisfactory after a re-test, then it is necessary to pressure test one additional dwelling of the same dwelling type.

16.6.5.5 Reduced criteria for re-testing following the failure of an air pressure test

In the period up to 31 October 2007, but not thereafter, reasonable provision after an unsatisfactory test result would be to carry out remedial work, re-test, and then treat the results of the re-test as follows:

- show that either:
  - the re-test result shows an improvement of 75% of the difference between the initial test result and the design air permeability; or

○ if less demanding, the re-test result is within 15% of the required design perme-
    ability; and
● in the calculation of TER, replace the reference value of the air permeability from Table
  16.7 with the measured re-test value, and show that the recalculated DER is less than
  or equal to the recalculated TER.

16.6.5.6   *Alternative to air pressure testing on small developments*

On a development where only one or at most two dwellings are to be erected, air pressure
testing can be avoided by using the following procedure:

● demonstrate that during the preceding 12-month period a dwelling of the same dwelling
  type constructed by the same builder was pressure tested and achieved an air permeability that did not exceed the limiting value of 10 $m^3/h.m^2$; and
● use a value of 15 $m^3/h.m^2$ for the air permeability when calculating the DER.

However, using 15 $m^3/h.m^2$ will result in an unusually high value of DER and hence an
increased difficulty of satisfying criterion 1, i.e. that the DER is less than or equal to the
TER. If this approach is used, it would be advisable to include compensatory measures
elsewhere in the design.

16.6.5.7   *Commissioning of heating and hot-water systems*

Heating and hot water systems, including their controls, must be left such that:

● they have been correctly commissioned;
● they are in full and proper working order;
● they can operate efficiently for the purposes of the conservation of fuel and power.

The correct commissioning procedure is set out in the Domestic Heating Compliance
Guide [20], and notification must be given to the local authority within the time limits
set by regulation 20C.

16.6.6   Section 3: Operating and maintenance instructions – criterion 5

Criterion 5 requires that the owner of a dwelling must be provided with sufficient information about the building, its fixed building services and their maintenance in order that the
building can be operated in an energy-efficient manner. The operating and maintenance
instructions should:

● relate directly to the particular dwelling and to the particular services installed within
  it;
● explain how to achieve economy in the use of fuel and power;
● be written in layman's language so that they are comprehensible to the householder;
● specifically, explain how to make adjustments to the time and temperature settings;

- specify the routine maintenance that is necessary to maintain optimum operating efficiency.

Criterion 5 also requires that an energy rating for the dwelling is prepared, using SAP 2005, and fixed to the dwelling in a conspicuous place.

16.6.7 Section 4: model designs

Some builders and building designers may wish to have access to standard dwelling designs that have been shown on previous occasions to be compliant with AD L1A. However, L1A is framed to be as flexible as possible in order to encourage new and imaginative approaches to the task of designing energy-efficient dwellings. Excessive reliance on a few model designs would not only be contrary to this objective but may also result in the stereotyping of dwelling designs. Builders, especially those who develop large sites, may, over a period of time, develop a number of standard designs that they know will be likely to pass. The small builder, doing one-off designs, will have to treat each property separately, and so one can only make very general recommendations, such as, for example:

- ensure that the external envelope U-values are low, preferably low enough to bring the area weighted average U-values (NOT the limit values) below the maxima given in Table 16.10;
- avoid excessive glazed areas or, if a large glazed area is being considered, adopt a compensatory measure in the overall design;
- choose heat-raising equipment with a boiler efficiency above the recommended minimum.

In particular cases, it may be useful to do a sensitivity analysis of the SAP calculation for a dwelling to discover the effect of changes in the input data on the calculated DER. This should be possible if the calculation is being carried out with a good software package which allows easy editing of the input data.

16.7 Part L1B – Conservation of fuel and power in existing dwellings

16.7.1 Section 0: General guidance

This section includes:

- types of work covered by AD L1B;
- technical risk – see section 16.5.1;
- historic buildings;
- calculation of U-values – see section 16.10.

16.7.1.1   Types of work covered by AD L1B

The guidance covers:

- extensions to dwellings;
- the creation of a new dwelling, or part of a dwelling, through material change of use of an existing structure;
- material alterations to existing dwellings;
- the provision of a controlled fitting;
- the provision or extension of a controlled service;
- the provision or renovation of a thermal element.

If the building work includes or affects areas that are not classified as a dwelling, for example because they are part of a mixed-use building, or because they are common areas in a multiple-dwelling building, then it will also be necessary to consult AD L2B.

The energy efficiency requirements of regulation 4A apply to work in existing dwellings, and so the BCB must, in most circumstances, be notified in the usual way, the exceptions being:

- Where the work is being carried out under the terms of an approved Competent Persons Scheme.
- Where the work involves an emergency repair (e.g. a failed boiler or leaking hot water cylinder), the repair work can be commenced without the need for advanced notification to the BCB. Nevertheless, the work must comply with the requirements. Either the BCB must be given notice at the earliest opportunity, or if the installer is registered under an approved Competent Persons Scheme, a completion certificate must be issued.
- Where the work is minor, as described in Schedule 2B (see Table 16.3). The work must still comply, but need not be notified to a BCB.

16.7.1.2   Historic buildings

These normally include:

- listed buildings;
- buildings situated in conservation areas;
- buildings of architectural and historical interest and which are referred to as a material consideration in a local authority development plan;
- buildings of architectural and historical interest within national parks, areas of outstanding natural beauty and world heritage sites.

Special considerations may apply, as any work on an historic building must balance the need to improve energy efficiency requirements against the following factors:

- the need to avoid prejudicing the character of the historic building;
- the danger of increasing the risk of long-term deterioration of the building fabric;

- the danger of increasing the risk of long-term deterioration of the building's fittings;
- the extent to which energy conservation measures are a practical possibility.

Advice on achieving the correct balance should be sought from the conservation officer of the local authority. The guidance given in the English Heritage publication [28] should be taken into account when determining appropriate energy performance standards. Advice from other published sources, e.g. PPG15 [29], BS 7913 [30] and SPAB Information sheet 4 [31], may also be appropriate. Particular consideration should be given to:

- the restoration of the historic character of a building that has been the subject of inappropriate alteration, such as the replacement of windows, doors or rooflights;
- the rebuilding of a former historic building, which may have been damaged or destroyed due to some mishap (such as a fire), or in-filling a gap in a terrace;
- the provision of a means for the fabric of a historic building to 'breathe' so that moisture movement may be controlled and the potential for long-term decay problems reduced.

16.7.2 Section 1: Guidance relating to building work, and section 2: Guidance on thermal elements

These two sections deal with:

- extensions to an existing dwelling;
- material change of use;
- material alteration;
- work on controlled fittings and services;
- guidance on thermal elements, i.e. U-values.

There is considerable cross-referencing between sections 1 and 2 in AD L1B, and so they are considered here as one. In all cases covered by AD L1B, one of the main criteria for compliance with the requirements of heat loss through the building envelope is the U-value. The requirements and guidance on U-values is therefore considered first.

16.7.2.1 The U-values of thermal elements and windows, etc.

The principal standards for maximum U-values for work on existing dwellings are collected together in Table 16.13. Additional criteria, intended to allow a measure of design flexibility, apply to newly constructed thermal elements, as shown in Table 16.14. The average U-value is the area-weighted value found from:

$$U_{av} = \frac{U_1 A_1 + U_2 A_2 + U_3 A_3 + \ldots}{A_1 + A_2 + A_3 + \ldots}$$

where U_1, A_1, etc. are the individual U-values and areas.

See section 16.10 regarding retained and renovated thermal elements and continuity of insulation.

Table 16.13   U-value standards.

Thermal element or controlled fitting	New thermal element or fitting in an extension	Replacement thermal element or fitting in an existing dwelling	Upgrade or renovation of a retained thermal element or fitting	
	Maximum U-value (Wm⁻²K⁻¹)	Maximum U-value (Wm⁻²K⁻¹)	Threshold U-value (Wm⁻²K⁻¹)	Improved U-value (Wm⁻²K⁻¹)
Column no.	1	2	3	4
Cavity wall	0.30	0.35	0.70	0.55
All other walls	0.30	0.35	0.70	0.35
Pitched roof – insulation at ceiling level	0.16	0.16	0.35	0.16
Pitched roof – insulation at rafter level	0.20	0.20	0.35	0.20
Flat roof or roof with integral insulation	0.20	0.25	0.35	0.25
Floors	0.22	0.25	0.70	0.25
Window, roof window or rooflight	1.80 **or** centre pane 1.2 **or** WER band D	2.00 **or** centre pane 1.2 **or** WER band E	n/a	n/a
Door, 50% or more glazed	2.20 **or** centre pane 1.2	2.20 **or** centre pane 1.2	n/a	n/a
Other doors	3.0	3.0	n/a	n/a

Notes:

1   In this table, cavity wall means a wall suitable for the installation of cavity insulation.
2   The roof parts of dormer windows are treated as a roof, and the wall parts (cheeks) are treated as a wall.
3   WER means Window Energy Rating, as described in reference [32].
4   In certain circumstances, for replacement walls and floors, a higher maximum U-value may be appropriate – see text.

Table 16.14   Limiting U-values.

Element	Maximum average U-value (Wm⁻²K⁻¹)	Maximum limiting U-value (Wm⁻²K⁻¹)
Wall	0.35	0.70
Floor	0.25	0.70
Roof	0.25	0.35
Windows, roof windows, rooflights and doors	2.20	3.30

16.7.2.2 A new extension to an existing dwelling – normal approach

When constructing a new extension to an existing dwelling, it is necessary to observe:

- minimum standards for the thermal insulation (i.e. maximum U-values) of the external fabric;
- a maximum area for the windows, doors and rooflights;
- minimum requirements for energy-efficient heating and lighting.

The external fabric includes walls, roofs, floors, windows and doors. For all newly constructed elements, the appropriate maximum U-value should be chosen from column 2 of Table 16.13, and the maximum area of windows and doors and rooflights is calculated from:

$$\text{Maximum area} = 0.25A_F + A_{old}$$

where A_F is the floor area of the extension and A_{old} is the area of any windows or doors that no longer exist or are no longer exposed as a result of the new extension. The BCB may allow some flexibility in the calculation of the maximum area if there is a problem in achieving sufficient daylight in the extension.

If a part of the existing dwelling is to become part of the external fabric of the extension, then columns 4 and 5 of Table 16.13 apply. If the existing thermal element has a U-value greater than the maximum (or threshold) value in column 4, then it would be sufficient to upgrade it to have a U-value equal to or less than the 'improved' value given in column 5. However, there may be cases where technical, functional or economic problems, make it difficult to reach the 'improved' value. These problems could include, for example:

- The necessary thickness of extra insulation reduces the usable floor area by more than 5%.
- The extra insulation creates difficulties with adjoining floor levels.
- The weight of additional insulation is too great for the existing structure.

If these problems cannot be resolved, then the element should be upgraded to the best standard possible, such that it achieves a simple payback of 15 years or less (see section 16.2 for a definition of simple payback).

Some design flexibility in the choice of U-values and opening areas is possible, provided that:

- the area-weighted U-value of all the elements in the extension is equal to or less than the area-weighted U-value for an extension of the same shape and size that complies with the standard limits on U-values and opening areas;
- the area-weighted U-value for each element type in the extension does not exceed the maximum average U-value in Table 16.14;
- the U-value of any individual element does not exceed the maximum limiting U-value in Table 16.14.

16.7.2.3   A new extension to an existing dwelling – SAP approach for maximum flexibility

If the normal approach to demonstrating compliance is found to be too restrictive, an alternative is to use the SAP 2005 calculation. This approach requires the following steps:

(1) Using SAP 2005, calculate the CO_2 emission rate from the dwelling and its proposed extension.
(2) Using SAP 2005, calculate the CO_2 emission rate from the dwelling with a notional extension, of the same size and shape as the proposed extension, built to the normal standards for U-values and opening areas.
(3) Show that the CO_2 emission rate for the dwelling plus proposed extension is equal to or less than that for the dwelling plus notional extension.
(4) If, in order to achieve (3) above, it is necessary to improve the thermal performance of parts of the existing dwelling which will be retained, then they should, after improvement, have a U-value that is equal to or less than the 'improved' value in column 4 of Table 16.13.

If the thermal performance of elements of the existing dwelling is not known, the data in section 16.10 or in Appendix S of SAP 2005 can be used as an estimate.

16.7.2.4   Conservatories and substantially glazed extensions

Conservatories and substantially glazed extensions must be thermally separated from the heated area of a dwelling. To be regarded as providing thermal separation, the walls, doors and windows between the conservatory or extension and the heated space must be insulated and draught-stripped to at least the same standard as the rest of the existing dwelling. If the construction is not thermally separated from the dwelling, then it must be considered as a conventional extension, and sections 16.7.2.1 and 16.7.2.2 apply. If a conservatory is built at ground level and has a floor area of 30 m² or less, it is exempt from the Building Regulations (apart from Part N). Otherwise, the requirements for conservatories are:

- The U-values of glazed elements must not exceed the relevant values for replacement glazed elements in column 2 of Table 16.13.
- The U-values of all other elements must not exceed the values for replacement thermal elements in column 2 of Table 16.13.
- Any heating system must have independent temperature and ON/OFF controls.
- Any heating appliance must satisfy the requirements described in section 16.7.2.7.

If the extension is thermally separated from the dwelling, but is not sufficiently glazed to qualify as a conservatory, it may be described as a substantially glazed space. The requirements are the same as for a conservatory, except that some flexibility may be achieved by demonstrating that the area-weighted U-value of all the external elements of the glazed space is the same or less than that of a notional conservatory of the same size and shape that complies.

16.7.2.5   A dwelling formed by material change of use

According to regulation 5, a dwelling is said to be formed by material change of use when:

- the building is used as a dwelling where previously it was not, or
- the building contains a flat where previously it did not, or
- the building, which contains at least one dwelling, contains a greater or lesser number of dwellings than it did previously.

In carrying out the material change, attention must be paid to:

- the provision of windows, roof windows, rooflights and doors (i.e. controlled fittings);
- the possible need to replace an existing window, roof window, rooflight or door;
- the provision of new or extended fixed building services;
- new, replaced, renovated or retained walls, roofs and floors (i.e. thermal elements).

New windows, roof windows, rooflights and doors within a newly constructed part of the building should have a U-value that does not exceed the maximum given in column 1 of Table 16.13. For a replacement or renovated window, roof window or rooflight in the existing structure, the maximum is given in column 2 of Table 16.13. However, any existing window, roof window or rooflight that separates a conditioned space from an unconditioned space or the external environment may be retained unaltered if its U-value is $3.3\,\mathrm{Wm^{-2}K^{-1}}$ or less. If it is more than $3.3\,\mathrm{Wm^{-2}K^{-1}}$, it must be replaced, and column 2 of Table 16.13 applies. In all cases, units should be suitably draught-proofed. See also section 16.10 regarding retained and renovated thermal elements and continuity of insulation.

The requirements for new or extended fixed building services are as described in sections 16.7.2.7, 16.7.2.8 and 16.7.2.9 below.

16.7.2.6   A dwelling subject to material alteration

Material alterations are defined in regulations 3(2) and 3(3). The requirements are essentially the same as for material change of use as described in section 16.7.2.5 above. There is an additional requirement to keep the glazed area within reasonable limits. The calculation of maximum glazed area for a new extension may be used as a means of demonstrating compliance. See also section 16.10 regarding retained and renovated thermal elements and continuity of insulation.

16.7.2.7   Heating and hot-water systems

If the heating and/or hot-water system of the existing dwelling is to be extended or altered or, if as a result of the building work, a new appliance or system is to be installed, then it is necessary to install an appliance and controls that meet minimum requirements, and to ensure that pipes and ducts are suitably insulated. The *Domestic Heating Compliance*

Guide [20] provides all the relevant criteria. The appliance itself must have an efficiency that satisfies both of the following:

- It must have an efficiency not less than that recommended for its type in the *Domestic Heating Compliance Guide*.
- If it is the primary heating service, it must have an efficiency not more than 2% below that of the appliance being replaced.

If the new appliance uses a different fuel, then before making the comparison with the old appliance, the efficiency of the new appliance must be adjusted according to the CO_2 emission factors of the new and old fuels, using the CO_2 emission factors from Table 16.18. The formula is:

$$\eta_{adjusted} = \eta_{new} \times \left(\frac{E_{old}}{E_{new}}\right)$$

For example, an old natural-gas-fired boiler with a SEDBUK of 65% is replaced by an oil-fired boiler with a SEDBUK of $\eta_{new} = 87\%$. The efficiency of the new boiler exceeds the minimum for a new dwelling, but it must also be compared with the old boiler. The CO_2 emission factors are:

- $E_{old} = 0.194$ kgCO_2 per kWh for gas and $E_{new} = 0.265$ kgCO_2 per kWh for oil.

The adjusted efficiency is therefore:

$$\eta_{adjusted} = 87 \times \left(\frac{0.194}{0.265}\right) = 64\%$$

This is less than the efficiency of the old boiler, but within the permissible margin of 2%. Therefore the new oil-fired boiler meets both requirements and is satisfactory.

Once completed, the systems and controls must be commissioned and left in working order and operating efficiently. The *Domestic Heating Compliance Guide* must be consulted for the correct procedure, and notice given to the BCB that commissioning has been successfully carried out. This could be done by a member of an approved Competent Persons Scheme.

16.7.2.8   *Mechanical ventilation and cooling systems*

The requirements are the same as for a new dwelling as described in section 16.6.3.5. If a mechanical ventilation system is provided, it must:

- be not worse than the systems described in General Practice Guide 268 [22]; and
- have specific fan powers and heat recovery efficiency not worse than those given in Table 16.11.

Note also that mechanical ventilation systems must satisfy the requirements of Part F.

In the case of mechanical cooling, if a fixed household air conditioner is installed, it should have an energy efficiency classification not worse than class C in Schedule 3 of the labelling scheme adopted under the Energy Information (Household Air Conditioners) (No. 2) Regulations 2005 [23].

16.7.2.9   Design limits for fixed internal and external lighting

The requirements for lighting are the same as for a new dwelling, as described in section 16.6.3.6. However, with respect to internal lighting, the calculation of the minimum number of energy efficient light fittings is based not on the floor area of the whole dwelling, but on the floor area affected by the building work or, if it is greater, the area affected by the newly installed lighting system.

16.7.3   Section 3: Providing information

On completion of the work, the owner of the dwelling must be provided with appropriate operating and maintenance information. The details are the same as for a new dwelling (see section 16.6.8 above). However, the need to prepare an energy rating for the dwelling (using SAP 2005) and to fix it to the dwelling in a conspicuous place is only a specific requirement if the dwelling is formed by material change of use.

16.7.4   Appendix A: Work on thermal elements

When carrying out any of the work to which AD L1B applies, the renovation or upgrading of thermal elements is a possibility or even a necessity. Table A1 in Appendix A of AD L1B gives numerous practical examples of cost-effective methods for renovating typical roofs, walls and floors in order to achieve a specified target U-value. See section 16.10 for details on thermal bridging, continuity of insulation, etc.

16.8   Part L2A – Conservation of fuel and power in new buildings other than dwellings

16.8.1   Section 0: General guidance

This section includes:

- types of work covered by AD L2A;
- technical risk – see section 16.5.1;
- demonstrating compliance;
- modular buildings.

16.8.1.1   Types of work covered by AD L2A

The guidance covers:

- the construction of new buildings other than dwellings;
- fit-out works where this is included as part of the construction of a new building;
- the construction of extensions to existing buildings that are not dwellings where the total useful floor area of the extension is greater than 100 m² and is also greater than 25% of total useful floor area of the existing building;
- the construction of buildings that contain rooms for residential purposes (e.g. hostels, student accommodation blocks, boarding houses, etc.) but do not contain a dwelling (see section 16.2 for definitions for rooms for residential purposes and for dwellings);
- where a building contains one or more dwellings, the construction of those parts of the building which are used for other purposes, including common areas between and providing access to dwellings (but see exception below).

AD L2A does not apply to a new building that contains both living accommodation and space used for industrial or commercial purposes, subject to all the following conditions:

- The commercial part could revert to domestic use on change of ownership.
- There is direct access between the living accommodation and the commercial space.
- Both are contained within the same thermal envelope.
- The living accommodation occupies a substantial portion of the total area of the building. (This rules out the case of a small manager's or caretaker's flat in a large non-domestic building.)

The implication of these conditions is that the building must be conceived of primarily as being a dwelling in which part of the space is used commercially, and so AD L1A applies.

Buildings which are exempt from Part L are listed in section 16.4.

16.8.1.2   Demonstrating compliance

In order to demonstrate compliance it is necessary to meet all of the following five criteria.

- **Criterion 1:** The predicted CO_2 emission rate from the building as constructed is less than or equal to the target emission rate from an equivalent notional building, i.e.:

$$BER \leq TER$$

- **Criterion 2:** The performance of the building fabric and the heating, hot water and fixed lighting systems meets or exceeds certain minimum standards.
- **Criterion 3:** Those parts of the building that are not provided with comfort cooling systems have appropriate passive control measures to limit solar gains.
- **Criterion 4:** The performance of the building as built is consistent with the predicted BER.

New non-dwellings

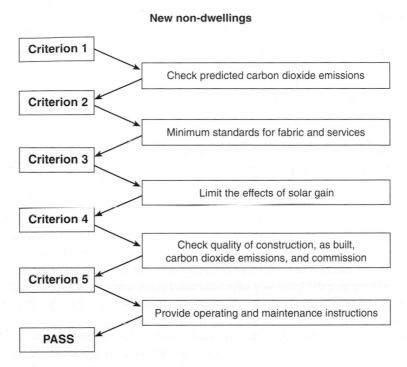

Fig. 16.2   New non-dwellings – the five criteria.

- **Criterion 5:** Information is provided to the building owner/operator to ensure energy-efficient operation of the building.

These criteria, which are illustrated in Fig. 16.2, are dealt with in turn in detail in subsequent sections of the AD L2A. However, it should be noted that they are not wholly independent of each other. In particular, air permeability and pressure testing come under criterion 4, but the results affect the calculations necessary for criterion 1.

16.8.1.3   *Modular buildings*

Temporary buildings that are intended to be used for two years or less are exempt from Part L. If the intended life is more than two years, they are not exempt, but special considerations may apply. For example, if more than 70% of the external envelope is created from sub-assemblies manufactured before the current Part L came into effect (6 April 2006), then it would be reasonable to refer for guidance to *Energy performance standards for modular and portable buildings* [33].

16.8.2   Section 1: Design standards – criterion 1

Section 1 of AD L2A discusses criteria 1, 2 and 3. We begin with criterion 1.

Table 16.15 Approved calculation tools for buildings other than dwellings.

Approved calculation tool	Source
Simplified Building Energy Model, SBEM and iSBEM	The National Calculation Method, www.ncm.bre.co.uk
IES Virtual Environment, version 5.5	IES Ltd (UK) www.iesve.com
EDSL TAS, version 9.0.9	Environmental Design Solutions Limited www.edsl.net

16.8.2.1   Criterion 1: The TER (Target Emission Rate) and the BER (Building Emission Rate)

The essential feature of criterion 1 is the calculation of CO_2 emission rates. The calculation must be carried out using an approved calculation tool. The calculation tools are normally available in the form of computer software, and new software may be approved from time to time. Currently approved calculation tools are shown in Table 16.15. The TER and BER calculations are carried out as follows:

(1) Calculate, using one of the approved calculation tools, the CO_2 emission rate, $C_{notional}$ from a notional building of the same size and shape as the actual building, and hence calculate the TER.
(2) Calculate, using the same calculation tool, a preliminary value of the BER from the actual building, based on the plans and specifications of the building as it is to be constructed. Although not essential, this should be done to enable both the builder and the BCB to discover whether or not the building is likely to be compliant.
(3) Calculate a final value of the BER from the actual building as constructed, allowing for any changes in performance specifications made during construction, and using measured values of air permeability, ductwork leakage and fan performance.
(4) Provide the BCB with either:
   (a) a notice specifying the TER and the final as constructed BER; or
   (b) a certificate stating that the building will satisfy criterion 1 if it is constructed according to an accompanying list of specifications. This certificate must be issued by a person who is registered for CO_2 emission rate calculations by FAERO Ltd [2] or by BRE Certification Ltd [3].

16.8.2.2   Calculating the TER

In order to calculate the TER, the construction of the notional building must be assumed to be according to a reference specification, the outline of which is given in Table 16.16. Further details of the specification are to be found within the user document for SBEM in the national calculation method. The notional building is evaluated using one of the approved calculation tools in order to obtain an initial value of the target emission rate, called $C_{notional}$. This is then adjusted to give TER using the formula:

$$TER = C_{notional} \times (1 - \text{improvement factor}) \times (1 - LZC_{benchmark})$$

Table 16.16 Reference specification for the notional building.

Element or system	Value
Size and shape	Same as actual building
Building fabric, and fixed building services	Comply with the energy performance values set out in the detailed definition of the notional building as stated in the SBEM.
Vehicle access doors and display windows	Same as the actual building
Any service that is not a fixed building service	To be excluded (e.g. vertical transport systems)
Building usage and building services	The same activity areas and classes of building services as in the actual building, selected from the predefined standard activity areas specified in the SBEM
Occupancy times, temperatures, ventilation rates, etc.	Selected in each activity area according to the standard data associated with the reference schedules
Lighting	The actual and notional buildings must be assumed to have the same lighting levels as that given in the reference schedules, even if the actual building uses a different level
External conditions	The assumed external climate must be that defined by the CIBSE test reference year for the location most appropriate to the actual building
Heating fuel	Assume mains gas if that is what is to be used in the actual building. For all other cases assume oil
Fuel for other all other building services	Assume grid mains electricity
CO_2 emission factors	Use the values given in Table 16.18
Fit-out works	Where fit-out works are excluded from the proposal, and the space is to be offered with a range of service options, assume the most energy-intensive of the specifications throughout. In addition, spaces that potentially may be fitted without air conditioning must comply with criterion 3, limiting solar gains, as if they were not air conditioned.

In this formula, the improvement factor depends on the class of building services strategy for controlling the indoor climate of the actual building. Values must be selected from Table 16.17. If different areas are served by different classes, then the improvement factor must be applied to each area individually. This will result in different values of the TER for each area, and it may, with the agreement of the BCB be acceptable to take the average, weighted according to floor area, as the TER of the whole building.

Table 16.17 Improvement factor and LZC benchmark.

Building services strategy for the actual building	Improvement factor	LZC benchmark
Heated and naturally ventilated	0.15	0.10
Heated and mechanically ventilated	0.20	0.10
Air conditioned	0.20	0.10

Note: Mechanical ventilation means a system that runs continuously during occupied hours.

The LZC benchmark is a factor to allow for low and zero carbon energy (LZC) sources. The single value of 0.10 in Table 16.17 represents an assumed or benchmark provision in the notional building. If the actual building has a greater or lesser contribution from LZC, this will be accounted for in the calculation of BER.

16.8.2.3   Calculating the BER

The BER must be calculated, using the same approved calculation tool that was used for the TER, but using data for the actual building as constructed, allowing for any changes made during construction to performance specifications, and using measured values of the air permeability, ductwork leakage and fan performances. The BER calculation requires the carbon emission factors of the fuels actually used, as shown in Table 16.18. In selecting a value from Table 16.18, the following should be taken into account:

- Grid displaced electricity comprises all electricity generated in or on the building premises by, for example, PV panels, wind-powered generators, CHP, etc. The associated CO_2 emissions are deducted from the total CO_2 emissions for the building before calculating the BER. However, the CO_2 emissions arising from fuels used by the building's power generation system (e.g. to power a CHP engine) must be included in the calculation of the building CO_2 emissions.
- Waste heat includes waste heat from industrial processes and from power stations rated at more than 10 MWe and with a power efficiency greater than 35%.
- For bio-mass fired systems rated at greater than 100 kW output but where there is an alternative appliance to provide standby, the CO_2 emission factor should be based on the fuel that is normally expected to provide the lead.

Table 16.18   The CO_2 emission factors for various fuels.

Fuel	CO_2 emission factor ($kgCO_2/kWh$)
Natural gas	0.194
LPG	0.234
Biogas	0.025
Oil	0.265
Coal	0.291
Anthracite	0.317
Smokeless fuel (including coke)	0.392
Dual fuel appliances (mineral + wood)	0.187
Biomass	0.025
Grid supplied electricity	0.422
Grid displaced electricity	0.568
Waste heat	0.018

- For systems rated at less than 100 kW output, where the same appliance is capable of burning both bio-fuel and fossil fuel, the CO_2 emission factor for dual-fuel appliances should be used, except that the figure for anthracite should be used when the building is in a Smoke Control Area.
- In all other cases where more than one fuel can be used, the highest of the emission factors of those fuels should be used.
- If thermal energy is supplied from a district or community heating or cooling system, the emission factor should be based on the annual average performance of the whole system, and it will be necessary to supply a report (authorised by a suitable qualified person) explaining how that emission factor has been calculated.

The result obtained for the BER may be reduced if certain management or control features are incorporated in the actual building. These and their adjustment factors are listed in Table 16.19. Each reduction and its appropriate adjustment factor only applies to:

- the CO_2 emissions from the system to which the feature is applied;
- in the case of the power factor, only if the whole-building power factor is corrected;
- one or other of the power factor adjustment factors (i.e. the most appropriate, not both) can be chosen.

For example, consider a building whose total BER is calculated to be 130 $kgCO_2$ per m^2 per year, of which 70 $kgCO_2$ per m^2 per year is due to electricity consumption. The provision of equipment to give a whole-building power factor correction of 0.95 allows an adjustment factor of 0.025, and the adjusted BER is:

$$BER = 130 - 0.025 \times 70 = 130 - 1.75 = 128.25 \text{ } kgCO_2 \text{ per } m^2 \text{ per year}$$

Reductions in the BER may also be achieved by using low- or zero-carbon energy sources. Some of the energy sources that may qualify in this category could include:

- solar hot water;
- photovoltaic power;
- bio-fuels such as wood fuels and oil blends;
- combined heat and power;
- heat pumps.

Information on the effect LZC sources on the BER calculation may be found in *Low or Zero Carbon Energy Sources – Strategic Guide* [18].

Table 16.19   Adjustment factors for management and control features.

Management or control feature	Adjustment factor
Automatic monitoring and targeting with alarms for out-of-range values	0.050
Power factor correction to achieve a whole-building power factor of at least 0.90	0.010
Power factor correction to achieve a whole-building power factor of at least 0.95	0.025

The final value of BER, incorporating data for the actual building as built, any applicable adjustment factors and the effects of any LZC sources, is compared with the TER, the requirement being:

$$BER \leq TER$$

16.8.2.4   *Strategies for achieving criterion 1*

Should the test for criterion 1 result in a fail, remedial work will be necessary. This would be a most unwelcome consequence. Nevertheless, it is unlikely that the final test for criterion 1 will fail, provided that:

- a preliminary calculation of BER was carried out at the design stage and found to be less than TER, preferably by a reasonable margin;
- the design was executed according to the original design and specifications with all due care to construction detail;
- the consequences of any changes to the design or to the specifications of any components were fully explored by recalculating the preliminary BER.

The design features and strategies which can help to bring the BER below the TER could include:

- adopting a highly insulating external envelope, i.e. low U-values for the roof, wall, floor, windows, etc.;
- adopting a design that minimises summer overheating and the need for comfort cooling;
- achieving, by attention to construction detailing, an air permeability significantly less than $10m^3/h.m^2$;
- adopting low- or zero-energy supply systems, as discussed in *Low or zero carbon energy sources – strategic guide* [18].

However, the effectiveness of each of these or any other strategies depends on the overall design of the building, making it difficult to formulate rules for general guidance.

16.8.3   Section 1: Design standards – criterion 2

Criterion 1 does not lay down specific minimum standards for the thermal performance of a building and its installed services (e.g. by specifying maximum permitted U-values). This is intended to allow a substantial measure of design flexibility which, in practice, allows some trade-off between some aspects of the design. However, in addition to criterion 1, and in order to prevent this flexibility being used unwisely, criterion 2 imposes limits on design flexibility. These limits apply to:

- insulation standards (U-values) of the external envelope;
- air permeability of the external envelope;
- controls and energy meters;

- heating and hot water service systems;
- cooling and air handling plant;
- insulation of pipes and ducts;
- lighting, its efficacy and controls.

The details are as follows.

16.8.3.1   *Design limits for the external envelope – U-values*

Table 16.20 gives limiting (i.e. maximum) U-values which should not be exceeded. In this table, U_{av} is the area-weighted average U-value of the particular element:

$$U_{av} = \frac{U_1 A_1 + U_2 A_2 + U_3 A_3 + \ldots}{A_1 + A_2 + A_3 + \ldots}$$

where U_1, A_1, etc. refer to all the individual elements of that type, and U_{part} refers to that part of a particular element which has an unusually high U-value. It should be noted that:

- Although these limiting maxima are described in the AD as reasonable, in practice it would be advisable to treat them as absolute.
- If one part of an element, for example part of the external wall, has a high U-value (say, close to the maximum of 0.70 Wm²K⁻¹), then the U-value of the other parts of the external wall would have to have a sufficiently low U-value to bring the average down to the permitted maximum of 0.35Wm²K⁻¹.
- The figures given in Table 16.20 for the U-values of roof windows and rooflights assume that they are in the vertical position; if they are not mounted vertically, an adjustment to the U-value must be applied (see section 16.10) before making a comparison with the limiting maximum value in the table.
- Display windows are *not* subject to any of the above limiting values, and therefore may have any U-value. However, they are included in the CO_2 emission calculations, and so a poor thermal performance (e.g. a high U-value and/or a large area) could lead to an unacceptably high value for the BER.

Table 16.20   Limiting U-values – buildings other than dwellings.

Element	Maximum U_{av} (Wm⁻²K⁻¹)	Maximum U_{part} (Wm⁻²K⁻¹)
Wall	0.35	0.70
Floor	0.25	0.70
Roof	0.25	0.35
Windows, roof windows, rooflights and curtain walling	2.2	3.3
Pedestrian doors	2.2	3.0
Vehicle access and similar large doors	1.5	4.0
High-usage entrance doors	6.0	6.0
Roof ventilators (including smoke vents)	6.0	6.0

The U-values themselves must be measured or calculated according to an approved method. Section 16.10 gives details.

16.8.3.2   *Design limits for the external envelope – air permeability*

The limiting (i.e. maximum) value for the air permeability is 10 m^3/h.m^2, which is the same as the reference value that must be used in the calculation of TER. Although this is not a mandatory maximum, in practice it would be advisable to treat it as such, particularly as it may be necessary to achieve a lower measured value in order to ensure that the final BER is less than the TER. Values of air permeability down to 5 m^3/h.m^2 or less should normally be achievable if the guidance given in *Limiting thermal bridging and air leakage* [19] is followed. A high degree of air tightness, i.e. a low value for the air permeability, is advantageous for buildings with mechanical ventilation and air conditioning.

16.8.3.3   *Design limits for building services – controls and energy meters*

In principle, all systems which consume energy should:

- be controlled by an appropriate control system capable of achieving a reasonable standard of energy efficiency;
- be capable of having their energy consumption monitored.

The control features considered appropriate for heating, ventilation and air conditioning systems should include all the following:

- separate control zones corresponding to each area of the building that has a significantly different solar exposure, or pattern of use, or type of use;
- in each separate control zone, independent controls for timing, temperature and (where appropriate) ventilation and air recirculation rate;
- the ability to respond to the requirements of the space (i.e. the control zone) being served – in particular, if both heating and cooling are provided, the control system should ensure that they do not operate simultaneously;
- the ability to switch off the central plant when it is not needed, and to switch it on only when one or more of the control zones require it, i.e. the default condition should be off.

More detailed requirements and controls for specific systems are discussed below.

Energy meters must be capable of monitoring energy consumption for each fuel. The consumption for each fuel must be further broken down according to its end-use category, so that the amount used for heating, lighting, etc. can be identified. Any LZC system is considered as a separate fuel and must be monitored separately. The metering system is considered reasonable if it can measure at least 90% of the estimated annual energy consumption of each fuel. Detailed guidance is given in CIBSE TM 39 [34].

16.8.3.4 Design limits for building services – heating and hot-water service systems

The essential, and also principal, source of information on heating and hot-water service systems is the *Non-domestic Heating Cooling and Ventilation Compliance Guide* [35]. This includes checklists to assist in demonstrating that reasonable provision has been made in providing:

- suitably efficient heating plant, and
- effective control systems.

16.8.3.5 Design limits for building services – cooling and air handling plant

The guidance and checklists in the *Non-domestic Heating Cooling and Ventilation Compliance Guide* [35] are essential as an aid in determining reasonable provision for:

- suitably efficient cooling plant;
- suitably efficient air handling plant;
- effective control systems.

It is important to note that the carbon emissions due to cooling systems are often very high compared to other energy consuming systems, and can therefore have a disproportionate effect on the calculated BER. For this reason, it is prudent to attempt to adopt design strategies to minimise cooling loads. These could include:

- reducing solar gains by a combination of building style, orientation, window design and shading;
- reducing internal heat gains;
- avoiding strong peaks in the cooling demand by, for example, the incorporation of high thermal mass in the building structure;
- night cooling;
- careful selection of systems and controls to match demand over the whole of the cooling season.

Air handling systems (except for systems used in abnormal circumstances such as smoke control fans) must also meet minimum standards with respect to fan power. The Approved Document requires:

$$SFP_{25} \leq SFP_{100}$$

where SFP_{25} is the specific fan power at 25% of design flow rate and SFP_{100} is the specific fan power at 100% of design flow rate.

If a ventilation fan is rated at more than 1100 watts, reasonable provision would be to equip it with a variable-speed drive.

It is also important to limit air leakage from ventilation ductwork, and so ductwork must be reasonably airtight. Adhering to the specifications given in HVCA DW/144 [36] would be one way of demonstrating compliance.

16.8.3.6   Design limits for building services – insulation of pipes and ducts

Standards for the insulation of pipes and ducts are given in the *Non-domestic Heating Cooling and Ventilation Compliance Guide* [35]. Meeting or exceeding these standards would be considered to be reasonable provision. The TIMSA Guide [37] also provides information on insulation standards and how they may be achieved.

16.8.3.7   Design limits for building services – lighting, its efficacy and controls

AD L2A does not impose any requirement or limitation on:

- the use of daylight;
- the level of artificial light.

It does, however, require that electric lighting systems should be reasonably efficient and have appropriate controls. The efficiency and the control of lighting systems are considered separately, and the guidance varies according to building type.

General lighting efficacy in office, industrial and storage areas in all building types
This category of building type includes all those spaces that have predominantly desk-based tasks, and includes classrooms, seminar rooms, conference rooms, etc. The electric lighting system should be provided with reasonably efficient lamp/luminaire combinations. This requirement can be met by an initial average luminaire efficacy, averaged over the whole area of spaces in this category, of not less than 45 luminaire-lumens per circuit watt. When interpreting this criterion, it should be noted that:

- The figure of 45 refers to the number of lumens emitted by the lamp and luminaire combination, and not just the lamp.
- The circuit-watts includes all the power consumed in the lighting circuits, including the lamps, their associated control gear and power-factor correction equipment.
- The effect of the above is to allow flexibility in the choice of lamp, luminaire and control gear. For example, a low-efficiency luminaire (chosen perhaps for reasons of appearance or low glare) can be compensated by a lamp of high efficacy and/or more efficient control gear.

The initial average luminaire efficacy, η_{lum}, in luminaire-lumens per circuit watt may be calculated by means of the equation:

$$\eta_{lum} = \frac{\Sigma(\text{Lamp lumens} \times \text{LOR})}{(\text{Total circuit watts of all luminaires in the space}) \times C_L}$$

where:

- lamp lumens is the sum of the average initial (100 hour) lumen output of all the lamp or lamps in a luminaire;
- LOR is the ratio of the total light output of a luminaire under stated practical conditions to the total light output of the bare lamp or lamps used in that luminaire under reference conditions;
- the product (lamp lumens × LOR) is summed for all luminaires in the space;
- C_L is the control factor – for all new buildings, $C_L = 1$.

General lighting efficacy in all other types of space
In other types of space, it may be appropriate to use luminaires for which photometric data is not available (i.e. the value of the LOR is not known) and/or luminaires that take low power and/or less efficient lamps. For these cases, the criterion is based on the light output of the lamps themselves, without the need to allow for the performance of the luminaire. The requirement is that the installed lighting has an initial (100 hour) lamp plus ballast efficacy of not less than 50 lamp-lumens per circuit-watt.

Display lighting in all types of space
Display lighting is defined as:

- lighting intended to highlight displays of exhibits or merchandise;
- lighting used in spaces for public entertainment (e.g. dance halls auditoria, conference halls and cinemas).

The special requirements of display lighting may make it necessary to accept lower standards of energy performance than for general lighting. Nevertheless, it is still necessary for display lighting to be energy efficient, and in order to demonstrate compliance it is necessary to ensure that the installed capacity of the display lighting has an initial (100 hour) efficacy of not less than 15 lamp-lumens per circuit-watt. The circuit-watts should include the power consumed by transformers and ballasts.

Spaces in which display lighting is installed will normally also have a general lighting system which may be used in conjunction with the display lighting or when the display lighting is switched off. This general lighting must conform to the appropriate requirements for general lighting for the type of space.

Emergency escape lighting and specialist process lighting
Lighting for these purposes is not subject to the requirements of Part L.

Lighting controls in all types of space
The aim of lighting controls should be to encourage the maximum use of daylight and to avoid unnecessary use of lighting when spaces are not occupied. This should not, however, create a situation in which the operation of an automatically switched lighting system endangers the passage of building occupants. Guidance on lighting controls is given in BRE Digest 498 [38]. The requirement can be met by providing local manual switches in

easily accessible positions within each working area or at boundaries between working areas and circulation routes. In this context, a local manual switch is defined as follows:

- **Local Manual Switch:** A switch whose distance from the luminaire it controls is not more than 6 m on plan, or twice the height of the luminaire above the floor if this is greater, and:
  - ○ is operated by deliberate action of the occupants (rocker switch, push button, pull cord, etc.), or
  - ○ is operated by remote control (infra-red or wireless transmitter, sonic or ultrasonic device, telephone handset control, etc.).

Dimmer switches are included, and dimming is taken as synonymous with switching. However, dimming should normally be achieved by reducing rather than diverting the energy supply. If a space is a daylit space (see definition in section 16.2) illuminated by side windows, then, whichever control system is used, it should be possible to switch the row of perimeter luminaires separately.

Occupant control of local switching can be supplemented by automatic systems that:

- switch the lighting off when there are no occupants in the space; or
- dim or switch off the artificial lighting when there is sufficient daylight.

Although automatic systems are not a requirement, their inclusion has an ameliorating effect on carbon emissions, and in some circumstances could contribute to a significant reduction in the BER.

Display lighting should be operated by dedicated circuits, and have controls that switch them off when they are not necessary, i.e. when people are not inspecting exhibits or merchandise, or when there is no-one present at an entertainment event. Typically, the necessary control could be achieved using timers linked to the opening hours and/or usage of the building. However, in the case of, for example, a retail store, display lighting may remain switched on for displays designed to be viewed from outside the building through a display window.

16.8.4   Section 1: Design standards – criterion 3

Criterion 3 is concerned with the limiting of solar gains in summer. The main concern is with spaces that are not served by an air conditioning system, that is, spaces in which refrigeration is not used to provide comfort cooling for the occupants. Spaces which are excluded from criterion 3 are:

- unoccupied atria intended to drive natural ventilation via buoyancy;
- stacks;
- spaces adjacent to display glazing (i.e. the glazing of a display window) that are not served by an air conditioning system.

The requirement is that provision should be made to limit solar gains in order to prevent an excessive rise in internal temperature in summer. There are three methods for demonstrating that reasonable provision has been made in an internal space:

- Show that the total gains, per unit floor area, due to solar energy, people, lighting and equipment, averaged over the period 06.30 to 16.30 solar time, do not exceed 35 Wm^{-2}. The calculation must use the solar irradiances for July as given in the table of design irradiances for the month of July in CIBSE Design Guide A [39]. Necessary supporting data with guidance is given in CIBSE TM 37 [40], as well as a factor to allow the criterion of 35 Wm^{-2} to be adjusted for building location. For a perimeter area with side windows it would be normal calculate the gains over the area within 6 m of the window wall
- Using the CIBSE Design Summer Year appropriate to the building location, show that the operative temperature (as defined in CIBSE Design Guide A) in the occupied space does not exceed a threshold value for more than a reasonable number of occupied hours per year. The Approved Document specifies neither the threshold temperature nor the reasonable number of hours, and leaves it to designers, clients and health and safety inspectors to agree suitable values according to the activities within the space. A threshold of 28°C exceeded for no more than 1% of the annual occupied period may be acceptable for offices and similar spaces
- For school buildings, consult Building Bulletin 101 [41] for suitable overheating criteria and for guidance on methods to demonstrate that reasonable provision has been made.

For spaces served by air conditioning systems, in which refrigeration is employed to cool the occupied space, there is no specific requirement. This is because the TER is calculated for a notional building with modest amounts of glazing, and therefore limited solar gains. If solar gains in the actual building are higher than in the notional building, it is likely to be difficult to prevent the BER exceeding the TER, and thereby infringing criterion 1. It is therefore necessary to control solar gains in order to keep BER below TER. Otherwise, if large solar gains are an inevitable consequence of the design of the actual building, compensating energy efficiency measures must be included elsewhere in the building. Whatever the circumstances, an excessive summer cooling load should always be avoided, as it leads to a larger than necessary installed capacity and to unnecessarily high running costs.

Strategies for limiting solar gains include:

- choosing window size and orientation to minimise solar gains;
- protecting windows with solar control measures such as shading, solar control glazing, etc.;
- providing the capability for adequate ventilation at night as well as during the day;
- choosing a structure (e.g. predominantly brick, concrete or similar masonry products) with high thermal capacity.

However, these strategies can conflict with other requirements, for example:

- a desire to take advantage of solar gains in winter;
- the need to provide adequate daylight, and hence to avoid the excessive use of electric lighting;
- if the building has a high thermal-capacity structure, the possible need to increase the capacity of the heating system to provide adequate performance during pre-heating from cold.

Further information relevant to criterion 3 is given in reference [42].

16.8.5   Section 2: Quality of construction and commissioning – criterion 4

This section of the Approved Document deals with criterion 4 and sets minimum standards for two aspects of the completed building, that is the quality of construction of the building fabric (i.e. the external envelope) as built, and the commissioning of the building services systems as installed. The first of these, the building fabric, is to ensure that:

- thermal insulation is reasonably continuous over the whole building envelope;
- the air permeability of the external envelope is within reasonable limits;
- it has been adequately demonstrated that both of these have been achieved.

Construction details for achieving both continuity of insulation and a low air permeability are given in two principal references:

- *Limiting thermal bridging and air leakage* [19] for buildings using similar design details to dwellings;
- MCRMA Technical note 17 [43] for cladding systems.

16.8.5.1   *Continuity of insulation*

At the design stage, if details given in reference [19] or any other similarly approved details have been used then the design will be considered reasonable. Otherwise it is necessary to demonstrate that the design details will be at least as good as the approved details. This requires an assessment of thermal bridging using the techniques given in BRE IP 1/06 [26]. See also section 16.10.

Whatever design details have been used, it is also necessary to demonstrate that site inspection of construction procedures is sufficient to ensure that:

- the design details have been correctly executed;
- reasonable continuity of insulation has been achieved; and
- reasonable avoidance of thermal bridging has been achieved.

This can be done by submitting a suitable report showing that construction checklists have been completed and show satisfactory results. Reference [19] provides checklists suitable for dwellings and similarly constructed buildings, and these checklists may, with suitable modification, be appropriate to other buildings.

Table 16.21 Possible exemptions from pressure testing.

Exempt building	Description and procedure
Buildings with a total useful floor area of less than 500 m²	Pressure testing may be avoided by using a value for the air permeability of 15 m³/h.m² at 50 Pa in the calculation of the BER. This will require compensating measures elsewhere in the design to ensure that criterion 1 is satisfied
Factory-made modular buildings where no site assembly work is needed	The particular module type must have been subjected to an in-situ test programme and certified by an approved pressure testing organisation as having satisfactory design air permeability and that this is routinely achieved on site. Testing is necessary to demonstrate that the building is strong enough to resist flexure during lifting and transportation
Large extensions, whose compliance is being assessed as if they were new (see AD L2B), where sealing off the extension from the existing building is impractical	Guidance on how extensions can be tested and where pressure tests are inappropriate is given in reference [44]. If the BCB agrees that testing is impractical, the extension should be treated as a large complex building, see below
Large complex buildings whose size and/or complexity make pressure testing of the whole building impractical. Reference [44] indicates the circumstances when this might apply	Before construction work commences, developers must produce, in accordance with the approved procedure, a detailed justification of the impracticality of pressure testing. This must be endorsed by a suitably qualified person, e.g. a Competent Person approved for pressure testing. Compliance may be demonstrated by appointing a suitable qualified person to carry out a detailed programme of design development, component testing and site supervision to ensure that a continuous air barrier will be achieved. In such cases, the air permeability cannot be taken to be less than 5 m³/h.m² at 50 Pa
Compartmentalised buildings made up of self-contained units with no internal connections, making it impractical to perform a whole-building pressure test	Reasonable provision would be to pressure test a representative area of the building as detailed in reference [44]. In the event of a test failure, remedial measures must be carried out on the whole building. It would then be reasonable to test the original area plus an additional representative area to confirm that the expected standard has been achieved throughout.

16.8.5.2 *Pressure testing to determine air permeability*

The approved procedure for carrying out pressure tests, collecting data and recording results is given in the ATTMA guide *Measuring air permeability of buildings* [44]. The procedure must be carried out by a suitably qualified person, such as an ATTMA member.

Pressure testing is a mandatory requirement for all buildings that are not dwellings (including extensions that are being treated as new buildings), with the possible exceptions listed in Table 16.21.

16.8.5.3 *Compliance conditions for an air pressure test*

Compliance is demonstrated if both:

- the measured air permeability is less than or equal to the limit value of 10 m³/h.m²; and

- when the measured air permeability is used to calculate the BER, the BER is less than or equal to the TER, i.e:

$$BER \leq TER$$

This means that even if the measured air permeability is less than the limit value of $10 \, m^3/h.m^2$, it will nevertheless count as a fail if, when used in the BER calculation, the resulting BER is greater than the TER. In some cases, the air permeability may need to be significantly less than $10 \, m^3/h.m^2$ in order to produce a low enough result for BER. Note that the design value chosen for the air permeability for the preliminary calculation of BER ceases to be relevant once a measured value from the completed building becomes available.

16.8.5.4   *Failure of an air pressure test and its consequences*

When a building fails an air pressure test, remedial work and a repeat test become necessary, but the remedial strategy will depend on the nature of the failure. An air pressure test may fail for one of the following three reasons:

(1) The measured air permeability exceeds the limit of $10 \, m^3/h.m^2$ but the resulting BER is satisfactory.
(2) The measured air permeability satisfies the limit value but the resulting BER exceeds the TER.
(3) The measured air permeability exceeds the limit of $10 \, m^3/h.m^2$ and also the resulting BER exceeds the TER.

For case (1), it is essential to improve the air tightness of the external envelope in order to get within the limiting permeability. This should be possible, because the target figure of $10 \, m^3/h.m^2$ is not excessive, and should be easily attainable with proper attention to design details and their execution on site. A permeability above the target is likely to indicate a major omission in sealing, draught proofing or membrane fixing, though it may not be easy to find the error. However, if remedial work and a repeat test show that the air permeability has been sufficiently reduced to satisfy the limit value, then the recalculated BER must also have been reduced and must therefore still be satisfactory.

Case (2) is most likely to arise if the design air permeability was chosen to be less than $10 \, m^3/h.m^2$ in order to ensure the predicted BER was below the TER. In this case remedial action to reduce the air permeability is possible, but may be more difficult. Other features of the building should also be considered for treatment. For example, improvements to the U-values of some components by adding insulation will reduce the BER and may bring it within the TER without making changes to reduce the air permeability.

Case (3) is similar to case (1), and improvements to the air tightness of the external envelope are essential. If, after re-test, it is found that the air permeability satisfies the limit value but the BER is still too high, case (3) then becomes an example of case (2) and must be treated accordingly.

16.8.5.5 *Reduced criteria for re-testing following the failure of an air pressure test*

This applies only to buildings with a total useful floor area of less than 1000 m^2. In the period up to 31 October 2007, but not thereafter, reasonable provision after an unsatisfactory test result would be to carry out remedial work, re-test, and then treat the results of the re-test as follows:

- show that either:
  - ○ the re-test result shows an improvement of 75% of the difference between the initial test result and the design air permeability; or
  - ○ if less demanding, the re-test result is within 15% of the required design permeability; and
- in the calculation of TER, replace the reference value of the air permeability, as given in the detailed definition of the notional building in the SBEM, with the measured re-test value, and show that the recalculated BER is less than or equal to the recalculated TER.

16.8.5.6 *Commissioning of the building services systems*

Following regulation 20C, AD L2A requires that the building services systems be commissioned so that at completion the systems and their controls are left in working order and can operate efficiently for the purposes of conserving fuel and power. The approved procedures are given in CIBSE Code M, Commissioning Management [45]. The notification which must be given to the BCB should include a declaration that:

- commissioning has been followed so that every system has been inspected and commissioned in an appropriate sequence and to a reasonable standard; and
- the results of tests confirm that the measured performance is reasonably in accordance with the design performance for the actual building, with written commentaries where excursions are proposed to be accepted.

Ductwork leakage testing should be carried out according to the procedures set out in HVCA DW/143 [46] for sections of ductwork where:

- the system is served by a fan with a design flow rate greater than 1 m^{3}s^{-1}; and
- the pressure class is such that HVCA DW/143 recommends testing; and
- the BER calculation assumes a leakage rate for a given section of ductwork that is lower than the standard defined in HVCA DW/143 for its particular pressure class. In such cases, any low pressure ductwork should be tested using the provisions for medium pressure ductwork in HVCA DW/143.

If a ductwork system fails to meet the leakage standard, sufficient remedial work must be carried out to achieve the standard in a re-test. Additionally, further sections of ductwork should be tested, as described in HVCA DW/143.

All commissioning and testing should be carried out by suitably qualified persons. The following are likely to be acceptable:

- for heating, ventilating and air conditioning systems, a member of the Commissioning Specialists Association [47] or of the Commissioning Group of HVCA [48];
- for leakage testing of ductwork, a member of the Ductwork Group of HVCA [48], or of the Association of Ductwork Contractors and Allied Services [49];
- for lighting systems, a member of the Lighting Industry Commissioning Scheme [50].

Persons with other relevant training and experience may also be acceptable to BCBs.

16.8.6   Section 3: Operating and maintenance instructions – criterion 5

Criterion 5 derives from Part L1(c), and requires that the owner of a building should be provided with sufficient information about the building, its fixed building services and their maintenance requirements so that the building can be operated so as to use no more fuel and power than is reasonable. The information should be provided in the form of a building log book, and a way of demonstrating compliance would be to follow the guidance in CIBSE TM 31 *Building log book toolkit* [51]. The information should be presented in the form of templates the same as or similar to those in TM 31. If convenient, the information may draw on or refer to information available in other documentation such as:

- operation and maintenance manuals;
- the Health and Safety file required by CDM regulations.

The log book should include the data used to calculate both TER and BER, and preferably an electronic copy of the input file for the energy calculation should be retained to facilitate any future analysis that may be required should the building be altered or improved.

16.9   Part L2B – Conservation of fuel and power in existing buildings other than dwellings

The principal objectives of AD L2B are:

- to ensure that all new work in an existing building meets the relevant current Part L standards for the conservation of fuel and power; and
- to oblige the owner (or developer or contractor) to carry out 'consequential improvements', that is to say to improve the rest of the building and bring it up to current Part L standards insofar as such improvements are technically, functionally and economically feasible.

Such is the variety and complexity of carrying out work on existing buildings that are not dwellings, the first consideration will normally be to determine:

- the extent to which AD L2B is applicable to the proposed works;
- the extent to which consequential improvements will have to be carried out;
- the possibility that some or all of the works may be more appropriately treated using one of the other Approved Documents, e.g. AD L1B or AD L2A;
- the possibility that some or all of the works may be exempt from Part L.

In many cases it may be prudent to obtain the opinion of the BCB at an early stage.

16.9.1   Section 0: General guidance

This section includes:

- types of work covered by AD L2B;
- technical risk – see section 16.5.1;
- historic buildings;
- calculation of U-values – see section 16.10.

16.9.1.1   *Types of work covered by AD L2B*

The guidance covers:

- consequential improvements;
- extensions;
- material change of use;
- material alteration;
- the provision or extension of a controlled service or fitting;
- the replacement or renovation of a thermal element.

Note that rooms for residential purposes are not dwellings, and so AD L2B does apply.

Certain types of work in or on an existing building fall more properly within the guidance of one of the other Approved Documents, usually either AD L1B or AD L2A. Some of the circumstances where this might or might or not occur are:

- Fit-out works: in buildings erected in compliance with the Building Regulations in force prior to 6 April 2006, fit-out works must comply with the current Part L requirements, and guidance given in AD L2B applies.
- Fit-out works: in buildings such as shell and core office buildings or business park units built in compliance with the Building Regulations in force prior to 6 April 2006, although fit-out works would normally need to comply with the current Part L, it may be necessary to adopt a different approach. It may be reasonable to provide fixed building services that have efficiencies no worse than those assumed in the calculations showing that the existing building complied with regulation 17C when originally completed; in such a case it would be reasonable to make no further improvement.
- Large extensions (see section 16.2 for definition): the extension itself is considered to be a new building and AD L2A applies. Nevertheless, the existing building is still subject to consequential improvements as specified in AD L2B.

- Extensions using sub-assemblies: if the extension is formed from sub-assemblies obtained from stock or from the disassembly or relocation of other buildings, the extension is considered to be a new building and AD L2A applies. Nevertheless, the existing building is still subject to consequential improvements as specified in AD L2B.
- Buildings containing dwellings: AD L2B does not apply to work on a building that either before the work is started or after the work is completed contains one or more dwellings. In such cases AD L1B would apply.
- A new free-standing building on an existing site (e.g. a new building on an existing hospital site or university campus): this would be a new building, and AD L1A or AD L2A applies as appropriate.

Regulation 17C, which requires that the buildings CO_2 emission rate (DER) does not exceed the target CO_2 emission rate (TER), does not apply to work in existing buildings. However, all the other energy efficiency requirements do apply, and so the BCB must, in most circumstances, be notified in the usual way either by deposit of full plans or by a building notice. The exceptions are:

- Where the work is being carried out under the terms of an approved Competent Persons Scheme (see regulation 16A and Schedule 2A). No advance notice to the BCB is needed, but at completion the Competent Person provides the building owner with a certificate confirming that the installation has been carried out in accordance with the relevant requirements, and the scheme operator notifies the local authority.
- Where the work involves an emergency repair (e.g. a failed boiler or leaking hot water cylinder), the repair work can be commenced without the need for advanced notification to the BCB. Nevertheless the work must comply with the requirements. Either the BCB must be given notice at the earliest opportunity, or if the installer is registered under an approved Competent Persons Scheme, a completion certificate must be issued.
- Where the work is minor, as described in Schedule 2B (see Table 16.3). The work must still comply, but need not be notified to a BCB.

16.9.1.2   Historic buildings

These normally include:

- listed buildings;
- buildings situated in conservation areas;
- buildings of architectural and historical interest and which are referred to as a material consideration in a local authority development plan;
- buildings of architectural and historical interest within national parks, areas of outstanding natural beauty, and world heritage sites.

Special considerations may apply, as any work on an historic building must balance the need to improve energy efficiency requirements against the following factors:

- the need to avoid prejudicing the character of the historic building;
- the danger of increasing the risk of long-term deterioration of the building fabric;
- the danger of increasing the risk of long-term deterioration of the building's fittings;
- the extent to which energy conservation measures are a practical possibility.

Advice on achieving the correct balance should be sought from the conservation officer of the local authority. The guidance given in the English Heritage publication [28] should be taken into account when determining appropriate energy performance standards. Advice from other published sources, e.g. PPG15 [29], BS 7913 [30] and SPAB Information sheet 4 [31], may also be appropriate. Particular consideration should be given to:

- the restoration of the historic character of a building that has been the subject of inappropriate alteration, such as the replacement of windows, doors or rooflights;
- the rebuilding of a former historic building, which may have been damaged or destroyed due to some mishap (such as a fire), or in-filling a gap in a terrace;
- the provision of a means for the fabric of an historic building to 'breathe' so that moisture movement may be controlled and the potential for long-term decay problems reduced.

16.9.2   Section 1: Consequential improvements

If an existing building has a total useful floor area over 1000 m^2, and if the proposed building work (i.e. the principal works) consists of or includes:

- an extension;
- the initial provision of any fixed building service; or
- an increase to the installed capacity of any fixed building services,

then in addition to the principal works, consequential improvements must be made to the existing building. The objective is to bring the building up to the energy efficiency standards of the current Part L, at least as far as the improvements are technically, functionally and economically feasible.

In most cases, the technical and functional feasibility of an improvement can be established. However, economic feasibility depends primarily on the payback resulting from the improvement, and this requires the selection of a suitable maximum payback period. AD L2B states that improvements with a simple payback (see definition in section 16.2) not exceeding 15 years will, in most ordinary circumstances, be economically feasible. An exception would be if a building had a remaining life of less than 15 years, in which case the payback period would have to be reduced accordingly.

16.9.2.1   *Consequential improvements when a building is extended*

Table 16.22 lists improvements that in most ordinary circumstances would be considered practical and economically feasible. However, it is also necessary to demonstrate that the

Table 16.22   Consequential improvements that are normally practical and feasible when a building is extended.

Item	Description of improvement
1	Upgrading heating systems that are more than 15 years old by the provision of new plant or improved controls
2	Upgrading cooling systems that are more than 15 years old by the provision of new plant or improved controls
3	Upgrading air-handling systems that are more than 15 years old by the provision of new plant or improved controls
4	Upgrading general lighting systems that have an average lamp efficacy of less than 40 lamp-lumens per circuit-watt and that serve areas greater than 100 m² by the provision of new luminaires or improved controls
5	Installing energy metering following the guidance given in CIBSE TM 39 [34]
6	Upgrading thermal elements which have U-values worse than those set out in column 3 of Table 16.25 following the guidance in section 16.9.3.1
7	Replacing existing windows, roof windows or rooflights (but excluding display windows) or doors (but excluding high usage doors) which have U-values worse than $3.3 \, \mathrm{Wm^{-2}K^{-1}}$ following the guidance in section 16.9.3.1
8	Increasing the on-site low and zero carbon (LZC) energy-generating systems if the existing on-site LZC systems provide less than 10% of the total on-site energy demand, subject to this improvement achieving a simple payback of seven years or less. The shorter payback is because this improvement is likely to be more capital intensive or more risky than items 1 to 7. Guidance on LZC systems is given in reference [18]

improvements are sufficient, and this can be done by showing that their total value is equal to or greater than 10% of the value of the principal works, i.e.:

$$\text{Total value of consequential improvements} \geq 0.1 \times \text{value of principal works}$$

The values should be established using prices current at the date the proposals are made known to the BCB, and submitted in a report signed by a suitably qualified person (e.g. a chartered quantity surveyor)

16.9.2.2   *Consequential improvements on installing building services*

This applies to:

- the installation of a fixed building service as a first installation; or
- an installation which increases the installed capacity per unit area of an existing service.

Installed capacity is taken to mean the design output of the distribution system output devices (the terminal units) serving the space in question divided by the total useful floor area of that space.

Reasonable provision would be to carry out the improvements in two stages.

(1) Improve those parts of the building served by the service, where this is economically feasible.

This normally means that the thermal performance of the fabric must be improved. The purpose is to ensure that the capacity and energy consumption of the new or extended installation is not excessive. The value of the principal works is the value of the work due to the new or expanded building service, excluding the value of the improvements. However, these improvements *cannot* contribute to the value of the consequential improvements made under this heading.

(2) Carry out consequential improvements to the rest of the building served by the service, where this is practical and economically feasible.

If these additional consequential improvements were not carried out, it is probable that the new or expanded system capacity would result in the building having a higher rate of CO_2 emissions. Improvements under this heading *do* count towards the value of consequential improvements.

The required improvements depend on the exact nature of the new or expanded service, there being three cases which cover nearly all possibilities.

(1) Increase in the size/capacity of the central plant solely for the purpose of serving a new extension whilst maintaining the same level of provision elsewhere.

The required improvements are as for a new extension, and so section 16.9.2.1 and Table 16.22 apply.

(2) A heating system whose installed capacity per unit area has been increased.

The required improvements are items 6 and 7 in Table 16.22. However, there is no 10% threshold on the value of the improvements.

(3) A cooling system whose installed capacity per unit area has been increased.

The required improvements are item 4 of Table 16.22, but applied to any lighting system within the area and regardless of the area it covers, plus item 6 in Table 16.22, plus, if the design solar load exceeds 25 Wm^{-2}, a requirement to improve solar control measures up to one of the targets specified in Table 16.23.

Table 16.23   Criteria for improved solar control measures if design solar load exceeds 25 Wm^{-2}.

Glazing (excluding display windows) in the space served by the cooling system	**Alternative targets for installed or upgraded solar control measures**
Either if the area of windows and roof windows within the area exceeds 40% of the façade area, **or** the area of rooflights exceeds 20% of the area of the roof	The design solar load must be reduced to 25 Wm^{-2} or less **or** the design solar load is reduced by a least 20% **or** the effective g-value (see CIBSE TM 37 [40]) must be no worse than 0.3

16.9.3   Section 2: Guidance relating to building work, and section 3: Guidance on thermal elements

Section 16.9.2 above specifies the consequential improvements that are necessary when work is carried out on buildings with a floor area greater than 1000 m². Sections 2 and 3 of AD L2B cover the requirements for the work itself, and apply to all non-domestic buildings regardless of size. They include:

- extensions, including conservatories;
- material change of use and change of energy status;
- material alteration;
- work on controlled services;
- work on controlled fittings, i.e. windows, etc.;
- guidance on thermal elements.

There is considerable cross-referencing between sections 2 and 3 in AD L2B, and so they are considered here as one. In all cases covered by AD L2B, one of the main criteria for compliance with the requirements of heat loss through the building envelope is the U-value. The requirements and guidance on U-values are therefore considered first.

16.9.3.1   *The U-values of thermal elements and windows, etc.*

The principal standards for maximum U-values for work on existing buildings are collected together in Table 16.24. Additional criteria, intended to allow a measure of design flexibility, are shown in Table 16.25. In this table, the average U-value, U_{av}, is the area-weighted value found from:

$$U_{av} = \frac{U_1 A_1 + U_2 A_2 + U_3 A_3 + \dots}{A_1 + A_2 + A_3 + \dots}$$

where U_1, A_1, etc. are the individual U-values and areas.

A higher U-value for glazing than that given in column 2 of Table 16.24 may be appropriate in some buildings formed by material change of use, change of energy status or material alteration. This would apply if the finished building has a high internal heat gain, and if higher glazing U-values could contribute to a reduction in the cooling load and hence a reduction in the CO_2 emissions. If this can be satisfactorily demonstrated, then higher values can be used, but nevertheless the area-weighted average U-value for windows doors and rooflights should not exceed 2.7 $Wm^{-2}K^{-1}$.

In the case of curtain walling, the maximum overall U-value is calculated from:

$$U_{max} = 0.9 + 1.3X$$

where X is the fraction of the curtain wall that is glazed.

See also section 16.10 regarding retained and renovated thermal elements and continuity of insulation.

Table 16.24 U-value standards.

Thermal element or controlled fitting	New thermal element or fitting in an extension — Maximum U-value ($\mathrm{Wm^{-2}K^{-1}}$)	Replacement thermal element or fitting in an existing building — Maximum U-value ($\mathrm{Wm^{-2}K^{-1}}$)	Upgrade of a retained thermal element or fitting — Threshold U-value ($\mathrm{Wm^{-2}K^{-1}}$)	Upgrade of a retained thermal element or fitting — Improved U-value ($\mathrm{Wm^{-2}K^{-1}}$)
Column no.	1	2	3	4
Cavity wall	0.30	0.35	0.70	0.55
All other walls	0.30	0.35	0.70	0.35
Pitched roof, insulation at ceiling level	0.16	0.16	0.35	0.16
Pitched roof, insulation at rafter level	0.20	0.20	0.35	0.20
Flat roof or roof with integral insulation	0.20	0.25	0.35	0.25
Floors	0.22	0.25	0.35	0.25
Window, roof window or rooflight	1.80 or centre pane 1.2	2.20 or centre pane 1.2	n/a	n/a
Alternative for windows in buildings essentially domestic in character	WER band D	WER band E	n/a	n/a
Pedestrian door, 50% or more of internal face area glazed	2.20	2.20	n/a	n/a
High-usage entrance doors	6.0	6.0	n/a	n/a
Vehicle access and similar large doors	1.5	1.5	n/a	n/a
Roof ventilators, including smoke extract ventilators	6.0	6.0	n/a	n/a

Notes:
1 Cavity wall means a wall suitable for the installation of cavity insulation.
2 The roof parts of dormer windows are treated as a roof, and the wall parts (cheeks) are treated as a wall.
3 WER means Window Energy Rating, as described in reference [32].
4 In certain circumstances, for replacement walls and floors, a higher maximum U-value may be appropriate – see text.

Table 16.25   Limiting U-values.

Element	Maximum average U-value (Wm⁻²K⁻¹)	Maximum limiting U-value (Wm⁻²K⁻¹)
Wall	0.35	0.70
Floor	0.25	0.70
Roof	0.25	0.35
Windows, roof windows, rooflights and doors	2.20	3.30

16.9.3.2   A new extension to an existing building (except large extensions)

This section does not apply to a large extension (see section 16.2 for definition). A large extension is considered to be and is treated as a new building to which AD L2A applies, with the added requirement that consequential improvements, as described in section 16.9.2.1, must be made to whatever is retained of the original building.

All extensions which are not large extensions must meet acceptable standards for the fabric and for the services. When constructing a new extension to an existing building, it is necessary to observe:

- minimum standards for the thermal insulation (i.e. maximum U-values) of the external fabric;
- a maximum area for the windows, doors and rooflights;
- minimum requirements for energy efficient operation of all services.

This section deals with fabric; services are discussed in section 16.9.3.7. The external fabric includes walls, roofs, floors, windows and doors. For all newly constructed elements, the appropriate maximum U-value should be chosen from column 1 of Table 16.24, and the maximum area of openings is selected as appropriate from Table 16.26.

If a part of the opaque fabric of the existing building is to become part of the thermal envelope whereas previously it was not, then the relevant maximum U-values are given in columns 3 and 4 of Table 16.24. If the existing thermal element has a U-value greater than

Table 16.26   Normal maximum areas of openings in an extension.

Building type	Windows and personnel doors Percentage of exposed wall area	Rooflights Percentage of roof area
Residential buildings where people temporarily or permanently reside	30	20
Places of assembly, offices and shops	40	20
Industrial and storage buildings	15	20
Vehicle access doors, display windows and similar glazing	As required	n/a
Smoke vents	n/a	As required

the maximum (or threshold) value in column 3, then it would be sufficient to upgrade it to have a U-value equal to or less than the 'improved' value given in column 4. However, there may be cases where technical, functional or economic problems, make it difficult to reach the 'improved' value. These problems could include, for example:

- The necessary thickness of extra insulation reduces the usable floor area by more than 5%.
- The extra insulation creates difficulties with adjoining floor levels.
- The weight of additional insulation is too great for the existing structure.

If these problems cannot be resolved, then the element should be upgraded to the best standard possible, such that it achieves a simple payback of 15 years or less.

16.9.3.3   Design flexibility

Some design flexibility in the choice of U-values and opening areas is possible, provided that:

- the area-weighted U-value of all the elements in the extension is equal to or less than the area-weighted U-value for an extension of the same shape and size that complies with the standard limits on U-values and opening areas;
- the area-weighted U-value for each element type in the extension does not exceed the maximum average U-value in Table 16.25;
- the U-value of any part of any individual element does not exceed the maximum limiting U-value in Table 16.25.

An additional consideration in some buildings is that rooflights are the primary source of useful daylight. A significant reduction in their area could lead to an increased use of electric lighting, and hence to an increase in energy consumption that outweighs any potential savings due to reduced heat loss or gain. The NARM publication [52] gives guidance on this point.

Where maximum flexibility of design is required, it is acceptable to use an approved calculation tool to demonstrate that the CO_2 emissions from the building and its proposed extension are reasonable. The procedure for this is:

(1) Ensure that, in the proposed extension, the limiting average and maximum U-values in Table 16.25 have been observed
(2) Set up a notional extension, of the same size and shape as the proposed extension, which complies with the fabric standards of Tables 16.24, 16.25 and 16.26 and with the minimum standards for fixed building services (section 16.9.3.7 below).
(3) Using an approved calculation tool, calculate the CO_2 emissions from the building with its notional extension.
(4) Using the same calculation tool, calculate the CO_2 emissions from the actual building with its proposed extension.
(5) For both calculations, ensure that the building incorporates all the necessary and proposed consequential improvements.

(6) Show that the emissions from the actual building and its proposed extension are less than or equal to the emissions from the building plus notional extension.

Whichever approach to achieving design flexibility is used, if additional upgrades to the actual building are proposed in order to compensate for lower performance in the extension, such upgrades, including those to retained thermal elements, must meet the standards set out in AD L2B, as shown in Table 16.24.

In general, the standards set out in AD L2B are expected to be cost effective, and so it is recommended that they be implemented in full even if this is more than is necessary to achieve compliance.

16.9.3.4   *Conservatories and substantially glazed extensions*

Conservatories and substantially glazed extensions must be thermally separated from the heated area of a building. To be regarded as providing thermal separation, the walls, doors and windows between the conservatory or extension and the heated space must be insulated and draught-stripped to at least the same standard as the rest of the existing building. If the construction is not thermally separated from the building, then it must be considered as a conventional extension, and section 16.9.3.2 applies. If a conservatory is built at ground level and has a floor area of 30 m^2 or less, it is exempt from the Building Regulations (apart from Part N). Otherwise, the requirements for conservatories are:

- The U-values of glazed elements must not exceed the relevant values for replacement glazed elements in column 2 of Table 16.24.
- The U-values of all other elements must not exceed the values for replacement thermal elements in column 2 of Table 16.24.
- Any heating system must have independent temperature and ON/OFF controls.
- Any heating appliance must satisfy the requirements for heating appliances described in section 16.9.3.7.

If the extension is thermally separated from the dwelling, but is not sufficiently glazed to qualify as a conservatory, it may be described as a substantially glazed space. The requirements are the same as for a conservatory, except that some flexibility may be achieved by demonstrating that the area-weighted U-value of all the external elements of the glazed space is the same as or less than that of a notional conservatory of the same size and shape that complies.

16.9.3.5   *Material change of use and change of energy status*

According to regulation 5, a building other than a dwelling is said to be formed by material change of use when:

- the building is used as a hotel or boarding house where previously it was not;
- the building is used as an institution where previously it was not;
- the building is used as a public building where previously it was not;

- the building is not a building described in Classes I to VI in Schedule 2 where previously it was;
- the building contains a room for residential purposes where previously it did not;
- the building, which contains at least one room for residential purposes, contains a greater or lesser number than it did previously;
- the building is used as a shop where previously it was not.

In carrying out the material change, attention must be paid to:

- the provision of windows, roof windows, rooflights and doors (i.e. controlled fittings);
- the possible need to replace an existing window, roof window, rooflight or door;
- new, replaced, renovated or retained walls, roofs and floors (i.e. thermal elements);
- the provision of new or extended fixed building services.

New windows, roof windows, rooflights and doors within a newly constructed part of the building should have a U-value that does not exceed the maximum given in column 1 of Table 16.24. For a replacement or renovated window, roof window, or rooflight in the existing structure, the maximum is given in column 2 of Table 16.24. However, any existing window, roof window, or rooflight that separates a conditioned space from an unconditioned space or the external environment may be retained unaltered if its U-value is $3.3 \text{W m}^{-2}\text{K}^{-1}$ or less. If it is more than $3.3 \text{ Wm}^{-2}\text{K}^{-1}$, it must, except in certain circumstances, be replaced, and column 2 of Table 16.24 applies. The exceptions are display windows and high-usage entrance doors, for which a lower standard (i.e. higher U-value) may be reasonable.

New and replacement thermal elements must meet the maximum U-value standards in column 1 of Table 16.24.

The requirements for services are discussed in section 16.9.3.7.

The requirements for renovated and retained thermal elements are discussed in section 16.9.3.1 above, and should satisfy the maximum U-value standards given in columns 3 and 4 of Table 16.24.

16.9.3.6   *A building subject to material alteration*

Material alterations are defined in regulations 3(2) and 3(3). The requirements for new, replacement, renovated and retained thermal elements and openings are essentially the same as for material change of use as described in sections 16.9.3.5 (material change of use) and 16.9.3.1 (U-values). There is, however, no specific limitation on the area of openings.

16.9.3.7   *Building services*

Where the work involves the provision or extension of fixed building services, the normal expectation is that new services will be provided. The requirements are therefore almost identical, with some minor differences, to those for a new building. The relevant sections are:

- 16.8.3.3 Controls and energy meters;
- 16.8.3.4 Heating and hot water service systems;
- 16.8.3.5 Cooling and air handling plant;
- 16.8.3.6 Insulation of pipes and ducts;
- 16.8.3.7 Lighting, its efficacy and controls;
- 16.8.5.6 Commissioning of the building services systems;
- 16.8.6   Operating and maintenance instructions – providing information.

The differences are as follows.

CENTRAL PLANT – boilers, chillers, main air handling plant
In addition to the minimum requirements for the efficiency of new appliances as set out for new buildings, if any of these appliances are replacing an existing appliance, its efficiency must be not less than the efficiency of the appliance being replaced by more than 2%. If the new appliance uses a different fuel, then before making the comparison, the efficiency of the new appliance must be adjusted according to the CO_2 emission factors of the new and old fuels, using the CO_2 emission factors from Table 16.18. The formula is:

$$\eta_{adjusted} = \eta_{new} \times \left(\frac{E_{old}}{E_{new}}\right)$$

For example, an old natural-gas-fired boiler of efficiency 66% is replaced by an oil-fired boiler of efficiency $\eta_{new} = 86\%$. This would meet the requirement for a new building, but in this case it must also be compared with the old boiler. The CO_2 emission factors are:

- $E_{old} = 0.194$ kgCO_2 per kWh for gas and $E_{new} = 0.265$ kgCO_2 per kWh for oil.

The adjusted efficiency is therefore:

$$\eta_{adjusted} = 86 \times \left(\frac{0.194}{0.265}\right) = 63\%$$

As this is more than 2% below the efficiency of the old boiler, it is unacceptable. The new oil-fired boiler would have to have an efficiency of 88% to raise $\eta_{adjusted}$ above 64% and hence bring it within 2% of the old boiler.

FIXED INTERNAL LIGHTING
There are two differences. The first is that, when the area covered by the new lighting system is less than 100 m², the work should still comply with the relevant standards, but need not be notified to the BCB.

The second difference concerns the Luminaire Control Factor C_L. In newly constructed buildings, $C_L = 1$ in all cases. For work in existing buildings, C_L is chosen from Table 16.27. The appropriate luminaire control factor is then substituted in the equation in section 16.8.3.7 to calculate the efficiency, η_{lum}, of the lighting installation. The calculated value must not be less than 45 luminaire-lumens per circuit-watt.

Table 16.27   Luminaire control factors, C_L.

Control function	C_L
The luminaire is in a daylit space and its light output is controlled by photoelectric switching or dimming control, with or without manual override	0.90
The luminaire is in a space that is likely to be unoccupied for a significant proportion of working hours and where a sensor switches off the lighting in the absence of occupants but switching on is done manually, except where this would be unsafe	0.90
Both of the above circumstances combined	0.85
None of the above	1.00

PROVIDING INFORMATION

The guidance in section 16.8.6 should be followed. If, however, a log book already exists in an acceptable form, it is permissible to add to it. The new or updated log book should provide details of:

- any newly provided, renovated or upgraded thermal elements;
- any newly provided fixed building services, their method of operation and maintenance;
- any newly installed energy meters;
- any other details that collectively enable the energy consumption of the building and building services comprising the works to be monitored and controlled.

16.9.4   Appendix A: Work on thermal elements

When any of the work to which AD L2B applies is carried out, the renovation or upgrading of thermal elements is a possibility or even a necessity. Table A1 in Appendix A of AD L1B, which is intended for dwellings, gives numerous practical examples of cost-effective methods for renovating typical roofs, walls and floors in order to achieve a specified target U-value. These examples may also, in some cases, be suitable for other buildings. See section 16.10 on U-values below.

16.10 U-values

Although U-values themselves are no longer a sufficient means for demonstrating compliance, they remain an essential part of the compliance procedures. It is therefore necessary to know the U-values of the external elements of a building. The usual sources are:

- measurement of an actual building element;
- tables or charts of U-values from an authoritative source;
- calculation from materials' properties.

Of these, the most reliable are those U-values obtained by measurement of the exact building element. However, such measurements are available for a relatively small number of constructions. If measured data is not available, the next most convenient source is pre-calculated data, usually in the form of a table or chart, provided the particular case for which a U-value is required is within the range of the data. If neither measured nor pre-calculated data is available, the U-value must be calculated from the properties of the constituent materials. If the U-value of a particular element is not already known, then the best approach to obtaining a value depends on the type of element, i.e. whether it is wall, roof window or floor.

16.10.1 Measured U-values

For a measured value to be acceptable, it must have been obtained in accordance with an approved method. Table 16.28 lists the most suitable methods. Where measured U-values are provided in trade (or any other) literature, they should be certified as having been obtained according to the appropriate standard.

16.10.2 Calculation methods for U-values

Elementary calculation methods for U-values are no longer acceptable. This is because the U-values which are necessary to achieve compliance are sufficiently low for thermal-bridging effects and geometrical details to be highly significant. Calculation methods must therefore be sufficiently sophisticated to take these effects into account. This is particularly true of windows, roof windows, rooflights and doors, where the U-value refers to the complete unit and therefore includes the frame. Consequently, in order to be acceptable, a calculated value must also have been obtained by an approved procedure. Table 16.29 lists the principal sources of information on approved calculation methods.

Table 16.28   Measurement methods for building elements.

Element	Title	Reference
Building materials and products	Thermal performance of building materials and products – Determination of thermal resistance by means of guarded hot plate and heat flow meter methods – Dry and moist products of medium and low thermal resistance	BS EN 12664
Building materials and products	Thermal performance of building materials and products – Determination of thermal resistance by means of guarded hot plate and heat flow meter methods – Products of high and medium thermal resistance	BS EN 12667
Building materials and products	Thermal performance of building materials and products – Determination of thermal resistance by means of guarded hot plate and heat flow meter methods – Thick products of high and medium thermal resistance	BS EN 12939
Windows and doors	Thermal performance of windows and doors – Determination of thermal transmittance by hot box method – Part 1: Complete windows and doors	BS EN ISO 12567–1
Windows and doors	Thermal performance of windows and doors – Determination of thermal transmittance by hot box method – Part 2: Roof windows and other projecting windows	BS EN ISO 12567–2

Table 16.29   Calculation methods for building components.

Element	Title	Reference
All building elements	Conventions for U-value calculations	BR 443, Building Research Establishment, 2006
Buildings and building components	Thermal performance of buildings and building components – Thermal resistance and thermal transmittance – Calculation method	BS EN ISO 6946
Windows, doors and shutters	Thermal performance of windows, doors and shutters – Calculation of thermal transmittance – Part 1: Simplified methods; Part 2: Numerical methods for frames	BS EN ISO 10077–1 and BS EN ISO 10077–2
Thermal bridges	Thermal bridges in building construction – Heat flows and surface temperatures – Part 1: General calculation methods; Part 2: Linear thermal bridges	BS EN ISO 10211–1 and BS EN ISO 10211–2
Ground floors	Thermal performance of buildings – Heat transfer via the ground – Calculation methods	BS EN ISO 13370
Buildings	Thermal performance of buildings – Transmission heat loss coefficient – Calculation method	BS EN ISO 13789
Windows, doors and shading devices	Thermal performance of windows, doors and shading devices – Detailed calculations	ISO 15099
Glass	Glass in building – Determination of thermal transmittance (U-value) – Calculation method	BS EN 673
Curtain walling	Thermal performance of curtain walling – Calculation of thermal transmittance	EN 13947

The first of these, i.e. BR 443, is specified in the Approved Documents as being the most suitable for the majority (but not all) of wall and some other constructions. It utilises the methods given in the British Standards, and a version of it in the form of a computer software programme is available from the Building Research Establishment [53]. The methods described here are suitable for many common situations.

16.10.3 U-values for walls and roofs

The lower the U-value of a construction element, the more significant is the effect of thermal bridging on the calculation of the U-value. Consequently it is usually necessary to include thermal bridging in the calculation method. The theory is based on the calculation of thermal resistances. For a single layer of material, the thermal resistance R is given by:

$$R = \frac{d}{\lambda}$$

where d is the thickness of the layer in metres, and λ is the thermal conductivity. The combined resistance of several materials depends on whether the heat flows through them sequentially, i.e. in series, or in parallel, as shown in Fig. 16.3.

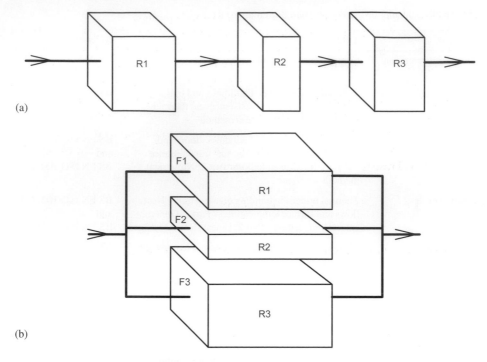

Fig. 16.3   Resistances in series and in parallel.

Resistances in series

For three materials in series (Fig. 16.3(a)), the total combined resistance R_{series} is given by:

$$R_{series} = R_1 + R_2 + R_3$$

Resistances in parallel

For three materials in parallel (Fig 16.3(b)), the total combined resistance $R_{parallel}$ is given by:

$$\frac{1}{R_{Parallel}} = \frac{F_1}{R_1} + \frac{F_2}{R_2} + \frac{F_3}{R_3}$$

where F_1, F_2 and F_3 are the cross-sectional areas of each material expressed as a fraction of the total.

If the structure consists solely of elements in series or solely of elements in parallel, then the U-value is found from:

$$U = \frac{1}{R_{Series}} \quad \text{or} \quad U = \frac{1}{R_{Parallel}}$$

Construction elements with materials in series and in parallel

Many practical construction elements consist of several layers through which heat passes in series, with some components embedded within them through which the heat passes in parallel (the thermal bridges). The total thermal resistance can be calculated in several ways, the method given here being one of the simplest and most direct. The method is suitable when the bridging material is timber or mortar, or some other material that is thermally similar. It is not suitable when the bridging material is metal, nor is it suitable for ground floors and basements. The method calculates an upper resistance limit for the construction element, R_{upper}, and a lower resistance limit, R_{lower}, and then finds the total combined resistance, R_T, by taking the average:

$$R_T = \frac{1}{2}(R_{upper} + R_{Lower})$$

The U-value for the total construction is then found from:

$$U_T = \frac{1}{R_T}$$

The U-value found in this way can then be corrected for a number of additional factors, including:

- air gaps that penetrate an insulation layer;
- mechanical fasteners that form a thermal bridge effect;
- inverted roofs;
- an unheated space on the exterior face of an element and the outside environment.

For the first three, a correction factor, ΔU, is added to U_T to give a corrected value U_c as follows:

$$\text{Corrected U-value: } U = U_T + \Delta U$$

where

$$\Delta U = \Delta U_g + \Delta U_f + \Delta U_r$$

ΔU_g = correction for air gaps, ΔU_f = correction for mechanical fasteners and ΔU_r = correction for inverted roofs.

Correction for air gaps

$$\Delta U_g = \Delta U^{11} \times \left(\frac{R_1}{R_T}\right)^2$$

where R_1 = thermal resistance of layer containing the gaps and R_T = total thermal resistance of the whole component. ΔU^{11} is obtained from Table 16.30.

Table 16.30   Correction factors for air gaps.

Level	ΔU^{11} (Wm^{-2}K^{-1})	Type of air gap
0	0.00	Insulation installed in such a way that no air circulation is possible on the warm side of the insulation. No air gaps penetrating the entire insulation layer
1	0.01	Insulation installed in such a way that no air circulation is possible on the warm side of the insulation. Air gaps may penetrate the insulation layer
2	0.04	Air circulation is possible on the warm side of the insulation. Air gaps may penetrate the insulation layer

Reproduced with permission from BS EN ISO 6946.

Table 16.31   Corrections for mechanical fasteners.

Type of fastener	α (m^{-1})
Wall tie between masonry leaves	6
Roof fixing	5

Reproduced with permission from BS EN ISO 6946.

Correction for mechanical fasteners

$$\Delta U_f = \alpha \lambda_f \, n_f A_f$$

where λ_f = thermal conductivity of the fastener, n_f = number of fasteners per square metre and A_f = cross-sectional area of one fastener. α is obtained from the Table 16.31. Corrections for fasteners must *not* be applied when:

- the wall ties are across an empty cavity;
- the wall ties are between a masonry leaf and timber studs;
- the thermal conductivity of the fastener, or part of it, is less than 1 Wm^{-1}K.

Correction for inverted roofs
The correction for an inverted roof may be obtained from BS EN ISO 6946 Annex D.

Correction for a U-value via an unheated space
The precise calculation of the heat flow through a building element, and then via an unheated space to the outside, requires complex procedures. These can be found in BS EN ISO 13789 (see Table 16.28). However, for the purposes of Part L, a simpler procedure, in which the unheated space is assumed to behave like an additional homogeneous plane layer, is usually adequate. With this assumption, the extra thermal resistance of an unheated space may be included in the calculation of the U-value of an element using the formula:

$$U = \cfrac{1}{\cfrac{1}{U_c} + R_{extra}}$$

where U is the U-value of the element including the effect of the unheated space, U_c is the U-value of the element between the heated and unheated spaces, calculated as if exposed directly to the outside and R_{extra} is the effective extra thermal resistance due to the unheated space.

This formula is acceptable provided R_{extra} is small compared to $1/U_c$, say if $R_{extra} < 0.3/U_c$. Values of R_{extra} are given in SAP 2005 for some typical unheated spaces attached to dwellings, including:

- single and double garages in various configurations;
- stairwells;
- access corridors;
- conservatories;
- roof spaces adjacent to a room in a roof.

These values, which are shown in Tables 16.32 and 16.33, can also be applied to similar situations in other buildings.

For other unheated spaces, it may be possible to calculate R_{extra} from:

$$R_{extra} = \frac{A_{INT}}{\Sigma(A_{EXT} \times U_{EXT}) + 0.33NV}$$

Table 16.32 Extra thermal resistance due to unheated spaces – garages.

	R_{extra} (m²KW⁻¹)	
Garage type and description	**Garage inside insulation layer of building**	**Garage outside insulation layer of building**
Single, fully integral, sharing side wall, end wall and floor with building	0.68	0.33
Single, fully integral, sharing side wall and floor with building	0.54	0.25
Single, partially integral, projecting forward, sharing part of side wall, part of floor and end wall with building	0.56	0.26
Single, adjacent, sharing side wall only with building	0.23	0.09
Double, fully integral, sharing side wall, end wall and floor with building	0.59	0.28
Double, half integral, sharing side wall, half of end wall and half of floor with building	0.34	0.17
Double, partially integral, projecting forward, sharing part of side wall, part of floor and end wall with building	0.28	0.13
Double, adjacent, sharing side wall only with building	0.13	0.05

Adapted with permission from SAP 2005.

Table 16.33 Extra thermal resistance due to unheated spaces – various.

Type and description of unheated space	R_{extra} (m²KW⁻¹)
Stairwell between heated space and external (exposed) wall	0.82
Stairwell between heated space and internal (not exposed) wall	0.90
Access corridor between heated space and external (exposed) wall, with another corridor above *or* below	0.31
Access corridor between heated space and external (exposed) wall, with another corridor above *and* below	0.28
Access corridor between heated space and internal (not exposed) wall, with another corridor above *or* below	0.43
Conservatory, double glazed, sharing one wall with heated space	0.06
Conservatory, double glazed, sharing two walls with heated space, i.e. in the angle between two walls	0.14
Conservatory, double glazed, sharing three walls with heated space, i.e. recessed	0.25
Conservatory, single glazed, sharing any number of walls with heated space	0.10
Loft space, between the roof covering and the wall of a heated room formed within a pitched roof above an insulated ceiling, for heat flow horizontally through the wall of the heated room	0.50
Loft space, between the roof covering and the wall of a heated room formed within a pitched roof above an insulated ceiling, for heat flow vertically through the insulated ceiling of the room below	0.50

Adapted with permission from SAP 2005

where A_{INT} is the total area of the elements separating the heated space from the unheated space, A_{EXT} is the total area of the elements separating the unheated space from the outside, excluding ground floors, U_{EXT} is the U-value of the external elements of the unheated space, excluding ground floors, N is the ventilation rate in air changes per hour of the unheated space (if this is not known, it is reasonable to assume N = 3) and V is the volume in m³ of the unheated space.

Calculation of U-values for pitched roofs

Flat roofs can be treated in the same way as walls, but for pitched roofs the calculation can be particularly difficult as it may be necessary to allow for:

- the irregular shape of loft spaces;
- the many different possible positions for the insulation layer;
- air leakage into the loft through hatches, access doors and light fittings;
- the flow of external air through the loft.

Because of this it is recommended that a calculation tool (e.g. the BRE U-value calculator [53]) is used rather than manual calculation. However, the following three examples give an indication of the U-values to be expected for the main styles of domestic roof construction.

Roof example 1 – pitched roof with insulated ceiling
- Tiled roof with felt or sarking boards;
- 100 mm ceiling joists with 100 mm mineral wool quilt ($\lambda = 0.040$) laid between the joists and 150 mm mineral wool quilt ($\lambda = 0.040$) laid across the joists;
- plasterboard ceiling;
- loft hatch with 50 mm of insulation;
- U-value = 0.16 Wm^{-1}K^{-1}.

Roof example 2 – pitched roof with insulation above loft and fixed to rafters
- Tiled roof with felt or sarking boards;
- 120 mm rafters with insulation board ($\lambda = 0.025$) between;
- 50 mm insulation board ($\lambda = 0.025$) below rafters;
- plasterboard ceiling;
- U-value = 0.21 Wm^{-2}K^{-1}.

Roof example 3 – pitched roof with insulated slope and sloping ceiling
- 15 mm clay tiles;
- 25 mm ventilated air layer;
- sarking felt;
- 50 mm ventilated air layer;
- 150 mm rafters with insulation board ($\lambda = 0.022$) between;
- plasterboard ceiling;
- U-value = 0.20 Wm^{-2}K^{-1}.

16.10.3.1 *Example calculations for walls*

The calculation method is most conveniently explained by means of examples.

Example 1 – cavity wall
Figure 16.4 shows a cavity wall consisting of external brickwork, cavity, lightweight blockwork, mineral wool insulation within a timber sub-frame, and internal plasterboard. The blockwork and the mineral wool are the main providers of thermal insulation in this construction, and both suffer from thermal bridging. The blockwork is bridged by the mortar joints, and the mineral wool by the timber frame. In each case, the proportion of the area bridged is:

- blockwork 93% of area, mortar joints 7% of area;
- mineral wool 88% of area, timber battens 12% of area.

Table 16.34 gives the thermal data for the wall.

The upper resistance limit, R$_{upper}$
Each possible heat flow path through the wall is considered separately, and in this case it can be seen that there are four such paths, as shown in Fig. 16.5. The resistance of each path is calculated on the basis that the materials are in series, and then the four paths are

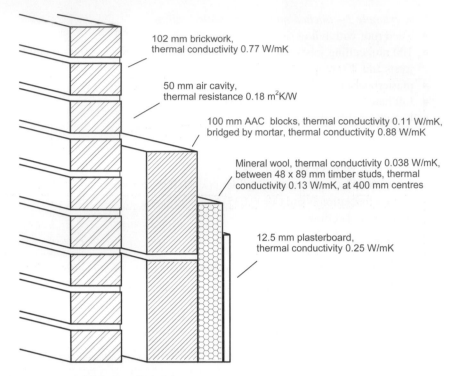

Fig. 16.4   Brick and blockwork cavity wall.

Table 16.34   Thermal data for cavity wall.

Material	Thickness (mm)	Thermal conductivity (Wm⁻¹K⁻¹)	Thermal resistance (m²KW⁻¹)
External surface			0.040
Outer brickwork	102	0.77	0.132
Cavity, unvented			0.180
AAC blocks	100	0.11	0.909
Mortar	100	0.88	0.114
Mineral wool insulation	89	0.038	2.342
Timber battens	89	0.13	0.685
Plasterboard	12.5	0.25	0.050
Internal surface			0.130

combined on the basis that they are in parallel. The first part of the calculation is illustrated in Table 16.35. When the resistance of each path has been found, the four paths are combined in parallel to find R_{upper}:

$$\frac{1}{R_{upper}} = \frac{F_1}{R_1} + \frac{F_2}{R_2} + \frac{F_3}{R_3} + \frac{F_4}{R_4} = \frac{0.818}{3.783} + \frac{0.062}{2.988} + \frac{0.112}{2.126} + \frac{0.008}{1.331} = 0.2957$$

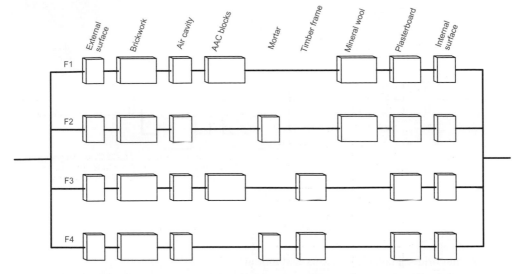

Fig. 16.5 Brick and blockwork cavity wall – upper resistance limit.

Table 16.35 Calculation of the upper resistance limit, cavity wall.

	Thermal resistance (m²KW⁻¹)			
	Path 1	**Path 2**	**Path 3**	**Path 4**
External surface resistance	0.040	0.040	0.040	0.040
Resistance of brickwork	0.132	0.132	0.132	0.132
Resistance of cavity	0.180	0.180	0.180	0.180
Resistance of AAC blocks	0.909		0.909	
Resistance of mortar		0.114		0.114
Resistance of mineral wool	2.342	2.342		
Resistance of timber			0.685	0.685
Resistance of plasterboard	0.050	0.050	0.050	0.050
Internal surface resistance	0.130	0.130	0.130	0.130
Total thermal resistance of path	3.783	2.988	2.126	1.331
Fractional area of path	93% × 88% = 0.818	7% × 88% = 0.062	93% × 12% = 0.112	7% × 12% =0.008

$$R_{upper} = 3.382 \, m^2 KW^{-1}$$

The lower resistance limit, R_{lower}

Each thermal bridge in the construction element is first converted to a single combined resistance, as shown in Fig. 16.6. Using these combined resistances, the construction can then be considered as a single heat flow path with all components in series. Thus in the present example:

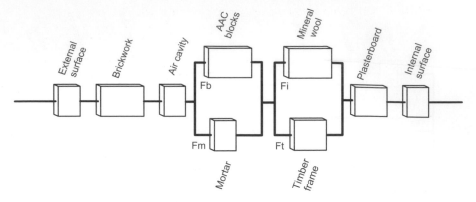

Fig. 16.6   Brick and blockwork cavity wall – lower resistance limit.

- first bridged layer (blockwork and mortar):

$$\frac{1}{R_{bm}} = \frac{F_{blocks}}{R_{blocks}} + \frac{F_{mortar}}{R_{mortar}} = \frac{0.93}{0.909} + \frac{0.07}{0.114} = 1.637$$

$$R_{bm} = 0.611 \, \text{m}^2 \text{KW}^{-1}$$

- second bridged layer (insulation and timber):

$$\frac{1}{R_{it}} = \frac{F_{insulation}}{R_{insulation}} + \frac{F_{timber}}{R_{timber}} = \frac{0.88}{2.342} + \frac{0.12}{0.685} = 0.5509$$

$$R_{it} = 1.815 \, \text{m}^2 \text{KW}^{-1}$$

These combined resistances may now be used to find the lower resistance limit, as shown in Table 16.36. Note that R_{upper} is an overestimate of the true resistance, whereas R_{lower} is an underestimate. The average of these is very close to the true value. Hence, the total resistance of the wall is found from:

$$R_T = \frac{1}{2}(R_{upper} + R_{Lower}) = \frac{1}{2}(3.382 + 2.958) = 3.170 \, \text{m}^2 \text{KW}^{-1}$$

and the U-value is:

$$U_T = \frac{1}{R_T} = \frac{1}{3.170} = 0.315 \, \text{Wm}^{-2}\text{K}^{-1}$$

Corrections to the U-value for air gaps and mechanical fixings
If there are small air gaps or mechanical fixings (such as wall ties) penetrating the insulation layer, it may be necessary to add a correction, ΔU_g, to the U-value. The correction is

Table 16.36   Calculation of the lower resistance limit, cavity wall.

	Thermal resistances (m²KW⁻¹)		
	Thermal bridges		
	Components	**Combined**	
External surface resistance			0.040
Resistance of brickwork			0.132
Resistance of cavity			0.180
Resistance of AAC blocks (93%)	0.909	0.611	0.611
Resistance of mortar (7%)	0.114		
Resistance of mineral wool (88%)	2.342	1.815	1.815
Resistance of timber (12%)	0.685		
Resistance of plasterboard			0.050
Internal surface resistance			0.130
Total thermal resistance, R_{lower}			2.958

required if ΔU_g is 3% or more of the uncorrected U-value, but may be ignored if it is less than 3%. The correction is calculated from:

$$\Delta U_g = \Delta U^{11} \times \left(\frac{R_1}{R_T}\right)^2$$

In this case, ΔU^{11} is 0.01, R_1 is 1.815 and R_T is 3.170, and so ΔU_g is 0.003 W/m²K. As this is less than 3% of U, it may be ignored. The final U-value is rounded to two decimal places, and so the result is:

$$U = 0.32\,Wm^{-2}K^{-1}$$

Example 2 – Timber framed wall

Figure 16.7 shows a timber-framed wall consisting of an outer layer of brickwork, a clear ventilated cavity, 10 mm plywood, 38 × 140 mm timber stud framing with 140 mm mineral-wool quilt insulation placed between the studs, and two sheets of 12.5 mm plasterboard with an integral vapour check. The timber studs account for 15% of the area, corresponding to 38 mm studs at 600 mm centres, with allowances for horizontal noggins and additional framing at junctions and around openings. Thermal data is given in Table 16.37.

The upper resistance limit, R_{upper}

Each possible heat flow path through the wall is considered separately, and in this case it can be seen that there are two such paths. This is illustrated in Fig. 16.8. The resistance of each path is calculated on the basis that the materials are in series, and then the two paths are combined on the basis that they are in parallel. The first part of the calculation is illustrated in Table 16.38. When the resistance of each path has been found, the two paths are combined in parallel to find R_{upper}:

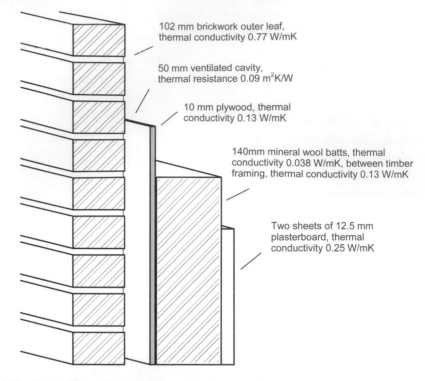

102 mm brickwork outer leaf,
thermal conductivity 0.77 W/mK

50 mm ventilated cavity,
thermal resistance 0.09 m²K/W

10 mm plywood, thermal
conductivity 0.13 W/mK

140mm mineral wool batts, thermal
conductivity 0.038 W/mK, between timber
framing, thermal conductivity 0.13 W/mK

Two sheets of 12.5 mm
plasterboard, thermal
conductivity 0.25 W/mK

Fig. 16.7   Timber frame wall.

Table 16.37   Thermal data for timber frame wall.

Material	Thickness (mm)	Thermal conductivity $(Wm^{-1}K^{-1})$	Thermal resistance (m^2KW^{-1})
External surface			0.040
Outer brickwork	102	0.77	0.132
Cavity, vented			0.090
Plywood	10	0.13	0.077
Mineral wool quilt insulation	140	0.038	3.684
Timber framing	140	0.13	1.077
Plasterboard	25	0.25	0.100
Internal surface			0.130

$$\frac{1}{R_{upper}} = \frac{F_1}{R_1} + \frac{F_2}{R_2} = \frac{0.85}{4.253} + \frac{0.15}{1.646} = 0.291$$

$$R_{upper} = 3.437\,m^2KW^{-1}$$

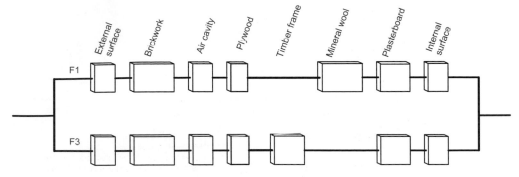

Fig. 16.8   Upper resistance limit – timber frame wall.

Table 16.38   Calculation of the upper resistance limit, timber frame wall.

	Thermal resistance (m²KW⁻¹)	
	Path 1	Path 2
External surface resistance	0.040	0.040
Resistance of brickwork	0.132	0.132
Resistance of cavity	0.090	0.090
Resistance of plywood	0.077	0.077
Resistance of mineral wool quilt	3.684	
Resistance of timber		1.077
Resistance of plasterboard	0.100	0.100
Internal surface resistance	0.130	0.130
Total thermal resistance of path	4.253	1.646
Fractional area of path	85% = 0.85	15% = 0.15

The lower resistance limit, R_{lower}

Each thermal bridge in the construction element is first converted to a single combined resistance, as shown in Fig. 16.9. Using these combined resistances, the construction can then be considered as a single heat-flow path with all components in series. In this example there is one bridged layer, insulation and timber:

$$\frac{1}{R_{it}} = \frac{F_{insulation}}{R_{insulation}} + \frac{F_{timber}}{R_{timber}} = \frac{0.85}{3.684} + \frac{0.15}{1.077} = 0.370$$

$$R_{it} = 2.703 \, \text{m}^2\text{KW}^{-1}$$

This combined resistance may now be used to find the lower resistance limit, as shown in Table 16.39. Note that R_{upper} is an overestimate of the true resistance, whereas R_{lower} is an underestimate. The average of these is very close to the true value. Hence, the total resistance of the wall is found from:

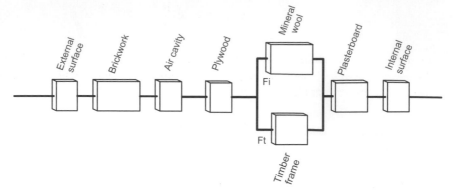

Fig. 16.9   Lower resistance limit – timber frame wall.

Table 16.39   Calculation of the lower resistance limit, timber frame wall.

| | Thermal resistances (m²KW⁻¹) | | |
| | Thermal bridges | | |
	Components	Combined	
External surface resistance			0.040
Resistance of brickwork			0.132
Resistance of cavity			0.090
Resistance of plywood			0.077
Resistance of mineral wool (85%)	3.684	2.703	2.703
Resistance of timber (15%)	1.077		
Resistance of plasterboard			0.100
Internal surface resistance			0.130
Total thermal resistance, R_{lower}			3.272

$$R_T = \frac{1}{2}(R_{upper} + R_{Lower}) = \frac{1}{2}(3.437 + 3.272) = 3.354 \, m^2KW^{-1}$$

and the U-value is:

$$U_T = \frac{1}{R_T} = \frac{1}{3.354} = 0.298 \, Wm^{-2}K^{-1}$$

Corrections to the U-value for air gaps and mechanical fixings
If there are small air gaps or mechanical fixings (such as wall ties) penetrating the insulation layer, it may be necessary to add a correction, ΔU_g, to the U-value. The correction is required if ΔU_g is 3% or more of the uncorrected U-value, but may be ignored if it is less than 3%. The correction is calculated from:

$$\Delta U_g = \Delta U^{11} \times \left(\frac{R_\perp}{R_T}\right)^2$$

In this case, ΔU^{11} is 0.01, R_1 is 2.703 and R_T is 3.354, and so ΔU_g is $0.006 \mathrm{Wm^{-2}K^{-1}}$. As this is less than 3% of U, it may be ignored. The final U-value is rounded to two decimal places, and so the result is:

$$U = 0.30 \mathrm{Wm^{-2}K^{-1}}$$

16.10.4 U-values for ground floors

The accurate calculation of the U-value of ground floors is difficult, and requires the rigorous procedures given in BS EN ISO 13370 or in CIBSE Guide A [39]. However, the full rigour of these methods may not be necessary, and the following simple approach may be adequate for most of the common constructions and ground conditions to be found in the UK. The method is based on pre-calculated tabulated values. There are several points to be noted:

- For solid ground floors, if the perimeter to area ratio is less than $0.12 \mathrm{m/m^2}$, the U-value will normally be $0.25 \mathrm{Wm^{-2}K^{-1}}$ or less without the need for insulation.
- For suspended ground floors, if the perimeter to area ratio is less than $0.09 \mathrm{m/m^2}$, the U-value will normally be $0.25 \mathrm{Wm^{-2}K^{-1}}$ or less without the need for insulation.
- For ground floors the U-value depends on the type of soil beneath the building; clay soil is the most typical in the UK and this is assumed to be the case in the following tables.
- Where the soil is neither clay nor silt, the U-value must be calculated in accordance with BS EN ISO 13370.

As the U-value of a ground floor depends on the ratio of the perimeter to the area, the rules for calculating this ratio must be observed. These are:

- Floor dimensions should be measured between the finished internal faces of the external elements of the building and must include any projecting bays.
- For semi-detached houses, terraced houses, blocks of flats and similar structures, the floor dimensions can be either those of the individual unit or the whole building.
- When considering extensions to existing buildings, the floor dimensions may be taken as those of the complete building including the extension.
- Unheated spaces outside the insulated fabric (e.g. attached garages and porches) are excluded from the perimeter and area calculation, but the length of common wall between them must be included in the perimeter.

In addition to meeting U-value requirements, it is also important that the floor design should prevent excessive thermal bridging at the floor edge. This is to reduce the risk of condensation and mould growth.

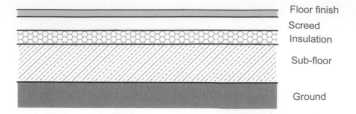

Fig. 16.10 Solid floor in contact with the ground.

16.10.4.1 Solid ground floors

For solid ground floors with all-over insulation, as shown in Fig. 16.10, the U-values in Table 16.40 apply.

16.10.4.2 Solid ground floors with edge insulation

It is often better to insulate a floor with horizontal or vertical edge insulation *instead of* all-over insulation. When this is done, as shown in Fig. 16.11, a correction factor is subtracted from the U-value for the equivalent uninsulated solid ground floor. The correction factor is a combination of the perimeter to area ratio and an edge insulation factor, Ψ. Thus:

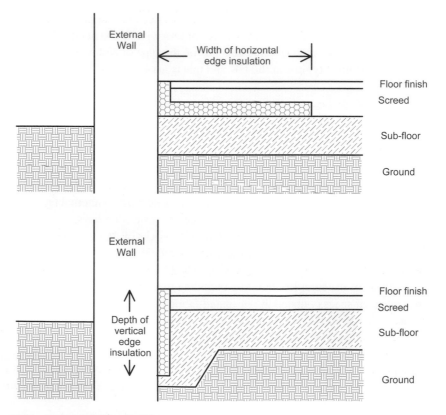

Fig. 16.11 Edge insulation of solid ground floors.

$$U = U_0 - \frac{P}{A}\Psi$$

where U_0 is the value for an uninsulated floor taken from Table 16.40 (the column for zero thermal resistance), and Ψ is obtained from Table 16.41.

16.10.4.3 Ground floors with both all-over insulation and edge insulation

There are no tables for floors with both types of insulation. For such cases, the calculation method of BS EN ISO 13370 must be used.

16.10.4.4 Uninsulated suspended ground floors

The U-values for uninsulated suspended ground floors are given in Table 16.42. These values can be used when:

Table 16.40   U-values for solid ground floors.

Perimeter to area ratio (m/m²)	Thermal resistance of all-over insulation (m²KW⁻¹)					
	0	0.5	1	1.5	2	2.5
	U-value of solid ground floor(Wm⁻²K⁻¹)					
0.05	0.13	0.11	0.10	0.09	0.08	0.08
0.10	0.22	0.18	0.16	0.14	0.13	0.12
0.15	0.30	0.24	0.21	0.18	0.17	0.15
0.20	0.37	0.29	0.25	0.22	0.19	0.18
0.25	0.44	0.34	0.28	0.24	0.22	0.19
0.30	0.49	0.38	0.31	0.27	0.23	0.21
0.35	0.55	0.41	0.34	0.29	0.25	0.22
0.40	0.60	0.44	0.36	0.30	0.26	0.23
0.45	0.65	0.47	0.38	0.32	0.27	0.23
0.50	0.70	0.50	0.40	0.33	0.28	0.24
0.55	0.74	0.52	0.41	0.34	0.28	0.25
0.60	0.78	0.55	0.43	0.35	0.29	0.25
0.65	0.82	0.57	0.44	0.35	0.30	0.26
0.70	0.86	0.59	0.45	0.36	0.30	0.26
0.75	0.89	0.61	0.46	0.37	0.31	0.27
0.80	0.93	0.62	0.47	0.37	0.32	0.27
0.85	0.96	0.64	0.47	0.38	0.32	0.28
0.90	0.99	0.65	0.48	0.39	0.32	0.28
0.95	1.02	0.66	0.49	0.39	0.33	0.28
1.00	1.05	0.68	0.50	0.40	0.33	0.28

Table 16.41 Edge insulation factors for solid ground floors.

Width of horizontal insulation (m)	Thermal resistance of insulation (m²KW⁻¹)			
	0.5	**1.0**	**1.5**	**2.0**
	Edge insulation factor Ψ (Wm⁻¹K⁻¹)			
0.50	0.13	0.18	0.21	0.22
1.00	0.20	0.27	0.32	0.34
1.50	0.23	0.33	0.39	0.42
Depth of vertical insulation (m)				
0.25	0.13	0.18	0.21	0.22
0.50	0.20	0.27	0.32	0.34
0.75	0.23	0.33	0.39	0.42
1.00	0.26	0.37	0.43	0.48

Table 16.42 U-values for uninsulated suspended ground floors.

Perimeter to area ratio (m per m²)	Ventilation opening area per unit perimeter of underfloor space	
	0.0015 m² per m	**0.0030 m² per m**
	U-value of suspended ground floor (Wm⁻²K⁻¹)	
0.05	0.15	0.15
0.10	0.25	0.26
0.15	0.33	0.35
0.20	0.40	0.42
0.25	0.46	0.48
0.30	0.51	0.53
0.35	0.55	0.58
0.40	0.59	0.62
0.45	0.63	0.66
0.50	0.66	0.70
0.55	0.69	0.73
0.60	0.72	0.76
0.65	0.75	0.79
0.70	0.77	0.81
0.75	0.80	0.84
0.80	0.82	0.86
0.85	0.84	0.88
0.90	0.86	0.90
0.95	0.88	0.92
1.00	0.89	0.93

- the floor deck is not more than 500 mm above the external ground level;
- the wall surrounding the underfloor space is uninsulated.

The U-values depend on the amount of ventilation which is provided to the underfloor space. This is expressed as the area in square metres of ventilation opening per unit length in metres of floor perimeter, and the table provides data for two typical values.

16.10.4.5 *Insulated suspended ground floors*

The U-value of an insulated suspended floor, Figs 16.12 and 16.13, is calculated from the parameters U_o, R_f and U_f where:

- U_o is the U-value of the equivalent uninsulated floor taken from Table 16.42;
- U_f is the U-value of the floor deck, including allowances for thermal bridging, and calculated according to the methods recommended in BS EN ISO 6946, or by a numerical modelling method; and
- R_f is the thermal resistance of the floor deck itself.

The procedure is to first obtain U_f, and then find R_f from the formula:

$$R_f = \frac{1}{U_f} - 0.17 - 0.17$$

The two values of 0.17 are the surface resistances. The U-value of the floor is then found from:

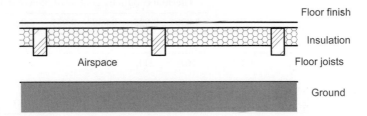

Floor finish

Insulation

Airspace   Floor joists

Ground

Fig. 16.12   Suspended timber ground floor.

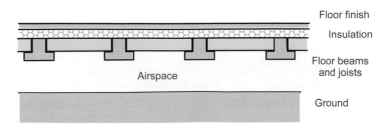

Floor finish

Insulation

Floor beams
and joists

Airspace

Ground

Fig. 16.13   Suspended concrete ground floor.

$$U = \frac{1}{[(1/U_o) - 0.2 + R_f]}$$

16.10.4.6   Determining the thickness of insulation for upper floors

For upper floors, i.e. floors above an external space, the U-values given in Table 16.43, which is derived from the Approved Document for Part L, 2002 edition, should be sufficient to give an indication of the minimum thickness of insulation required to achieve a given U-value. In practice, the minimum thicknesses in this table may be reduced by 5 to 10 mm because of the floor and ceiling finishes above and below the floor joists. Note also that in the table it is assumed that the proportion by area of structural timber in the timber floor construction is 12%, corresponding to 48 mm wide timber joists at 400 mm centres. The U-value for other proportions of timber must be calculated using the method of section 16.10.3.

16.10.4.7   Example calculations for floors

Example 1 – Solid ground floor over clay subsoil
Figure 16.14 illustrates the floor plan of a detached house. The subsoil is clay, and the floor is uninsulated. First, calculate the perimeter to area ratio:

$$P = 2 \times (10.2 + 6.7) = 33.8 \text{ m}$$

Table 16.43   Recommended thickness of insulation for upper floors.

| | U-value $(\text{Wm}^{-2}\text{K}^{-1})$ | Thermal conductivity of insulation material $(\text{Wm}^{-1}\text{K}^{-1})$ | | | | | | |
| | | 0.020 | 0.025 | 0.030 | 0.035 | 0.040 | 0.045 | 0.050 |
		Thickness of insulation layer (mm)						
Timber construction	0.20	167	211	256	298	341	383	426
	0.25	109	136	163	193	225	253	281
	0.30	80	100	120	140	160	184	208
Concrete construction	0.20	95	119	142	166	190	214	237
	0.25	75	94	112	131	150	169	187
	0.30	62	77	92	108	123	139	154

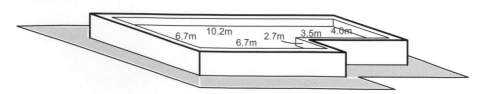

Fig. 16.14   Solid ground floor.

$$A = 10.2 \times 6.7 - 3.5 \times 2.7 = 58.89 \text{ m}^2$$

$$P/A = 33.8/58.89 = 0.57 \text{ m/m}^2$$

The U-value is found from Table 16.40, using the column for zero thermal resistance. The Approved Document recommends that the row nearest to the actual P/A value is used; in this case the nearest row to 0.57 is P/A = 0.55. It is not necessary to interpolate between rows because the change in U-value between rows is not large enough to warrant the extra work. The U-value of this floor is therefore:

$$U = 0.74 \text{ Wm}^{-2}\text{K}^{-1}$$

Example 2, Solid ground floor over clay subsoil with all-over insulation

The floor in Fig. 16.14 is now provided with all-over insulation between the screed and the structural floor. The insulation layer is 75 mm thick and has a thermal conductivity of 0.040 W/mK. The resistance of the insulation layer is:

$$R_{ins} = 0.075/0.040 = 1.875 \text{ m}^2\text{KW}^{-1}$$

In Table 16.40 we again use the row for P/A = 0.55, but this time we must interpolate between the columns for $R_{ins} = 1.5$ and $R_{ins} = 2.0$. This interpolation is necessary because the change in U-value between columns is more significant than the change between rows. Thus:

$$\text{At } R_{ins} = 1.875, \ U = 0.34 - \left(\frac{1.875 - 1.5}{2.0 - 1.5}\right) \times (0.34 - 0.28) = 0.295 \text{ Wm}^{-2}\text{K}^{-1}$$

Example 3 – Solid ground floor over clay subsoil with vertical edge insulation

The floor in Fig. 16.14 is provided with vertical edge insulation instead of all-over insulation. The insulation is to a depth of 750 mm, and the insulation is 75 mm thick with a thermal conductivity of 0.040 W/mK. The resistance of the insulation layer is:

$$R_{ins} = 0.075/0.040 = 1.875 \text{ m}^2\text{KW}^{-1}$$

From example 1, the perimeter to area ratio of this floor is 0.57, and its uninsulated U-value is 0.74 Wm^{-2}K^{-1}. We require to obtain the edge insulation factor from Table 16.41, and it is necessary to interpolate between the columns for $R_{ins} = 1.5$ and $R_{ins} = 2.0$. Hence:

$$\Psi = 0.39 + \left(\frac{1.875 - 1.5}{2.0 - 1.5}\right) \times (0.42 - 0.39) = 0.413 \text{ Wm}^{-2}\text{K}^{-1}$$

The U-value of the floor is now:

$$U = 0.74 - 0.57 \times 0.413 = 0.50 \text{ Wm}^{-2}\text{K}^{-1}$$

Example 4 – Uninsulated, suspended ground floor

Now assume that the floor in Fig. 16.14 is an uninsulated suspended timber ground floor. The floor deck is less than 500 mm above external ground level, and the under-floor ventilation openings amount to approximately 0.0015 m² per metre of floor perimeter. The perimeter to area ratio is 0.57, and taking 0.55 as the nearest value in Table 16.42, the U-value is:

$$U = 0.69 \, \text{Wm}^{-2}\text{K}^{-1}$$

Example 5 – Insulated, suspended ground floor

Now let the floor in example 4 be insulated, with insulation fitted between the floor joists. The U-value of this floor deck may be calculated in the same way as the U-value of a wall, using the method of Chapter 5. Assuming the result of this calculation is a U-value of 0.45 Wm⁻²K⁻¹, the U-value of the floor is found as follows:

$$R_f = \frac{1}{U_f} - 0.17 - 0.17 = \frac{1}{0.45} - 0.34 = 2.22 - 0.34 = 1.88 \, \text{m}^2\text{KW}^{-1}$$

The uninsulated U-value from example 4 is $U_o = 0.69$, and so:

$$U = \frac{1}{[(1/U_o) - 0.2 + R_f]} = \frac{1}{1.45 - 0.2 + 1.88} = 0.32 \, \text{Wm}^{-2}\text{K}^{-1}$$

16.10.5 U-values of windows, roof windows and rooflights

The U-values of windows, doors and rooflights must normally include the effects of heat flow through the frame, the calculation of which requires advanced techniques. It is therefore usual to select values from pre-prepared data. When available, manufacturer's certified U-values (by approved methods of measurement or calculation) should be used. If these are not available, values for single, double and triple glazing may be taken from Tables 16.44, 16.45 and 16.46, modified where necessary for metal frames according to Table 16.47. Low emissivity (Low-E) coatings are of two main types, 'hard' and 'soft'. If the exact value of the emissivity, ε_n, is not known, then for hard coatings or where the

Table 16.44 Single glazing U-values for windows, rooflights and doors.

Single glazing type	Wm⁻²K⁻¹
Windows in wood or PVC-U frames	4.8
Rooflights in dwellings in wood or PVC-U frames	5.1
Rooflights in buildings other than dwellings in wood or PVC-U frames	4.8
Windows in metal frames (4 mm thermal break)	5.7
Solid wood door	3.0

Table 16.45   Double glazing U-values for windows and rooflights (Wm^{-2}K^{-1}).

Double glazing description	Gap between panes			Adjustment for rooflights in dwellings
	6 mm	12 mm	16 mm or more	
Wood or PVC-U frames				
Air filled	3.1	2.8	2.7	
Low-E, $\varepsilon_n = 0.2$	2.7	2.3	2.1	
Low-E, $\varepsilon_n = 0.15$	2.7	2.2	2.0	For dwellings only,
Low-E, $\varepsilon_n = 0.1$	2.6	2.1	1.9	add 0.2 for all wood
Low-E, $\varepsilon_n = 0.05$	2.6	2.0	1.8	and PVC-U frames
Argon filled	2.9	2.7	2.6	
Low-E, $\varepsilon_n = 0.2$, argon filled	2.5	2.1	2.0	
Low-E, $\varepsilon_n = 0.1$, argon filled	2.3	1.9	1.8	
Low-E, $\varepsilon_n = 0.05$, argon filled	2.3	1.8	1.7	
Metal frames, 4 mm thermal break				
Air filled	3.7	3.4	3.3	
Low-E, $\varepsilon_n = 0.2$	3.3	2.8	2.6	
Low-E, $\varepsilon_n = 0.1$	3.2	2.6	2.5	For metal frames, see
Low-E, $\varepsilon_n = 0.05$	3.1	2.5	2.3	Table 16.47
Argon filled	3.5	3.3	3.2	
Low-E, $\varepsilon_n = 0.2$, argon filled	3.1	2.6	2.5	
Low-E, $\varepsilon_n = 0.1$, argon filled	2.9	2.4	2.3	
Low-E, $\varepsilon_n = 0.05$, argon filled	2.8	2.3	2.1	

type of coating is unknown use the data for $\varepsilon_n = 0.2$, and for soft coatings use the data for $\varepsilon_n = 0.1$.

For doors that are half-glazed, the U-value is the average of the non-glazed door and the appropriate U-value for the glazing.

For windows and rooflights with metal frames where the thermal break differs from 4 mm, the corrections in Table 16.47 should be applied. Note that if corrections for thermal break *and* rooflight are applicable, both should be made.

16.10.5.1   *Minimum specifications for windows*

Inspection of the U-values in Tables 16.45 and 16.46 reveals the required design specification for a window to be less than the permitted maximum U-values of 1.8, 2.0 or 2.2 W/m^{2}K given in Tables 16.13 and 16.24. For double glazing with a 6 mm air gap in a wood or PVC-U frame, it is not possible to satisfy the U-value requirement for any of the listed types of glass. For example, double glazing units in wood or PVC-U frames which may be acceptable can be listed as follows.

Table 16.46   Triple glazing U-values for windows and rooflights, $Wm^{-2}K^{-1}$

Triple glazing description	Gap between panes			Adjustment for rooflights in dwellings
	6 mm	**12 mm**	**16 mm or more**	
Wood or PVC-U frames				
Air filled	2.4	2.1	2.0	
Low-E, $\varepsilon_n = 0.2$	2.1	1.7	1.6	For dwellings only,
Low-E, $\varepsilon_n = 0.1$	2.0	1.6	1.5	add 0.2 for all wood
Low-E, $\varepsilon_n = 0.05$	1.9	1.5	1.4	and PVC-U frames
Argon filled	2.2	2.0	1.9	
Low-E, $\varepsilon_n = 0.2$, argon filled	1.9	1.6	1.5	
Low-E, $\varepsilon_n = 0.1$, argon filled	1.8	1.4	1.3	
Low-E, $\varepsilon_n = 0.05$, argon filled	1.7	1.4	1.3	
Metal frames, 4mm thermal break				
Air filled	2.9	2.6	2.5	
Low-E, $\varepsilon_n = 0.2$	2.6	2.2	2.0	
Low-E, $\varepsilon_n = 0.1$	2.5	2.0	1.9	For metal frames,
Low-E, $\varepsilon_n = 0.05$	2.4	1.9	1.8	see Table 16.47
Argon filled	2.8	2.5	2.4	
Low-E, $\varepsilon_n = 0.2$, argon filled	2.4	2.0	1.9	
Low-E, $\varepsilon_n = 0.1$, argon filled	2.2	1.9	1.8	
Low-E, $\varepsilon_n = 0.05$, argon filled	2.2	1.8	1.7	

Table 16.47   Corrections for metal frames with various thermal breaks.

Thermal break (mm)	Correction to U-value ($Wm^{-2}K^{-1}$)	
	Window, or rooflight in buildings other than dwellings	**Rooflight in dwellings**
0 (no break)	+0.3	+0.7
4	0	+0.3
8	-0.1	+0.2
12	-0.2	+0.1
16	-0.2	+0.1

- For all windows in all new buildings or newly constructed extensions:
  Maximum U-value 1.80 $Wm^{-2}K^{-1}$ – double glazed in wood or UPVC frames:
  - 12 mm air gap   Argon filled   low-E, $\varepsilon_n = 0.05$;
  - 16 mm air gap   Air filled   low-E, $\varepsilon_n = 0.05$;
  - 16 mm air gap   Argon filled   low-E, $\varepsilon_n = 0.1$;
  - 16 mm air gap   Argon filled   low-E, $\varepsilon_n = 0.05$.

- For windows in existing dwellings:
  Maximum U-value 2.00 $Wm^{-2}K^{-1}$ – double glazed in wood or UPVC frames:
  All the above plus:
  - 12 mm air gap   Air filled   low-E, $\varepsilon_n = 0.05$;
  - 12 mm air gap   Argon filled   low-E, $\varepsilon_n = 0.1$;
  - 16 mm air gap   Air filled   low-E, $\varepsilon_n = 0.1$;
  - 16 mm air gap   Air filled   low-E, $\varepsilon_n = 0.15$.
  - 16 mm air gap   Argon filled   low-E, $\varepsilon_n = 0.2$.
- For windows in existing buildings other than dwellings:
  Maximum U-value 2.20 $Wm^{-2}K^{-1}$ – double glazed in wood or UPVC frames:
  All the above plus:
  - 12 mm air gap   Air filled   low-E, $\varepsilon_n = 0.15$;
  - 12 mm air gap   Air filled   low-E, $\varepsilon_n = 0.1$;
  - 12 mm air gap   Argon filled   low-E, $\varepsilon_n = 0.2$;
  - 16 mm air gap   Air filled   low-E, $\varepsilon_n = 0.2$;

For triple glazing, even though there is a much higher number of glass/frame combinations that satisfy the U-value standard, a significant number do not, especially metal frames with a small air gap.

A particular problem arises with U-values for windows, roof windows and rooflights. The values quoted are normally correct for the component mounted in a vertical plane, i.e. at 90° to the horizontal. If the component is not mounted vertically, then the U-value must be increased to take account of its increased exposure to the sky. The increase above the normally quoted vertical U-value is given in Table 16.48. Although not precise, the corrections in Table 16.48 are normally adequate.

16.10.7 Thermal bridging, continuity of insulation, and replacement renovated and retained thermal elements

16.10.7.1 *Thermal bridging and continuity of insulation*

Thermal bridges in insulation layers can seriously reduce the effectiveness of the insulation. Every care should be taken to avoid such bridges and to ensure continuity of insulation

Table 16.48   Correction for inclination to glazing U-values.

Inclination of component from the horizontal	Correction to U-value (Wm^2K^{-1})	
	Twin skin or double glazed	Triple skin or triple glazed
70° or over (taken as vertical)	0.0	0.0
69° to 61°	+0.2	+0.1
60° to 41°	+0.3	+0.2
39° to 21°	+0.4	+0.2
20° to 0° (taken as horizontal)	+0.5	+0.3

Adapted with permission from BR443

over the building envelope, and this can be most conveniently done by adopting accredited Robust Details [19]. There are two main types of thermal bridge:

- repeating thermal bridges, which repeat in regular fashion throughout a thermal element, for example the roof joists that interrupt the insulation laid in a roof space, or the studding in a timber frame wall;
- non-repeating thermal bridges such as lintels, door posts, etc.

The effect of repeating bridges is usually accounted for in the calculation of the U-value, and no further allowance is necessary. On the other hand, the effect of non-repeating bridges must be calculated, or an approximate allowance must be made. For dwellings, the SAP calculation makes an automatic allowance. For other buildings and for dwellings if it is thought to confer an advantage, the heat loss coefficient, H, due to non-repeating bridges may be calculated from:

$$H = \Sigma(L \times \Psi)$$

where L is the length of the thermal bridge and Ψ is linear thermal transmittance. Values for Ψ are given in SAP 2005, IP 1/06 and in the technical sections of many trade catalogues. Metal roofing and cladding systems can present particular problems with thermal bridging, and guidance specifically relevant to Part L is given in MCRMA Technical Paper 17 [43].

If the details of thermal bridges are not known, their effect may be found from:

$$H = k\Sigma A_e$$

where k is a constant and ΣA_e is the sum of the areas (m^2) of all the exposed elements. For dwellings, k = 0.08 if accredited construction details have been used, or if not k = 0.15.

Testing of the continuity of insulation may be carried out on the completed building. This is usually done by measuring the temperature of the external surface, and the most widely used technique is infra-red photography.

16.10.7.2 Replacement, renovated and retained thermal elements

The U-value standards for replacement, renovated and retained thermal elements are given in Table 16.13 for dwellings and in Table 16.24 for buildings other than dwellings. Part L applies to a retained thermal element when:

- an existing thermal element is part of a building subject to material change of use; or
- an existing element is to become part of the building envelope and be upgraded.

There are however several circumstances where the normal U-value requirements can be relaxed:

- In the case of a renovated thermal element, if the work applies to less than 25% of its surface area, it is permissible to make no improvement to it.

● For both renovated and retained thermal elements, if the upgrade is not technically, functionally or economically feasible, it is acceptable to upgrade the element to the best standard possible provided the upgrade has a simple payback of 15 years or less.

16.10.7 Thermal conductivity and density of building materials

Table 16.49 lists the thermal conductivities and densities of some common building materials.

Table 16.49   Thermal conductivity and density of common building materials.

	Density (kgm^{-3})	Thermal conductivity (Wm^{-1}K^{-1})
Walls		
Brickwork (outer leaf)	1700	0.77
Brickwork (inner leaf)	1700	0.56
Lightweight aggregate concrete block	1400	0.57
Autoclaved aerated concrete block	600	0.18
Concrete, medium density (inner leaf)	1800	1.13
	2000	1.33
	2200	1.59
Concrete, high density	2400	1.93
Reinforced concrete, 1% steel	2300	2.30
Reinforced concrete, 2% steel	2400	2.50
Mortar, protected	1750	0.88
Mortar, exposed	1750	0.94
Gypsum	600	0.18
	900	0.30
	1200	0.43
Gypsum plasterboard	900	0.25
Sandstone	2600	2.30
Limestone, soft	1800	1.10
Limestone, hard	2200	1.70
Fibreboard	400	0.10
Plasterboard	900	0.25
Tiles, ceramic	2300	1.30
Timber, softwood	500	0.13
Timber, hardwood	700	0.18
Timber, plywood and chipboard	500	0.13
Wall ties, stainless steel	7900	17.0

(Contd).

Table 16.49   (*Contd*).

	Density (kgm⁻³)	Thermal conductivity (Wm⁻¹K⁻¹)
Surface finishes		
External rendering	1300	0.57
Plaster, dense	1300	0.57
Plaster, lightweight	600	0.18
Roofs		
Aerated concrete slab	500	0.16
Asphalt	2100	0.70
Felt/bitumen layers	1100	0.23
Screed	1200	0.41
Stone chippings	2000	2.00
Tiles, clay	2000	1.00
Tiles, concrete	2100	1.50
Wood wool slab	500	0.10
Floors		
Cast concrete	2000	1.35
Metal tray, steel	7800	50.00
Screed	1200	0.41
Timber, softwood	500	0.13
Timber, hardwood	700	0.18
Timber, plywood and chipboard	500	0.13
Insulation		
Expanded polystyrene (EPS) board	15	0.040
Mineral wool quilt	12	0.042
Mineral wool batt	25	0.038
Phenolic foam board	30	0.025
Polyurethane board	30	0.025

16.11 Checklists

Many of the first tier and second tier documents provide a number of checklists as an aid to achieving compliance with Part L. There is insufficient space to include them here, but the following is a list of the most important.

AD L1A   Appendix A
Checklist for a dwelling as designed or as built.

AD L1A   Appendix B
Checklist of unusual values reported by SAP 2005.

AD L1B   Appendix A
Cost-effective U-value targets when undertaking renovation works to thermal elements.

AD L2A   Appendix A
Checklist for a new building that is not a dwelling.

Domestic Heating Compliance Guide   Appendix A
Assessing the case for a non-condensing boiler.

Non-domestic Heating, Cooling and Ventilation Compliance Guide   Appendix 2
Compliance checklist.

16.12 References

1   British Institute of Non-destructive testing: www.bindt.org
2   FAERO Limited: www.faero.co.uk
3   BRE Certification Limited: www.brecertification.co.uk
4   Gas Safety (Installation and Use) Regulations 1998
5   CORGI Services Limited: www.corgi-gas-safety.com
6   Oil Firing Technical Association Limited: www.oftec.co.uk
7   NAPIT Certification Limited: www.napit.org.uk
8   Building Engineering Services Competence Accreditation Limited: www.besca.org.uk
9   HETAS Limited: www.hetas.co.uk
10 The Electrical Contractors Association Limited: www.eca.co.uk
11 British Standards Institution: www.bsi-global.com
12 ELECSA Limited: www.elecsa.org.uk
13 NICEIC Group Limited: www.niceic.org.uk
14 Fenestration Self-Assessment Scheme by Fensa Limited: www.fensa.co.uk
15 CERTASS Limited: www.certass.co.uk
16 Electrical Equipment (Safety) Regulations 1994
17 *Thermal insulation: avoiding risks.* ISBN 1 86081 515 4. BR 262, Building Research Establishment 2001
18 *Low or Zero Carbon Energy Sources – Strategic Guide.* ISBN 1 85946 224 3. NBS, 2006
19 *Limiting thermal bridging and air leakage: robust construction details for dwellings and similar buildings.* ISBN 0 11753 631 8. TSO (Energy Saving Trust) 2002
20 *Domestic Heating Compliance Guide.* ISBN 1 85946 225 1. NBS, 2006
21 Domestic Heating Design Guide, DHDG/1 2004. Heating and Ventilating Contractors Association: www.hvca.org.uk
22 *Energy efficient ventilation in dwellings – a guide for specifiers*, GPG 268. Energy Saving Trust 2006

23   The Energy Information (Household Air Conditioners) (No. 2) Regulations 2005 SI 2005/1726

24   *Low energy domestic lighting*, GIL 20. Energy Saving Trust 2006

25   *Reducing overheating – a designer's guide*, CE129. Energy Saving Trust 2006

26   *Assessing the effects of thermal bridging at junctions and around openings in the external elements of buildings*, IP01/06. ISBN 1 86081 904 4. BRE 2006

27   *Checklists for thermal bridging.* Amendment 1 to *Limiting thermal bridging and air leakage: robust construction details and similar buildings*, 2002. ISBN 0 11753 631 8. Department of Transport, Local Government and the Regions (DLTR)

28   *Buildings regulations and historic buildings.* English Heritage 2002 (revised 2004)

29   *Planning policy guidance: Planning and the historic environment.* PPG 15, Departments of Environment and National Heritage 1994, Amended by DCLG Circulars 01/2001 and 09/2005

30   *Guidance to the principles of the conservation of historic buildings.* BS 7913, 1998

31   *The need for old buildings to 'Breathe'.* Society for the Protection of Ancient Buildings. SPAB Information Sheet 4, 1993

32   *Windows for new and existing houses*, CE66. Energy Saving Trust 2006

33   *Energy performance standards for modular and portable buildings.* The Modular & Portable Buildings Association, 2006: www.mpba.biz

34   *Building energy metering*, TM 39. ISBN 1 90328 707 7. CIBSE, 2006

35   *Non-domestic Heating Cooling and Ventilation Compliance Guide.* ISBN 1 85946 226 X. NBS, 2006

36   Specification for Sheet Metal Ductwork – Low, Medium & High Pressure/Velocity Air Systems. DW/144. ISBN 0 90378 327 4. Heating and Ventilating Contractors Association, 1998: www.hvca.org.uk

37   *Guidance for achieving compliance with Part L of the Building Regulations.* Thermal Insulation Manufacturers & Suppliers Association, 2006: www.timsa.org.uk

38   BRE Digest 498, *Selecting lighting controls.* BRE, 2006

39   *Design Guide A – Environmental Design.* ISBN 1 90328 766 8. CIBSE, 2006

40   *Design for improved solar shading control*, TM 37. ISBN 1 90328 757 X. CIBSE 2006

41   *Ventilation of school buildings.* Building Bulletin 101, 2005, Department for Education and Skills (DfES): www.teachernet.gov.uk/iaq

42   BR 364 *Solar shading of buildings.* BRE 1999 (reprinted 2001)

43   *Design of metal roofing and cladding systems: guidance to complement AD L2A and AD L2B.* MCRMA Technical Paper 17, The Metal Cladding and Roof Manufacturers Association, 2006

44   *Measuring air permeability of building envelopes.* ATTMA, 2006

45   *Commissioning code M: Commissioning management.* ISBN 1 90328 733 2. CIBSE, 2003

46   *A practical guide to ductwork leakage testing*, DW/143. ISBN 0 90378 330 4. Heating and Ventilating Contractors Association, 2000: www.hvca.org.uk

47   Commissioning Specialists Association: www.csa.org.uk

48   Commissioning group and ductwork group of HVCA: www.hvca.org.uk

49   Association of Ductwork Contractors and Allied Services: www.adcas.co.uk

50   Lighting industry commissioning scheme. NICEIC Group Limited: www.niceic.org.uk

51 *Building log book tool kit*, TM 31. ISBN 1 90328 771 5. CIBSE, 2006

52 *Use of rooflights to satisfy the 2002 Building Regulations for the conservation of fuel and power*. National Association of Rooflight Manufacturers: www.narm.org.uk

53 BRE U-value Calculator. Available at www.brebookshop.com

Other useful websites

54 BRE (The Building Research Establishment): www.bre.co.uk

55 EST (Energy Savings Trust): www.est.org.uk

56 NBS (National Building Specification): www.thenbs.com

57 English Heritage: www.english-heritage.org.uk

58 SPAB (Society for the Protection of Ancient Buildings): www.spab.org.uk

59 MCRMA (Metal Cladding and Roof Manufacturer's Association): www.mcrma.co.uk

60 ATTMA (Air Tightness Testing and Measurement Association): www.attma.org

17 Access to and use of buildings (Part M)

K.T. Bright

17.1 Introduction

The law concerning access for disabled people to buildings has a relatively short history. The first provisions were contained in the Chronically Sick and Disabled Persons Act 1970. These provisions were mostly advisory and were only applied if it was reasonably practicable to do so. There were no enforcement powers contained in the Act and it proved to be rather ineffective. It was clear that some form of legislation with 'teeth' was required.

Therefore, it was considered that the Building Regulations were the most suitable medium for any future legislation. This resulted in the fourth amendment to the 1976 Regulations, which introduced Part T, Facilities for disabled people, in August 1985. Following that the former Part T was recast in format as Part M, supported by Approved Document M (AD M), and this has been the subject of four further revisions to extend its scope and coverage. The 1999 revision to Part M and the AD M (1999) extended the requirements of M1 to M3 to new dwellings as well as to other types of buildings. That revision came into force on 25 October 1999.

The latest edition to Part M and its AD M (2004) was introduced in early 2004 and came into effect on 1 May 2004. For non-domestic buildings, there are several changes from the recommendations contained in the 1999 edition, although the guidance for domestic properties remains largely the same. Part M now applies generally to material alterations of and extensions to existing non-domestic buildings. It also applies to material changes of use to some non-domestic uses. Historic buildings are also covered by the changes and the AD M (2004).

The guidance on dwellings is unchanged from the 1999 edition of the AD M (1999), although some changes to reference numbers for diagrams and text references have been altered to ensure consistency with the amended document.

Perhaps the most obvious change in the 2004 Part M can be seen in the title 'Access to and use of buildings', which no longer segregates 'disabled people' as a separate user group (the 1999 edition was entitled 'Access and Facilities for disabled people'), and now emphasises the usability of an environment for all users, not just the provision of facilities.

When dealing with the provision of equal access to services, employment opportunities, transport and education for disabled people, reference should also be made to the Disability Discrimination Act 1995 (DDA) and the subsequent amendments and extensions to it.

Since coming into effect on the 1 May 2004, some amendments have been made to the AD M (2004) and clarification offered on some of the guidance it contains. Details can be found at www.communities.gov.uk/index.asp?id=1130902.

17.1.1 The Access Statement – the concept

In undertaking new building work or major refurbishment of an existing one, the AD M 2004 offers guidance on how the minimum requirements of Part M may be met. However, it is important to emphasis that in meeting the requirements of Part M, the guidance offered in the AD M is not prescriptive. It is one potential way of meeting the requirements, not the only way. Indeed the section entitled 'Use of Guidance' states:

> 'However, there may well be alternative ways of achieving compliance with the requirements. Thus there is no obligation to adopt any particular solution contained in an Approved Document if you prefer to meet the relevant requirement in some other way.'

For new buildings, which will often enjoy the benefits of being built on an vacant plot of land, will be freely conceived as a design concept and will be drawn onto a 'blank piece of paper', there should be no reason why full accessibility should not be achieved. This may be done by either following the guidance in the AD M (2004) or by following other, well researched and established guidance. For refurbishments, constraints within the existing fabric and design of the existing building may preclude accessibility being provided precisely as described in the AD M (2004), and compromises may be needed. For either of these situations, the question that needs to be addressed is that, if the AD M (2004) is not to be, or cannot be, followed prescriptively, how can designers justify their design proposals as being equivalent to or more suitable than those described in the AD M.

To address this issue, Part M (2004) introduces the important and exciting concept of 'The Access Statement' as a means of recording and justifying design decisions in all Building Regulation applications.

The Access Statement provides an opportunity for developers, designers, product providers and managers of environments to demonstrate their commitment to ensuring accessibility in the work they undertake. It allows them to demonstrate how they:

- are meeting, or will meet, the various obligations placed on them by legislation; and
- will continue to manage accessibility throughout the delivery of the services they provide or the employment opportunities they create.

Therefore, if a design proposal does not precisely follow the suggested solutions in the AD M, the Statement can be used to demonstrate that the proposal is equivalent (in terms of accessibility) to that described in the AD M (2004) or, perhaps, offers a more suitable solution to a given situation. The use of a Statement expands considerably the opportunity for creative solutions to address accessibility issues and the range of good-practice guidance on which the appropriateness of alternative solutions can be based. However, in being used to justify a proposed alternative, the use of a Statement will also increase the responsibility of the proposer to make sure that the solution actually works in practice.

17.1.2   Developing an Access Statement

An Access Statement should be viewed as a document which 'grows' with the project.

Starting at the strategic level, the Statement should record, and explain, decisions on accessibility associated with the planning, design and ongoing management of a project. The client or project sponsor should take ownership of the early stages by developing a strategy document to demonstrate how access issues will be considered from the initial inception stage through to, and including, the occupancy phases of the project.

The precise form of an Access Statement will depend upon the size, nature and complexity of the proposed building or the space. However, whilst Statements may be project specific, each one should, insofar as it is relevant, contain the following information:

- a brief explanation of the Project Sponsor's policy and approach to access;
- a description of how the sources of advice on accessibility and technical issues will be, or have been, followed;
- details of any consultations undertaken or planned;
- details of any professional advice that has been followed, or will be sought, including recommendations from access audits or appraisals;
- details of access solutions adopted to overcome any access issues, including those that deviate from recognised sources of good practice;
- details of the management and maintenance management policies adopted, or to be adopted, to maintain the features enhancing accessibility.

Where good practice cannot or may not be met, the Access Statement should say why, what the implications are for the users and what other steps are being taken to lessen any adverse effects on accessibility. It should also state why, in the view of the designer, the proposed design feature is considered to be equivalent to the guidance in the AD M (2004) in terms of its effective usability by disabled people.

Importantly, consultation with disabled people and other interested groups should be seen as a crucial element in the preparation of an Access Statement and undertaken as early as possible in the development process.

By developing a document that passes to those who will undertake the long-term management of the facility, the Access Statement process will assist in ensuring that the 'evolving duty' placed on service providers, employers and educators under the DDA can be better addressed.

As a document that 'captures' the good decisions on accessibility made at the design stage, the Statement can also help to ensure that those decisions are not lost or reversed by those responsible for the management of the building throughout its life. Further guidance from the Disability Rights Commission on the suggested content and format of an Access Statement can be found at www.drc-gb.org/businessandservices/bizdetails.asp?id=97&title=bs.

17.1.3   The Access Statement at Part M application stage

An Access Statement for building control applications should be seen as a document that complements the submission and forms part of the application, rather than as a separate document. It should identify the philosophy and approach to the design adopted, the key issues of the scheme and the sources of advice and guidance used. It should also record deviations from the AD M (2004) and the designer's reasoning as to the appropriateness of the alternative proposals.

For extensions to existing buildings or for major refurbishments or work on historic buildings, the Access Statement can be used to identify any constraints imposed by the existing structure and areas where prescriptively following the AD M (2004) would be impracticable or unreasonable.

17.2   Interpretation

DISABLED PEOPLE – Unlike the 1999 edition, M1 of the 2004 amendments does not contain a definition of 'disabled people'. Reference is now made to the need for an environment to be suitable for people 'regardless of disability, age or gender'.

A number of other terms are defined in AD M, that apply throughout the document as follows:

ACCESS – approach, entry or exit.

ACCESSIBLE – with respect to buildings or parts of buildings means that people, regardless of disability, age or gender, are able to gain access.

BUILDING – in addition to dwellings (see section 17.12) the rules in AD M apply to the following buildings:

- shops, offices, factories and warehouses;
- schools, other educational establishments and student residential accommodation in traditional halls of residence;
- institutions;
- premises which admit the public, whether on immediate payment, subscription, fee or otherwise.

The rules apply to the whole building and equally to any parts of the building that comprise separate individual premises.

CONTRAST VISUALLY – the term used to identify the contrast of one element of a building with another in terms of the light reflectance value between the two surfaces.

DWELLING – a house or a flat. New block of flats built as student accommodation should be treated as hotel/motel accommodation in respect of space requirements and internal facilities.

INDEPENDENT ACCESS – access to part of a building (from outside, and therefore from the site boundary and from any car park on site) that does not pass through the rest of the building.

LEVEL – surfaces of a level approach, access routes and landings associated with steps, stairs and ramps means predominantly level, but with a maximum gradient along the direction of travel of 1:60.

OPENING/CLOSING FORCE – the effort (measured in Newtons) required to either open a door from the closed position or to close a door.

Note that the AD M (2004) recommends that the force required to open a door should be no greater than 20 N at the leading edge. This guidance has subsequently been amended to:

'For disabled people to have independent access through single or double swing doors, the opening force, when measured at the leading edge of the door, should be not more than 30 N from 0° (the door in the closed position) to 30° open, and not more than 22.5 N from 30° to 60° of the opening cycle.'

Further details on door opening forces can be found on the web page identified in section 17.1.

PRINCIPAL ENTRANCE (NON-DOMESTIC BUILDING) – the entrance which a visitor not familiar with the building would normally expect to approach.

PRINCIPAL ENTRANCE (DOMESTIC) – the entrance which a visitor not familiar with the building would normally expect to approach or the common entrance to a block of flats.

SUITABLE – means of access and facilities that are designed for use by people regardless of disability, age or gender, but subject to the usual gender-related conventions regarding sanitary accommodation.

17.3   Application

Part M of Schedule 1 to the 2000 Regulations applies to the following:

- a newly erected non-domestic building or a newly erected dwelling;
- an extension to an existing non-domestic building or an existing building that is undergoing a material alteration;
- an existing building or part of a building that is undergoing a material change of use to become a hotel or boarding house, institution, public building or shop.

The requirements of Part M (2004) do not apply to areas that are used solely to gain access for inspecting, repairing or maintaining parts of a building or its services.

17.3.1   Dwellings

Part M does not apply to an extension or material alteration to a dwelling. However, if a dwelling is extended or undergoes a material alteration, any work carried out must not make the building less satisfactory in terms of access than it was before the work or material alteration was carried out. A dwelling that was previously constructed to meet Part M must not be extended or altered in any way which lessens or negates the previously approved work.

17.3.2   Alterations and extensions

A material alteration is the term applied to work being undertaken to a building:

- that currently complies with Part M but, as a result of the new work, no longer will; or
- that does not comply with Part M and, as a result of the work, will be less satisfactory in relation to Part M than it currently is.

In this respect, any alteration or extension that has the potential to reduce the compliance of the building as a whole with Part M must be undertaken in a manner that ensures the existing level of compliance is not lessened.

If a non-domestic building is to be extended, the extension should be treated as a new building and should fully meet the requirements of Part M. If the extension has its own independent access from the outside, the site boundary or, if provided, the car parking area and has its own sanitary accommodation, there may be no requirement to alter the existing part of the building. However, if access to the extension is through the existing building, or the sanitary facilities in the existing part are to be used to support activities in the extension, the routes through the existing building to the extension and/or the sanitary accommodation should be accessible to current Part M (2004) standards.

17.4   Buildings other than dwellings

17.4.1   The main provisions

Reasonable provision must be made in buildings for people, regardless of disability, age or gender to:

- reach the principal entrance to a building and its other entrances from the site boundary, the car parking area and between other buildings (if any) on the same site;
- have safe access into, and within, any storey within the building, and to the facilities available;
- have access to appropriate sanitary accommodation;
- expect suitable accommodation to be provided for wheelchair users, or people with other disabilities, in audience or spectator seating;
- expect aids to communication to be present in auditoria, meeting rooms, reception areas, ticket offices and at information points.

These access provisions should be available to all people who use the building, including employees and visitors.

17.5   Means of access

17.5.1   The approach to the building

All people, regardless of disability, age or gender, should be able to reach the principal entrance into the building or any other entrances that are provided (see section 17.6).

Access should be provided from the entrance into the site curtilage or from any car parking that is provided within the building site.

Routes between buildings on a site should be accessible to all users.

The following recommendations are given in AD M (2004) regarding the approach to a building.

Level approach

- Wherever possible, the approach should be level and not steeper than 1 in 20.
- Approaches should be well lit.
- A surface width of 1800 mm will allow any amount of non-vehicular traffic to use the approach without the need for passing places. A surface width of 1500 mm may be acceptable if appropriately sized passing places (1800 mm wide and 2000 mm long) are also provided. If this width is not possible, perhaps on a restricted site, 1200 mm may be acceptable if the case for the narrowness of the approach is addressed in an Access Statement (see 17.1.1).
- People should be able to use an approach without excessive effort and without the risk of tripping or falling. Surfaces should be firm, durable and slip-resistant, especially when wet. Where different surface materials are used in an approach, there may be an increased risk of users stumbling. To reduce this risk, the frictional characteristics of adjacent materials (or materials used in close proximity) should be similar.
- To minimise the danger of inadvertently walking into a vehicle access route, pedestrian routes should be appropriately separated. Where crossing points for pedestrians are necessary across the vehicular route, a tactile method of warning of the potential danger (blister pattern) should be provided (see Fig. 17.1).
  Note that the colour of the tactile warning surface at controlled crossings should be red. At uncontrolled crossings, the colour of the surface should be buff or any colour that visually contrasts with the adjacent pavement – except red.
- If the approach needs to be steeper than 1 in 20, an appropriately designed ramped approach should be provided (see section 17.5.3 and Fig. 17.3).
- Some people find it easier to negotiate steps than ramps, therefore easy-going steps should complement a ramped approach (see section 17.5.4 and Fig. 17.3).
- To advise visually impaired people of an imminent hazard, tactile warning surfaces (corduroy pattern) should be provided at the top of and bottom of flights of steps. Examples of appropriate designs for tactile hazard surfaces are shown in Fig. 17.1.
- The area up to 2100 mm above ground level should always be kept clear to reduce the risk of head collisions for all users, but especially for those with restricted vision (see section 17.5.6).

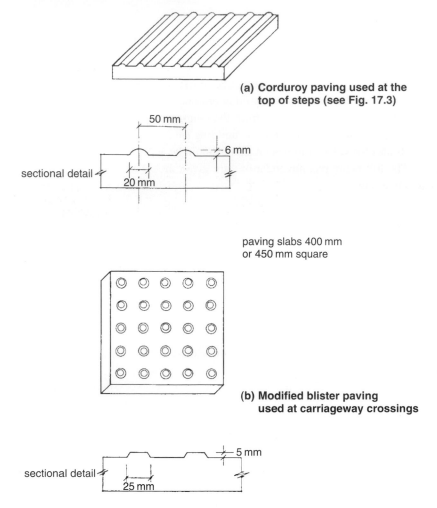

(a) Corduroy paving used at the top of steps (see Fig. 17.3)

50 mm

6 mm

sectional detail

20 mm

paving slabs 400 mm or 450 mm square

(b) Modified blister paving used at carriageway crossings

5 mm

sectional detail

25 mm

Fig. 17.1   Tactile pavings.

17.5.2   On site car-parking and setting down

People who travel to a building by car should be provided with car-parking facilities appropriate to their needs. This includes the position of the parking spaces within the site, the design of the bays and the approach from the car-parking area to the entrances into the building.

Generally, the surfaces to car parks should be firm, and the marking out of bays should allow sufficient space for motorists to exit and enter their vehicle comfortably and without undue hazard.

If barriers are used to control entry into and/or out of the car park, or payment is required to use the car park, the facilities provided should be usable by all people using the car-park.

For people who arrive at a building as passengers, a setting-down point to allow them to exit the vehicle close to the entrance should also be provided.

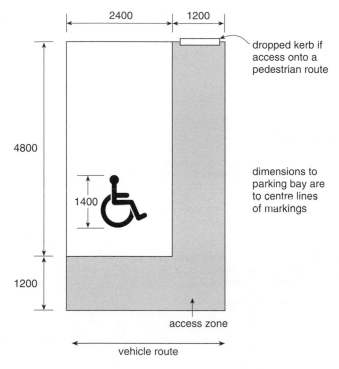

Fig. 17.2 Designated parking bay.

The following recommendations are given in AD M (2004) regarding the designated disabled car-parking provision for a building:

- At least one designated and appropriately identified parking bay should be provided close to the principal entrance of the building for use by disabled people.
- The location of the parking bay should be clearly sign-posted both at the bay itself and from the entrance to the car-park.
- The car-park surface should be a firm surface which is level and durable. The area within the access zones to the side and rear of the designated parking bay should be slip-resistant (see Fig. 17.2).
- Ticket machines, if they are required to be used by disabled people, should be accessible in terms of the approach to the machine and the height of the controls.

17.5.3   Ramped provision

If ramps are necessary, perhaps because constraints within a site necessitate the use of approaches at 1:20 or steeper, they should be as shallow as practicable, of sufficient width and provided with surfaces that are slip resistant, especially when wet. Appropriately designed handrails, visually contrasted with the background, should be provided on both sides of the ramp, and lighting to the ramp should be appropriate in daylight and under artificial lighting conditions (see Fig. 17.3).

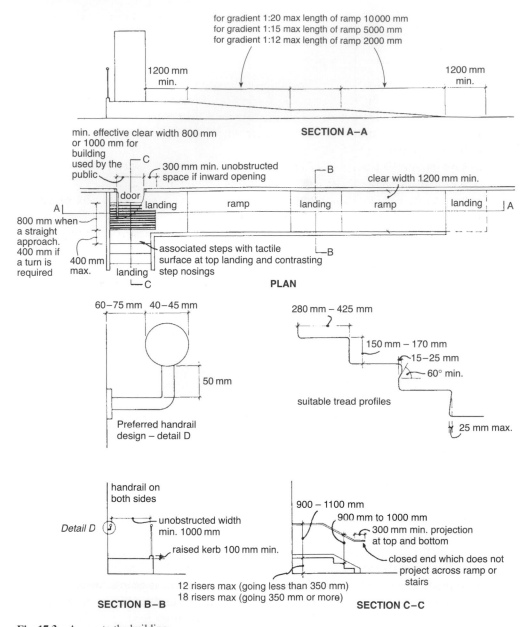

Fig. 17.3 Access to the building.

Steep ramps can be difficult for many users of buildings to negotiate, but especially so for older people, people with restricted mobility and wheelchair users. Equally, ramps can present difficulties for disabled people and their helpers, who may need to make frequent rest stops when negotiating long ramps. Therefore, adequate space to allow people to pass on ramps and the slip resistance and nature of the surface finish are also design and use issues.

The recommendations in AD M (2004) regarding the provision of ramps on an approach can be summarised as follows.

Ramps should:

- be readily identifiable or their location clearly sign-posted;
- at 1:20, have no flight with a going greater than 10 m or a rise of more than 500 mm;
- at 1:15, have no flight with a going greater than 5 m or a rise of more than 333 mm;
- at 1:12, have no flight with a going greater than 2 m or a rise of more than 166 mm;
- have no gradient steeper than 1:12;
- have intermediate landings of 1800 mm x 1800 mm where a wheelchair users cannot see along the whole length of the ramp before starting to use it;
- have a landing at top and bottom at least 1200 mm long, clear of any door swing;
- have intermediate landings at least 1500 mm long, clear of any door swing;
- have on any open side of a flight or landing a raised kerb at least 100 mm high, appropriately visually contrasted with the ramp or landing;
- have a surface width of at least 1200 mm;
- have an appropriately designed and visually contrasted handrail on both sides;
- have a slip-resistant surface with adjacent surface finishes having similar frictional characteristics;
- have clearly signed alternative steps where the rise of the ramp is more than 300 mm.

17.5.4   Stepped access

Many of the design factors that apply to ramps also apply to steps (see Fig. 17.3).

However, there are a few additional factors to be considered. For example, sudden changes of level marked by steps may create dangers for people:

- with impaired sight; or
- with impaired hearing who may be communicating with companions or others when walking along and using their less sensitive peripheral vision to identify hazards in front of them.

Therefore, the presence of each individual step in a flight should be made apparent with appropriate visual contrast. Single steps are not permitted in an access route.

Some ambulant disabled people may have stiffness in their hip or knee joints or may need to wear calipers. If the steps have projected nosings or open risers, there is a risk that they may catch their feet or calipers under nosings or treads on the staircase. Therefore, sharply projecting nosings and open risers should not be used on an access route.

People with physical weakness on one side or the other or who have sight impairments will need to place their feet squarely onto the treads. Therefore, steps should:

- have a level landing with an unobstructed length of not less than 1200 mm at the top and bottom of each flight;

- be provided with a 'corduroy' hazard warning surface to the top and bottom landings.

In essence, the surface of the steps/stairs should:

- have a clear width at surface level of 1200 mm and a clear width between handrails of 1000 mm;
- have a rise and going which is consistent throughout the flight, and risers that are not open;
- have uniform rises between 150 mm and 170 mm high and uniform goings between 280 mm and 425 mm long; in schools, the preferred dimension for the riser is 150 mm and for the going is 280 mm;
- if provided, have projecting nosings that blend smoothly with the tread and riser (see Fig. 17.3);
- have no more than 12 risers between landings if the going of each step is less than 350 mm;
- have no more than 18 risers between landings if the going of each steps is 350 mm or greater;
- have nosings that are 55 mm wide on both the riser and the tread and which visually contrast with the surface of the tread and the riser.

17.5.5   Handrails to steps and ramps

Handrails can have a major impact on the accessibility and usability of a building. That can be a positive impact if they are appropriately designed, installed and visually contrasting, and a negative impact if they are not. For many users, the provision of appropriate handrails is critical to their safe and independent use of an environment.

The AD M (2004) makes several recommendations about the provision of handrails to steps and ramps. In essence, handrails should:

- extend horizontally at least 300 mm beyond the top and bottom of a ramped or stepped access; the design should ensure that the extension does not project hazardously into a circulation route and the end of the handrail should be designed so that the risk of catching clothing is minimised;
- be placed between 900 mm and 1000 mm above the surface of a ramp or pitch-line of stairs, and between 900 mm and 1100 mm above the surface of any landings; if a second lower handrail is provided, the top of the handrail should be 600 mm above surface level or pitch-line;
- be easy to grip; the profile of the handrail can be circular or oval, with oval being preferred – if round, the diameter should be between 40 mm and 45 mm; if oval, the width should be 50 mm and the depth 38 mm with a 15 mm radius at the corners (see Fig. 17.3);
- visually contrast with the background against which they will be viewed;
- be continuous on each side of a flight and around landings.

Other recommendations are described in Fig. 17.3.

17.5.6   Hazards on access routes

Where a circulation route passes close to a building, care should be taken to avoid projections that might be a hazard to building users.

If there is a projection of more than 100 mm into an access route or if doors or windows can swing outwards into the route, the area of projection should be protected with a permanent barrier to minimise the risk of collision and injury to the building users. The barrier should include a kerb or solid barrier at low level to assist people using a mobility cane.

Areas under stairs or ramps (including landings) that are less than 2100 mm above ground or floor level should be protected by a permanent guardrail to minimise the potential danger of collision for all users, but especially those with restricted vision.

17.6   Access to the building

In general, buildings should be designed so that there is convenient access into and out of the building, and that all entrances and exits are suitable for the people who use the building. This includes the doors and any lobbies that may be incorporated in the design.

In a limited number of cases, space restriction, congestion or sloping ground may make the provision of a single accessible main entrance impossible. Similarly, physical constraints may prevent the provision of car parking spaces close to the principal entrance. In such cases it could be appropriate to provide an additional accessible entrance, but it is important that this entrance is designed and used as an additional entrance for general use, rather than a segregated facility for disabled people. It should also be located in a position whereby suitable access to all internal areas, including the principal entrance, is ensured.

The entrance into the building should not be obstructed with hazards such as supports for canopies, entry phones or controls for powered opening devices. Thresholds should be level. Any weather protection provided must be suitable for use by wheelchair users and not provide a tripping hazard for other users. Coir matting should not be used.

If it is necessary to provide a separate accessible entrance, its location should be clearly identifiable. All accessible entrances should be appropriately signed using the International Symbol of Access.

If an entrance door is opened manually (i.e. without the assistance of a manually activated door control or an automatic opening device) a means of protecting the users from the weather should be provided:

- a level landing at least 1500 mm x 1500 mm immediately in front of the entrance. The area should be clear of any door swings and the surface should be of a material that is suitable for wheelchair users;
- a threshold that is level or presents a maximum upstand of 15 mm with the surrounding area; any upstand in excess of 5 mm should be chamfered or rounded.

If an entry system is provided it must be suitable for people with vision, hearing and speech impairments. If a mat well is provided, the upper level of the mat should be level with the surrounding floor area.

17.6.1   Entrance doors and lobbies

Some building users, but especially wheelchair users, people carrying luggage, people with assistance dogs, and parents with pushchairs and small children, will need sufficient space when negotiating an entrance door and entrance lobby. For all new buildings, the entrance door should, when open, provide an effective clear width that is sufficient to allow all users to pass through with relative ease. If, in an existing building, this is not possible, the reasons should be made in an Access Statement, together with an indication of what other measures will be taken to lessen the impact of the reduced level of accessibility.

For an entrance door in a building to be used by the general public the effective clear width should be 1000 mm. For entrance doors into other buildings the minimum effective clear width is 800 mm, although if the door is approached from an access route less than 1200 mm wide, the minimum effective clear width is 825 mm. These dimensions apply to a door which is a single leaf, or for one leaf of a double leaf door.

Whichever type of door is used, it is important to allow sufficient room for the door to be opened by a person in a wheelchair. Therefore, for manually opened doors there should be an unobstructed space of at least 300 mm on the pull side of the door between the leading edge of the door and the return wall.

To assist all users of a building (including children), vision panels should be provided in all door leaves and side panels over 450 mm wide, unless it can be argued in an Access Statement that a panel is not appropriate (i.e. for reasons of security, etc). The vision panel should be located towards the leading edge of the door and should cover a minimum visibility zone extending between 500 mm and 1500 mm above floor level (see Fig. 17.4). If necessary, this zone can be interrupted between 800 mm and 1150 mm above floor level to accommodate a horizontal rail to the door.

If it is not possible to achieve a door opening force less than the maximum permitted (see section 17.2), assisting the user by providing power operated opening and closing systems should be considered.

If glass doors are provided their presence should be made clear with manifestation that visually contrasts with the background against which it will be viewed. That applies to how the users will view the manifestation when both entering and exiting the building. The manifestation should be provided at two levels, one between 850 mm and 1000 mm and another between 1400 mm and 1600 mm above floor level. Glazed doors that can be held in the open position represent a potential collision hazard for all users, but especially for those with restricted vision. Therefore, on doors that can be held open, visual contrast should be provided to the leading edge of the door. In situations where visual contrast may not be effective, some other method of minimising the hazard such as, for example, protective guarding, could be considered.

17.6.2 Powered entrance doors

If powered entrance doors are provided they should be provided with controls that are easy to use, reachable by all users, and incorporate safety features that ensure the safety of all users of the building. Features such as the dwell times for the doors when open, and the speed at which they open and close, should be carefully designed and managed.

The position of manual controls for powered door systems should be logical and clearly identified using signage and visual contrast. Any manual controls should be located between 750 mm and 1000 mm above ground level and set back 1400 mm from the leading edge of the door when in the fully open position.

When open, doors should not project into an access route. If the doors swing towards people approaching the building, visual and audible warnings should also be provided.

17.6.3 Revolving doors

Revolving doors can create particular difficulties for some disabled people and risks for assistance dogs and people using mobility aids such as canes. They are also difficult to use for parents with pushchairs and or small children. Therefore, they are not considered to be accessible.

If a revolving door is provided, there should also be:

- an additional accessible manually operated non-powered door with:
  ○ a required opening force of less than 30 N;
  ○ appropriate visual contrast;
  ○ door furniture that can be operated using one hand or a closed fist;
  ○ a 300 mm clear space to the leading edge; or
- a sliding, swinging or folding action door that is controlled by a push pad, card swipe, coded entry, automatic sensor or other proximity sensor.

17.6.4 Entrance lobbies

The design of a lobby should allow for assistance to be given if required and to allow people to avoid others who may be passing in the opposite direction. It should also be possible to move clear of the swing of one door when opening the next. For this reason entrance doors and lobbies need to be built to certain minimum dimensions. Figure 17.4 shows the design principles stated in the AD M (2004).

The force required to open doors to lobbies should not exceed 30 N (see section 17.1) and any glazing used in the doors should be of an appropriate size (see section17.6.1) and not create distracting reflections.

17.7   Access around the building

Once inside the building, users must be able to identify the location of, travel to, and use, the facilities provided.

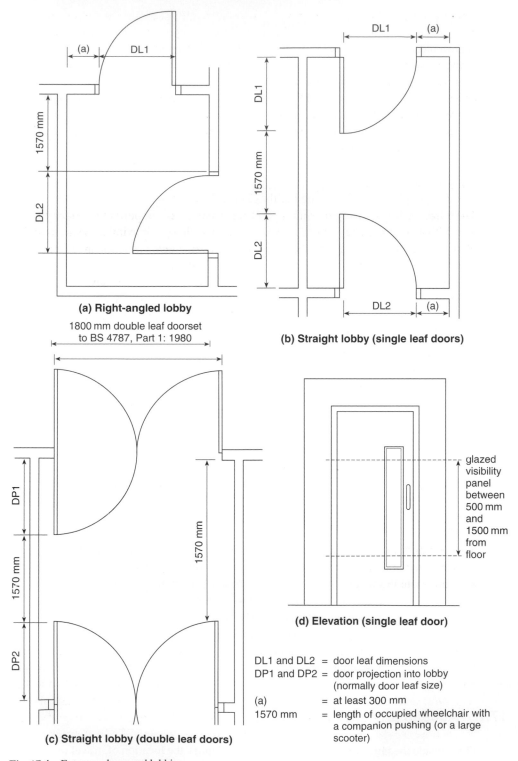

(a) Right-angled lobby

1800 mm double leaf doorset
to BS 4787, Part 1: 1980

(b) Straight lobby (single leaf doors)

(c) Straight lobby (double leaf doors)

glazed
visibility
panel
between
500 mm
and
1500 mm
from
floor

(d) Elevation (single leaf door)

DL1 and DL2 = door leaf dimensions
DP1 and DP2 = door projection into lobby
                        (normally door leaf size)

(a)   = at least 300 mm
1570 mm   = length of occupied wheelchair with
                        a companion pushing (or a large
                        scooter)

Fig. 17.4   Entrance doors and lobbies.

Different building types will have a variety of facilities, and some will be unique to a particular use. The AD M (2004) does not attempt to provide exhaustive guidance on all the facilities that may be relevant. It divides guidance into design considerations and provisions for both vertical and horizontal circulation. In general terms, however, it is clear that the objective of any building design should be to allow all people to travel vertically and horizontally within the building to reach and use facilities, and for that to be done conveniently and without undue discomfort.

The AD M then considers the journey around a building as a sequential horizontal and vertical route, starting with the entrance hall and reception areas.

17.7.1   Entrance hall and reception area

The entrance hall and any reception point represent a very important interface for any user with a building. They can be critical for those wishing to maximise the accessibility of an environment and it is necessary to provide a sufficient space to manoeuvre for the comfortable and safe use of all people visiting the building.

If a reception point is provided, its position should be logical and easily identifiable for anyone entering the building. There should be sufficient space for all users to manoeuvre around the point with relative ease and the design of the desk should allow easy and comfortable communication between the visitor and those staffing the reception desk. The design of the reception desk should also allow ease of use by standing and seated visitors for activities such as, if required, signing-in. There is not a requirement to have a knee recess at a reception desk, but this would clearly be needed if such management activities as signing-in were adopted.

Generally a reception desk should have:

- a clear manoeuvring space of 1200 mm deep by 1800 mm wide in front of the desk (if a knee recess of at least 500 mm deep is provided);
- a clear manoeuvring space of 1400 mm deep by 2200 mm wide (if no knee recess is provided);
- one section of the desk that is at least 1500 mm wide and with a writing surface between 700 mm and 760 mm above finished floor level;
- a hearing enhancement system;
- slip-resistant floors adjacent to the desk;
- appropriate illuminance.

17.7.2   Internal doors

All doors are potential barriers to free and easy access into and around buildings. Therefore, their use should be restricted to places where their use is essential to address the activities taking place in the building. If doors are required, they should have a suitable effective clear width and an unobstructed space at the leading edge (minimum 300 mm). Internal doors should be suitable for all users to open and close with relative ease, or provided with assisted openers or, where appropriate, hold-open devices that will minimise the impact of their presence in everyday use.

Recognising the guidance laid down in BS 8300:2001 on the installation and use of internal doors, the AD M (2004) recommends that internal doors are:

- able to be opened by the user with a force applied at the leading edge of no greater than 30 N;
- provided with appropriate visual contrast to the door furniture, the leading edge of the door (if it could present a collision hazard) and between the door frame and the surrounding wall;
- provided with a vision panel as described in section 17.6.1;
- appropriately provided at two levels with manifestation as described in section 17.6.1;
- designed not to open into circulation routes (except doors to minor utility facilities). The door to a unisex wheelchair accessible WC may open onto a non-major access route or escape route providing that a minimum clear width of 1800 mm exists in the corridor when the door is open.

If double doors are provided across a corridor that forms a major access route or an escape route, and the doors within the set are of unequal width, the wider leaf should be located on the same side throughout the length of the corridor.

17.7.3   Corridors

Corridors and passageways generally should have a clear width of at least 1800 mm or, where this is not possible, they should have passing places at least 1800 mm long by 1800 mm wide at reasonable intervals along the route. In school buildings where lockers may be provided in corridors the preferred minimum width is 2700 mm.

Floors should be level (maximum gradient 1:60). Where that is not possible, a gradient of up to 1:20 can be provided as long as the overall vertical change in level does not exceed 500 mm without an appropriately designed level rest area being provided (minimum 1500 mm long). Any sloping floor section should extend for the full width of the corridor or, where this is not possible, the exposed edge should be identified with visual contrast and/or a protective barrier.

Gradients steeper than 1:20 should be designed as a ramp (see sections 17.5.3 and 17.5.6).

Floor surfaces should be firm and slip resistant, and finishes should not be visually confusing in terms of the pattern or design used. In general, the materials and surface finishes used in corridors should, wherever possible, contribute to a good acoustic design for the space, rather than detract from it.

If glazed screens are provided alongside a corridor, the potential hazard should be identified using appropriate manifestation (see section 17.6.1).

17.7.4   Internal lobbies

The principles of design that apply to the provision of internal lobbies are the same as described in section 17.6.1 for entrance lobbies. Therefore, internal lobbies should meet the minimum dimensions shown in Figs 17.4.

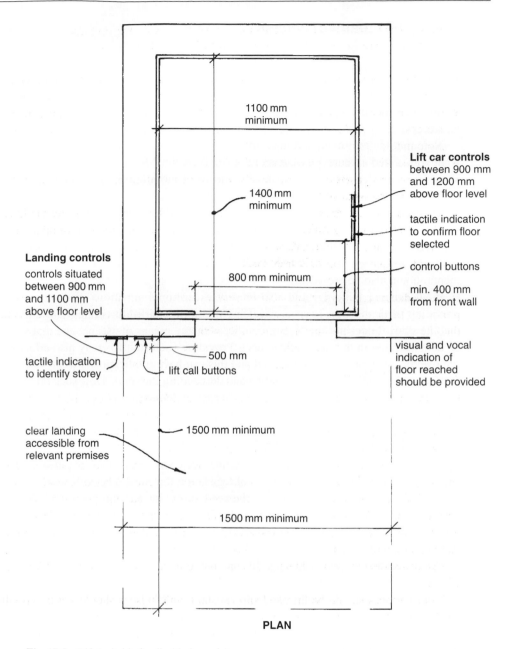

Fig. 17.5 Lift (suitable for disabled people).

17.8   Vertical means of access

17.8.1   Generally

A lift is the recommended way of providing vertical access for people moving from one storey to another in all buildings. New developments should have a passenger lift serving all storeys.

Note that the minimum nett floor areas that were described in the AD M (1999) for what constituted a 'storey' have been removed from the AD M (2004). Therefore, a lift should be considered as the most suitable means of moving people from one storey to another, regardless of the floor area of the storey.

If, because of site or other constraints, it is not possible to provide a passenger lift, an appropriately designed platform lift may be suitable. However, the case for providing it in preference to a passenger lift should be argued in an Access Statement. Whichever one is chosen, a staircase suitable for the needs of people with ambulant and sensory impairments should also be provided.

If, in existing buildings and also in very exceptional situations, the only way of providing vertical access is via a wheelchair platform stairlift, it is important to ensure that the stairlift does not impinge or conflict with any means of escape, and that should be demonstrated in the Access Statement. Providing such stairlifts may also effectively remove the availability of one handrail and this can have implications for any user who requires a handrail when both ascending and descending the stairs. How such effects can be minimised should also be argued in the Access Statement.

17.8.2   General requirements for lifting devices

The AD M (2004), requires any installed lifting device to be fit for its intended purpose, and identifies a series of relevant pieces of legislation that should be addressed.

Other general requirements relate to the need to provide appropriate illuminance and general lighting design on the approach to the lift and in the lift car, the provision of controls that allow relatively easy and safe use by all users, and the importance of providing surfaces that will not contribute to visual confusion for users of the lift.

Recommended ways of achieving this include:

- The routes to any lift should be accessible and it should be possible to access the lift from all areas of the storey it serves.
- There should be an unobstructed area to manoeuvre of at least 1500 mm x 1500 mm outside the lift doors, or a straight approach to the doors at least 900 mm wide.
- Tactile lift call buttons on the landings that visually contrast with their surroundings should be provided. These should be located between 900 mm and 1100 mm above finished floor level, and at least 500 mm from any return wall.
- Light-coloured floor coverings in the lift car should be provided.
- Floor coverings should display similar frictional characteristics to the surfaces provided to the landings.
- There should be a handrail in the lift car on at least one side of the car located with the top surface 900 mm above the finished floor level in the lift car.

- If the size of the lift does not allow a wheelchair user to turn around within the car, a mirror should be provided to the rear wall to assist them when reversing out of the lift. Full-height floor-to-ceiling mirrors can cause confusion for visually impaired people and present an impact hazard for wheelchair users. It is important that mirrors on the rear wall of a lift car do not extend below handrail height.
- There should be an emergency call system. It is important that the system provided addresses the needs of all users of the lift including those of people with sensory impairments (vision and hearing).

17.8.3   Passenger lifts

The number and size of the lifts provided should be suitable to address the needs of all users of the building. This will depend on anticipated demand and the number of disabled people who may be using the building.

In addition to the general recommendations described above, the AD M (2004) identifies the following issues that should also be addressed in the provision of passenger lifts (see Fig. 17.5):

- The minimum dimensions specified in the AD M (2004) as being suitable for a passenger lift are 1100 mm wide and 1400 mm deep. This size is suitable to accommodate a wheelchair user and an accompanying person.
- A larger lift with internal dimensions of 2000 mm wide by 1400 mm deep will accommodate a wheelchair user and several other passengers, as well as people using larger wheelchairs. A lift of this size will also allow a wheelchair user or someone using mobility aids to move through 180° and exit the lift travelling forwards.
- The landings and the doors to lift cars should contrast visually with the surrounding walls and the minimum effective clear open width for the doors should be 800 mm.
- Dwell times of doors must allow adequate time for all users, including those with assistance dogs, to enter and leave the lift without undue hazard.
- Within the lift car, controls should be located between 900 mm and 1200 mm above finished floor level (1100 mm preferred). They should not be closer than 400 mm to any return wall.
- Audible and visual information should be provided to indicate the location of the lift, its arrival and the floor level it has reached.
- On a landing-call buttons should be reachable and provided between 900 mm and 1100 mm above finished floor level.
- The lift car should be designed as a visually and acoustically accessible environment that can be used by all users, including people with sensory impairments. If glass is provided in the enclosure to the lifts its presence should be clearly identifiable, perhaps with appropriate manifestation (see section 17.6.1).
- Lifts that are designed to evacuate people in an emergency should conform to the recommendations in BS 5588–8.

It is necessary to ensure that the door controls can be overridden in order that lift passengers do not get caught in the closing doors. A door re-opening activator that uses a photo-eye or infra-red detector or similar non-contact sensor should be provided.

17.8.4   Lifting platform

The availability of lifting platforms in buildings should be restricted to use by wheelchair users, ambulant disabled people and their companions/carers. Lifting platforms are not suitable as a means of providing vertical access for other users and should not be installed for that purpose.

Any controls should be installed such that they can be used independently by a wheelchair user or by their carer. Clear instructions on how to use the platform should be given. Instructions should be provide in a way that meets the communication needs of all disabled people likely to require them, including those with sensory impairments.

In general, platforms are operated by applying continuous pressure to the call button and the movement controls. The means of providing continuous control should take into account the needs of users who experience restricted mobility or limited dexterity and those who are easily fatigued. The position of the controls differs slightly from that described earlier for passenger lifts in that the controls should be located between 800 mm and 1100 mm above finished floor level, although the distance of 400 mm from any return wall is the same. The position of landing call controls for lifting platforms is identical to that for passenger lifts (see section 17.8.3).

The speed of movement of the platform should not exceed 0.15 m/s.

The minimum clear dimensions of a lifting platform also vary from that specified for a passenger lift with two smaller sized platforms being identified as acceptable in certain circumstances. Recommendations for clear dimensions are generally as follows:

(a)   800 mm wide by 1200 mm deep where the platform is not enclosed and where the disabled person using the lift will be unaccompanied when doing so;

(b)   900 mm wide by 1400 mm deep where the platform is enclosed and provision is being made for the disabled person to be accompanied by a companion or carer;

(c)   1100 mm wide by 1400 mm deep if the platform is as (b) above, but the doors used to enter and exit the platform are not opposite each other (requiring the disabled person to turn through 90° to exit the platform).

Entrance doors should contrast visually with the surrounding wall and, if glass doors or enclosures are provided, their presence should be identified using appropriate manifestation. The minimum effective clear width for doors to the platform should be 800 mm for (a) and (b) and 900 mm for (c).

As with passenger lifts, people using the lifting platform should be given audible and visual advice on the position of the lift, in terms of the arrival of the lift and the floor level reached.

The vertical rise of a lifting platform without a lift-way enclosure should not exceed 2 m, although this can be exceeded if an enclosure is provided.

17.8.5   Wheelchair platform stairlifts

In *existing buildings* that are being converted or altered, some small floor areas may exist that accommodate unique facilities such as, for example, staff rest areas, training rooms or small galleried libraries. In such areas, suitable access for wheelchair users must be provided but it may not be practical to provide a full passenger lift or a lifting platform as described above. In such situations the provision of a wheelchair platform stairlift may be considered. However, this facility is only appropriate in those situations where users can be given instructions in their use and where appropriate management supervision can be ensured.

Therefore, they would not be suitable for situations where general independent, unmanaged access by users is possible, and management practices to ensure this should also be in place.

The minimum clear dimensions of the platform should be 800 mm wide by 1250 mm long and access to the platform should be a minimum of 800 mm effective clear width.

If a wheelchair platform lift is provided, similar issues relating to the controls provided such as, for example, continuous pressure controls, user fatigue, visual contrast, speed of movement, etc. that apply to lifting platforms also apply here.

17.8.6   Internal stairways

The design considerations for internal stairs, including the provision of handrails, are the same as those for external stairs with the following exceptions.

Hazard warning surfaces are not required at the top and bottom of a flight of steps internally. In general, this is because it is difficult to control the difference in the slip resistance factors of adjacent materials that may be used internally for the hazard warning area and the adjacent flooring (timber, carpet, ceramics, vinyl, etc.), and the permutations in which they will be used. Whilst hazard warnings are required at the top and bottom of external stairs, the materials used in such areas are more limited (concrete, tarmac, etc.) and difference in slip resistance factors are more easily determined and controlled in use. However, the AD M (2004) does advise designers to consider other methods of reducing the potential risk posed by stairs by, for example, providing appropriate visual contrast and good design and ensuring the stairs are not sited in the direct line of an access route.

In normal circumstances:

- a flight should contain no more than 12 risers (each between 150 mm and 170 mm deep), but, in exceptional circumstances, for example in small buildings, this may be increased to 16 risers;
- the going of each step should be at least 250 mm although 300 mm is preferred. However, for school buildings, the going should not exceed 280 mm.

17.8.7   Internal ramps

The design considerations for internal ramps, including the provision of handrails, are the same as those for external ramps with the following exceptions:

- for changes in level of less than 300 mm, a ramp can be provided instead of a single step;
- landings must be level and have a maximum gradient along their length not greater than 1:60.

17.9   Facilities in buildings other than dwellings

All people, regardless of disability, age or gender should be able to get access to, and use, all of the facilities within buildings. This includes facilities used in enjoying educational, business, leisure and social activities, and regardless of whether the person using the facility is a spectator, a participant or a member of staff.

17.9.1   Access to restaurant and bar facilities

Facilities should be located in areas that have suitable access by all users and, if changes in level are incorporated into a design, all users, including wheelchair users, should be able to gain access to where the facility or activity is located. If changes in level are provided in refreshment areas such as, for example, restaurants, cafeterias and bars, suitable access to all floor areas should be provided.

The provision of permanent or temporary seating should always allow for disabled people to have a choice of position from which to participate in or view an activity taking place, and be located in a position that affords a clear view without obstructing the views of others.

The full range of services offered should be accessible including, where provided, bars and self-service counters. Counters should be designed to allow use by wheelchair users, either for the whole facility or in a section of it.

17.9.2   Audience and spectator facilities

In addition to the guidance already described for reception counters (see section 17.7.1), consideration should also be given to how all people, but especially those with restricted mobility and sensory impairments, can view, listen to, communicate during or participate in activities as a member of an audience.

In general, routes to and from all seating areas and associated facilities should be accessible, and the seating areas for disabled people should, wherever possible, be integrated into the auditoria, not segregated into special areas. The seating areas should be provided with appropriate space for a wheelchair user to manoeuvre and some should have sufficient space adjacent to them (in front of, or under, the seat) to accommodate an assistance dog. Some seats should also be provided with detachable or lift-up arms. Wherever possible, wheelchair users, people who have difficulty in using seats with fixed arms and those with assistance dogs should also have the opportunity of sitting next to a non-disabled companion or carer.

Auditoria provided with sections of seating that are removable will offer a greater choice of seating location for disabled people, and also improve flexibility in terms of the number of disabled people who can be accommodated at any one time. Some sections, perhaps at the back of a block of seats or at the end of rows, may be provided with extra legroom to accommodate users of large stature.

The AD M (2004) suggests that for auditoria with a seating capacity of up to 600, at least six seats should be permanently available for wheelchair users, with a further six seats capable of being made available (removable seating) if required. For auditoria with between 600 and 10 000 seats, at least 1% of the total seating capacity is required. The provision of any additional removable seating in addition to this amount is left to the discretion of the designer or the venue providers. For auditoria with over 10 000 seats the AD M draws attention to the guidance contained in a Football Foundation and Football Licensing Authority document entitled *Accessible Stadia: a good practice guide to the design of facilities to meet the needs of disabled spectators and other users*.

17.9.3   Lecture and conference facilities

In lecture and conference facilities, consideration should always be given to ensuring that sight-lines to the screen, lectern, etc. are appropriate, and to the lighting, acoustics and general décor within the area. People who are deaf or hard of hearing should be able to participate fully in conferences, committee meetings and associated groups.

Podiums and presentation facilities (where provided) should be accessible to all users.

17.9.4   Entertainment, leisure and social facilities

In places of entertainment such as, for example, cinemas and theatres, where seating may be more closely sited, the location of accessible seating should always ensure that all users can enjoy the atmosphere of the area. In all cases, reference should be made to *Technical standards for places of entertainment*, published by the District Surveyors Association and the Association of British Theatre Technicians (ISBN 1–90403105–6).

17.9.5   Sports facilities

The provision of facilities in sports stadia should also take into account the guidance documents specified in the AD M (2004).

17.9.6   Refreshment facilities

Refreshment areas should be accessible to all users, whether they are using them independently or with companions. If the design of an area includes the provision of different levels, perhaps to designate different functions or enhance aesthetics, all levels should be accessible. This applies to public and staff areas.

Any facilities provided such as toilets, telephones and external terraces should also be accessible to all users. If the refreshment area includes self-service and served areas, all customers should have access to both.

Bars and serving counters should include a lowered section (not more than 850 mm above finished floor level) for use by wheelchair users. Any worktops provided in shared refreshment facilities should include a lowered section (not more than 850 mm above finished floor level) and a clear space underneath of not less than 700 mm above finished floor level (see also Fig. 17.6).

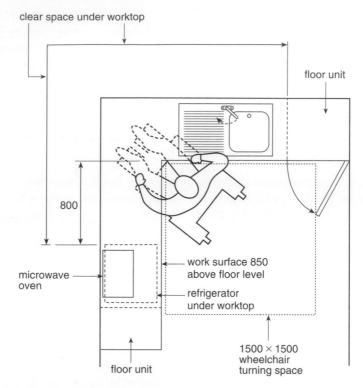

clear space under worktop

floor unit

800

microwave oven

work surface 850 above floor level

refrigerator under worktop

1500 × 1500 wheelchair turning space

floor unit

Fig. 17.6   Refreshment facilities (suitable for disabled people).

17.9.7   Sleeping accommodation

The AD M (2004) contains a considerable number of recommendations related to the design of accessible bedrooms, associated sanitary facilities and the provision of switches, outlets and controls.

In essence, where sleeping accommodation is provided, for example in hotels, motels and student accommodation, it should be designed to be convenient for all users (see Fig. 17.7). To accommodate the increased spatial needs of wheelchair users, people with restricted mobility and those with assistance dogs, some rooms should be designed and located to address these needs. However, all facilities within the building should be accessible to all people using it.

The AD M (2004) recommends that at least one wheelchair-accessible bedroom should be provided for every 20 bedrooms, or part thereof, in the building. Therefore, if 21 guest bedrooms were provided, two of them would have to be suitably accessible for a wheelchair user. Accessible bedrooms should be of a standard similar to other bedrooms within the building and provided such that a wheelchair user has a reasonable choice of location.

Within bedrooms, sufficient space should be provided to enable a wheelchair user independent or assisted transfer to at least one side of the bed. There should be sufficient manoeuvring space within the room to allow wheelchair users to easily access en-suite facilities and operate switches, controls and facilities such as wide-angle viewers in doors. An appropriate emergency assistance alarm and associated reset button should be provided in each wheelchair accessible bedroom and en-suite facility.

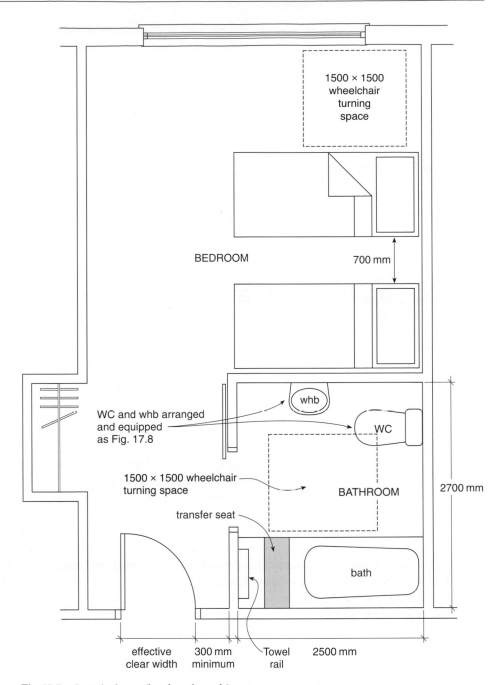

Fig. 17.7 Guest bedroom (hotels and motels).

Facilities for hanging or stacking clothes and other belongings should be provided at accessible locations within the room and be convenient to use. Consideration should be given to providing automatic or remotely triggered controls for facilities such as curtains and blinds. Other facilities such as, for example, telephones and televisions should also be accessible in both their location and the manner of operation.

Wheelchair users may wish to visit companions in other bedrooms when on holiday or attending residential conferences/courses, and the design of the building should allow for this. In rooms that are not specifically designed for independent use by wheelchair users, the space available to manoeuvre may be insufficient. Therefore, the main entrance door into all bedrooms should be appropriate in terms of width, maximum opening forces and position to allow access into the room by wheelchair users (see section 17.7.2 and/or AD M (2004) – Table 2).

17.9.8   Aids to communication

All people will benefit from environments where the design of the building or the provision of enhancement systems allows for an appropriate standard of communication to be achieved.

For all users, the type and quality of public address systems can be critical in their ability to gather information. Therefore, systems should be carefully chosen to suit the environment and the purpose for which they will be needed. In addition, good lighting design, visual contrast and acoustics can all play an important part in a person's ability to communicate within an environment, and this is especially so for people with sensory impairments (vision and hearing).

In that respect, recommendations included the following:

- Avoid shiny, reflective surfaces and designs that do not adequately control glare;
- Avoid large or highly visually contrasting patterns on walls and floors;
- Where used, artificial lighting should not create glare or strong pools of bright light and strong shadow;
- Avoid lighting regimes that provide uplighters at floor or low level;
- Where appropriate, such as, for example, at reception areas and in meeting rooms, provide lighting that illuminates the face of the person speaking;
- Avoid acoustically 'live' environments within a design. This can be done by careful allocation of the amount of 'hard' surfaces (floor, wall and ceiling finishes) in areas where communication is important;
- Provide induction loops or infra-red hearing enhancement systems in rooms or spaces designed for meetings, lectures, classes, performances, spectator sports, films, etc.;
- Provide induction loops or infra-red hearing enhancement systems at service or reception desks if they are in noisy areas or provided with glazed screens;
- Consider the guidance in BS 8300:2001 when selecting surface finishes, visual, audible and tactile signs, and the use of enhancement systems.

17.10 Sanitary accommodation in buildings other than dwellings

It is necessary to provide sanitary conveniences for all users of a building, whether they are visitors, customers or staff.

The guidance given in the AD M (2004) is much more detailed than that given in the previous AD (1999), and it is not possible to describe all of the new guidance here. However, the following issues are some of those that need to be considered.

17.10.1 Sanitary accommodation generally

The following general design principles should be considered:

- Suitable sanitary conveniences should be provided in a new or extended building. In an extended building the sanitary accommodation can be either in the extension or, providing the approach is accessible, in the existing part of the building being extended;
- Sanitary accommodation that has been suitably designed to meet the needs of all disabled and non-disabled users should be provided. It should be sited in logical, consistent locations within the building. Sanitary accommodation should include facilities for wheelchair and ambulant disabled users, as well as people of either gender who may be using the facility whilst encumbered with children, luggage, etc.;
- The number of toilets provided will depend on the size and nature of the building and on the ease of access to the facility. Travel distances should always reflect the fact that some disabled people may need to reach a WC quickly;
- Facilities such as taps, door locks, flush controls and light controls should be suitable to meet the needs of all users, including those with restricted dexterity or strength. The needs of users with visual or hearing impairments should also be taken into account in terms of, for example, visual contrast, lighting, surface finishes and the provision of visual and audible alarm systems;
- At least one unisex wheelchair accessible facility should be provided at every location where facilities are provided for non-disabled staff, visitors or customers;
- Where there is sufficient space within a building for only one toilet, it should be a unisex wheelchair accessible facility. However, the internal width should be increased from 1500 mm to 2000 mm to accommodate a standing height wash basin in addition to the finger rinse basin provided adjacent to the WC;
- Within separate-sex toilets at least one cubicle should be designed and equipped to meet the needs of ambulant disabled people (see Fig. 17.8);
- If more than four cubicles are provided, one cubicle should be wider to accommodate the needs of people who need extra space. Such users would include, for example, parents with children, people with luggage or shopping etc., and people with guide dogs.

17.10.2 Wheelchair-accessible unisex sanitary accommodation

A wheelchair user should not have to travel more than 40 m:

- to reach a suitable sanitary facility on the same floor, unless the reasons can be argued in an Access Statement (see section 17.1.1);
- if the route includes the need to move between floors; however, the route must incorporate an appropriately accessible passenger or platform lift and not require a travel journey of more than one storey.

Where more than one WC compartment is provided in a building, both left-hand and right-hand transfer layouts should be provided and a disabled person should not be required to travel more than one storey to reach an appropriate layout.

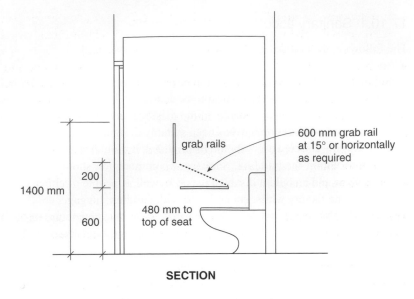

SECTION

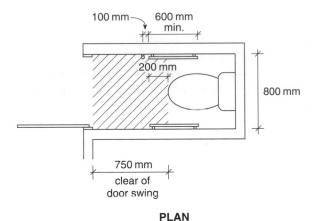

PLAN

Fig. 17.8   WC for ambulant disabled people.

Individual wheelchair accessible toilets should be designed for ease of access and use and should:

- Provide adequate space for wheelchair to manoeuvre and for the presence of a carer to assist with frontal, lateral, diagonal or backward transfer onto the WC;
- Have hand washing and drying facilities that may be reached from the WC before transfer back to the wheelchair;
- Have either a left-hand or right-hand transfer layout and, whether unisex or integral within WC compartments, be consistent in layout and in the provision of facilities;
- Be located in a way that ensures the privacy of users.

- Be provided with an audible and visual alarm emergency call system that can be reached from the WC and the area close to the WC. A reset control should also be provided close to the WC or reachable whilst seated in a wheelchair;
- Be provided with outward opening doors. If there is sufficient space to manoeuvre internally to allow the door to open inwards, it is necessary to provide an emergency release system to allow the door to open outwards in an emergency;
- Be provided with lever action or automatically controlled flush mechanisms located on the open or transfer side of the space;
- Have appropriate visual contrast between the walls and the sanitary fittings, and at wall/floor junction, and appropriate lighting;
- Not also be used as a baby changing facility.

The AD M (2004) indicates a preference for unisex wheelchair-accessible WC accommodation and lists the following advantages over integral facilities:

- The approach to it is separated from other sanitary accommodation.
- Direct-access unisex accessible toilet facilities allow a partner or carer of a different sex to enter the toilet and offer assistance if required.
- The facility is more easily identified.
- There is more chance of it being available when needed.
- Less space is needed overall, since duplication of facilities is necessary with integral accommodation for the same level of provision.

Visitors to hotels and motels will often find that guest bedrooms have en-suite sanitary accommodation. If this is the case, then there should also be bedrooms suitable with en-suite facilities for disabled people. Where en-suite facilities are not provided in the general sanitary arrangement, then unisex facilities should be provided within easy reach. There is still, of course, the necessity to provide additional sanitary accommodation for staff and for daytime or non-resident visitors.

Figure 17.9 illustrates a typical layout for a WC compartment suitable for wheelchair users in either unisex or integral facilities.

17.10.3 Wheelchair-accessible changing and shower facilities

The AD M (2004) contains several recommendations for the provision of suitable wheelchair-accessible changing and shower facilities, and it is not be possible to describe all of them here. However, the following represents an example of the type of recommendations made:

- In areas where more than one individual changing compartment or shower compartment is provided, a choice of layouts of seats and/or WCs in terms of left- and right-hand transfer should be provided.
- Wall-mounted drop-down rails and a wall-mounted, slip-resistant (not spring-loaded) seat should be provided.
- An alarm system should be provided.
- Suitable storage for artificial limbs should be provided.

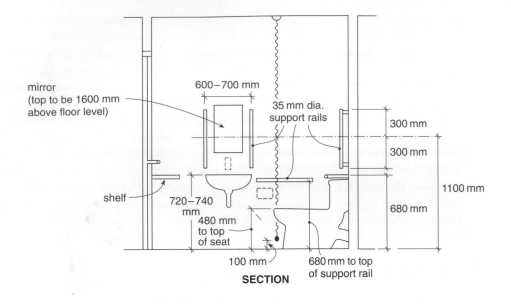

SECTION

PLAN

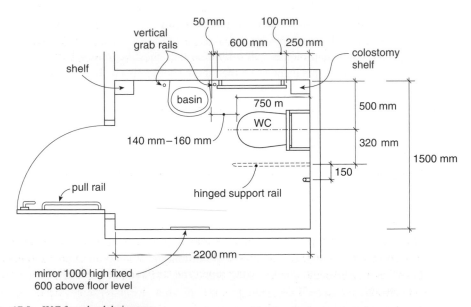

Fig. 17.9   WC for wheelchair users.

- In front of lockers there should be a space to manoeuvre at least 1500 mm deep.
- Floors to showers should be slip resistant and self draining.
- In commercial properties where showers are provided for the use of staff, at least one wheelchair-accessible shower compartment should be provided.

17.10.4 Wheelchair-accessible bathrooms

The AD M (2004) makes recommendations for wheelchair-accessible bathing facilities provided in buildings such as, for example, hotels, motels, student accommodation and sports facilities.

It suggests that the facilities provided in a bathroom should allow wheelchair users and ambulant disabled people to be able to wash or bath either independently or with assistance from a carer. There should be sufficient manoeuvring space and the layout of the facilities should not compromise the ability of a disabled person to use them independently and safely. There are several detailed recommendations relating to minimum dimensions of the space, choice of layouts, slip-resistant floor coverings, emergency alarms and direction of door swing, and there are additional recommendations regarding the provision of seating to assist transfer into the bath.

17.11 Means of escape in case of fire

Part M and the AD M (2004), are primarily aimed at the needs of users, regardless of disability, age and gender, when accessing and using a building and its surroundings. They do not extend to provisions for escape in an emergency, although, clearly, appropriate consideration of such issues should be an inherent part of the design and management of any building or environment. Means of escape for disabled people in the event of a fire is described in the Approved Document B, *Fire safety*, and BS 5588, Part 8: *Code of practice for means of escape for disabled people* (see also Chapter 7).

17.12 Access and facilities in dwellings

17.12.1 Interpretation

The following definitions of terms used in AD M (2004) apply only to the provisions concerning dwellings:

DWELLINGS – This means a house, flat or maisonette.
Note that in the AD M (1999), the term dwelling included purpose-built student accommodation, other than traditional halls of residence providing bedrooms and not equipped as self-contained accommodation. In the AD M (2004) new blocks of flats built as student accommodation should be treated as though they are hotel/motel accommodation for sleeping accommodation and internal facilities. Details are covered in the AD M (2004) in the section entitled 'Buildings other than dwellings'. Other issues about the building should be designed in accordance with the requirements for flats, as described in the section entitled 'Dwellings'.

COMMON – Serving more than one dwelling.

HABITABLE ROOM – When used to define the principal storey in a dwelling, it means a room intended to be used for dwelling purposes. This includes a kitchen but not a utility room or bathroom.

MAISONETTE – A self-contained dwelling that occupies more than one storey in a building, but which is not a dwelling-house.

POINT OF ACCESS – The place where a person would alight from a vehicle when visiting a dwelling, before they approached the dwelling. This could be inside or outside the plot boundary.

PRINCIPAL ACCESS – The entrance that would normally be used by a visitor who was not familiar with the dwelling. In a block of flats it would be the common entrance.

PLOT GRADIENT – The slope measured between the point of access and the finished floor level of the dwelling.

STEEPLY SLOPING PLOT – A plot gradient that exceeds 1 in 15.

USABLE – Convenient for independent use.

17.12.2 The main provisions

Reasonable provision must be made:

- for disabled people to gain access to and use the facilities in the dwelling;
- for sanitary conveniences suitable for disabled people in the entrance storey of the dwelling.

The 'entrance storey' is the storey that contains the principal entrance to the dwelling (see above for definition) and in the design of some dwellings the entrance storey may not contain any habitable rooms. If this is the case, then the option is given of placing the sanitary conveniences in either the entrance storey or a principal storey. Regulation M4 defines 'principal storey' as 'the storey nearest to the entrance storey which contains a habitable room, or where there are two such storeys equally near, either such storey'.

Although this legal terminology may seem confusing, the intention of the regulation is simply to require sanitary conveniences suitable for disabled people to be placed in the entrance storey of the dwelling, if this contains any habitable rooms. Otherwise the sanitary conveniences may be placed in either the entrance storey or the nearest storey to the principal entrance that contains habitable rooms.

Therefore, in dwellings, it should be reasonably safe and convenient for disabled people to:

- approach the principal (or other suitable alternative) entrance to the dwelling from the edge of the site or from car parking within the site;
- gain access into the building;

- gain access within the building;
- use sanitary accommodation.

It should be noted that the provision of access and facilities for disabled people in dwellings is to enable them to visit new dwellings and use the principal storey. The intention is that disabled occupants will be able to cope better with reducing mobility and will be able to remain living in their own homes longer than would otherwise be the case. On the other hand, the provisions are not intended to facilitate fully independent living for all disabled people.

17.12.3 Means of access

In general terms, a disabled person should be able to gain access into a dwelling from the point of leaving a vehicle (which is parked either within or outside the plot boundary).

In most cases it should be possible to provide a safe and convenient level or ramped approach, thereby permitting wheelchair users to gain access to the dwelling. Clearly, there will be situations on steeply sloping plots (i.e. where the plot gradient exceeds 1 in 15), where it will only be practicable to provide a stepped approach. Where this is the case the approach should be suitable for an ambulant disabled person using a walking aid.

17.12.4 The approach to the dwelling

The choice of a suitable approach to the dwelling from the point of access to the plot will be influenced by the topography and available area of the plot, and by the distance to the dwelling from the point of access. Account may also need to be taken of local planning requirements, especially for new developments in conservation areas. Developers are advised to discuss the access requirements of Part M with their local planning authority, in conjunction with their building control supervisor (local authority or approved inspector) at an early stage in the design process, to avoid later conflicts.

It may be possible to reduce the effect of a steeply sloping plot by means of a suitable driveway. This could allow for the parking space within the plot boundary to be at a sufficiently high level to permit a level or ramped approach from the parking space to the dwelling.

The surface material of the approach to the dwelling should be firm enough to support a wheelchair and user, and smooth enough to allow satisfactory manoeuvre. Ambulant disabled people using walking aids also need to be considered. Therefore, loose surfacing materials such as gravel or shingle are unlikely to be satisfactory and the approach should be sufficiently wide (in addition to the width of the parking space) to allow safe and convenient passage.

In practical design terms the provisions can be summarised in the following paragraphs:

- Provide a suitable approach:
  - from a reasonably level point of access (i.e. from the vehicle parking position) to the dwelling entrance (i.e. the principal entrance or a suitable alternative entrance if it is not possible to access the principal entrance);

○ with cross falls that do not exceed 1 in 40;

○ that may consist in whole or in part of a vehicle driveway.

- A level approach will have:
  ○ a gradient not exceeding 1 in 20; a firm and even surface and a minimum width of 900 mm.
- A ramped approach will be needed where the overall plot gradient exceeds 1 in 20 but does not exceed 1 in 15.
- A ramped approach should have:
  ○ a firm and even surface;
  ○ minimum unobstructed flight widths of 900 mm;
  ○ individual flights no longer than 10 m in length with a maximum gradient of 1 in 15 (although gradients not exceeding 1 in 12 are allowed where the individual flight does not exceed 5 m in length);
  ○ top, bottom and, if necessary, intermediate landings at least 1200 mm long, clear of any door or gate swinging across it.
- A stepped approach will be needed where the overall plot gradient exceeds 1 in 15. In this case the stepped approach should have:
  ○ minimum unobstructed flight widths of 900 mm;
  ○ a maximum rise of 1800 mm between landings;
  ○ top, bottom and, if necessary, intermediate landings at least 900 mm long;
  ○ a suitable tread profile as illustrated in Fig. 17.10;
  ○ uniform risers between 75 and 150 mm high and goings at least 280 mm long (measured 270 mm in from the narrow edge of the tread for tapered steps);
  ○ where the flight consists of three or more risers, a suitable handrail should be provided on one side; the handrail should have a profile that can be gripped, be positioned between 850 and 1000 mm above the pitch line of the flight and project at least 300 mm beyond the top and bottom nosings (see Fig. 17.10).

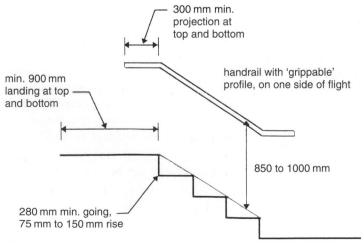

Fig. 17.10   Stepped approach to dwelling.

17.12.5 Access into the dwelling

In general, the entrance into a dwelling or a block of flats from outside should be provided with an accessible threshold irrespective of whether the approach to the entrance is level, ramped or stepped. Exceptionally, if the approach is stepped and, for practical reasons, a step into the dwelling is unavoidable, it should not exceed 150 mm in height.

The design of accessible thresholds should follow the guidance contained in The Stationery Office publication *Accessible thresholds in new housing* (ISBN 011 702333 7).

The entrance door to an individual dwelling and/or a block of flats should be wide enough to accommodate a person using a wheelchair. This requirement can be satisfied if such a door has a minimum clear opening width of 775 mm.

17.12.6 Access within the dwelling

The requirements of regulation M4 mean that access must be facilitated to habitable rooms and to a WC (which may be in a bathroom) in the entrance storey or the principal storey of the dwelling, as appropriate.

Where it is not possible to make the principal entrance accessible and an alternative is provided instead, the route to the remainder of the entrance storey from the alternative entrance must be carefully considered, especially if the route passes through other rooms. Therefore, corridors and passageways should be wide enough to allow convenient access for a person in a wheelchair, whilst at the same time allowing for manoeuvre past local obstructions such as radiators and other fixtures.

Doors to rooms need to be wide enough to cater for both head-on approach and right-angled approach from a person in a wheelchair. The rules for access within the entrance or principal storey of a dwelling are illustrated in Fig. 17.11.

17.12.7 Vertical circulation

Steps within the entrance storey of a dwelling should be avoided wherever possible. Sometimes, such as in the case of severely sloping plots, it may not be possible to avoid putting a change of level involving steps in the entrance storey. In these circumstances a stair should be provided that is wide enough to be negotiated by an ambulant disabled person with assistance and with handrails on both sides. Therefore, any stair provided in the entrance storey which gives access to habitable rooms should have:

- a minimum clear width of 900 mm;
- a continuous handrail on each side, and on any intermediate landings, where the
  - flight consists of three or more risers; and
  - rise and going are in accordance with the guidance for private stairs in Approved Document K (see Chapter 15).

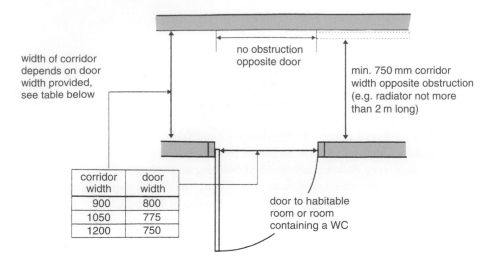

Right-angled approach to door

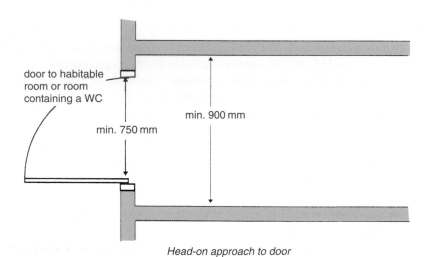

Head-on approach to door

Fig. 17.11   Corridor and door widths in dwellings.

17.12.8 Access to socket outlets and switches in dwellings

Ambulant disabled people and people who use wheelchairs are less mobile and likely to have more limited reach than able-bodied people. Therefore switches and socket outlets for such things as electrical appliances, lighting, television aerials, telephone jack points, etc. should be mounted at suitable heights so that they can be easily reached. Essentially, this means locating sockets and switches in habitable rooms between 450 mm and 1200 mm from finished floor level.

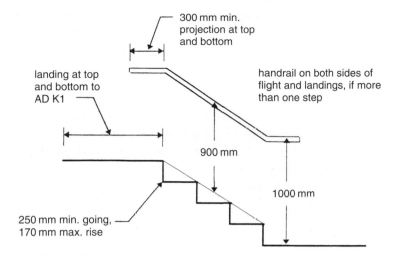

300 mm min.
projection at top
and bottom

landing at top
and bottom to
AD K1

handrail on both sides of
flight and landings, if more
than one step

900 mm

250 mm min. going,
170 mm max. rise

1000 mm

Fig. 17.12   Common-access stairs in blocks of flats.

17.12.9 Common-access stairs in blocks of flats

It should be possible for a disabled person to visit an occupant on any storey in a build-ing containing flats. The most suitable means of vertical access for a disabled person is a lift; however, AD M (2004) recognises that lifts are not always provided (and does not, at present, recommend that lifts should be installed). Therefore, where there are no passenger lifts, the common-access stairs should be suitable for use by ambulant disabled people, (as well as being suitable for people with impaired sight) and be designed to have:

- step nosings that are clearly visible by the use of contrasting brightness;
- top and bottom landings that follow the guidance contained in Part K1 of Approved Document K (see Chapter 15);
- uniform risers not exceeding 170 mm in height;
- uniform goings not less than 250 mm in length (measured 270 mm in from the narrow edge of the flight for tapered steps);
- a suitable handrail on each side of flights and landings if comprising more than one riser;
- a suitable tread nosing profile (e.g. as illustrated in Fig. 17.3 but with 170 mm maximum rise and 250 mm minimum going); and
- risers that are not open.

Some of this guidance is illustrated in Fig. 17.12.

17.12.10 Passenger lifts in blocks of flats

If passenger lift access is to be provided to flats above the entrance storey in a building, it should be suitable for both unaccompanied wheelchair users and people with sensory impairments. It should also contain suitable delay systems to enable disabled people more time to enter and leave the car and lessen the risk of contact with the closing doors.

A suitable passenger lift should have:

- a minimum load capacity of 400 kg;
- an unobstructed, accessible landing space at least 1500 mm square in front of the lift doors;
- a door or doors with a clear opening width of 800 mm;
- a car with minimum dimensions of 900 mm wide by 1250 mm deep (although other dimensions may be suitable if it can be demonstrated by test evidence or experience in use, etc. that they are suitable for an unaccompanied wheelchair user);
- landing and car controls between 900 mm and 1400 mm from landing or car floor levels and at least 400 mm from the front wall;
- tactile floor-level indication on each landing next to the lift-call button to identify the storey in question, and on or adjacent to the lift buttons in the lift car to confirm the floor selected;
- visual and audible floor indication where the lift serves more than three floors;
- a signalling system which gives a visual warning that the lift is about to stop at a floor and once stopped, a minimum of five seconds before the doors begin to close after being fully open.

It is necessary to ensure that the door controls can be overridden in order that lift passengers do not get caught in the closing doors. This should be a suitable electronic system (such as a photo-eye or infra-red detector), but not a door-edge pressure system, which might cause a disabled person to lose their balance. The door should remain fully open for at least three seconds.

17.12.11 Provision of accessible sanitary conveniences in dwellings

For dwellings that contain more than one storey, sanitary conveniences suitable for disabled people should be provided in the entrance storey, if this contains any habitable rooms. Otherwise, the sanitary conveniences may be placed in either the entrance storey or the nearest storey to the principal entrance that contains habitable rooms. There should be no need to negotiate a stairway to reach the WC from the habitable rooms in that storey. (Obviously, for single-storey dwellings and individual flats the sanitary conveniences can only be provided in the entrance storey).

Additionally, the following provisions apply to the design and location of the sanitary accommodation:

- The WC may be located in a bathroom if there is one available in the relevant storey of the dwelling.
- It is accepted in AD M that it may not always be practical for a wheelchair to be fully accommodated inside the WC compartment.
- The WC door should open outwards and should be positioned so as to allow access to the WC for a person in a wheelchair.
- The WC door should have a minimum width as shown in Fig. 17.13 (and be wider if possible, so as to allow easier access and manoeuvring by wheelchair users).
- There should be sufficient space in the WC compartment to allow wheelchair users to access the WC.

- The position of the washbasin should not impede access to the WC.

Typical, suitable WC compartments are illustrated in Fig. 17.13.

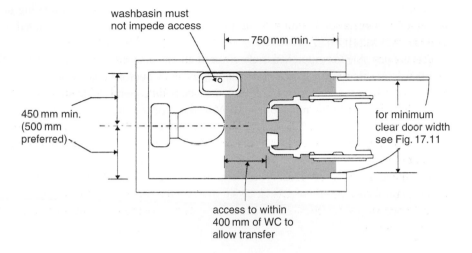

Typical WC compartment with frontal access

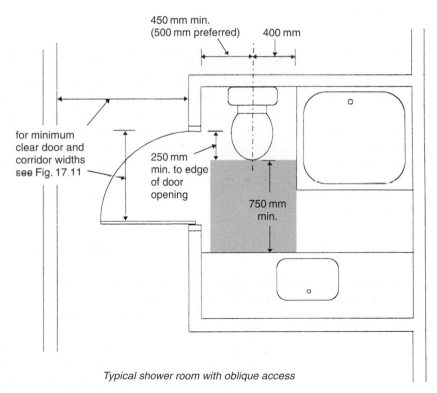

Typical shower room with oblique access

Fig. 17.13   Accessible sanitary conveniences in dwellings.

17.13 In conclusion

The AD M (2004) is a document that is substantially different in terms of scope and guidance contained in AD M (1999). Certain elements of the AD M (2004) are very similar, such as the comments relating to housing, but the guidance it contains relating to the design of non-domestic buildings is far more comprehensive than that contained in the previous AD M (1999).

It is not possible to comprehensively cover of all the changes in the new AD M (2004), and that is not the purpose of an informative and introductory chapter contained here. Therefore, whilst several issues have been identified and general guidance given, designers are advised to always refer to the AD M (2004) for the full extent of the areas now covered.

18 Glazing (Part N)

18.1 Introduction

This chapter deals with safety issues associated with glazing in terms of protecting people against the risks of impact and making sure that glazed elements may be opened, closed and cleaned safely.

The 1998 edition of Approved Document N introduced new provisions governing the safe use and cleaning of glazed elements mainly in order to ensure compliance with the Workplace (Health, Safety and Welfare) Regulations 1992. Therefore, compliance with the revised regulations in Part N prevents action being taken against the occupier under the Workplace Regulations when the building is eventually in use. This involves considering aspects of design which will affect the way a building is used and applies only to workplaces.

Although dwellings are excluded from the changes, in mixed use developments the requirements for the non-domestic part of the use would apply to any shared parts of the building (such as common access staircases and corridors). With flats the situation may be less clear for although certain sections of Part N do not apply to dwellings, it is still necessary for people such as cleaners, wardens and caretakers to work in the common parts. Therefore, the requirements of the Workplace Regulations may still apply even though the Building Regulations do not.

18.2 Application

Part N of Schedule 1 to the 2000 Regulations applies to all new glazing used in the erection, extension or material alteration of a building.

It does not apply to exempt buildings (except small conservatories or porches in Class VII of Schedule 2, see section 2.5), or to replacement glazing although this may be the subject of consumer protection legislation in the future.

Requirement N1 applies to all building types whereas requirements N2, N3 and N4 apply to all building types *except* dwellings (i.e. dwelling-houses and flats).

18.3 Protection against impact

People are likely to come into contact with glazing in certain critical locations, when they are moving in or about a building. Accordingly, the glazing must:

- if broken on impact, break so that the dangers of injury are minimised (i.e. break safely); or
- resist impact without breaking; or
- be protected or shielded from impact.

Critical locations under the terms of paragraph N1 are shown in Fig. 18.1.

Two main areas may be identified where an accident may result in cutting or piercing injuries.

- Doors and door side panels between finished floor level and 1500 mm above.
- Internal and external walls and partitions between finished floor level and 800 mm above.

In doors and door side panels the main risk is in the area of door handles and push plates especially since doors are prone to stick. Also it is possible that an initial impact above waist level may result in a fall through the glass. In low level glazing away from doors the main risks are to children.

18.3.1   Possible solutions to N1

Approved Document N lists a number of solutions which can be adopted in order to minimise the risks of injury in these critical areas as follows:

(1)   If breakage occurs the glazing should break safely.

The concept of safe breakage is taken from BS 6206: 1981 *Specification for impact performance requirements for flat safety glass and safety plastics for use in buildings:* clause 5.3.

A test is carried out using a leather bag filled with lead shot. This is swung pendulum-fashion to impact a sheet of safety glazing material and the results are noted.

The test material is required to remain unbroken or it may break safely as defined in one of the following ways:

(a)   Cracks and fissures are allowed to develop provided it is not possible to pass a sphere of 76 mm diameter through any openings and any detached particles are of limited size; or

(b)   Disintegration is allowed to occur provided the particles are of small size; or

(c)   Breakage is allowed to occur provided the pieces are not sharp or pointed.

In essence a glazing material in a critical location will be satisfactory if it can be classified under the requirements of BS 6206 as Class C (i.e. it remains unbroken or breaks safely when impacted from a height of 305 mm). Additionally, if it is installed in a door or door side panel and has a pane width which exceeds 900 mm, then it should meet the requirements of Class B of BS 6206 (i.e. it should remain unbroken or break safely when impacted from heights of 305 mm and 457 mm).

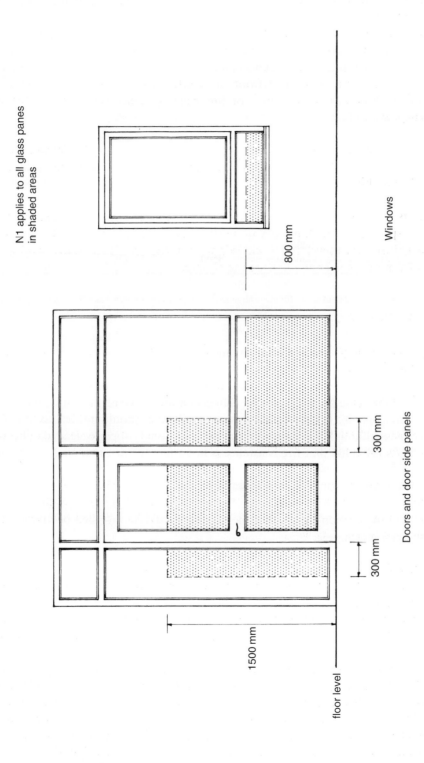

N1 applies to all glass panes in shaded areas

800 mm

Windows

1500 mm

300 mm

300 mm

Doors and door side panels

floor level

Fig. 18.1 Internal and external walls – critical locations.

(2)   The glazing should be robust.

Robustness refers to the strength of the glazing material. Some materials such as glass blocks or polycarbonates are inherently strong. Annealed glass gains its strength through increased thickness and AD N1 describes the use of this material for large glazed areas forming fronts in shops, showrooms, factories, offices or public buildings. The dimensions of these glazed areas and their related glass thicknesses are shown in Table 18.1.

Table 18.1   Annealed glass – thickness/dimension limits.

| Height (mm) | | Length (mm) | | Thickness (mm) |
From	To	From	To	
0	1100	0	1100	8
1100	2250	1100	2250	10
2250	3000	2250	4500	12
3000	any	4500	any	15

Note
Annealed glass sizes and thicknesses for use in large areas to shopfronts, showrooms, offices, factories and public buildings.

(3)   The glazing should be in small panes.

This relates to the use of a single pane or one of a number of panes within glazing bars, in either case having a smaller dimension not exceeding 250 mm and an area not greater than $0.5\,m^2$. Annealed glass in small panes should not be less than 6 mm in thickness although it is possible to install traditional copper or leaded lights using 4mm glass provided that fire resistance is not a factor. (See Fig. 18.2.)

(4)   The glazing should be permanently protected.

Permanent protection means that the glazing should be installed behind a permanent screen which:

- prevents a sphere of 75 mm diameter touching the glazing,
- is itself robust; and
- is difficult to climb in cases where the glazing forms part of protection from falling.

Where permanent screen protection is provided then the glazing itself does not need to comply with requirement N1. (See Fig. 18.2.)

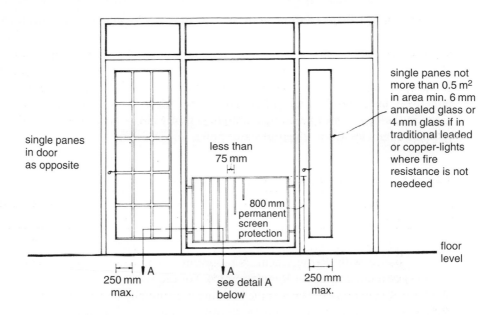

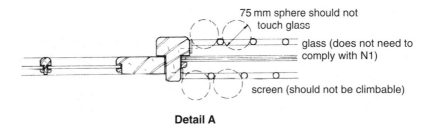

Detail A

Fig. 18.2   Small panes and permanent screen protection.

18.4   Manifestation of glazing

If there is a risk that people may come into contact with large, uninterrupted areas of transparent glazing while moving in or about a building, then paragraph N2 of Schedule 1 requires that such areas must incorporate features which make the glazing apparent. As mentioned above, this requirement does not apply to dwellings.

The risk of collision and consequent injury is most serious where parts of a building or its surroundings are separated by transparent glazing and the impression is given that direct access is possible through the area without interruption. In these critical locations (i.e. internal or external walls of shops, showrooms, offices, factories, public or other non-domestic buildings) it is necessary to adopt some means of making the glazing more apparent.

This is termed 'manifestation' of the glazing in AD N2 and it may take the form of patterns, company logos, broken or solid lines, etc. marked on the glazing at appropriate heights and intervals. This is illustrated in Fig. 18.3.

It is, of course, possible to indicate the presence of glazing by other means. Such features as mullions, transoms, door frames or large push or pull handles can be effectively used and AD N2 acknowledges that where these features are incorporated in the design then permanent manifestation will not be necessary.

Some examples of these features are shown in Fig. 18.3.

18.5   Safe use of windows, skylights and ventilators

Windows, skylights and ventilators must be constructed or equipped so that they can be opened, closed or adjusted safely, if they are so positioned as to be operable by people in or about the building. This requirement does not apply to dwellings.

Compliance with Requirement N3 prevents action being taken against the occupier of a building under Regulation 15(1) of the Workplace (Health, Safety and Welfare) Regulations 1992 when the building is eventually in use. (Regulation 15(1) relates to requirements for opening, closing or adjusting windows, skylights and ventilators).

In order to meet the performance standard for safe operation of windows, skylights and ventilators controls should typically be located as follows:

- Not more than 1.9 m above the floor or other stable surface where there is unobstructed access (ignoring small recesses, such as window reveals).
- At a lower level where there is an obstruction (e.g. 1.7 m from floor level if there is an obstruction 600 mm deep (including any recess) and not more than 900 mm high).

If controls cannot be positioned at a safe distance from a permanent stable surface, then it may be necessary to install either manual or electrical remote controls.

Above ground level there may be a danger of the operator or other person falling through a window. Where this is the case suitable opening limiters should be fitted or the window should be guarded as described in Approved Document K (see Chapter 15).

18.6   Safe access for cleaning glazed surfaces

Provision must be made for glazed surfaces to be safely accessible for cleaning. This includes:

- windows and skylights; and
- translucent walls, ceilings or roofs.

Regulation N4 does not apply to dwellings, or to any of the above transparent or translucent elements if their surfaces are not intended to be cleaned. In this case, compliance with Requirement N4 prevents action being taken against the occupier of a building under Regulation 16 of the Workplace (Health, Safety and Welfare)

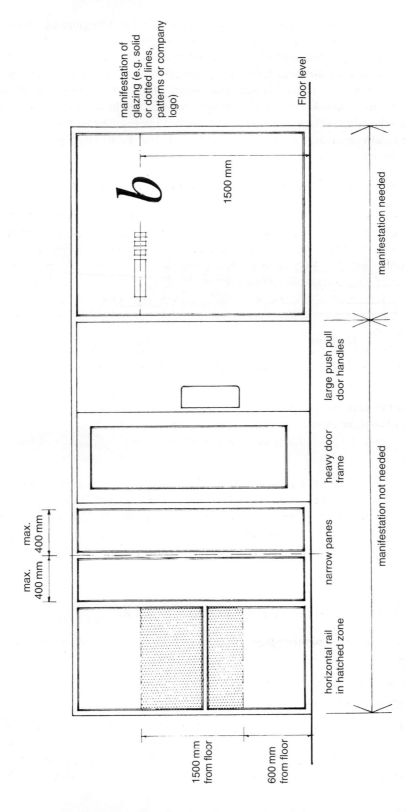

Fig. 18.3 Manifestation of large glazed areas (internal or external).

Regulations 1992 when the building is eventually in use. (Regulation 16 relates to requirements for cleaning glazed surfaces in buildings.)

In the context of Regulation N4, it will be necessary to make provision for safe means of access for cleaning *both* sides of any glazed surfaces which are positioned so that there is a danger of falling more than 2 m. Furthermore, glazed surfaces which cannot be cleaned safely by a person standing on the ground, a floor or other permanent stable surface will need to be catered for in other ways.

The following arrangements (illustrated in Figs 18.4 and 18.5) are typical examples of how it may be possible to satisfy Regulation N4.

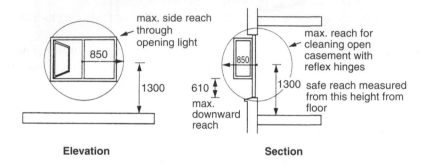

Fig. 18.4   Typical safe reaches for cleaning windows.

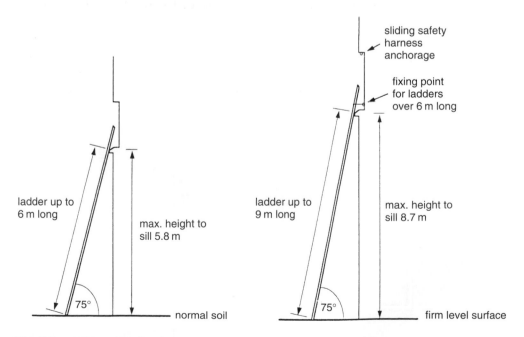

Fig. 18.5   Ladder access for cleaning windows. (**Note:** Since designers will need to relate the length of access ladders to window sill heights, the maximum sill heights from ground level are shown for both 6 m and 9 m ladders.)

- Install windows of a suitable design and size so that they can be cleaned from inside the building. Reversible type windows should be capable of being fixed in the reversed position for cleaning purposes (see Fig. 18.4). For additional information on windows see BS 8213: Part 1 *Windows, doors and rooflights,* and Approved Document K (see Chapter 15) for guidance on minimum sill heights.
- Use portable ladders up to 9 m long if there is an adequate area of firm, level ground situated in a safe place for siting the ladders. Ladders up to 6 m long may be sited on normal soil. (For ladders over 6 m long provide permanent tying or fixing points.) (See Fig. 18.5).
- Provide catwalks at least 400 mm wide with either 1100 mm high guarding or provision for anchorages for sliding safety harnesses.
- Provide access equipment, e.g. suspended cradles or travelling ladders with safety harness attachments.
- Provide adequate anchorage points for safety harnesses or abseiling hooks.
- In exceptional circumstances where other means of access cannot be used, provide suitably located space for scaffold towers.

18.7   Further references to glazing

Attention is drawn to the following Approved Documents where further information regarding glazing may be found:

- Approved Document B: *Fire safety* – guidance on fire resisting glazing and the reaction of glass to fire (see Chapter 7).
- Approved Document K: *Protection from falling, collision and impact* – guidance on glazing which forms part of protection from falling from one level to another, and which needs to provide containment as well as limiting the risk of injuries through contact. Recommendations are given concerning the heights of guarding and the means for achieving containment.

19 Electrical safety (Part P)

19.1 Introduction

Part P of Schedule 1 to the Building Regulations 2000 was added by virtue of the Building (Amendment) (No. 3) Regulations 2004 (SI 2004/3210). Originally, it consisted of two requirements related to:

- design, installation, inspection and testing of electrical installations; and
- provision of information.

This was later amended by the Building and Approved Inspectors (Amendment) Regulations 2006 (SI 2006/652) to a single requirement covering design and installation of electrical installations.

A revised Approved Document to support Part P entitled *Electrical safety – dwellings* came into effect on 6 April 2006 and, interestingly, it still contains advice on inspection and testing of installations.

Although Part P has only recently been introduced into the building regulations, other Parts of the regulations have long been concerned with requirements affecting electrical installations.

Some examples include:

- Part A (Structure) – depth of chases in walls, and size of holes and notches in floor and roof joists to receive services such as electrical trunking and cabling;
- Part B (Fire safety) – fire safety of certain electrical installations, such as lifts and mechanical ventilation and air conditioning systems, provision of fire alarm and fire detection systems, fire resistance of penetrations through floors and walls;
- Part C (Site preparation and resistance to moisture and contaminants) – moisture resistance of cable penetrations through external walls;
- Part E (Resistance to the passage of sound) – penetrations through floors and walls;
- Part L (Conservation of fuel and power) – energy efficient lighting, reduced current carrying capacity of cables in insulation;
- Part M (Access to and use of buildings) – wall-mounted socket outlets, switches and consumer units need to be located so as to be easily reachable.

It should be noted that, although Part P covers the safety of fixed electrical installations, it does not cover the functionality of systems such as fire alarms and fans used in ventilation systems. Details of the way in which such systems are required to function may be found in the other Parts of the regulations referred to above and in other legislation.

19.2   Design and installation

Requirement P1 states that 'Reasonable provision shall be made in the design and installation of electrical installations in order to protect persons operating, maintaining or altering the installations from fire or injury'.

The requirement is qualified in the limits on application so that it only applies to electrical installations that are intended to operate at low or extra-low voltage (see section 19.5 for definitions) and are:

- in or attached to a dwelling;
- in the common parts of a building serving one or more dwellings, but excluding power supplies to lifts;
- in a building that receives its electricity from a source located within or shared with a dwelling; and
- in a garden or in or on land associated with a building where the electricity is from a source located within or shared with a dwelling.

19.3   Application

Part P applies to electrical installations in buildings or parts of buildings which comprise:

- dwelling-houses and flats;
- dwellings and business premises sharing a common supply, e.g. shops and public houses with a flat above;
- common access areas in blocks of flats, e.g. corridors and staircases;
- amenities that are shared in blocks of flats, e.g. laundries and gymnasia.

Part P also applies to parts of the electrical installations mentioned above, such as:

- in or on land associated with the buildings, e.g. fixed lighting and pond pumps in gardens;
- in outbuildings like sheds, detached greenhouses and garages.

In many cases, when electrical work to buildings is carried out, it will not be necessary to give a building notice or deposit full plans with the local authority or notify an approved inspector, since most electrical work will be covered by regulation 12(5), which enables the work to be certified by a competent person under the provisions of Schedule 2A to the Building Regulations 2000. This means that when the work has been completed, the person ordering the work should receive a signed Building Regulations compliance certificate (for rented properties a copy should also be sent to the occupier) and the relevant building control body should either receive a copy of the information contained in the certificate or be informed within 30 days of completion of the work by the person carrying it out. In practice, the competent person merely sends the information on completion of the work to the Competent Person Scheme of which he or she is a member, and the task of informing the local authority or client is carried out by the Scheme administrators (see Chapter 5 for full details). In addition to this, the person ordering the work should be given a completed

Electrical Installation Certificate by the installer. There are two standard forms of certificate illustrated in IEE Guidance Note 3 (2002 edition) and these are reproduced in Appendix B of AD P as Forms 1 and 2, together with notes and guidance (see below). Form 1 is a short form for use when the design, construction, inspection and testing of an installation is the responsibility of a single person. Form 2 is the full form designed for use when the design, construction, and inspection and testing are carried out by different individuals.

For minor works (i.e. works not involving the replacement of consumer units or similar items) it is possible for a Minor Electrical Installation Works Certificate to be issued (see Appendix B of AD P Form 5 reproduced below). BS 7671 does not require this certificate to be issued for the replacement of equipment such as accessories and luminaries but does advise that the certificate should be issued where appropriate inspection and testing has been carried out.

Each Electrical Installation Certificate must be completed and signed by the competent person or persons in respect of the design, construction, inspection and testing of the work. The competent person is required to:

- have a sound knowledge and experience relevant to the nature of the work undertaken and to the technical standards set down in BS 7671;
- be fully versed in the inspection and testing procedures contained in BS 7671; and
- employ adequate testing equipment.

References to the following notes are included in Forms 1 and 2 and should be read in conjunction with those Forms:

(1) The Electrical Installation Certificate must be used only for the initial certification of a new installation or for an alteration or addition to an existing installation where new circuits have been introduced. It must not be used for a Periodic Inspection for which a Periodic Inspection Report form should be used. For an alteration or addition which does not extend to the introduction of new circuits, a Minor Electrical Installation Works Certificate may be used. The original Certificate must be given to the person ordering the work. A duplicate should be retained by the contractor.

(2) The Certificate is only valid if accompanied by the Schedule of Inspections and the Schedule(s) of Test Results.

(3) The signatures appended are those of the persons authorised by the companies executing the work of design, construction and inspection and testing respectively. A signatory authorised to certify more than one category of work should sign in each of the appropriate places.

(4) The time interval recommended before the first periodic inspection must be inserted (see IEE Guidance Note 3 for guidance).

(5) The page numbers for each of the Schedules of Test Results should be indicated, together with the total number of sheets involved.

(6) The maximum prospective fault current recorded should be the greater of either the short-circuit current or the earth fault current.

(7) The proposed date for the next inspection should take into consideration the frequency and quality of maintenance that the installation can reasonably be expected to receive during its intended life, and the period should be agreed between the designer, installer and other relevant parties.

AD P Copies of BS 7671 and IEE Model Forms

Form 1   Form No   /1

ELECTRICAL INSTALLATION CERTIFICATE (notes 1 and 2)

(REQUIREMENTS FOR ELECTRICAL INSTALLATIONS - BS 7671 [IEE WIRING REGULATIONS])

DETAILS OF THE CLIENT (note 1)

....

....

....

INSTALLATION ADDRESS

....

....

....Postcode ....

DESCRIPTION AND EXTENT OF THE INSTALLATION Tick boxes as appropriate

Description of installation: .... New installation ☐

Extent of installation covered by this Certificate: ....

.... Addition to an

.... existing installation ☐

.... Alteration to an

.... existing installation ☐

FOR DESIGN, CONSTRUCTION, INSPECTION & TESTING

I being the person responsible for the Design, Construction, Inspection & Testing of the electrical installation (as indicated by my signature below), particulars of which are described above, having exercised reasonable skill and care when carrying out the Design, Construction, Inspection & Testing, hereby CERTIFY that the said work for which I have been responsible is to the best of my knowledge and belief in accordance with BS 7671 : ...., amended to .... (date) except for the departures, if any, detailed as follows:

> Details of departures from BS 7671 (Regulations 120-01-03, 120-02):

The extent of liability of the signatory is limited to the work described above as the subject of this Certificate.

Name (IN BLOCK LETTERS):...   Position: ....

Signature (note 3): ....   Date:..

For and on behalf of: ....

Address: ....

....

....Postcode ....   Tel No: ....

NEXT INSPECTION

I recommend that this installation is further inspected and tested after an interval of not more than .... years/months (notes 4 and 7)

SUPPLY CHARACTERISTICS AND EARTHING ARRANGEMENTS Tick boxes and enter details, as appropriate

Earthing arrangements	Number and Type of Live Conductors		Nature of Supply Parameters	Supply Protective Device Characteristics
TN-C ☐	a.c. ☐	d.c ☐	Nominal voltage, U/Uo[1] V	
TN-S ☐	1-phase, 2-wire ☐	2-pole ☐	Nominal frequency, f [1] ....Hz	Type:.........................
TN-C-S ☐	1-phase, 3-wire ☐	3-pole ☐	Prospective fault current, Ipf [2] (note 6) ....kA	....
TT ☐	2-phase, 3-wire ☐	other ☐	External loop impedance, Ze [2] ....Ω	
IT ☐	3-phase, 3-wire ☐		(Note: (1) by enquiry, (2) by enquiry or by measurement)	Nominal current rating
	3-phase, 4-wire ☐			...................................A
Alternative source ☐ of supply (to be detailed on attached schedules)				

PARTICULARS OF INSTALLATION REFERRED TO IN THE CERTIFICATE Tick boxes and enter details, as appropriate

Means of Earthing

Distributor's facility ☐

Maximum Demand

Maximum demand (load)...Amps per phase

Details of Installation Earth Electrode (where applicable)

Installation
earth electrode ☐

Type
(e.g. rod(s), tape etc)

....

Location

....

Electrode resistance to earth

....Ω

Main Protective Conductors

Earthing conductor: material .... csa ....mm²   connection verified ☐

Main equipotential
bonding conductors   material .... csa ....mm²   connection verified ☐

To incoming water and/or gas service ☐   To other elements ....

Main Switch or Circuit-breaker

BS, Type ....   No. of poles ....   Current rating ....A   Voltage rating V

Location ....   Fuse rating or setting ....A

Rated residual operating current I ∆n =...................... mA, and operating time of.............ms (at I ∆n)
(applicable only where an RCD is suitable and is used as a main circuit-breaker)

COMMENTS ON EXISTING INSTALLATION: (In the case of an alteration or additions see Section 743)

....

....

....

....

....

....

....

....

SCHEDULES (note 2)

The attached Schedules are part of this document and this Certificate is valid only when they are attached to it.

.... Schedules of Inspections and .... Schedules of Test Results are attached.

(Enter quantities of schedules attached)

GUIDANCE FOR RECIPIENTS

This safety Certificate has been issued to confirm that the electrical installation work to which it relates has been designed, constructed and inspected and tested in accordance with British Standard 7671 (The IEE Wiring Regulations).

You should have received an original Certificate and the contractor should have retained a duplicate Certificate. If you were the person ordering the work, but not the user of the installation, you should pass this Certificate, or a full copy of it including the schedules, immediately to the user.

The "original" Certificate should be retained in a safe place and be shown to any person inspecting or undertaking further work on the electrical installation in the future. If you later vacate the property, this Certificate will demonstrate to the new owner that the electrical installation complied with the requirements of British Standard 7671 at the time the Certificate was issued. The Construction (Design and Management) Regulations require that for a project covered by those Regulations, a copy of this Certificate, together with schedules is included in the project health and safety documentation.

For safety reasons, the electrical installation will need to be inspected at appropriate intervals by a competent person. The maximum time interval recommended before the next inspection is stated on Page 1 under "Next Inspection".

This Certificate is intended to be issued only for a new electrical installation or for new work associated with an alteration or addition to an existing installation. It should not have been issued for the inspection of an existing electrical installation. A "Periodic Inspection Report" should be issued for such a periodic inspection.

AD P Copies of BS 7671 and IEE Model Forms

Form 2   Form No   /2

ELECTRICAL INSTALLATION CERTIFICATE (notes 1 and 2)

(REQUIREMENTS FOR ELECTRICAL INSTALLATIONS - BS 7671 [IEE WIRING REGULATIONS])

DETAILS OF THE CLIENT (note 1) ....
....
....

INSTALLATION ADDRESS
....
....
....Postcode ....

DESCRIPTION AND EXTENT OF THE INSTALLATION Tick boxes as appropriate
(note 1)

Description of installation:	New installation ☐
Extent of installation covered by this Certificate:	
....	Addition to an existing installation ☐
....	
....	Alteration to an existing installation ☐

FOR DESIGN

I/We being the person(s) responsible for the design of the electrical installation (as indicated by my/our signatures below), particulars of which are described above, having exercised reasonable skill and care when carrying out the design, hereby CERTIFY that the design work for which I/we have been responsible is to the best of my/our knowledge and belief in accordance with BS 7671 : ...., amended to .... (date) except for the departures, if any, detailed as follows:

> Details of departures from BS 7671 (Regulations 120-01-03, 120-02):

The extent of liability of the signatory or the signatories is limited to the work described above as the subject of this Certificate.

For the DESIGN of the installation: **(Where there is mutual responsibility for the design)

Signature: .... Date .... Name (BLOCK LETTERS): ....Designer No 1

Signature: .... Date .... Name (BLOCK LETTERS): ....Designer No 2**

FOR CONSTRUCTION

I/We being the person(s) responsible for the construction of the electrical installation (as indicated by my/our signatures below), particulars of which are described above, having exercised reasonable skill and care when carrying out the construction, hereby CERTIFY that the construction work for which I/we have been responsible is to the best of my/our knowledge and belief in accordance with BS 7671 : ...., amended to .... (date) except for the departures, if any, detailed as follows:

> Details of departures from BS 7671 (Regulations 120-01-03, 120-02):

The extent of liability of the signatory is limited to the work described above as the subject of this Certificate.

For CONSTRUCTION of the installation:

Signature: .... Date ....

Name (BLOCK LETTERS): .... Constructor

FOR INSPECTION & TESTING

I/We being the person(s) responsible for the inspection & testing of the electrical installation (as indicated by my/our signatures below), particulars of which are described above, having exercised reasonable skill and care when carrying out the inspection & testing, hereby CERTIFY that the work for which I/we have been responsible is to the best of my knowledge and belief in accordance with BS 7671 : ...., amended to .... (date) except for the departures, if any, detailed as follows:

> Details of departures from BS 7671 (Regulations 120-01-03, 120-02):

The extent of liability of the signatory is limited to the work described above as the subject of this Certificate.

For INSPECTION & TEST of the installation: **(Where there is mutual responsibility for the design)

Signature: .... Date ....

Name (BLOCK LETTERS): .... Inspector

NEXT INSPECTION (notes 4 and 7)

I/We the designer(s) recommend that this installation is further inspected and tested after an interval of not more than .... years/months

PARTICULARS OF THE SIGNATORIES TO THE ELECTRICAL INSTALLATION CERTIFICATE (note 3)

Designer (No 1)
Name: .... Company: ....

Address: ....

.... Postcode: .... Tel No: ....

Designer (No 2)
(if applicable) Name: .... Company: ....

Address: ....

.... Postcode: .... Tel No: ....

Constructor
Name: .... Company: ....

Address: ....

.... Postcode: .... Tel No: ....

Inspector
Name: .... Company: ....

Address: ....

.... Postcode: .... Tel No: ....

SUPPLY CHARACTERISTICS AND EARTHING ARRANGEMENTS Tick boxes and enter details, as appropriate

Earthing arrangements	Number and Type of Live Conductors	Nature of Supply Parameters	Supply Protective Device Characteristics
TN-C ☐	a.c. ☐ d.c ☐	Nominal voltage, $U/U_o^{(1)}$ V	
TN-S ☐	1-phase, 2-wire ☐ 2-pole ☐	Nominal frequency, $f^{(1)}$ ....Hz	Type: ....
TN-C-S ☐	1-phase, 3-wire ☐ 3-pole ☐	Prospective fault current, Ipf $^{(2)}$ (note 6) ....kA	....
TT ☐	2-phase, 3-wire ☐ other ☐	External loop impedance, Ze $^{(2)}$ ....Ω	
IT ☐	3-phase, 3-wire ☐	(Note: (1) by enquiry, (2) by enquiry or by measurement)	Nominal current rating
	3-phase, 4-wire ☐		...................................A
Alternative source ☐ of supply (to be detailed on attached schedules)			

PARTICULARS OF INSTALLATION REFERRED TO IN THE CERTIFICATE Tick boxes and enter details, as appropriate

Means of Earthing

Distributor's facility ☐

Maximum Demand

Maximum demand (load) ....Amps per phase

Details of Installation Earth Electrode (where applicable)

Installation earth electrode ☐

Type (e.g. rod(s), tape etc) ....

Location ....

Electrode resistance to earth ....Ω

Main Protective Conductors

Earthing conductor: material .... csa ....mm² connection verified ☐

Main equipotential bonding conductors material .... csa ....mm² connection verified ☐

To incoming water and/or gas service ☐   To other elements ....

Main Switch or Circuit-breaker

BS, Type............................... No. of poles .... Current rating ....A Voltage rating V

Location .... Fuse rating or setting ....A

Rated residual operating current I ∆n = ... mA, and operating time of ....ms (at I ∆n)
(applicable only where an RCD is suitable and is used as a main circuit-breaker)

COMMENTS ON EXISTING INSTALLATION: (In the case of an alteration or additions see Section 743)

....

....

....

....

SCHEDULES (note 2)
The attached Schedules are part of this document and this Certificate is valid only when they are attached to it.
.... Schedules of Inspections and .... Schedules of Test Results are attached. (Enter quantities of schedules attached)

ELECTRICAL INSTALLATION CERTIFICATE - GUIDANCE FOR RECIPIENTS (to be appended to the Certificate)

This safety Certificate has been issued to confirm that the electrical installation work to which it relates has been designed, constructed and inspected and tested in accordance with British Standard 7671 (The IEE Wiring Regulations).

You should have received an original Certificate and the contractor should have retained a duplicate Certificate. If you were the person ordering the work, but not the user of the installation, you should pass this Certificate, or a full copy of it including the schedules, immediately to the user.

The "original" Certificate should be retained in a safe place and be shown to any person inspecting or undertaking further work on the electrical installation in the future. If you later vacate the property, this Certificate will demonstrate to the new owner that the electrical installation complied with the requirements of British Standard 7671 at the time the Certificate was issued. The Construction (Design and Management) Regulations require that for a project covered by those Regulations, a copy of this Certificate, together with schedules is included in the project health and safety documentation.

For safety reasons, the electrical installation will need to be inspected at appropriate intervals by a competent person. The maximum time interval recommended before the next inspection is stated on Page 1 under "Next Inspection".

This Certificate is intended to be issued only for a new electrical installation or for new work associated with an alteration or addition to an existing installation. It should not have been issued for the inspection of an existing electrical installation. A "Periodic Inspection Report" should be issued for such a periodic inspection.

The Certificate is only valid if a Schedule of Inspections and Schedule of Test Result are appended.

For the minor electrical alterations covered by Schedule 2B to the Building Regulations 2000 (see Chapter 5) the person carrying out the electrical work does not even have to notify the local authority, although the work must still comply with Part P. This raises an interesting question as to how the person carrying out the work can ensure that the non-notifiable work complies with the regulations. If this work is to be undertaken by a DIY worker, one way of showing compliance would be to follow the IEE guidance or guidance in other authoritative manuals that are based on this, and to have the work inspected and tested by a qualified person who could then supply a BS 7671 Periodic Inspection Report or similar. This would indicate that the installation had been tested and shown to comply with the BS 7671 criteria, although such a person would not be able to supply an installation certificate since this can only be done by the person actually installing the work. The competent person does not necessarily need to be registered with an electrical self-certification scheme but, as required by BS 7671, must be competent in respect of the inspection and testing of an installation. Nevertheless, non-notifiable works should always be drawn to the attention of a person carrying out subsequent work or periodic inspections. It should be noted that although work may be non-notifiable it must still comply with the regulations and local authorities are empowered to take enforcement action if electrical work is found to be unsafe or non-compliant.

Where the work falls outside the scope of Schedules 2A and 2B it will be necessary to either serve a building notice or deposit full plans with the local authority (see Chapter 3) or engage the services of an Approved Inspector (see Chapter 4).

In an emergency the relevant building control body should be notified as soon as possible. The building control body then becomes responsible for ensuring that the work is safe and complies with the regulations.

Where a relevant building control body is used for ensuring compliance, installers should provide a copy of the BS 7671 installation certificate if they are qualified to carry out inspection and testing. The building control body will take the certificate into account when considering if the installation is in compliance with the regulations and may ask for evidence that the installer is sufficiently qualified.

AD P Copies of BS 7671 and IEE Model Forms

Form 5

MINOR ELECTRICAL INSTALLATION WORKS CERTIFICATE

(REQUIREMENTS FOR ELECTRICAL INSTALLATIONS - BS 7671 [IEE WIRING REGULATIONS])

To be used only for minor electrical work which does not include the provision of a new circuit

---

PART 1 : Description of minor works

1.   Description of the minor works : ....

2.   Location/Address :..

3.   Date minor works completed : ....

4.   Details of departures, if any, from BS 7671

     ....

     ....

     ....

PART 2 : Installation details

1.   System earthing arrangement: TN-C-S ☐   TN-S ☐   TT ☐

2.   Method of protection against indirect contact:...

3.   Protective device for the modified circuit: Type BS ....   Rating ....A

4.   Comments on existing installation, including adequacy of earthing and bonding arrangements:
     (see Regulation 130-07)..

     ....

     ....

     ....

PART 3 : Essential Tests

1.   Earth continuity : satisfactory ☐

2.   Insulation resistance:

                              Phase/neutral ....MΩ

                              Phase/earth ....MΩ

                              Neutral/earth ....MΩ

3.   Earth fault loop impedance   ....Ω

4.   Polarity : satisfactory   ☐

5.   RCD operation (if applicable): Rated residual operating current $I_{\Delta n}$ ....mA and operating time of ....ms (at $I_{\Delta n}$)

PART 4 : Declaration

1.   I/We CERTIFY that the said works do not impair the safety of the existing installation, that the said works have been designed, constructed, inspected and tested in accordance with BS 7671 : .... (IEE Wiring Regulations), amended to .... and that the said works, to the best of my/our knowledge and belief, at the time of my/our inspection, complied with BS 7671 except as detailed in Part 1.

2.   Name: ....   3. Signature: ....

     For and on behalf of: ....   Position: ....

     Address:..

     ....   Date: ....

     ....

     ....Postcode ....

MINOR ELECTRICAL INSTALLATION WORKS CERTIFICATE

GUIDANCE FOR RECIPIENTS (to be appended to the Certificate)

This Certificate has been issued to confirm that the electrical installation work to which it relates has been designed, constructed and inspected and tested in accordance with British Standard 7671 (The IEE Wiring Regulations).

You should have received an original Certificate and the contractor should have retained a duplicate. If you were the person ordering the work, but not the owner of the installation, you should pass this Certificate, or a copy of it, to the owner. A separate Certificate should have been received for each existing circuit on which minor works have been carried out. This Certificate is not appropriate if you requested the contractor to undertake more extensive installation work, for which you should have received an Electrical Installation Certificate.

The Certificate should be retained in a safe place and be shown to any person inspecting or undertaking further work on the electrical installation in the future. If you later vacate the property, this Certificate will demonstrate to the new owner that the minor electrical installation work carried out complied with the requirements of British Standard 7671 at the time the Certificate was issued.

It may be the case that an installer is not qualified to complete the BS 7671 certificates. Where the work is notifiable it will be the responsibility of the building control body to carry out the necessary tests at their own expense. This may be done by the building control body itself or may be sub-contracted to a specialist body.

It should be noted that a building control body cannot issue a BS 7671 certificate as these can only be issued by the person carrying out the work.

Approved Document P contains, in Tables 1 and 2, additional guidance on what constitutes non-notifiable work (Table 1) and Special Locations and Installations (Table 2). This augments Schedule 2B and should be read in conjunction with that Schedule. Tables 1 and 2 are reproduced below.

The following additional examples of notifiable and non-notifiable work are given in AD P and should be read in conjunction with Table 1:

- **Notifiable work:**
    - new circuits back to consumer unit;
    - extensions to circuits in kitchens and special locations;
    - extensions to circuits associated with special installations (e.g. garden lighting and power installations, etc.;
    - replacement of consumer unit;
    - work to outdoor lighting and power installations that involves crossing a garden or is in a garden;
    - installation of a socket outlet on an external wall;
    - new central heating control wiring installations wherever located.
- **Non-notifiable work:**
    - replacement, repair and maintenance jobs (except replacement of consumer unit);
    - work in conservatories and attached garages unless it involves the installation of a new circuit or the extension of a circuit in a kitchen or special location or associated with a special installation;
    - work in detached garages and sheds not involving new outdoor lighting;

AD P, Section 0

Table 1   Work that need not be notified to building control bodies.

Work consisting of:
Replacing any fixed electrical equipment (for example, socket-outlets, control switches and ceiling roses) which does not include the provision of any new fixed cabling
Replacing the cable for a single circuit only, where damaged, for example, by fire, rodent or impact[a]
Re-fixing or replacing the enclosures of existing installation components[b]
Providing mechanical protection to existing fixed installations [c]
Installing or upgrading main or supplementary equipotential bonding [d]
Work that is not in a kitchen or special location and does not involve a special installation [e] and consists of:
Adding lighting points (light fittings and switches) to an existing circuit[f]
Adding socket-outlets and fused spurs to an existing ring or radial circuit[f]
Work not in a special location, on:
Telephone or extra-low voltage wiring and equipment for the purposes of communications, information technology, signalling, control and similar purposes
Prefabricated equipment sets and associated flexible leads with integral plug and socket connections

Notes:
(a) On condition that the replacement cable has the same current-carrying capacity and follows the same route.
(b) If the circuit's protective measures are unaffected.
(c) If the circuit's protective measures and current-carrying capacity of conductors are unaffected by increased thermal insulation.
(d) Such work will need to comply with other applicable legislation, such as the Gas Safety (Installation and Use) Regulations.
(e) Special locations and installations are listed in Table 2.
(f) Only if the existing circuit protective device is suitable and provides protection for the modified circuit, and other relevant safety provisions are satisfactory.

Table 2   Special locations and installations[a]

Special locations
Locations containing a bath tub or shower basin
Swimming pools or paddling pools
Hot air saunas
Special installations
Electric floor or ceiling heating systems
Outdoor lighting or power installations
Solar photovoltaic (PV) power supply systems
Small scale generators such as microCHP units
Extra-low voltage lighting installations, other than pre-assembled, CE-marked lighting sets

Note:
(a) See IEE Guidance Note 7 which gives more guidance on achieving safe installations where risks to people are greater.
NB. In large bathrooms, the location containing a bath or shower is defined by the walls of the bathroom

○ the installation of fixed equipment unless it involves fixed wiring and the installation of a new circuit or the extension of a circuit in a kitchen or special location or associated with a special installation;

○ the installation of equipment attached to the outside wall of a house (e.g. security lighting, air conditioning equipment and radon fans) unless there are exposed outdoor connections or the work involves the installation of a new circuit or the extension of a circuit in a kitchen or special location or associated with a special installation;

○ the installation of prefabricated, 'modular' systems (e.g. kitchen lighting systems and armoured garden cabling) linked by plug and socket connectors provided that the products are CE-marked and that final connections in kitchens and special locations are made to existing connection units or points (possibly a 13A socket outlet).

19.4   Performance

The requirements of Part P can be met by adhering to the 'Fundamental Principles' for achieving safety given in BS 7671: 2001 Chapter 13 (incorporating Amendments No 1: 2002 and No 2: 2004), *Requirements for electrical installations (IEE Wiring Regulations 16th Edition)*, published by The Institution of Electrical Engineers, 2004.

Electrical installations can achieve these requirements by being:

● designed and installed so that they do not present electric shock and fire hazards to people, and afford appropriate protection against mechanical and thermal damage; and

● inspected and tested in a suitable manner so as to verify that they meet the relevant standards for equipment and installation.

The 'Fundamental Principles' referred to above can be met by following:

● the technical rules described in BS 7671: 2001, or an equivalent standard approved by a member of the EEA (this should include issuing an electrical installation certificate to the person ordering the work); and

● guidance given in installation manuals that are consistent with BS 7671: 2001, such as:

○ the IEE (Institution of Electrical Engineers) On-Site Guide, or

○ the series of Guidance Notes Nos 1 to 7 published by the IEE as follows:
   – *IEE Guidance Note 1: Selection and erection of equipment.* 4th Edition. 2002;
   – *IEE Guidance Note 2: Isolation and switching.* 4th Edition. 2002;
   – *IEE Guidance Note 3: Inspection and testing.* 4th Edition. 2002;
   – *IEE Guidance Note 4: Protection against fire.* 4th Edition. 2003;
   – *IEE Guidance Note 5: Protection against electric shock.* 4th Edition. 2002;
   – *IEE Guidance Note 6: Protection against overcurrent.* 4th Edition. 2003;
   – *IEE Guidance Note 7: Special locations.* 2nd Edition (incorporating the 1st and 2nd amendments). 2003.

19.5   Interpretation

The following definitions apply throughout Approved Document P:

ELECTRICAL INSTALLATION – BS 7671 defines this as 'an assembly of associated electrical equipment supplied from a common origin to fulfill a specific purpose and having certain coordinated characteristics'. A simpler and more specific definition is given for the purposes of Building Regulations as – 'fixed electrical cables or fixed electrical equipment located on the consumer's side of the electricity supply meter'.

EXTRA-LOW VOLTAGE is defined in BS 7671 as 'normally not exceeding 50 volts alternating current or 120 volts ripple free direct current, whether between conductors or to earth'.

LOW VOLTAGE is defined in BS 7671 as 'normally exceeding extra-low voltage but not exceeding 1000 volts alternating current or 1500 volts direct current between conductors, or 600 volts alternating current or 900 volts direct current between conductors and earth'.

These last two definitions are phrased slightly differently to those given in regulation 2 (see Chapter 2) but are essentially the same.

KITCHEN – A room or part of a room which contains a sink and food preparation facilities. In open plan areas a kitchen zone can be considered to extend from the edge of the sink for a distance of 3 m or to a dividing wall if this is nearer.

19.6   Guidance on design, installation, inspection and testing

19.6.1   Electricity at Work Regulations

In general, electrical installation work that is carried out by a qualified electrician should comply with the Electricity at Work Regulations 1989 (as amended by SI 1996/192, 1997/1993 and 1999/2024). The Regulations impose duties on employers, employees and the self-employed and regulation 3(2)(b) places duties on employees equivalent to those placed on employers and self-employed persons where there are matters within their control.

The full text of the Electricity at Work Regulations and guidance on how to comply with them may be found in the Health and Safety guidance document *Memorandum of guidance on the electricity at work regulations 1989* – HSR25. Important elements of the Regulations include:

- Electrical work should only be carried out by persons that are competent to prevent danger and injury while doing it, or who are appropriately supervised (regulation 16).
- Regulations 4(1), 5, 6, 7, 8, 9, 10, 11, 12 set general requirements for the design, construction and suitability of equipment for its intended use.

19.6.2   Public safety, quality and continuity of supply

From the viewpoint of an electrical installation, it is most important that measures are provided so that people are protected against the risks of electric shock, burns or fire injuries. In order to achieve these aims the installation should be suitably:

- designed and installed;
- enclosed; and
- separated by appropriate distances to provide mechanical and thermal protection.

One way of complying with the above would be to follow the technical rules in BS 7671: 2001 or an equivalent standard. Another example of a suitable approach may be found in the Electricity Safety, Quality and Continuity Regulations 2002 (SI 2002/2665) (detailed guidance on the Regulations may be obtained from www.dti.gov.uk/electricity-regulations).

Amongst many other things, these regulations make it mandatory for the electricity distributor to install the incoming main supply cut-out and the electricity meter in a safe location (i.e. where they can be mechanically protected and safely maintained). It may also be necessary in certain circumstances for the electricity distributor and installer to take account of the risk of flooding (see the DCLG publication *Preparing for floods*, available from www.communities.gov.uk for further guidance). Where there are proposals for new electrical installations or it is intended to make significant alterations to existing ones, these same regulations (together with the mains electricity supply contract) mean that agreement over the proposals must be reached with the electricity distributor.

The Electricity Safety, Quality and Continuity Regulations 2002 (which came into force on 31 January 2003) were introduced not only to improve standards in public safety but also to align requirements to modern electricity markets. Therefore, they also specify power quality and supply continuity requirements so that an efficient and economic electricity supply service can be ensured for consumers.

Under the regulations, duties are imposed on:

- electricity generators, distributors and suppliers; and
- meter operators, consumers and specified persons.

However, the majority of the duties apply to distributors who own or operate networks used to supply consumers' installations, street furniture or other networks. Some of the principal duties of electricity distributors are as follows:

- Unless inappropriate for safety reasons, an earthing facility must be provided for new connections to maintain the supply within defined tolerance limits.
- Certain technical and safety information must be provided to consumers to enable them to design their installations.
- Together with meter operators, distributors must make sure that their equipment on consumers' premises is:
  - o suitable for its purpose;
  - o safe in its particular environment; and
  - o the polarity of conductors is clearly indicated.

Additionally, under the provisions of the regulations the Secretary of State may issue safety enforcement notices to consumers where their installations outside buildings present a danger to the public.

19.6.3   Embedded generation

One area covered by the Electricity Safety, Quality and Continuity Regulations 2002 concerns the generation of electricity by the public from, for example, renewable sources (such as windmills and photovoltaic cells) and combined heat and power (where, for example, an industrial process that generates heat can be used to run a small privately owned power station). This so-called 'embedded' generation ('embedded' generators are those connected to the distribution networks of public electricity suppliers rather than directly to the National Grid) must be controlled in order to prevent a parallel connection occurring with the distributor's network and must comply with BS 7671. Sources of energy that operate in parallel with the distributor's network must meet certain additional safety standards (for example, the equipment must not be a source of danger or cause interference with the distributor's network) and persons installing domestic combined heat and power equipment must advise the local distributor of their intentions before or at the time of commissioning the source.

The regulations prevent distributors from connecting installations to their networks which do not comply with BS 7671. Other persons may connect installations to distributors' networks provided that they obtain the prior consent of the distributor. The distributor may require evidence that:

- the installation complies with BS 7671; and
- the connection itself will meet safety and operational requirements.

Distributors may disconnect consumers' installations that are a source of danger or cause interference with their networks or other installations.

19.6.4   Inspection and testing before taking into service

Before they are taken into service, electrical installations should be inspected and tested both during and at the end of installation. This is to verify that they comply with BS 7671: 2001 (i.e. they are reasonably safe to use maintain and alter) and that they comply with Part P and any other relevant parts of the regulations.

Chapters 71 and 74 of BS 7671 contain procedures that may be used to demonstrate compliance by supplying:

- a declaration to the relevant building control body that compliance with Part P of the Building Regulations has been achieved, where the work has been carried out by a competent person registered with an electrical self certification scheme (see Chapter 5); and
- copies of the forms called for by BS 7671, to the person ordering the work, signed by a person qualified to do so ('qualified' means having appropriate qualifications, knowledge and experience to carry out the inspection and testing procedures and to complete the relevant electrical installation certificate).

The forms should show that the electrical installation work has been:

- inspected during installation and on completion to verify that the components used:
  - comply with appropriate British Standards or harmonised European Standards;
  - have been selected and installed to the requirements of BS 7671;
  - have not been visibly damaged and are not defective as to be unsafe;
- tested to check satisfactory performance in relation to:
  - conductor's continuity;
  - insulation resistance;
  - circuits separation;
  - polarity;
  - earthing and bonding arrangements;
  - earth fault loop impedance;
  - functionality of all protective devices including residual current devices.

A list of all the necessary inspections may be found in section 712 of BS 7671 (although not all the elements may be relevant in all cases). Appendix 6 of BS 7671 contains a schedule of inspections and this has been reproduced in Approved Document P in Appendix B as Form 3 and in this chapter (see below).

A list of all the necessary tests may be found in section 713 of BS 7671 (although not all the elements may be relevant in all cases). Appendix 6 of BS 7671 contains a blank schedule of test results, and this has also been reproduced in Approved Document P in Appendix B as Form 4 and in this chapter (see below).

Test results should be recorded on forms similar to Form 4 and the tests themselves should be carried out using appropriate and accurate instruments under the BS 7671 conditions. Compliance should then be confirmed by comparing the test results with relevant performance criteria.

The Minor Works Certificate (see Form 5 above) lists a series of essential tests for additions and alterations that do not include the provision of a new circuit. The nature of the work will determine which tests are appropriate in the circumstances.

Electrical installation certificates based on those in BS 7671 are available from the Institution of Electrical Engineers (go to the IEE website on www.iee.org/Publish/WireRegs/forms.cfm) and other sources, and model forms are given in Appendix B of Approved Document P. Where appropriate they have been reproduced from AD P in this chapter. The scope of work covered by the certificates ranges from minor works to large projects such as blocks of flats. The most appropriate form should be used in each particular case and it should be signed by the person responsible for carrying out the design, construction, inspection and testing of the works.

19.6.5   Provision of information

In order that an electrical installation can be operated, maintained and altered with reasonable safety the following information (called for by BS 7671 or an equivalent standard) should be left with the occupant of the premises in which the installation is located:

AD P Copies of BS 7671 and IEE Model Forms

Form 3

Form No   /3

SCHEDULE OF INSPECTIONS

Methods of protection against electric shock

(a) Protection against both direct and indirect contact:

☐ (i) SELV (note 1)

☐ (ii) Limitation of discharge of energy

(b) Protection against direct contact: (note 2)

☐ (i) Insulation of live parts

☐ (ii) Barriers or enclosures

☐ (iii) Obstacles (note 3)

☐ (iv) Placing out of reach (note 4)

☐ (v) PELV

☐ (vi) Presence of RCD for supplementary protection

(c) Protection against indirect contact:

(i) EEBADS including:

☐ Presence of earthing conductor

☐ Presence of circuit protective conductors

☐ Presence of main equipotential bonding conductors

☐ Presence of supplementary equipotential bonding conductors

☐ Presence of earthing arrangements for combined protective and functional purposes

☐ Presence of adequate arrangements for alternative source(s), where applicable

☐ Presence of residual current device(s)

☐ (ii) Use of Class II equipment or equivalent insulation (note 5)

☐ (iii) Non-conducting location: (note 6) Absence of protective conductors

☐ (iv) Earth-free equipotential bonding: (note 7) Presence of earth-free equipotential bonding conductors

☐ (v) Electrical separation (note 8)

Prevention of mutual detrimental influence

☐ (a) Proximity of non-electrical services and other influences

☐ (b) Segregation of band I and band II circuits or band II insulation used

☐ (c) Segregation of safety circuits

Identification

☐ (a) Presence of diagrams, instructions, circuit charts and similar information

☐ (b) Presence of danger notices and other warning notices

☐ (c) Labelling of protective devices, switches and terminals

☐ (d) Identification of conductors

Cables and conductors

☐ (a) Routing of cables in prescribed zones or within mechanical protection

☐ (b) Connection of conductors

☐ (c) Erection methods

☐ (d) Selection of conductors for current-carrying capacity and voltage drop

☐ (e) Presence of fire barriers, suitable seals and protection against thermal effects

General

☐ (a) Presence and correct location of appropriate devices for isolation and switching

☐ (b) Adequacy of access to switchgear and other equipment

☐ (c) Particular protective measures for special installations and locations

☐ (d) Connection of single-pole devices for protection or switching in phase conductors only

☐ (e) Correct connection of accessories and equipment

☐ (f) Presence of undervoltage protective devices

☐ (g) Choice and setting of protective and monitoring devices for protection against indirect contact and/or overcurrent

☐ (h) Selection of equipment and protective measures appropriate to external influences

☐ (I) Selection of appropriate functional switching devices

Inspected by ....   Date   ....

Notes:

T   to indicate an inspection has been carried out and the result is satisfactory
C   to indicate an inspection has been carried out and the result was unsatisfactory
N/A   to indicate the inspection is not applicable
LIM   to indicate that, exceptionally, a limitation agreed with the person ordering the work prevented the inspection or test being carried out

1.   SELV An extra-low voltage system which is electrically separated from Earth and from other systems. The particular requirements of the Regulations must be checked (see Regulations 411-02 and 471-02)

2.   Method of protection against direct contact - will include measurement of distances where appropriate

3.   Obstacles - only adopted in special circumstances (see Regulations 412-04 and 471-06)

4.   Placing out of reach - only adopted in special circumstances (see Regulations 412-05 and 471-07)

5.   Use of Class II equipment - infrequently adopted and only when the installation is to be supervised (see Regulations 413-03 and 471-09)

6.   Non-conducting locations - not applicable in domestic premises and requiring special precautions (see Regulations 413-04 and 471-10)

7.   Earth-free local equipotential bonding - not applicable in domestic premises, only used in special circumstances (see Regulations 413-05 and 471-11)

8.   Electrical separation (see Regulations 413-06 and 471-12)

AD P Copies of BS 7671 and IEE Model Forms

Form 4

SCHEDULE OF TEST RESULTS

Form No .... /4

Contractor: ....

Test Date: ....

Signature ....

Method of protection against indirect contact: ....

Equipment vulnerable to testing: ....

Address/Location of distribution board: ....

* Type of Supply: TN-S/TN-C-S/TT ....
* Ze at origin: ....ohms
* PFC: ....kA

Instruments

loop impedance: ....

continuity: ....

insulation: ....

RCD tester: ....

Description of Work: ....

Circuit Description	Overcurrent Device				Wiring Conductors		Continuity			Insulation Resistance		Polarity	Earth Loop Impedance	Functional Testing		Remarks
	*Short-circuit capacity: ....kA	type	Rating I_n	live	cpc	$(R_1 + R_2)$*	R_2*	Ring	Live/ Live	Live/ Earth		Zs	RCD time	Other	
			A	mm²	mm²	Ω	Ω		MΩ	MΩ		Ω	ms		
1	2	3		4	5	*6	*7	*8	*9	*10	*11	*12	*13	*14	15

Test Results

Deviations from Wiring Regulations and special notes:

* See notes on schedule of test results

NOTES ON SCHEDULE OF TEST RESULTS (to be read in conjunction with Form 4)

* **Type of supply** is ascertained from the supply company or by inspection.
* **Ze at origin.** When the maximum value declared by the electricity supplier is used, the effectiveness of the earth must be confirmed by a test. If measured the main bonding will need to be disconnected for the duration of the test.
* **Short-circuit capacity** of the device is noted, see Table 7.2A of the On-Site Guide or 2.7.15 of GN3
* **Prospective fault current (PFC).** The value recorded is the greater of either the short-circuit current or the earth fault current. Preferably determined by enquiry of the supplier.

The following tests, where relevant, shall be carried out in the following sequence:

Continuity of protective conductors, including main and supplementary bonding

Every protective conductor, including main and supplementary bonding conductors, should be tested to verify that it is continuous and correctly connected.

6   Continuity

Where Test Method 1 is used, enter the measured resistance of the phase conductor plus the circuit protective conductor (R1+ R2).

See 10.3.1 of the On-Site Guide or 2.7.5 of GN3.

During the continuity testing (Test Method 1) the following polarity checks are to be carried out:

(a)   every fuse and single-pole control and protective device is connected in the phase conductor only

(b)   centre-contact bayonet and Edison screw lampholders have outer contact connected to the neutral conductor

(c)   wiring is correctly connected to socket-outlets and similar accessories.

Compliance is to be indicated by a tick in polarity column 11.

(R1 + R2) need not be recorded if R2 is recorded in column 7.

7   Where Test Method 2 is used, the maximum value of R2 is recorded in column 7.

Where the alternative method of Regulation 413-02-12 is used for shock protection, the resistance of the circuit protective conductor R2 is measured and recorded in column 7.

See 10.3.1 of the On-Site Guide or 2.7.5 of GN3.

8   Continuity of ring final circuit conductors

A test shall be made to verify the continuity of each conductor including the protective conductor of every ring final circuit.

See 10.3.2 of the On-Site Guide or 2.7.6 of GN3.

9, 10 Insulation Resistance

All voltage sensitive devices to be disconnected or test between live conductors (phase and neutral) connected together and earth.

The insulation resistance between live conductors is to be inserted in column 9.

The minimum insulation resistance values are given in Table 10.1 of the On-Site Guide or Table 2.2 of GN3.

See 10.3.3(iv) of the On-Site Guide or 2.7.7 of GN3.

All the preceding tests should be carried out before the installation is energised.

11   Polarity

A satisfactory polarity test may be indicated by a tick in column 11.

Only in a Schedule of Test Results associated with a Periodic Inspection Report is it acceptable to record incorrect polarity.

12   Earth fault loop impedance Zs

This may be determined either by direct measurement at the furthest point of a live circuit or by adding (R1 + R2) of column 6 to Ze. Ze is determined by measurement at the origin of the installation or preferably the value declared by the supply company used.

Zs = Ze + (R1 + R2). Zs should be less than the values given in Appendix 2 of the On-Site Guide or App 2 of GN3.

13   Functional testing

The operation of RCDs (including RCBOs) shall be tested by simulating a fault condition, independent of any test facility in the device.

Record operating time in column 13. Effectiveness of the test button must be confirmed.

See Section 11 of the On-Site Guide or 2.7.16 of GN3.

14   All switchgear and controlgear assemblies, drives, control and interlocks, etc must be operated to ensure that they are properly mounted, adjusted and installed.

Satisfactory operation is indicated by a tick in column 14.

Earth electrode resistance

The earth electrode resistance of TT installations must be measured, and normally an RCD is required.

For reliability in service the resistance of any earth electrode should be below 200 Ω. Record the value on Form 1, 2 or 6, as appropriate.

See 10.3.5 of the On-Site Guide or 2.7.13 of GN3.

- electrical installation certificates which describe the installation and give details of the work carried out;
- permanent labels on, for example:
  - o earth connections and bonds, and
  - o items of electrical equipment such as consumer units and RCDs;
- log books and operating instructions;
- detailed plans (only if the installation is unusually large or complex).

19.7   Work on existing electrical installations

19.7.1   Extensions, material alteration and material changes of use

With reference to Part P, the Building Regulations apply not only to new electrical installations, but also to the extension and material alteration of an existing installation and to electrical work required when a building undergoes a material change of use.

Specifically, electrical work comes under the definition of 'controlled service or fitting' in regulation 2 and also under the definition of 'building work' as defined by regulation 3(1) in the following sub-paragraphs:

- '(b) the provision or extension of a controlled service or fitting in or in connection with a building';
- '(c) the material alteration of a building, or a controlled service or fitting';
- '(d) work required by regulation 6 (requirements relating to material change of use)'.

Furthermore, regulation 6 makes it clear that Part P applies to all classes of material change of use.

All the above terms are defined and explained in detail in Chapter 2.

Clearly, any new electrical installation work must comply fully with Part P; however the extent to which compliance may be required of the existing installation (as affected by the new works), may not be so clear cut. Guidance on the way in which the requirements of Part P, paragraph P1 may be met when carrying out work to existing electrical installations is given in section 2 of Approved Document P.

In such cases it is necessary to ensure that:

- the mains supply equipment is suitable for the works of addition and alteration; and
- work is carried out on the existing electrical installation to the extent that the additions and alterations, the circuits which feed them, the protective measures and the relevant earthing and bonding systems will meet the requirements of Part P.

One way of showing compliance would be to follow the guidance given in section 19.6 above regarding design and installation and to show that for the altered circumstances brought about by the works of addition and alteration:

- the rating and the condition of the existing equipment belonging to both the consumer and to the electricity distributor either can carry the additional loads being allowed for, or can be improved to do so; and
- the correct protective measures are used; and
- satisfactory earthing and equipotential bonding arrangements are in place.

19.7.2   Obsolete practice found in alteration work

Electricians often encounter old and/or obsolete practice when carrying out alteration work on existing electrical installations and some of the features encountered will differ considerably from those found in modern installations. If they are suitably experienced they will know what to expect and will be able to advise the client on the suitability of the existing installation for extension or alteration. This may not be the case with others who have not been specifically trained as electricians and DIY workers.

Electrical installations have been a common feature in domestic dwellings since the 1920s. However, in the intervening period considerable changes have taken place, particularly with regard to the types of wiring materials and other equipment being installed, and in the ways that electrical installations are structured. From the viewpoint of electrical safety, the two principal causes of these changes have been:

- technological advances; and
- amendments to the Wiring Regulations published by the Institution of Electrical Engineers (now issued as British Standard BS 7671).

Appendix C of AD P provides examples of some of the features mentioned above and which may present a safety hazard if encountered by people who are unfamiliar with them. These are discussed below.

Insufficient means of earthing for the electrical installation

All electrical installations must be properly earthed in order to provide a safe route to earth should a fault occur in the installation. Up until 1966 it was common for metal service pipes (e.g. water pipes) to be used as a means of earthing. Gas pipes have never been permitted to be used in this way. Current practice is to earth the installation to an electricity distributor's earthing terminal, provided for this purpose near the electricity meter.

Equipotential bonding conductors

Equipotential bonding conductors are wires used to connect metal pipes and metal appliance casings to earth to avoid the risk of electric shock should a live connecting wire accidentally come into contact with the metal body of an appliance, etc.

Since 1966 the installation of *main* equipotential bonding conductors on pipes that run underground, such as water service pipes, oil supply pipes and gas installation pipes, has been a requirement of the Wiring Regulations. In the 1980s the minimum size of main equipotential bonding conductors was increased to a cross sectional area of 10 mm^2.

Since 1981 the installation of ***supplementary*** equipotential bonding conductors has been required where there is deemed to be an increased risk of electric shock. For example, electrical appliances (such as light fittings) in bathrooms and shower rooms would come into this category of risk. The minimum sizing for supplementary equipotential bonding conductors was increased during the 1980s and 1990s. In most dwellings the minimum permissible size that can be installed without mechanical protection is 4 mm².

Changes in cabling

Circuit protective conductors (more commonly referred to as earth wires) are required in most installed cabling. Some of the changes that affect circuit protective conductors (CPCs) are as follows:

- Prior to 1977 the regulations accepted the single colour green for the identification of CPCs. Since that date a yellow–green coding has been required. Whenever the older, green sleeving is encountered when connections are remade, it should be replaced with the new yellow–green striped sleeving.
- In older electrical installations where 2.5 mm² twin & earth pvc/pvc cables are encountered, it may be the case that the CPC has a cross-sectional area of only 1 mm² rather than the 1.5 mm² which is current practice. Such cables are manufactured to BS 6004 and the size of the CPC was increased because it was apparent that where the cable was used in a ring final circuit protected by a 30 amp semi-enclosed (rewirable) fuse, the 1 mm² CPC might not always be properly protected against thermal effects in the event of an earth fault. Where such cables are encountered, consideration should be given to upgrading the cables and/or the consumer unit.
- In lighting circuits installed before 1966, if no metal accessories were included in the design it was common practice for the CPCs to be omitted. Where additions or alterations are made to such installations, it is most important that any new or replacement switches, light fittings or other components are of a non-metallic variety not requiring earthing, or it will be necessary to install new CPCs. Since 1966 all lighting circuits (other than certain extra-low voltage circuits) have been required to include CPCs.

Imperial sized cables – cables having imperial (rather than metric) sized conductors – could be obtained up until the beginning of the 1970s, and can still be found in older installations. Conductors in these imperial sized cables may be:

- single-stranded (as in 1/.044);
- three-stranded (as in 3/.029); or
- contain seven or more strands (as in 7/.029 and 19/.044).

It takes experience to recognise such cables unless they are compared with metric cables and it should be appreciated that their current carrying capacity and voltage drop characteristics are likely to be different from those that may at first be expected. Where it is intended to install a new appliance with a higher rating (such as a washing machine, dishwasher, towel rail or an appliance rated at more than 2 kW), it would be prudent to engage a competent electrician to establish whether the performance limits of the cables are being exceeded.

Cable sheathing – in the 1960s pvc-insulated cables were introduced and soon became universally used especially in domestic dwellings. Prior to the use of cables with pvc sheathing a number of other materials were used and may still be met with when carrying out alterations or additions to older installations as follows:

- Tough rubber sheathed (TRS) vulcanised rubber insulation was used for cables in dwellings prior to the introduction of pvc-insulated cables. TRS rubber-insulated cables are immediately recognisable by their black exterior. Unfortunately, they are prone to deterioration in service especially if they have been subjected to overloading and/or excessive temperature, or the rubber has been exposed to direct sunlight. Deterioration can be recognised by evidence of dryness and inflexibility in the rubber (often causing it to crumble), resulting in loss of insulating properties. When identified, wiring installations of this type should be tested at the earliest opportunity by a competent person. Otherwise they should be left undisturbed until replacement, since they are beyond their normal expected safe working life.
- Lead-sheathed cables are much more rarely encountered than rubber-sheathed insulated cables and may sometimes be found in unimproved properties dating from before about 1948. They are characterised by an outer sheath of lead enclosing rubber insulated tinned copper conductors. Since the outer sheath is a conductor of electricity and could lead to electric shock if the cable was damaged it is essential that the sheath is (and remains) properly earthed. As is the case for TRS cables, the conductor insulation (being made of rubber) is also prone to deterioration. Again, such wiring installations should be tested at the earliest opportunity by a competent person since they are beyond their normal expected safe working life.

The changes to cables mentioned above are illustrated in Fig. 19.1.

Concealed cables – permitted cable zones in walls

There is always a possibility that concealed cables may be struck when work is carried out on existing walls. In the latter part of the 1980s the Wiring Regulations were changed to contain specific requirements for the positioning of cables concealed in walls and partitions (see regulation 522–06–06 in BS 7671), and these are illustrated in Fig. 19.2.

In any event, it is good practice to take extreme care when carrying out any activity that involves penetrating a wall or partition (e.g. by using a cable and stud detector before attempting to drill into walls, floors or ceilings) even when it is known that concealed cables were only recently installed. This is especially true where the cables were installed prior to 1980 as they are particularly likely to be found outside of the zones illustrated in Fig. 19.2.

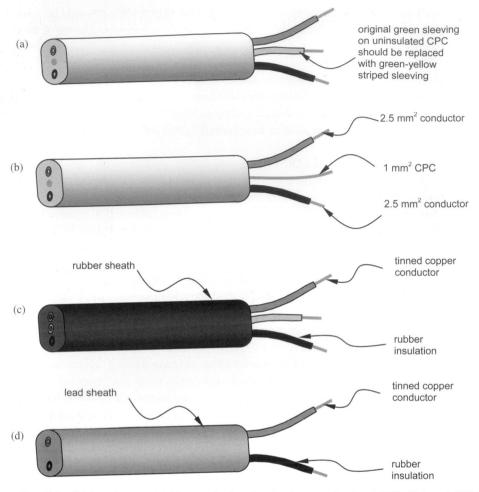

(a)

original green sleeving
on uninsulated CPC
should be replaced
with green-yellow
striped sleeving

2.5 mm² conductor

(b)

1 mm² CPC

2.5 mm² conductor

rubber sheath

tinned copper
conductor

(c)

rubber
insulation

lead sheath

tinned copper
conductor

(d)

rubber
insulation

Fig. 19.1 Cabling changes. (a) Change of colour for sleeving on uninsulated CPC. (b) 1 mm² CPC in 2.5 mm² twin and earth cables. (c) Tough rubber sheathed rubber insulated cables. (d) Lead-sheathed cables.

Socket-outlets and accessories

Obsolete socket-outlets – the current 13 amp square pinned socket-outlets were mainly introduced in the 1950s. Before that, two common types of socket-outlet were prevalent – the 15 amp outlet, which accepted fused and non-fused plugs with three round pins, and the 5 amp outlet, which accepted unearthed non-fused plugs with two pins. These older socket outlets will cause danger if connected to a ring main. Additionally, the socket-outlets that accept unearthed non-fused plugs with two pins must never be used to supply equipment that needs to be earthed. These obsolete socket outlets are rarely encountered today since it is relatively easy to change them for modern 13 amp outlets and the plugs that fit them are no longer available. Where encountered they should always be changed.

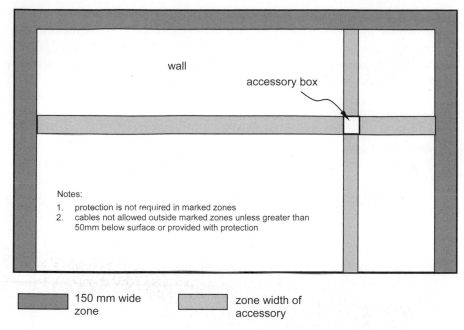

Fig. 19.2   Cable zones for concealed cables in walls.

Wooden mounting blocks – up until the mid 1960s wooden mounting blocks were often used for accessories such as ceiling roses, light switches and socket outlets. Often the wooden mounting block formed the rear enclosure of the accessory and was in the immediate proximity of unsheathed cores and connection terminals. The current Wiring Regulations contain ignitability requirements that the mounting block materials are unlikely to satisfy.

Fuses and circuit-breakers

Double-pole fusing – up to the 1950s, it was common practice for the circuits to have a fuse in the neutral conductor as well as in the phase (i.e. live) conductor. This is a potentially dangerous practice for alternating current (ac) installations, since, if a short-circuit occurs, there is a 50/50 chance that the fuse in the neutral conductor will operate. When this happens, the phase (live) conductor is not automatically disconnected from the faulty circuit (as would now normally be expected), thereby leaving a danger for the unwary when the fault is repaired. This practice was banned by the Wiring Regulations in about 1955.

Voltage-operated earth-leakage circuit-breakers – earth-leakage circuit-breakers are used to protect circuits should a fault arise by rapidly cutting off the flow of electricity. Up until 1981 the Wiring Regulations recognised two basic types of earth-leakage circuit-breaker: those operated by current, and those operated by voltage. Today, only those operated by current are recognised and they are now referred to as residual current devices (or RCDs).

Where encountered, the voltage-operated type of earth-leakage circuit-breaker can be recognised by its two separate earthing terminals, one of which is used as an earthing connection to the load and the other as an earthing connection to a means of earthing (such as a driven rod). This type of device suffers from the major drawback that a parallel earth path can render it disabled.

Inadequate RCD protection for socket-outlets likely to supply portable equipment outdoors – a person using portable electrical equipment outdoors can, if they receive an electric shock, be at great risk of death or serious injury. This risk can be significantly reduced if the supply to the equipment is provided with sensitive RCD protection which can be fitted either at the socket-outlet itself or at the consumer unit.

No such protection was provided by the Wiring Regulations prior to 1981 and even after that date the required minimum standard of provision was for only one protected socket-outlet. This minimum provision was found to be inadequate and was amended so that sensitive RCD protection is now required for all socket-outlets which are installed having a rating of 32 A or less, and which may reasonably be expected to supply portable equipment for use outdoors. It should be noted that the rated residual operating current of the RCD should not exceed 30 mA.

19.7.3   Harmonised cable identification colours

It has already been mentioned under the notes on changes in cabling in section 19.7.2 that the colours of the various conductors in a cable have been changed over the years. When additions and alterations are made to existing installations, the older cable colours will often be encountered. Therefore, Appendix D of Approved Document P offers guidance on applying the harmonised cable identification system agreed throughout the European Union. On 31 March 2004 Amendment No. 2 to BS 7671: 2001 was published. This document, which specifies new (harmonised) cable core colours for all new fixed wiring in electrical installations in the UK, includes guidance for alterations and additions to installations wired in the old cable colours.

Table 19.1 gives details of the new cable core colours for ac power circuits and those that existed prior to 31 March 2004 for both single and three phase cables.

Further information, including cable identification colours for extra-low voltage and direct current power circuits, is available from the following sources:

- *New wiring colours.* Leaflet published by the IEE, 2004 (may be downloaded from the IEE website at www.iee.org/cablecolours).
- *ECA comprehensive guide to harmonised cable colours, BS 7671: 2001 Amendment No 2.* Published by the Electrical Contractors' Association, March 2004.
- *New fixed wiring colours – A practical guide.* Published by the National Inspection Council for Electrical installation Contracting (NICEIC), Spring 2004.

Table 19.1   Identification of conductors in ac power and lighting circuits – comparison of current and previous practice.

Conductor	New BS 7671 (harmonised) cable core colour (see notes 1 to 3)	Cable core colours in use prior to 31 March 2004 (but see notes 1 to 3)
Protective conductor (earth)	Green and yellow	Green
Neutral	Blue	Black
Phase of single phase circuit (live)	Brown	Red
Phase 1 of three-phase circuit	Brown	Red
Phase 2 of three-phase circuit	Black	Green or Yellow
Phase 3 of three-phase circuit	Grey	Blue

Notes:

(1) It has been permissible to use the new (harmonised) colour cables on site from 31 March 2004.

(2) For new installations or alterations to existing installations, it was permissible to use either new or old colours, but not both, between 31 March 2004 and 31 March 2006.

(3) Only the new colours have been permitted since 31 March 2006.

19.8   Electrical installation diagrams

Approved Document P provides, in Appendix A, a number of diagrams that are intended to: 'give an indication of the sorts of electrical services encountered in dwellings, some of the ways they can be connected and the complexity of the wiring and protective systems necessary to supply them'. This sounds like useful advice. However, rather curiously, the Approved Document adds (in bold type) a warning that 'they are intended as an indication of the scope of Part P for those who are not electricians; they must not be used for installation purposes' and this warning is reinforced in Appendix A where a number of notes (some of them containing highly technical language) are produced which show the limitations of the use of the diagrams.

The diagrams given relate to:

- an indication of the many electrical appliances that can be found in the home and how they might be supplied (Diagram 1(a));
- the earthing and bonding arrangements that can be necessary (Diagram 1(b));
- an indication of the earthing arrangements as might be provided by electricity distributors (Diagram 2(a));
- an indication of the earthing arrangement as might need to be provided by the consumer (Diagram 2(b)).

This information is readily available in a number of excellent textbooks designed for both electricians and DIY workers where the technical terminology is properly explained and the Wiring Regulations are considered. Readers wanting a better understanding of electrical installation work would be better off consulting such texts rather than the much qualified, simplified examples found in Approved Document P.

For the sake of completeness we reproduce Diagram 1 and its key below as an example. All the diagrams and accompanying notes may be obtained from Approved Document P which is obtainable free on-line at www.communities.gov.uk.

AD P Examples of electrical installation diagrams (Diagram 1(a))

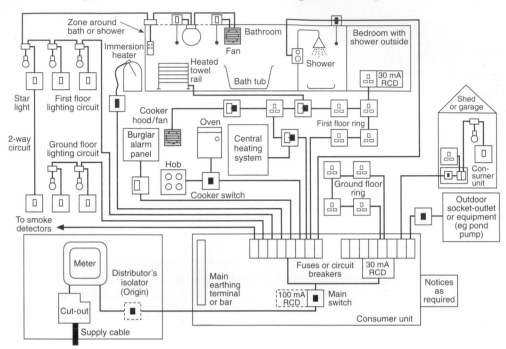

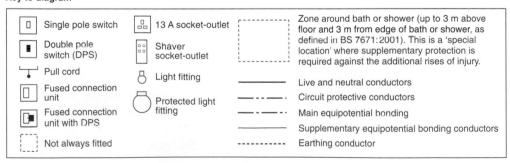

Notes:
1. See the general rules in BS 7671:2001.
2. The RCD component in the main switch is required for TT systems (see Diagram 2(b)). Individual circuit 30 mA RCD's may be required to avoid unnecessary tripping.
3. The notices include advice on periodic testing and regular test operation of the RCDs.
4. The zone shown around the bath or shower corresponds to zone 3 in Section 601 of BS 7671:2001.

   The socket-outlet shown in the bedroom with the shower cublicle must be outside zone 3.

19.9   Further guidance on electrical installations

Further guidance on Part P is available from:

- the website of the Institution of Electrical Engineers (IEE) at www.iee.org/Publish/WireRegs/IEE_Building_Regs.pdf;
- the websites of the National Inspection Council for Electrical Installation Contracting (NICEIC) at www.niceic.org.uk and the Electrical Contractors' Association (ECA) at www.eca.co.uk. These organisations have jointly published the *Electrical Installers' Guide to the Building Regulations*.

III | Appendix

Appendix
Local Acts of Parliament

The following list of Local Acts of Parliament is reproduced by kind permission of Carillion Specialist Services Ltd, which is a corporate approved inspector under the Building Regulations.

The sections marked with an asterisk (*) in column 3 of the list are applicable where there are matters to be satisfied by a developer either as part of a submission under Building Regulations or which otherwise need to be addressed in parallel with that submission.

County Acts usually, but not inevitably, apply across a whole county. References to district authorities in the following list however merely illustrate those which may have been consulted, and do not necessarily constitute a complete list of districts within that county. However, where it is known that a particular section applies only in a particular district this is indicated.

By virtue of the Building (Repeal of Provisions of Local Acts) Regulations 2003 (SI 2003/3030) certain sections in Local Acts of Parliament dealing with separate systems of drainage were repealed with effect from 1 March 2004. This list was last updated on 30 September 2006.

(1) *Local Act*	(2) *Relevant sections*	(3) *Apply?*
County of Avon Act 1982		
• Bath City Council	s.7 Parking places, safety requirements	*
• Bristol City Council	s.35 Hot springs, excavations in certain areas of Bath	
Berkshire Act 1986		
• Newbury District Council	s.28 Safety of stands	*
	s.32 Access for fire brigade	*
	s.36 Parking places, safety requirements	*
	s.37 Fire precautions in large storage buildings	*
	s.38 Fire precautions in high buildings	
• Reading Borough Council		
Bournemouth Borough Council Act 1985	s.15 Access for fire brigade	*
	s.16 Parking places, safety requirements	*
	s.17 Fire precautions in certain large buildings	*
	s.18 Fire precautions in high buildings	*
	s.19 Amending s.72 Building Act 1984	*
Cheshire County Council Act 1980	s.48 Parking places, safety requirements	*
• Chester City Council	s.50 Access for fire brigade	*
• Warrington BC	s.49 Fireman switches	
	s.54 Means of escape, safety requirements	

(1) *Local Act*	(2) *Relevant sections*	(3) *Apply?*
County of Cleveland Act 1987		
• Stockton on Tees Borough Council	s.5 Access for fire fighting	*
	s.6 Parking places, safety requirements	*
	s.15 Safety of stands	
• Middlesborough Borough Council		
Clwyd Act 1985		
• Borough of Rhuddlan	s.19 Parking places, safety requirements	*
• Colwyn District Council	s.20 Access for fire brigade	*
Cornwall Act 1984		
• Caradon DC		
• North Cornwall		
Croydon Corporation Act 1960	s.93/94 Buildings of excess cubic capacity	*
	s.95 Buildings used for trade and for dwellings	
	s.79 Separate drainage systems	
Cumbria Act 1982		
• Barrow Borough Council	s.23 Parking places, safety requirements	*
• Carlisle City Council	s.25 Access for fire brigade	*
	s.28 Means of escape from certain buildings	
Derbyshire Act 1981		
• Borough of High Peak	s.16 Safety of stands	*
	s.23 Access for fire brigade	*
	s.24 MoE from certain buildings	*
	s.28 Parking places; safety requirements	*
	s.25 Fireman switches	
Dyfed Act 1987		
• South Pembrokeshire District Council	s.46 Safety of stands	*
	s.47 Parking places; safety requirements	*
	s.51 Access for fire brigade	
• Carmarthen District Council		
East Ham Corporation Act 1957		
• Newham London Borough Council	s.54 Separate access to tenements	*
	s.61 Access for fire brigade	*
East Sussex Act 1981	s.34 Fireman switches	
• Hastings Borough Council	s.35 Access for fire brigade	*
Essex Acts 1952 & 1958 (GLC areas formerly in Essex)		
Essex Act 1987		
• Uttlesford District Council	s.13 Access for fire brigade	*
Exeter Act 1987		
Greater Manchester Act 1981		
• Trafford Metropolitan Borough Council	s.58 Safety of stands	
	s.61 Parking places; safety requirements	*
• Manchester City Council	s.62 Fireman switches	
	s.63 Access for fire brigade	*
	s.64 Fire precautions in high buildings	*
	s.65 Fire precautions in large storage buildings	*
	s.66 Fire and safety precautions in public and other buildings	*

(1) *Local Act*	(2) *Relevant sections*	(3) *Apply?*
Hampshire Act 1983		
• Southampton City Council	s.11 Parking places; safety requirements	*
	s.12 Access for fire brigade	*
	s.13 Fire precautions in certain large buildings	*
Hastings Act 1988		
Hereford City Council Act 1985	s.17 Parking places; safety requirements	*
	s.18 Access for fire brigade	*
Humberside Act 1982		
• Gt.Grimsby BC	s.12 Parking places; safety requirements	*
	s.13 Fireman switches	
	s.14 Access for the fire brigade	*
	s.15 Means of escape in certain buildings	
	s.17 Temporary structures, byelaws	
Isle of Wight Act 1980	s.32 Access for fire brigade	*
(Part VI)	s.31 Fireman switches	
	s.30 Parking places; safety requirements	*
Kent 1958 (GLC areas formerly in Kent)		
County of Kent Act 1981		
• Canterbury City Council	s.51 Parking places; safety requirements	*
• Rochester City Council	s.52 Fireman switches	
	s.53 Access for fire brigade	*
	s.78 Annulment of plans approvals	*
Lancashire Act 1984	s.31 Access for fire brigade	*
Leicestershire Act 1985		
• Leicester City Council	s.21 Safety of stands	
North West Leicestershire District	s.49 Parking places; safety requirements	*
Council	s.50 Access for fire brigade	*
	s.52 Fire precautions in high buildings	*
	s.53 Fire precautions in large storage buildings	
	s.54 Means of escape, safety requirements	
London Building Acts 1930–1939	s.20 Buildings of excess height or cubic capacity	*
	N.B.: By definition (BA 1984, s.88 & Sch.3), these Acts do not constitute Local Acts but otherwise have a similar effect.	
• Corporation of London		
• City of Westminster London Borough Council		
• Royal Borough of Kensington & Chelsea		
County of Merseyside Act 1980	s.20 Safety of stands	
• Liverpool City Council	s.48/49 Means of escape from fire	
	s.50 Parking places; safety requirements	*
• Borough of Wirral	s.51 Fire and safety precautions in public and other buildings	*
	s.52 Fire precautions in high buildings	*
	s.53 Fire precautions in large storage buildings	*
	s.54 Fireman switches	
	s.55 Access for fire brigade	*

(1) *Local Act*	(2) *Relevant sections*	(3) *Apply?*
Middlesex Act 1956 (GLC areas formerly in Middlesex)	s.33 Access for fire brigade	*
Mid Glamorgan County Council Act 1987 • Merthyr Tydfil • Taff-Ely BC	s.9 Access for fire brigade	*
Nottinghamshire Act 1985 • City of Nottingham		
Plymouth Act 1987		
Poole Act 1986	s.10 Parking places; safety requirements s.11 Access for fire brigade s.14 Fire precautions in certain large buildings S.15 Fire precautions in high buildings	* * *
County of South Glamorgan Act 1976 • Cardiff City Metropolitan District Council	s.27 Safety of stands s.48/50 Underground parking places s.51 Means of escape for certain buildings s.52 Fireman switches s.53 Precautions against fire in high buildings s.54 Byelaws for temporary structures	 * *
South Yorkshire Act 1980 • City of Sheffield Metropolitan District Council • Barnsley Metropolitan Borough Council • Rotherham Metropolitan Borough Council • Doncaster Metropolitan Borough Council	s.53 Parking places; safety precautions s.54 Fireman switches s.55 Access for fire brigade	* *
Staffordshire Act 1983 • Staffordshire Moorlands District Council	s.25 Parking places; safety precautions s.26 Access for fire brigade	* *
Surrey Act 1985 • Guildford Borough Council • Spelthorne Borough Council	s.18 Parking places; safety requirements s.19 Fire precautions in large storage buildings s.20 Access for fire brigade	* * *
Tyne & Wear 1980	s.24 Access for fire brigade	*
West Glamorgan Act 1987 • City of Swansea • Neath Borough Council	s.43 Parking places; safety precautions	*
West Midlands Act 1980 • Birmingham City Council	s.39 Safety of stands s.44 Parking places; safety requirements s.45 Fireman switches s.46 Access for fire brigade s.49 Means of escape from certain buildings	 * *
West Yorkshire Act 1980 • Kirklees Metropolitan District Council • Bradford City Council	s.9 Culverting water courses s.51 Fireman switches	*
Worcester City Council Act 1985		

Index